Second Edition

Essentials of Biology

Sylvia S. Mader

with significant contributions by

Andrew Baldwin
Mesa Community College

Rebecca Roush
Sandhills Community College

Stephanie Songer
North Georgia College and State University

Michael Thompson
Middle Tennessee State University

Mc Graw Hill **Higher Education**

Boston   Burr Ridge, IL   Dubuque, IA   New York   San Francisco   St. Louis
Bangkok   Bogotá   Caracas   Kuala Lumpur   Lisbon   London   Madrid   Mexico City
Milan   Montreal   New Delhi   Santiago   Seoul   Singapore   Sydney   Taipei   Toronto

Higher Education

ESSENTIALS OF BIOLOGY, SECOND EDITION

1 2 3 4 5 6 7 8 9 0 VNH/VNH 0 9

ISBN 978–0–07–340342–7
MHID 0–07–340342–3

Publisher: *Janice Roerig-Blong*
Executive Editor: *Michael S. Hackett*
Director of Development: *Kristine Tibbetts*
Senior Developmental Editor: *Rose M. Koos*
Marketing Manager: *Tamara Maury*
Senior Project Manager: *Jayne L. Klein*
Lead Production Supervisor: *Sandy Ludovissy*
Senior Media Project Manager: *Jodi K. Banowetz*
Designer: *Laurie B. Janssen*
Cover/Interior Designer: *Elise Lansdon*
(USE) Cover Image: *©Tim Davis/Corbis*
Senior Photo Research Coordinator: *Lori Hancock*
Photo Research: *Evelyn Jo Johnson*
Supplement Producer: *Mary Jane Lampe*
Typeface: *10/13 Times LT Std*
Printer: *R. R. Donnelley, Jefferson City, MO*

The credits section for this book begins on page C-1 and is considered an extension of the copyright page.

Library of Congress Cataloging-in-Publication Data

Mader, Sylvia S.
   Essentials of biology / Sylvia S. Mader. -- 2nd ed.
      p. cm.
   Includes index.
   ISBN 978-0-07-340342-7 — ISBN 0-07-340342-3 (hard copy : alk. paper) 1. Biology. I. Title.
   QH308.2.M24 2010
   570--dc22
                                                              2008031863

www.mhhe.com

Brief Contents

Preface

Essentials of Biology was written to engage non-major students in the study of biology and inspire them to learn more. It continues to be traditional in its approach and organized around the major concepts of biology. However, the book is succinct enough to be appropriate for a one semester course.

Three primary goals guided the revision of *Essentials of Biology,* Second Edition:

1. build upon the strengths of the first edition,
2. increase student awareness of evolution and why it is a central concept of biology,
3. add more applications to enhance the relevancy of the content for students.

Strengths of *Essentials of Biology*

Content Organization Writing Style Illustrations Pedagogy

The content and organization of *Essentials of Biology* continues to be solid—accurate and up-to-date, and logical. Equally important, the text reaches out to the student because it has a conversational writing style and an exciting illustration program. The design and illustration program of the book is appropriate for those who are accustomed to being visually stimulated, and instructors can easily use these same great illustrations in lecture or as part of assignments.

The *Essentials of Biology* Learning System also helps beginning biology students by guiding them through each chapter. The chapter-opening page lists the key concepts to give students an overview of the chapter. The opening vignette captures students' interest and encourages them to begin their study of the chapter. Distributed throughout each chapter are *Check Your Progress* questions designed to foster confidence as readers proceed through the chapter. The chapter end matter gives students an opportunity to review the chapter and test themselves on how well they understand the concepts. *Thinking Scientifically* questions allow students to apply their knowledge to concrete scientific explorations. Most chapters now end with a *Bioethical Issue* that can be used as a basis for class discussion.

Four contributors joined me in revising *Essentials of Biology,* and their varied backgrounds and extensive teaching experience

> "*Essentials of Biology* is an outstanding textbook for the non-major. The material is clear and specific, but not intimidating. The text contains numerous graphics, diagrams, etc. to enhance student-learning. This text invites students to open it and read!"
>
> *Barbara J. Salvo*
> *Carthage College*

further enhanced the strengths of the book. Andrew Baldwin, *Mesa Community College,* revised the evolution and ecology chapters. Rebecca Roush, *Sandhills Community College,* revised Chapters 22-26 of the animal structure and function chapters. Stephanie Songer, *North Georgia College and State University,* revised the diversity of life and several animal structure and function chapters. Michael Thompson, *Middle Tennessee State University,* revised the cell, genetics, and plant chapters.

We worked collaboratively to improve this new edition, and through my review of every chapter, the book retains the hallmark features of a Mader text: accurate and current content, logical organization of chapters, clear writing, well-developed visuals, and a strong pedagogical system. For more information on the features of this book, see the Guided Tour on pages vii-xi.

Increased Evolution Coverage

For several years now, biology has been undergoing a revolutionary change. Molecular biology is indeed wedded to evolution and the binding together of these two fields of biology has opened up a vista into the evolutionary relationships of organisms never before possible. With the help of nucleic acid sequencing, we can now more confidently reconstruct the evolutionary history of life. *Essentials of Biology* makes full use of our newfound knowledge and relates the evolutionary consequences for the broadest categories, the domains, and for each of the kingdoms.

Being good educators we realized that students need a firm foundation in the principles of evolution in order to fully appreciate the phylogenic data now available to biologists. Reviewers will find a renewed emphasis on evolution that begins in the first chapter with a new section devoted to evolutionary principles and a generalized tree of life. The evolution chapters in Part III, *Evolution,* received special attention before Part IV, *Diversity of Life,* specifically

> "This text clearly presents topics in biology in such an interesting fashion that it is sure to intrigue and fascinate students. The text is also written in such as way as to be quite understandable by the non-science student."
>
> *Kathleen Pelkki,*
> *Saginaw Valley State University*

discusses the evolution of various groups and presents new evolutionary trees for protists, plants, fungi, and animals. Increased evolution coverage is also weaved throughout this edition, and a *Thinking Scientifically* question that pertains to evolution is presented at the end of each chapter (identified with an icon, *left*) to help students realize that evolution is pertinent to all aspects of biology.

Additional examples of the increased evolution coverage are noted below in the overview of changes.

Relevancy of Biology

More applications have been added to the second edition of *Essentials of Biology* to increase the relevancy of the text. In reviews, instructors emphasize the need to relate the concepts of biology to students' lives. Such relevant examples help students understand the concepts, and increase their chances of remembering the concepts.

Each chapter opens with a brief story that relates the biological concepts of that chapter to something of interest to students. For example, Chapter 2 introduces chemistry by discussing how exposure to extreme levels of radiation can cause sunburn, hair loss, or cancer. Chapter 25 discusses saturated, unsaturated, and trans fats to open the nutrition chapter.

> "I really like this book because it is student friendly. Content has real world application and is easy to read and understand. [It] has beautiful illustrations that stimulate interest and excellent end of chapter materials."
> *Mary Ruth Hood, Tomball College*

Most chapters also end with a Bioethical Issue; a brief reading about an ethical issue related to biology that confronts our society. Examples of topics covered include cloning, global climate change, water pollution, health risks associated with certain plastics, gene therapy, and blood doping.

Additional examples of relevancy are given below in the overview of changes.

Overview of Changes to *Essentials of Biology,* Second Edition

Chapter 1   *A View of Life*

This chapter previews the text by discussing the characteristics of life, principles of evolution, organization of the biosphere, and the scientific process. An expanded section on evolution including a generalized tree of life, is new to this edition. The scientific process includes a new example of a controlled experiment that is more applicable to students' lives.

Part I   *The Cell*

This part was revised to emphasize the dynamic nature of cells. Chapter 3, *The Organic Molecules of Life,* has been enriched with new references to health concerns including high-fructose drinks, trans fats in processed foods and the abuse of steroids to gain muscle mass. Chapter 5, *The Dynamic Cell,* has been reorganized to place enzymatic reactions next to the discussion of metabolic pathways. In Chapter 6, *Energy for Life,* the diversity of photosynthetic process is explained as adaptations to different environments.

New bioethical issues occur in this part including *Human Cloning, Arsenic in Drinking Water,* and *Are Carbon Offset Credits Effective?*

Part II   *Genetics*

This part has been updated to be more applicable to students' everyday lives. In Chapter 9, *Sexual Reproduction,* the importance of meiosis for the generation of variations so necessary to evolution is stressed. Chapter 11, *DNA Biology and Technology,* now contains a new illustration (Fig. 11.9) to introduce the topic of protein synthesis. The discussion of genomics and bioinformatics has been moved from Chapter 13 to this chapter. The Human Genome Project is reviewed, and the current emphasis on genomic variation between individuals is also addressed. Further, the contributions of genomics to our knowledge of evolutionary ties between species is discussed. In Chapter 12, *Gene Regulation and Cancer,* an improved section on the genetics of cancer is supported by new Figure 12.15.

New bioethical issues occur in this part including *Selecting the Sex of Your Child, Should BPA be Banned?,* and *Gene Therapy Trials.*

Part III   *Evolution*

This part is much improved by current examples of the evolutionary process. In Chapter 14, *Darwin and Evolution,* Alfred Russell Wallace's contribution to Darwin's theory is emphasized, and Darwin's theory is better explained. In Chapter 15, *Evolution on a Small Scale,* an improved discussion of a Hardy-Weinberg equilibrium is supported by a new figure. Examples of genetic mutations have been updated by a discussion of SNPs (single nucleotide polymorphisms). Figures 15.10-15.11, which describe the different types of natural selection, have been clarified and revised. Chapter 16, *Evolution on a Large Scale,* contains two new sections: 16.2 is now a discussion of

> "I think the revisions made so far and the process of updating this textbook will continue to make it one of the sought after texts for general biology."
> *Bo Sosnicki, Piedmont Virginia Community College*

the fossil record, and 16.3 now gives a review of systematics that includes an improved discussion of cladistics and its relationship to Linnaean classification.

Part IV   *Diversity of Life*

This part has been updated to include a more comprehensive look at the evolutionary relationships of organisms. In Chapter 17, *The First Forms of Life,* both the origin of the first cell and the origin of the eukaryotic cell are now included. Chapter 18, *Land Environment: Plants and Fungi,* has been thoroughly revised to include a new evolutionary tree of plants, and a discussion that traces their evolution from an algal ancestor and emphasizes five significant events as they became fully adapted to land. Chapter 19, *Both Water and Land: Animals,* now includes a new evolutionary tree of animals and a discussion that traces their evolution from an ancestral protist. The tree emphasizes number of germ layers, type of symmetry, and type of embryonic development. Human evolution, which is in this chapter, was updated to include the new designation, hominins, which also includes the chimpanzees.

Part V   *Plant Structure and Function*

Chapter 20, *Plant Anatomy and Growth,* and Chapter 21, *Plant Responses and Reproduction,* were revised to make the material more student friendly and to emphasize the specific adaptations of flowering plants that have resulted in their great success on land.

Part VI   *Animal Structure and Function*

Various applications and analogies have been inserted and more material was added to selected topics. For example, in Chapter 24, *The Maintenance Systems,* digestive enzymes are reviewed in a new section. In Chapter 26, *Defenses Against Disease,* the new terms "acquired immunity" and "innate immunity" are introduced. The emphasis on homeostasis is maintained in this part, and new applications have been provided to enliven the discussion. The bioethical issues have been updated and expanded.

Part VII   Ecology

The principles of ecology are more thoroughly reviewed in this section. The interplay between ecology and evolution is emphasized because adaptations are often a response to ecological factors. In Chapter 32, *Human Impact on the Biosphere,* the negative human impacts on the environment are counter balanced by an expanded section that shows how humans can foster sustainability.

What's New to This Edition

Increased Evolution Coverage

- Chapter 1 now introduces students to the principles of evolution and includes a generalized tree of life. The process of science contains a more applicable example of a controlled study.
- Plant and animal biology now begin with evolutionary trees based on nucleic acid sequencing and emphasizing the innovations that have occurred since they evolved from protistan ancestors.
- Thinking Scientifically questions that pertain to all aspects of evolution have been added and are identified with an icon.

Improved Genetics

The relationship between genetics and cancer is highlighted and a reorganization has resulted in expanded coverage of genomics and bioinformatics as they apply to student lives.

Expanded Ecology

The negative impact of human activities on the environment is balanced by a new section on sustainability that gives examples of ways humans can and do contribute to the sustainability of the planet.

Additional Applications

Many new and/or improved applications and analogies have been added to the text. They will help students relate all aspects of biology to their lives and the world around them.

 CONNECT

McGraw-Hill CONNECT Biology is a web-based assignment and assessment platform. Instructors can deliver assignments, quizzes, and tests easily online. Students can practice important skills at their own pace and on their own schedule.

Enhanced Instructor Presentation Tools

All line art figures in *Essentials of Biology,* Second Edition are available to instructors for presentation purposes in an easy-to-use table of assets. Included are enhanced image PowerPoints, (including editable art, tables, and photos), lecture PowerPoints with animations, animation PowerPoints, as well as labeled and unlabeled JPEG images. With the help of a content expert, figures have been split or stepped out as necessary to ensure optimal quality during presentations.

Electronic Books

If you or your students are ready for an alternative version of the traditional textbook, an innovative and inexpensive electronic textbook is available for *Essentials of Biology,* Second Edition. Smart, interactive, searchable, and portable, the electronic book comes with a powerful suite of built-in tools that allow detailed searching, highlighting, note taking, and student-to-student or instructor-to-student note sharing. In addition, relevant animations and videos are integrated into the textbook content for a true multimedia learning experience.

Guided Tour

Increased Evolution Coverage

Provides a firm foundation in the principles of evolution

Chapter 1

introduces you to the principles of evolution and a generalized tree of life.

Part IV   Diversity of Life

includes a more comprehensive look at the evolutionary relationships of organisms.

Thinking Scientifically Question

related to evolution is presented at the end of each chapter (and identified with an icon) to help you understand how evolution is pertinent to all aspects of biology.

Thinking Scientifically

1. Recently, three fossil skulls of *Homo sapiens*, dating to about 160,000 years BP, were discovered in eastern Africa. These skulls fill a gap between the 100,000-year-old *H. sapiens* skulls found in Africa and Israel and 500,000-year-old skulls of archaic *H. sapiens* found in Ethiopia. Supporters of the replacement model (out-of-Africa hypothesis) argue that this discovery strengthens their position by documenting the succession of human ancestors from 6 MYA through this most recent group. Does this finding prove that the replacement model is correct? If not, what fossil evidence might yet be found to support the multiregional continuity hypothesis?

2. Think of the animals in this chapter that are radially symmetrical (cnidarians, many adult echinoderms). How is their lifestyle different from that of bilaterally symmetrical animals? How does their body plan complement their lifestyle?

1.2 | Evolution: The Core Concept of Biology

From bacteria to bats, toadstools to trees, whip-poor-wills to whales—the diversity of the living world boggles the mind. Yet all organisms are united by a common heritage that began during the early years of our planet. Just as you are descended from your parents, grandparents, and so forth from generation to generation back in time, all species can trace their history to the original living thing. **Evolution**, the process by which species have changed and diversified since life arose, explains the unity—all living things share the same characteristics—and the diversity of life. If you have ever studied a family tree, then you know it is a diagram used to show how individuals are related to one another and descended from common ancestors. An evolutionary tree is very much like a family tree, only it depicts relationships between groups of organisms. The evolutionary tree in **Figure 1.4** illustrates the lineages of the major groups of living things. The tree summarizes the history of life on Earth over the approximately 3.5 billion years that have passed since the first, ancestral cell. All the different groups of organisms on our planet are related to one another, and are represented by branches on the same "family tree of life."

The Diversity of Life

Think of an enormous department store, offering thousands of different items for sale. The various items are grouped in departments—electronics, apparel, furniture, and so on—to make them easy for customers to find. Because life is so diverse, it is helpful to have a system that groups organisms into categories (see Appendix B). **Taxonomy** is the discipline of identifying and classifying organisms according to certain rules. Taxonomy makes sense out of the bewildering variety of life on Earth by classifying organisms according to their presumed evolutionary relationship. As more is learned about evolutionary

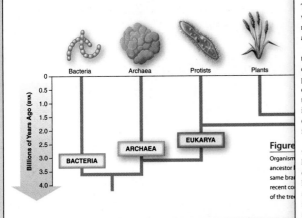

17.4 | The Protists

Like ancient creatures from another planet, the **protists** inhabit the oceans and other watery environments of the world. Their morphological diversity is their most outstanding feature—unicellular diatoms are encrusted in silica "petri dishes"; dinoflagellates have plates of armor; and ciliates shaped like slippers have complex structures.

Many protists are unicellular, but all are eukaryotes with a nucleus and a wide range of organelles. It is widely held that the organelles of eukaryotic cells arose from close symbiotic associations between bacteria and primitive eukaryotes (**Fig. 17.19**). This so-called *endosymbiotic hypothesis* is supported by the presence of double membranes around mitochondria and chloroplasts. Also, these organelles have their own genomes, although incomplete, and their ribosomal genes point to bacterial origins. The mitochondria appear closely related to certain bacteria, and the chloroplasts are most closely related to cyanobacteria.

To explain the diversity of protists, we can well imagine that once the eukaryotic cell arose, it provided the opportunity for many different lineages to begin. Some unicellular protists have organelles not seen in other eukaryotes. For example, food is digested in food vacuoles, and excess water is expelled when contractile vacuoles discharge their contents.

Protists also possibly give us insight into the evolution of a multicellular organism with differentiated tissues. Some protists are a colony of single cells, with certain cells specialized to produce eggs and sperm, and others are multicellular, with tissues specialized for various purposes. Perhaps the first type of organization preceded the second in a progression toward multicellular organisms.

General Biology of Protists

The complexity and diversity of protists make it difficult to classify them. The variety of protists is so great that it's been suggested they could be split into more than a dozen kingdoms. Due to limited space, this text groups the protists according to modes of nutrition.

Traditionally, the term **algae** means aquatic photosynthesizer. At one time, botanists classified algae as plants because they contain chlorophyll *a* and carry on photosynthesis. In aquatic environments, algae are a part of the phytoplankton, photosynthesizers that lie suspended in the water. They are producers, which serve as a source of food for other organisms and pour oxygen into the environment. In terrestrial systems, algae are found in soils, on rocks, and in trees. One type of alga is a symbiote of animals called corals, which depend on them for food as they build the coral reefs of the world. Others partner with fungi in lichens capable of living in harsh terrestrial environments.

The definition of a **protozoan** as a unicellular chemoheterotroph explains why protozoans were originally classified with the animals. Often a protozoan has some form of locomotion, either by flagella, pseudopods, or cilia. In aquatic environments, protozoans are a part of the zooplankton, suspended microscopic heterotrophs that serve as a food source for animals. While most are free-living, some protozoans are human pathogens, often causing diseases of the blood or intestines. In many cases, their complex life cycles inhibit the development of suitable treatments.

There is still one other group of protists: the slime molds and water molds. These microrganisms are chemoheterotrophs, but the slime molds

Figure 17.19 Evolution of the eukaryotic cell.

Invagination of the plasma membrane could account for the formation of the nucleus and certain other organelles. The endosymbiotic hypothesis suggests that mitochondria and chloroplasts are derived from prokaryotes that were taken up by a much larger eukaryotic cell.

The Visuals

A brilliant visuals program brings biology to life!

Multi-Level Perspective

Illustrations depicting complex structures show macroscopic and microscopic views to help you see the relationships between increasingly detailed drawings.

phospholipid

polar head

nonpolar tail

Outside of cell

carbohydrate chain

glycoprotein

external membrane surface

phospholipid bilayer

hydrophilic

hydrophobic

hydrophilic

internal membrane surface

protein molecule

cholesterol

cytoskeleton filaments

Inside of cell

Combination Art

Drawings of structures are paired with micrographs to give you the best of both perspectives: the realism of photos and the explanatory clarity of line drawings.

cuticle

upper epidermis

Water and minerals enter leaf through xylem.

palisade mesophyll

Sugar exits leaf through phloem.

leaf vein

spongy mesophyll

lower epidermis

×94

nucleus

mitochondrion

central vacuole

chloroplast

guard cell

O₂ and H₂O exit leaf through stoma.

CO₂ enters leaf through stoma.

epidermal cell

Leaf cell

Stoma and guard cells

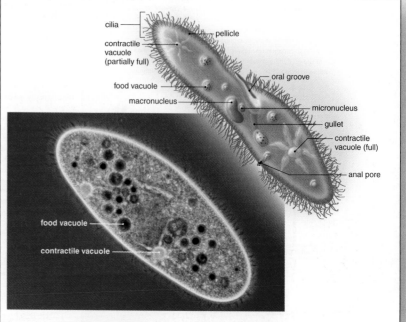

cilia

pellicle

contractile vacuole (partially full)

oral groove

food vacuole

macronucleus

micronucleus

gullet

contractile vacuole (full)

anal pore

food vacuole

contractile vacuole

Process Figures

Complex processes are broken down into a series of smaller steps that are easy-to-follow. Numbers guide you through the process.

Evaporation of water (transpiration) creates tension that pulls the water column from the roots to the leaves. H_2O

Cohesion and adhesion of water molecules keeps the water column intact within xylem.

Water enters a plant at root cells.

H_2O

Nucleus

Cytoplasm

DNA

mRNA

nuclear pore

4 At termination, the polypeptide becomes a protein. The ribosomal subunits disengage, and the mRNA is released.

small subunit

large subunit

2 In the cytoplasm, the mRNA and ribosomal subunits join, and polypeptide synthesis begins.

3 If a ribosome attaches to a receptor on the ER, the polypeptide enters the lumen of the ER.

1 mRNA is produced in the nucleus but moves through a nuclear pore into the cytoplasm.

ribosome

ribosome

protein

receptor

polypeptide

lumen of the ER

ER membrane

ribosome

Endoplasmic reticulum

amino acid

complementary base pairing

tRNA binding sites

P site   A site

large subunit

mRNA binding site

small subunit

b. Ribosome

U G G

anticodon

Transcription is taking place— the nucleotides of mRNA are joined by the enzyme RNA polymerase in an order complementary to a strand of DNA.

DNA template strand

RNA polymerase

peptide

anticodon

mRNA

ribosome

c. tRNA–amino acid at ribosome

DNA coding strand

This mRNA transcr... ready to be proces...

amino acid methionine

initiator tRNA

mRNA

small ribosomal subunit

start codon

P site

A site

large ribosomal subunit

mRNA

mRNA

to processing

Color Consistency

Consistent use of color organizes information and clarifies concepts.

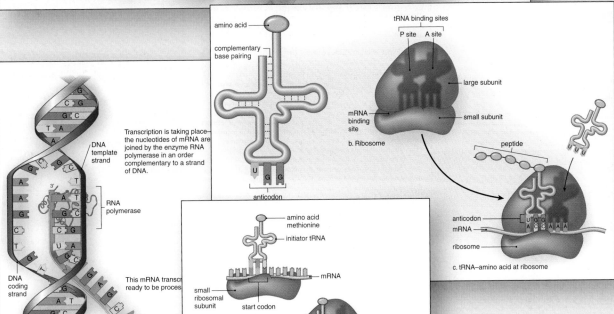

The Learning System

Pedagogical features facilitate your understanding of biology

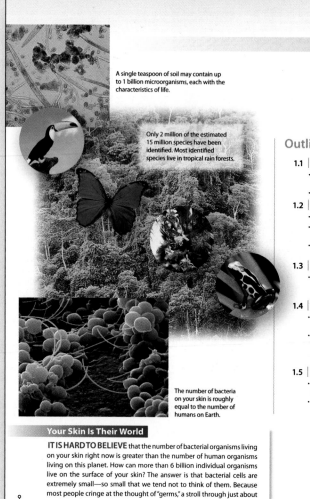

A single teaspoon of soil may contain up to 1 billion microorganisms, each with the characteristics of life.

Only 2 million of the estimated 15 million species have been identified. Most identified species live in tropical rain forests.

The number of bacteria on your skin is roughly equal to the number of humans on Earth.

Your Skin Is Their World

IT IS HARD TO BELIEVE that the number of bacterial organisms living on your skin right now is greater than the number of human organisms living on this planet. How can more than 6 billion individual organisms live on the surface of your skin? The answer is that bacterial cells are extremely small—so small that we tend not to think of them. Because most people cringe at the thought of "germs," a stroll through just about any aisle of your local grocery store will reveal a slew of products that claim to be antibacterial. Advertisers tend to make us think that all

—Continued next page

1

A View of Life

Outline

1.1 | The Characteristics of Life
- Although life is quite diverse, it can be defined by certain common characteristics. 2
- Living things exhibit levels of organization. 2

1.2 | Evolution: The Core Concept of Biology
- Evolution explains the unity and diversity of life. 5
- Living things are grouped (classified) according to their evolutionary relationships. 6
- An evolutionary tree is a pictorial representation of how groups of organisms are related. 8

1.3 | How the Biosphere Is Organized

1.4 | S

1.5 | S

Detailed Outline

focuses your attention on the major concepts.

Numbered Sections

help you organize the content.

Captivating Illustrations and Story

show how the concepts relate to your life.

Check Your Progress

as you read the chapter. Maximize your study time by identifying concepts that you don't understand before you move on.

---

promoter. The lac operon is normally turned off in this way because lactose is usually absent (Figure 12.5a).

What turns the operon on when lactose is present? Lactose binds to the repressor and it changes shape. Now the repressor is unable to bind to the operator and RNA polymerase is able to bind to the promoter. (Figure 12.5b). Transcription of the genes needed for lactose metabolism occurs. When lactose is present and glucose, the preferred sugar, is absent, a protein called CAP (not shown) assists the binding of RNA polymerase to the promoter. This further ensures that the lactose-metabolizing enzymes are transcribed when they are needed.

Other bacterial operons, such as those that control amino acid synthesis, are usually turned on. For example, in the *trp* operon, the regulatory gene codes for a repressor that ordinarily is unable to attach to the operator. Therefore, the genes needed to make the amino acid tryptophan are ordinarily expressed. When tryptophan is present, it binds to the repressor. A change in shape activates the repressor and allows it to bind to the operator. Now the operon is turned off.

Gene Expression in Eukaryotes

In bacteria, a single promoter serves several genes that make up a transcription unit, while in eukaryotes, each gene has its own promoter where RNA polymerase binds. Bacteria rely mostly on transcriptional control, but eukaryotes employ a variety of mechanisms to regulate gene expression. These mechanisms affect whether the gene is expressed, the speed with which it is expressed, and how long it is expressed.

Some mechanisms of gene expression occur in the nucleus, and others occur in the cytoplasm (**Fig. 12.6**). In the nucleus, chromatin condensation, DNA transcription, and mRNA processing all play a role in determining which genes are expressed in a particular cell type.

In the cytoplasm, mRNA translation into a polypeptide at the ribosomes can occur right away or be delayed. The mRNA can last a long time or be destroyed immediately, and the same holds true for a protein. These mechanisms control the quantity of gene product and/or how long it is active.

Chromatin Condensation   Eukaryotes utilize chromatin condensation as a way to keep genes turned on or off. The more tightly chromatin is compacted,

Check Your **Progress**

1. Contrast the activity of the *lac* operon when lactose is absent to its activity when lactose is present.
2. List the various levels at which eukaryotes regulate gene expression.

Answers: 1. When lactose is absent, the repressor is active. The active repressor binds the operator and inhibits transcription of the Lactose-metabolizing genes X2. When lactose is present, lactose binds and inactivates the repressor, allowing transcription of the DNA.   2. Eukaryotes may regulate transcription at the level of chromatin condensation, transcription, mRNA processing, mRNA translation, and protein activity.

Figure 12.6   Control of gene expression in eukaryotic cells.

Gene expression which can be turned on by an external signal (red) is controlled at various levels in eukaryotic cells.

The Chapter in Review

Summary

7.1 | Cellular Respiration

During cellular respiration, glucose from food is oxidized to CO_2, which we exhale. Oxygen (O_2), which we breathe in, is reduced to H_2O. When glucose is oxidized, energy is released. Cellular respiration captures the energy of oxidation and uses it to produce ATP molecules. The following equation gives an overview of these events:

$$C_6H_{12}O_6 \ + \ 6\ O_2 \ \longrightarrow \ 6\ CO_2 \ + \ 6\ H_2O \ + \ \boxed{ATP}$$
glucose

7.2 | Outside the Mitochondria: Glycolysis

Glycolysis, the breakdown of glucose to 2 molecules of pyruvate, is a series of enzymatic reactions that occur in the cytoplasm. During glycolysis:

- Glucose is oxidized by removal of hydrogen atoms ($H^+ + e^-$).
- When NAD^+ accepts electrons, NADH results.

Breakdown releases enough energy to immediately give a net gain of 2 ATP by substrate-level ATP synthesis. The inputs and outputs of glycolysis are summarized here:

Glycolysis	
inputs	outputs
glucose	2 pyruvate
2 NAD$^+$	2 NADH
2 ATP	2 ADP
4 ADP + 4 P	4 ATP
	2 ATP net

The Citric Acid Cycle

Acetyl groups enter the citric acid cycle, a series of reactions occurring in the mitochondrial matrix. During one turn of the cycle, oxidation results in 2 CO_2, 3 NADH, and 1 $FADH_2$. One turn also produces 1 ATP. The cycle turns twice and doubles these amounts per glucose.

The Electron Transport Chain

The final stage of cellular respiration involves the electron transport chain located in the cristae of the mitochondria. The chain is a series of electron carriers that accept high-energy electrons (e^-) from NADH and $FADH_2$ and pass electrons along until they are finally low-energy electrons received by oxygen, which combines with H^+ to produce water.

The carriers of the electron transport chain are located in molecular complexes on the cristae of mitochondria. These carriers capture energy from the passage of electrons and use it to pump H^+ into the intermembrane space of the mitochondrion. When H^+ flows down its gradient into the matrix through ATP synthase complexes, energy is released and used to form ATP molecules from ADP and P.

Energy Yield from Glucose Metabolism

Of the maximum 38 ATP formed by complete glucose breakdown, 4 are the result of substrate-level ATP synthesis, and the rest are produced as a result of the electron transport chain and ATP synthase:

Alternative Metabolic Pathways

Besides carbohydrates, glycerol and fatty acids from fats, and amino acids from proteins can undergo cellular respiration by entering glycolysis and/or the citric acid cycle. These metabolic pathways also provide substrates for the synthesis of fats and proteins.

Key Terms

acetyl-CoA 104	fermentation 109
citric acid cycle 101	glycolysis 101
coenzyme 101	intermembrane space 107
coenzyme A (CoA) 101	oxygen deficit 109
electron transport chain 101	preparatory (prep) reaction 101

Testing Yourself

Choose the best answer for each question.

1. During cellular respiration, _____ is oxidized and _____ is reduced.
   a. glucose, oxygen
   b. glucose, water
   c. oxygen, water
   d. water, oxygen
   e. oxygen, carbon dioxide

2. Cellular respiration requires _____ and _____ are produced.
   a. water and carbon dioxide; oxygen and glucose
   b. oxygen and carbon dioxide; glucose and water
   c. oxygen and glucose; carbon dioxide and water
   d. water and glucose; oxygen and carbon dioxide
   e. ATP and glucose; oxygen and water

3. During the energy-harvesting steps of glycolysis, which are produced?
   a. ATP and NADH   c. ATP and NAD$^+$
   b. ADP and NADH   d. ADP and NAD$^+$

Oxidation results in NADH and CO_2.
ATP is produced by substrate-level ATP synthesis.
Oxidation produces more NADH and $FADH_2$.

11. The strongest and final electron acceptor in the electron transp... chain is

The Chapter in Review

provides an excellent overview of the chapter concepts using concise, bulleted summaries and key illustrations.

Thinking Scientifically

questions give you an opportunity to reason as a scientist. An icon identifies questions that are related to evolution. (See Appendix A for answers.)

Thinking Scientifically

1. The compound malonate, a substrate of the citric acid cycle, can be poisonous at high concentrations because it can block cellular respiration. How might a substrate of the citric acid cycle block respiration when it is present in great excess?

2. Insulin resistance can lead to diabetes. Insulin normally causes cells to take in glucose from the blood, but insulin resistance causes cells to not respond properly to this signal. The pancreas then compensates by secreting even more insulin into the body. Researchers have found a connection between a high-fat diet, decreased mitochondrial function, and insulin resistance in some elderly patients. What might be the connection between these three events?

3. You observe that prokaryotic cells sometimes have invaginations of the plasma membrane where metabolic pathways take place. How could this observation be used to support your hypothesis that mitochondria evolved when a prokaryote cell was engulfed by a nucleated cell?

Bioethical Issue

Regulating Dietary Supplements

Americans spend millions of dollars every year on products and diets that claim to boost metabolic rates and assist in weight loss. These and other supplements are available in supermarkets, by mail order, and through the Internet. Drug makers are required to prove that a drug is safe and effective before it can be sold to the public. But the Food and Drug Administration (FDA) cannot legally supervise the supplement industry because their products are not defined as drugs. Thus, regulators must prove that a supplement is dangerous before it can be removed from the market. Although some of the more dangerous supplements, such as ephedra, have been banned, many questionable and dangerous products remain freely available to anyone who wants them. Since the FDA does not have the time or resources to perform the necessary research, most dietary supplements receive scant supervision.

Many scientists and other concerned individuals contend that FDA regulation of the supplement industry is needed to protect public safety. They argue that FDA regulation would protect the public from potentially dangerous products and ensure the efficacy of dietary supplements and diet products. However, some are concerned that regulation of the supplement industry would undermine the public interest by reducing choices in the marketplace. Furthermore, they argue that FDA regulation of supplements would create the need for additional taxes and divert resources that would be better spent regulating prescription and over-the-counter drugs.

Do you believe that the government should increase scrutiny of dietary and weight-loss supplements? Or do you believe that FDA regulation of these products would be a waste of time and money?

Testing Yourself

questions help you review material and prepare for tests. (See Appendix A for answers.)

Bioethical Issue

readings discuss a variety of controversial topics that confront our society so you can begin to understand those issues in the context of biology.

Essentials of Biology Website

provides practice tests, animations, and videos organized and integrated by chapter to help you succeed in your study of biology.

Essentials of Biology Website

The companion website for *Essentials of Biology* provides a wealth of information organized and integrated by chapter. You will find practice tests, animations, videos, and much more that will complement your learning and understanding of general biology.

http://www.mhhe.com/maderessentials2

Acknowledgments

While I guided the development of both editions of *Essentials of Biology*, many instructors have lent their talents to ensuring its increasing success. This edition, I want to particularly thank Andrew Baldwin of Mesa Community College, Rebecca Roush of Sandhills Community College, Stephanie Songer of North Georgia College and State University, and Michael Thompson of Middle Tennessee State University. My involvement ensured that each of these chapters was written and illustrated in the familiar Mader style.

Many dedicated and talented individuals assisted in the development of *Essentials of Biology*. I am very grateful for the help of so many professionals at McGraw-Hill who were involved in bringing this book to fruition. In particular, let me thank Rose Koos, the developmental editor who lent her talents and advice to all those who worked on this text. The biology publisher and editor are Janice Roerig-Blong and Michael Hackett, who steadfastly encouraged and supported this project. The project manager, Jayne Klein, faithfully and carefully steered the book through the publication process. Tamara Maury, the marketing manager, tirelessly promoted the text and educated the sales reps on its message.

The design of the book is the result of the creative talents of Laurie Janssen and many others who assisted in deciding the appearance of each element in the text. Electronic Publishing Services followed their guidelines as they created and reworked each illustration, emphasizing pedagogy and beauty to arrive at the best presentation on the page. Lori Hancock and Evelyn Jo Johnson did a superb job of finding just the right photographs and micrographs.

My office assistant, Beth Butler, worked faithfully as she helped proof the chapters and made sure all was well before the book went to press. As always, my family was extremely patient with me as I remained determined to meet every deadline on the road to publication. My husband, Arthur Cohen, is also a teacher of biology. The many discussions we have about the minutest detail to the gravest concept are invaluable to me.

The second edition of *Essentials of Biology* would not have the same excellent quality without the input of these contributors and those of the many contributors and reviewers listed below.

McGraw-Hill's 360° Development Process is an ongoing, never-ending, market-oriented approach to building accurate and innovative print and digital products. It is dedicated to continual large-scale and incremental improvement driven by multiple customer feedback loops and checkpoints. This is initiated during the early planning stages of our new products, and intensifies during the development and production stages, then begins again upon publication in anticipation of the next edition.

This process is designed to provide a broad, comprehensive spectrum of feedback for refinement and innovation of our learning tools, for both student and instructor. The 360° Development Process includes market research, content reviews, course- and product-specific symposia, accuracy checks, and art reviews. We appreciate the expertise of the many individuals involved in this process.

Ancillary Authors

Instructor's Manual
Farshad Tamari, *Kean University*

Practice Tests
Josh Dobkins and Denise Slayback-Barry, *Keiser University*

Media Asset Correlations
Jennifer Burtwistle, *Northeast Community College*

Test Bank
Andrew Baldwin, *Mesa Community College*

Laurie Faber-Foster, *Grand Rapids Community College*

Sarah Krajewski, *Grand Rapids Community College*

Rebecca Roush, *Sandhills Community College*

Cara Schillington, *Eastern Michigan University*

Stephanie Songer, *North Georgia College and State University*

Michael Thompson, *Tennessee State University*

Lecture Outlines/Image PowerPoints
Brenda Leady, *University of Toledo*

BioInteractive Questions
Jennifer Herzog, Herkimer County Community College

Second Edition Reviewers

Eugene L. Bass
University of Maryland Eastern Shore

Bill Radley Bassman
Stern College

Loren A. Bertocci
Salem International University

Barbara Boss
Keiser University

Patty Bostwick Taylor
Florence Darlington Technical College

Anthony Botyrius
York College of Pennsylvania

Ben F. Brammell
Morehead State University

Mimi Bres
Prince George's Community College

Carol Britson
University of Mississippi

Nikolle L. Brown
State Center Community College District, Madera Center

Matthew Burnham
Jones County Junior College

Claire M. Carpenter
Yakima Valley Community College

Steven Christenson
Brigham Young University Idaho

Julie Clements
Keiser University

Anthony Cooper
Albany State University

Laurie-Ann Crawford
Hawkeye Community College

Peter A. Daempfle
SUNY College of Technology at Delhi

Debbie Dalrymple
Thomas Nelson Community College

Chris Davison
Long Beach City College

Jean DeSaix
University of North Carolina at Chapel Hill

Josh Dobkins
Keiser University

Li'anne Drysdale
Ozarks Technical Community College, Richwood Valley Facility

Pamela Faber, D.A.
Heidelberg College

Farshad Tamari
Kean University

Jeffry Fasick
Kean University

Lynn Firestone
Brigham Young University, Idaho

Diane Fritz
Northern Kentucky University

Wendy Garrison
University of Mississippi

Norma Golovoy
Schoolcraft College

David M. Gordman
Pittsburg State

Tamar Goulet
University of Mississippi

Colleen Hatfield
California State University, Chico

Jennifer A. Herzog
Herkimer County Community College

Elizabeth Hodgson
York College of Pennsylvania

Mary Ruth Hood
Tomball College

George A. Jacob
Xavier University

Ashok Jain
Albany State University

Scott Johnson
Wake Technical Community College

Gregory A. Jones
Santa Fe Community College

Mary King Kananen
Penn State Altoona

Brenda Kelley
Alpena Community College

David Knowles
East Carolina University

Eliot Krause
Seton Hall University

Ruhul H. Kuddus
Utah Valley University

Brenda Leady
University of Toledo

Debabrata Majumdar
Norfolk State University

Lisa Maranto
Prince George's Community College

Rob McCandless
Methodist University

Michael McCarthy
Eastern Arizona College

Quintece M. McCrary
University of Maryland–Eastern Shore

John Mishler
Delaware Valley College

Sarah E. Myer
Oral Roberts University

Joanna Padolina
Virginia Commonwealth University

Kathleen Pelkki
Saginaw Valley State University

Mary Perry
Keiser University

Patrick K. Pfaffle
Carthage College

Megan Pickering Stringer
Jones County Junior College

Rongsun Pu
Kean University

Denice Robertson
Northern Kentucky University

Carolina Ross
Laramie County Community College

Barbara J. Salvo
Carthage College

Susan Sawyer
Kellogg Community College

John Richard Schrock
Emporia State University

James M. Schwartz
Front Range Community College

Marilyn Shopper
Johnson County Community College

William Simcik
Tomball College, NHMCCD

Robert Singer
New Jersey City University

Denise Slayback-Barry
Keiser University

Kathryn Sloan Ponnock
Delaware Valley College

Bo Sosnicki
Piedmont Virginia Community College

Chebium Subrahmanyam
Florida A & M University

Marshall D. Sundberg
Emporia State University

Karen S. Teitelbaum
State University of NY College of Technology at Delhi

Kip R. Thompson
Ozarks Technical Community College

Kevin R. Toal
St. Louis Community College at Florissant Valley

Kate Warpeha
University of Illinois at Chicago

Stan Willenbring
Dabney S Lancaster Community College

Heather Wilson-Ashworth
Utah Valley University

W. Brooke Yeager III
Luzerne County Community College

First Edition Reviewers

Sylvester Allred
Northern Arizona University

Paul E. Arriola
Elmhurst College

Tammy Atchison
Pitt Community College

James S. Backer
Daytona Beach Community College

Gail F. Baker
LaGuardia Community College

Sirakaya Beatrice
Pennsylvania State University

Carla Bundrick Benejam
California State University–Monterey Bay

Charles L. Biles
East Central University

Donna H. Bivans
Pitt Community College

Steven G. Brumbaugh
Greenriver Community College

Neil Buckley
SUNY–Plattsburgh

Nancy Butler
Kutztown University

Michelle Cawthorn
Georgia Southern University

Van D. Christman
Brigham Young University

Genevieve C. Chung
Broward Community College

Kimberly Cline-Brown
University of Northern Iowa

Mary C. Colavito
Santa Monica College

Mark A. Coykendall
College of Lake County

Don C. Dailey
Austin Peay State University

Cathy A. Davison
Empire State College

Bonnie L. Dean
West Virginia State University

William R. DeMott
Indiana-Purdue University–Fort Wayne

Amy Stinnett Dewald
Eureka College

Lee C. Drickamer
Northern Arizona University

Marie D. Dugan
Broward Community College

James W. DuMond, Jr.
Texas State University

Kathryn A. Durham
Lorain County Community College

Andrew R. Dyer
University of South Carolina–Aiken

Steven E. Fields
Winthrop University

Lynn Firestone
Brigham Young University–Idaho

Susan Fisher
Ohio State University

Edison R. Fowlks
Hampton University

Dennis W. Fulbright
Michigan State University

Ron Gaines
Cameron University

John R. Geiser
Western Michigan University

Beatriz Gonzalez
Santa Fe Community College

Andrew Goyke
Northland College

Richard Gringer
Augusta State University

Lonnie J. Guralnick
Western Oregon University

William F. Hanna
Massasoit Community College

Lisa K. Johansen
University of Colorado at Denver

Ragupathy Kannan
University of Arkansas–Fort Smith

Arnold Karpoff
University of Louisville

Darla E. Kelly
Orange Coast College

Elaine B. Kent
California State University–Sacramento

Scott L. Kight
Montclair State University

Kristin Lenertz
Black Hawk College

Melanie Loo
California State University–Sacramento

Michelle Malott
Minnesota State University–Moorhead

Paul Mangum
Midland College

Mara Manis
Hillsborough Community College

Karen Benn Marshall
Montgomery College–Takoma Park

Cynthia Conaway Mauroidis
Northwest State Community College

Elizabeth McPartlan
De Anza College

Dwight Meyer
Queensborough Community College

Rod Nelson
University of Arkansas–Fort Smith

Donald J. Padgett
Bridgewater State College

Tricia L. Paramore
Hutchinson Community College

Brian K. Paulson
California University of Pennsylvania

Debra K. Pearce
Northern Kentucky University

Lisa Rapp
Springfield Technical Community College

Jill Raymond
Rock Valley Community College

Cara Shillington
Eastern Michigan University

Lee Sola
Glendale Community College

John D. Sollinger
Southern Oregon University

Andrew Storfer
Washington State University

Janis G. Thompson
Lorain County Community College

Briana Timmerman
University of South Carolina

James R. Triplett
Pittsburg State University

Paul Twigg
University of Nebraska–Kearney

Garland Rudolph Upchurch, Jr.
Texas State University, San Marcos

James A. Wallis II
St. Petersburg College, Tarpon Springs Campus

Cosima B. Wiese
College Misericordia

Melissa Zwick
Longwood University

General Biology Symposia

Every year McGraw-Hill conducts several General Biology Symposia, which are attended by instructors from across the country. These events are an opportunity for editors from McGraw-Hill to gather information about the needs and challenges of instructors teaching non-majors level biology courses. They also offer a forum for the attendees to exchange ideas and experiences with colleagues they might not have otherwise met. The feedback we have received has been invaluable, and has contributed to the development of *Essentials of Biology*, Second Edition and its ancillaries.

Norris Armstrong
University of Georgia

David Bachoon
Georgia College and State University

Sarah Bales
Moraine Valley Community College

Lisa Bellows
North Central Texas College

Joressia Beyer
John Tyler Community College

James Bidlack
University of Central Oklahoma

Mark Bloom
Texas Christian University

Paul Bologna
Montclair University

Linda Brandt
Henry Ford Community College

Marguerite Brickman
University of Georgia

Bradford Boyer
Suffolk County Community College

Art Buikema
Virginia Polytechnic Institute

Sharon Bullock
Virginia Commonwealth University

Raymond Burton
Germanna Community College

Nancy Butler
Kutztown University of Pennsylvania

Jane Caldwell
West Virginia University

Carol Carr
John Tyler Community College

Kelly Cartwright
College of Lake County

Rex Cates
Brigham Young University

Sandra Caudle
Calhoun Community College

Genevieve Chung
Broward Community College

Jan Coles
Kansas State University

Marian Wilson Comer
Chicago State University

Renee Dawson
University of Utah

Lewis Deaton
University of Louisiana at Lafayette

Jody DeCamilo
St. Louis Community College

Jean DeSaix
University of North Carolina at Chapel Hill

JodyLee Estrada-Duek
Pima Community College, Desert Vista

Laurie Faber-Foster
Grand Rapids Community College

Susan Finazzo
Broward Community College

Theresa Fischer
Indian River Community College

Theresa Fulcher
Pellissippi State Technical College

Dennis Fulbright
Michigan State University

Steven Gabrey
Northwestern State University

Cheryl Garett
Henry Ford Community College

Farooka Gauhari
University of Nebraska–Omaha

John Geiser
Western Michigan University

Cindy Ghent
Towson University

Julie Gibbs
College of DuPage

William Glider
University of Nebraska, Lincoln

Christopher Gregg
Louisiana State University

Carla Guthridge
Cameron University

Bob Harms
St. Louis Community College, Meramec

Wendy Hartman
Palm Beach Community College

Tina Hartney
California State Polytechnic University

Kelly Hogan
University of North Carolina, Chapel Hill

Eva Horne
Kansas State University

David Huffman
Texas State University, San Marcos

Shelley Jansky
University of Wisconsin, Stevens Point

Sandra Johnson
Middle Tennessee State University

Tina Jones
Shelton State Community College

Arnold Karpoff
University of Louisville

Jeff Kaufmann
Irvine Valley College

Kyoungtae Kim
Missouri State University

Michael Koban
Morgan State University

Todd Kostman
University of Wisconsin Oshkosh

Steven Kudravi
Georgia State University

Nicki Locascio
Marshall University

Dave Loring
Johnson County Community College

Janice Lynn
Alabama State University

Phil Mathis
Middle Tennessee State University

Mary Victoria McDonald
University of Central Arkansas

Susan Meiers
Western Illinois University

Daryl Miller
Broward Community College, South Campus

Marjorie Miller
Greenville Technical College

Meredith Norris
University of North Carolina at Charlotte

Mured Odeh
South Texas College

Nathan Olia
Auburn University, Montgomery

Rodney Olsen
Fresno City College

Alexander Olvido
Virginia State University

Clark Ovrebo
University of Central Oklahoma

Forrest Payne
University of Arkansas at Little Rock

Nancy Pencoe
University of West Georgia

Murray P. "Pat" Pendarvis
Southeastern Louisiana University

Jennie Plunkett
San Jacinto College

Scott Porteous
Fresno City College

David Pylant
Wallace State Community College

Fiona Qualls
Jones County Junior College

Eric Rabitoy
Citrus College

Karen Raines
Colorado State University

Kirsten Raines
San Jacinto College

Jill Reid
Virginia Commonwealth University

Darryl Ritter
Okaloosa-Walton College

Chris Robinson
Bronx Community College

Robin Robison
Northwest Mississippi Community College

Vickie Roettger
Missouri Southern State University

Bill Rogers
Ball State University

Vicki Rosen
Utah State University

Kim Sadler
Middle Tennessee State University

Cara Shillington
Eastern Michigan University

Greg Sievert
Emporia State University

Jimmie Sorrels
Itawamba Community College

Judy Stewart
Community College of Southern Nevada

Julie Sutherland
College of DuPage

Bill Trayler
California State University–Fresno

Linda Tyson
Santa Fe Community College

Eileen Underwood
Bowling Green State University

Heather Vance-Chalcraft
East Carolina University

Marty Vaughan
IUPUI, Indianapolis

Paul Verrell
Washington State University

Thomas Vogel
Western Illinois University

Brian Wainscott
Community College of Southern Nevada

Jennifer Warner
University of North Carolina, Charlotte

Scott Wells
Missouri Southern State University

Lan Xu
South Dakota State University

Robin Whitekiller
University of Central Arkansas

Allison Wiedemeier
University of Illinois, Columbia

Michael Windelspecht
Appalachian State University

Mary Wisgirda
San Jacinto College, South Campus

Tom Worcester
Mount Hood Community College

Frank Zhang
Kean University

Michelle Zjhra
Georgia Southern University

Victoria Zusman
Miami Dade College

Supplements

Dedicated to providing high-quality and effective supplements for instructors and students, the following supplements were developed for *Essentials of Biology,* Second Edition.

For Instructors

Laboratory Manual

The ***Essentials of Biology Laboratory Manual*** is written by Dr. Sylvia Mader. With few exceptions, each chapter in the text has an accompanying laboratory exercise in the manual. Every laboratory has been written to help students learn the fundamental concepts of biology and the specific content of the chapter to which the lab relates, and to gain a better understanding of the scientific method.

 ### Connect

McGraw-Hill Connect Biology is a web-based assignment and assessment platform that gives students the means to better connect with their coursework, with their instructors, and with the important concepts that they will need to know for success now and in the future. With Connect Biology, instructors can deliver assignments, quizzes and tests easily online. Students can practice important skills at their own pace and on their own schedule. With Connect Biology Plus, students also get 24/7 online access to an eBook—an online edition of the text—to aid them in successfully completing their work, wherever and whenever they choose.

Companion Website

www.mhhe.com/maderessentials2

The companion website contains the following resources for instructors:

- ***Presentation Tools*** Everything you need for outstanding presentations in one place! This easy-to-use table of assets includes
  - *Enhanced Image PowerPoints*—including every piece of art that has been sized and cropped specifically for superior presentations as well as labels that you can edit. Also included are tables, photographs, and unlabeled art pieces.
  - *Animation PowerPoints*—Numerous full-color animations illustrating important processes are also provided. Harness the visual impact of concepts in motion by importing these files into classroom presentations or online course materials.
  - *Lecture PowerPoints*—with animations fully embedded

- *Labeled and Unlabeled JPEG Images*—Full-color digital files of all illustrations that can be readily incorporated into presentations, exams, or custom-made classroom materials.

- ***Presentation Center*** In addition to the images from your book, this online digital library contains photos, artwork, animations, and other media from an array of McGraw-Hill textbooks that can be used to create customized lectures, visually enhanced tests and quizzes, compelling course websites, or attractive printed support materials. All assets are copyrighted by McGraw-Hill Higher Education, but can be used by instructors for classroom purposes.

- ***Instructor's Manual*** The instructor's manual contains chapter outlines, lecture enrichment ideas, and critical thinking questions.

- ***Computerized Test Bank*** A comprehensive bank of test questions is provided within a computerized test bank powered by McGraw-Hill's flexible electronic testing program EZ Test Online. EZ Test Online allows you to create paper and online tests or quizzes in this easy to use program! A new tagging scheme allows you to sort questions by difficulty level, topic, and section. Imagine being able to create and access your test or quiz anywhere, at any time, without installing the testing software. Now, with EZ Test Online, instructors can select questions from multiple McGraw-Hill test banks or author their own, and then either print the test for paper distribution or give it online.

Test Creation
- Author/edit questions online using the 14 different question type templates
- Create question pools to offer multiple versions online— great for practice
- Export your tests for use in WebCT, Blackboard, PageOut, and Apple's iQuiz
- Sharing tests with colleagues, adjuncts, TAs is easy

Online Test Management
- Set availability dates and time limits for your quiz or test
- Assign points by question or question type with dropdown menu
- Provide immediate feedback to students or delay feedback until all finish the test
- Create practice tests online to enable student mastery
- Your roster can be uploaded to enable student self-registration

Online Scoring and Reporting

- Automated scoring for most of EZ Test's numerous question types
- Allows manual scoring for essay and other open-response questions
- Manual rescoring and feedback are also available
- EZ Test's grade book is designed to easily export to your grade book
- View basic statistical reports

Support and Help

- Flash tutorials for getting started on the support site
- Support Website: www.mhhe.com/eztest
- Product specialist available at 1-800-331-5094
- Online Training: http://auth.mhhe.com/mpss/workshops

Go to www.mhhe.com/maderessentials2 to learn more.

McGraw-Hill: Biology Digitized Video Clips

ISBN (13) 978-0-312155-0
ISBN (10) 0-07-312155-X

McGraw-Hill is pleased to offer an outstanding presentation tool to text adopting instructors—digitized biology video clips on DVD! Licensed from some of the highest-quality science video producers in the world, these brief segments range from about five seconds to just under three minutes in length and cover all areas of general biology from cells to ecosystems. Engaging and informative, McGraw-Hill's digitized videos will help capture students' interest while illustrating key biological concepts and processes such as mitosis, how cilia and flagella work, and how some plants have evolved into carnivores.

Student Response System

Wireless technology brings interactivity into the classroom or lecture hall. Instructors and students receive immediate feedback through wireless response pads that are easy to use and engage students. This system can be used by instructors to take attendance, administer quizzes and tests, create a lecture with intermittent questions, manage lectures and student comprehension through the use of the grade book, and integrate interactivity into their PowerPoint presentations.

For Students

Companion Website

www.mhhe.com/maderessentials2

The Mader: *Essentials of Biology* companion website is an electronic study system that offers students a digital portal of knowledge. Students can readily access a variety of digital learning objects that include:

- Chapter-level quizzing with pretest and posttest
- Bio Tutorial animations with quizzing
- Vocabulary flashcards
- Virtual Labs

Biology Prep, also available on the companion website, helps students to prepare for their upcoming coursework in biology. This website enables students to perform self assessments, conduct self-study sessions with tutorials, and perform a post-assessment of their knowledge in the following areas:

- Introductory Biology Skills
- Basic Math Review I and II
- Chemistry
- Metric System
- Lab Reports and Referencing

Electronic Books

Electronic books are an innovative way for students to save money and help create a greener environment at the same time! By purchasing eBooks from McGraw-Hill, students can save on selected titles delivered on the most advanced eBook platform available.

They also save time—every eBook comes with a powerful search engine that allows students to quickly find the topics needed to study. Contact your McGraw-Hill sales representative to discuss eBook packaging options.

How to Study Science

ISBN (13) 978-0-07-234693-0
ISBN (10) 0-07-234693-0

This workbook offers students helpful suggestions for meeting the considerable challenges of a science course. It gives practical advice on such topics as how to take notes, how to get the most out of laboratories, and how to overcome science anxiety.

Photo Atlas for General Biology

ISBN (13) 978-0-07-284610-2
ISBN (10) 0-07-284610-0

Atlas was developed to support our numerous general biology titles. It can be used as a supplement for a general biology lecture or laboratory course.

Contents

PART VII Ecology

1

A View of Life

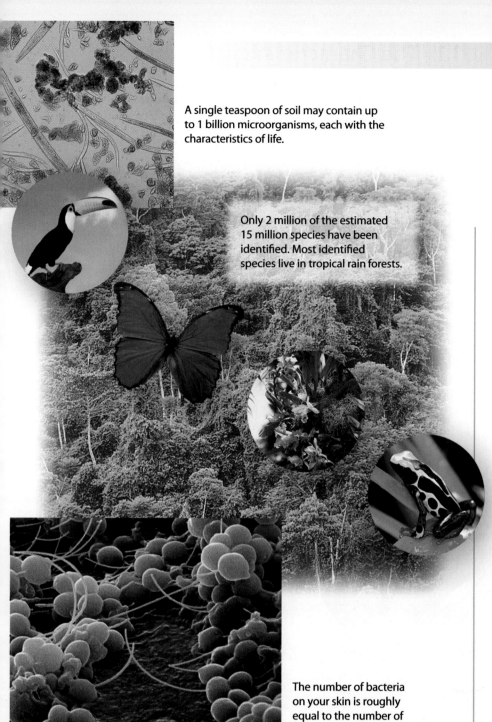

A single teaspoon of soil may contain up to 1 billion microorganisms, each with the characteristics of life.

Only 2 million of the estimated 15 million species have been identified. Most identified species live in tropical rain forests.

The number of bacteria on your skin is roughly equal to the number of humans on Earth.

Outline

Your Skin Is Their World

IT IS HARD TO BELIEVE that the number of bacterial organisms living on your skin right now is greater than the number of human organisms living on this planet. How can more than 6 billion individual organisms live on the surface of your skin? The answer is that bacterial cells are extremely small—so small that we tend not to think of them. Because most people cringe at the thought of "germs," a stroll through just about any aisle of your local grocery store will reveal a slew of products that claim to be antibacterial. Advertisers tend to make us think that all bacteria

—Continued next page

Continued—

are harmful. But only a handful of bacterial species are actually dangerous to humans, while the overwhelming majority are beneficial. The billions of bacteria that live on your skin right now are termed normal flora (or resident bacteria), and they essentially occupy all the available space on your skin surface. Should a detrimental bacterium come along, it will literally have no room to settle and multiply. Because the normal flora protect you in this way, we certainly don't want to do away with them.

The bacterial cells occupying your skin have a lot in common with the cells that make up your body. This chapter introduces the common characteristics, the diversity, and the organization of living things, as well as explaining how scientists perform their studies.

1.1 | The Characteristics of Life

From huge menacing sharks to minuscule exotic orchids, life is very diverse. Despite life's diversity, all living things, also called **organisms,** share certain characteristics, and these characteristics give us insight into the nature of life and help distinguish organisms from nonliving things.

Living Things Are Organized

The complex organization of living things begins with small molecules that join to form larger molecules within a **cell,** the smallest, most basic unit of life. Although a cell is alive it is made from nonliving molecules.

Some organisms are **unicellular,** being only a single cell. Plants and animals are **multicellular** and are composed of many types of cells. In multicellular organisms, similar cells combine to form **tissues.** Tissues make up **organs,** as when various tissues combine to form a heart or a leaf. Organs work together in **organ systems;** for example, the heart and blood vessels form the cardiovascular system. Various organ systems work together within complex organisms (**Fig. 1.1**).

Later in this chapter, we will see that levels of biological organization extend beyond the individual organism. For example, all the members of one species in a particular area belong to a population. Zebras form one of the many populations on an African plain. All the populations make up a community. The community of populations interacts with the physical environment to form an ecosystem. Finally, all the Earth's ecosystems make up the biosphere.

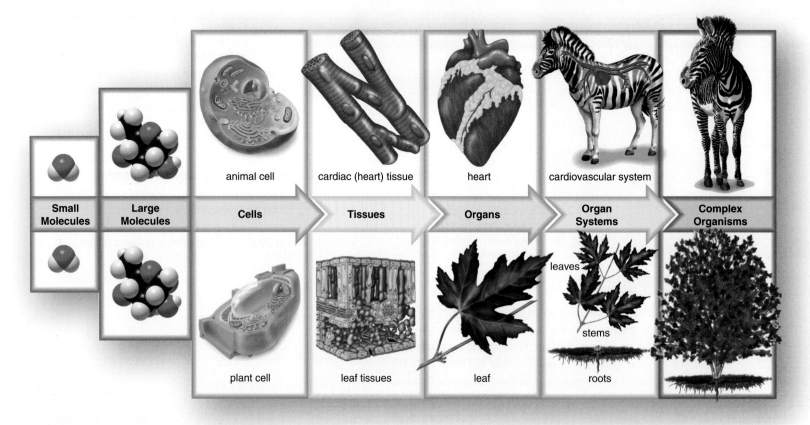

Figure 1.1 **Levels of biological organization.**

Living Things Acquire Materials and Energy

Living things cannot maintain their organization or carry on life's activities without an outside source of materials and energy (**Fig. 1.2**). Food provides nutrient molecules, which are used as building blocks or energy sources. **Energy** is the capacity to do work, and it takes work to maintain the organization of the cell and the organism. When cells use nutrient molecules to make their parts and products, they carry out a sequence of chemical reactions. The term **metabolism** encompasses all the chemical reactions that occur in a cell.

The ultimate source of energy for nearly all life on Earth is the sun. Plants and certain other organisms are able to capture solar energy and carry on **photosynthesis,** a process that transforms solar energy into the chemical energy of nutrient molecules. Animals and plants get energy by metabolizing nutrient molecules made by photosynthesizers.

Remaining Homeostatic

For metabolic processes to continue, living things need to keep themselves stable in temperature, moisture level, acidity, and other factors critical to maintain life. This is **homeostasis**—the maintenance of internal conditions within certain boundaries.

Many organisms depend on behavior to regulate their internal environment. A chilly lizard may raise its internal temperature by basking in the sun on a hot rock. When it starts to overheat, it scurries for cool shade. Other organisms have control mechanisms that do not require any conscious activity. When a student is so engrossed in her textbook that she forgets to eat lunch, her liver releases stored sugar to keep the blood sugar level within normal limits. Hormones regulate sugar storage and release, but in other instances the nervous system is involved in maintaining homeostasis.

Living Things Respond

Living things find energy and/or nutrients by interacting with their surroundings. Even unicellular organisms can respond to their environment. The beating of microscopic hairs or the snapping of whip-like tails moves them toward or away from light or chemicals. Multicellular organisms can manage more complex responses. A monarch butterfly can sense the approach of fall and begin its flight south where resources are still abundant. A vulture can smell meat a mile away and soar toward dinner.

The ability to respond often results in movement: The leaves of a plant turn toward the sun, and animals dart toward safety. Appropriate responses help ensure survival of the organism and allow it to carry on its daily activities. Altogether, we call these activities the *behavior* of the organism.

Living Things Reproduce and Develop

Life comes only from life. Every type of living thing can **reproduce,** or make another organism like itself. Bacteria and other types of unicellular organisms simply split in two. In multicellular organisms, the reproductive process usually begins with the pairing of a sperm from one

Figure 1.2 Acquiring nutrient materials and energy.

a. Osprey, a type of hawk, eating a fish. **b.** Humans harvesting crops.

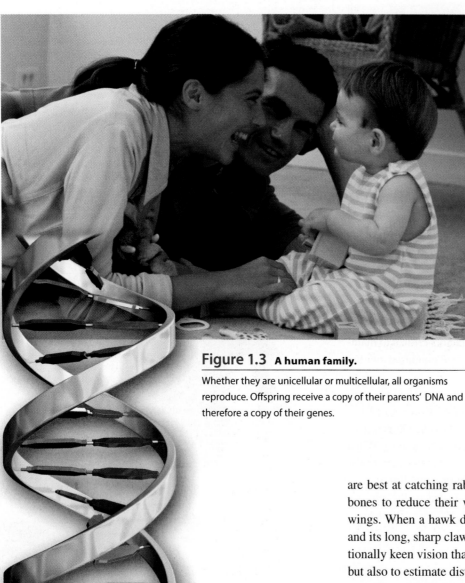

DNA

Figure 1.3   A human family.

Whether they are unicellular or multicellular, all organisms reproduce. Offspring receive a copy of their parents' DNA and therefore a copy of their genes.

partner and an egg from the other partner. The union of sperm and egg, followed by many cell divisions, results in an immature individual, which grows and develops through various stages to become an adult.

An embryo develops into a whale or a yellow daffodil or a human being because of the specific set of genes inherited from its parents (**Fig. 1.3**). In all organisms, the **genes** are made of long DNA (deoxyribonucleic acid) molecules, but even so the genes are different between species and individuals. Genetic comparisons are the basis of paternity testing. Although an individual's overall genetic makeup is unique, it consists of DNA from both parents. Thus, the DNA profiles of a child and his or her biological parents will be measurably similar. DNA provides the blueprint or instructions for the organization and metabolism of the particular organism. All cells in a multicellular organism contain the same set of genes, but only certain ones are turned on in each type of specialized cell.

Living Things Have Adaptations

Adaptations are modifications that make organisms suited to their way of life. Some hawks have the ability to catch fish (see Fig. 1.2), and others are best at catching rabbits. Hawks can fly in part because they have hollow bones to reduce their weight and flight muscles to depress and elevate their wings. When a hawk dives, its strong feet take the first shock of the landing, and its long, sharp claws reach out and hold onto the prey. Hawks have exceptionally keen vision that enables them not only to spot prey from great heights but also to estimate distance and speed.

Penguins look very different from hawks, although they are both birds. Penguins are adapted to an aquatic existence in the Antarctic. While birds like a hawk have forelimbs proportioned for flying, penguins have stubby flattened wings suitable for swimming. Their feet and tails serve as rudders in the water, but the flat feet also allow them to walk on land. Penguins live where it is cold. An extra layer of downy feathers is covered by short, thick feathers that form a waterproof coat. Layers of blubber also keep the birds warm in cold water.

The manner in which species become adapted to their environment is discussed in the next section of this chapter. The members of a species have a similar structure and, if sexually reproducing, are capable of interbreeding.

Check Your Progress

1. Define unicellular and multicellular.
2. List the levels of organization of a multicellular organism, beginning with molecules.
3. List five features common to organisms.

Answers: 1. Unicellular refers to one cell, and multicellular indicates many cells.   2. molecules–cells–tissues–organs–organ systems–organism.   3. Organisms are organized, acquire materials and energy, respond to external stimuli, reproduce and develop, and have adaptations.

1.2 | Evolution: The Core Concept of Biology

From bacteria to bats, toadstools to trees, whip-poor-wills to whales—the diversity of the living world boggles the mind. Yet all organisms are united by a common heritage that began during the early years of our planet. Just as you are descended from your parents, grandparents, and so forth from generation to generation back in time, all species can trace their history to the original living thing. **Evolution,** the process by which species have changed and diversified since life arose, explains the unity—all living things share the same characteristics—and the diversity of life. If you have ever studied a family tree, then you know it is a diagram used to show how individuals are related to one another and descended from common ancestors. An evolutionary tree is very much like a family tree, only it depicts relationships between groups of organisms. The evolutionary tree in **Figure 1.4** illustrates the lineages of the major groups of living things. The tree summarizes the history of life on Earth over the approximately 3.5 billion years that have passed since the first, ancestral cell. All the different groups of organisms on our planet are related to one another, and are represented by branches on the same "family tree of life."

The Diversity of Life

Think of an enormous department store, offering thousands of different items for sale. The various items are grouped in departments—electronics, apparel, furniture, and so on—to make them easy for customers to find. Because life is so diverse, it is helpful to have a system that groups organisms into categories (see Appendix B). **Taxonomy** is the discipline of identifying and classifying organisms according to certain rules. Taxonomy makes sense out of the bewildering variety of life on Earth by classifying organisms according to their presumed evolutionary relationship. As more is learned about evolutionary

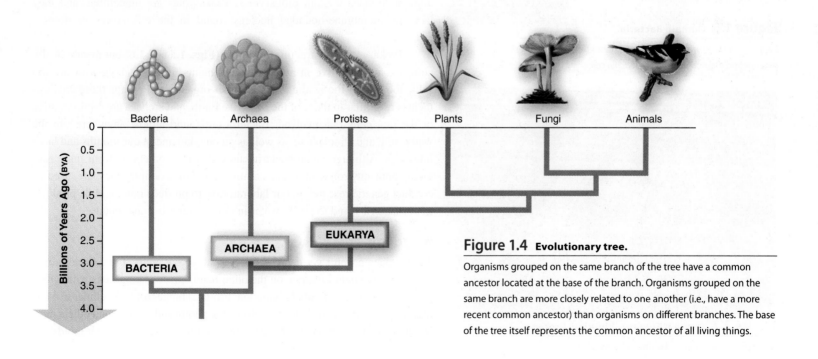

Figure 1.4 Evolutionary tree.

Organisms grouped on the same branch of the tree have a common ancestor located at the base of the branch. Organisms grouped on the same branch are more closely related to one another (i.e., have a more recent common ancestor) than organisms on different branches. The base of the tree itself represents the common ancestor of all living things.

×20,000

×64,000

Figure 1.5   **Domain Archaea.**

Archaea are capable of living in extreme environments. These are exterior and interior views of *Methanosarcina mazei.*

×6,600

×65,000

Figure 1.6   **Domain Bacteria.**

Bacteria are structurally simple but metabolically diverse. These are exterior and interior views of *Escherichia coli.*

×150

Figure 1.7   **Domain Eukarya, kingdom Protista.**

Many protists are unicellular. This is *Euglena.*

relationships between species, taxonomy changes. Taxonomists are even now making observations and performing experiments that will one day bring about changes in the classification system adopted by this text.

Categories of Classification

The classification categories, going from least inclusive to most inclusive, are **species, genus, family, order, class, phylum, kingdom,** and **domain.** Each successive category above species contains more types of organisms than the preceding one. Species placed within one genus share many specific characteristics and are the most closely related, while species placed in the same domain share only general characteristics. For example, all species in the genus *Pisum* look pretty much the same—that is, like pea plants—but species in the plant kingdom can be quite varied, as is evident when we compare grasses to trees. By the same token, only modern humans are in the genus *Homo,* but many types of species, from tiny hydras to huge whales, are members of the animal kingdom. Species placed in different domains are the most distantly related.

Scientific Naming

Biologists give each living thing a two-part scientific name called a **binomial name.** For example, the scientific name for the garden pea is *Pisum sativum;* our own species is *Homo sapiens.* The first word is the genus, and the second word is the specific epithet of a species within a genus. The genus may be abbreviated (e.g., *P. sativum* or *H. sapiens*). Scientific names are universally used by biologists to avoid confusion. Common names tend to overlap, and often they are in the language of the people who use that particular name. But scientific names are based on Latin, a universal language that not too long ago was well known by most scholars.

Domains

Biochemical evidence suggests that there are only three domains: **domain Bacteria, domain Archaea,** and **domain Eukarya.** Both domain Bacteria and domain Archaea contain prokaryotes. Prokaryotes are unicellular, and they lack the membrane-bounded nucleus found in the eukaryotes of domain Eukarya.

Prokaryotes are structurally simple (**Figs. 1.5** and **1.6**) but metabolically complex. Archaea live in aquatic environments that lack oxygen or are too salty, too hot, or too acidic for most other organisms. Perhaps these environments are similar to those of the primitive Earth, and archaea are representative of the first cells that evolved. Bacteria are found almost anywhere—in the water, soil, and atmosphere, as well as on our skin and in our mouths and large intestines. Although some bacteria cause diseases, others perform many services, both environmental and commercial. For example, they are used to conduct genetic research in our laboratories, to produce innumerable products in our factories, and to purify water in our sewage treatment plants.

Kingdoms

Taxonomists haven't yet decided how to categorize archaea and bacteria into kingdoms. Domain Eukarya, on the other hand, has four kingdoms with which you may be familiar. **Protists** (kingdom Protista) range from unicellular to a few multicellular organisms (**Fig. 1.7**). Some are photosynthesizers, and others must ingest their food. Among the **fungi** (kingdom Fungi) are the familiar molds and mushrooms that, along with many types of bacteria, help decompose dead organisms (**Fig. 1.8**). **Plants** (kingdom Plantae) are well known as multicellular photosynthesizers (**Fig. 1.9**). **Animals** (kingdom Animalia) are multicellular organisms that ingest their food (**Fig. 1.10**). The three domains and the four kingdoms within the domain Eukarya are depicted in the evolutionary tree in Figure 1.4. This tree which is largely based on the DNA of organisms says that domain Bacteria was the first to arise in the history of life, followed by the domain Archaea and finally domain Eukarya. The domain Archaea is more closely related to domain Eukarya than either is to domain Bacteria. Among the Eukarya, the protists gave rise to the kingdoms plants, fungi, and animals. Later in this text we will return to this evolutionary tree and the evolution of each kingdom will be discussed separately. Now that we know how scientists group organisms, let's return to our discussion of evolution and the process by which diversity arises.

Natural Selection

In the nineteenth century, two naturalists—Charles Darwin and Alfred Russell Wallace—came to the conclusion independently that evolution occurs by means of a process called **natural selection.** Charles Darwin is the more famous of the two because he wrote a book called *On the Origin of Species,* which presented much data to substantiate the occurrence of evolution by natural selection. Since that time evolution has become the core concept of biology because the theory explains so many different types of observations in every field of biology. The study of evolution encompasses all levels of biological organization. Indeed, much of today's evolution research is carried out at the molecular level, comparing the DNA of different groups of organisms to determine how they are related. Looking at how life has changed over time, from its origin to the current day, helps us understand why there are so many different kinds of organisms, and why they have the characteristics they do. Our growing knowledge of evolution by natural selection also has practical applications including the prevention and treatment of disease.

Natural selection is a process in which steps 1–3 result in a population adapted to the environment (4):

1. The members of a population have heritable variations: differences in their DNA that can be passed from one generation to the next.
2. The population produces more offspring than the resources of an environment can support.
3. There is competition for resources, and as a result the more adapted individuals survive and reproduce to a greater extent than those that lack the adaptations.
4. Across generations, a larger proportion of the population becomes adapted to the environment.

One example that Darwin used concerned giraffes, which are browsers feeding on tree leaves. Imagine an early population of giraffes in which the neck length varied from short to long. Giraffes with long necks had an advantage over those with short necks because they could reach more food. The ability to capture more food allowed long-neck giraffes to have more offspring than short-neck giraffes. Therefore, slowly over many generations giraffes became exclusively long-necked animals.

Figure 1.8 **Domain Eukarya, kingdom Fungi.**

Fungi are multicellular and break down organic debris. This is an *Amanita* mushroom.

Figure 1.9 **Domain Eukarya, kingdom Plantae.**

Plants are multicellular photosynthesizers. These are roses.

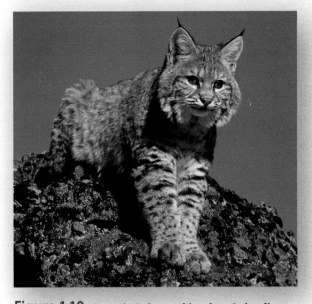

Figure 1.10 **Domain Eukarya, kingdom Animalia.**

Animals are multicellular and ingest their food. This is a bobcat.

Today, we know that because of selection, resistance to antibiotic drugs has become increasingly common in a number of bacterial species, including those that cause tuberculosis, gonorrhea, and staph infections. Antibiotic drugs such as penicillin kill susceptible bacteria. However, some bacteria in the body of a patient undergoing antibiotic treatment may be unharmed by the drug. Bacteria can survive antibiotic drugs in many different ways. For example, certain bacteria can endure treatment with penicillin because they break down the drug, rendering it harmless. If even one bacterial cell lives because it is antibiotic-resistant, then its descendants will inherit this drug-defeating ability. The more antibiotic drugs are used, the more natural selection favors resistant bacteria, and the more often antibiotic-resistant infections occur.

Descent with Modification

Darwin said that evolution is descent with modification. One species can be a common ancestor to several species, each adapted to a particular set of environmental conditions. Specific adaptations allow species to play particular roles in an ecosystem. The diversity of life-forms is best understood in terms of the many different ways in which organisms carry on their life functions within the ecosystem where they live, acquire energy, and reproduce.

The Hawaiian honeycreepers are a remarkable example of evolution (**Fig. 1.11**). The various species of honeycreepers all evolved from one species of finch which likely originated in North America and arrived in the Hawaiian Islands millions of years ago. Honeycreepers have an assortment of different bill shapes adapted to different types of food. Some honeycreeper species have curved, elongated bills used for drinking flower nectar. Others have strong, hooked bills suited to digging in tree bark and seizing wood-boring insects; or short, straight, finchlike bills for feeding on small seeds and fruits. Even with such dramatic differences in feeding habits and bill shapes, honeycreepers still share certain characteristics which stem from their common finch ancestor. The various honeycreeper species are similar in body shape and size, as well as mating and nesting behavior.

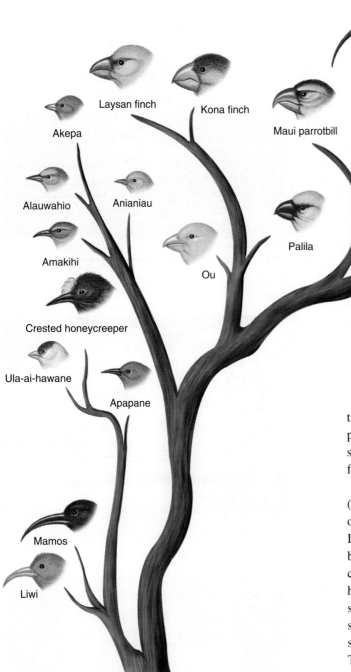

Figure 1.11 **Evolution of Hawaiian honeycreepers.**

Hawaiian honeycreepers, descendants of a single ancestral species, display an amazing diversity of bill shapes and sizes.

Check Your **Progress**

1. List the eight classification categories, from least to most inclusive.
2. List the four kingdoms in domain Eukarya.
3. Define evolution.

Answers: 1. species, genus, family, order, class, phylum, kingdom, domain.   2. Protista, Fungi, Plantae, Animalia.   3. Evolution is descent of species from common ancestors, with genetic modifications that make each species more suited to its environment.

1.3 | How the Biosphere Is Organized

The organization of life extends beyond the individual to the **biosphere,** the zone of air, land, and water at the surface of the Earth where living organisms are found. Individual organisms belong to a **population,** all the members of a species within a particular area. The populations within a **community** interact among themselves and with the physical environment (soil, atmosphere, etc.), thereby forming an **ecosystem.**

Ecosystem

One example of an ecosystem is a North American grassland, which is inhabited by populations of rabbits, hawks, and various types of grasses, among many others. These populations interact with each other by forming food chains in which one population feeds on another. For example, rabbits feed on grasses, while hawks feed on rabbits and other organisms.

As **Figure 1.12** shows, ecosystems are characterized by chemical cycling and energy flow, both of which begin when photosynthesizers, such as grasses, take in solar energy and inorganic nutrients to produce food (organic nutrients) by photosynthesis. Chemical cycling (aqua arrows) occurs as chemicals move from one population to another in a food chain, until death and decomposition allow inorganic nutrients to be returned to the photosynthesizers once again. Energy (red arrows), on the other hand, flows from the sun through plants and the other members of the food chain as they feed on one another. The energy gradually dissipates and returns to the atmosphere as heat. Because energy does not cycle, ecosystems could not stay in existence without solar energy and the ability of photosynthesizers to absorb it.

Biosphere

Climate largely determines where different ecosystems are found in the biosphere. For example, deserts exist in areas of minimal rain, while forests require much rain. The two most biologically diverse ecosystems—tropical rain forests and coral reefs—occur where solar energy is most abundant. The human population tends to modify these and all ecosystems for its own purposes. Humans clear forests or grasslands in order to grow crops; later, they build houses on what was once farmland; and finally, they convert small towns into cities. As coasts are developed, humans send sediments, sewage, and other pollutants into the sea and onto coral reefs.

Tropical rain forests, coral reefs, and most ecosystems are severely threatened as the human population increases in size. Some coral reefs are 50 million years old, and yet in just a few decades, human activities have destroyed 10% of all reefs and seriously degraded another 30%. At this rate, nearly three-quarters could be destroyed within 50 years. Similar statistics are available for tropical rain forests.

It has long been clear that human beings depend on healthy ecosystems for food, medicines, and various raw materials. We are only now beginning to realize that we depend on them even more for the services they provide. The workings of ecosystems ensure that environmental conditions are suitable for the continued existence of humans.

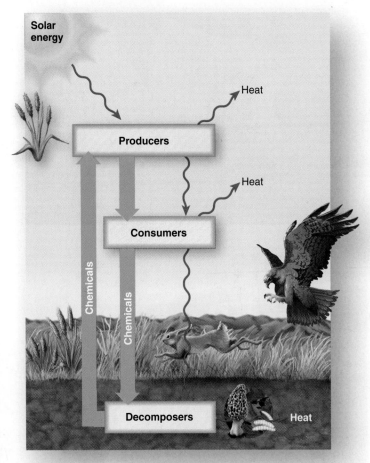

Figure 1.12   A grassland, a terrestrial ecosystem.

In an ecosystem, chemical cycling (aqua arrows) and energy flow (red arrows) begin when plants use solar energy and inorganic nutrients to produce their own food. Chemicals and energy are passed from one population to another in a food chain. Eventually, energy dissipates as heat. With the death and decomposition of organisms, chemicals are returned to living plants once more.

Check Your Progress

1. Contrast a community with an ecosystem.

Answer: 1. A community consists of different populations that interact in a particular place. An ecosystem is a community together with the physical environment with which it interacts.

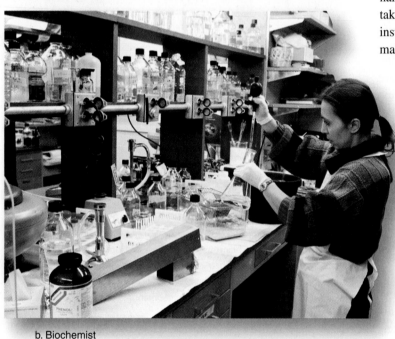

a. Botanist

b. Biochemist

c. Ecologists

1.4 | Science: A Way of Knowing

Biology, with its numerous branches, is the scientific study of life (**Fig. 1.13**). Religion, aesthetics, ethics, and science are all ways that human beings have of finding order in the natural world. Science differs from the other fields by its process, which often involves the use of the scientific method (**Fig. 1.14**).

Observation

The scientific method begins with **observations.** We can observe with our noses that dinner is almost ready, observe with our fingertips that a surface is smooth and cold, and observe with our ears that a piano needs tuning. Scientists also extend the ability of their senses by using instruments; for example, the microscope enables them to see objects that could never be seen by the naked eye. Finally, scientists may expand their understanding even further by taking advantage of the knowledge and experiences of other scientists. For instance, they may look up past studies on the Internet or at the library, or they may write or speak to others who are researching similar topics.

Hypothesis

After making observations and gathering knowledge about a phenomenon, a scientist uses inductive reasoning. **Inductive reasoning** occurs whenever a person uses creative thinking to combine isolated facts into a cohesive whole. Chance alone can help a scientist arrive at an idea. The most famous case pertains to the antibiotic penicillin, which was discovered in 1928. While examining a petri dish of bacteria that had accidentally become contaminated with the mold *Penicillium*, Alexander Fleming observed an area around the mold that was free of bacteria. Fleming had long been interested in finding cures for human diseases caused by bacteria, and was very knowledgeable about antibacterial substances. So when Fleming saw the dramatic effect of *Penicillium* mold on bacteria, he reasoned that the mold might be producing an antibacterial substance. We call such a possible explanation for a natural event a **hypothesis.** A hypothesis is based on existing knowledge, so it is much more informed than a mere guess. Fleming's hypothesis was supported by further study. Sometimes a hypothesis is not supported, and must be either modified and subjected to additional study, or rejected.

All of a scientist's past experiences, no matter what they might be, may influence the formation of a hypothesis. But a scientist only considers hypotheses that can be tested by experiments or further observations. Moral and religious beliefs, while very important to our lives, differ between cultures and through time, and are not always testable.

Figure 1.13 Biologists.

Biologists work in a variety of settings. For example, (**a**) some botanists work in greenhouses; (**b**) some biologists, such as this biochemist, work in laboratories; and (**c**) many ecologists and environmentalists collect data in the field.

Experiments/Further Observations

Scientists often perform an **experiment,** a series of procedures to test a hypothesis. The manner in which a scientist intends to conduct an experiment is called the **experimental design.** A good experimental design ensures that scientists are testing what they want to test and that their results will be meaningful. When an experiment is done in a laboratory, all conditions can be kept constant except for an **experimental variable,** which is deliberately changed. One or more **test groups** are exposed to the experimental variable, but one other group, called the **control group,** is not. If, by chance, the control group shows the same results as the test group, the experimenter knows the results are invalid.

Scientists often use a **model,** a representation of an actual object. Modeling occurs when scientists use software to decide how human activities will affect climate, or when they use mice instead of humans for, say, testing a new drug. Ideally, a medicine that is effective in mice should still be tested in humans. And whenever it is impossible to study the actual phenomenon, a model remains a hypothesis in need of testing. Someday, a scientist might devise a way to test it.

Data

The results of an experiment are referred to as the **data.** Mathematical data are often displayed in the form of a graph or table. Sometimes studies rely on statistical data. Let's say an investigator wants to know if eating onions can prevent women from getting osteoporosis (weak bones). The scientist conducts a survey asking women about their onion-eating habits and then correlates these data with the condition of their bones. Other scientists critiquing this study would want to know: How many women were surveyed? How old were the women? What were their exercise habits? What criteria were used to determine the condition of their bones? And what is the probability that the data are in error? The greater the variance in the data, the greater the probability of error. In any case, even if the data do suggest a correlation, scientists would want to know if there is a specific ingredient in onions that has a direct biochemical or physiological effect on bones. After all, correlation does not necessarily mean causation. It could be that women who eat onions eat lots of vegetables, and have healthier diets overall than women who do not eat onions. In this way scientists are skeptics who always pressure one another to keep investigating.

Conclusion

Scientists must analyze the data in order to reach a **conclusion** about whether a hypothesis is supported or not. Because science progresses, the conclusion of one experiment can lead to the hypothesis for another experiment (Fig. 1.14). In other words, results that do not support one hypothesis can often help a scientist formulate another hypothesis to be tested. Scientists report their findings in scientific journals so that their methodology and data are available to other scientists. Experiments and observations must be *repeatable*—that is, the reporting scientist and any scientist who repeats the experiment must get the same results, or else the data are suspect.

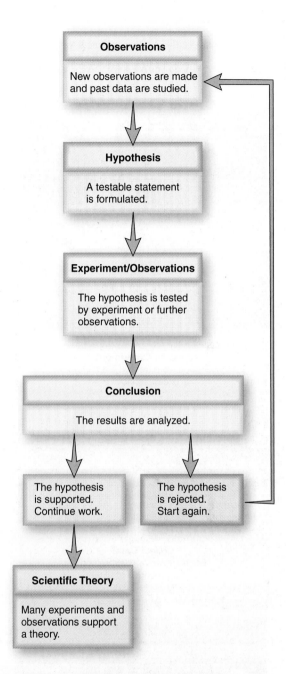

Figure 1.14 **Flow diagram for the scientific method.**

On the basis of new and/or previous observations, a scientist formulates a hypothesis. The hypothesis is tested by further experiments and/or observations, and new data either support or do not support the hypothesis. Following an experiment, a scientist often chooses to retest the same hypothesis or to test a related hypothesis. Conclusions from many different but related experiments may lead to the development of a scientific theory. For example, studies pertaining to development, anatomy, and fossil remains all support the theory of evolution.

Scientific Theory

The ultimate goal of science is to understand the natural world in terms of **scientific theories,** which are accepted explanations for how the world works. Some of the basic theories of biology are the cell theory, which says that all organisms are composed of cells; the gene theory, which says that inherited information dictates the form, function, and behavior of organisms; and the theory of evolution, which says that all organisms have a common ancestor, and each one is adapted to a particular way of life.

The theory of evolution is considered the unifying concept of biology because it pertains to many different aspects of organisms. For example, the theory of evolution enables scientists to understand the history of life, the variety of organisms, and the anatomy, physiology, and development of organisms. The theory of evolution has been a very fruitful scientific theory, meaning that it has helped scientists generate new testable hypotheses.

Because this theory has been supported by so many observations and experiments for over 100 years, some biologists refer to the **principle** of evolution, a term sometimes used for theories that are generally accepted by an overwhelming number of scientists. Others prefer the term **law** instead of principle.

How to Do a Controlled Study

Most stomach and intestinal ulcers (open sores) are caused by the bacterium *Helicobacter pylori.* Let's say investigators wanted to determine which of two antibiotics was best for the treatment of an ulcer. When clinicians do an experiment, they try to vary just the experimental variables, in this case, the medications being tested. A control group is not given the medications, but one or more test groups are given the medications. If by chance, the control group shows the same results as a test group, the investigators immediately know the results of their study are invalid because it would mean the medications may have nothing to do with the results.

The study depicted in **Figure 1.15** shows how investigators may study this hypothesis:

Hypothesis Newly discovered antibiotic B is a better treatment for ulcers than antibiotic A, which is in current use.

Next, the investigators might decide to use three experimental groups: one control group and two test groups. It is important to reduce the number of possible variables (differences) such as sex, weight, other illnesses, and so forth between the groups. Therefore, the investigators *randomly* divide a very large group of volunteers equally into the three groups. The hope is any differences will be distributed evenly among the three groups. This is possible only if the investigators have a large number of volunteers.

Figure 1.15   A controlled laboratory experiment to test the effectiveness of a medication in humans.

In this study, a large number of people were divided into three groups. The control group received a placebo and no medication. One of the test groups received medication A, and the other test group received medication B. The results are depicted in a graph, and it shows that medication B was a more effective treatment than medication A for the treatment of ulcers.

The three groups are to be treated like this:

Control group: Subjects with ulcers are not treated with either antibiotic.
Test group 1: Subjects with ulcers are treated with antibiotic A.
Test group 2: Subjects with ulcers are treated with antibiotic B.

After the investigators have determined that all volunteers do have ulcers, they will want the subjects to think they are all receiving the *same* treatment. This is an additional way to protect the results from any influence other than the medication. To achieve this end, the subjects in the control group can receive a **placebo,** a treatment that appears to be the same as that administered to the other two groups but actually contains no medication. In this study, the use of a placebo would help ensure the same dedication by all subjects to the study.

The Results

After two weeks of administering the same amount of medication (or placebo) in the same way, the stomach and intestinal lining of each subject is examined to determine if ulcers are still present. Endoscopy, a procedure depicted in the photograph, is one possible way to examine a patient for the presence of ulcers. This procedure is performed under sedation and involves inserting an endoscope—a small, flexible tube with a tiny camera on the end—down the throat and into the stomach and the upper part of the intestine. Now, the doctor can see the lining of these organs, and check for possible ulcers. Tests performed during an endoscopy can also determine if *Helicobacter pylori* is present.

Because endoscopy is somewhat subjective, it is probably best if the examiner is not aware of which group the subject is in, otherwise the prejudice of the examiner may influence the examination. When neither the patient nor the technician is aware of the specific treatment, it is called a *double-blind* study.

In this study, the investigators may decide to determine effectiveness of the medication by the percentage of people who no longer have ulcers. So, if 20 people out of 100 still have ulcers, the medication is 80% effective. The difference in effectiveness is easily read in the graph portion of Figure 1.15.

Conclusion On the basis of their data, the investigators conclude that their hypothesis has been supported.

Publication of Scientific Studies

Scientific studies are customarily published in a scientific journal (**Fig. 1.16**), so that all aspects of a study are available to the scientific community. Before information is published in scientific journals, it is typically reviewed by experts who ensure the research is credible, accurate, unbiased, and well executed. Another scientist should be able to read about an experiment in a scientific journal, repeat the experiment in a different location, and get the same (or very similar) results. Some articles are rejected for publication by reviewers when they believe there is something questionable about the design or manner in which an experiment was conducted.

As mentioned previously, the conclusion of one experiment often leads to another experiment. The need for scientists to expand on findings explains why science changes and the findings of yesterday may be improved upon tomorrow.

Figure 1.16   Scientific periodicals.

Some scientific periodicals, such as *Nature, Nature Medicine,* and *Science & Medicine,* are scholarly journals in which researchers share their findings with other scientists. Journals are different from magazines such as *Scientific American* and *Science News.* Articles in science magazines are usually written by reporters for a broad audience.

Check Your Progress

1. Explain how a scientist uses the scientific method to create a hypothesis.
2. Compare and contrast the control group with the test groups in a scientific study.

Answers: 1. First, a scientist makes observations about some phenomenon in the natural world. Then, s/he uses inductive reasoning to formulate a hypothesis, a possible explanation for the phenomenon. Finally, s/he uses the information gained to create an explanation for the phenomenon.
2. In an experiment, the control group is not exposed to the experimental variable, but the test groups are exposed to the experimental variable. Otherwise, if the control groups and the test group were exposed to identical conditions, there would be no basis for comparing the results.

1.5 | Science and Bioethical Issues

Many scientists work in the field or the laboratory collecting data and coming to conclusions that sometimes seem remote from our everyday lives. Other scientists are interested in using the findings of past and present scientists to produce a product or develop a technique that does affect our lives. The application of scientific knowledge for a practical purpose is called **technology.**

Who should decide how, and even whether, a technology is put to use? Making value judgments is not a part of science. Ethical and moral decisions must be made by all people. Therefore, the responsibility for how to use the fruits of science must reside with people from all walks of life, and not with scientists alone. **Bioethics** is the branch of ethics that is concerned with the development and consequences of biological technology. Scientists should provide the public with as much information as possible, but all citizens, including scientists, should make decisions about the use of technologies.

Biodiversity is perhaps the single most significant bioethical issue that we face today. Biodiversity may be defined as variation in life on Earth, and usually refers to the number of different species. The present number of species has been estimated to be as high as 15 million, but so far, fewer than 2 million have been identified, named, and classified. The biodiversity that surrounds us is the result of approximately 3.5 billion years of evolution. Unfortunately, technology-enabled human activities present a major threat to biodiversity. As the human population grows, more and more of the biosphere is exploited for human needs such as food and housing.

Extinction is the death of a species or a larger taxonomic group. Although it is natural for species extinctions to occur over time, human activities have increased the extinction rate by a factor of 100 to 1,000. It is estimated that as many as 400 species per day are lost because of human activities. Many biologists are alarmed about the present rate of extinction and the resulting loss of biodiversity worldwide.

Our own species may be in peril if biodiversity loss continues, because we are interconnected with all of life on Earth in ways that we do not yet fully understand. We depend on other living things to supply us with food, clothing, medicines, and construction materials. We need other species to help clean up our pollution, preserve soil fertility, and maintain the balance of gases in the atmosphere. We can protect biodiversity through the preservation of ecosystems. Just as a native fisherman who assists in overfishing a reef is doing away with his own food source, we as a society are contributing to the destruction of our home, the biosphere, when we destroy ecosystems. If instead we adopt a conservation ethic that preserves ecosystems, we are helping to ensure the continued existence of all species, including our own.

The Chapter in Review

■■■ Summary

1.1 | The Characteristics of Life

Living things, often called organisms, share several common characteristics. Organisms
- are organized (have levels of organization).
- acquire materials and energy.
- respond to external stimuli.
- reproduce and develop.
- have adaptations.

1.2 | Evolution: The Core Concept of Biology

Evolution explains the unity and diversity of life. Descent from a common ancestor explains why organisms share some characteristics, and adaptation to various ways of life explains the diversity of life-forms. An evolutionary tree diagram may be used to describe how groups of organisms are related to one another.

According to the rules of taxonomy, species belong to the following categories (from the most inclusive to the least inclusive):

Domains:
- Archaea: prokaryotes
- Bacteria: prokaryotes
- Eukarya: eukaryotes

Kingdoms in domain Eukarya:
- Protista: unicellular to multicellular organisms with various modes of nutrition
- Fungi: live on debris; molds and mushrooms
- Plantae: multicellular photosynthesizers
- Animalia: multicellular organisms that ingest food

To classify a particular organism, scientists use a consistent sequence of categories. A binomial name consists of the genus and the specific epithet. For example, here is how the names *Homo sapiens* and *Zea mays* are derived:

LEVELS OF CLASSIFICATION		
Category	**Human**	**Corn**
Domain	Eukarya	Eukarya
Kingdom	Animalia	Plantae
Phylum	Chordata	Anthophyta
Class	Mammalia	Liliopsida
Order	Primates	Commelinales
Family	Hominidae	Poaceae
Genus	Homo	Zea
*Species**	H. sapiens	Z. mays

* To specify an organism, you must use the full binomial name, such as *Homo sapiens*.

1.3 │ How the Biosphere Is Organized

Populations within a community interact with one another and with the physical environment, forming an ecosystem. With an ecosystem, chemicals cycle, but energy flows unidirectionally and eventually becomes heat.

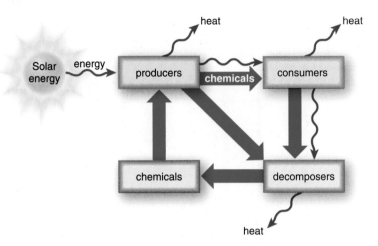

1.4 │ Science: A Way of Knowing

The scientific process includes a series of systemic steps known as the scientific method:

- Observations, which use the senses and may also include studies done by others
- A hypothesis (a statement to be tested)
- New observations and experiments
- A conclusion reached by analyzing data to determine whether the results support or do not support the hypothesis

A hypothesis confirmed by many different studies becomes known as a theory. Examples are the cell theory, the gene theory, and the theory of evolution, which is the unifying concept of biology.

1.5 │ Science and Bioethical Issues

Scientific findings often lead to the development of a technology that can be of service to human beings. Technologies have both benefits and drawbacks. Bioethics is concerned with the consequences of using biological technology. Every member of society needs to be prepared to participate in making bioethical decisions. Human activities, many of which are made possible by advances in technology, are damaging ecosystems and causing a worldwide loss of biodiversity; this is arguably the most important bioethical dilemma of our time.

Key Terms

adaptation   4
animal   7
binomial name   6
biodiversity   14
bioethics   14
biology   10
biosphere   9
cell   2
class   6
community   9
conclusion   11

control group   11
data   11
domain   6
domain Archaea   6
domain Bacteria   6
domain Eukarya   6
ecosystem   9
energy   3
evolution   5
experiment   11
experimental design   11

experimental variable   11
extinction   14
family   6
fungus   6
gene   4
genus   6
homeostasis   3
hypothesis   10
inductive reasoning   10
kingdom   6
law   12
metabolism   3
model   11
multicellular   2
natural selection   7
observation   10
order   6
organ   2

organism   2
organ system   2
photosynthesis   3
phylum   6
placebo   13
plant   7
population   9
principle   12
protist   6
reproduce   3
scientific theory   12
species   6
taxonomy   5
technology   14
test group   11
tissue   2
unicellular   2

Testing Yourself

Choose the best answer for each question.

1. The region of the Earth's surface where all organisms are found is called the
   a. ecosystem.
   c. community.
   b. population.
   d. biosphere.

2. In the following diagram, label the levels of biological organization:

3. All of the chemical reactions that occur in a cell are called
   a. homeostasis.
   c. heterostasis.
   b. metabolism.
   d. cytoplasm.

4. Which of the following statements is not true?
   a. Darwin described evolution as descent with modification.
   b. Evolution is an explanation for both the unity and the diversity of life.
   c. Evolution occurs through a process called natural selection.
   d. Evolution is just one man's idea.

5. A modification that helps equip organisms for their way of life is a(n)
   a. homeostasis.
   c. adaptation.
   b. natural selection.
   d. extinction.

6. The process of turning solar energy into chemical energy is called
   a. work.
   c. photosynthesis.
   b. metabolism.
   d. respiration.

7. Place the steps in the process of natural selection in the correct order:
   (1) Over generations, a larger proportion of the population becomes adapted to the environment.
   (2) The population produces more offspring than the environment can support.
   (3) Members of a population have heritable variations.
   (4) The more adapted individuals survive and reproduce to a greater extent than those individuals lacking adaptations.
   a. 2, 4, 3, 1   d. 2, 3, 1, 4
   b. 4, 1, 3, 2   e. 3, 2, 4, 1
   c. 3, 1, 4, 2   f. 1, 2, 4, 3

For questions 8–12, choose one answer from the following key.

Key:

a. unicellular organism
b. multicellular organism
c. neither
d. both

8. Capable of responding to the environment.
9. Reproduces by uniting an egg with a sperm.
10. Contains tissues.
11. Contains genes.
12. Creates energy.
13. A group of interbreeding individuals is a(n)
    a. ecosystem.   c. community.
    b. species.   d. phylum.
14. Information collected from a scientific experiment is known as
    a. a scientific theory.   c. data.
    b. a hypothesis.   d. a conclusion.
15. In an ecosystem,
    a. energy flows and nutrients cycle.
    b. energy cycles and nutrients flow.
    c. energy and nutrients flow.
    d. energy and nutrients cycle.
16. Taxonomy is a biological discipline in which organisms are grouped according to their
    a. geographic location.   c. common ancestries.
    b. size.   d. gene number.
17. The first cells to evolve in the primitive Earth were most likely members of the domain
    a. Archaea.   c. Eukarya.
    b. Bacteria.
18. Which of the following categories is the most inclusive?
    a. domain   c. class
    b. family   d. species
19. A scientist's suggested explanation for a natural event is a(n)
    a. observation.   c. principle.
    b. experiment.   d. hypothesis.
20. Choose the correct way to represent the scientific name of corn.
    a. Zea Mays   e. *Zea Mays*
    b. zea mays   f. *zea mays*
    c. Zea mays   g. *Zea mays*
    d. zea Mays   h. *zea Mays*
21. A representation of a natural phenomenon studied by scientists when the real object is impossible to study is a(n)
    a. model.   c. control group.
    b. experimental variable.   d. scientific theory.

22. The smallest living unit is a(n)
    a. atom.   c. cell.
    b. molecule.   d. tissue.

Thinking Scientifically

1. Based on the evolutionary tree below, which prokaryotic domain gave rise to the domain Eukarya? Which kingdom in domain Eukarya gave rise to plants, animals, and fungi?

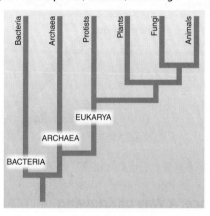

2. Several years before Alexander Fleming discovered penicillin, he made another exciting discovery. Fleming had a bad cold but kept working in his lab. He decided to put some of his nasal mucus in a petri dish of bacteria to see what would happen. Fleming observed that, over time, the mucus-covered area of the petri dish became clear of bacteria. What hypothesis might he have formulated when he saw the effect of his nasal mucus on the bacteria? Fleming soon became notorious for squeezing lemon juice into people's eyes to make them produce tears. He would then place the tears in petri dishes of bacteria. What hypothesis do you think he was testing?

Bioethical Issue

Placebo Use in Clinical Drug Trials

Clinical trials on drug efficacy are typically carried out on two groups of patients. The test group receives the drug, while the control group receives a placebo. The control group determines the extent to which psychology affects people who believe they are taking the new drug.

Suppose a new cancer drug has been developed. Many people sign up to participate in a clinical trial to test the drug's ability to slow the progression of the disease. These people know that they may receive a placebo in place of the real drug, but they participate because they have run out of options. Is it right to give the placebo to some of the participants, knowing that their disease is going to progress during the course of the trial? What if the drug was one that alleviated pain in chronic pain sufferers? In this case, no disease progression would occur during the course of the trial, but participants taking the placebo would have to endure serious pain.

Essentials of Biology Website

The companion website for *Essentials of Biology* provides a wealth of information organized and integrated by chapter. You will find practice tests, animations, videos, and much more that will complement your learning and understanding of general biology.

http://www.mhhe.com/maderessentials2

2

The Chemical Basis of Life

The pH of your stomach is acidic enough to dissolve steel.

The chemical composition of human blood closely resembles that of seawater.

UV radiation can cause skin cancer.

Outline

Radiation Causes Chemical Damage to Cells

YOU ARE PROBABLY AWARE that the effects of exposure to extreme levels of radiation in humans can range from sunburns, to hair loss, to cancer, and maybe even death. The former spy Alexander Litvinenko died of poisoning by the radioactive element polonium-210, for example. But what is it about exposure to radioactivity that causes such damage to living cells? When we refer to a "radioactive atom," we really mean that the atom is not stable and has a tendency to release subatomic particles. If subatomic particles are released from an atom, they can cause major damage. When organisms are exposed to radioactivity, loose subatomic

—Continued next page

Radiation causes cancer, but it can also be used to cure cancer.

Continued—

particles penetrate their cells and cause damage to cellular molecules, such as the genetic material (DNA). When DNA is damaged, a mutation may result. Many mutations may cause cancer; thus, exposure to high levels of radioactivity often causes cancer in humans. Radioactivity can also cause so much damage that many cells (and the individual) die.

Chemistry is the very essence of all organisms. You may have never thought of your body as simply a pile of chemicals, how the chemistry of your body changes when you eat or drink certain foods, or the changes that occur in cells when they are exposed to radiation. But more and more we see the boundaries between chemistry and biology becoming blurred. Scientists have made much progress in medicine, agriculture, and ecology by understanding the chemical reactions that occur in cells. Knowledge of the basic principles of chemistry will greatly enhance your ability to understand biology and provide a foundation on which to build further competency.

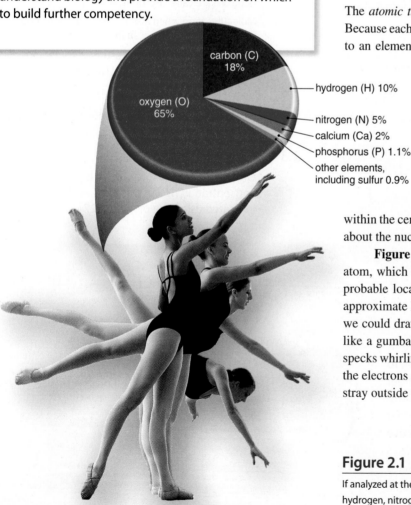

2.1 | The Nature of Matter

Kiss your sweetheart, pat your dog, catch a bus, mow your lawn—everything you touch is matter. **Matter** refers to anything that takes up space and has mass. Matter can exist as a solid, a liquid, or a gas. Then we can realize that not only are we matter, but so are the water we drink and the air we breathe.

All matter, both nonliving and living, is composed of elements. Formally speaking, an **element** is a substance that cannot be broken down into another substance by ordinary chemical means. There are only 92 naturally occurring elements, and each of these differs from the others in its properties. (A *property* is a physical or chemical characteristic, such as density, solubility, melting point, and reactivity.)

Both the Earth's crust and all organisms are made up of elements, but these elements occur in different proportions (**Fig. 2.1**). Six elements—carbon, hydrogen, nitrogen, oxygen, phosphorus, and sulfur—are of special significance to us. They make up about 98% of the body weight of most organisms. The acronym CHNOPS helps us remember these six elements, whose properties are essential to the uniqueness of living things, from simple one-celled life-forms to complex multicellular organisms.

Atomic Structure

The *atomic theory* states that elements consist of tiny particles called **atoms.** Because each element consists of only one kind of atom, the same name is given to an element and its atoms. This name is represented by one or two letters, called the **atomic symbol.** For example, the symbol H stands for a hydrogen atom, and the symbol Na (for *natrium* in Latin) stands for a sodium atom.

From our discussion of elements, you might expect each atom to have a certain mass. The **atomic mass** of an atom is dependent upon the presence of three types of subatomic particles: **neutrons,** which have no electrical charge; **protons,** which have a positive charge; and **electrons,** which have a negative charge. Protons and neutrons are located within the center of an atom, which is called its **nucleus,** while electrons move about the nucleus.

Figure 2.2 shows the arrangement of the subatomic particles in a helium atom, which has only two electrons. In Figure 2.2*a*, the stippling shows the probable location of electrons, and in Figure 2.2*b*, the circle represents the approximate location of electrons. Most of an atom is empty space. In fact, if we could draw an atom the size of a baseball stadium, the nucleus would be like a gumball in the center of the stadium, and the electrons would be tiny specks whirling about in the upper stands. Usually, we can only indicate where the electrons are expected to be. In our analogy, the electrons might very well stray outside the stadium at times.

Figure 2.1 Elements in living organisms.

If analyzed at the level of atoms, human beings are mostly composed of oxygen, carbon, hydrogen, nitrogen, calcium, and phosphorus.

In effect, the atomic mass of an atom is just about equal to the sum of its protons and neutrons. Protons and neutrons are assigned one atomic mass unit each. Electrons are so small that their mass is assumed to be zero in most calculations. The term atomic mass is used, rather than atomic weight, because mass is constant while weight changes according to the gravitational force of a planet.

All atoms of an element have the same number of protons. This is called the atom's **atomic number.** The number of protons makes an atom unique. In **Figure 2.3,** the atomic number is written above the atomic symbol. The mass number is written below the atomic symbol. For example, the carbon atom is shown in this way:

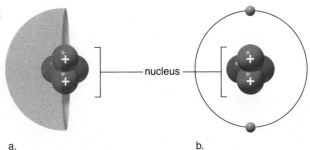

Figure 2.2 **Two models of helium (He).**

Atoms contain subatomic particles, which are located as shown in these two simplified models of helium. Protons are positively charged, neutrons have no charge, and electrons are negatively charged. Protons and neutrons are found within the nucleus, and electrons are outside the nucleus. **a.** This model shows electrons as a negatively charged cloud around the nucleus. **b.** In this model, the average location of electrons is sometimes represented by a circle.

The Periodic Table

Once chemists discovered a number of the elements, they began to realize that the elements' chemical and physical characteristics recur in a predictable manner. The periodic table (Fig. 2.3) was developed as a way to display the elements, and therefore the atoms, according to these characteristics. Every atom is in a particular period (the horizontal rows) and in a particular group (the vertical columns). The atoms in group 8 are called the *noble gases* because they are gases that rarely react with another atom for reasons we will discuss later. Notice that helium is a noble gas.

The atomic number written above the atomic symbol tells you the number of positively charged protons, and also the number of negatively charged electrons if the atom is electrically neutral. To determine the usual number of neutrons, subtract the number of protons from the mass number, written below the atomic symbol, and take the closest whole number.

Isotopes

Isotopes are atoms of the same element that differ in the number of neutrons. In other words, isotopes have the same number of protons, but they have different mass numbers. A nucleus with excess neutrons is unstable and may decay and emit radiation. The radiation given off by radioactive isotopes can be detected in various ways. Most people are familiar with the use of a Geiger counter to detect radiation. However, other methods to detect radiation exist that are useful in medicine and science.

Figure 2.3 **A portion of the periodic table.**

In the periodic table, the elements, and therefore the atoms that compose them, are in the order of their atomic numbers but arranged in periods (horizontal rows) and groups (vertical columns). All the atoms in a particular group have certain chemical characteristics in common. The six elements previously mentioned (CHNOPS) are highlighted in red.

Check Your Progress

1. Contrast atomic mass with atomic number.

Answer: 1. Atomic mass is approximately the sum of the protons and neutrons in an atom. Atomic number is the number of protons in an atom.

thyroid gland

Uses of Radioactive Isotopes   The importance of chemistry to biology and medicine is nowhere more evident than in the many uses of radioactive isotopes. For example, radioactive isotopes can be used as tracers to detect molecular changes, or to destroy abnormal or infectious cells.

The chemical behavior of a radioactive isotope is essentially the same as that of the stable isotopes of an element. This means that you can put a small amount of radioactive isotope in a sample and it becomes a **tracer** by which to detect molecular changes because it behaves like the stable isotope. Positron-emission tomography (PET) utilizes tracers to determine the comparative activity of tissues. A radioactively labeled glucose tracer that emits a subatomic particle known as a positron is injected into the body. The radiation given off is detected by sensors and analyzed by a computer. The result is a color image that shows which tissues took up glucose and are thus metabolically active (**Fig. 2.4***a,b*). Although not shown in Figure 2.4, a PET scan can help diagnose a malfunctioning thyroid, a brain tumor, Alzheimer disease, epilepsy, or whether a stroke has occurred.

Radioactive substances in the environment can cause harmful chemical changes in cells, damage DNA, and cause cancer. The release of radioactive particles following a nuclear power plant accident can have far-reaching and long-lasting effects on human health. Even relatively small doses of radioactivity can be fatal—former KGB spy Alexander Litvinenko was fatally poisoned with the radioactive element polonium-210 (**Fig. 2.5***a*). But the effects of radiation can also be put to good use. Packets of radioactive isotopes can be placed in the body so that the subatomic particles emitted destroy only cancer cells, with little risk to the rest of the body (**Fig. 2.5***b*).

Radiation from radioactive isotopes has been used for many years to sterilize medical and dental equipment. Now the possibility exists that it can also be used to sterilize the U.S. mail to free it of possible pathogens, such as anthrax spores introduced by terrorists.

a.

Figure 2.4   PET scans.

In a PET scan, red indicates areas of greatest metabolic activity, and blue means areas of least activity. Computers analyze the data from different sections of an organ. **a.** A longitudinal scan of the thyroid resembles its overall shape. **b.** A cross section of the brain is a powerful diagnostic tool for the physician.

83 MM

b.

Arrangement of Electrons in an Atom

Electrons in atoms are much like the blades of a ceiling fan. When the fan is moving, it is difficult to see the individual blades, and all you see is a blur in the area where the blades are. When the fan is stopped, the blades have a specific location and can be seen. Likewise, the electrons of an atom are constantly moving. Although it is not possible to determine the precise location of an individual electron at any given moment, it is useful to construct models of atoms that show electrons at discrete energy levels about the nucleus (**Fig. 2.6**). It seems reasonable to suggest that negatively charged electrons are attracted to the positively charged nucleus, and therefore it takes an increasing amount of energy to push them farther away from the nucleus. Electrons in outer shells, therefore, contain more energy than those in inner shells.

Each energy level contains a certain number of electrons. In the models shown in Figure 2.6, the energy levels (**electron shells**) are drawn as concentric rings about the nucleus. For atoms up through number 20 (i.e., calcium), the first shell closest to the nucleus can contain two electrons; thereafter, each additional shell can contain eight electrons. For these atoms, each lower level is filled with electrons before the next higher level contains any electrons.

a.   b.

Figure 2.5   High levels of radiation.

a. High levels of radiation can cause damage to the body. Former KGB spy Alexander Litvinenko was fatally poisoned with a high dose of radioactivity. **b.** Radiation can be also be used to heal. Physicians treat patients with high levels of radioactive isotopes to kill cancer cells.

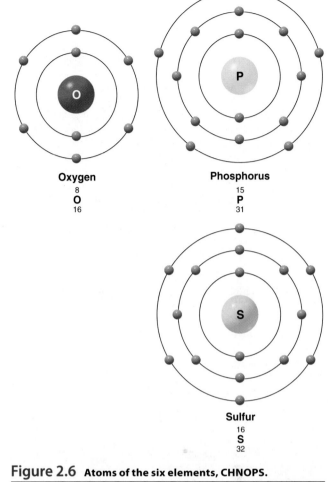

Figure 2.6   **Atoms of the six elements, CHNOPS.**

Electrons orbit the nucleus at particular energy levels (electron shells): The first shell contains up to two electrons, and each shell thereafter can contain up to eight electrons (until we consider atoms with an atomic number above 20). Each shell is to be filled before electrons are placed in the next shell.

The sulfur atom, with an atomic number of 16, has two electrons in the first shell, eight electrons in the second shell, and six electrons in the third, or outer, shell. Revisit the periodic table (see Fig. 2.3), and note that sulfur is in the third period. In other words, the period tells you how many shells an atom has. Also note that sulfur is in group 6. The group tells you how many electrons an atom has in its outer shell.

If an atom has only one shell, the outer shell is complete when it has two electrons. If an atom has two or more shells, the outer shell is most stable when it has eight electrons; this is called the **octet rule.** As mentioned previously, atoms in group 8 of the periodic table are called the noble gases because they do not ordinarily undergo reactions. Atoms with fewer than eight electrons in the outer shell react with other atoms in such a way that each has a completed outer shell after the reaction. Atoms can give up, accept, or share electrons in order to have eight electrons in the outer shell. In other words, the number of electrons in an atom's outer shell, called the **valence shell,** determines its chemical reactivity. The size of an atom is also important. Both carbon (C) and silicon (Si) atoms are in group 4, and therefore they have four electrons in their valence shells. This means they can bond with as many as four other atoms in order to achieve eight electrons in their outer shells. But carbon in period 2 has two shells, and silicon in period 3 has three shells. The smaller atom, carbon, can bond to other carbon atoms and form long-chained molecules, while the larger silicon atom is unable to bond to other silicon atoms. Still, the chemical properties of atoms—that is, how they react—are largely determined by the arrangement of their electrons.

Types of Chemical Bonds

A group of atoms bonded together is called a **molecule.** When a molecule contains atoms of more than one element, it can be called a **compound.** Compounds and molecules contain two types of chemical bonds: ionic and covalent. The type of bond that forms depends on whether two bonded atoms share electrons or whether one has given electrons to the other. For example, in hydrogen gas (H_2) the two hydrogen atoms are sharing electrons in order to fill the valence shells of both atoms. When sodium chloride (NaCl) forms, however, the sodium atom (Na) gives an electron to the chlorine (Cl) atom, and in that way each atom has eight electrons in the outer shell.

Check Your Progress

1. **List some uses of radioactive isotopes in biology and medicine.**
2. **Explain how the octet rule affects an element's reactivity.**

Answers: 1. Uses of radioactive isotopes include imaging of body parts (e.g., PET scans, thyroid imaging), sterilization of medical equipment, and cancer therapy.   2. Elements with two or more electron shells will give up, accept, or share electrons with other elements to obtain eight electrons in their valence (outer) shells.

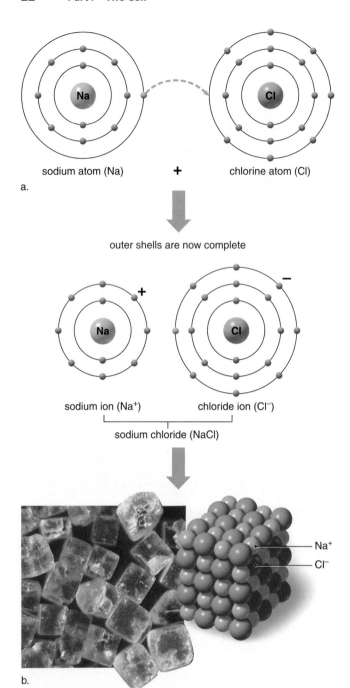

a.

outer shells are now complete

sodium chloride (NaCl)

b.

Figure 2.7 **Formation of sodium chloride.**

a. During the formation of sodium chloride, an electron is transferred from the sodium atom to the chlorine atom. At the completion of the reaction, each atom has eight electrons in the outer shell, but each also carries a charge as shown. **b.** In a sodium chloride crystal, ionic bonding between Na^+ and Cl^- causes the atoms to assume a three-dimensional lattice shape.

Ionic Bonding

An **ionic bond** forms when two atoms are held together by the attraction between opposite charges. The reaction between sodium and chlorine atoms is an example of how an ionic bond is formed. Consider that sodium (Na), with only one electron in its third shell, usually gives up an electron (**Fig. 2.7**a). Once it does so, the second shell, with eight electrons, becomes its outer shell. Chlorine (Cl), on the other hand, tends to take on an electron, because its outer shell has seven electrons. If chlorine gets one more electron, it has a completed outer shell. So, when a sodium atom and a chlorine atom react, an electron is transferred from sodium to chlorine. Now both atoms have eight electrons in their outer shells.

This electron transfer causes these atoms to become **ions,** or charged atoms. The sodium ion has one more proton than it has electrons; therefore, it has a net charge of $+1$ (symbolized by Na^+). The chloride ion has one more electron than it has protons; therefore, it has a net charge of -1 (symbolized by Cl^-). Negatively charged ions often have names that end in "ide," and thus Cl^- is called a chloride ion. In the periodic table, atoms in groups 1 and 2 and groups 6 and 7 become ions when they react with other atoms. Atoms in groups 2 and 6 always transfer two electrons. For example, calcium becomes Ca^{2+}, while oxygen becomes O^{2-}.

Ionic compounds are often found as **salts.** Sodium chloride is the salt we use at the table. Salts can exist as a dry solid or in a dissociated (ionized) form. A sodium chloride crystal illustrates the solid form of a salt (Fig. 2.7b). When salts are placed in water, they release ions as they dissolve. NaCl separates into Na^+ and Cl^-. Ionic compounds are most commonly found in this dissociated (ionized) form in biological systems because these systems are 70–90% water.

Covalent Bonding

A **covalent bond** results when two atoms share electrons in order to have a completed outer shell. In a hydrogen atom, the outer shell is complete when it contains two electrons. If hydrogen is in the presence of a strong electron acceptor, it gives up its electron to become a hydrogen ion (H^+). But if this is not possible, hydrogen can share with another atom, and thereby have a completed outer shell. For example, one hydrogen atom can share with another hydrogen atom. In this case, the two orbitals overlap and the electrons are shared between them—that is, you count the electrons as belonging to both atoms:

Hydrogen gas (H_2)

Rather than drawing an orbital model like the above, scientists often use simpler ways to indicate molecules. A *structural formula* uses straight lines, as in H—H. The straight line is used to indicate a pair of shared electrons. A *molecular formula* omits the lines that indicate bonds and simply shows the number of atoms involved, as in H_2.

Sometimes, atoms share more than two electrons to complete their octets. A double covalent bond occurs when two atoms share two pairs of electrons, as in this molecule of oxygen gas:

Methane (CH₄)

Oxygen gas (O₂)

In order to show that oxygen gas (O₂) contains a double bond, the structural formula would be written as O═O to indicate that two pairs of electrons are shared between the oxygen atoms.

It is also possible for atoms to form triple covalent bonds, as in nitrogen gas (N₂), which can be written as N≡N. Single covalent bonds between atoms are quite strong, but double and triple bonds are even stronger.

A single atom may form bonds with more than one other atom. For example, the molecule methane results when carbon binds to four hydrogen atoms (**Fig. 2.8**). In methane, each bond actually points to one corner of a tetrahedron. The best model to show this arrangement is a ball-and-stick model (Fig. 2.8c). Space-filling models (Fig. 2.8d) come closest to showing the actual shape of a molecule. The shapes of molecules help dictate the functional roles they play in organisms.

Chemical Reactions

Chemical reactions are very important to organisms. For example, we have already noted that the process of photosynthesis enables plants to make molecular energy available to themselves and other organisms. An overall equation for photosynthesis indicates that some bonds are broken and others are formed:

$$6\ CO_2\ +\ 6\ H_2O\ \longrightarrow\ C_6H_{12}O_6\ +\ 6\ O_2$$
| carbon dioxide | water | glucose | oxygen |

This equation says that six molecules of carbon dioxide react with six molecules of water to form one glucose molecule and six molecules of oxygen. The **reactants** (molecules that participate in the reaction) are shown on the left of the arrow, and the **products** (molecules formed by the reaction) are shown on the right. Notice that the equation is "balanced"—that is, the same number of each type of atom occurs on both sides of the arrow.

Note the glucose molecule in the equation above. Glucose is more complex than any molecule discussed so far. The arrangement of atoms in a glucose molecule is shown below:

one molecule

$$C_6H_{12}O_6$$

| indicates 6 atoms of carbon | indicates 12 atoms of hydrogen | indicates 6 atoms of oxygen |

a. Electron model showing covalent bonds

hydrogen

covalent bond

carbon

109°

c. Ball-and-stick model

d. Space-filling model

b. Structural model

Figure 2.8   Shapes of covalently bonded molecules.

An electron model (**a**) and a structural model (**b**) show that methane (CH₄) contains one carbon atom bonded to four hydrogen atoms. **c.** The ball-and-stick model shows that when carbon bonds to four other atoms, as in methane, each bond actually points to one corner of a tetrahedron. **d.** The space-filling model is a three dimensional representation of the molecule.

Check Your Progress

1. **Contrast an ionic bond with a covalent bond.**
2. **Ammonia is formed by covalent bonding between nitrogen and hydrogen. How many hydrogen atoms would be bonded to the nitrogen atom?**

Answers: 1. An ionic bond is created when one atom gives up electron(s) and another gains electron(s) so that both atoms have their outer shells filled. A covalent bond is formed when two atoms share electrons to fill their outer shell.   2. Carbon requires four electrons to fill its outer shell, so it bonds with four hydrogen atoms.

2.2 | Water's Importance to Life

Life began in water, and it is the single most important molecule on Earth. All organisms are 70–90% water; their cells consist of membranous compartments enclosing aqueous solutions. Water has unique properties that make it a life-supporting substance. These properties of water stem from the structure of the molecule.

The Structure of Water

The electrons shared between two atoms in a covalent bond are not always shared equally. Atoms differ in their **electronegativity**—that is, their affinity for electrons in a covalent bond. Atoms that are more electronegative tend to hold shared electrons more tightly than those that are not. This causes the bond to become **polar,** meaning that the atoms on either side of the bond are partially charged, even though the overall molecule itself bears no net charge. For example, in water, oxygen shares electrons with two hydrogen atoms. Oxygen is more electronegative than hydrogen, so the two bonds are polar. The shared electrons spend more time orbiting the oxygen nucleus than the hydrogen nuclei, and this unequal sharing of electrons makes water a polar molecule. The covalent bonds are angled, and the molecule is bent roughly into a Λ shape. The point of the Λ (oxygen) is the negative (−) end, and the two hydrogens are the positive (+) end (**Fig. 2.9a**).

The polarity of water molecules causes them to be attracted to one another. The positive hydrogen atoms in one molecule are attracted to the negative oxygen atoms in other water molecules. This attraction is called a **hydrogen (H) bond,** and each water molecule can engage in as many as four H bonds (Fig. 2.9b). The covalent bond is much stronger than an H bond, but the large number of H bonds in water makes the H bond strong overall. The properties of water are due to its polarity and its ability to form hydrogen bonds.

Properties of Water

Water is so familiar that we take it for granted, but without water, life as we know it would not exist. The properties of water that support life are solvency, cohesion and adhesion, high surface tension, high heat capacity, high heat of vaporization, and varying density.

Oxygen is slightly negative (−)

H   O   H

Hydrogens are slightly positive (+)

a.

hydrogen bond

b.

Figure 2.9   The structure of water.

a. The space-filling mode shows the ∧ shape of a water molecule. Oxygen attracts the shared electrons more than hydrogen atoms do, and this causes the molecule to be polar: The oxygen carries a slight negative charge and the hydrogens carry a slight positive charge. **b.** The positive hydrogens form hydrogen bonds with the negative oxygen in nearby molecules. Each water molecule can be joined to other water molecules by as many as four hydrogen bonds.

Water is a solvent Because of its polarity and H-bonding ability, water dissolves a great number of substances. Molecules that are attracted to water are said to be **hydrophilic** (*hydro*, water; *phil*, love). Nonionized and nonpolar molecules that are not attracted to water are said to be **hydrophobic** (*hydro*, water; *phob*, fear).

When a salt such as sodium chloride (NaCl) is put into water, the negative ends of the water molecules are attracted to the sodium ions, and the positive ends of the water molecules are attracted to the chloride ions. This attraction causes the sodium ions and the chloride ions to break up, or dissociate, in water:

The salt NaCl dissociates in water.

Water dissolves many polar nonionic substances, such as glucose, by forming H bonds with them. When ions and molecules disperse in water, they move about and collide, allowing reactions to occur. As mentioned, cellular fluids are aqueous solutions of various substances, and so are the oceans. Without the dissolving power of water, aquatic organisms could not take up the substances they need from the water.

Water molecules are cohesive and adhesive *Cohesion* refers to the ability of water molecules to cling to each other due to hydrogen bonding. Because of cohesion, water exists as a liquid under ordinary conditions of temperature and pressure. The strong cohesion of water molecules is apparent because water flows freely, and yet the molecules do not separate from each other. *Adhesion* refers to the ability of water molecules to cling to other polar surfaces. This is because of water's polarity. The positive and negative poles of water molecules cause them to adhere to other polar surfaces.

Liquid water is an excellent transport system, both outside and within living organisms, due to adhesion and cohesion. Multicellular animals often contain internal vessels in which water assists the transport of nutrients and wastes because the cohesion and adhesion of water allows blood to fill the tubular vessels of the cardiovascular system. For example, the liquid portion of our blood, which transports dissolved and suspended substances about the body, is 90% water.

Cohesion and adhesion also contribute to the transport of water in plants. Plants have their roots anchored in the soil, where they absorb water, but the leaves are uplifted and exposed to solar energy. Water evaporating from the leaves is immediately replaced with water molecules from transport vessels that extend from the roots to the leaves (**Fig. 2.10**). Because water molecules are cohesive, a tension is created that pulls a water column up from the roots. Adhesion of water to the walls of the vessels also helps prevent the water column from breaking apart.

Water has a high surface tension Because the water molecules at the surface are more strongly attracted to each other than to the air above, water molecules at the surface cling tightly to each other. Thus we say that water exhibits surface tension. The stronger the force between molecules in a liquid, the greater the surface tension. Hydrogen bonding is the main force that causes water to have a high surface tension. This property makes it possible for humans to skip rocks across the surface of a body of water. An insect called a water strider can even walk on the surface of a pond without breaking through (**Fig. 2.11**).

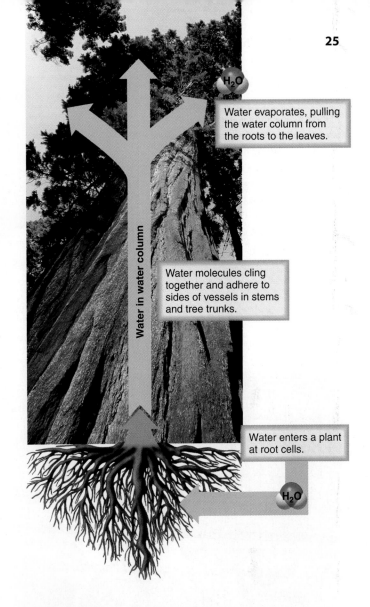

Water evaporates, pulling the water column from the roots to the leaves.

Water molecules cling together and adhere to sides of vessels in stems and tree trunks.

Water enters a plant at root cells.

Figure 2.10 **Cohesion and adhesion of water molecules.**

How does water rise to the top of all trees? Water-filled vessels extend from the roots to the leaves. When water evaporates from the leaves, this water column is pulled upward due to the cohesion of water molecules with one another and the adhesion of water molecules to the sides of the vessel.

Figure 2.11 **Surface tension of water.**

A water strider can flit across a pond on widespread legs because the surface tension of water is strong enough to support it.

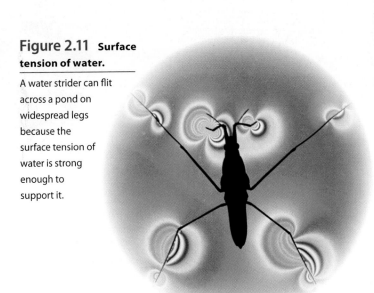

Figure 2.12 Heat of vaporization.

At room temperature, water is a liquid. **a.** Water boils at 100°C, vaporizes, and takes up a large amount of heat. **b.** It takes much body heat to vaporize sweat, which is mostly liquid water, and this helps keep our bodies cool when the temperature rises.

a.

b.

Water has a high heat capacity The many hydrogen bonds that link water molecules allow water to absorb heat without greatly changing in temperature. Water's high heat capacity is important not only for aquatic organisms but for all organisms. Because the temperature of water rises and falls slowly, terrestrial organisms are better able to maintain their normal internal temperatures and are also protected from rapid temperature changes.

Water also has a high heat of vaporization: It takes a great deal of heat to break the hydrogen bonds in water so that it becomes gaseous and evaporates into the environment. If the heat given off by our metabolic activities were to go directly into raising our body temperature, death would follow. Instead, the heat is dispelled as sweat evaporates (**Fig. 2.12**).

Because of water's high heat capacity and high heat of vaporization, temperatures along the Earth's coasts are moderate. During the summer, the ocean absorbs and stores solar heat, and during the winter, the ocean releases it slowly. In contrast, the interior regions of the continents experience abrupt changes in temperature.

Water is less dense as ice Unlike other substances, water expands as it freezes, which explains why cans of soda burst when placed in a freezer and how roads in northern climates become bumpy because of "frost heaves" in the winter. Since water expands as it freezes, it is less dense than liquid water, and therefore ice floats on liquid water (**Fig. 2.13**).

If ice were more dense than water, it would sink, and ponds, lakes, and perhaps even the ocean would freeze solid, making life impossible in the water and also on land. Instead, bodies of water always freeze from the top down. When a body of water freezes on the surface, the ice acts as an insulator to prevent the water below it from freezing. This protects aquatic organisms so that they can survive the winter. As ice melts in the spring, it draws heat from the environment, helping to prevent a sudden change in temperature that might be harmful to life.

a. Ice

Figure 2.13 Properties of ice.

a. The geometric requirements of hydrogen bonding of water molecules cause ice to be less dense than liquid water. **b.** Therefore, bodies of water freeze from the top down, and organisms in ponds and lakes are protected during the winter.

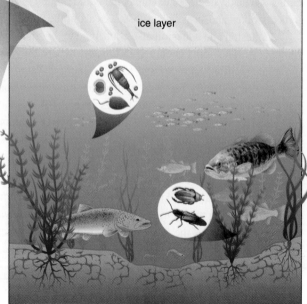

ice layer

b. Pond

2.3 | Acids and Bases

Figure 2.14 shows that when water dissociates (breaks apart), it releases an equal number of **hydrogen ions (H⁺)** and **hydroxide ions (OH⁻)** as in the following reaction:

$$H-O-H \rightleftarrows H^+ + OH^-$$

water   hydrogen   hydroxide
ion   ion

Acidic Solutions (High H⁺ Concentration)

Lemon juice, vinegar, tomato juice, and coffee are all acidic solutions. What do they have in common? Acidic solutions have a sharp or sour taste, and therefore we sometimes associate them with indigestion. To a chemist, **acids** are substances that dissociate in water, releasing hydrogen ions (H⁺). For example, an important acid is hydrochloric acid (HCl), which dissociates in this manner:

$$HCl \rightarrow H^+ + Cl^-$$

The acidity of a substance depends on how fully it dissociates in water. HCl dissociates almost completely; therefore, it is called a strong acid. If hydrochloric acid is added to a beaker of water, the number of hydrogen ions (H⁺) increases greatly (**Fig. 2.15**).

Basic Solutions (Low H⁺ Concentration)

Milk of magnesia and ammonia are common basic solutions that most people are familiar with. Basic solutions have a bitter taste and feel slippery when in water. To a chemist, **bases** are substances that either take up hydrogen ions (H⁺) or release hydroxide ions (OH⁻). For example, an important base is sodium hydroxide (NaOH), which dissociates in this manner:

$$NaOH \rightarrow Na^+ + OH^-$$

Like acids, the strength of a base is determined by how fully it dissociates. Dissociation of sodium hydroxide is almost complete; therefore, it is called a strong base. If sodium hydroxide is added to a beaker of water, the number of hydroxide ions increases (**Fig. 2.16**).

Many strong bases, such as ammonia, are useful household cleansers. Ammonia has a poison symbol and carries a strong warning not to ingest the product. Neither acids nor bases should be tasted because they are quite destructive to cells.

▬ Check
Your Progress

1. **Contrast cohesion and adhesion of water.**
2. **What properties of water protect organisms that live underwater?**
3. **Contrast an acidic solution with a basic solution.**

Answers: 1. Cohesion refers to the tendency of water molecules to cling to each other, whereas adhesion refers to the ability of water molecules to cling to polar surfaces.   2. Water expands as it freezes, so that bodies of water freeze from the top down and insulate the water below. Water's high heat capacity protects organisms from sudden changes in temperature.   3. An acidic solution contains more H⁺ ions than OH⁻ ions; a basic solution contains more OH⁻ ions than H⁺ ions.

Figure 2.14 **Dissociation of water molecules.**

Dissociation of water molecules produces an equal number of hydrogen ions (H⁺) and hydroxide ions (OH⁻). (These illustrations are not meant to be mathematically accurate.)

Figure 2.15 **Addition of hydrochloric acid (HCl).**

HCl releases hydrogen ions (H⁺) as it dissociates. The addition of HCl to water results in a solution that is acidic (more H⁺ than OH⁻).

Figure 2.16 **Addition of sodium hydroxide (NaOH), a base.**

NaOH releases OH⁻ as it dissociates. The addition of NaOH to water results in a solution that is basic (more OH⁻ than H⁺).

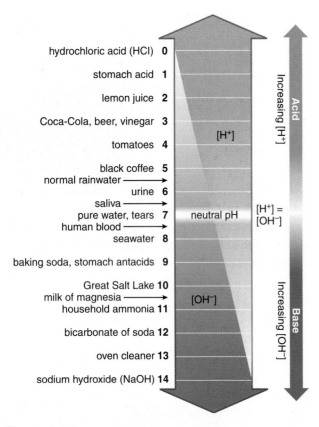

Figure 2.17 **The pH scale.**

The proportionate amount of hydrogen ions to hydroxide ions is indicated by the diagonal line. Any solution with a pH above 7 is basic, while any solution with a pH below 7 is acidic.

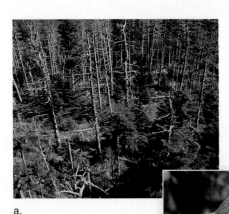

a.

Figure 2.18 **Effects of acid deposition.**

The burning of gasoline derived from oil, a fossil fuel, leads to acid deposition, which causes (**a**) trees to die and (**b**) statues to deteriorate.

b.

pH and the pH Scale

pH[1] is a mathematical way of indicating the number of hydrogen ions in a solution. The **pH scale** is used to indicate the acidity or basicity (alkalinity) of a solution. The pH scale ranges from 0 to 14 (**Fig. 2.17**). A pH of 7 represents a neutral state in which the hydrogen ion and hydroxide ion concentrations are equal, as in pure water. A pH below 7 is acidic because the hydrogen ion concentration, commonly expressed in brackets as $[H^+]$, is greater than the hydroxide concentration, $[OH^-]$. A pH above 7 is basic because $[OH^-]$ is greater than $[H^+]$. Further, as we move down the pH scale from pH 7 to pH 0, each unit has 10 times the acidity $[H^+]$ of the previous unit. As we move up the scale from 7 to 14, each unit has 10 times the basicity $[OH^-]$ of the previous unit.

The pH scale was devised to eliminate the use of cumbersome numbers. For example, the hydrogen ion concentrations of these solutions are on the left, and the pH is on the right:

		$[H^+]$ (moles per liter)	**pH**
0.000001	=	1×10^{-6}	6 (acid)
0.0000001	=	1×10^{-7}	7 (neutral)
0.00000001	=	1×10^{-8}	8 (base)

The effect of pH on organisms is dramatically illustrated by the phenomenon known as acid deposition. When fossil fuels are burned, sulfur dioxide and nitrogen oxides are produced, and they combine with water in the atmosphere to form acids. These acids then come in contact with organisms and objects, leading to damage or even death (**Fig. 2.18**).

Buffers and pH

In the human body, pH needs to be kept within a narrow range in order to maintain homeostasis. A **buffer** is a chemical or a combination of chemicals that keeps pH within normal limits. Buffers resist pH changes because they can take up excess hydrogen ions (H^+) or hydroxide ions (OH^-). Many commercial products, such as Bufferin, shampoos, and deodorants, are buffered as an added incentive for us to buy them.

---

[1] pH is defined as the negative log of the hydrogen ion concentration $[H^+]$. A log is the power to which 10 must be raised to produce a given number.

Check Your Progress

1. What information is given by the pH of a solution?
2. Describe how buffers can help regulate pH in the human body?

Answers: 1. The pH of a solution tells you the concentration of H^+ in a solution. At a high pH (above 7), the H^+ concentration is low, and the solution is basic. At a lower level pH (below 7), the H^+ concentration is higher, and the solution is more acidic.   2. Buffers can keep the pH between normal levels by taking up excess H^+ or OH^- ions.

The Chapter in Review

Summary

2.1 | The Nature of Matter

Definitions to remember:

- **Matter:** Takes up space, has mass, and is composed of elements
- **Element:** A fundamental constituent of matter that cannot be broken down into another substance; composed of atoms
- **Nucleus:** The center of an atom, contains protons (+) and neutrons
- **Atomic mass:** Protons plus neutrons
- **Isotopes:** Atoms of same type that differ in the number of neutrons
- **Atomic number:** Determined by the number of protons in an atom; equals number of electrons when atom is neutral
- **Electrons** (−)

  First shell: Complete with two electrons

  Other shells: Complete with eight electrons, assuming an atomic number of 20 or below.

  Valence shell: Number of electrons in outer shell determines reactivity of atom.

- **Octet rule:** Atoms react with one another in order to have a completed outer shell with eight electrons.

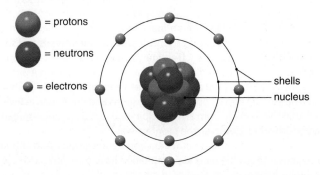

= protons

= neutrons

= electrons

shells

nucleus

- **Ionic bond:** Attraction between oppositely charged ions. Ions form when atoms lose or gain one or more electrons to achieve a completed outer shell.
- **Covalent bond:** Sharing of electrons between two atoms. There are single covalent bonds (sharing one pair of electrons), double (sharing two pairs of electrons), and triple (sharing three pairs of electrons).

2.2 | Water's Importance to Life

The two covalent O—H bonds of water are polar because the electrons are not shared equally between oxygen and hydrogen. The positively and negatively charged ends of the molecules are attracted to each other to form hydrogen bonds. The polarity and hydrogen bonding in water account for its unique properties, which can be summarized as follows:

Properties	Chemical Reasons	Effect
Water is a solvent.	Polarity	Water facilitates chemical reactions.
Water is cohesive and adhesive.	Hydrogen bonding; polarity	Water serves as a transport medium.
Water has a high surface tension.	Hydrogen bonding	The surface tension of water is hard to break.
Water has a high heat capacity.	Hydrogen bonding	Water protects organisms from rapid changes in temperature.
Water has a high heat of vaporization.	Hydrogen bonding	Water helps organisms resist overheating.
Water is less dense as ice.	Hydrogen bonding changes	Ice floats on liquid water.

2.3 | Acids and Bases

Water dissociates to produce an equal number of hydrogen ions and hydroxide ions. This is neutral pH. This illustration shows the range of acidic and basic solutions:

Cells are sensitive to pH changes. Biological systems contain buffers that help keep the pH within a normal range.

Key Terms

acid   27
atom   18
atomic mass   18
atomic number   19
atomic symbol   18
base   27
buffer   28
compound   21
covalent bond   22
electron   18
electronegativity   24
electron shell   20
element   18
hydrogen (H) bond   24
hydrogen ion (H⁺)   27
hydrophilic   25
hydrophobic   25
hydroxide ion (OH⁻)   27

ion   22
ionic bond   22
isotope   19
matter   18
molecule   21
neutron   18
nucleus   18
octet rule   21
pH   28
pH scale   28
polar   24
product   23
proton   18
reactant   23
salt   22
tracer   20
valence shell   21

Testing Yourself

Choose the best answer for each question.

1. The atomic mass of an atom depends primarily on the number of
   a. protons and neutrons.   c. neutrons and electrons.
   b. positrons.   d. protons and electrons.
2. The most abundant element by weight in the human body is
   a. carbon.   c. oxygen.
   b. hydrogen.   d. nitrogen.
3. This rule states that the outer electron shell is most stable when it contains eight electrons.
   a. stability rule   c. octet rule
   b. atomic rule   d. shell rule
4. How many more electrons does nitrogen require to fill its outer shell?
   a. 0   c. 2
   b. 4   d. 3

Nitrogen

5. A substance with a radioactive tag attached that allows a researcher to track it in a living system is called a
   a. neutron.   c. positron.
   b. tracer.   d. noble gas.
6. Which type of chemical bond is formed when atoms share electrons?
   a. covalent   c. hydrogen
   b. ionic   d. nitrogen
7. A covalent bond in which electrons are not shared equally is called
   a. polar.   c. nonpolar.
   b. normal.   d. neutral.
8. Water flows freely, but does not separate into individual molecules because water is
   a. cohesive.   c. hydrophobic.
   b. hydrophilic.   d. adhesive.
9. Compounds having an affinity for water are said to be
   a. cohesive.   c. hydrophobic.
   b. hydrophilic.   d. adhesive.
10. Water freezes from the top down because
    a. water has a high surface tension.
    b. ice has a high heat capacity.
    c. ice is less dense than water.
    d. ice is less cohesive than water.
11. Water can absorb a large amount of heat without much change in temperature because it has a high
    a. surface tension.
    b. heat capacity.
    c. hydrogen ion (H^+) concentration.
    d. hydroxide ion (OH^-) concentration.

Thinking Scientifically

1. If you watch a bird dive into the water, you will see that the surface of the water is smooth and continuous as the bird enters. However, when the bird flies back into the air, many drops of water fly into the air as well (see Fig. 2.9). Explain this observation based on properties of water.

2. Like carbon, silicon has four electrons in its outer shell. Yet life evolved to be carbon-based. What is there about silicon's structure that might prevent it from sharing with four other elements and prevent it from forming the many varied shapes of carbon molecules?

Bioethical Issue

Acid Precipitation

Acid precipitation is produced when atmospheric water is polluted by sulfur dioxide and nitrous oxide emissions. These compounds are converted to sulfuric and nitric acids, absorbed into water droplets in the atmosphere, and then fall back to Earth as acid precipitation. These emissions are mostly produced by the burning of fossil fuels, particularly coal. Because the emissions are airborne, the acid precipitation may occur far from where the fossil fuels are being burned. For example, most of the acid precipitation in Scandinavian countries comes from emissions produced in Great Britain, central Europe, and Russia. Similarly, much of the acid precipitation in southeastern Canada results from industrial activity in the midwestern and eastern United States.

Acid precipitation harms aquatic ecosystems and affects the balance of species within ecosystems on land. It also corrodes metals and some types of building material such as limestone and marble. However, not all scientists are as concerned. The National Acidic Precipitation Assessment Program (NAPAP) in 1990 concluded that although acid precipitation has had a noticeable impact on many ecosystems and on human activities as well, that the damage from acid precipitation was not as widespread as had been feared. However, the program was reauthorized and continues to monitor the situation, as do other environmental groups. And although the emissions responsible for acid precipitation have been linked to respiratory problems in humans, no studies to date have shown acid precipitation to be harmful to human health.

To address these problems, many industrialized countries, including the United States and Canada, signed the Convention on Long-Range Transboundary Air Pollution (CLRTAP) between 1979 and 1983. While this and other agreements are a positive first step in addressing the problem, acid precipitation persists, and has become a large problem in other parts of the world, particularly in India and southwest China.

Do you think acid precipitation is a significant environmental problem, and why or why not? Do you think the countries producing pollution should compensate their neighbors for the damage being done? Are countries at least obligated to minimize the effects of their air pollutants on other countries? And since much of the emissions that cause acid precipitation are due to the burning of coal for electricity, would you support the use of nuclear energy to reduce the use of coal-burning plants?

Essentials of Biology Website

The companion website for *Essentials of Biology* provides a wealth of information organized and integrated by chapter. You will find practice tests, animations, videos, and much more that will complement your learning and understanding of general biology.

http://www.mhhe.com/maderessentials2

3

The Organic Molecules of Life

Some human cells can produce up to 10,000 different types of proteins.

Dietary fiber, found in green vegetables, grains, and supplements, keeps us regular and helps protect us from cancer.

Cholesterol found in foods such as egg yolks has many important functions, including the production of sex hormones.

fat
plaque
cholesterol crystals

Outline

Not All Cholesterol Is Bad

WITHOUT CHOLESTEROL YOUR body would not function well. What is it about cholesterol that can cause problems in our bodies? Although blood tests now distinguish between "good" cholesterol and "bad" cholesterol, in reality there is only one molecule called cholesterol. Cholesterol is absolutely essential to the normal functioning of our cells. It is a part of the cell's outer membrane, and it also serves as a precursor to hormones such as the sex hormones testosterone and estrogen. Without cholesterol, your cells would be in major trouble.

—Continued next page

Continued—

So when does cholesterol become "bad"? Your body packages cholesterol in high-density lipoproteins (HDLs) or low-density lipoproteins (LDLs), depending on such factors as genetics, your diet, your activity level, and whether you smoke. Cholesterol packaged as HDL is primarily on its way from the tissues to the liver for recycling. This reduces the likelihood of deposits called plaques forming in the arteries; thus, HDL cholesterol is "good." LDL is carrying cholesterol to the tissues; high levels of LDL-cholesterol contribute to the development of plaque that can result in heart disease, making LDL the "bad" form. When cholesterol levels are measured, the ratio of LDL to HDL is determined. Exercising, improving the diet, and not smoking can lower LDL. In addition, a variety of prescription medications can help lower LDL levels and raise HDL levels.

The molecules that make up our bodies are largely carbohydrates, lipids, proteins, and nucleic acids. The building blocks for nearly all of these molecules come from our diet. Cells join the building blocks together to form the molecules that allow them to perform their many complex duties. In this chapter you will learn about the structure and function of the major molecules of cells.

3.1 | Organic Molecules

If you decide to take a chemistry course, you will most likely have a choice of either inorganic chemistry or organic chemistry. **Inorganic chemistry** is often thought of as the chemistry of the nonliving world, and **organic chemistry** is the chemistry of the living world (**Fig. 3.1**). But water, the molecule that makes up 70–90% of a cell, is an inorganic molecule. Why should that be? For a molecule to be **organic,** it must contain carbon and hydrogen; otherwise, it is an inorganic molecule. The chemistry of carbon accounts for the molecules that make up the structure of cells and organisms.

The Carbon Atom

A bacterial cell contains some 5,000 different organic molecules, and a plant or animal cell has twice that number. What is there about carbon that makes organic molecules so diverse and also so complex? Carbon, with a total of six electrons, has four electrons in the outer shell. In order to acquire four electrons to complete its outer shell, a carbon atom almost always shares electrons with—you guessed it—CHNOPS, the elements that make up most of the weight of living things (see p. 18).

Because carbon is small and needs to acquire four electrons, carbon can bond with as many as four other elements. Carbon atoms most often share electrons with other carbon atoms. The C—C bond is stable, and the result is that carbon chains can be quite long. Hydrocarbons are

Figure 3.1 **Organic molecules as structural materials.**

Some organic molecules are purely structural in nature. Related organic molecules (**a**) help hold a plant stem erect, (**b**) make the shell of a crab hard, and (**c**) support the wall of a bacterium.

b.

c.   ×38,000   a.

Carbon chains can vary in length, and/or have double bonds, and/or be branched.

Carbon chains can form rings of different sizes and have double bonds.

Figure 3.2 Hydrocarbons are highly versatile.

Hydrocarbons contain only hydrogen and carbon. Even so, they can be quite varied according to the number of carbons, the placement of any double bonds, possible branching, and possible ring formation.

Functional Groups		
Group	**Structure**	**Found in**
Hydroxyl	R—O—H	Alcohols, sugars
Carboxyl	R—C(=O)—O—H	Amino acids, fatty acids
Amino	R—N with H, H	Amino acids, proteins
Sulfhydryl	R—S—H	Amino acid cysteine, proteins
Phosphate	R—O—P(=O)(O)—O—H	ATP nucleic acids
R = remainder of molecule		

Figure 3.3 Functional groups.

Molecules with the same carbon skeleton can still differ according to the type of functional group attached. Many of these functional groups are polar, helping to make the molecule soluble in water. In this illustration, the remainder of the molecule, the hydrocarbon chain, is represented by an *R*.

chains of carbon atoms that are bonded only to hydrogen atoms. Any carbon atom of a hydrocarbon molecule can start a branch chain, and a hydrocarbon can also turn back on itself to form a ring compound (**Fig. 3.2**). Carbon can form double bonds with itself and other atoms. **Isomers** are molecules that have the same number and kinds of atoms but different chemical properties because the atoms occur in different arrangements. It is apparent, then, that the chemistry of carbon leads to a huge diversity of organic molecules.

The Carbon Skeleton and Functional Groups

The carbon chain of an organic molecule is called its skeleton, or backbone. This terminology is appropriate because just as your skeleton accounts for your shape, so does the carbon skeleton of an organic molecule account for its shape. The reactivity of an organic molecule is largely dependent on the attached functional groups (**Fig. 3.3**). A **functional group** is a specific combination of bonded atoms that always has the same chemical properties and therefore always reacts in the same way, regardless of the particular carbon skeleton to which it is attached. As in Figure 3.3, we often use an *R* to stand for the rest of the molecule to save space, because only the functional group is involved in the reaction.

The functional groups of an organic molecule therefore help determine its chemical properties. For example, sugars contain a carbon skeleton with many attached polar —OH groups. Thus, although hydrocarbons are nonpolar and **hydrophobic** (not soluble in water), glucose with several —OH groups is actually **hydrophilic** (soluble in water; see Chapter 2). Because cells are mainly composed of water, the ability to interact with and be soluble in water profoundly affects the activity of organic molecules in cells.

Organic molecules containing carboxyl groups (—COOH) are both polar (hydrophilic) and weakly acidic. They partially ionize and release hydrogen ions in solution:

$$—COOH \longrightarrow —COO^- + H^+$$

Check Your Progress

1. Describe the attributes of a carbon atom that allow it to form varied molecules.

Answer: 1. The carbon atom can bond with up to four different atoms. Carbon-to-carbon bonds are stable, so long chains can be built. These chains can be variable in length and branching patterns. Some organic molecules can form isomers.

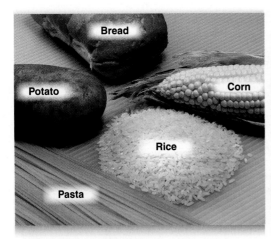

Figure 3.4 **Carbohydrates.**

The familiar carbohydrates, sugar and starch, are present in the foods shown. A diet loaded with these carbohydrates may make you prone to diabetes type 2 and perhaps to other illnesses as well. In contrast, moderate amounts of multigrain foods are better for you and provide fiber to keep you regular.

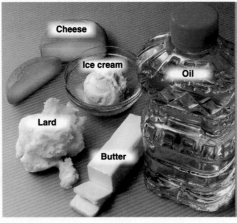

Figure 3.5 **Lipid foods.**

A diet rich in fat can indeed make you fat. However, you do need some fat in your diet. Choose vegetable oils over the animal fat in lard, butter, and other dairy products. The fats in vegetable oils lower your risk of cardiovascular disease if they haven't been hydrogenated, as they are in processed foods.

Figure 3.6 **Protein foods.**

It's surprising how little protein is needed in the diet. Just 6 ounces a day will provide you with all the amino acids you need. Beef, but not chicken, can be loaded with animal fat. Fish, however, contains beneficial oils that lower the incidence of cardiovascular disease. Also, the diet can include an egg a day, usually with no ill effects.

Figure 3.7 **Synthesis and breakdown of polymers.**

a. In cells, synthesis often occurs when monomers bond during a dehydration reaction (removal of H_2O). **b.** Breakdown occurs when the monomers in a polymer separate because of a hydrolysis reaction (the addition of H_2O).

3.2 | The Biological Molecules of Cells

Despite their great diversity, organic molecules in cells, called **biological molecules,** are grouped into only four categories: carbohydrates, lipids, proteins, and nucleic acids. You are very familiar with these molecules because certain foods are known to be rich in carbohydrates, lipids, and proteins, as illustrated in **Figures 3.4, 3.5,** and **3.6.** When you digest these foods, they break down into subunit molecules. Your body then takes these subunits and builds from them the large macromolecules that make up your cells. Many different foods also contain nucleic acids, the type of molecule that forms your genes.

Certain biological molecules in cells are composed of a large number of the same type of subunits, called **monomers.** When monomers join the result is a **polymer.** A protein can contain hundreds of amino acid monomers, and a nucleic acid can contain hundreds of nucleotide monomers. How can polymers get so large? Just as a train increases in length when boxcars are hitched together one by one, so a polymer gets longer as monomers bond to one another.

A cell uses the same type of reaction to synthesize any type of biological molecule. It is called a **dehydration reaction** because the equivalent of a water molecule—that is, an —OH (hydroxyl group) and an —H (hydrogen atom), is removed as the reaction occurs (**Fig. 3.7**a). To break down a biological molecule, a cell and also the digestive tract uses an opposite type of reaction: During a **hydrolysis reaction,** an —OH group from water attaches to one monomer, and an —H from water attaches to the other monomer (Fig. 3.7b). In other words, water is used to break the bond holding monomers together.

Carbohydrates

Carbohydrates are almost universally used as an immediate energy source in living things, but they also play structural roles in a variety of organisms. Carbohydrates are polymers of individual monomers called saccharides, or sugars. Typically, the sugar glucose is a monomer of carbohydrate polymers. The term carbohydrate may refer to either a single sugar molecule (monosaccharide), two sugar molecules (disaccharide), or many sugar molecules bonded together (polysaccharide).

Monosaccharides: Ready Energy

Because **monosaccharides** have only a single sugar molecule they are called simple sugars. A simple sugar can have a carbon backbone of three to seven carbons. The word *carbohydrate* might make you think that every carbon atom is bonded to an H and an —OH. This is not strictly correct, as you can see by examining the structural formula for glucose (**Fig. 3.8**). Still, sugars do have many polar —OH groups, which make them soluble in water.

   Glucose, with six carbon atoms, has a molecular formula of $C_6H_{12}O_6$. Glucose has two important isomers, called *fructose* and *galactose*, but even so, we usually think of glucose when we see the formula $C_6H_{12}O_6$. That's because glucose has a special place in the chemistry of organisms. Glucose is transported in the blood of animals, and it is also the molecule that is broken down in nearly all types of organisms as an immediate source of energy. In other words, cells use glucose as the energy source of choice.

   Ribose and **deoxyribose,** with five carbon atoms, are significant because they are found in the nucleic acids RNA and DNA, respectively. RNA and DNA are discussed later in this chapter.

Disaccharides: Varied Uses

A **disaccharide** (*di*, two; *saccharide*, sugar) contains two monosaccharides bonded together. The brewing of beer relies on maltose, a disaccharide usually derived from barley. During the production of beer, yeast breaks down maltose to two units of glucose and then uses glucose as an energy source. As the yeast use glucose for energy, they use a process called *fermentation*, which produces ethyl alcohol (**Fig. 3.9**).

   Sucrose, a disaccharide acquired from sugar beets and sugarcane, is of special interest because we use it at the table and in baking as a sweetener. Our body digests sucrose to its components, glucose and fructose. (Later, the fructose is changed to glucose, our usual energy source.) If the body doesn't need more energy at the moment, the glucose is metabolized to fat! While glucose is the energy source of choice for animal cells, fat is the body's primary energy storage form. That's why eating lots of sugary desserts can make you gain weight.

   When you make lemonade at home, you add sucrose. But drinks made commercially often contain high-fructose corn syrup (HFCS). In the 1980s, a commercial method was developed for converting the glucose in corn syrup to the much sweeter-tasting fructose. Nutritionists are not in favor of eating highly *processed* foods that are rich in sucrose, HFCS, and white starches. They say these foods provide "empty" calories, meaning that although they supply energy, they don't supply any of the vitamins, minerals, and fiber needed in the diet. In contrast, minimally processed foods provide glucose, starch, and many other types of nutritious molecules.

Figure 3.8 **Glucose.**

Each of these structural formulas represents glucose. **a.** The carbon skeleton (with carbon atoms numbered) and all attached groups are shown. **b.** The carbon skeleton is omitted. **c.** The carbon skeleton and attached groups are omitted. **d.** Only the ring shape of the molecule remains.

Figure 3.9 **Breakdown of maltose, a disaccharide.**

Maltose is the energy source for yeast during the production of beer. Yeasts differ as to the amount of maltose they convert to alcohol, so selection of the type of yeast is important.

starch granule in potato cell

nonbranched

branched

×57

a. Starch structure

glycogen granules in liver cell

highly branched

×59,400

b. Glycogen structure

cellulose fibers in plant cell wall

— H bond

×20

c. Cellulose structure

Figure 3.10   Starch and glycogen structure and function.

a. Glucose is stored in plants as starch. The electron micrograph shows the location of starch in plant cells. Starch is a chain of glucose molecules that can be nonbranched or branched. **b.** Glucose is stored in animals as glycogen. The electron micrograph shows glycogen deposits in a portion of a liver cell. Glycogen is a highly branched polymer of glucose molecules. **c.** In plant cell walls, each cellulose fiber contains several microfibrils. Each microfibril contains many polymers of glucose hydrogen-bonded together. Three such polymers are shown.

Polysaccharides as Energy Storage Molecules

Polysaccharides are polymers of monosaccharides. Some types of polysaccharides function as short-term energy storage molecules because they are much larger than a sugar and are relatively insoluble. Polysaccharides cannot easily pass through the plasma membrane and are kept (stored) within the cell.

Plants store glucose as **starch.** The cells of a potato contain granules where starch resides during winter until energy is needed for growth in the spring. Notice in **Figure 3.10***a* that starch exists in two forms—one is nonbranched and the other is branched. Animals store glucose as **glycogen,** which is more highly branched (Fig. 3.10*b*). Branching subjects a polysaccharide to more attacks by hydrolytic enzymes; therefore, branching makes a polysaccharide easier to break down. Spiraling makes these polysaccharides more compact.

The storage and release of glucose from liver cells is controlled by hormones. After we eat, the pancreas releases the hormone insulin, which promotes the storage of glucose as glycogen.

Polysaccharides as Structural Molecules

Some types of polysaccharides function as structural components of cells. **Cellulose** is the most abundant of all the carbohydrates, which in turn are the most abundant of all the organic molecules on Earth. Plant cell walls contain cellulose, and therefore cellulose is plentiful in the wood of tree trunks. The seeds of cotton plants have long fibers composed mostly of cellulose. Humans use wood for construction, and they use cotton fibers to make a cloth of the same name.

The bonds joining the glucose subunits in cellulose are different from those found in starch and glycogen (Fig. 3.10*c*). As a result, the molecule does not spiral nor have branches. The long glucose chains are held parallel to each other by hydrogen bonding to form strong microfibrils and then fibers. The fibers crisscross within plant cell walls for even more strength.

The digestive juices of animals can't hydrolyze cellulose, but some microorganisms can digest it. Cows and other ruminants (cud-chewing animals) have a special internal pouch where microorganisms break down cellulose to glucose. In humans, cellulose has the benefit of serving as dietary fiber, which maintains regular elimination and digestive system health.

Chitin, which is found in the exoskeleton of crabs and related animals such as lobsters and insects, is also a polymer of glucose. However, each glucose subunit has an amino group ($-NH_2$) attached to it. Since the functional groups attached to organic molecules determine their properties, chitin is chemically different from cellulose, even though the linkage between the glucose molecules is like that found in cellulose. Even though chitin, like cellulose, is not digestible by humans but still has many good uses. Seeds are coated with chitin, and this protects them from attack by soil fungi. Because chitin also has antibacterial and antiviral properties, it is processed and used in medicine as a wound dressing and suture material. Chitin is even useful during the production of cosmetics and various foods.

Lipids

Although molecules classified as **lipids** are quite varied, they have one characteristic in common: They are all insoluble in water due to their long nonpolar hydrocarbon chains and their relative lack of hydrophylic functional groups. It is said that oil and water do not mix. For example, salad dressings are rich in vegetable oils. Even after shaking, the vegetable oil will separate out from the water.

Lipids are very diverse and they have varied structures and functions. **Fats** (such as bacon fat, lard, and butter), and **oils** (such as corn oil, olive oil, and coconut oil) are some well-known lipids. In animals, fats are used for both insulation and long-term energy storage. Fat below the skin of marine mammals is called blubber; around the waist of humans, it is often referred to as a "spare tire." Instead of fat, plants use oils for long-term energy storage. In animals, the secretions of oil glands help waterproof skin, hair, and feathers (**Fig. 3.11**).

Fats and Oils: Long-Term Energy Storage

Fats and oils contain two types of subunit molecules: glycerol and fatty acids (**Fig. 3.12**). **Glycerol** is a compound with three —OH groups. The —OH groups are polar; therefore, glycerol is soluble in water. A **fatty acid** has a long chain of carbon atoms bonded only to hydrogen, with a carboxyl group at one end. A fat or oil forms when the carboxyl portions of three fatty acids react with the —OH groups of glycerol. This is a dehydration reaction because in addition to a fat molecule, three molecules of water result. Fats and oils are degraded during a hydrolysis reaction, in which water is added to the molecule (see Fig. 3.12). Because three long fatty acids are attached to each glycerol molecule, fats and oils are called **triglycerides** and pack a lot of energy in one molecule. Thus it is logical that fats and oils are the body's primary long-term energy storage molecules.

Fatty acids are the primary components of fats and oils. Most of the fatty acids in cells contain 16 or 18 carbon atoms per molecule, although smaller or larger ones are also found. Fatty acids are either saturated or unsaturated (**Fig. 3.13**).

Figure 3.11   Preening in birds.

As a bird preens, it transfers oil to its feathers from a gland at the base of its tail. Because the oil repels water, the feathers become waterproof.

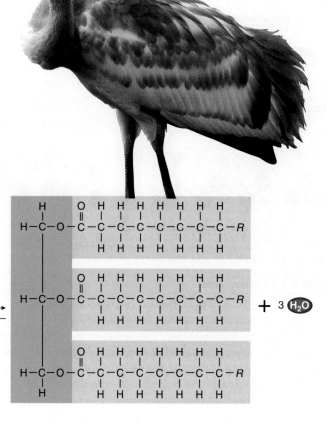

Glycerol   **3 Fatty acids**   **Fat**   **3 Waters**

Figure 3.12   Synthesis and breakdown of fat.

Following a dehydration reaction, glycerol is bonded to three fatty acid molecules, and water is given off. Following a hydrolysis reaction, the bonds are broken due to the addition of water. *R* represents the remainder of the molecule, which in this case is a continuation of the hydrocarbon chain, composed of 16 or 18 carbons.

canola oil

carboxyl group

$C_{18}H_{34}O_2$

bend caused by double bond

a. Oleic acid, a monounsaturated fatty acid (one double bond) found in canola oil.

butter

carboxyl group

$C_{18}H_{36}O_2$

b. Stearic acid, a saturated fatty acid (no double bonds) found in butter.

donut

carboxyl group

$C_{18}H_{34}O_2$

c. Elaidic acid, a *trans* fatty acid (one double bond) found in many snack foods.

Figure 3.13 Fatty acids.

A fatty acid has a carboxyl group attached to a long hydrocarbon chain. **a.** If there is one double bond between adjacent carbon atoms in the chain, the fatty acid is monounsaturated. **b.** If there are no double bonds, the fatty acid is saturated. A diet high in saturated fats appears to contribute to disease of the heart and blood vessels. **c.** Hydrogenation of oils is not always complete. Further, at some remaining double bonds, the attached hydrogens are in the trans configuration. If so, the fatty acid is a trans fatty acid. Trans fats are also linked to cardiovascular disease.

Unsaturated fatty acids have double bonds in the carbon chain wherever the number of hydrogens is less than two per carbon atom (Fig 3.13*a*). **Saturated fatty acids** have no double bonds between the carbon atoms (Fig 3.13*b*). The carbon chain is saturated, so to speak, with all the hydrogens it can hold. Saturation or unsaturation of a fatty acid determines its chemical and physical properties.

In general, oils are liquids at room temperature because they contain unsaturated fatty acids. Notice in Figure 3.13*a* that the double bond creates a bend in the fatty acid chain. Such kinks prevent close packing between the hydrocarbon chains and account for the fluidity of oils. On the other hand, butter, which contains mostly saturated fatty acids, is a solid at room temperature. The saturated fatty acid chains can pack together more tightly because they have no kinks (Fig. 3.13*b*).

Trans fats contain fatty acids that have been partially hydrogenated to make them more saturated, and thus more solid (Fig. 3.13*c*). Trans fats are often found in processed foods—margarine, baked goods, and fried foods in particular. Saturated fats and trans fats have been linked to a cardiovascular disease called atherosclerosis, in which lipid material, called plaque, accumulates inside blood vessels. Plaque contributes to high blood pressure and heart attacks. Unsaturated oils, particularly *monounsaturated* (one double bond) oils but also *polyunsaturated* (many double bonds) oils, have been found to be protective against atherosclerosis. These healthy oils are found in abundance in olive oil, canola oil, and certain fish.

Phospholipids: Membrane Components

Phospholipids, as implied by their name, contain a phosphate group. Essentially, a phospholipid is constructed like a triglyceride, except that in place of the third fatty acid attached to glycerol, there is a charged phosphate group. The phosphate group is usually bonded to another polar functional group, indicated by *R* in **Figure 3.14***a*. Thus, one end of the molecule is water-soluble. This portion of the molecule is the polar head. The hydrocarbon chains of the fatty acids, or the nonpolar tails, are not water-soluble.

Because phospholipids have both hydrophilic (polar) heads and hydrophobic (nonpolar) tails, they tend to arrange themselves so that only the polar heads interact with the watery environment outside and the nonpolar tails crowd inward away from the water. Between two compartments of water, such as the outside and inside of a cell, phospholipids become a bilayer in which the polar heads project outward and the nonpolar tails project inward. The bulk of the plasma membrane that surrounds cells consists of a fairly fluid phospholipid bilayer, as do all the other membranes in the cell (Fig. 3.14*b*). A plasma membrane is absolutely essential to the structure and function of a cell, and thus phospholipids are vitally important to humans and other organisms.

Steroids: Four Fused Rings

Steroids are lipids that possess a unique carbon skeleton made of four fused rings, shown in red in **Figure 3.15.** Unlike other lipids, steroids do not contain fatty acids, but they are similar to other lipids because they are insoluble in water. Steroids are also very diverse. Each type of steroid differs primarily in the types of functional groups attached to the carbon skeleton.

Figure 3.14 **Phospholipids from membranes.**

a. Phospholipids are constructed like fats, except that in place of the third fatty acid, they have a charged phosphate group. The hydrophilic (polar) head group is soluble in water, whereas the two hydrophobic (nonpolar) tail groups are not. **b.** This causes the molecules to arrange themselves as a bilayer in the plasma membrane that surrounds a cell.

a. Phospholipid structure

b. Plasma membrane of a cell

a. Cholesterol

b. Testosterone

c. Estrogen

Figure 3.15 **Steroid diversity.**

a. All the steroids are derived from cholesterol, an important component of the plasma membrane. Cholesterol and all steroid molecules have four adjacent rings, but their effects on the body largely depend on the attached groups indicated in red. The different effects of (**b**) testosterone and (**c**) estrogen on the body are due to different groups attached to the same carbon skeleton.

Cholesterol (Fig. 3.15a) is a component of an animal cell's plasma membrane, and it is the precursor of other steroids, such as the sex hormones testosterone and estrogen. The male sex hormone, testosterone, is formed primarily in the testes, and the female sex hormone, estrogen, is formed primarily in the ovaries (Fig. 3.15b,c). Testosterone and estrogen differ only by the functional groups attached to the same carbon skeleton, and yet they have a profound effect on the body and the sexuality of humans and other animals.

Most likely you are aware that some people take anabolic steroids, such as synthetic testosterone, to increase muscle mass. The result is usually unfortunate. The presence of the steroid in the body upsets the normal hormonal balance: The testes atrophy, and males may develop breasts; females tend to grow facial hair and lose hair on their heads. Because steroid use gives athletes an unfair advantage and destroys their health—heart, kidney, liver, and psychological disorders are common—they are banned by professional athletic associations.

Check Your Progress

1. **Compare the structure of a saturated fatty acid with an unsaturated fatty acid.**
2. **Explain why phospholipids form a bilayer in a watery medium.**
3. **Describe how the structure of fats and oils differs from that of steroids.**

Answers: 1. A saturated fatty acid contains no double bonds between carbon atoms, while an unsaturated fatty acid contains one or more double bonds. 2. Phospholipids arrange themselves so that their hydrophilic heads are adjacent to water, while the hydrophobic tails point inward toward each other. 3. Fats and oils contain fatty acids and glycerol, while steroids have four fused carbon rings.

Proteins

Proteins are of primary importance in the structure and function of cells. Here are some of their many functions:

Support Some proteins are structural proteins. Examples include the protein in spiderwebs; keratin, the protein that makes up hair and fingernails; and collagen, the protein that lends support to skin, ligaments, and tendons (**Fig. 3.16***a*).

Metabolism Many proteins are **enzymes**. They bring reactants together and thereby speed chemical reactions in cells. They are specific for one particular type of reaction and can function at body temperature.

Transport Channel and carrier proteins in the plasma membrane allow substances to enter and exit cells. Other proteins transport molecules in the blood of animals—for example, **hemoglobin** is a complex protein that transports oxygen (Fig. 3.16*b*).

Defense Some proteins, called antibodies, combine with disease-causing agents to prevent them from destroying cells and upsetting homeostasis, the relative constancy of the internal environment.

Regulation Hormones are regulatory proteins. They serve as intercellular messengers that influence the metabolism of cells. For example, the hormone insulin regulates the content of glucose in the blood and in cells, while growth hormone determines the height of an individual.

Motion The contractile proteins actin and myosin allow parts of cells to move and cause muscles to contract (Fig. 3.16*c*). Muscle contraction enables animals to move from place to place.

a. Structural proteins

b. Transport proteins

c. Contractile proteins

Figure 3.16 Types of proteins.

a. The protein in hair, fingernails, and spiderwebs (keratin) is a structural protein, as is collagen. Collagen injections can reduce wrinkles and scarring by providing extra support.
b. Hemoglobin, a major protein in red blood cells, is involved in transporting oxygen. **c.** Contractile proteins, actin and myosin, cause muscles to move.

The structures and functions of cells differ according to the type of protein they contain. Muscle cells contain actin and myosin; red blood cells contain hemoglobin; support cells produce the collagen they secrete.

Amino Acids: Subunits of Proteins

Proteins are polymers and their monomers are called **amino acids.** Amino acids have an unique carbon skeleton in which a central carbon atom bonds to a hydrogen atom, two functional groups, and a side chain, or *R* group (**Fig. 3.17***a*). The name *amino acid* is appropriate because one of the two functional groups is an —NH$_2$ (amino group) and another is a —COOH (an acid group as discussed on page 33).

There are 20 different amino acids, which differ according to their particular *R* group. The *R* groups range in complexity from a single hydrogen atom to a complicated ring compound. The unique chemical properties of an amino acid depend on those of the *R* group. For example, some *R* groups are polar and some are not. Also, the amino acid cysteine has an *R* group that ends with a sulfhydryl (—SH) group, which can serve to connect one chain of amino acids to another by a disulfide bond, —S—S—. Four amino acids commonly found in cells are shown in Figure 3.17*b*.

Peptides

Figure 3.18 shows how two amino acids join by a dehydration reaction between the carboxyl group of one and the amino group of another. The resulting covalent bond between two amino acids is called a **peptide bond.** The atoms associated with the peptide bond share the electrons unevenly because oxygen is more electronegative than nitrogen. Therefore, the hydrogen attached to the nitrogen has a slightly positive charge (δ^+), while the oxygen has a slightly negative charge (δ^-):

Therefore, the peptide bond is polar, and hydrogen bonding is possible between the C=O of one amino acid and the N—H of another amino acid in a polypeptide.

Figure 3.17   **Amino acids.**

a. Structure of an amino acid. **b.** Proteins contain 20 different kinds of amino acids, four of which are shown here. Amino acids differ by the particular *R* group (blue) attached to the central carbon. Some *R* groups are nonpolar and hydrophobic, some are polar and hydrophilic, and some are ionized and hydrophilic.

Figure 3.18   **Synthesis and degradation of peptide.**

Following a dehydration reaction, a peptide bond joins two amino acids, and water is given off. Following a hydrolysis reaction, the bond is broken due to the addition of water.

A **peptide** is two or more amino acids covalently bonded together, and a **polypeptide** is a chain of many amino acids joined by peptide bonds. Polypeptides containing more than 100 amino acids are generally referred to as proteins, but a protein may contain one or more polypeptide chains. Therefore, you can see why most proteins are composed of a very large number of amino acids. Ribonuclease, a small protein that breaks down RNA, contains barely more than 100 amino acids. But some proteins may be very large, like titin, which contains over 33,000 amino acids! Titin is an integral part of the structure of your muscles; without it, your muscles would not function properly.

Each polypeptide has its own normal sequence. The amino acid sequence determines the final three-dimensional shape of the protein, and thus its function. Proteins that have an abnormal sequence of amino acids often have the wrong shape and cannot function properly.

Shape of Proteins

The function of a protein is determined by its three-dimensional shape. All proteins have multiple levels of structure. These levels are called the primary, secondary, tertiary, and quaternary structures (**Fig. 3.19**). A protein's sequence of amino acids is called its *primary structure*. Consider that an almost infinite number of words can be constructed by changing the number and sequence of the 26 letters in our alphabet. In the same way, many different proteins can result by varying the number and sequence of just 20 amino acids.

The *secondary structure* of a protein occurs when portions of the amino acid chain take on a certain orientation in space depending on the number and identity of the amino acids present in the chain. Portions of the polypeptide can have the spiral shape of an alpha helix, or it can turn back on itself, forming a pleated sheet. Hydrogen bonds between nearby peptide bonds maintain the secondary structure of a protein. Any polypeptide may contain one or more secondary structures within the same chain.

Proteins also have a *tertiary structure*. The tertiary structure of a protein is its overall three-dimensional shape that results from the folding and twisting of its secondary structure. The tertiary structure is held in place by interactions between the *R* groups of amino acids making up the helices and pleated sheets within the polypeptide. Several types of interactions are possible. For example,

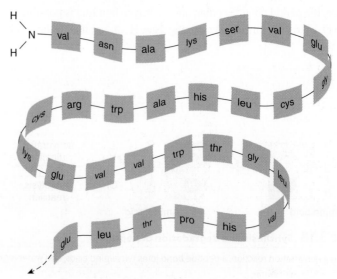

Figure 3.19   Levels of protein organization.

All proteins have a primary structure. Fibrous proteins have a secondary structure; they are either helices (e.g., keratin, collagen) or pleated sheets (e.g., silk). Globular proteins always have a tertiary structure, and most have a quaternary structure (e.g., hemoglobin, enzymes).

Primary structure: sequence of amino acids

ionic bonds, hydrogen bonds, and even covalent bonds can occur between *R* groups. Also, a tight packing of side chains can occur when hydrophobic *R* groups come together at the same location to avoid contact with water.

The tertiary structure of a protein determines its function, and this structure can be affected by the environmental conditions such as the pH and temperature. Acids and bases may interrupt interactions between *R* groups and affect a protein's structure. Likewise, a rise in temperature can also disrupt the shape of an enzyme. Frying an egg disrupts the tertiary structure of the proteins in it, causing the protein to become solid and change color. The protein has been **denatured,** and has lost its function.

Some proteins, like hemoglobin and insulin have a *quaternary structure* because they contain more than one polypeptide chain. Each polypeptide chain in the protein has its own primary, secondary, and tertiary structure. The quaternary structure is determined by how the individual polypeptide chains interact. The quaternary structure can also affect a protein's function. If hemoglobin's quaternary structure is disrupted, it can no longer carry oxygen in the blood.

The overall shapes of many proteins may be classified as fibrous or globular. **Fibrous proteins** adopt a rodlike structure. *Keratin,* the fibrous protein in hair, fingernails, horns, reptilian scales, and feathers, has many helical regions. *Collagen,* the protein that gives shape to the skin, tendons, ligaments, cartilage, and bones of animals, also contains many helical secondary structures, and adopts a rodlike shape. Thus, fibrous proteins most often have structural roles. **Globular proteins** have a rounded or irregular three-dimensional tertiary shape. Enzymes are globular proteins, as is hemoglobin.

Check Your Progress

1. What is the primary structure of a protein?
2. Describe the interactions that create secondary and tertiary structures in proteins.

Answers: 1. The protein's sequence of amino acids is its primary structure.   2. The secondary structure is created by hydrogen bonds that occur between peptide bonds of a polypeptide chain. The tertiary structure comes into being because of interactions between *R* groups, such as ionic bonds, hydrogen bonds, and covalent bonds.

Secondary structure: alpha helix and pleated sheet   **Tertiary structure:** globular shape   **Quaternary structure:** more than one polypeptide

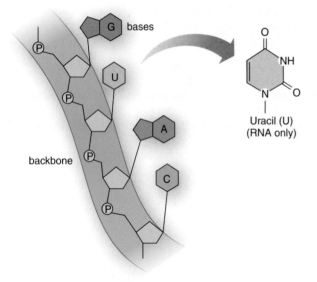

b. DNA structure with base pairs: G with C and A with T

c. RNA structure with bases G, U, A, C

Figure 3.20   DNA and RNA structure.

a. Structure of a nucleotide. **b.** Structure of DNA and its bases **c.** Structure of RNA in which uracil replaces thymine.

Nucleic Acids

DNA (*deoxyribonucleic acid*) and **RNA** (*ribonucleic acid*) are the **nucleic acids** in cells. Early investigators called them nucleic acids because they were first detected in the nucleus. As already stated, DNA stores genetic information. Each DNA molecule contains many genes, and genes specify the sequence of the amino acids in proteins. RNA is the helper that takes this information to the site of protein synthesis.

Nucleic acids are polymers in which the monomer is called a **nucleotide**. All nucleotides are composed of three parts: a phosphate, a 5-carbon sugar, and a nitrogen-containing base (**Fig. 3.20***a*). The phosphate is simply —PO_4^-. The sugar is deoxyribose in DNA and ribose in RNA, which accounts for their names. (Deoxyribose has one less oxygen than does ribose.) Each nucleotide of DNA contains one of four different bases: adenine (A), guanine (G), cytosine (C), or thymine (T).

As you can see in Figure 3.20*b*, DNA is a double helix, meaning that two strands spiral around one another. In each strand, the backbone of the molecule is composed of phosphates bonded to sugars, and the bases project to the inside. Interestingly, the base guanine (G) is always paired with cytosine (C), and the base adenine (A) is always paired with thymine (T). This is called *complementary base pairing*. Complementary base pairing holds the two strands together and is very important when DNA makes a copy of itself, a process called replication.

RNA differs from DNA not only by its sugar but also because it uses the base uracil (U) instead of thymine. Whereas DNA is double-stranded, RNA is single-stranded (Fig. 3.20*c*). Complementary base pairing also allows DNA to pass genetic information to RNA. The information is stored in the sequence of bases. DNA has a triplet code, and every three bases stands for one of the 20 amino acids in cells. Once you know the sequence of bases in a gene, you know the sequence of amino acids in a polypeptide.

Many scientists around the world were involved in the Human Genome Project. Not long ago, members of this project determined the sequence of all the base pairs in all of our 20,500 genes. They hope this information will lead to cures for many human illnesses that are caused by proteins that don't function as they should.

Figure 3.21   Sickle cell disease.

One of the amino acid chains in hemoglobin is 146 amino acids long. Sickle cell disease, characterized by sickled red blood cells, results when valine occurs instead of glutamate at the sixth amino acid.

Relationship Between Proteins and Nucleic Acids

We have learned that the functional group of an amino acid determines its behavior, and that the order of the amino acids within a polypeptide determines its shape. The shape of a protein determines its function. The structure and function of cells are determined by the types of proteins they contain. The same is true for organisms—that is, they differ with regard to their proteins. Next, we learned that DNA, the genetic material, bears instructions for the sequence of amino acids in polypeptides. The proteins of organisms differ because their genes differ.

Sometimes very small changes in a gene may cause large changes in the protein encoded by the gene, and can cause illness. This is illustrated by sickle cell disease. In sickle cell disease, the individual's red blood cells are sickle-shaped because in one particular spot the amino acid valine (Val) appears where the amino acid glutamate (Glu) should be in the protein (**Fig. 3.21**). This substitution makes red blood cells lose their normally round, flexible shape and become hard and jagged. When these abnormal red blood cells go through the small blood vessels, they may clog the flow of blood. This condition can cause pain, organ damage, and a low red blood cell count (called anemia), and can be fatal if untreated.

What's the root of the problem? The individual inherited a faulty DNA code for an amino acid in one of the hemoglobin's polypeptides.

Check Your Progress

1. List the three components of a nucleotide.
2. How are DNA and RNA chemically similar and how are they different?
3. When is complementary base pairing utilized to copy a strand of DNA?

Answers: 1. Phosphate, 5-carbon sugar, and nitrogen-containing base.   2. Both DNA and RNA are polymers of nucleotides. DNA contains the sugar deoxyribose; RNA contains ribose. DNA is double-stranded, while RNA is single stranded. In place of thymine (T) in DNA, RNA contains the base uracil (U).   3. Complementary base pairing is utilized when DNA replicates and when genetic information is passed from DNA to RNA.

The Chapter in Review

Summary

3.1 | Organic Molecules

In order to be organic, a molecule must contain carbon and hydrogen. The chemistry of carbon contributes to the diversity of organic molecules. Organic molecules have different chemical properties depending on the size and shape of their carbon skeleton and its attached functional groups. Some functional groups are nonpolar (hydrophobic), and others are polar (hydrophilic).

3.2 | The Biological Molecules of Cells

Biological molecules are synthesized by dehydration reactions and degraded by hydrolysis reactions.

Living things are composed of four types of biological molecules: carbohydrates, lipids, proteins, and nucleotides. Polysaccharides, proteins, and nucleic acids are polymers constructed when their particular monomer forms long chains. Lipids are varied and do not have a particular monomer. The characteristics of these molecules are summarized in the following chart:

Organic molecules	Examples	Monomers	Functions
Carbohydrates	Monosaccharides, disaccharides, polysaccharides	Glucose	Immediate energy and stored energy; structural molecules
Lipids	Fats, oils, phospholipids, steroids	Glycerol / Fatty acid	Long-term energy storage; membrane components
Proteins	Structural, enzymatic, carrier, hormonal, contractile	Amino acid	Support, metabolic, transport, regulation, motion
Nucleic acids	DNA, RNA	Nucleotide	Storage of genetic information

Carbohydrates

- Glucose is a monosaccharide that serves as blood sugar and as a monomer of starch, glycogen, and cellulose. Its isomers are fructose and galactose.
- Sucrose is a disaccharide (composed of glucose and fructose) that we know as table sugar.
- Starch, a polymer of glucose, stores energy in plants.
- Glycogen, a polymer of glucose, stores energy in animals.
- Maltose is a disaccharide that results from the digestion of starch and glycogen.
- Cellulose is a polymer of glucose that makes up the structure of plant cell walls.

Lipids

Fats and oils, which are composed of glycerol and fatty acids, are called triglycerides. Triglycerides made of saturated fatty acids (having no double bonds) are solids and are called fats. Triglycerides composed of unsaturated fatty acids (having double bonds) are liquids and are called oils.

Phospholipids have the same structure as triglycerides, except that a group containing phosphate takes the place of one fatty acid. Phospholipids make up the plasma membrane as well as other cellular membranes.

Steroids are lipids with a unique structure composed of four fused hydrocarbon rings. Cholesterol, a steroid, is a component of the plasma membrane. The sex hormones testosterone and estrogen are steroids.

Proteins

Proteins are polymers of amino acids. A peptide is composed of two amino acids; a polypeptide contains many amino acids. A peptide bond is a polar covalent bond.

A protein has levels of structure:
- Primary structure is the primary sequence of amino acids.
- Secondary structure is a helix or pleated sheet.
- Tertiary structure forms due to folding and twisting of the secondary structure.
- Quaternary structure occurs when a protein has more than one polypeptide.

Nucleic Acids

Cells and organisms differ because of their proteins, which are coded for by genes composed of nucleic acids. Nucleic acids are polymers of nucleotides.

DNA (deoxyribonucleic acid) contains deoxyribose sugar and is the chemical that makes up our genes. DNA is double-stranded and shaped like a double helix. The two strands of DNA are joined by complementary base pairing: Adenine (A) pairs with thymine (T), and guanine (G) pairs with cytosine (C).

RNA (ribonucleic acid) serves as a helper to DNA during protein synthesis. Its sugar is ribose, and it contains the base uracil in place of thymine. RNA is single-stranded and thus does not form a double helix. **Table 3.1** compares the structure of DNA and RNA.

Table 3.1 |

DNA STRUCTURE COMPARED TO RNA STRUCTURE		
	DNA	**RNA**
Sugar	Deoxyribose	Ribose
Bases	Adenine, guanine, thymine, cytosine	Adenine, guanine, uracil, cytosine
Strands	Double-stranded with base pairing	Single-stranded
Helix	Yes	No

Key Terms

amino acid 41	isomer 33
biological molecules 34	lipid 37
carbohydrate 35	monomer 34
cellulose 36	monosaccharide 35
chitin 36	nucleic acid 44
cholesterol 39	nucleotide 44
dehydration reaction 34	oil 37
denatured 43	organic 32
deoxyribose 35	organic chemistry 32
disaccharide 35	peptide 42
DNA (deoxyribonucleic acid) 44	peptide bond 41
enzyme 40	phospholipid 38
fat 37	polymer 34
fatty acid 37	polypeptide 42
fibrous protein 43	polysaccharide 36
functional group 33	protein 40
globular protein 43	ribose 35
glucose 35	RNA (ribonucleic acid) 44
glycerol 37	saturated fatty acid 38
glycogen 36	starch 36
hemoglobin 40	steroid 38
hydrolysis reaction 34	trans fat 38
hydrophilic 33	triglyceride 37
hydrophobic 33	unsaturated fatty acid 38
inorganic chemistry 32	

Testing Yourself

Choose the best answer for each question.

1. Which of the following is an organic molecule?
    a. CO_2
    b. H_2O
    c. $C_6H_{12}O_6$
    d. O_2
    e. More than one of these are correct.

2. Carbon requires how many electrons to complete its outer shell?
    a. 2
    b. 3
    c. 4
    d. Any one of these is correct.

3. An amino acid is to a protein as a _____ is to a nucleic acid.
    a. nucleotide
    b. fatty acid
    c. monosaccharide
    d. triglyceride

4. Carbon chains can vary in
    a. length.
    b. number of double bonds.
    c. branching pattern.
    d. All of these are correct.

5. Organic molecules containing carboxyl groups are
    a. nonpolar.
    b. acidic.
    c. basic.
    d. More than one of these are correct.

6. Biological macromolecules are polymers that are formed when _____ are joined by a _____ reaction.
    a. monomers, dehydration
    b. multimers, dehydration
    c. subunits, reduction
    d. monomers, hydrolysis

7. Which of the following is a monosaccharide?
    a. glucose
    b. cellulose
    c. lactose
    d. sucrose

For questions 8–15, match the items to those in the key. Answers can be used more than once.

Key:

    a. carbohydrate
    b. lipid
    c. protein
    d nucleic acid

8. Cellulose, the major component of plant cell walls.
9. Keratin, found in hair, fingernails, horns, and feathers.
10. Steroids such as cholesterol and sex hormones.
11. Composed of nucleotides.
12. Insoluble in water due to hydrocarbon chains.
13. Sometimes undergoes complementary base pairing.
14. May contain pleated sheets and helices.
15. May be a ring of six carbon atoms attached to hydroxyl groups.
16. A triglyceride contains
    a. glycerol and three fatty acids.
    b. glycerol and two fatty acids.
    c. protein and three fatty acids.
    d. a fatty acid and three sugars.
17. Variations in three-dimensional shapes among proteins are due to bonding between the
    a. amino groups.
    b. ion groups.
    c. R groups.
    d. H atoms.
18. The following picture illustrates a
    a. saturated fatty acid.
    b. monounsaturated fatty acid.
    c. polyunsaturated fatty acid.
    d. All of these are correct.

19. Which of the following interactions help to form the tertiary structure of a protein?
    a. hydrogen bonding
    b. folding and twisting of secondary structures within the protein chain
    c. interactions between amino acid R groups
    d. All of these are correct.

For questions 20–24, match the structure in the following diagram with the name of the functional group.

20. phosphate
21. carboxyl
22. hydroxyl
23. sulfhydryl
24. amino

Functional Groups

Structure

a. R—O—H

b. R—C(=O)(O—H)

c. R—N(H)(H)

d. R—S—H

e. R—O—P(=O)(O—H)(O—H)

R = remainder of molecule

25. Carbon and oxygen can be found in
    a. amino acids.
    b. glucose.
    c. starch.
    d. All of these are correct.

26. Animals store glucose as
    a. maltose.
    b. glycogen.
    c. starch.
    d. None of these are correct.

27. The polysaccharide found in plant cell walls is
    a. glucose.   c. maltose.
    b. starch   d. cellulose.

28. Phosphates can be found in
    a. DNA.
    b. RNA.
    c. glucose.
    d. Both a and b are correct.

29. All _____ are
    _____.
    a. proteins, enzymes   c. enzymes, proteins
    b. sugars, monosaccharides d. sugars, polysaccharides

30. _____ is the precursor of _____.
    a. Cholesterol, testosterone and estrogen
    b. Cholesterol, glucose
    c. Testosterone, cholesterol
    d. Estrogen, cholesterol

31. An example of a hydrocarbon would be
    a. heptane.   c. maltose.
    b. glucose.   d. None of these are correct.

32. A 5-carbon sugar is associated with
    a. glucose.   b. a protein.   c. DNA.   d. lipids.

33. Nucleotides
    a. contain a fatty acid, a nitrogen-containing base, and a phosphate molecule.
    b. are the monomers for fats and polysaccharides.
    c. join together by covalent bonding between the bases.
    d. are found in DNA, RNA, and proteins.
    e. None of these are correct.

34. Amino acids in a protein are joined by
    a. peptide bonds.   c. ionic bonds.
    b. hydrogen bonds.   d. phosphodiester bonds.

35. Which of the following pertains to an RNA nucleotide and not to a DNA nucleotide?
    a. contains the sugar ribose
    b. contains uracil (U) instead of thymine (T)
    c. contains a phosphate molecule

d. becomes bonded to other nucleotides by condensation
    e. Both a and b are correct.

36. Saturated fatty acids and unsaturated fatty acids differ in
    a. the number of carbon-to-carbon bonds.
    b. their consistency at room temperature.
    c. the number of hydrogen atoms present.
    d. All of these are correct.

37. Which of the following lipids would most likely be solid at room temperature?
    a. palmitic acid, a saturated fatty acid
    b. oleic acid, an unsaturated fatty acid
    c. cholesterol, a steroid
    d. elaidic acid, a trans fat

Thinking Scientifically

1. Nutritionists advise us to avoid consuming oils from tropical plants because, compared to oils from temperate plants, they contain high levels of saturated fatty acids. Let's try to determine a physiological reason for this difference in fatty acid composition.

   Look at the following graph and answer questions a–d.

Change in plasma membrane linolenic acid content of seedlings during exposure to cold temperatures.

Percent Linolenic Acid → (y-axis: 0, 10, 20, 30, 40, 50, 60)
Weeks of Exposure to Chilling (5°C) (x-axis: 1, 2, 3, 4, 5, 6)

   a. What plant cell structure are we considering here?
   b. Linolenic acid is an unsaturated fatty acid. What happens to its level following exposure of plants to cold temperatures?
   c. If you put a product with a high level of unsaturated fats (such as canola oil) and a product with a high level of saturated fats (such as butter) in your refrigerator (chilling temperature), how would they differ in consistency?
   d. How might fatty acid composition relate to chilling tolerance in plants?

2. In order to understand the relationship between enzyme structure and function, researchers often study mutations that swap one amino acid for another. In one enzyme, function is retained if a particular amino acid is substituted with one that has a nonpolar *R* group, but function is lost if the amino acid is replaced by one with a polar *R* group. Why might that be?

3. Scientists have observed that while two species might have the same protein, the more distantly related they are, the more likely the sequence of the amino acids has changed. What might account for this observation?

Essentials of Biology Website

The companion website for *Essentials of Biology* provides a wealth of information organized and integrated by chapter. You will find practice tests, animations, videos, and much more that will complement your learning and understanding of general biology.

http://www.mhhe.com/maderessentials2

4

Inside the Cell

A bone cell encased within bone.

Cone cell in eye accounts for color vision.

The human body is composed of up to 100 trillion cells.

A nerve cell within the brain.

An egg cell protected within an ovarian follicle.

A cell caught in the act of committing suicide, a necessary part of development.

Red blood cells coursing through a blood vessel.

Outline

The Diversity and Unity of Cells

RED BLOOD CELLS are the most abundant cell type in your body, and they perform a critical job—delivering oxygen throughout your body and picking up carbon dioxide waste from its cells. Along with other types of blood cells, new red blood cells are continually produced from precursor cells, known as stem cells, located in the marrow of your bones. Red blood cells do not resemble a typical animal cell in structure because they are chock full of hemoglobin, the pigment that makes them red and enables them to carry oxygen. Because red blood cells do not have

—Continued next page

49

Continued—

the usual contents of an animal cell, they only live about 120 days before being destroyed by the liver and the spleen. Thus, the body continually faces the daunting task of replacing the 3 million red blood cells it loses every day. If this process of replacing the red blood cells is impaired, as occurs in a condition called aplastic anemia, a person can quickly die.

Cells are the fundamental units of all living things. Although the diversity of organisms is incredible, the cells of all organisms share many similarities. In fact, there are only a few structural differences between most types of cells. In this chapter you will learn about the two main types of cells and how they are similar as well as different. Learning about the structure of cells will later help you understand their complex functions and interactions.

4.1 | Cells Under the Microscope

Cells are extremely diverse, but nearly all require a microscope to see them (**Fig. 4.1**). When we speak of "a cell" or "the cell," we mean a cell that can be discussed in general terms. Our own bodies are composed of several hundred cell types, and each type is present billions of times over. There are nerve cells, muscle cells, gland cells, and bone cells, to name a few. Each type of cell is specialized to perform a particular function—a nerve cell conducts, a muscle cell contracts, a gland cell secretes, and a bone cell supports.

Cells are complex and yet tiny. The light microscope, invented in the seventeenth century, allows us to see cells but

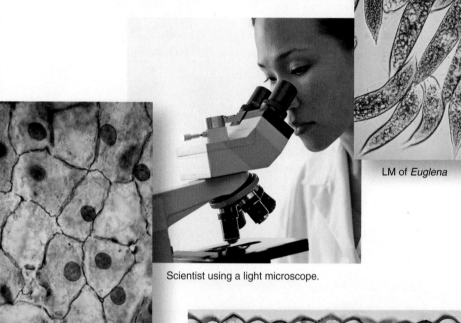
LM of *Euglena*

TEM of human lymphocyte

LM of human epithelial cells

Scientist using a light microscope.

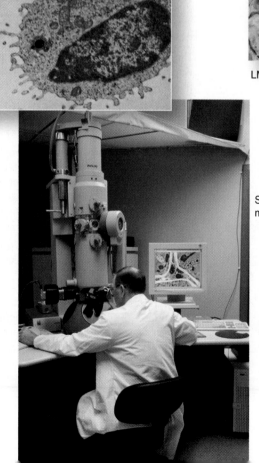
Scientist using an electron microscope.

LM of leaf cells

Figure 4.1 **Diversity of cells.**

All organisms are composed of cells, but it takes a microscope to see most of them. Today, three basic types of microscopes are in use: the light microscope, the transmission electron microscope, and the scanning electron microscope. The light microscope and the transmission electron microscope reveal the insides of cells, while the scanning electron microscope shows three-dimensional surface features. Pictures resulting from the use of the light microscope are called light micrographs (LM), and those that result from the use of the transmission electron microscope are called transmission electron micrographs (TEM). Use of the scanning electron microscope results in scanning electron micrographs (SEM).

Figure 4.2   **Relative sizes of some living things and their components.**

This diagram not only gives you an idea of the relative sizes of organisms, cells, and their components, but it also shows which of them can be seen with an electron microscope, a light microscope, and the unaided eye. These sizes are given in metric units, with each higher unit ten times greater than the lower unit. See Appendix B for the complete metric system.

not much of their complexity. That's because the properties of light limit the amount of detail a light microscope can reveal. Electron microscopes, invented in the 1930s, overcome this limit by using beams of electrons instead of beams of light as their source of illumination. An electron microscope enables us to see the surface features and fine details of cells, and even some of the larger molecules within them. **Figure 4.2** compares the visual ranges of the electron microscope, the light microscope, and the unaided eye.

Why are cells so small? To answer this question, consider that a cell needs a surface area large enough to allow adequate nutrients to enter and wastes to exit. If you were to cut a large cube into smaller cubes, the smaller cube would have a lot more surface area per volume than the large cube. For example, a 4-cm cube has a **surface-area-to-volume ratio** of only 1.5:1, but a 1-cm cube has a surface-area-to-volume ratio of 6:1. Therefore, small cells, not large cells, are most likely to have an adequate surface area for exchanging wastes and nutrients.

Many cells also possess adaptations that increase the surface-area-to-volume ratio. For example, animal cells that specialize in absorption have ways to increase their surface-area-to-volume ratio. Cells in your intestine have tiny fingerlike projections called microvilli (sing., microvillus), which increase the surface area available for absorbing nutrients. Without this adaptation, your intestine would have to be hundreds of meters long!

Check Your Progress

1. Explain why a large surface-area-to-volume ratio is needed for the proper functioning of cells.

Answer: 1. A cell needs a relatively large surface area for efficient absorption of nutrients and secretion of wastes.

4.2 | The Two Main Types of Cells

All cells have an outer membrane called a plasma membrane. The plasma membrane encloses a semifluid substance called the **cytoplasm** and the cell's genetic material. The plasma membrane regulates what enters and exits cells; the cytoplasm carries on chemical reactions; and the genetic material provides the information needed for growth and reproduction. When a cell reproduces, it produces more cells. The concepts that all organisms are composed of cells, and that cells come only from preexisting cells, are the two central tenets of the **cell theory.**

Cells are divided into two main types according to the way their genetic material is organized. **Prokaryotic cells** are so named because they lack a membrane-bounded nucleus. Their DNA is located in a region of the cytoplasm called the nucleoid. The other type of cell, called a **eukaryotic cell,** has a nucleus that houses its DNA. Eukaryotic cell structure is discussed in Section 4.4.

Prokaryotic Cells

Organisms from the domains Bacteria and Archaea are prokaryotic cells (see Figs. 1.5 and 1.6) whose evolutionary history dates back to the first cells on Earth. Prokaryotic cells are generally much smaller in size and simpler in structure than eukaryotic cells (**Fig. 4.3**). Their small size and simple structure allows them to reproduce very quickly and effectively; they exist in great numbers in the air, in bodies of water, in the soil, and even on you. Prokaryotes are an extremely successful group of organisms.

Bacteria are well known because they cause some serious diseases, including tuberculosis, throat infections, and gonorrhea. Even so, the biosphere would not long continue without bacteria. Many bacteria decompose dead remains and contribute to ecological cycles. Bacteria also assist humans in another way—we use them to manufacture all sorts of products, from industrial chemicals to foodstuffs and drugs. The active cultures found in a container of yogurt are bacteria that are beneficial to us. There are also billions of bacteria teeming within your intestines! We are dependent on them for many things, including the synthesis of some vitamins we can't make ourselves.

Our knowledge about how DNA specifies the sequence of amino acids in proteins was much advanced by doing experiments utilizing *E. coli*, a bacterium that lives in the human intestine. The fact that this information applies to all prokaryotes and eukaryotes, even ourselves, reveals the remarkable unity of living things and gives evidence that all organisms share a common descent.

Bacterial Structure

In bacteria, the cytoplasm is surrounded by a plasma membrane, a cell wall, and sometimes a capsule (**Fig. 4.4**). The cytoplasm contains a variety of different enzymes. These organic catalysts speed many types of chemical reactions that are required to maintain an organism. Both prokaryotes and eukaryotes have the same type of plasma membrane; its structure is discussed on page 54. The **cell wall** maintains the shape of the cell, even if the cytoplasm should happen to take up an abundance of water. The **capsule** is a protective layer of polysaccharides lying outside the cell wall.

The DNA of a bacterium is located in a single coiled chromosome that resides in a region called the **nucleoid.** The many proteins specified by bacte-

cell

flagella

Prokaryotic cell: simple
internal structure

Eukaryotic cell: complex
internal structure

Figure 4.3   Comparison of prokaryotic cells and eukaryotic cells.

Prokaryotic cells are much smaller than eukaryotic cells. They are also less complex in structure, as we shall see.

rial DNA are synthesized on tiny particles called **ribosomes.** A bacterial cell contains thousands of ribosomes.

The appendages of a bacterium, namely flagella, fimbriae, and conjugation pili, are made of protein. Some bacteria have **flagella** (sing., flagellum), which are tail-like appendages that allow bacteria to propel themselves. The bacterial flagellum has a filament, a hook, and a basal body, which is a series of rings anchored in the cell wall and plasma membrane. The hook rotates 360° within the basal body, and this rotary motion propels bacteria—the bacterial flagellum does not move back and forth like a whip. Sometimes flagella occur only at the ends of a cell, and other times they are dispersed randomly over the surface.

Fimbriae are small, bristle-like fibers that sprout from the cell surface. They don't have anything to do with motility, but they do help bacteria attach to a surface, such as a tabletop, or even your skin! **Conjugation pili** are rigid tubular structures used by bacteria to pass DNA from cell to cell. Bacteria can also take up DNA directly from the external medium or by way of viruses. All three of these methods allow bacteria to pass information from cell to cell, including drug resistance genes.

Check Your Progress

1. **What is the major distinction between a prokaryotic cell and a eukaryotic cell?**
2. **List the three types of bacterial appendages. What do they have in common?**

Answers: 1. A prokaryotic cell lacks a membrane-bounded nucleus, while a eukaryotic cell has one.
2. The three bacterial appendages—flagella, fimbriae, and conjugation pili—are all made of protein.

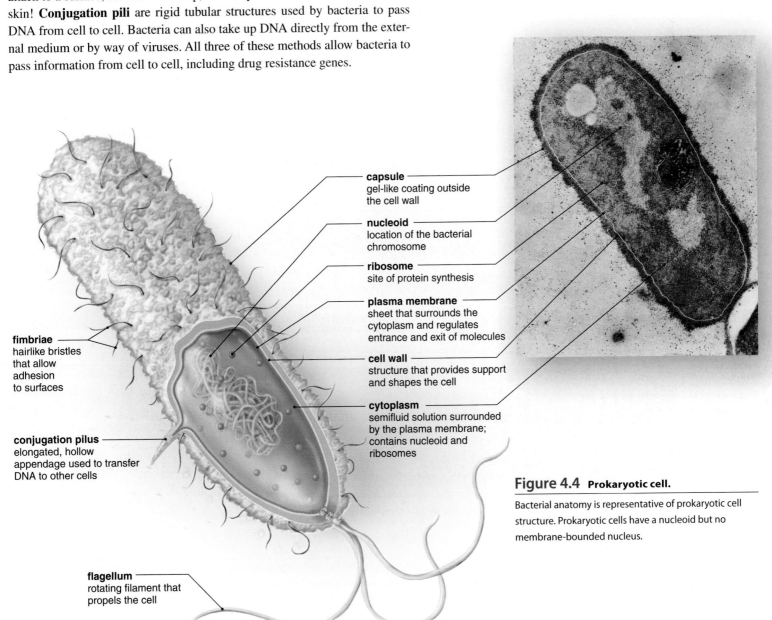

capsule
gel-like coating outside the cell wall

nucleoid
location of the bacterial chromosome

ribosome
site of protein synthesis

plasma membrane
sheet that surrounds the cytoplasm and regulates entrance and exit of molecules

cell wall
structure that provides support and shapes the cell

cytoplasm
semifluid solution surrounded by the plasma membrane; contains nucleoid and ribosomes

fimbriae
hairlike bristles that allow adhesion to surfaces

conjugation pilus
elongated, hollow appendage used to transfer DNA to other cells

flagellum
rotating filament that propels the cell

Figure 4.4   Prokaryotic cell.

Bacterial anatomy is representative of prokaryotic cell structure. Prokaryotic cells have a nucleoid but no membrane-bounded nucleus.

4.3 │ The Plasma Membrane

The **plasma membrane** marks the boundary between the outside and inside of a cell. Its integrity and function are vital to a cell because it acts much like a gatekeeper, regulating the passage of molecules and ions into and out of the cell.

In both prokaryotes and eukaryotes, the plasma membrane is a phospholipid bilayer with embedded proteins (see Fig. 3.14). The polar heads of the phospholipids face toward the outside of the cell and toward the inside of the cell, where there is a watery medium. The nonpolar tails face inward toward each other, where there is no water. Cholesterol molecules, present in the plasma membrane of eukaryotic cells, lend support to the membrane, which has the consistency of olive oil.

The fluid-mosaic model states that the protein molecules embedded in the membrane have a pattern (form a mosaic) within the phospholipid bilayer (**Fig. 4.5**). The pattern varies according to the particular membrane and also within the same membrane at different times. When you consider that the plasma membrane of a red blood cell contains over 50 different types of proteins, you can see why the membrane is said to be a mosaic. Short chains of sugars are attached to the outer surface of some of these proteins, forming glycoproteins. The sugar chain helps a protein perform its particular function.

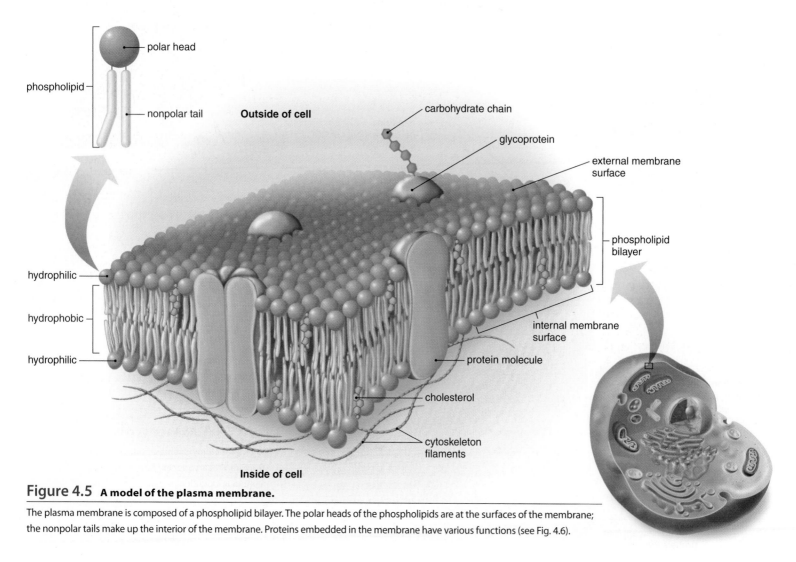

Figure 4.5   **A model of the plasma membrane.**

The plasma membrane is composed of a phospholipid bilayer. The polar heads of the phospholipids are at the surfaces of the membrane; the nonpolar tails make up the interior of the membrane. Proteins embedded in the membrane have various functions (see Fig. 4.6).

Functions of Membrane Proteins

The proteins that occur within a membrane have varied specific functions:

Channel Proteins

Channel proteins form a tunnel across the entire membrane that allows only one or a few types of specific molecules to simply move across the membrane (**Fig. 4.6***a*) For example, aquaporins are channel proteins that allow water to enter or exit a cell. Without aquaporins in the kidney your body would soon dehydrate.

Transport Proteins

Transport proteins are also involved in the passage of molecules and ions through the membrane. They often combine with a substance and help it move across the membrane, with an input of energy (Fig. 4.6*b*). For example, a transport protein conveys sodium and potassium ions across a nerve cell membrane. Without this transport protein, nerve conduction would be impossible.

Cell Recognition Proteins

Cell recognition proteins are glycoproteins (Fig. 4.6*c*). Among other functions, these proteins enable our bodies to distinguish between our own cells and the cells of other organisms. Without this distinction, pathogens would be able to freely invade the body.

Receptor Proteins

A receptor protein has a shape that allows a specific molecule, called a signal molecule, to bind to it (Fig. 4.6*d*). The binding of a signal molecule causes the receptor protein to change its shape, and thereby bring about a cellular response. For example, the hormone insulin binds to a receptor protein in liver cells, and thereafter these cells store glucose.

Enzymatic Proteins

Some plasma membrane proteins are enzymatic proteins that directly participate in metabolic reactions (Fig. 4.6*e*). Without enzymes, some of which are attached to the various membranes of the cell, a cell would never be able to perform the degradative and synthetic reactions that are important to its function.

Junction Proteins

As discussed on page 67, proteins are also involved in forming various types of junctions between cells (Fig. 4.6*f*). The junctions assist cell-to-cell adhesion and communication. The adhesion junctions in your bladder keep the cells bound together as the bladder swells with urine.

a. Channel protein   b. Transport protein

c. Cell recognition protein   d. Receptor protein

e. Enzymatic protein   f. Junction proteins

Figure 4.6   Membrane protein diversity.

These are some of the types of proteins embedded in the plasma membrane.

Check Your **Progress**

1. Why is the plasma membrane described as a fluid mosaic?
2. List six types of proteins found in the plasma membrane.

Answers: 1. The membrane is fluid because it has the consistency of oil; it is a mosaic because the many proteins form different patterns at various times in various membranes.   2. Channel, transport, cell recognition, receptor, enzymatic, and junction proteins.

4.4 | Eukaryotic Cells

As discussed in Chapter 1, protists, fungi, plants, and animals are all composed of eukaryotic cells. Unlike prokaryotic cells, eukaryotic cells have a membrane-bounded **nucleus** that houses their DNA. The DNA forms the genes that are located on chromosomes.

Eukaryotic cells are much larger than prokaryotic cells, and as mentioned they often have projections that increase their surface area. They are also compartmentalized and contain small structures called **organelles** that differ in structure and function. All the molecules necessary to perform a particular function are concentrated inside the organelle.

In this chapter, the organelles are divided into four categories:

1. The nucleus and the ribosomes. The nucleus communicates with the ribosomes, which are the small particles that synthesize proteins coded for by DNA located within the chromosomes.

Figure 4.7   Structure of a typical animal cell.

a. False-colored TEM of a animal cell. **b.** Generalized drawing.

2. Organelles of the endomembrane system (ES). Each of these organelles has its own particular set of enzymes and produces its own products. The products are carried between the ES organelles by **transport vesicles,** small membranous sacs that keep the products from entering the cytoplasm.
3. The energy-related organelles. The energy-related organelles—chloroplasts in plant cells and mitochondria in plant and animal cells—are more self-contained than those of the ES. They even contain DNA and possess ribosomes that resemble those of prokaryotic cells.
4. The cytoskeleton. The **cytoskeleton** is a lattice of protein filaments and tubules that maintains the shape of the cell and assists in the movement of organelles. Without an efficient means of moving organelles and their products, eukaryotic cells could not exist.

Each structure in the typical animal and plant cells shown in **Figures 4.7** and **4.8** has been given a particular color, and this color will be used to represent the same structure throughout this text.

Figure 4.8   Structure of a typical plant cell.

a. False-colored TEM of a plant cell. **b.** Generalized drawing.

Nucleus and Ribosomes

The nucleus stores genetic information, and the ribosomes in the cytoplasm use this information to carry out protein synthesis.

The Nucleus

Because of its large size, the nucleus is a noticeable structure in the eukaryotic cell (**Fig. 4.9**). The nucleus contains **chromatin** within a semifluid nucleoplasm. Chromatin looks grainy, but actually it is a network of strands. Just before the cell divides, the chromatin condenses and coils into rodlike structures called **chromosomes.** All the cells of an organism contain the same number of chromosomes except for the egg and sperm, which have half this number.

Chromatin (and therefore chromosomes) is composed of DNA, protein, and some RNA. The DNA is organized into genes, each of which has a specific sequence of bases that codes for a polypeptide. The coded information is relayed to the ribosomes by a type of RNA called messenger RNA (mRNA). This mRNA has a sequence of bases that mirrors the sequence of bases in a gene. This sequence specifies the order of amino acids that are to occur in a particular polypeptide. A protein contains one or more polypeptides. Because the proteins of a cell determine its structure and functions, the nucleus may be thought of as the command center of the cell.

The **nucleolus** is a dark structure within the nucleus where another type of RNA, called ribosomal RNA (rRNA), is produced. Proteins join with rRNA to form the subunits of ribosomes. The assembled ribosomal subunits are then sent out of the nucleus into the cytoplasm, where

nuclear envelope — outer membrane
— inner membrane

nucleolus

chromatin

nucleoplasm

ER lumen

ribosome

endoplasmic reticulum

nuclear pores

SEM of freeze-fractured nuclear envelope

Figure 4.9   Structure of the nucleus.

The nuclear envelope contains pores that allow substances to pass from the nucleus to the cytoplasm.

Nucleus

DNA

mRNA

nuclear pore

Cytoplasm

1 mRNA is produced in the nucleus but moves through a nuclear pore into the cytoplasm.

2 In the cytoplasm, the mRNA and ribosomal subunits join, and polypeptide synthesis begins.

ribosome

receptor

3 If a ribosome attaches to a receptor on the ER, the polypeptide enters the lumen of the ER.

ribosome

polypeptide

4 At termination, the polypeptide becomes a protein. The ribosomal subunits disengage, and the mRNA is released.

small subunit

large subunit

protein

lumen of the ER

ER membrane

ribosome

Endoplasmic reticulum

they join and assume their role in protein synthesis.

The nucleus is separated from the cytoplasm by a double membrane known as the **nuclear envelope.** Even so, the nucleus communicates with the cytoplasm. The nuclear envelope has **nuclear pores** of sufficient size (100 nm) to permit the passage of ribosomal subunits and mRNA out of the nucleus into the cytoplasm, and the passage of proteins from the cytoplasm into the nucleus.

Ribosomes

Ribosomes are found in both prokaryotes and eukaryotes. In both types of cells, ribosomes are composed of two subunits, one large and one small. Each subunit has its own mix of proteins and rRNA. As mentioned, ribosomes are sites of protein synthesis. They receive mRNA from the nucleus, which carries a coded message from DNA. Each mRNA codes for the correct sequence of amino acids in a polypeptide and, as mentioned, a protein contains one or more polypeptides.

In eukaryotic cells, some ribosomes occur freely within the cytoplasm, either singly or in groups called **polyribosomes.** Proteins synthesized by polyribosomes are used in the cytoplasm. Many ribosomes are attached to the **endoplasmic reticulum (ER)**, an organelle composed of many saccules and channels. After the ribosome binds to a receptor at the ER, the polypeptide being synthesized enters the lumen of the ER, where it may be further modified (**Fig. 4.10**). A protein takes its final shape inside the lumen of the ER.

Figure 4.10   The nucleus, ribosomes, and endoplasmic reticulum (ER).

After mRNA leaves the nucleus, it attaches itself to a ribosome, and polypeptide synthesis begins. When a ribosome combines with a receptor at the endoplasmic reticulum (ER), the polypeptide enters the lumen of the ER through a channel in the receptor. Exterior to the ER, the ribosome splits, releasing the mRNA while a protein takes shape inside the ER lumen.

Check Your **Progress**

1. List the components of the nucleus and give a function for each.
2. Where are ribosomal subunits produced, and where do ribosomes form in the cell?

Answers: 1. nuclear envelope—defines the nucleus; nuclear pore—allows substances to move into and out of nucleus; nucleolus—produces ribosomal subunits; chromatin—becomes chromosomes and contains DNA, organized into genes.   2. Ribosomal subunits are produced in the nucleolus, and these subunits are assembled to form ribosomes in the cytoplasm. Many ribosomes are attached to the endoplasmic reticulum, and others occur freely within the cytoplasm.

Endomembrane System

The **endomembrane system** consists of the nuclear envelope, the membranes of the endoplasmic reticulum, the Golgi apparatus, and many small membranous sacs called **vesicles.** This system helps compartmentalize the cell, so that particular enzymatic reactions are restricted to specific regions. Transport vesicles carry molecules from one part of the system to another.

Endoplasmic Reticulum

The endoplasmic reticulum (ER) consists of a complicated system of membranous channels and saccules (flattened vesicles). It is physically continuous with the outer membrane of the nuclear envelope (**Fig. 4.11**).

Rough ER is studded with ribosomes on the side of the membrane that faces the cytoplasm; therefore, rough ER is able to synthesize polypeptides. It also modifies the polypeptides after they have entered the central enclosed region of the ER, called the lumen, where a protein takes shape. The rough ER forms transport vesicles which take proteins to other parts of the cell. Often transport vesicles are on their way to the plasma membrane or the Golgi apparatus (described below).

Smooth ER, which is continuous with rough ER, does not have attached ribosomes. Smooth ER synthesizes lipids, such as phospholipids and steroids. The functions of smooth ER are dependent on the particular cell. In the testes, it produces testosterone, and in the liver, it helps detoxify drugs. Regardless of any specialized function, smooth ER also forms transport vesicles that carry molecules to other parts of the cell, notably the Golgi apparatus.

Golgi Apparatus

The **Golgi apparatus,** named for its discoverer, Camillo Golgi, consists of a stack of three to twenty slightly curved, flattened saccules resembling pancakes (**Fig. 4.12**). The Golgi apparatus may be thought of as a transfer station. First, it receives transport vesicles sent to it by rough and smooth ER. The molecules within the vesicles are modified as they move between saccules. For example, sugars may be added to or removed from proteins within the saccules of the Golgi. Finally, the Golgi apparatus sorts the modified molecules and packages them into new transport vesicles according to their particular destinations. Outgoing transport vesicles may return to the ER or proceed to the plasma membrane, where they discharge their contents during **secretion.** In animal cells, some of the vesicles that leave the Golgi are lysosomes, which are discussed next.

Lysosomes

Lysosomes are vesicles produced by the Golgi apparatus that digest molecules and even portions of the cell itself. Sometimes, after engulfing molecules outside the cell, a vesicle formed at the plasma membrane fuses with a lysosome. Lysosomal enzymes then digest the contents of the vesicle. In Tay-Sachs disease, a genetic disorder, lysosomes are missing an enzyme for a particular lipid molecule. The cells become so full of storage bodies that the individual dies, usually in childhood. Someday it may be possible to provide the missing enzyme for these children.

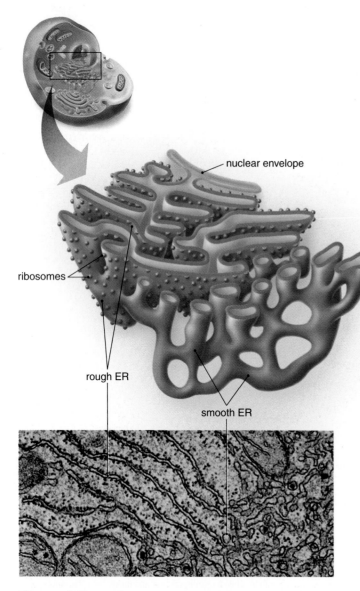

Figure 4.11 **Endoplasmic reticulum (ER).**

Ribosomes are present on rough ER, which consists of flattened saccules, but not on smooth ER, which is more tubular. Proteins are synthesized by ribosomes at the surface of the rough ER, then enter the rough ER for modification. Lipids are synthesized by the smooth ER, which can have other functions as well.

rough ER
synthesizes proteins and
packages them in vesicles.

smooth ER
synthesizes lipids and
performs other functions.

transport vesicles
from rough ER

transport vesicles
from smooth ER

Golgi apparatus
modifies lipids and proteins;
sorts them and packages
them in vesicles.

lysosomes
digest molecules
or old cell parts.

secretory vesicles
fuses with the plasma
membrane as secretion
occurs.

incoming vesicles
bring substances into the
cell.

Figure 4.12 Endomembrane system.

The organelles in the endomembrane system work together to carry out the functions noted. Plant cells do not have lysosomes.

Vacuoles

Vacuoles, like vesicles, are membranous sacs, but vacuoles are larger than vesicles. The vacuoles of some protists are quite specialized; they include contractile vacuoles for ridding the cell of excess water, and digestive vacuoles for breaking down nutrients (**Fig. 4.13**a). Vacuoles usually store substances, such as nutrients or ions. Plant vacuoles contain not only water, sugars, and salts, but also pigments and toxic molecules (Fig. 4.13b). The pigments are responsible for many of the red, blue, or purple colors of flowers and some leaves. The toxic substances help protect a plant from herbivorous animals. Few animal cells contain vacuoles, but fat cells, or adipocytes, contain a very large lipid-engorged vacuole that takes up nearly two-thirds of the volume of the cell (Fig. 4.13c)!

vacuoles

a.   ×800

c.   ×400

Figure 4.13 Vacuoles.

a. Contractile vacuoles of a
protist. **b.** Plant vacuole.
c. Fat cell vacuole.

b.   ×7,700

Check Your Progress

1. Contrast rough endoplasmic reticulum with smooth endoplasmic reticulum.
2. Describe the function of the Golgi apparatus.

Answers: 1. Rough ER contains ribosomes, while smooth ER does not. Rough ER synthesizes the polypeptides of proteins and modifies them, while smooth ER synthesizes lipids and detoxifies various chemicals, among other activities.   2. The Golgi apparatus receives transport vesicles from the ER, modifies their contents if necessary, and repackages them in new transport vesicles that take them to their destinations.

Energy-Related Organelles

Chloroplasts and mitochondria are the two eukaryotic membranous organelles that specialize in energy conversion. **Chloroplasts** use solar energy to synthesize carbohydrates. **Mitochondria** (sing., mitochondrion) break down carbohydrates to produce **adenosine triphosphate (ATP)** molecules. The production of ATP is of great importance because ATP serves as a carrier of energy in cells. The energy of ATP is used whenever a cell synthesizes molecules, transports molecules, or carries out a special function such as muscle contraction or nerve conduction. Without a constant supply of ATP, no cell could exist for long.

Chloroplasts

The chloroplast, a plant cell organelle, is quite large, being twice as wide and as much as five times the length of a mitochondrion. Chloroplasts have a three-membrane system. They are bounded by a double membrane, which includes an outer membrane and an inner membrane. The large inner space, called the **stroma,** contains a concentrated mixture of enzymes and thylakoids. The **thylakoids** are disclike sacs formed from the third membrane. A stack of thylakoids is called a **granum.** The lumens of thylakoid sacs are believed to form a large internal compartment called the thylakoid space (**Fig. 4.14**). The pigments that capture solar energy are located in the thylakoid membrane, and the enzymes that synthesize carbohydrates are in the stroma. The carbohydrates produced by chloroplasts serve as organic nutrient molecules for plants and indeed for all living things on Earth.

The discovery that chloroplasts have their own DNA and ribosomes supports an accepted theory that says chloroplasts are derived from photosynthetic bacteria that entered a eukaryotic cell in the distant past. As shown in Figure 4.8, plant cells contain *both* mitochondria and chloroplasts.

a.   ×35,000

b.

double membrane — { outer membrane / inner membrane }

granum

stroma

thylakoid space

thylakoid membrane

thylakoid

Figure 4.14   Chloroplast structure.

a. Electron micrograph of a chloroplast, which carries out photosynthesis. **b.** Generalized drawing in which the outer and inner membranes have been cut away to reveal the grana, each of which is composed of a stack of membranous sacs called thylakoids. In some grana, but not all, it is obvious that the thylakoid spaces are interconnected.

Mitochondria

Mitochondria are much smaller than chloroplasts, and are usually visible only under an electron microscope. We think of mitochondria as having a shape like that shown in **Figure 4.15**, but actually they often change shape, becoming longer and thinner or shorter and broader. Mitochondria can form long, moving chains, or they can remain fixed in one location—often where energy is most needed. For example, they are packed between the contractile elements of cardiac cells and wrapped around the interior of a sperm's flagellum.

Like chloroplasts, mitochondria are bounded by a double membrane. The inner membrane is highly convoluted into folds called **cristae** that project into the matrix. Cristae increase the surface area of the inner membrane so much that in a liver cell they account for about one-third of the total membrane in the cell.

Mitochondria are often called the powerhouses of the cell because they produce most of the ATP the cell utilizes. The inner membrane encloses the **matrix,** which contains a highly concentrated mixture of enzymes that assists the breakdown of carbohydrates and other nutrient molecules. These reactions supply the chemical energy that permits ATP synthesis to take place on the cristae. The complete breakdown of carbohydrates, which also involves the cytoplasm, is called **cellular respiration** because oxygen is needed and carbon dioxide is given off.

The matrix also contains mitochondrial DNA and ribosomes. The presence of mitochondrial DNA and ribosomes are evidence that mitochondria have a similar origin to chloroplasts and are also derived from bacteria that took up residence in an early eukaryotic cell. Today, with a few rare exceptions, all eukaryotic cells have mitochondria.

Check Your **Progress**

1. Compare the structure and function of chloroplasts and mitochondria.

Answer: 1. Structure: The two main parts of a chloroplast are the thylakoids and the stroma; the two main parts of a mitochondrion are the cristae and the matrix. Function: Chloroplasts capture energy from the sun to build carbohydrates. Mitochondria break down mainly carbohydrates to release energy for ATP production.

×70,000

Figure 4.15   **Mitochondrion structure.**

a. Generalized drawing in which the outer membrane and portions of the inner membrane have been cut away to reveal the cristae. **b.** Electron micrograph of a mitochondrion, which is involved in cellular respiration.

Figure 4.16 **Microtubules.**

Microtubules (tubulin dimer cylinders) radiate out from the centrosome.

Figure 4.17 **Actin filaments.**

Actin filaments (helical actin polymers) are organized into bundles or networks just under the plasma membrane, where they lend support to the shape of a cell.

a. Myosin

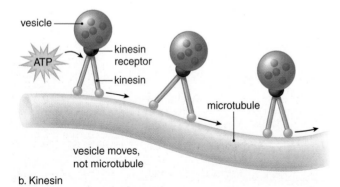

b. Kinesin

Figure 4.18 **Motor proteins.**

a. Myosin creates movement by detaching and reattaching to actin to pull the actin filament along the myosin filament. **b.** The motor molecule kinesin carries organelles along microtubule tracks. One end binds to an organelle, and the other end attaches, detaches, and reattaches to the microtubule. ATP supplies the energy for both myosin and kinesin.

The Cytoskeleton and Motor Proteins

The **cytoskeleton** is a network of interconnected protein filaments and tubules that extends from the nucleus to the plasma membrane in eukaryotic cells. Much as bones and muscles give an animal structure and produce movement, the elements of the cytoskeleton maintain cell shape and, along with **motor proteins,** allow the cell and its organelles to move. But unlike an animal's skeleton, the cytoskeleton is highly dynamic—its elements can be quickly assembled and disassembled as appropriate. The cytoskeleton includes microtubules, intermediate filaments, and actin filaments. The major motor proteins are myosin, kinesin, and dynein.

Microtubules

Microtubules are small, hollow cylinders composed of 13 long chains of tubulin dimers (two tubulin molecules at a time). The chains surround what appears to be an empty central core in electron micrographs (**Fig. 4.16**). Microtubule assembly is controlled by the **centrosome,** a microtubule organizing center which lies near the nucleus. Microtubules radiating from the centrosome help maintain the shape of the cell and act as tracks along which organelles and other materials can move.

Intermediate Filaments

Intermediate filaments are intermediate in size between actin filaments and microtubules. They are ropelike assemblies of proteins that typically run between the nuclear envelope and the plasma membrane. The network they form supports both the nucleus and the plasma membrane.

The protein making up intermediate filaments differs according to the cell type. Intermediate filaments made of the protein keratin give great mechanical strength to skin cells.

Actin Filaments

Each **actin filament** consists of two chains of globular actin monomers twisted about one another in a helical manner to form a long filament. Actin filaments support the cell, forming a dense, complex web just under the plasma membrane (**Fig. 4.17**). Actin filaments also support projections of the plasma membrane, such as microvilli.

Motor Proteins

Motor proteins associated with the cytoskeleton are instrumental in allowing cellular movements. The major motor proteins include those called myosin, kinesin, and dynein.

Myosin often interacts with actin filaments when movement occurs. For example, myosin is interacting with actin filaments when cells move in an amoeboid fashion and/or engulf large particles. During animal cell division, actin, in conjunction with myosin, pinches the original cell into two new cells. When a muscle cell contracts, myosin pulls actin filaments toward the middle of the cell (**Fig. 4.18***a*).

Kinesin and dynein move along microtubules much as a car travels along a highway. First, an organelle, perhaps a vesicle, combines with the motor molecule, and then the motor molecule attaches, detaches, and reattaches further along the microtubule. In this way, the organelle moves from one place to another in the cell (Fig. 4.18*b*). Kinesin and dynein are acting similarly when transport vesicles take materials from the Golgi apparatus to their final destinations.

one pair of centrioles in a
centrosome

one microtubule
triplet

centrosome

Figure 4.19 **Centrioles.**

A pair of centrioles lies to one side of the nucleus.
Their exact function is unknown; however,
possibilities do exist.

Centrioles

Centrioles are short cylinders with a 9 + 0 pattern of microtubule triplets—that is, nine sets of triplets occur in a ring, and none are in the middle of the cylinder (**Fig. 4.19**). In animal cells and most protists, two centrioles lie at right angles to one another in the middle of a centrosome. A centrosome, as mentioned previously, is the major microtubule organizing center for the cell. Therefore, it is possible that centrioles are also involved in the process by which microtubules assemble and disassemble.

Before an animal cell divides, the centrioles replicate, and the members of each pair are at right angles to one another. Then each pair becomes part of a separate centrosome. During cell division, the centrosomes move apart and most likely play a role in organizing a microtubular apparatus (the spindle), which assists the movement of chromosomes. In any case, each new cell eventually has its own centrosome and pair of centrioles.

Plant and fungal cells have the equivalent of a centrosome, but this structure does not contain centrioles, suggesting that centrioles are not necessary for the assembly of cytoplasmic microtubules.

Cilia and Flagella

In eukaryotes, **cilia** and **flagella** are hairlike projections that can move either stiffly, like an oar, or in an undulating fashion, like a whip (**Fig. 4.20***a*). Cells that have these organelles are capable of movement. Cilia are much shorter than flagella, but they have a similar construction. Both are membrane-bounded cylinders with a basal body lying in the cytoplasm. They have a matrix that contains nine microtubule doublets arranged in a circle around two central microtubules. This is called the 9 + 2 pattern of microtubules (Fig. 4.20*b*). Cilia and flagella move when the microtubule doublets slide past one another.

■ Check
Your **Progress**

1. **List the components of the cytoskeleton.**
2. **What do motor proteins do?**
3. **Compare and contrast the internal structure of centrioles with cilia and flagella.**

Answers: 1. Microtubules, intermediate filaments, and actin filaments.   2. Motor proteins move vesicles and organelles from one location to another along microtubules or can create movement by interacting with actin filaments.   3. Centrioles, cilia, and flagella are all composed of microtubules arranged in a particular pattern and enclosed by the plasma membrane. In centrioles, the pattern is 9 triplets with none in the center, while the arrangement of cilia and flagella is 9 doublets with two microtubules in the center.

cilia in bronchial wall
flagella of sperm

a.

Flagellum

**Flagellum
cross section**

central
microtubules

microtubule
doublet

dynein side arms

TEM ×350,000

plasma
membrane

TEM ×101,000

triplets

Basal body

**Basal body
cross section**

b.

Figure 4.20 **Cilia and flagella.**

a. Cilia in the bronchial wall and the flagella of sperm are organelles capable of movement. The cilia in our bronchi sweep mucus and debris back up into the throat where it can be swallowed or ejected. The flagella of sperm allow them to swim to the egg. **b.** Cilia and flagella have a distinct pattern of microtubules bounded by a plasma membrane. Both have a basal body with the same pattern of microtubules as a centriole. Therefore, it is believed that centrioles give rise to basal bodies.

Figure 4.21   Plasmodesmata.

Plant cells are joined by membrane-lined channels called plasmodesmata that contain cytoplasm. Through these channels, water and small molecules can pass from cell to cell.

Figure 4.22   Animal cell extracellular matrix.

The extracellular matrix supports an animal cell and affects its behavior.

4.5 | Outside the Eukaryotic Cell

Now that we have completed our discussion of the plasma membrane and the cell contents, you might think we have finished our tour of the cell. However, most cells also have extracellular structures formed from materials the cell produces and transports across its plasma membrane.

Plant Cell Walls

All plant cells have a **cell wall.** A *primary cell wall* contains cellulose fibrils and noncellulose substances, and these allow the wall to stretch when the cell is growing. Adhesive substances are abundant outside the cell wall in the middle lamella, a layer that holds two plant cells together.

For added strength, some plant cells have a *secondary cell wall* that forms inside the primary cell wall. The secondary wall has a greater quantity of cellulose fibrils, which are laid down at right angles to one another. Lignin, a substance that adds strength, is a common ingredient of secondary cell walls.

In a plant, living cells are connected by **plasmodesmata** (sing., plasmodesma), numerous narrow, membrane-lined channels that pass through the cell wall (**Fig. 4.21**). Cytoplasmic strands within these channels allow direct exchange of some materials between adjacent plant cells and eventually among all the cells of a plant. The plasmodesmata allow only water and small solutes to pass freely from cell to cell.

Exterior Cell Surfaces in Animals

Animal cells do not have a cell wall, but they do have two other exterior surface features of interest: (1) An extracellula r matrix exists outside the cell, and (2) various junctions occur between some cell types.

Extracellular Matrix

An **extracellular matrix (ECM)** is a meshwork of fibrous proteins and polysaccharides in close association with the cell that produced them (**Fig. 4.22**).

Collagen and elastin are two well-known proteins in the extracellular matrix. Collagen resists stretching and elastin gives it resilience. The polysaccharides play a dynamic role by directing the migration of cells along collagen fibers during development. Other ECM proteins bind to receptors in a cell's plasma membrane, permitting communication between the extracellular matrix and the cytoskeleton within the cytoplasm of the cell.

The extracellular matrices of various tissues vary greatly. They may be quite flexible, as in cartilage, or rock solid, as in bone. The rigidity of the extracellular matrix is influenced mainly by the number and types of protein fibers present, and how they are arranged. The extracellular matrix of bone is very hard because in addition to the components mentioned, mineral salts, notably calcium salts, are deposited outside the cell.

Junctions Between Cells

Three types of junctions are found between certain cells: adhesion junctions, tight junctions, and gap junctions. The type of junction found between two cells depends on whether or not the cells need to be able to exchange materials and if the cells need to be joined together very tightly.

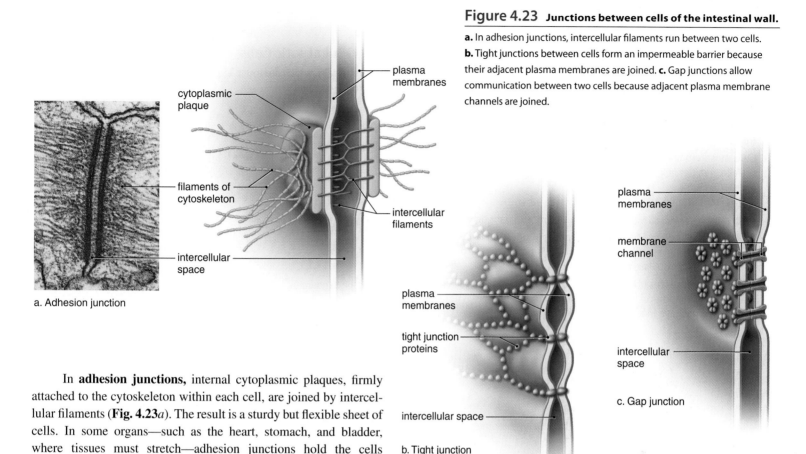

Figure 4.23 **Junctions between cells of the intestinal wall.**
a. In adhesion junctions, intercellular filaments run between two cells.
b. Tight junctions between cells form an impermeable barrier because their adjacent plasma membranes are joined. **c.** Gap junctions allow communication between two cells because adjacent plasma membrane channels are joined.

a. Adhesion junction

b. Tight junction

c. Gap junction

In **adhesion junctions,** internal cytoplasmic plaques, firmly attached to the cytoskeleton within each cell, are joined by intercellular filaments (**Fig. 4.23***a*). The result is a sturdy but flexible sheet of cells. In some organs—such as the heart, stomach, and bladder, where tissues must stretch—adhesion junctions hold the cells together.

Adjacent cells are even more closely joined by **tight junctions,** in which plasma membrane proteins actually attach to each other, producing a zipperlike fastening (Fig. 4.23*b*). The cells of tissues that serve as barriers are held together by tight junctions; for example, urine stays within kidney tubules because the cells of the tubules are joined by tight junctions.

A **gap junction** allows cells to communicate. A gap junction is formed when two identical plasma membrane channels join (Fig. 4.23*c*). The channel of each cell is lined by six plasma membrane proteins that allow the junction to open and close. A gap junction lends strength to the cells, but it also allows small molecules and ions to pass between them. Gap junctions are important in heart muscle and smooth muscle because they permit the flow of ions that is required for the cells to contract as a unit.

Check Your Progress

1. **Contrast the primary cell wall with the secondary cell wall in plants.**
2. **Describe the composition of the extracellular matrix of an animal cell.**
3. **Compare adhesion junctions with tight junctions.**
4. **Describe the functions of gap junctions.**

Answers: 1. All plant cells have a primary wall, which is capable of stretching during growth and is composed of cellulose and other compounds. The secondary cell wall has added strength and contains lignin in addition to cellulose. Lignin adds rigidity but also strength to the cell wall.   2. The extracellular matrix notably contains the fibrous proteins collagen and elastin. It also contains polysaccharides.   3. Adhesion junctions are flexible because intercellular filaments join cytoplasmic plaques. Tight junctions produce an impermeable barrier because plasma membrane proteins are joined to each other.   4. In gap junctions, plasma membrane proteins form channels that allow small molecules and ions to pass between two adjacent cells.

The Chapter in Review

Summary

4.1 | Cells Under the Microscope

- Cells are microscopic in size. Although a light microscope allows you to see cells, it cannot make out the detail that an electron microscope can.
- Cells must remain small in order to have an adequate amount of surface area per cell volume.

4.2 | The Two Main Types of Cells

- All cells have a plasma membrane, cytoplasm, and genetic material.
- Prokaryotic cells do not have a membrane-bounded nucleus. Eukaryotic cells have a membrane-bounded nucleus as well as various membranous organelles.
- Bacteria are representative of the prokaryotes. They have a cell wall and capsule, in addition to a plasma membrane. Their DNA is in the nucleoid. They have many ribosomes and three possible appendages.

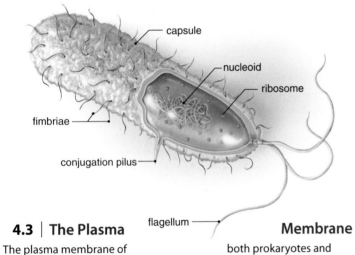

4.3 | The Plasma Membrane

The plasma membrane of both prokaryotes and eukaryotes is a phospholipid bilayer.

- The phospholipid bilayer regulates the passage of molecules and ions into and out of the cell.
- The fluid-mosaic model of membrane structure shows that the embedded proteins form a mosaic (varying) pattern.
- The types of embedded proteins include channel, transport, cell recognition, receptor, and enzymatic proteins.

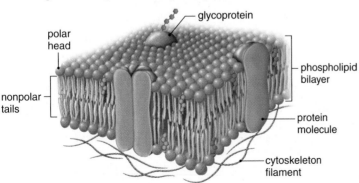

4.4 | Eukaryotic Cells

Eukaryotic cells, which are much larger than prokaryotic cells, have the following organelles:

Nucleus and Ribosomes

- The nucleus houses chromatin, which contains DNA, the genetic material. During division, chromatin becomes condensed into chromosomes.
- The nucleolus is an area within the nucleus where ribosomal RNA is produced. Proteins are combined with the rRNA to form the subunits of ribosomes, which exit the nucleus through nuclear pores.
- Ribosomes in the cytoplasm synthesize polypeptides, which later become proteins with the help of mRNA, a DNA intermediary.

Endomembrane System

- Rough ER contains ribosomes, which produce polypeptides. These polypeptides enter the ER, are modified, and become proteins that are then packaged in transport vesicles.
- Smooth ER synthesizes lipids, but also has various metabolic functions depending on the cell type. It can also form transport vesicles.
- The Golgi apparatus is a transfer station that receives transport vesicles and modifies, sorts, and repackages proteins into transport vesicles that fuse with the plasma membrane as secretion occurs.
- Lysosomes are produced by the Golgi apparatus. They contain enzymes that carry out intracellular digestion.

Vacuoles and Vesicles

- Vacuoles are large membranous sacs specialized for storage, contraction, digestion, and other functions.
- Vesicles are small membranous sacs.

Energy-Related Organelles

- Chloroplasts capture the energy of the sun and carry on photosynthesis, which produces carbohydrates.
- Mitochondria break down carbohydrate-derived products as ATP is produced during cellular respiration.

The Cytoskeleton and Motor Proteins

- The cytoskeleton maintains cell shape and allows the cell and the organelles to move.
- Microtubules radiate from the centrosome and are present in cytoplasm. They also occur in centrioles, cilia, and flagella.
- Intermediate filaments support the nuclear envelope and the plasma membrane. They give mechanical strength to skin cells, for example.
- Actin filaments are long, thin helical filaments that support the plasma membrane and projections of the cell.
- Motor proteins allow cellular movements to occur and move vesicles and organelles within the cell. The motor protein myosin is often interacting with actin when movement occurs, and kinesin interacts with microtubules to move vesicles and other organelles from place to place.

Centrioles

- Centrioles are short cylinders with a 9 + 0 pattern of microtubule triplets.
- In animal cells and those of most protists, two centrioles lie at right angles just outside the nucleus.
- Centrioles are not present in plant cells.

Cilia and Flagella

- Cilia and flagella are hairlike projections that allow some cells to move.
- Both have a cylindrical construction with an internal 9 + 2 pattern of microtubules.
- Their basal bodies resemble centrioles.

4.5 │ Outside the Eukaryotic Cell

Plant cells have a freely permeable cell wall, with cellulose as its main component.

- Some plant cells have a secondary cell wall containing lignin and cellulose that give them rigidity
- Small membrane-lined channels called plasmodesmata span the cell wall and contain strands of cytoplasm that allow materials to pass from one cell to another.
- The extracellular matrix contains proteins and polysaccharides produced by the cell that helps support cells and aids in communication between cells.
- Adhesion junctions and tight junctions, if present, help hold cells together.
- Gap junctions allow passage of small molecules between cells.

Key Terms

actin filament 64	lysosome 60
adenosine triphosphate (ATP) 62	matrix 63
adhesion junction 67	microtubule 64
capsule 52	mitochondrion 62
cell 50	motor protein 64
cell theory 52	nuclear envelope 59
cellular respiration 63	nuclear pore 59
cell wall 52, 66	nucleoid 52
centriole 65	nucleolus 58
centrosome 64	nucleus 56
chloroplast 62	organelle 56
chromatin 58	plasma membrane 54
chromosome 58	plasmodesmata 66
cilium (pl., cilia) 65	polyribosome 59
conjugation pili 53	prokaryotic cell 52
cristae 63	ribosome 53
cytoplasm 52	rough ER 60
cytoskeleton 57, 64	secretion 60
endomembrane system 60	smooth ER 60
endoplasmic reticulum (ER) 59	stroma 62
eukaryotic cell 52	surface-area-to-volume ratio 51
extracellular matrix (ECM) 66	thylakoid 62
fimbriae 53	tight junction 67
flagellum (pl., flagella) 53, 65	transport vesicle 57
gap junction 67	vacuole 61
Golgi apparatus 60	vesicle 60
granum 62	

Testing Yourself

Choose the best answer for each question.

1. Which of the following is not found in a prokaryotic cell?
   a. cytoplasm
   b. ribosome
   c. plasma membrane
   d. mitochondrion
   e. nucleoid

2. Which of the following can only be viewed with an electron microscope?
   a. virus
   b. chloroplast
   c. bacteria
   d. human egg

3. The plasma membrane has the consistency of
   a. butter.
   b. peanut butter.
   c. olive oil.
   d. water.

4. The plasma membrane is called a fluid mosaic because it contains
   a. waxes suspended within a mosaic of phospholipids.
   b. a mosaic of proteins suspended within a phospholipid bilayer.
   c. a polysaccharide mosaic suspended within a protein bilayer.
   d. a mosaic of phospholipids suspended within a protein bilayer.

5. Eukaryotic cells compensate for a low surface-to-volume ratio by
   a. taking up materials from the environment more efficiently.
   b. lowering their rate of metabolism.
   c. compartmentalizing their activities into organelles.
   d. reducing the number of activities in each cell.

6. Ribosomal subunits are produced in which organelle?
   a. mitochondria
   b. endoplasmic reticulum
   c. cytoplasm
   d. nucleolus

For questions 7–10, match the items to those in the key.

Key:

  a. channel proteins   d. receptor proteins
  b. transport proteins   e. enzymatic proteins
  c. cell recognition proteins   f. tight junction protein

  7. Assist with cell-to-cell communication.

  8. Combine with molecules to carry them across plasma membranes.

  9. Combine with signaling molecules.

 10. Participate in metabolic reactions.

For questions 11–14, match the items to those in the key.

Key:

  a. single membrane
  b. double membrane
  c. no membrane

 11. Ribosome

 12. Nucleus

 13. Lysosome

 14. Mitochondrion

 15. The organelle that sorts proteins and sends them to their final destinations is the
  a. ribosome.   c. Golgi apparatus.
  b. vacuole.   d. lysosome.

 16. Plant vacuoles may contain
  a. flower color pigments.
  b. toxins that protect plants against herbivorous animals.
  c. sugars.
  d. All of these are correct.

 17. The interior of a mitochondrion is called the
  a. cristae.   c. matrix.
  b. thylakoid.   d. granum.

 18. Label the parts of the following cell.

 19. Which of these structures is involved in protein synthesis?
  a. ribosomes   c. mitochondria
  b. plasma membrane   d. microtubules

Thinking Scientifically

1. The 1992 movie *Lorenzo's Oil* recounts the story of six-year-old Lorenzo Odone and his battle with adrenoleukodystrophy (ALD), which is caused by a malfunctioning organelle called a peroxisome.

A peroxisome is an organelle that ordinarily contains enzymes capable of breaking down long-chain fatty acids. In ALD, the enzyme is missing, and when long-chain fatty acids accumulate, muscle weakness and loss of control result.
  a. How might a dietitian be able to help a person with ALD?
  b. What organelle discussed in this chapter is most similar in structure and function to a peroxisome?

2. Protists of the phylum Apicomplexa cause malaria and contribute to infections associated with AIDS. These parasites are unusual because they contain plastids. (Chloroplasts are a type of plastid, but most plastids store materials.) Scientists have discovered that an effective antibiotic that kills the parasite inhibits the functioning of the plastid. Can you conclude that the plastid is necessary to the life of the parasite? Why or why not?

3. Scientists reason that a cell could not evolve until it had an outer plasma membrane. Why does that seem logical?

Bioethical Issue

Human Cloning

Cloning animals, and the specter of cloning humans, is an issue that has caused much concern among the scientific community, governments, and the general public. The use of a fertilized egg, from which the nucleus has been removed, is in disfavor because it requires the destruction of a human embryo.

Recent technological advances may sidestep this concern, but will create others. In 2007, scientists in the United Kingdom applied for licenses to create human-nonhuman cell lines, and this avoids the current prohibition on using human embryos in scientific research. Instead of using a human egg, the scientists proposed using fertilized eggs from other animal species, such as rabbits, and adding a human cell nucleus to it to produce a human-animal hybrid cell, which could then be implanted into a surrogate mother.

Proponents of this measure point out that this technology avoids the destruction of human embryos. However, there is still the moral issue over whether or not it is acceptable to clone individual people. Some insist that such technology could be abused to create clones of unsuspecting individuals. While those who oppose the measure insist that cloning human beings is still wrong, and that even the destruction of animal embryos is questionable, the procedure does satisfy the letter of the law prohibiting the destruction of human embryos for such research. Consider that the nucleus, and thus the DNA, of the resulting clone is definitely human, but the other contents of the cell, including mitochondria, endoplasmic reticulum, and cytoskeleton, for example, are that of another species. Since mitochondria contain DNA, not all of the DNA of the resulting cell would be human.

Do you believe that it is acceptable or unacceptable to clone human beings, even if human embryos are not destroyed to do so? Are the fears of unauthorized use of the technology real or unfounded? And should resulting clones be considered to be fully human? Would you be willing for taxpayer funds to be available for this type of research? Why or why not?

Essentials of Biology Website

The companion website for *Essentials of Biology* provides a wealth of information organized and integrated by chapter. You will find practice tests, animations, videos, and much more that will complement your learning and understanding of general biology.

http://www.mhhe.com/maderessentials2

Body temperature is maintained by energy conversions in cells.

Penicillin, a life-saving antibiotic, works by inhibiting the activity of certain enzymes.

Drinking excessive amounts of water can have a toxic effect on cells.

5

The Dynamic Cell

Outline

Can Water Be Poisonous?

ALTHOUGH WE NORMALLY think of all the wonderful things water does for our bodies, drinking an excessive amount of it during a short period of time can be toxic or even fatal. In 2006, a young mother in California died after drinking several gallons of water in an attempt to win a video game system. Drinking too much water has also been reported in people who are under the influence of drugs such as Ecstasy.

In humans, water intake typically equals water output, primarily by the kidneys. When huge amounts of water are consumed very quickly,

—Continued next page

Continued—

the kidneys sometimes can't keep up, and thus cells swell. This swelling, especially in the brain, causes symptoms similar to those associated with intoxication, and the condition is called *water intoxication.*

What makes cells swell? As you will learn in this chapter, the process of osmosis causes water to move from the side of a cell containing more water to the side containing less water. In the case of water intoxication, first there is much more water outside the cells than inside the cells. Then, when this water moves into the cells to achieve equilibrium, the cells swell and may literally explode and die.

This chapter also explores how other molecules move in and out of cells, basic properties of energy, and the functions of enzymes. Only after understanding these principles will you appreciate the processes of photosynthesis and cellular respiration. These processes are vital for maintaining life and will be described in Chapters 6 and 7.

5.1 | What Is Energy?

Energy is defined as the capacity to do work—to make things happen. Without a source of energy, we humans would not be here on Earth—nor would any other living thing. Our biosphere gets its energy from the sun, and thereafter, one form of energy is changed to another form as life processes take place.

The two basic forms of energy are potential energy and kinetic energy. **Potential energy** is stored energy, and **kinetic energy** is the energy of motion. Potential energy is constantly being converted to kinetic energy, and vice versa. Let's look at the example in **Figure 5.1.** The food a cross-country skier has for breakfast contains chemical energy, which is a form of potential energy. When our skier hits the slopes, she may have to ascend a hill. During her climb, the potential energy of food is converted to the kinetic energy of motion, a type of mechanical energy. Once she reaches the hilltop, kinetic energy has been converted to the potential energy of location. As she skis down the hill, this potential energy is converted to the kinetic energy of motion again. But with each conversion, some energy is lost as heat and other unusable forms.

Measuring Energy

Chemists use a unit of measurement called the joule to measure energy, but it is common to measure food energy in terms of calories. A **calorie** is the amount of heat required to raise the temperature of 1 gram of water by one degree Celsius. This isn't much energy, so the caloric value of food is listed in nutrition labels and in diet charts in terms of **kilocalories** (1,000 calories). In this text, we will use *Calorie* to mean 1,000 calories.

Two Energy Laws

Two **energy laws** govern energy flow and help us understand the principles of energy conversion. The first law, called the law of conservation of energy, tells us: *Energy cannot be created or destroyed, but it can be changed from one form to another.* Relating the law to our

Figure 5.1 **Potential energy versus kinetic energy.**

Food contains potential energy that a skier can convert to kinetic energy in order to climb a hill. Height is potential energy due to location, which the skier converts to kinetic energy as she descends the hill. With every conversion to kinetic energy, some potential energy is lost as heat.

previous example, we know that the skier had to acquire energy by eating food before she could climb the hill, and that energy conversions occurred before she reached the bottom of the hill again.

The second energy law tells us: *Energy cannot be changed from one form to another without a loss of usable energy.* Many forms of energy are usable, such as the energy of the sun, food, and ATP. **Heat** is diffuse energy and the least usable form. Every energy conversion results in a loss of usable energy in the form of heat. For example, much of the potential energy stored in gasoline is lost as heat as the car converts it into the kinetic energy of motion, which is why a car's engine heats up as it is driven. This heat represents a loss of usable energy.

Entropy

The second energy law can be stated another way: Every energy transformation leads to an increase in the amount of disorganization or disorder. The term **entropy** refers to the relative amount of disorganization. The only way to maintain or bring about order is to add more energy to a system. To take an example from your own experience, you know that a tidy room is more organized and less stable than a messy room, which is disorganized and more stable (**Fig. 5.2***a*). In other words, your room is much more likely to stay messy than it is to stay tidy. Why? Unless you continually add energy to keep your room organized and neat, it will inevitably become less organized and messy.

Because our universe is a *closed system* in which energy cannot be created, all energy transformations, including those in cells, increase the total entropy of the universe. Figure 5.2*b* shows a process that occurs in cells because it proceeds from a more ordered state to a more disordered state. Just as a tidy room tends to become messy, hydrogen ions (H^+) that have accumulated on one side of a membrane tend to move to the other side unless they are prevented from doing so by the addition of energy. Why? Because when H^+ ions are distributed equally on either side of the membrane, no additional energy is needed to keep them that way, and the entropy, or disorder, of their arrangement has increased. The result is a more stable arrangement of H^+ ions.

What about reactions in cells that apparently proceed from disorder to order? For example, we know that plant cells can make glucose out of carbon dioxide and water. How do they do it? In order to overcome the natural tendency toward disorder, energy input is required, just as it requires energy to organize the messy room. Likewise, energy provided by the sun allows plants to make glucose, a highly organized molecule, from the more disorganized water and carbon dioxide. Even this process, however, involves a loss of some potential energy. When light energy is converted to chemical energy in plant cells, some of the sun's energy is always lost as heat. In other words, the organization of a cell has a constant energy cost that also results in an increase in the entropy of the universe.

Entropy: Second Law of Thermodynamics

Tidy bedroom	Messy bedroom

a.

- more organized
- more potential energy
- less stable
- less entropy

- less organized
- less potential energy
- more stable
- more entropy

Unequal distribution of hydrogen ions	Equal distribution of hydrogen ions

b.

Figure 5.2 **Cells and entropy.**

The second energy law tells us that entropy (disorder) always increases. Therefore (**a**) a tidy room tends to become messy and disorganized, and (**b**) hydrogen ions (H^+) on one side of a membrane tend to move to the other side so that the ions are equally distributed. Both processes result in a loss of potential energy and an increase in entropy.

Check Your Progress

1. **Contrast potential energy with kinetic energy.**
2. **How do cells avoid entropy (disorder) and maintain their organization?**

Answers: 1. Potential energy is stored energy, while kinetic energy is energy of motion.   2. Cells carry out many energy requiring processes and reactions to maintain their organization.

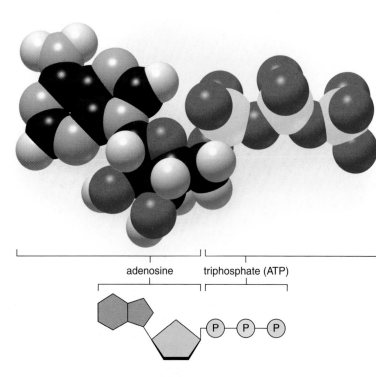

adenosine | triphosphate (ATP)

Figure 5.3 ATP.

ATP, the universal energy currency of cells, is composed of adenosine and three phosphate groups (called a triphosphate).

Figure 5.4 The ATP cycle.

When ATP is used as an energy source, a phosphate group is removed by hydrolysis. ATP is regenerated in mitochondria as cellular respiration occurs.

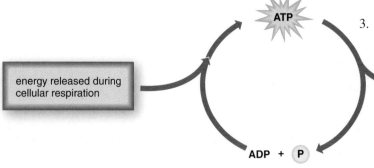

5.2 | ATP: Energy for Cells

ATP (adenosine triphosphate) is the energy currency of cells. Just as you use coins to purchase all sorts of products, a cell uses ATP to carry out nearly all of its activities, including synthesizing proteins, transporting ions across the plasma membranes, and causing organelles and cilia to move. Cells use the readily accessible energy provided by ATP to provide energy wherever it is needed.

Structure of ATP

ATP is a nucleotide, the type of molecule that serves as a monomer for the construction of DNA and RNA. ATP's name, adenosine triphosphate, means that it contains the sugar ribose, the nitrogen-containing base adenine, and three phosphate groups (**Fig. 5.3**). The three phosphate groups are negatively charged and repel one another. It takes energy to overcome their repulsion, and thus these phosphate groups make the molecule unstable. ATP easily loses the last phosphate group because the breakdown products, ADP (adenosine diphosphate) and a separate phosphate group symbolized as Ⓟ, are more stable than ATP. This reaction is written as ATP → ADP + Ⓟ and energy is released as ATP breaks down. ADP can also lose a phosphate group to become AMP (adenosine monophosphate).

Use and Production of ATP

The continual breakdown and regeneration of ATP is known as the ATP cycle (**Fig. 5.4**). ATP holds energy for only a short period of time before it is used in a reaction that requires energy. Then ATP is rebuilt from ADP + Ⓟ. Each ATP molecule undergoes about 10,000 cycles of synthesis and breakdown every day. Our bodies use some 40 kg (about 88 lb) of ATP daily, and the amount on hand at any one moment is sufficiently high to meet only current metabolic needs.

ATP's instability, the very feature that makes it an effective energy carrier, keeps it from being an energy storage molecule. Instead, the many H—C bonds of carbohydrates and fats make them the energy storage molecules of choice. Their energy is extracted during cellular respiration and used to rebuild ATP, mostly within mitochondria. You will learn in Chapter 7 that the breakdown of one molecule of glucose permits the buildup of some 38 molecules of ATP. During cellular respiration, only 39% of the potential energy of glucose is converted to the potential energy of ATP; the rest is lost as heat.

The production of ATP is still worthwhile for the cell for the following reasons:

1. ATP releases energy quickly, which facilitates the speed of enzymatic reactions.
2. When ATP becomes ADP + Ⓟ, the amount of energy released is usually just enough for a biological purpose. Breaking down an entire carbohydrate or fat molecule would be wasteful since it would release much more energy than is needed.
3. The structure of ATP allows its breakdown to be easily *coupled* to an energy-requiring reaction, as described next.

Coupled Reactions

Many metabolic reactions require energy. Energy can be supplied when a reaction that requires energy (e.g., building a protein) occurs in the vicinity of a reaction that gives up energy (e.g., ATP breakdown). A **coupled reaction** brings them together in such a way that the energy-releasing reaction can drive the energy-requiring reaction. Usually the energy-releasing reaction is ATP breakdown, which generally releases more energy than the amount consumed by the energy-requiring reaction. This increases entropy, but both reactions will proceed. The simplest way to represent a coupled reaction is like this:

Coupling

This reaction tells you that coupling occurs, but it does not show how coupling is achieved. A cell has two main ways to couple ATP breakdown to an energy-requiring reaction: ATP is used to energize a reactant or to change its shape. Both are often achieved by transferring a phosphate group to the reactant. For example, when polypeptide synthesis occurs at a ribosome, an enzyme transfers a phosphate group from ATP to each amino acid in turn, and this transfer activates the amino acid, causing it to bond with another amino acid.

Figure 5.5 shows how ATP breakdown provides the necessary energy for muscular contraction. During muscle contraction, myosin filaments pull actin filaments to the center of the cell, and the muscle shortens. First, myosin combines with ATP, and only then does ATP break down to ADP + P. The release of ADP + P from the molecule causes myosin to change shape and pull on the actin filament.

Check Your **Progress**

1. Explain why ATP is an effective energy carrier in cells.
2. Briefly explain how ATP functions in coupled reactions.

Answers: 1. The last phosphate group of ATP is easily and quickly removed. ATP holds an appropriate amount of energy for most cellular reactions so that energy is not wasted.   2. ATP can donate a phosphate to energize a compound for a reaction. Alternatively, it causes a molecule to change its shape.

Figure 5.5   Coupled reaction.

Muscle contraction occurs only when it is coupled to ATP breakdown. Myosin combines with ATP prior to its breakdown. Release of ADP + P causes myosin to change position and pull on an actin filament.

Energy Conversions

heat   solar energy

chloroplast

O_2

CO_2 and H_2O **Chemical energy** (carbohydrate)

mitochondrion

heat

ATP   **Chemical work Transport work Mechanical work**

The Flow of Energy

In the biosphere, the activities of chloroplasts and mitochondria enable energy to flow from the sun through all living things. During photosynthesis, the chloroplasts in plants capture solar energy and use it to convert water and carbon dioxide into carbohydrates, which serve as food for themselves and for other organisms. During cellular respiration, mitochondria complete the breakdown of carbohydrates and use the released energy to build ATP molecules.

Notice in **Figure 5.6** that cellular respiration requires oxygen and produces carbon dioxide and water, the very molecules taken up by chloroplasts. It is actually the cycling of molecules between chloroplasts and mitochondria that allows a flow of energy from the sun through all living things. This flow of energy maintains the levels of biological organization from molecules to organisms to ecosystems. In keeping with the energy laws, useful energy is lost with each chemical transformation, and eventually the solar energy captured by plants is lost in the form of heat. In this way, living things are dependent upon an input of solar energy.

Human beings are also involved in the cycling of molecules between plants and animals and in the flow of energy from the sun. We inhale oxygen and eat plants and their stored carbohydrates, or we eat other animals that have eaten plants. Oxygen and nutrient molecules enter our mitochondria, which produce ATP and release carbon dioxide and water. Without a supply of energy-rich foods, we could not produce the ATP molecules needed to maintain our bodies and carry on activities (**Fig. 5.7**).

Figure 5.6   Flow of energy.

Chloroplasts convert solar energy to the chemical energy of nutrient molecules. Mitochondria convert this chemical energy to ATP molecules, which cells use to perform chemical, transport, and mechanical work.

Potato

Fruit

Grain

Meat

Potatoes contain about 90 Calories per 3.5 oz (100 g)

Potatoes contain starch, which stores energy for the growing plant. When a potato is eaten, starch molecules are broken down to simple sugars in the body before their energy is released.

Apples contain about 60 Calories per 3.5 oz (100 g)

Fruits such as apples contain natural sugars that provide energy. The body converts excess sugar into fat, which is stored energy that can be used later.

Grain contains about 300 Calories per 3.5 oz (100 g)

Grain is an efficient energy source because it directly stores the energy of photosynthesis. Grains are a good source of energy for humans and animals.

Pork chops, grilled or broiled, contain about 215 Calories per 3.5 oz (100 g)

Meat is a good energy source; much of its energy is stored in fat. However, meat isn't an efficient energy source because only a small amount of the energy taken in by a living pig, for example, is converted to energy usable by the human or animal that eats the pork chop.

Figure 5.7   Energy of life.

Our food provides all the energy our bodies need. The energy in food is measured in Calories. To drive a car requires about 61 Calories per hour. To run at a speed of 1 mile in 7 minutes takes 865 Calories per hour. If you take in more Calories than you burn, you will gain weight.

5.3 | Metabolic Pathways and Enzymes

Reactions do not occur haphazardly in cells. Just as a house is built in a series of ordered steps, chemical reactions in a cell usually occur in a particular order. They are part of a **metabolic pathway,** a series of linked reactions. Metabolic pathways begin with a particular reactant and terminate with an end product. For example, glucose is broken down by a series of consecutive reactions until the end products CO_2 and H_2O result. In the pathway, one reaction leads to the next reaction, which leads to the next reaction, and so forth in an organized, highly structured manner.

A metabolic pathway can be represented by the following diagram:

$$E_1 \qquad E_2 \qquad E_3 \qquad E_4 \qquad E_5 \qquad E_6$$
$$A \rightarrow B \rightarrow C \rightarrow D \rightarrow E \rightarrow F \rightarrow G$$

In this diagram, the letters A–F are reactants and the letters B–G are products. In other words, the product from the previous reaction becomes the reactant of the next reaction. In the first reaction, A is the reactant and B is the product. Now B becomes the reactant in the next reaction of the pathway, and C is the product. This process continues until the final product (G) forms.

In the above diagram, the letters E_1–E_6 represent **enzymes,** which are protein molecules that function as organic catalysts to speed chemical reactions. Enzymes can only speed reactions that are possible to begin with. In the cell, an enzyme is analogous to a mutual friend who causes two people to meet and interact, because an enzyme brings together particular molecules and causes them to react with one another.

The molecules acted on by an enzyme are called its **substrates.** An enzyme converts substrates into products. The substrates and products of an enzymatic reaction vary greatly (**Fig. 5.8**). Many enzymes facilitate the breakdown of a substrate into multiple products. Or an enzyme may convert a single substrate into a single product. Still others may take two or more substrates and combine them into a single product.

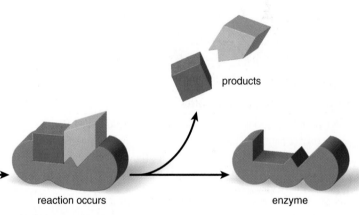

Figure 5.8 Enzymatic action.

An enzyme has an active site where the substrates and enzyme fit together in such a way that the substrates are oriented to react. Following the reaction, the products are released, and the enzyme is free to act again.

An Enzyme's Active Site

In most instances, only one small part of the enzyme, called the **active site,** accommodates the substrate(s). At the active site, the substrate fits into the enzyme seemingly like a key fits a lock, and thus most enzymes can only fit one substrate. However, the active site undergoes a slight change in shape in order to accommodate the substrate(s). This is called the **induced fit model** because the enzyme is induced to undergo a slight alteration to achieve optimal fit (Fig. 5.8).

The change in the shape of the active site facilitates the reaction that next occurs. After the reaction has been completed, the product(s) is released, and the active site returns to its original state, ready to bind to another substrate molecule. Only a very small amount of each enzyme is actually needed in a cell because enzymes are not used up by the reaction.

Check Your **Progress**

1. Explain the benefit of metabolic pathways in cells.
2. Describe how the induced fit model explains the binding of a substrate to an enzyme's active site.

Answers: 1. Various reactions of a pathway produce a product that is useable in one or more other pathways. Also, the reactions of a metabolic pathway allow energy to be used or generated in small increments.   2. A slight change in shape of the active site of an enzyme allows the substrate to fit more easily into the active site.

Active enzyme and active pathway

Feedback inhibition

Inactive enzyme and inactive pathway

Figure 5.9 **Feedback inhibition.**

This type of feedback inhibition occurs when the end product (P) of an active enzyme pathway is plentiful and binds to the first enzyme (E_1) of the pathway at a site other than the active site. This changes the shape of the active site so that the substrate (S) can no longer bind to the enzyme. Now the entire pathway becomes inactive.

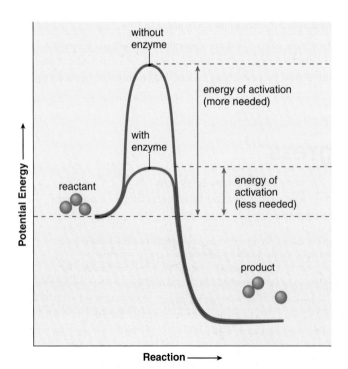

Figure 5.10 **Energy of activation (E_a).**

Enzymes speed the rate of reactions because they lower the amount of energy required for the reactants to react. Even reactions like this one, in which the energy of the product is less than the energy of the reactant, speed up when an enzyme is present.

Enzyme Inhibition

Enzyme inhibition occurs when an active enzyme is prevented from combining with its substrate. Enzyme inhibitors are often poisonous to certain organisms. Cyanide, for example, is an inhibitor of the enzyme cytochrome *c* oxidase, which performs a vital function in cells because it is involved in making ATP. Cyanide is a poison because it binds the enzyme, blocking its activity. But some enzyme inhibitors are useful drugs. For example, penicillin is a poison for bacteria, but not humans, because it blocks the active site of an enzyme unique to bacteria. Many other antibiotic drugs also act as enzyme inhibitors.

The activity of almost every enzyme in a cell is regulated by **feedback inhibition.** In the simplest case, when a product is in abundance, it competes with the substrate for the enzyme's active site. As the product is used up, inhibition is reduced, and then more product can be produced. In this way, the concentration of the product always stays within a certain range.

Most metabolic pathways are regulated by more complex types of feedback inhibition (**Fig. 5.9**). In these instances, when the end product is plentiful, it binds to a site other than the active site of the first enzyme in the pathway. This binding changes the shape of the active site, preventing the enzyme from binding to its substrate. Without the activity of the first enzyme, the entire pathway shuts down.

Energy of Activation

Molecules frequently do not react with one another unless they are activated in some way. In the lab, activation is very often achieved by heating a mixture to increase the number of effective collisions between molecules. The added heat causes the molecules to move more quickly, increasing the likelihood of effective collisions between molecules. The energy needed to cause molecules to react with one another is called the **energy of activation (E_a). Figure 5.10** compares the E_a without an enzyme to the E_a with an enzyme, illustrating that the enzyme lowers the amount of energy required for a reaction to occur. Enzymes lower the energy of activation by bringing the substrates into contact and even by participating in the reaction at times.

Check Your **Progress**

1. What is the benefit of using feedback inhibition to control metabolic pathways?

Answer: 1. It ensures that the enzyme is only active when product levels are low and more product is needed.

5.4 | Cell Transport

The plasma membrane regulates the passage of molecules into and out of the cell. This function is crucial because the life of the cell depends on maintenance of its normal composition. The plasma membrane can carry out this function because it is selectively permeable, meaning that certain substances can freely pass through the membrane, some are transported across, and others are prohibited from entering or leaving. Basically, substances enter a cell in one of three ways: passive transport, active transport, and bulk transport. Although there are different types of passive transport, in all cases substances move from a higher to a lower concentration, and no energy is required. Active transport moves substances against a concentration gradient and requires both a transport protein and lots of ATP. Bulk transport requires energy, but movement of the large substances involved is independent of concentration gradients.

Passive Transport: No Energy Required

Actually, no membrane is required for simple diffusion. During **simple diffusion,** molecules move down their concentration gradient until equilibrium is achieved and they are distributed equally. Simple diffusion occurs because molecules are in motion, but it is a *passive* form of transport because a cell does not need to expend energy for it to happen. Some small, noncharged molecules, such as oxygen and carbon dioxide, are able to slip between the phospholipid molecules making up the plasma membrane. Therefore, these molecules can diffuse across the membrane.

Figure 5.11 demonstrates simple diffusion to cross a membrane. Water is present on two sides of a membrane, and dye is added to one side. The dye particles move in various directions, but since there are more of them on the right side of the membrane, the net movement is toward the left side of the membrane. Eventually, the dye is dispersed, with equal numbers of dye particles on both sides of the membrane, and there is no net movement of dye in either direction. A **solution** contains both a solute and a solvent. In this case, the dye is called the **solute,** and the water is called the **solvent.** Solutes are usually solids or gases, and solvents are usually liquids.

Facilitated diffusion occurs when an ion or molecule diffuses across a membrane faster than expected, either by way of a specific channel protein or with the assistance of a carrier protein. Water, a polar molecule, moves across the membrane through channels called **aquaporins.** This is why water can cross a membrane much more quickly than expected. Glucose and amino acids are assisted across the membrane by carrier proteins that undergo a change in shape as they pass through. Channel proteins and carrier proteins are very specific to the molecule they assist across the membrane.

Osmosis

Osmosis is the diffusion of water from the lower to the higher concentration of solute across a semipermeable membrane. (Today we know that water diffuses through channel proteins called aquaporins.) To illustrate osmosis, a tube containing a 5% salt solution and covered at one end by a membrane is placed in a beaker that contains water only (**Fig. 5.12***a*). Water can cross the membrane, but the solute cannot. Therefore, there will be a net movement of water across the membrane from the beaker to the inside of the tube. Theoretically, the solution inside the tube will rise until there is an equal concentration of water on both sides of the membrane (Fig. 5.12*b,c*). What would happen if the beaker contained a 2% salt solution? Water would still diffuse into the tube, because the tube at 5% salt would still contain a lower concentration of water molecules than the beaker at 2%.

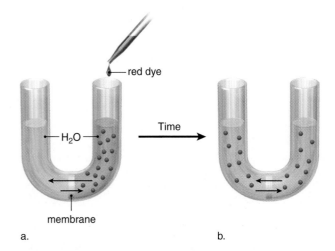

a.   b.

Figure 5.11   **Simple diffusion demonstration.**

Simple diffusion is spontaneous, and no energy is required to bring it about. **a.** Red dye is added to water separated by a membrane. The dye molecules can pass through the membrane. The dye molecules move randomly about, but over time the net movement of dye is toward the region of lower concentration. **b.** Eventually, the dye molecules are equally distributed throughout the container and there is no net movement of dye in either direction.

a.   b.   c.

Figure 5.12   **Osmosis demonstration.**

a. A tube covered at the broad end by a differentially permeable membrane contains a 5% salt solution, and the beaker contains only water. The salt ions are unable to pass through freely, but the water molecules can pass through the membrane. **b.** Therefore, there is a net movement of water toward the inside of the tube, where the percentage of water molecules is lower than the outside. **c.** The level of the solution rises in the tube because of the incoming water.

Red blood cells

normal cells   cells swell, burst   shriveled cells

| Isotonic solution | Hypotonic solution | Hypertonic solution |

normal cell   normal turgid cell   cytoplasm shrinks from cell wall

Plant cells

Figure 5.13   Osmosis in animal and plant cells.

In an isotonic solution, cells neither gain nor lose water. In a hypotonic solution, cells gain water. Red blood cells swell to bursting, and plant cells become turgid. In a hypertonic solution, cells lose water. Red blood cells shrivel, and plant cell cytoplasm shrinks away from the cell wall.

The Effect of Osmosis on Cells

When solutes are added to a cell's surroundings, its differentially permeable membrane often limits movement of the solute back and forth. Instead, water moves back and forth across the membrane by osmosis. This can affect the size and shape of cells, as shown in **Figure 5.13**. In the laboratory, cells are normally placed in **isotonic solutions** (*iso,* same as) in which the cell neither gains nor loses water—that is, the concentration of water is the same on both sides of the membrane. In medical settings, a 0.9% solution of the salt sodium chloride (NaCl) is known to be isotonic to red blood cells; therefore, intravenous solutions usually have this concentration.

Cells placed in a **hypotonic solution** (*hypo,* less than) gain water. Outside the cell, the concentration of solute is less, and the concentration of water is greater, than inside the cell. Animal cells placed in a hypotonic solution expand and sometimes burst. The term *lysis* refers to disrupted cells; *hemolysis,* then, is disrupted red blood cells.

When a plant cell is placed in a hypotonic solution, the large central vacuole gains water, and the plasma membrane pushes against the rigid cell wall as the plant cell becomes *turgid.* The plant cell does not burst because the cell wall does not give way. Turgor pressure in plant cells is extremely important in maintaining their erect position.

Cells placed in a **hypertonic solution** (*hyper,* more than) lose water. Outside the cell, the concentration of solute is more, and the concentration of water is less, than inside the cell. Animal cells placed in a hypertonic solution shrink. For example, meats are sometimes preserved by being salted. Bacteria are killed not by the salt, but by the lack of water in the meat.

When a plant cell is placed in a hypertonic solution, the plasma membrane pulls away from the cell wall as the large central vacuole loses water. This is an example of **plasmolysis,** shrinking of the cytoplasm due to osmosis. Cut flowers placed into salty water will wilt due to plasmolysis. The dead plants you may see along a roadside could have died due to exposure to a hypertonic solution during the winter, when salt was used on the road.

Active Transport: Energy Required

During **active transport,** molecules or ions move through the plasma membrane, accumulating on one side of the cell (**Fig. 5.14**). For example, iodine collects in the cells of the thyroid gland; glucose is completely absorbed from the digestive tract by the cells lining the digestive tract; and sodium can be almost completely withdrawn from urine by cells lining the kidney tubules. In these instances, molecules have moved against their concentration gradients, a situation that requires both a transport protein and much ATP. Therefore, cells involved in active transport, such as kidney cells, have many mitochondria near their plasma membranes to generate the ATP that is needed.

Check Your Progress

1. **Describe the relationship between a solute, a solvent, and a solution.**
2. **How is osmosis useful to cells?**

Answers: 1. A solution is composed of a solute dissolved in a solvent.   2. Osmosis passively moves water into cells.

transport protein

ATP

Outside   Inside

P

P

Figure 5.14   Active transport.

During active transport, a transport protein uses energy to move a solute across the plasma membranes toward a higher concentration. Note that the transport protein changes shape during the process.

The passage of salt (NaCl) across a plasma membrane is of primary importance in cells because the salt causes water to move to that side of the plasma membrane. First, sodium ions are actively transported across a membrane, and then chloride ions simply diffuse through channels that allow their passage. Chloride ion channels malfunction in persons with cystic fibrosis, leading to the symptoms of this inherited disorder. Proteins engaged in active transport are often called *pumps*. The **sodium-potassium pump,** vitally important to nerve conduction, undergoes a change in shape that allows it to combine alternately with sodium ions and potassium ions.

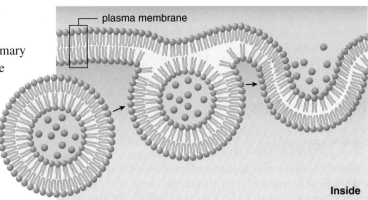

a. Exocytosis

Bulk Transport

Macromolecules, such as polypeptides, polysaccharides, or polynucleotides, are often too large to be moved by transport proteins. Instead, vesicle formation takes them into or out of a cell. For example, digestive enzymes and hormones use molecules transported out of the cell by **exocytosis (Fig. 5.15***a*). In cells that synthesize these products, secretory vesicles accumulate near the plasma membrane. These vesicles release their contents when the cell is stimulated to do so.

When cells take in substances by vesicle formation, the process is known as **endocytosis** (Fig. 5.15b). If the material taken in is large, such as a food particle or another cell, the process is called **phagocytosis.** Phagocytosis is common in unicellular organisms, such as amoebas. It also occurs in humans. Certain types of human white blood cells are amoeboid—that is, they are mobile like an amoeba, and are able to engulf debris such as worn-out red blood cells or bacteria. When an endocytic vesicle fuses with a lysosome, digestion occurs. In Chapter 26, we will see that this process is a necessary and preliminary step toward the development of immunity to bacterial diseases.

Pinocytosis occurs when vesicles form around a liquid or around very small particles. White blood cells, cells that line the kidney tubules and the intestinal wall, and plant root cells all use pinocytosis to ingest substances.

During **receptor-mediated endocytosis,** receptors for particular substances are found at one location in the plasma membrane. This location is called a coated pit because there is a layer of protein on its intracellular side (**Fig. 5.16**). Receptor-mediated endocytosis is selective and much more efficient than ordinary pinocytosis. It is involved when substances move from maternal blood into fetal blood at the placenta, for example.

receptor protein

molecule

b. Endocytosis

Figure 5.15 Bulk transport.

During exocytosis **(a)** and endocytosis **(b)**, vesicle formation transports substances out of or into a cell, respectively.

coated pit

coated vesicle

Figure 5.16 Receptor-mediated endocytosis.

During receptor-mediated endocytosis, molecules first bind to specific receptor proteins that are in a coated pit. The vesicle that forms contains the molecules and their receptors.

Courtesy of SynCardia Systems, Inc.

Check Your Progress

1. Compare and contrast pinocytosis and phagocytosis.

Answer: 1. Both use vesicles to transport materials across the plasma membrane into a cell. Very small molecules are transported by pinocytosis and larger ones by phagocytosis. Endocytic vesicles often fuse with lysosomes to digest the particles they contain.

The Chapter in Review

Summary

5.1 | What Is Energy?

Potential energy can be converted to kinetic energy, and vice versa. Solar and chemical energy are forms of potential energy; mechanical energy is a form of kinetic energy. Two energy laws hold true universally:
- Energy cannot be created or destroyed, but can be transferred or transformed.
- Energy may be converted from one form to another, but some is lost as unusable forms. Therefore, the entropy of the universe is increasing, and only a constant input of energy maintains the organization of living things.

5.2 | ATP: Energy for Cells

Energy flows from the sun through chloroplasts and mitochondria, which produce ATP. Because ATP has three adjoining negative phosphate groups, it is a high-energy molecule that tends to break down to ADP + P , releasing energy. ATP breakdown is coupled to various energy-requiring cellular reactions, including protein synthesis, active transport, and muscle contraction. Cellular respiration provides the energy for the production of ATP. The following diagram summarizes the ATP cycle:

5.3 | Metabolic Pathways and Enzymes

A metabolic pathway is a series of reactions that proceed in an orderly, step-by-step manner. Each reaction requires an enzyme that is specific to its substrate. Enzymes bring substrates together at an enzyme's active site, and speed reactions by lowering the energy of activation. The activity of most enzymes and metabolic pathways is regulated by feedback inhibition.

5.4 | Cell Transport

The plasma membrane is differentially permeable; some substances can freely cross the membrane, and some must be assisted across if they are to enter the cell.

Passive transport requires no metabolic energy and moves substances from a higher to a lower concentration.
- In simple diffusion, molecules move from an area of higher concentration to the area of lower concentration until the concentration of molecules at both sites are the same. Some molecules, such as dissolved gases, cross plasma membranes by simple diffusion.
- In facilitated diffusion, molecules diffuse across a plasma membrane through a channel protein (aquaporins are channel proteins for water) or with the assistance of carrier proteins.
- Osmosis is the simple diffusion of water through aquaporins across a membrane toward the area of lower water concentration (higher solute concentration).

Active transport requires metabolic energy (ATP) and moves substances from a lower to a higher concentration across a membrane.
- A transport protein acts as a pump that causes a substance to move against its concentration gradient. For example, the sodium-potassium pump carries Na^+ to the outside of the cell and K^+ to the inside of the cell.

Bulk transport requires vesicle formation and metabolic energy. It occurs independent of concentration gradients.
- Exocytosis transports macromolecules out of a cell via vesicle formation and often results in secretion.
- Endocytosis transports macromolecules into a cell via vesicle formation.
- Phagocytosis is a type of endocytosis that transports large particles; pinocytosis transports liquids or small particles; and receptor-mediated endocytosis is a form of pinocytosis, which makes use of receptor proteins in the plasma membrane.

The following diagram illustrates the different types of passive and active transport:

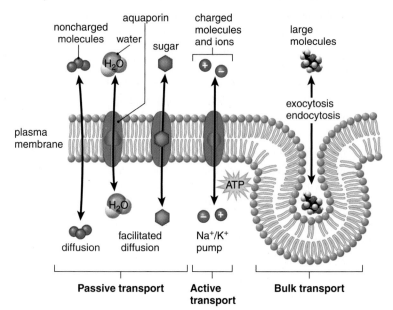

Key Terms

active site   77
active transport   80
aquaporin   79
calorie   72
coupled reaction   75
endocytosis   81
energy   72
energy laws   72
energy of activation (E_a)   78
entropy   73
enzyme   77
enzyme inhibition   78
exocytosis   81
facilitated diffusion   79
feedback inhibition   78
heat   73
hypertonic solution   80
hypotonic solution   80

induced fit model   77
isotonic solution   80
kilocalorie   72
kinetic energy   72
metabolic pathway   77
osmosis   79
phagocytosis   81
pinocytosis   81
plasmolysis   80
potential energy   72
receptor-mediated endocytosis   81
simple diffusion   79
sodium-potassium pump   81
solute   79
solution   79
solvent   79
substrate   77

Testing Yourself

Choose the best answer for each question.

1. Which of the following is not a fundamental law of energy?
   a. Energy cannot be created or destroyed.
   b. Energy can be changed from one form to another.
   c. Energy is lost when it is converted from one form to another.
   d. Potential energy cannot be converted to kinetic energy.

2. As a result of energy transformation,
   a. entropy increases.
   b. entropy decreases.
   c. potential energy increases.
   d. energy is lost in the form of heat.
   e. Both a and d are correct.

3. Which of the following is not a component of ATP?
   a. adenine   c. phosphate
   b. glucose   d. ribose

4. ATP is a good source of energy for a cell because
   a. it is versatile—able to be used in many types of reactions.
   b. its breakdown is easily coupled with energy-requiring reactions.
   c. it provides just the right amount of energy for cellular reactions.
   d. All of these are correct.

5. Compared to carbon dioxide and water, glucose
   a. is less organized.   c. is more stable.
   b. has less potential energy. d. exhibits lower entropy.

For questions 6–9, match the items to the correct answers in the key. Each answer may include more than one item.

Key:
   a. oxygen
   b. carbon dioxide
   c. water
   d. carbohydrates

6. Consumed by cellular respiration.
7. Produced by cellular respiration.
8. Consumed by photosynthesis.
9. Produced by photosynthesis.
10. In the following figure, one curve illustrates the energy of activation required for a reaction "with an enzyme," while the other illustrates the energy of activation "without an enzyme." Label each curve.

11. In the induced fit model of enzyme catalysis,
    a. the substrate exactly fits multiple enzymes.
    b. many substrates fit every enzyme.
    c. the enzyme changes shape slightly to accommodate the substrate.
    d. the enzyme doesn't fit the substrate until it is induced by another substrate.

For questions 12–18, match the items to the correct answers in the key. Each answer may include more than one item.

Key:
   a. simple diffusion
   b. facilitated diffusion
   c. osmosis
   d. active transport

12. Movement of molecules from high concentration to low concentration.
13. Requires a membrane.
14. Requires energy input.
15. Requires a protein pump.
16. Mechanism by which oxygen enters cells.
17. Mechanism by which glucose enters cells.
18. Movement of water across a membrane.
19. Cells involved in active transport have a large number of _____ near their plasma membrane.
    a. vacuoles   c. actin filaments
    b. mitochondria   d. aquaporins
20. The movement of water across the plasma membrane of a cell is facilitated by
    a. aquaporins.
    b. substrate-mediated endocytosis.
    c. receptor-mediated diffusion.
    d. channel proteins.
    e. Both a and d are correct.

21. Energy is used for
    a. simple diffusion.
    b. osmosis.
    c. active transport.
    d. None of these are correct.

22. A cell specializing in endocytosis would likely have many
    a. lysosomes.
    b. nuclei.
    c. mitochondria.
    d. aquaporins.

23. Seawater is nearly _____ to blood because it contains a/an _____ concentration of solutes.
    a. isotonic, higher
    b. hypotonic, higher
    c. hypertonic, lower
    d. isotonic, equal

24. A coated pit is associated with
    a. simple diffusion.
    b. osmosis.
    c. receptor-mediated endocytosis.
    d. pinocytosis.

25. Pumps used in active transport are made of
    a. sugars.   c. lipids.
    b. proteins.   d. cholesterol.

26. Which of these is correct?
    a. Energy can be transformed into another type as shown by the production of ATP in mitochondria.
    b. Energy can be created as shown by the production of ATP in mitochondria.
    c. Energy transformations cannot occur; therefore, ecosystems need a constant input of solar energy.
    d. Energy lost as heat is recaptured as decreased entropy.

27. Coupled reactions
    a. create energy through the hydrolysis of ATP.
    b. couple favorable reactions with unfavorable reactions.
    c. liberate energy trapped in enzymes.
    d. always result in decreased entropy.

28. The amount of energy needed to get a chemical reaction started is known as the
    a. starter energy.
    b. energy of activation.
    c. reaction energy.
    d. product energy.

29. Entropy is a term used to indicate the relative amount of
    a. organization.
    b. disorganization.
    c. enzyme action.
    d. None of these are correct.

30. Enzymes catalyze reactions by
    a. bringing the reactants together.
    b. lowering the activation energy.
    c. Both a and b are correct.
    d. Neither a nor b is correct.

31. The active site of an enzyme
    a. is identical to that of any other enzyme.
    b. is the part of the enzyme where its substrate can fit.
    c. can be used only once.

    d. is not affected by environmental factors such as pH and temperature.

Thinking Scientifically

1. Some coupled reactions in cells, including many involved in protein synthesis, use the nucleotide GTP as an energy source instead of ATP. What would be the advantage of using GTP instead of ATP as an energy source for these cellular reactions?

2. Cystic fibrosis is a genetic disorder caused by a defective membrane transport protein. The defective protein closes chloride channels in membranes, preventing chloride from being exported out of cells. This results in the development of a thick mucus on the outer surfaces of cells. This mucus clogs the ducts that carry digestive enzymes from the pancreas to the small intestine, clogs the airways in the lungs, and promotes lung infections. How do you think the defective protein results in a thick, sticky mucus outside the cells, instead of a loose, fluid covering?

3. Some prokaryotes, especially archaea, are capable of living in extreme environments, such as deep sea heat vents, where temperatures can reach 80°C (176°F). Few organisms can survive at this temperature. What adaptations might archaea possess that allow them to survive in such extreme heat?

Bioethical Issue

Arsenic in Drinking Water

Arsenic is chemically similar to phosphorus, and may form arsenate ions (AsO_4^{3-}) within the body that can substitute for phosphate (PO_4^{3-}) in many enzymatic reactions. Arsenate blocks some of the processes in the cell that are responsible for producing ATP, resulting in cell death. Metabolically active cells are particularly vulnerable to this toxin.

Arsenic occurs naturally in the water and soil; thus, arsenic levels in drinking water are a concern in many parts of the world. In 2000, the Clinton administration directed the EPA to reset the standard for maximal allowable levels in water to 10 parts per billion (ppb) from the previous level of 50 ppb set in 1942. These rules were suspended by the Bush administration in 2001, citing the high cost of compliance by utilities. Supporters asserted the costs of compliance would be significant and that smaller utilities serving rural communities would be particularly hard hit. Other critics objected on the basis that the cost-to-benefit ratio of the new standards did not justify the new rules. But many health and environmental organizations and many members of Congress blasted the decision to rescind the new rules, citing studies demonstrating health problems associated with arsenic levels in water below 50 ppb. Proponents also pointed out that rural water customers would be the primary beneficiaries of the new rules. The Bush administration ultimately relented and reinstated the newer, lower standard.

Do you think that the government did the right thing in reinstating the tougher standard? Why or why not? And should politics play a role in making scientific decisions, especially when health and environmental safety are at risk?

Essentials of Biology Website

The companion website for *Essentials of Biology* provides a wealth of information organized and integrated by chapter. You will find practice tests, animations, videos, and much more that will complement your learning and understanding of general biology.

http://www.mhhe.com/maderessentials2

The oceans teem with photosynthesizers.

Plants grow fastest when exposed to red and blue light.

Solar energy

H_2O
H_2O
H_2O

O_2

Plants produce oxygen by splitting water molecules.

6

Energy for Life

Outline

Photosynthesis Is Essential to Life on Earth

MOST PEOPLE ARE aware that plants produce oxygen gas. Further, they often think plants produce this oxygen as some sort of altruistic act to benefit humans and other animals! But the facts don't bear this out. During the process of photosynthesis, plants are actually making food for themselves. To do this, they reduce the carbon dioxide you exhale to carbohydrates after converting energy from the sun into chemical energy. This conversion process is complex, and depends on the energizing of electrons removed from water. In the process, water splits and releases

—Continued next page

Continued—

oxygen. This oxygen is the gas we and other animals depend upon. Thus, plants provide the oxygen you inhale and the glucose you use as a source of energy only because they are in the process of photosynthesizing to keep themselves alive. Plants are solar powered, and indirectly so are all other living things on Earth, including ourselves.

Understanding the process of photosynthesis and its critical importance to nearly all living organisms will help you appreciate just how dependent we humans are on plants. Scientists are now studying the complex process of photosynthesis to better understand it and how the process affects life, and even the environment. In fact, engineers are even exploring the possibility of using photosynthesis to produce electricity and reduce our dependence on fossil fuels! In this chapter, you will learn about basic plant cell structure and how plants perform photosynthesis.

6.1 | Overview of Photosynthesis

Photosynthesis transforms solar energy into the chemical energy of a carbohydrate. Photosynthetic organisms, including plants, algae, and cyanobacteria, produce an enormous amount of carbohydrate (**Fig. 6.1**). If the amount of carbohydrate were instantly converted to coal, and the coal loaded into standard railroad cars (each car holding about 50 tons), the photosynthesizers of the biosphere would fill more than 100 cars *per second* with coal.

Euglena

garden plants

trees

mosses

diatoms

kelp

cyanobacteria

Figure 6.1 Photosynthetic organisms.

Photosynthetic organisms include plants, such as trees, flowers, and mosses, which typically live on land; photosynthetic protists, such as *Euglena*, diatoms, and kelp, which typically live in water; and cyanobacteria, a type of bacterium that lives everywhere.

It is no wonder, then, that photosynthetic organisms are able to sustain themselves and, with a few exceptions,[1] all of the other living things on Earth. To appreciate this, consider that most food chains lead back to plants. In other words, producers, which have the ability to synthesize carbohydrates, feed not only themselves but also consumers, which must take in preformed organic molecules.

Our analogy about photosynthetic products becoming coal is apt because the bodies of plants formed the coal we burn today. This process occurred hundreds of thousands of years ago, and that is why coal is called a fossil fuel. The wood of trees is also commonly used as fuel. In addition, the fermentation of plant materials produces alcohol, which can be used to fuel automobiles directly or as a gasoline additive.

Flowering Plants as Photosynthesizers

The green portions of plants, particularly the leaves, carry on photosynthesis. Carbon dioxide in the air enters the many spaces of mesophyll tissue through small openings called **stomata** (sing., stoma). These spaces ensure that carbon dioxide, a raw material for photosynthesis, is efficiently delivered to mesophyll cells. The roots of a plant absorb water, which then moves in vascular tissue up the stem to the leaves where it exits at leaf veins. Carbon dioxide and water diffuse into mesophyll cells and then into **chloroplasts,** the organelles that carry on photosynthesis (**Fig. 6.2**).

In a chloroplast, a double membrane surrounds a fluid called the **stroma.** A third membrane system within the stroma forms flattened sacs called **thylakoids,** which in some places are stacked to form **grana** (sing., granum), so called because early microscopists thought they looked like piles of seeds. The space within each thylakoid is believed to be connected to the space within every other thylakoid, thereby forming an inner compartment within chloroplasts called the *thylakoid space.*

Chlorophyll and other pigments reside within the membranes of the thylakoids. These pigments are capable of absorbing solar energy, the energy that drives photosynthesis. A thylakoid membrane also contains complexes that convert solar energy into a chemical form useable by the enzymes in the stroma. The stroma is an enzyme-rich solution in which carbon dioxide is first attached to an organic compound and then reduced to a carbohydrate using the chemical energy provided by the thylakoid membranes.

This carbohydrate is the only source of energy for most organisms on Earth. When they metabolize the carbohydrate produced by chloroplasts, they release carbon dioxide into the air. This same carbon dioxide enters leaves and is converted to a carbohydrate. The relationship between producers (e.g., plants) and consumers (e.g., animals) is an intrinsic part of the intricate web of life on planet Earth.

Figure 6.2 Leaves and photosynthesis.

The raw materials for photosynthesis are carbon dioxide and water. Water enters a leaf by way of leaf veins. Carbon dioxide enters a leaf by way of the stomata. Chloroplasts have two major parts. The grana are made up of thylakoids, membranous disks that contain photosynthetic pigments such as chlorophyll. These pigments absorb solar energy. The stroma is a fluid-filled space where carbon dioxide is enzymatically reduced to a carbohydrate such as glucose.

[1] A few types of bacteria are chemosynthetic organisms, which obtain the necessary energy to produce their own organic nutrients by oxidizing inorganic compounds.

Figure 6.3 **The photosynthetic process.**

The process of photosynthesis consists of the light reactions and the Calvin cycle reactions. The light reactions, which produce ATP and NADPH, occur in a thylakoid membrane. These molecules are used by the Calvin cycle in the stroma to reduce carbon dioxide to a carbohydrate.

Check Your **Progress**

1. How do water and carbon dioxide for photosynthesis enter a leaf?
2. What molecules are produced by photosynthesis?

Answers: 1. Water flows from the roots through vascular tissue and enters the leaf from a leaf vein. Carbon dioxide enters the leaf through open stomata.   2. Carbohydrate and oxygen are produced by photosynthesis.

The Photosynthetic Process

The product of photosynthesis, carbohydrate (CH_2O), contains a lot more energy than the reactants carbon dioxide (CO_2) and water (H_2O). Why? Because electrons are removed from H_2O and energized by solar energy before they are given to CO_2. These electrons are accompanied by hydrogen ions (H^+), and that's why hydrogen atoms are apparently transferred during photosynthesis:

When a molecule gains hydrogen atoms ($e^- + H^+$), **reduction** occurs and the molecule is reduced. Therefore, carbon dioxide is reduced when it becomes CH_2O. When a molecule gives up hydrogen atoms ($e^- + H^+$), **oxidation** occurs and the molecule is oxidized. Therefore, water is oxidized when it splits and releases O_2. In metabolism, reduction goes along with oxidation and the reaction is called a **redox reaction.**

Notice that if you multiplied (CH_2O) by 6 you would get $C_6H_{12}O_6$, which is the formula for glucose. Some prefer to emphasize that glucose is the final product, although it is actually G3P (glyceraldehyde-3-phosphate) that is the end product of photosynthesis.

Two Sets of Reactions

An overall equation for photosynthesis tells us the beginning reactant and the end products of the pathway. But much goes on in between. The word photosynthesis suggests that the process requires two sets of reactions. *Photo*, which means light, refers to the reactions that capture solar energy, and *synthesis* refers to the reactions that produce a carbohydrate. The two sets of reactions are called the **light reactions** and the **Calvin cycle reactions** (**Fig. 6.3**).

Light Reactions   The following events occur in the thylakoid membrane during the light reactions:

- Chlorophyll within the thylakoid membranes absorbs solar energy and energizes electrons.
- ATP is produced from ADP + (P) with the help of an electron transport chain.
- [1]$NADP^+$, an enzyme helper, accepts electrons and becomes NADPH.

Calvin Cycle Reactions   The following events occur in the stroma during the Calvin cycle reactions:

- CO_2 is taken up by one of the molecules in the cycle.
- ATP and NADPH from the light reactions reduce CO_2 to a carbohydrate.

In the following sections, we discuss details of these two reactions.

---
[1] NADP = Nicotinamide adenine dinucleotide phosphate.

6.2 | Light Reactions

During the light reactions, the pigments within the thylakoid membranes absorb solar (radiant) energy. Solar energy can be described in terms of its wavelength and its energy content. **Figure 6.4** lists the different types of radiant energy, from the shortest wavelength, gamma rays, to the longest, radio waves. *Visible light* is only a small portion of this spectrum.

Visible light contains various wavelengths of light, as can be proven by passing it through a prism; the different wavelengths appear to us as the colors of the rainbow, ranging from violet (the shortest wavelength) to blue, green, yellow, orange, and red (the longest wavelength). The energy content is highest for violet light and lowest for red light.

Only about 42% of the solar radiation that hits the Earth's atmosphere ever reaches its surface, and most of this radiation is within the visible-light range. Higher-energy wavelengths are screened out by the ozone layer in the atmosphere, and lower-energy wavelengths are screened out by water vapor and CO_2 before they reach the Earth's surface. Both the organic molecules within organisms and certain life processes, such as vision and photosynthesis, are adapted to utilize the radiation that is most prevalent in the environment.

Photosynthetic Pigments

The pigments found within most types of photosynthesizing cells are **chlorophylls** and **carotenoids.** These pigments are capable of absorbing portions of visible light. Both chlorophyll *a* and chlorophyll *b* absorb violet, blue, and red wavelengths better than those of other colors. Because green light is reflected and only minimally absorbed, leaves appear green to us. Accessory pigments such as the carotenoids appear yellow or orange because they are able to absorb light in the violet-blue-green range, but not the yellow-orange range. These pigments and others become noticeable in the fall when chlorophyll breaks down and the other pigments are uncovered (**Fig. 6.5**).

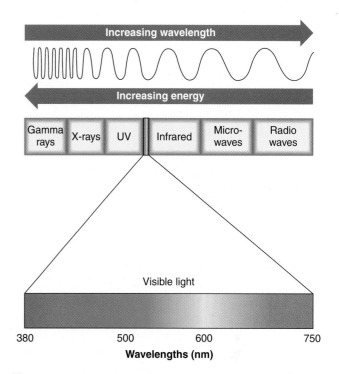

Figure 6.4   Radiant energy.

Radiant energy exists in a range of wavelengths that extends from the very short wavelengths of gamma rays through the very long wavelengths of radio waves. Visible light, which drives photosynthesis, is expanded to show its component colors. The components differ according to wavelength and energy content.

Figure 6.5   Leaf colors.

During the summer, leaves appear green because the chlorophylls absorb other portions of the visible spectrum better than they absorb green. During the fall, chlorophyll breaks down, and the carotenoids that remain cause leaves to appear yellow to orange because they do not absorb these colors.

The Electron Pathway of the Light Reactions

Much as a solar energy panel captures the sun's energy and stores it in a battery, we will see that the light reactions capture the sun's energy and store it in the form of a hydrogen ion (H^+) gradient. This hydrogen ion gradient will be used to produce ATP molecules. The light reactions also produce NADPH (**Fig. 6.6**).

The light reactions use two **photosystems,** called photosystem I (PS I) and photosystem II (PS II). The photosystems are named for the order in which they were discovered, not for the order in which they participate in the photosynthetic process. Here is a summary of how the pathway works:

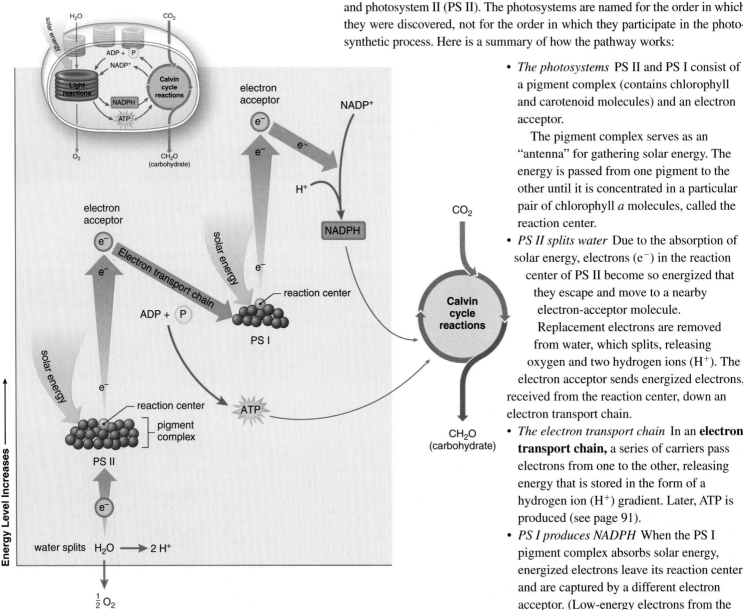

Figure 6.6   The electron pathway.

Energized electrons leave PS II and pass down an electron transport chain, leading to the formation of ATP. Energized electrons (replaced from PS II) leave PS I and pass to NADP$^+$, which then combines with H$^+$ to become NADPH. The electrons from PS II are replaced with electrons from water as it splits to release oxygen.

- *The photosystems* PS II and PS I consist of a pigment complex (contains chlorophyll and carotenoid molecules) and an electron acceptor.

  The pigment complex serves as an "antenna" for gathering solar energy. The energy is passed from one pigment to the other until it is concentrated in a particular pair of chlorophyll *a* molecules, called the reaction center.

- *PS II splits water* Due to the absorption of solar energy, electrons (e^-) in the reaction center of PS II become so energized that they escape and move to a nearby electron-acceptor molecule.

  Replacement electrons are removed from water, which splits, releasing oxygen and two hydrogen ions (H^+). The electron acceptor sends energized electrons, received from the reaction center, down an electron transport chain.

- *The electron transport chain* In an **electron transport chain,** a series of carriers pass electrons from one to the other, releasing energy that is stored in the form of a hydrogen ion (H^+) gradient. Later, ATP is produced (see page 91).

- *PS I produces NADPH* When the PS I pigment complex absorbs solar energy, energized electrons leave its reaction center and are captured by a different electron acceptor. (Low-energy electrons from the electron transport chain adjacent to PS II replace those lost by PS I.) The electron acceptor in PS I passes its electrons to a NADP$^+$ molecule. NADP$^+$ accepts electrons and a hydrogen ion to become NADPH.

Organization of the Thylakoid Membrane

PS I, PS II, and the electron transport chain are molecular complexes located within the thylakoid membrane (**Fig. 6.7**). The molecular complexes are densely clustered to promote efficient electron transfer between carriers. Also present is an ATP synthase complex.

ATP Production

During photosynthesis, the thylakoid space acts as a reservoir for hydrogen ions (H^+). First, each time water is split, two H^+ remain in the thylakoid space. Second, as the electrons move from carrier to carrier down the electron transport chain, the electrons give up energy. This energy is used to pump H^+ from the stroma into the thylakoid space. Therefore, there are many more H^+ ions in the thylakoid space than in the stroma, and an H^+ gradient has been established. This gradient contains a large amount of potential energy.

The H^+ ions now flow down their concentration gradient, across the thylakoid membrane at the ATP synthase complex, and energy is released. This causes the enzyme **ATP synthase** to change its shape and produce ATP from ADP + (P). The production of ATP captures the released energy.

NADPH Production

Enzymes often have nonprotein helpers called **coenzymes**. $NADP^+$ is a coenzyme that accepts electrons, becoming NADPH. When it gives up electrons to a substrate, the substrate is reduced.

We have seen that during the light reactions, $NADP^+$ receives electrons at the end of the electron pathway in the thylakoid membrane, and then it picks up a hydrogen ion to become NADPH (see page 90).

Figure 6.7   Organization of a thylakoid.

Molecular complexes of the electron transport chain within the thylakoid membrane pump hydrogen ions from the stroma into the thylakoid space. When hydrogen ions flow back out of the space into the stroma through the ATP synthase complex, ATP is produced from ADP + (P). $NADP^+$ accepts two electrons and joins with H^+ to become NADPH.

Check Your Progress

1. Trace the path of electrons during the light reactions.
2. What two molecules contain energy captured during the light reactions?

Answers: 1. Electrons are removed from water as it is split by PS II. They are passed from PS II to PS I through an electron transport chain, and are transferred to $NADP^+$ as it becomes NADPH to capture energy.   2. ATP and NADPH.

6.3 | Calvin Cycle Reactions

The ATP and NADPH generated by the light reactions power the Calvin cycle reactions. The Calvin cycle reactions occur in the stroma of chloroplasts and consist of a series of reactions that produce carbohydrate before returning to the starting point (**Fig. 6.8**). The end product of the Calvin cycle is often considered to be glucose, as we shall see. The cycle is named for Melvin Calvin, who, with colleagues, used the radioactive isotope ^{14}C as a tracer to discover the reactions that make up the cycle.

This series of reactions utilizes carbon dioxide from the atmosphere and consists of these three steps: (1) carbon dioxide fixation; (2) carbon dioxide reduction; and (3) regeneration of the first substrate, RuBP (ribulose 1,5-bisphosphate). The last step also results in a molecule of G3P (glyceraldehyde-3-phosphate) that is used by plants to produce glucose and other types of organic molecules. The Calvin cycle requires a lot of energy, which is supplied by ATP and NADPH from the light reactions.

Fixation of Carbon Dioxide

Carbon dioxide (CO_2) fixation is the first step of the Calvin cycle. During this reaction, CO_2 from the atmosphere is attached to RuBP, a 5-carbon molecule. The enzyme for this reaction is called **RuBP carboxylase (rubisco)**, and the result is a 6-carbon molecule that splits into two 3-carbon molecules.

Reduction of Carbon Dioxide

Reduction of CO_2 is the sequence of reactions that uses NADPH and some of the ATP from the light reactions. Electrons are added, and carbon dioxide is reduced to a carbohydrate as R–CO_2 becomes R–CH_2O. NADPH and ATP supply the needed electrons and energy for CO_2 reduction, respectively.

Figure 6.8 **The Calvin cycle reactions.**

The Calvin cycle is divided into three portions: CO_2 fixation, CO_2 reduction, and regeneration of RuBP. Because five G3P are needed to re-form three RuBP, it takes three turns of the cycle to achieve a net gain of one G3P. Two G3P molecules are needed to form glucose. Each ● is a carbon atom (C).

Regeneration of RuBP

The product of the Calvin cycle is actually G3P, which is used to form glucose. Notice that the Calvin cycle reactions in Figure 6.8 are multiplied by three because it takes three turns of the Calvin cycle to allow one G3P to exit for making glucose and other carbohydrates. Why? Because, for every three turns of the Calvin cycle, five molecules of G3P are used to re-form three molecules of RuBP, a C_5 molecule. These reactions also utilize some of the ATP produced by the light reactions.

The Fate of G3P

Compared to animal cells, algae and plants have enormous biochemical capabilities. From a G3P molecule, they can make all the molecules they need (**Fig. 6.9**). A plant can utilize the hydrocarbon skeleton of G3P to form fatty acids and glycerol, which are combined in plant oils. We are all familiar with corn oil, sunflower oil, and olive oil used in cooking. Also, when nitrogen is added to the hydrocarbon skeleton derived from G3P, amino acids are formed.

Notice also that glucose phosphate is among the organic molecules that result from G3P metabolism. Glucose is the molecule that plants and animals most often metabolize to produce ATP molecules to meet their energy needs. Glucose phosphate can be combined with fructose (and the phosphate removed) to form sucrose, the molecule that plants use to transport carbohydrates from one part of the plant to another. Glucose phosphate is also the starting point for the synthesis of starch and cellulose. Starch is the storage form of glucose. Some starch is stored in chloroplasts, but most starch is stored in amyloplasts in plant roots. Cellulose is a structural component of plant cell walls; it serves as fiber in our diet because we are unable to digest it.

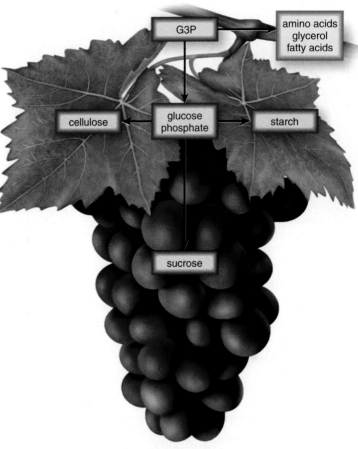

Figure 6.9 **The fate of G3P.**

G3P is the first reactant in a number of plant cell metabolic pathways such as those that lead to products other than carbohydrates (blue box) and those that are carbohydrates (green boxes). Two G3Ps are needed to form glucose phosphate; glucose is often considered the end product of photosynthesis. Sucrose is the transport sugar in plants; starch is the storage form of glucose; and cellulose is a major constituent of plant cell walls.

Check Your **Progress**

1. What are the three major steps of the Calvin cycle?
2. Where does the Calvin cycle get the NADPH and the ATP it uses to reduce carbon dioxide to a carbohydrate?
3. Describe carbon dioxide fixation in the Calvin cycle. What enzyme is responsible for this task?
4. In what part of a chloroplast do the light reactions occur? In what part do the Calvin cycle reactions occur?

Answers: 1. Carbon dioxide fixation, carbon dioxide reduction, and regeneration of RuBP. 2. From the light reactions. 3. Carbon dioxide is fixed by attaching it to RuBP to make a 6-carbon molecule. Rubisco performs this reaction. 4. The light reactions occur in the thylakoid membrane, and the Calvin cycle reactions occur in the stroma.

CO$_2$ fixation in a C$_3$
plant, blue columbine

Figure 6.10 **Carbon dioxide fixation in C$_3$ plants.**

In C$_3$ plants, such as columbines, CO$_2$ is taken up by the Calvin cycle directly in mesophyll cells, and the first detectable molecule is a C$_3$ molecule (red).

Figure 6.11 **Comparison of C$_3$ and C$_4$ plant anatomy.**

In C$_3$ plants, mesophyll cells are arranged in rows and contain chloroplasts. The Calvin cycle, including CO$_2$ fixation, occurs in mesophyll cells. In C$_4$ plants, mesophyll cells are arranged in rings around bundle sheath cells, and both contain chloroplasts. CO$_2$ fixation occurs within mesophyll cells, and the Calvin cycle occurs in the bundle sheath cells.

6.4 | Other Types of Photosynthesis

Plants are physically adapted to their environments. In the cold, windy climates of the north, evergreen trees have small, narrow leaves that look like needles. In the warm, wet climates of the south, some evergreen trees have large, flat leaves to catch the rays of the sun.

In the same way, plants are metabolically adapted to their environments. Where temperature and rainfall tend to be moderate, plants carry on C$_3$ photosynthesis, and are therefore called C$_3$ plants. In a **C$_3$ plant,** the first detectable molecule after CO$_2$ fixation is a C$_3$ molecule composed of three carbon atoms (**Fig. 6.10**). Look again at the Calvin cycle (see Fig. 6.8), and notice that the C$_6$ molecule formed when RuBP combines with carbon dioxide immediately breaks down to two C$_3$ molecules.

If the weather is hot and dry, stomata close, preventing the loss of water. (Water loss might cause the plant to wilt and die.) But this also prevents CO$_2$ from entering the leaf and traps O$_2$, a by-product of photosynthesis, within the leaf spaces where it can diffuse back into mesophyll cells. In C$_3$ plants, this O$_2$ competes with CO$_2$ for the active site of *rubisco*, the first enzyme of the Calvin cycle, and less C$_3$ is produced. Such yield decreases are of concern because many food crops are C$_3$ plants.

Successful plants adapt to hot, dry conditions by carrying on C$_4$ photosynthesis instead of C$_3$ photosynthesis. In a **C$_4$ plant,** the first detectable molecule following CO$_2$ fixation is a C$_4$ molecule composed of four carbon atoms. C$_4$ plants are able to avoid the uptake of O$_2$ by rubisco. Let's explore how C$_4$ plants do this.

C$_4$ Photosynthesis

The anatomy of a C$_4$ plant is different from that of a C$_3$ plant (**Fig. 6.11**). In a C$_3$ leaf, mesophyll cells, arranged in parallel rows, contain well-formed chloroplasts. The Calvin cycle, including CO$_2$ fixation, occurs in the chloroplasts of mesophyll cells. In a C$_4$ leaf, chloroplasts are located in the mesophyll cells, but they are also located in bundle sheath cells, which surround the leaf vein. Further, the mesophyll cells are arranged concentrically around the bundle sheath cells, shielding the bundle sheath cells from O$_2$ in the leaf spaces.

In C$_4$ plants, the Calvin cycle occurs only in the bundle sheath cells and not in the mesophyll cells. Because the bundle sheath cells are not accessible to leaf spaces, the CO$_2$ needed for the Calvin cycle is not directly taken from the air. Instead, CO$_2$ is fixed in mesophyll cells by a C$_3$ molecule, and a C$_4$ molecule forms. The C$_4$ molecule is modified and then pumped into bundle sheath cells (**Fig. 6.12**). Now CO$_2$ enters the Calvin cycle. This represents partitioning of CO$_2$ fixation and the Calvin cycle in space.

It takes energy to pump molecules, and you would think that the C$_4$ pathway would be disadvantageous. Yet in hot, dry climates, the net photosynthetic rate of C$_4$ plants such as sugarcane, corn, and Bermuda grass is two to three times that of C$_3$ plants such as wheat, rice, and oats. Why do C$_4$ plants enjoy such an advantage? The answer is that when the weather is hot and dry and stomata close, rubisco is not exposed to O_2, and yield is maintained.

When the weather is moderate, C_3 plants ordinarily have the advantage, but when the weather becomes hot and dry, C_4 plants have their chance to take over, and we can expect them to predominate. In the early summer, C_3 plants such as Kentucky bluegrass and creeping bent grass predominate in lawns in the cooler parts of the United States, but by midsummer, crabgrass, a C_4 plant, begins to take over.

CAM Photosynthesis

Another type of photosynthesis is called **CAM,** which stands for crassulacean-acid metabolism. It gets its name from the Crassulaceae, a family of flowering succulent (water-containing) plants that live in warm, arid regions of the world. CAM was first discovered in these plants, but now it is known to be prevalent among most succulent plants that grow in desert environments, including cactuses.

Whereas a C_4 plant represents partitioning in space—that is, CO_2 fixation occurs in mesophyll cells, and the Calvin cycle occurs in bundle sheath cells—CAM is partitioning by the use of time. During the night, CAM plants use C_3 molecules to fix CO_2, forming C_4 molecules. These molecules are stored in large vacuoles in mesophyll cells. During the day, the C_4 molecules release CO_2 to the Calvin cycle when NADPH and ATP are available from the light reactions (**Fig. 6.13**).

The primary advantage of this partitioning again relates to the conservation of water. CAM plants open their stomata only at night, and therefore only at that time is atmospheric CO_2 available. During the day, the stomata close. This conserves water, but CO_2 cannot enter the plant.

Photosynthesis in a CAM plant is minimal because of the limited amount of CO_2 fixed at night, but it does allow CAM plants to live under stressful conditions.

Evolutionary Trends

C_4 plants most likely evolved in, and are adapted to, areas of high light intensities, high temperatures, and limited rainfall. However, C_4 plants are more sensitive to cold, and C_3 plants probably do better than C_4 plants below 25°C.

CAM plants, on the other hand, compete well with either type of plant when the environment is extremely arid. Surprisingly, CAM is quite widespread and has evolved in 30 families of flowering plants, including cactuses, stonecrops, orchids, and bromeliads. It is also found among nonflowering plants, such as some ferns and cone-bearing trees.

CO$_2$ fixation in a
C$_4$ plant, corn

Figure 6.12 **Carbon dioxide fixation in C$_4$ plants.**

In C$_4$ plants, such as corn, CO$_2$ fixation results in C$_4$ molecules (red) in mesophyll cells. The C$_4$ molecule releases CO$_2$ to the Calvin cycle in bundle sheath cells.

CO$_2$ fixation in a
CAM plant, pineapple

Figure 6.13 **Carbon dioxide fixation in a CAM plant.**

CAM plants, such as pineapple, fix CO$_2$ at night, forming a C$_4$ molecule (red). The C$_4$ molecule releases CO$_2$ to the Calvin cycle during the day.

Check Your Progress

1. Describe how C$_4$ and CAM photosynthesis separate CO$_2$ fixation from the Calvin cycle.
2. Explain why C$_4$ photosynthesis is advantageous in hot, dry conditions.

Answers: 1. C$_4$ plants separate CO$_2$ fixation and the Calvin cycle into different cells, while CAM photosynthesis separates CO$_2$ fixation and the Calvin cycle reactions by time.   2. Stomata close under hot, dry conditions, increasing the concentration of oxygen relative to carbon dioxide in the leaf. The spatial separation of carbon fixation and the Calvin cycle reactions in C$_4$ plants prevents oxygen from competing with carbon dioxide for an active site on the enzyme rubisco.

The Chapter in Review

■ Summary

6.1 | Overview of Photosynthesis

Cyanobacteria, algae, and plants carry on photosynthesis, a process in which water is oxidized (split) and carbon dioxide is reduced using solar energy. The end products of photosynthesis include carbohydrate and oxygen:

$$\text{solar energy} + CO_2 + H_2O \longrightarrow (CH_2O) + O_2$$

In plants, photosynthesis takes place in chloroplasts. A chloroplast is bounded by a double membrane and contains two main components: the liquid stroma, and the membranous grana made up of thylakoids. During photosynthesis, the light reactions take place in the thylakoid membrane, and the Calvin cycle reactions take place in the stroma:

6.2 | Light Reactions

The light reactions use solar energy in the visible-light range. Pigment complexes in two photosystems (PS II and PS I) absorb various wavelengths of light. The collected solar energy results in high-energy electrons at the reaction centers of the photosystems.

- Solar energy enters PS II, and energized electrons are picked up by an electron acceptor. The oxidation (splitting) of water replaces these electrons in the reaction center. O_2 is released to the atmosphere, and hydrogen ions (H^+) remain in the thylakoid space.
- As electrons pass down from one acceptor to another in an electron transport chain (ETC), the release of energy allows the carriers to pump H^+ into the thylakoid space. The buildup of H^+ establishes an electrochemical gradient.
- When solar energy is absorbed by PS I, energized electrons leave and are ultimately received by $NADP^+$, which also combines with H^+ from the stroma to become NADPH. Electrons from PS II replace those lost by PS I.
- When H^+ flows down its concentration gradient through the channel present in ATP synthase complexes, ATP is synthesized from ADP and ℗ by ATP synthase.

6.3 | Calvin Cycle Reactions

The Calvin cycle consists of the following stages:
- **CO_2 fixation** The enzyme rubisco fixes CO_2 to RuBP, producing a 6-carbon molecule that immediately breaks down to two C_3 molecules.
- **CO_2 reduction** CO_2 (incorporated into an organic molecule) is reduced to carbohydrate (CH_2O). This step requires the NADPH and some of the ATP from the light reactions.
- **Regeneration of RuBP** For every three turns of the Calvin cycle, the net gain is one G3P molecule; the other five G3P molecules are used to re-form three molecules of RuBP. This step also requires the energy of ATP. G3P is then converted to all the organic molecules a plant needs. It takes two G3P molecules to make one glucose molecule.

6.4 | Other Types of Photosynthesis

C_4 Photosynthesis

When the weather is hot and dry, C_3 plants are at a disadvantage because stomata close and O_2 from photosynthesis competes with CO_2 for the active site of rubisco. C_4 plants avoid this drawback. In C_4 plants, CO_2 fixation occurs in mesophyll cells, and the Calvin cycle reactions occur in bundle sheath cells. In mesophyll cells, a C_3 molecule fixes CO_2, and the result is a C_4 molecule, for which this type of photosynthesis is named. A modified form of this molecule is pumped into bundle sheath cells, where CO_2 is released to the Calvin cycle. This represents partitioning of CO_2 fixation and the Calvin cycle in space.

CAM Photosynthesis

CAM plants live in hot, dry environments. At night when stomata remain open, a C_3 molecule fixes CO_2 to produce a C_4 molecule. The next day, CO_2 is released and enters the Calvin cycle within the same cells. This represents a partitioning of CO_2 fixation and the Calvin cycle in time: Carbon dioxide fixation occurs at night, and the Calvin cycle occurs during the day.

Key Terms

ATP synthase 91
C$_3$ plant 94
C$_4$ plant 94
Calvin cycle reaction 88
CAM 95
carbon dioxide (CO$_2$) fixation 92
carotenoid 89
chlorophyll 87, 89
chloroplast 87
coenzyme 91
electron transport chain 90

grana (sing., granum) 87
light reaction 88
oxidation 88
photosynthesis 86
photosystem 90
redox reaction 88
reduction 88
RuBP carboxylase (rubisco) 92
stomata (sing., stoma) 87
stroma 87
thylakoid 87

Testing Yourself

Choose the best answer for each question.

1. The raw materials for photosynthesis are
    a. oxygen and water.
    b. oxygen and carbon dioxide.
    c. carbon dioxide and water.
    d. carbohydrates and water.
    e. carbohydrates and carbon dioxide.
2. During photosynthesis, carbon dioxide is _____ and water is _____.
    a. reduced, oxidized
    b. oxidized, reduced
    c. reduced, reduced
    d. oxidized, oxidized
3. When electrons in the reaction center of PS I are passed to an energy-acceptor molecule, they are replaced by electrons that have been given up by
    a. oxygen.
    b. glucose.
    c. carbon dioxide.
    d. water.
4. PS I, PS II, and the electron transport chain are located in the
    a. thylakoid membrane.
    b. stroma.
    c. outer chloroplast membrane.
    d. cell's nucleus.
5. During the light reactions of photosynthesis, ATP is produced when hydrogen ions move
    a. down a concentration gradient from the thylakoid space to the stroma.
    b. against a concentration gradient from the thylakoid space to the stroma.
    c. down a concentration gradient from the stroma to the thylakoid space.
    d. against a concentration gradient from the stroma to the thylakoid space.
6. Oxygen is generated by the
    a. light reactions.
    b. Calvin cycle.
    c. light reactions and the Calvin cycle reactions.
    d. closing of stomata.
7. Energy for the Calvin cycle is supplied by _____ from the light reactions.
    a. carbon dioxide and water
    b. ATP and NADPH
    c. carbon dioxide and ATP
    d. ATP and water
    e. electrons and hydrogen ions

8. The product(s) of the Calvin cycle and a reactant(s) in many plant metabolic pathways is/are
    a. ribulose 1,5-bisphosphate, RuBP.
    b. adenosine diphosphate, ADP.
    c. fructose and glucose.
    d. glyceraldehyde-3-phosphate, G3P.
    e. carbon dioxide, CO$_2$.
9. Use these terms to label the following diagram of the light reactions.
    ATP
    thylakoid membrane
    electron transport chain
    stroma
    thylakoid space
    NADPH
    ATP synthase complex

10. In C$_4$ plants, CO$_2$ fixation occurs
    a. by adding electrons and a hydrogen ion when splitting water in bundle sheath cells.
    b. by joining CO$_2$ to a C$_3$ molecule and pumping it into bundle sheath cells.
    c. by splitting water only in bundle sheath cells.
    d. by oxidizing a C$_4$ molecule with O$_2$ derived from bundle sheath cells.
11. A leaf of a C$_4$ plant differs from that of a C$_3$ plant because
    a. stomata cannot close.
    b. the mesophyll has no leaf spaces.
    c. bundle sheath cells are protected from leaf spaces by a ring of mesophyll cells.
    d. C$_4$ plant leaves are narrower to allow more O$_2$ to diffuse away.

For questions 12–15, match the items to those in the key. Answers can be used more than once, and each question can have more than one answer.

Key:
    a. C$_3$ plant
    b. C$_4$ plant
    c. CAM plant

12. CO$_2$ fixation occurs in mesophyll cells.
13. The Calvin cycle occurs in bundle sheath cells.
14. CO$_2$ fixation and the Calvin cycle are separated by space.
15. CO$_2$ fixation and the Calvin cycle are separated by time.

16. Carbon dioxide fixation occurs when CO_2 combines with
    a. ATP.
    b. NADHP.
    c. G3P.
    d. RuBP.
17. Photosynthesis occurs best at wavelengths that are
    a. blue.
    b. gamma.
    c. infrared.
    d. ultraviolet.
18. Which of these descriptions is NOT true of chlorophyll?
    a. absorbs solar energy
    b. located in the grana
    c. located in the thyakoid membranes
    d. passes electrons directly to NADP
    e. passes electrons to an acceptor molecule
19. The two major sets of reactions involved in photosynthesis are
    a. the cyclic and noncyclic electron pathways.
    b. glycolysis and the citric acid cycle.
    c. the Calvin and citric acid cycles.
    d. the Calvin cycle and the electron transport chain.
    e. the light reactions and the Calvin cycle reactions.
20. Which of the following is NOT a stage in the Calvin cycle?
    a. carbon dioxide fixation
    b. carbon dioxide oxidation
    c. carbon dioxide reduction
    d. RuBP regeneration
21. The enzyme that produces ATP from ADP + (P) in the thylakoid is
    a. RuBP carboxylase.
    b. G3P.
    c. ATPase.
    d. ATP synthase.
    e. coenzyme A.
22. Which statement is NOT true?
    a. H^+ concentration is higher in the stroma than in the thlakloid space.
    b. The electron transport chain pumps H^+ from the stroma into the thylakoid space.
    c. The ATP synthase complex is present in the thylakoid membrane.
    d. All of these are true.
23. Each of the following is a product of light reactions EXCEPT
    a. ATP.
    b. NADHP.
    c. oxygen.
    d. sugar.

Thinking Scientifically

1. In 1882, T. W. Engelmann carried out an ingenious experiment to demonstrate that chlorophyll absorbs light in the blue and red portions of the spectrum. He placed a single filament of a green alga in a drop of water on a microscope slide. Then he passed light through a prism and onto the string of algal cells. The slide also contained aerobic bacterial cells. After some time, he peered into the microscope and saw the bacteria clustered around the regions of the algal filament that were receiving blue light and red light, as shown in the following illustration. Why do you suppose the bacterial cells were clustered in this manner?

Oxygen-seeking bacteria

Filament of green alga

2. Photosynthesis makes use of light in the visual range, the same range that is suitable to our vision. What explanation can you give to account for this observation?

Bioethical Issue

Are Carbon Offset Credits Effective?

With the prospect of climate change looming, many leading corporations, public figures, and concerned citizens have made an effort to fight global climate change by reducing the impact of their activities on the environment. Many have turned to the planting of trees and other plants, which can remove excess carbon dioxide from the atmosphere by photosynthesis, as a way to offset lifestyle choices, such as driving a car or flying on a plane, that cause carbon dioxide emissions. In response, many new companies and organizations have begun selling "carbon offset credits" in return for planting trees to offset the purchaser's carbon dioxide–emitting activities. But are carbon offset credits effective?

Advocates of carbon offset credits contend that such activities are effective at removing excess carbon dioxide from the atmosphere. They point out that carbon offset credits are effective at discouraging behaviors that result in carbon dioxide emissions, such as excessive travel or energy use, by attaching a monetary penalty to the activity. Indeed, such credits systems are encouraged by the 1997 Kyoto Protocol, and usage-based taxes are generally supported by politicians of all stripes. Furthermore, many environmentally conscious citizens insist that it is their duty to ensure that they do whatever is necessary to minimize their impact on the environment.

However, not everyone believes that carbon offset credits are an effective way to address the problem. Environmentalist George Monbiot compares the largely unregulated carbon offset credit system to the indulgences sold by the Catholic Church in the Middle Ages. Many skeptics view offsets as snake oil and ridicule purchasers as people trying to assuage their guilt while failing to address the real problem of their carbon dioxide–emitting activities. Many also point out that the current system has failed to reduce carbon dioxide emissions in countries where the credits are being sold. Furthermore, some environmentalists have become concerned that the planting of large stands of nonnative trees in some tropical areas by carbon offset credit suppliers may reduce biodiversity and disrupt native ecosystems.

Do you believe that carbon offset credits are an effective way to reduce excess carbon dioxide emissions and discourage excess energy consumption? Or are they, as some environmentalists contend, simply an excuse to pollute?

Essentials of Biology Website

The companion website for *Essentials of Biology* provides a wealth of information organized and integrated by chapter. You will find practice tests, animations, videos, and much more that will complement your learning and understanding of general biology.

http://www.mhhe.com/maderessentials2

7 Energy for Cells

Lack of oxygen for cellular respiration makes muscles ache.

Carbon monoxide and cyanide kill by stopping cellular repiration.

If a diet sounds too good to be true, it probably is.

Outline

Can You Lose Weight Without Dieting?

AN OBESITY EPIDEMIC rages unfettered in the United States. Not surprisingly, an entire industry has emerged, hawking weight-loss supplements, "miracle" fat burners, and fad diets. Advertisements for all types of weight-loss pills and dietary supplements fill the airwaves, claiming to speed up fat breakdown and burn off excess energy without dieting or exercising. But have you ever considered the science behind how these products claim to work? When you do, you will see that many of these claims simply can't be verified. Yet millions of people spend

—*Continued next page*

Continued—

a fortune hoping for a miracle, and many of them suffer health problems as a result.

Cells break down glucose in a process called cellular respiration to extract the energy from your food to create the ATP they need. However, if you continue to eat too much, your body will continue to produce the metabolites and ATP your body can use to build fat. In this chapter, you will learn how cells use glucose, fat, and protein to produce ATP. You will also learn how the body uses metabolites to build glucose, protein, and fat. Understanding these processes will give you the information you need to make informed decisions about weight loss, diets, and nutritional supplements.

7.1 | Cellular Respiration

Whether you go skiing, take an aerobics class, or just hang out, ATP molecules provide the energy needed for your muscles to contract. ATP molecules are produced during cellular respiration, a process that requires the participation of mitochondria. Cellular respiration is aptly named because just as you take in oxygen (O_2) and give off carbon dioxide (CO_2) during breathing, so do the mitochondria in your cells (**Fig. 7.1**). In fact, cellular respiration, which occurs in all cells of the body, is the reason you breathe.

Oxidation of substrates is a fundamental part of cellular respiration. In living things, oxidation doesn't occur by the addition of oxygen (O_2). Instead, oxidation is the removal of hydrogen atoms from a molecule. As cellular respiration occurs, hydrogen atoms are removed from glucose (and glucose products) and transferred to oxygen atoms, forming carbon dioxide (CO_2) and water (H_2O):

$$\overbrace{C_6H_{12}O_6}^{} + 6\,O_2 \longrightarrow 6\,CO_2 + 6\,H_2O + \text{energy}$$

Oxidation

Reduction

glucose

Note that respiration is essentially the reverse of photosynthesis. Carbohydrates are oxidized to release energy. Oxygen is consumed, and carbon dioxide and water are formed as by-products. In photosynthesis, carbon dioxide is reduced to make carbohydrate; in respiration, carbohydrate is oxidized and carbon dioxide is formed.

oxygen

H_2O

CO_2

glucose

double membrane
— outer membrane
— inner membrane

matrix

intermembrane space

cristae

Figure 7.1 Cellular respiration.

Glucose from our food and the oxygen we breathe are requirements for cellular respiration, a process completed within the mitochondria.

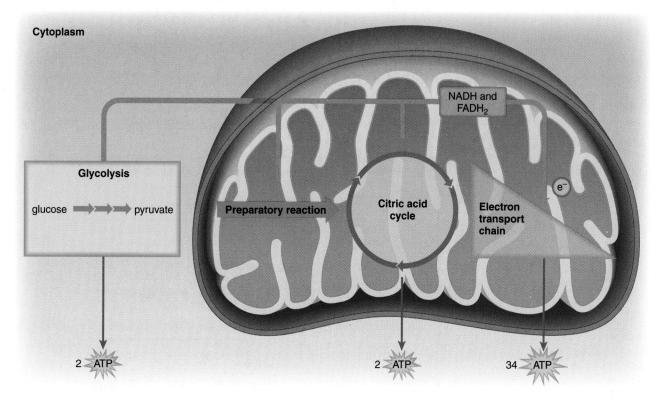

Cytoplasm

Glycolysis

glucose ➡➡➡ pyruvate

Preparatory reaction

Citric acid cycle

Electron transport chain

NADH and $FADH_2$

e^-

2 ATP   2 ATP   34 ATP

Phases of Complete Glucose Breakdown

The breakdown of glucose releases a lot of energy. If you mistakenly burn sugar in a skillet, the energy escapes into the atmosphere as heat. A cell is more sophisticated than that. In a cell, glucose is broken down slowly—not all at once—and the energy given off isn't all lost as heat. Hydrogen atoms are removed bit by bit, and this allows energy to be captured and used to make ATP molecules.

The enzymes that carry out oxidation during cellular respiration are assisted by nonprotein helpers called **coenzymes.** As glucose is oxidized, the coenzymes NAD^+ and FAD[1] receive hydrogen atoms ($H^+ + e^-$) and become NADH and $FADH_2$, respectively (**Fig. 7.2**).

During these phases, notice where CO_2 and H_2O are produced.

- **Glycolysis,** which occurs in the cytoplasm outside the mitochondria, is the breakdown of glucose to 2 molecules of pyruvate. Energy is invested to activate glucose, 2 ATP are gained, and oxidation results in NADH, which can be used for further ATP production.
- The **preparatory (prep) reaction** takes place in the matrix of mitochondria. Pyruvate is broken down to a 2-carbon acetyl group carried by **coenzyme A (CoA).** Oxidation of pyruvate results in not only NADH, but also CO_2.
- The **citric acid cycle** also takes place in the matrix of mitochondria. As oxidation occurs, NADH and $FADH_2$ result and more CO_2 is released. The citric acid cycle is able to produce 2 ATP per glucose molecule.
- The **electron transport chain** is a series of electron carriers in the cristae of mitochondria. NADH and $FADH_2$ give up electrons to the chain. Energy is released and captured as the electrons move from a higher energy to a lower energy state. Later, this energy will be used for the production of ATP. Oxygen (O_2) is the final electron acceptor. It then combines with hydrogen ions (H^+) to produce water (H_2O).

[1] NAD = Nicotinamide adenine dinucleotide; FAD = Flavin adenine dinucleotide

Figure 7.2 **The four phases of complete glucose breakdown.**

The enzymatic reactions of glycolysis take place in the cytoplasm. The preparatory reaction, the citric acid cycle, and the electron transport chain occur in mitochondria.

Check Your **Progress**

1. **Why is breathing necessary to cellular respiration?**
2. **List the four phases of complete glucose breakdown.**
3. **Identify the phases during which CO_2 and H_2O are formed.**

Answers: 1. Breathing takes in oxygen needed for cellular respiration and rids the body of carbon dioxide, a waste product of cellular respiration. 2. Glycolysis, the preparatory reaction, the citric acid cycle, and the electron transport chain. 3. CO_2 is released during the preparatory reaction and citric acid cycle. H_2O is the end product of the electron transport chain.

RuBP	ribulose 1,5-bisphosphate
3PG	3-phosphoglycerate
BPG	1,3-bisphosphoglycerate
G3P	glyceraldehyde-3-phosphate

Figure 7.3 Glycolysis.

This metabolic pathway begins with glucose (C_6) and ends with 2 pyruvate (C_3) molecules. A net gain of 2 ATP can be calculated by subtracting those expended during the energy-investment step from those produced during the energy-harvesting steps. Each ● is a carbon atom (C).

7.2 | Outside the Mitochondria: Glycolysis

In eukaryotes, such as plants and animals, glycolysis takes place within the cytoplasm outside the mitochondria. During glycolysis, glucose, a C_6 molecule, is broken down to 2 molecules of pyruvate, a C_3 molecule. Glycolysis is divided into (1) the energy-investment step when ATP is used; and (2) the energy-harvesting steps, when NADH and ATP are produced (**Fig. 7.3**).

Energy-Investment Step

During the energy-investment steps, 2 ATP transfer phosphate groups to substrates, and 2 ADP + Ⓟ result. In other words, ATP has been broken down, not built up. However, the phosphate groups activate the substrates so that they can undergo reactions.

Energy-Harvesting Steps

During the energy-harvesting steps, substrates are oxidized by the removal of hydrogen atoms, and 2 NADH result.

Oxidation produces substrates with energized phosphate groups, which are used to synthesize 4 ATP. As a phosphate group is transferred to ADP, ATP results. The process is called substrate-level ATP synthesis (**Fig. 7.4**).

What is the net gain of ATP from glycolysis? Confirm that 2 ATP are used to get started, and 4 ATP are produced by substrate-level ATP synthesis. Therefore, there is a net gain of 2 ATP from glycolysis.

If oxygen is available, pyruvate, the end product of glycolysis, enters mitochondria, where it undergoes further breakdown. If oxygen is not available, pyruvate remains in the cytoplasm and undergoes reduction. In humans, if oxygen is not available, pyruvate is reduced to lactate, as discussed on page 109.

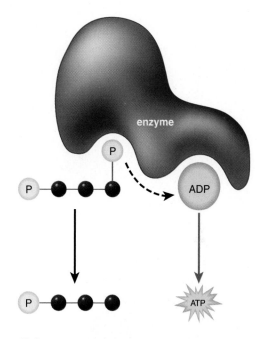

Figure 7.4 **Substrate-level ATP synthesis.**

Substrate level ATP synthesis is shown in this diagram. At an enzyme's active site, ADP acquires an energized phosphate group from a substrate of a metabolic pathway, and ATP results. The net gain of 2 ATP from glycolysis is the result of substrate-level ATP synthesis.

Glycolysis	
inputs	outputs
glucose	2 pyruvate
2 NAD⁺	2 NADH
2 ATP	2 ADP
4 ADP + 4 Ⓟ	4 ATP
	2 ATP net

Check Your **Progress**

1. Where in the cell does glycolysis occur?
2. Contrast the energy-investment step of glycolysis with the energy-harvesting steps.
3. What is the net yield of ATP per glucose molecule as a result of glycolysis?

Answers: 1. Glycolysis occurs within the cytoplasm of the cell.   2. During the energy-investment step, ATP breakdown provides the phosphate groups that activate substrates. During the energy-harvesting steps, NADH and ATP are produced.   3. 2 ATP.

cytoplasm
location of glycolysis

outer membrane   matrix   cristae

matrix
location of the prep reaction and the citric acid cycle

inner membrane
forms cristae

intermembrane
space

cristae
location of the electron transport chain

Figure 7.5 Mitochondrion structure and function.

A mitochondrion is bounded by a double membrane. The inner membrane folds inward to form the shelflike cristae. Glycolysis takes place in the cytoplasm outside the mitochondria. The preparatory reaction and the citric acid cycle occur within the mitochondrial matrix. The electron transport chain is located on the cristae of a mitochondrion.

7.3 | Inside the Mitochondria

The other three phases of cellular respiration occur inside the mitochondria (**Fig. 7.5**).

Preparatory Reaction

Occurring in the matrix, the preparatory (prep) reaction is so called because it produces a substrate that can enter the citric acid cycle. The preparatory reaction occurs twice per glucose molecule because glycolysis results in 2 pyruvate molecules.

During a prep reaction:

- Pyruvate is oxidized, and a CO_2 molecule is given off. This is part of the CO_2 we breathe out!
- NAD^+ accepts a hydrogen atom, and NADH results.
- The product, a C_2 acetyl group, is attached to coenzyme A (CoA), forming **acetyl-CoA.**

Therefore,

- Per glucose molecule, the total is 2 CO_2, 2 NADH, and 2 acetyl CoA.

The Citric Acid Cycle

The citric acid cycle is a cyclical metabolic pathway located in the matrix of mitochondria (**Fig. 7.6**). It was originally called the *Krebs cycle* to honor the man who first studied it. At the start of the citric acid cycle, the C_2 acetyl group carried by CoA joins with a C_4 molecule, and a C_6 citrate molecule results. The CoA returns to the preparatory reaction to be used again.

During the citric acid cycle:

- The acetyl group is oxidized, and the rest of the CO_2 we breathe out per glucose molecule is released.
- Both NAD^+ and FAD accept hydrogen atoms, resulting in NADH and $FADH_2$.
- Substrate-level ATP synthesis occurs (see Fig. 7.4), and an ATP results.

Because the citric acid cycle turns twice for each original glucose, the inputs and outputs of the citric acid cycle per glucose are as follows:

Citric acid cycle	
inputs	outputs
2 acetyl-CoA	4 CO_2
6 NAD^+	6 NADH
2 FAD	2 $FADH_2$
2 ADP + 2 P	2 ATP

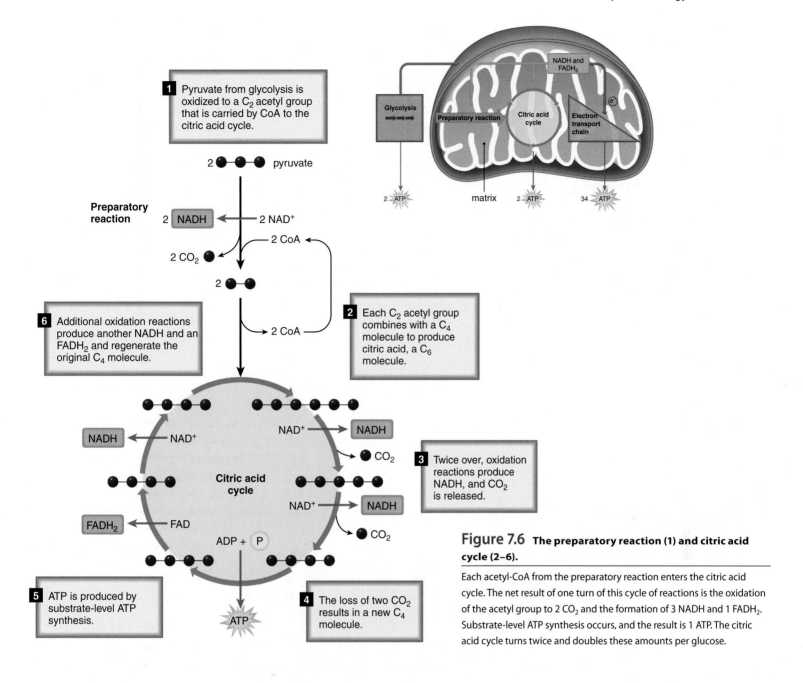

Figure 7.6 The preparatory reaction (1) and citric acid cycle (2–6).

Each acetyl-CoA from the preparatory reaction enters the citric acid cycle. The net result of one turn of this cycle of reactions is the oxidation of the acetyl group to 2 CO_2 and the formation of 3 NADH and 1 $FADH_2$. Substrate-level ATP synthesis occurs, and the result is 1 ATP. The citric acid cycle turns twice and doubles these amounts per glucose.

The labels within the figure read:

1 Pyruvate from glycolysis is oxidized to a C_2 acetyl group that is carried by CoA to the citric acid cycle.

2 Each C_2 acetyl group combines with a C_4 molecule to produce citric acid, a C_6 molecule.

3 Twice over, oxidation reactions produce NADH, and CO_2 is released.

4 The loss of two CO_2 results in a new C_4 molecule.

5 ATP is produced by substrate-level ATP synthesis.

6 Additional oxidation reactions produce another NADH and an $FADH_2$ and regenerate the original C_4 molecule.

Preparatory reaction: 2 pyruvate → 2 NADH (2 NAD^+), 2 CO_2, 2 CoA

Citric acid cycle: NADH (NAD^+), $FADH_2$ (FAD), ADP + P → ATP, NADH (NAD^+), CO_2, NADH (NAD^+), CO_2

Mitochondrion diagram labels: Glycolysis, Preparatory reaction, Citric acid cycle, Electron transport chain, NADH and $FADH_2$, matrix, e^-, 2 ATP, 2 ATP, 34 ATP

Check Your Progress

1. How many NADH are produced from each glucose molecule as a result of the preparatory reaction?
2. What are the products of the citric acid cycle as a result of further breakdown of glucose?

Answers: 1. As a result of the preparatory reaction, one NADH is produced for each pyruvate oxidized. Therefore, two NADH are produced per glucose.   2. The citric acid cycle turns twice, producing 4 CO_2, 6 NADH, 2 $FADH_2$, and 2 ATP per glucose molecule.

The Electron Transport Chain

The electron transport chain located in the cristae of mitochondria is a series of carriers that pass electrons from one to the other. NADH and $FADH_2$ deliver electrons to the chain. Consider that the hydrogen atoms attached to NADH and $FADH_2$ consist of an electron (e^-) and a hydrogen ion (H^+). The members of the electron transport chain accept only e^- and not H^+.

In **Figure 7.7,** high-energy electrons enter the chain, and low-energy electrons leave the chain. When NADH gives up its electrons, the next carrier gains the electrons and is reduced. This oxidation-reduction reaction starts the process, and each of the carriers in turn becomes reduced and then oxidized as the electrons move down the system. As the pair of electrons is passed from carrier to carrier, energy is released and captured for ATP production. The final acceptor of electrons is oxygen (O_2), the very O_2 we breathe in. It's remarkable to think that the role of oxygen in cellular respiration is to keep the electrons moving from the first to the last carrier. Why can oxygen play this role? Because oxygen attracts electrons to a greater degree than the carriers of the chain. Once oxygen accepts electrons, it combines with H^+ to form a water molecule, the other end product of cellular respiration (see the equation on page 100).

When each NADH delivers electrons to the first carrier of the electron transport chain, enough energy is captured by the time the electrons are received by O_2 to permit the production of three ATP molecules. When each $FADH_2$ delivers electrons to the electron transport chain, only 2 ATP are produced.

Once NADH has delivered electrons to the electron transport chain, NAD^+ is regenerated and can be used again. In the same manner, FAD is regenerated and can be used again. The recycling of coenzymes, and for that matter ADP, increases cellular efficiency since it does away with the need to synthesize NAD^+, FAD, and ADP anew.

The Cristae of a Mitochondrion

The carriers of the electron transport chain are located in molecular complexes within the inner mitochondrial membrane. ATP synthesis is carried out by ATP synthase complexes also located in this membrane (**Fig. 7.8**).

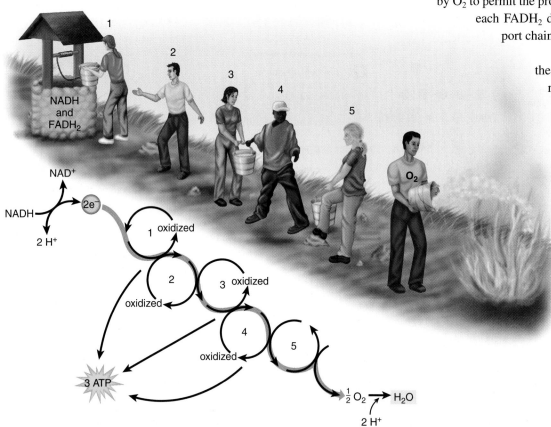

Figure 7.7 **The electron transport chain.**

An electron transport chain operates like a bucket brigade. Each electron carrier (1–5) is alternatively reduced (orange) and oxidized as if the electrons were a bucket being passed from person to person. As oxidation-reduction occurs, energy is released that will be used to make ATP.

a. Electron transport chain

b. ATP synthesis

Figure 7.8   **The organization of cristae.**

Molecular complexes that contain the electron transport carriers are located in the cristae as are ATP synthase complexes. **a.** As electrons move from one carrier to the other, hydrogen ions (H⁺) are pumped from the matrix into the intermembrane space. **b.** As hydrogen ions flow back down a concentration gradient through an ATP synthase complex, ATP is synthesized by the enzyme ATP synthase.

The carriers of the electron transport chain accept electrons from NADH or $FADH_2$ and then pass them from one to the other by way of two additional mobile carriers (orange arrow). What happens to the hydrogen ions (H⁺) carried by NADH and $FADH_2$? The complexes use the energy released by oxidation-reduction to pump H⁺ from the mitochondrial matrix into the **intermembrane space** located between the outer and inner membrane of a mitochondrion. The pumping of H⁺ into the intermembrane space establishes an unequal distribution of H⁺; in other words, there are many H⁺ in the intermembrane space and few in the matrix of a mitochondrion.

Just as in photosynthesis, the H⁺ gradient contains a large amount of stored energy that can be used to drive forward ATP synthesis. The cristae of mitochondria (like the thylakoid membrane of chloroplasts) contain an *ATP synthase complex* that allows H⁺ to return to the matrix. The flow of H⁺ through the ATP synthase complex brings about a conformational change, which causes the enzyme ATP synthase to synthesize ATP from ADP + Ⓟ. ATP leaves the matrix by way of a channel protein. This ATP remains in the cell and is used for cellular work.

Check Your **Progress**

1. Explain how the electron transport chain results in the synthesis of ATP.

Answer: 1. As electrons move from one carrier to another in the cristae, energy is released, and this energy is used to pump hydrogen ions from the matrix to the intermembrane space. The flow of hydrogen ions back down the concentration gradient into the matrix drives the synthesis of ATP by ATP synthase.

Phase	NADH	FADH$_2$	ATP Yield
Glycolysis	2	–	2
Prep reaction	2	–	–
Citric acid cycle	6	2	2
Electron transport chain	10 ⟶	2 ⟶	30 / 4
Total ATP			38

Figure 7.9 **Calculating ATP energy yield per glucose molecule.**

Substrate-level ATP synthesis during glycolysis and the citric acid cycle accounts for 4 ATP. The electron transport chain produces a maximum of 34 ATP, and the maximum total is 38 ATP. Some cells, however, produce only 36 ATP per glucose or even less.

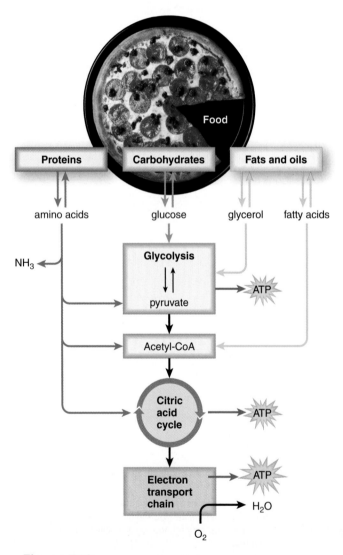

Figure 7.10 **Alternative metabolic pathways.**

All the types of food in a pizza can be used to generate ATP.

Energy Yield from Glucose Metabolism

Figure 7.9 calculates the ATP yield for the complete breakdown of glucose to CO_2 and H_2O. Per glucose, there is a net gain of 2 ATP from glycolysis, which takes place in the cytoplasm. The citric acid cycle, which occurs in the matrix of mitochondria, accounts for 2 ATP per glucose. This means that a total of 4 ATP form due to substrate-level ATP synthesis outside the electron transport chain.

Most of the ATP produced comes from the electron transport chain and the ATP synthase complex. Per glucose, 10 NADH and 2 FADH$_2$ take electrons from glycolysis and the citric acid cycle to the electron transport chain. The maximum number of ATP produced by the chain is therefore 34 ATP, and the maximum number produced by both the chain and substrate-level ATP synthesis is 38. However, for reasons beyond the scope of this book, the maximum number of ATP produced per glucose in some cells is only 36 ATP or lower. A yield of 36–38 ATP represents about 40% of the available energy in a glucose molecule. The rest of the energy is lost in the form of heat.

Alternative Metabolic Pathways

Let's say you are on a low-carbohydrate diet. Will you then run out of ATP? No, because your cells can also utilize other energy sources—the components of fats and oils, namely glycerol and fatty acids, and amino acids, which are derived from proteins (**Fig. 7.10**).

Because glycerol is a carbohydrate, it enters the process of cellular respiration during glycolysis. Fatty acids can be metabolized to acetyl groups, which enter the citric acid cycle. A fatty acid with a chain of 18 carbons can make three times the number of acetyl groups as does glucose. For this reason, fats are an efficient form of stored energy. The complete breakdown of glycerol and fatty acids results in many more ATP per fat molecule than does the breakdown of glucose.

Only the hydrocarbon backbone of amino acids can be used by the cellular respiration pathways. The amino group becomes ammonia (NH_3), which becomes part of urea, the primary excretory product of humans. Just where the hydrocarbon backbone from an amino acid begins degradation to produce ATP depends on its length. Figure 7.10 shows that the hydrocarbon backbone from an amino acid can enter cellular respiration pathways at pyruvate, at acetyl-CoA, or during the citric acid cycle.

The smaller molecules in Figure 7.10 can also be used to synthesize larger molecules. In such instances, ATP is used instead of generated. For example, some substrates of the citric acid cycle can become amino acids through the addition of an amino group, and these amino acids can be employed to synthesize proteins. Similarly, substrates from glycolysis can become glycerol, and acetyl groups can be used to produce fatty acids. When glycerol and three fatty acids join, a fat results. This explains why you can gain weight from eating carbohydrate-rich foods.

Check Your Progress

1. Explain why fats and oils are more efficient energy storage molecules than are carbohydrates.

Answer: 1. Each fat or oil molecule contains three long-chain fatty acids and one glycerol. Many more acetyl-CoA result from the breakdown of the three fatty acids than from the breakdown of one glucose. Therefore, many more ATP result from the breakdown of a fat than from glucose.

7.4 | Fermentation

As you have learned, oxygen is required for the complete breakdown of glucose. But what happens when oxygen is not available and the complete breakdown of glucose is not possible? **Fermentation,** which is the anaerobic breakdown of glucose resulting in the production of 2 ATP per glucose, ensures that ATP is available for cellular processes (**Fig. 7.11**).

During fermentation in animal cells, the pyruvate formed by glycolysis accepts 2 hydrogen atoms and is reduced to lactate. Notice in Figure 7.11 that 2 NADH pass hydrogen atoms to pyruvate, reducing it. Why is it beneficial for pyruvate to be reduced to lactate when oxygen is not available? The answer is that this reaction regenerates NAD^+, which can then pick up more electrons during the earlier reactions of glycolysis. This keeps glycolysis going, during which ATP is produced by substrate-level ATP synthesis.

The 2 ATP produced by fermentation represent only a small fraction of the potential energy stored in a glucose molecule. Following fermentation, most of this potential energy is still waiting to be released. Despite its low yield of only 2 ATP, fermentation is essential. It can provide a rapid burst of ATP, and muscle cells are more apt than other cells to carry on fermentation. When our muscles are working vigorously over a short period of time, as when we run, fermentation is a way to produce ATP even though oxygen is temporarily in limited supply.

When we stop running, our bodies are in **oxygen deficit,** as signified by the fact that we continue to breathe very heavily for a time. Recovery is complete when enough oxygen is present to completely break down glucose. Blood carries away the lactate formed in muscles and transports it to the liver, where it is reconverted to pyruvate. Some of the pyruvate is oxidized completely, and the rest is converted back to glucose.

Microorganisms and Fermentation

Bacteria utilize fermentation to produce an organic acid, such as lactate, or an alcohol and CO_2, depending on the type of bacterium.

Yeasts are good examples of microorganisms that generate ethyl alcohol and CO_2 when they carry out fermentation. When yeast is used to leaven bread, the CO_2 makes the bread rise. When yeast is used to ferment grapes for wine production or to ferment wort—derived from barley—for beer production, ethyl alcohol is the desired product. However, the yeast are killed by the very product they produce.

The inputs and outputs of fermentation are as follows:

Fermentation	
inputs	outputs
glucose	2 lactate or 2 alcohol and 2 CO_2
2 ATP	2 ADP
4 ADP + 4 P	4 ATP
	2 ATP net

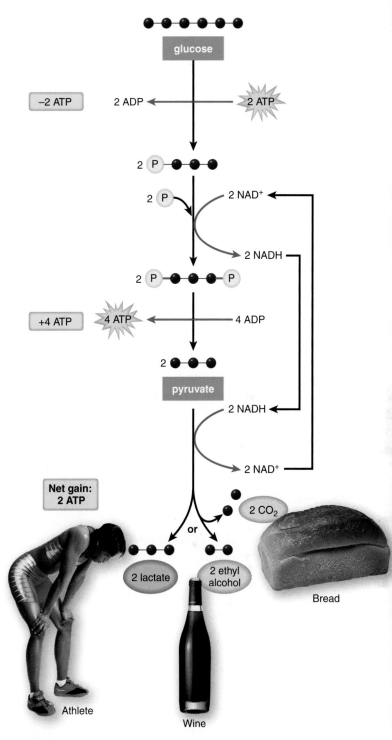

Figure 7.11 Fermentation.

Fermentation consists of glycolysis followed by a reduction of pyruvate by NADH. This regenerates NAD^+, which returns to the glycolytic pathway to pick up more hydrogen atoms.

Check Your Progress

1. How is the reduction of pyruvate beneficial when oxygen is not available?
2. What are the products of fermentation by humans and yeast?

Answers: 1. NAD^+ must be regenerated in order to continue glycolysis.   2. Humans produce lactate; yeast produce ethyl alcohol and CO_2.

The Chapter in Review

Summary

7.1 | Cellular Respiration

During cellular respiration, glucose from food is oxidized to CO_2, which we exhale. Oxygen (O_2), which we breathe in, is reduced to H_2O. When glucose is oxidized, energy is released. Cellular respiration captures the energy of oxidation and uses it to produce ATP molecules. The following equation gives an overview of these events:

$$C_6H_{12}O_6 + 6\,O_2 \longrightarrow 6\,CO_2 + 6\,H_2O + \text{ATP}$$
glucose

7.2 | Outside the Mitochondria: Glycolysis

Glycolysis, the breakdown of glucose to 2 molecules of pyruvate, is a series of enzymatic reactions that occur in the cytoplasm. During glycolysis:

- Glucose is oxidized by removal of hydrogen atoms ($H^+ + e^-$).
- When NAD^+ accepts electrons, NADH results.

Breakdown releases enough energy to immediately give a net gain of 2 ATP by substrate-level ATP synthesis. The inputs and outputs of glycolysis are summarized here:

Glycolysis	
inputs	outputs
glucose	2 pyruvate
2 NAD$^+$	2 NADH
2 ATP	2 ADP
4 ADP + 4 (P)	4 ATP
	2 ATP net

When oxygen is available, pyruvate from glycolysis enters a mitochondrion.

7.3 | Inside the Mitochondria

Preparatory Reaction

During the preparatory reaction in the matrix:

- Oxidation occurs as CO_2 is removed from pyruvate.
- NAD^+ accepts hydrogen atoms, and NADH results.
- An acetyl group, the end product, combines with CoA.

This reaction takes place twice per glucose.

The Citric Acid Cycle

Acetyl groups enter the citric acid cycle, a series of reactions occurring in the mitochondrial matrix. During one turn of the cycle, oxidation results in 2 CO_2, 3 NADH, and 1 $FADH_2$. One turn also produces 1 ATP. The cycle turns twice and doubles these amounts, per glucose.

The Electron Transport Chain

The final stage of cellular respiration involves the electron transport chain located in the cristae of the mitochondria. The chain is a series of electron carriers that accept high-energy electrons (e^-) from NADH and $FADH_2$ and pass electrons along until they are finally low-energy electrons received by oxygen, which combines with H^+ to produce water.

The carriers of the electron transport chain are located in molecular complexes on the cristae of mitochondria. These carriers capture energy from the passage of electrons and use it to pump H^+ into the intermembrane space of the mitochondrion. When H^+ flows down its gradient into the matrix through ATP synthase complexes, energy is released and used to form ATP molecules from ADP and (P).

Energy Yield from Glucose Metabolism

Of the maximum 38 ATP formed by complete glucose breakdown, 4 are the result of substrate-level ATP synthesis, and the rest are produced as a result of the electron transport chain and ATP synthase:

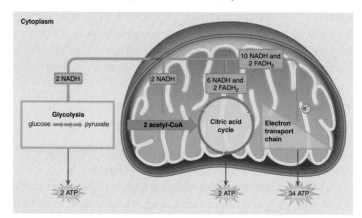

Alternative Metabolic Pathways

Besides carbohydrates, glycerol and fatty acids from fats, and amino acids from proteins can undergo cellular respiration by entering glycolysis and/or the citric acid cycle. These metabolic pathways also provide substrates for the synthesis of fats and proteins.

7.4 | Fermentation

Fermentation involves glycolysis followed by the reduction of pyruvate by NADH, either to lactate or to alcohol and CO_2. The reduction of pyruvate regenerates NAD^+, which can accept more hydrogen atoms during glycolysis.

- Although fermentation results in only 2 ATP, it still provides a quick burst of ATP energy for short-term, strenuous muscular activity.
- The accumulation of lactate puts the individual in oxygen deficit, which is the amount of oxygen needed when lactate is completely metabolized to CO_2 and H_2O.

Key Terms

acetyl-CoA   104
citric acid cycle   101
coenzyme   101
coenzyme A (CoA)   101
electron transport chain   101

fermentation   109
glycolysis   101
intermembrane space   107
oxygen deficit   109
preparatory (prep) reaction   101

Testing Yourself

Choose the best answer for each question.

1. During cellular respiration, _____ is oxidized and _____ is reduced.
   a. glucose, oxygen
   b. glucose, water
   c. oxygen, water
   d. water, oxygen
   e. oxygen, carbon dioxide

2. Cellular respiration requires _____ and _____ are produced.
   a. water and carbon dioxide; oxygen and glucose
   b. oxygen and carbon dioxide; glucose and water
   c. oxygen and glucose; carbon dioxide and water
   d. water and glucose; oxygen and carbon dioxide
   e. ATP and glucose; oxygen and water

3. During the energy-harvesting steps of glycolysis, which are produced?
   a. ATP and NADH   c. ATP and NAD$^+$
   b. ADP and NADH   d. ADP and NAD$^+$

4. During the energy-investment step of glycolysis, which is consumed?
   a. ATP   c. ADP + P
   b. NADH   d. NAD$^+$

5. The end product of glycolysis is
   a. phosphoenol pyruvate.   c. phosphoglyceraldehyde.
   b. glucose.   d. pyruvate.

6. Acetyl-CoA is the end product of
   a. glycolysis.   c. the citric acid cycle.
   b. the preparatory reaction. d. the electron transport chain.

7. The citric acid cycle results in the release of
   a. carbon dioxide.   c. oxygen.
   b. pyruvate.   d. water.

8. The energy needed for ATP synthesis by ATP synthase is obtained from which of the following?
   a. electrons   c. H$^+$ gradient in intermembrane space
   b. NADH   d. electrons from FADH$_2$

9. The following reactions occur in the matrix of the mitochondria:
   a. glycolysis and the preparatory reaction
   b. the preparatory reaction and the citric acid cycle
   c. the citric acid cycle and the electron transport chain
   d. the electron transport chain and glycolysis

10. Match the descriptions below to the lettered events in the preparatory reaction and citric acid cycle:
    Pyruvate is broken down to an acetyl group.
    Acetyl group is taken up and a C$_6$ molecule results.

Oxidation results in NADH and CO$_2$.
ATP is produced by substrate-level ATP synthesis.
Oxidation produces more NADH and FADH$_2$.

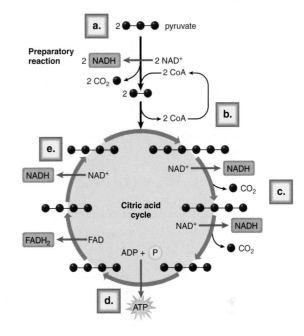

11. The strongest and final electron acceptor in the electron transport chain is
    a. NADH.   c. oxygen.
    b. FADH$_2$.   d. water.

For questions 12–18, match the items to those in the key. Answers can be used more than once, and each question can have more than one answer.

Key:

    a. glycolysis
    b. preparatory reaction
    c. citric acid cycle
    d. electron transport chain

12. produces ATP
13. uses ATP
14. produces NADH
15. uses NADH
16. produces carbon dioxide
17. occurs in cytoplasm
18. occurs in mitochondria

19. The carriers in the electron transport chain undergo
    a. oxidation only.
    b. reduction only.
    c. oxidation and reduction.
    d. the loss of hydrogen ions.
    e. the gain of hydrogen ions.

20. The final acceptor for hydrogen atoms during fermentation is
    a. O$_2$.   c. FAD.
    b. acetyl CoA.   d. pyruvic acid.

21. Which of the following do not enter the cellular respiration pathways?
    a. fats
    b. amino acids
    c. nucleic acids
    d. carbohydrates

22. When animals carry out fermentation, they produce _____, while yeasts produce _____.
    a. lactate, NADH
    b. lactate, ethyl alcohol and CO_2
    c. NADH, ethyl alcohol and CO_2
    d. $FADH_2$, lactate
    e. ethyl alcohol and CO_2, lactate

23. Fermentation does not yield as much ATP as cellular respiration does because fermentation
    a. generates mostly heat.
    b. makes use of only a small amount of the potential energy in glucose.
    c. creates by-products that require large amounts of ATP to break down.
    d. creates ATP molecules that leak into the cytoplasm and are broken down.

24. Which type of human cell carries on the most fermentation?
    a. muscle
    b. fat
    c. nerve
    d. bone

25. The metabolic process that produces the most ATP molecules is
    a. glycolysis.
    b. the citric acid cycle.
    c. the electron transport chain.
    d. fermentation.

26. The greatest contributor of electrons to the electron transport chain is
    a. oxygen.
    b. glycolysis.
    c. the citric acid cycle.
    d. the preparatory reaction.
    e. fermentation.

27. Substrate-level ATP synthesis takes place in
    a. glycolysis and the citric acid cycle.
    b. the electron transport chain and the preparatory reaction.
    c. glycolysis and the electron transport chain.
    d. the citric acid cycle and the preparatory reaction.

28. Match the terms to their definitions. Only four of these terms are needed:

    anaerobic   oxygen deficit
    citric acid cycle   pyruvate
    fermentation   preparatory reaction

    a. occurs in mitochondria and produces CO_2, ATP, NADH, and $FADH_2$
    b. growing or metabolizing in the absence of oxygen
    c. end product of glycolysis
    d. Anaerobic breakdown of glucose that results in a gain of 2 ATP and end products such as alcohol and lactate.

29. Which of the following is not true of fermentation? Fermentation
    a. has a net gain of only 2 ATP.
    b. occurs in the cytoplasm.
    c. donates electrons to the electron transport chain.
    d. is carried on by humans.
    e. is carried on by yeast.

Thinking Scientifically

1. The compound malonate, a substrate of the citric acid cycle, can be poisonous at high concentrations because it can block cellular respiration. How might a substrate of the citric acid cycle block respiration when it is present in great excess?

2. Insulin resistance can lead to diabetes. Insulin normally causes cells to take in glucose from the blood, but insulin resistance causes cells to not respond properly to this signal. The pancreas then compensates by secreting even more insulin into the body. Researchers have found a connection between a high-fat diet, decreased mitochondrial function, and insulin resistance in some elderly patients. What might be the connection between these three events?

3. You observe that prokaryotic cells sometimes have invaginations of the plasma membrane where metabolic pathways take place. How could this observation be used to support your hypothesis that mitochondria evolved when a prokaryote cell was engulfed by a nucleated cell?

Bioethical Issue

Regulating Dietary Supplements

Americans spend millions of dollars every year on products and diets that claim to boost metabolic rates and assist in weight loss. These and other supplements are available in supermarkets, by mail order, and through the Internet. Drug makers are required to prove that a drug is safe and effective before it can be sold to the public. But the Food and Drug Administration (FDA) cannot legally supervise the supplement industry because their products are not defined as drugs. Thus, regulators must prove that a supplement is dangerous before it can be removed from the market. Although some of the more dangerous supplements, such as ephedra, have been banned, many questionable and dangerous products remain freely available to anyone who wants them. Since the FDA does not have the time or resources to perform the necessary research, most dietary supplements receive scant supervision.

Many scientists and other concerned individuals contend that FDA regulation of the supplement industry is needed to protect public safety. They argue that FDA regulation would protect the public from potentially dangerous products and ensure the efficacy of dietary supplements and diet products. However, some are concerned that regulation of the supplement industry would undermine the public interest by reducing choices in the marketplace. Furthermore, they argue that FDA regulation of supplements would create the need for additional taxes and divert resources that would be better spent regulating prescription and over-the-counter drugs.

Do you believe that the government should increase scrutiny of dietary and weight-loss supplements? Or do you believe that FDA regulation of these products would be a waste of time and money?

Essentials of Biology Website

The companion website for *Essentials of Biology* provides a wealth of information organized and integrated by chapter. You will find practice tests, animations, videos, and much more that will complement your learning and understanding of general biology.

http://www.mhhe.com/maderessentials2

8
Cellular Reproduction

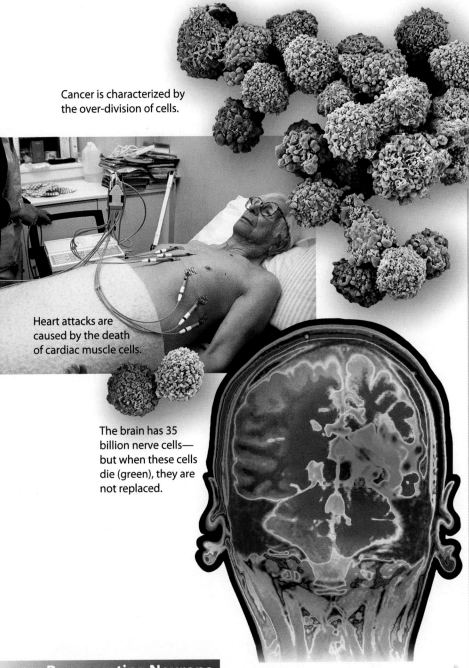

Cancer is characterized by the over-division of cells.

Heart attacks are caused by the death of cardiac muscle cells.

The brain has 35 billion nerve cells—but when these cells die (green), they are not replaced.

Outline

Regenerating Neurons

MANY OF OUR BODY tissues and organs can repair themselves if they become injured or damaged. This is because their cells are capable of undergoing mitosis, a process of nuclear division that enables cells and tissues to grow and be repaired. Exceptions are nerve cells, the cells that make up the nervous system. Injuries to the brain and spinal cord are very serious because nerve cells normally last a lifetime and seldom undergo mitosis. When nerve

—Continued next page

Chromosomes

An important event that occurs in preparation for cell division is **DNA replication,** when a cell copies its DNA. Then, a full copy of all the DNA is passed to both daughter cells. This step is critical because cells cannot continue to live without a copy of the genetic material. But a human cell contains an astonishing 2 meters of DNA, and moving this large quantity of DNA is no easy task. However, DNA and associated proteins are packaged into a set of **chromosomes,** which allow it to be distributed to the daughter cells. We shall see that proteins are involved in packaging the DNA, and these proteins are associated with still other proteins involved in DNA replication and function.

Check Your **Progress**

1. **In what ways is cellular reproduction necessary to the continued existence of a mature organism and the production of a new organism?**
2. **What two processes usually occur when a cell reproduces?**
3. **Contrast the structure of chromatin when the cell is not dividing to that of a chromosome just before cellular reproduction occurs.**

Answers: 1. Mature organisms have to replenish worn-out cells and repair injuries. Unicellular organisms divide in order to reproduce, and multicellular organisms develop from a single cell.   2. Growth and cell division usually occur.   3. Chromatin consists of strands of DNA that are organized around histone proteins. Chromatin forms a zigzag that is folded into loops. When cellular reproduction occurs, the chromatin is further condensed into larger loops that produce a greatly compacted chromosome.

Chromatin to Chromosomes

When a eukaryotic cell is not undergoing cell division, the DNA and associated proteins have the appearance of thin threads called **chromatin.** Closer examination reveals that chromatin is periodically wound around a core of eight protein molecules so that it looks like beads on a string. The protein molecules are **histones,** and each bead is called a **nucleosome (Fig. 8.2).** Chromatin normally adopts a zigzag structure, and then it is folded into loops for further compaction. This looped chromatin more easily fits within the nucleus.

Just before cell division occurs, the chromatin condenses multiple times into large loops that produce highly compacted chromosomes. Each species has a characteristic number of chromosomes; a human cell has 46 chromosomes. We can easily see chromosomes with a light microscope because just before division occurs a chromosome is over 10,000 times more compact than is chromatin.

By the time we can clearly see the chromosomes, they are duplicated. A duplicated chromosome is composed of two identical halves called **sister chromatids** held together at a constricted region called a **centromere.** Each sister chromatid contains an identical DNA double helix.

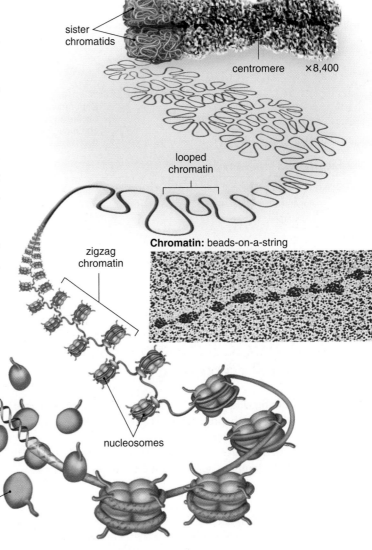

Chromosome: duplicated and condensed

sister chromatids

centromere   ×8,400

looped chromatin

Chromatin: beads-on-a-string

zigzag chromatin

nucleosomes

histones

Figure 8.2   Chromosome compaction.

Histones are responsible for packaging chromatin so that it fits into the nucleus. Prior to cell division, chromatin is further condensed into chromosomes so that the genetic material can be easily divided between the daughter cells.

Growth occurs as organelles double.

G₀

cytokinesis
telophase
anaphase
metaphase
prophase

M

Mitosis and cytokinesis occur.

Interphase

G₁

G₂

S

Growth occurs as cell prepares to divide.

DNA replication occurs as chromosomes duplicate.

Figure 8.3   The cell cycle.

Cells go through a cycle that consists of four stages: G_1, S, G_2, and M. The major activity for each stage is given. Some cells can exit G_1 and enter a G_0 stage.

8.2 | The Cell Cycle

We have already indicated that cellular reproduction involves duplication of cell contents followed by cell division. For cellular reproduction to be orderly, you would expect the first event to occur before the second event, and that's just what happens during the cell cycle. The **cell cycle** is an orderly sequence of stages that takes place between the time a new cell has arisen from division of a parent cell to the point when it has given rise to two daughter cells.

Interphase

As **Figure 8.3** shows, most of the cell cycle is spent in **interphase.** This is the time when a cell performs its usual functions, depending on its location in the body. The amount of time the cell takes for interphase varies widely. Embryonic cells complete the entire cell cycle in just a few hours. A rapidly dividing mammalian cell, such as an adult stem cell, typically takes about 24 hours to complete the cell cycle and spends 22 hours in interphase.

DNA replication occurs in the middle of interphase and serves as a way to divide interphase into three stages: G_1, S, and G_2. G_1 is the stage before DNA replication, and G_2 is the stage following DNA synthesis. Originally, G stood for "gap," but now that we know how metabolically active the cell is, it is better to think of G as standing for "growth." Protein synthesis is very much a part of these growth stages.

During G_1, a cell doubles its organelles (such as mitochondria and ribosomes) and accumulates materials that will be used for DNA replication. At this point, the cell integrates internal and external signals and "decides" whether to continue with the cell cycle. Some cells, such as muscle cells, typically remain in interphase and cell division is permanently arrested. These cells are said to have entered a G_0 stage. When an injury occurs, many cells in the G_0 stage can reenter the cell cycle and divide again to repair the damage. But a few cell types, such as nerve cells, almost never divide again once they enter G_0.

Following G_1, the cell enters the S stage. The *S* stands for synthesis, and certainly DNA synthesis is required for DNA replication. At the beginning of the S stage, each chromosome has one **chromatid** consisting of a single DNA double helix. At the end of this stage, each chromosome is composed of two *sister chromatids,* each having one double helix. The two chromatids of each chromosome remain attached at the centromere. We say that DNA replication results in duplicated chromosomes.

The G_2 stage extends from the completion of DNA replication to the onset of mitosis. During this stage, the cell synthesizes proteins that will be needed for cell division, such as the protein found in microtubules. The role of microtubules in cell division is described in a later section.

M (Mitotic) Stage

Cell division occurs during the M stage, which encompasses both division of the nucleus and division of the cytoplasm. The type of nuclear division associated with the cell cycle is called **mitosis,** which accounts for why this stage is called the M stage. As a result of mitosis, the daughter nuclei are identical to the parent cell and to each other—they all have the same number and kinds of chromosomes. Division of the cytoplasm, which starts even before mitosis is finished, is called **cytokinesis.**

■ **Check**
Your Progress

1. During what stage of the cell cycle is the DNA replicated?
2. Compare the structure of a chromosome from before and after DNA replication.
3. List the major events that occur during the G_1 stage of the cell cycle.
4. What happens during the M stage of the cell cycle?

Answers: 1. DNA replication occurs during the S stage of the cell cycle.   2. Before replication, each chromosome consists of a single chromatid. Following replication, each chromosome consists of two sister chromatids held together at a centromere.   3. During the G_1 stage, the cell increases in size and doubles its organelles in preparation for cell division. The decision whether to divide again or not is also made during G_1.   4. Mitosis (nuclear division in which the chromosome number stays constant) and cytokinesis (division of cytoplasm) occur.

8.3 | Mitosis and Cytokinesis

During mitosis, the duplicated nuclear contents of the parent cell are distributed equally to the daughter cells. Recall that during the S stage, the DNA of each chromosome is replicated to produce a duplicated chromosome that contains two identical sister chromatids that remain attached at the centromere (**Fig. 8.4**). Each chromatid is a single DNA double helix containing the same sequence of base pairs as the original chromosome. During mitosis, the sister chromatids of each chromosome separate and are now called *daughter chromosomes*. Because each original chromosome goes through the same process of DNA replication followed by separation of the sister chromatids, the daughter nuclei produced by mitosis are genetically identical to each other and to the parent nucleus. Thus, if the parent nucleus has four chromosomes, each daughter nucleus also has four chromosomes of exactly the same type. One way to keep track of the number of chromosomes in drawings is to count the number of centromeres, because every chromosome has a centromere.

Every animal has an even number of chromosomes because each parent contributed half of the chromosomes to the new individual. In drawings of mitosis, some chromosomes are colored red and some are colored blue to illustrate that half of the chromosomes are derived from one parent and the other half are derived from the other parent.

The Spindle

While it may seem easy to separate the chromatids of only 4 duplicated chromosomes, imagine the task when there are 46 chromosomes, as in humans, or 78, as in dogs. Certainly it is helpful that chromosomes be highly condensed before the task begins, but clearly some mechanism is needed to complete separation in an organized manner. Most eukaryotic cells rely on a **spindle,** a cytoskeletal structure, to pull the chromatids apart. A spindle has spindle fibers made of microtubules that are able to assemble and disassemble. First, the microtubules assemble to form the spindle that takes over the center of the cell and separates the chromatids. Later, they disassemble.

A **centrosome** is the primary microtubule organizing center of a cell. Centrosome duplication occurs at the start of the S phase of the cell cycle and is completed by G_2. During the first part of the M stage, the centrosomes separate and move to opposite sides of the nucleus, where they form the poles of the spindle. As the nuclear envelope breaks down, spindle fibers take over the center of the cell. Certain ones overlap at the **spindle equator,** which is midway between the poles. Others attach to duplicated chromosomes in a way that ensures the separation of the sister chromatids and their proper distribution to the daughter cells. Whereas the chromosomes will be inside the newly formed daughter nuclei, a centrosome will be just outside.

Traditionally, mitosis is divided into a sequence of events, even though it is a continuous process. We will describe mitosis as having four phases: **prophase, metaphase, anaphase,** and **telophase.** These phases are labeled in **Figure 8.5** for a dividing plant cell nucleus. Plant cells have centrosomes, but they are not clearly visible because they lack centrioles. In an animal cell, each centrosome has two centrioles and an array of microtubules called an **aster.** Otherwise the descriptions on the next two pages apply to both animal and plant cells.

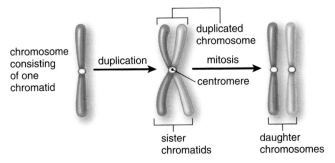

Figure 8.4   Overview of mitosis.

Counting the number of centromeres tells you the number of chromosomes.

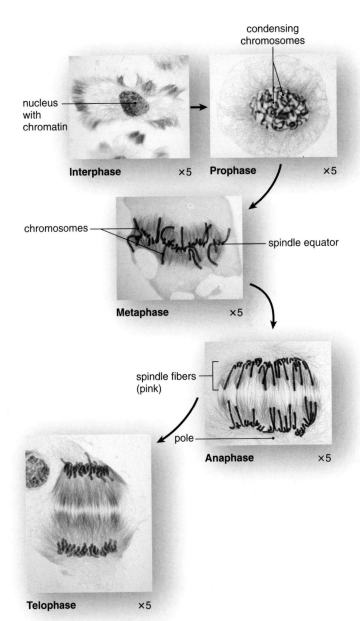

Figure 8.5   Mitosis in a plant cell.

The chromosomes are stained blue, and microtubules of the spindle fibers are stained pink in these photos of a dividing African blood lily. For a description of these phases see Figure 8.6.

Phases of Mitosis in Animal and Plant Cells

Mitosis requires the phases described previously for plant cells in Figure 8.5 and described for animal cells in **Figure 8.6.** Although mitosis is divided into these phases, it is a continuous process. Before studying the descriptions of these phases in Figure 8.6, recall that before mitosis begins, DNA has been replicated. Each double helix is located in a chromatid; therefore, each chromosome consists of two sister chromatids attached at a centromere. Notice that some chromosomes are colored red and some are colored blue. The red chromosomes were inherited from one parent and the blue chromosomes from the other parent.

During mitosis, a spindle arises that will separate the sister chromatids of each duplicated chromosome. Once separated, the sister chromatids become

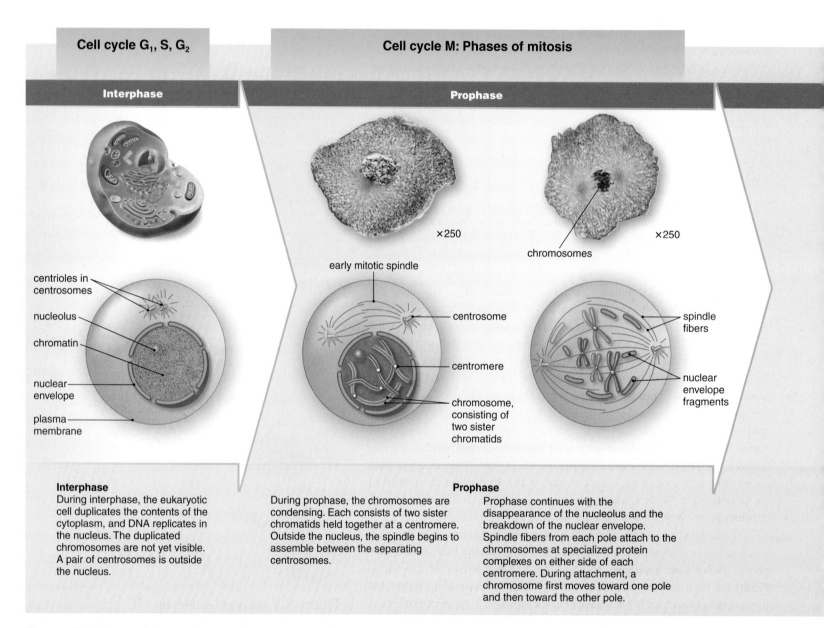

Interphase
During interphase, the eukaryotic cell duplicates the contents of the cytoplasm, and DNA replicates in the nucleus. The duplicated chromosomes are not yet visible. A pair of centrosomes is outside the nucleus.

Prophase
During prophase, the chromosomes are condensing. Each consists of two sister chromatids held together at a centromere. Outside the nucleus, the spindle begins to assemble between the separating centrosomes.

Prophase continues with the disappearance of the nucleolus and the breakdown of the nuclear envelope. Spindle fibers from each pole attach to the chromosomes at specialized protein complexes on either side of each centromere. During attachment, a chromosome first moves toward one pole and then toward the other pole.

Figure 8.6   Phases of mitosis in animal cells.

The red chromosomes were inherited from one parent and the blue chromosomes from the other parent.

daughter chromosomes. In this way, the resulting daughter nuclei are identical to each other and to the parental nucleus. In Figure 8.6, the daughter nuclei not only have four chromosomes; they each have one red short, one blue short, one red long, and one blue long, the same as the parent cell had.

Mitosis is usually followed by division of the cytoplasm, or cytokinesis. Cytokinesis begins during telophase and continues after the nuclei have formed in the daughter cells. The cell cycle is now complete, and the daughter cells enter interphase, during which the cell will grow and DNA will replicate once again. In rapidly dividing mammalian stem cells, the cell cycle lasts for 24 hours. Mitosis and cytokinesis require about two hours of this time period.

Check Your Progress

1. During what phase of mitosis are duplicated chromosomes aligned at the spindle equator?
2. During what phase of mitosis are daughter chromosomes moving toward the spindle poles?
3. Contrast the centrosomes of plant cells and animal cells.

Answers: 1. Metaphase.   2. Anaphase.   3. Animal cell centrosomes contain centrioles, whereas plant cell centrosomes lack centrioles and are not clearly visible.

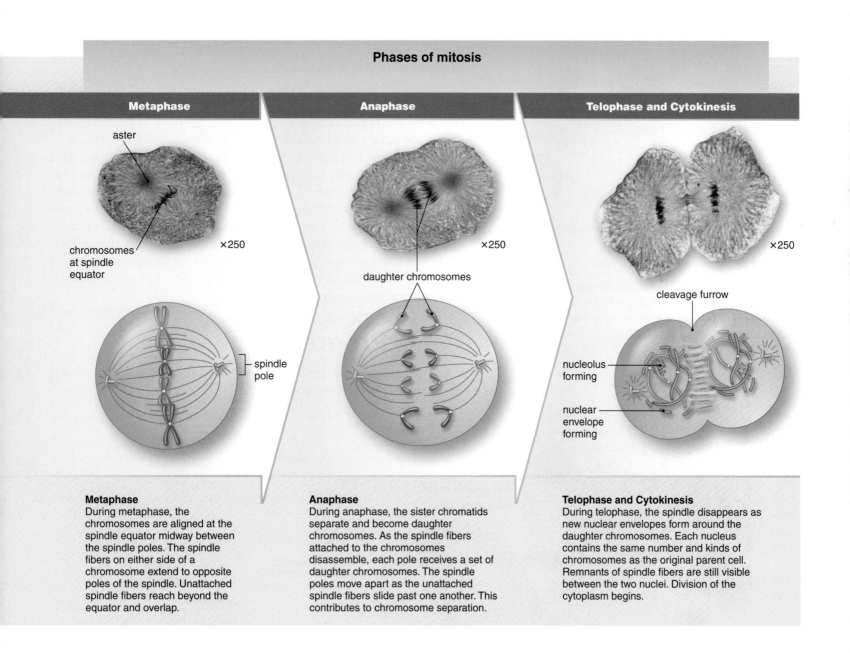

Phases of mitosis

Metaphase | **Anaphase** | **Telophase and Cytokinesis**

aster
chromosomes at spindle equator
×250
spindle pole

daughter chromosomes
×250

cleavage furrow
×250
nucleolus forming
nuclear envelope forming

Metaphase
During metaphase, the chromosomes are aligned at the spindle equator midway between the spindle poles. The spindle fibers on either side of a chromosome extend to opposite poles of the spindle. Unattached spindle fibers reach beyond the equator and overlap.

Anaphase
During anaphase, the sister chromatids separate and become daughter chromosomes. As the spindle fibers attached to the chromosomes disassemble, each pole receives a set of daughter chromosomes. The spindle poles move apart as the unattached spindle fibers slide past one another. This contributes to chromosome separation.

Telophase and Cytokinesis
During telophase, the spindle disappears as new nuclear envelopes form around the daughter chromosomes. Each nucleus contains the same number and kinds of chromosomes as the original parent cell. Remnants of spindle fibers are still visible between the two nuclei. Division of the cytoplasm begins.

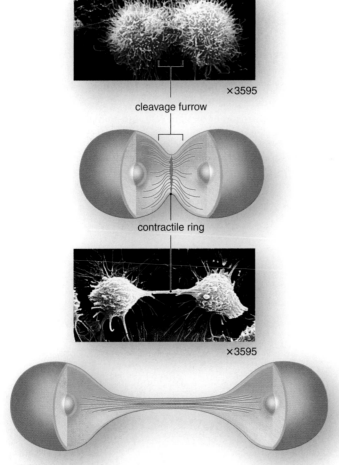

×3595

cleavage furrow

contractile ring

×3595

Figure 8.7 Cytokinesis in animal cells.

A single cell becomes two cells by a furrowing process. A cleavage furrow appears as early as anaphase, and a contractile ring, composed of actin filaments, gradually gets smaller until there are two cells.

Copyright by R. G. Kessel and C. Y. Shih, *Scanning Electron Microscopy in Biology: A Students' Atlas on Biological Organization*, Springer-Verlag, 1974.

Cytokinesis in Animal and Plant Cells

Cytokinesis accompanies mitosis in most cells but not all. When mitosis occurs, but cytokinesis doesn't occur, the result is a multinucleated cell. For example, skeletal muscle cells in vertebrate animals and the embryo sac in flowering plants are multinucleated.

Cytokinesis in Animal Cells

In animal cells, a cleavage furrow, which is an indentation of the membrane between the two daughter nuclei, begins as anaphase draws to a close. The cleavage furrow deepens when a band of actin filaments, called the contractile ring, slowly forms a circular constriction between the two daughter cells. The action of the contractile ring can be likened to pulling a drawstring ever tighter about the middle of a balloon. A narrow bridge between the two cells is visible during telophase, and then the contractile ring continues to separate the cytoplasm until there are two independent daughter cells (**Fig. 8.7**).

Cytokinesis in Plant Cells

Cytokinesis in plant cells occurs by a process different from that seen in animal cells (**Fig. 8.8**). The rigid cell wall that surrounds plant cells does not permit cytokinesis by furrowing. Instead, cytokinesis in plant cells involves the building of new plasma membrane and cell walls between the daughter cells.

Cytokinesis is apparent when a small, flattened disk appears between the two daughter plant cells. In electron micrographs, it is possible to see that the disk is composed of vesicles. The Golgi apparatus produces these vesicles, which move along microtubules to the region of the disk. As more vesicles arrive and fuse, a cell plate can be seen. The **cell plate** is simply newly formed plasma membrane that expands outward until it reaches the old plasma membrane and fuses with it. The new membrane releases molecules that form the new plant cell walls. These cell walls are later strengthened by the addition of cellulose fibrils.

Check Your Progress

1. What explanation can you give for the differences in cytokinesis between plant cells and animal cells?

Answers: 1. The rigid cell wall of plant cells does not permit cytokinesis by furrowing as in animal cells.

daughter cells

nucleoli

daughter nucleus

cell plate formation

daughter nucleus

vesicles containing membrane components fusing to form cell plate

Figure 8.8 Cytokinesis in plant cells.

During cytokinesis in a plant cell, a cell plate forms midway between two daughter nuclei and extends to the original plasma membrane.

8.4 | The Cell Cycle Control System

In order for a cell to reproduce successfully, the cell cycle must be controlled. The importance of cell cycle control can be appreciated by comparing the cell cycle to the events that occur in an automatic washing machine. The washer's control system starts to wash only when the tub is full of water, doesn't spin until the water has been emptied, and so forth. Similarly, the cell cycle's control system ensures that the G_1, S, G_2, and M stages occur in order and only when the previous stage has been successfully completed.

Cell Cycle Checkpoints

Just as a washing machine will not begin to agitate the load until the tub is full, the cell cycle has **checkpoints** that can delay the cell cycle until all is well. The cell cycle has many checkpoints, but we will consider only three: G_1, G_2 and the mitotic checkpoint (**Fig. 8.9**).

The *G_1 checkpoint* is especially significant because if the cell cycle passes this checkpoint, the cell is committed to divide. If the cell does not pass this checkpoint, it can enter G_0, during which it performs its normal functions but does not divide. The proper growth signals must be present for a cell to pass the G_1 checkpoint. Additionally, the integrity of the cell's DNA is also checked. If the DNA is damaged, the protein p53 can stop the cycle at this checkpoint and initiate DNA repair. If that is not possible, it brings about the death of the cell by apoptosis (see Fig. 8.11).

The cell cycle halts momentarily at the *G_2 checkpoint* until the cell verifies that DNA has replicated. This prevents the initiation of the M stage unless the chromosomes are duplicated. Also, if DNA is damaged, as from exposure to solar radiation or X-rays, arresting the cell cycle at this checkpoint allows time for the damage to be repaired so that it is not passed on to daughter cells.

Another cell cycle checkpoint occurs during the mitotic stage. The cycle hesitates between metaphase and anaphase to make sure the chromosomes are properly attached to the spindle and will be distributed accurately to the daughter cells. The cell cycle does not continue until every chromosome is ready for the nuclear division process.

Internal and External Signals

The checkpoints of the cell cycle are controlled by internal and external signals. A **signal** is a molecule that stimulates or inhibits an event. Signals outside the cell are called external signals and those inside the cell are called internal signals. Inside the cell, enzymes called **kinases** remove phosphate from ATP and add it to another molecule. Just before the S stage, a protein called S-cyclin combines with a kinase called S-kinase, and synthesis of DNA takes place. Just before the M stage, M-cyclin combines with a kinase called M-kinase, and mitosis occurs. **Cyclins** are so named because they are present only during certain stages of the cell cycle. They increase in amount until they combine with a kinase that not only activates a protein that drives the cell cycle, but also activates an enzyme that destroys the cyclin (**Fig. 8.10**). Destruction of the cyclin at the appropriate time is necessary for the cell cycle to progress normally.

Some external signals, such as growth factors and hormones, stimulate cells to go through the cell cycle. Growth factors also stimulate repair of tissues.

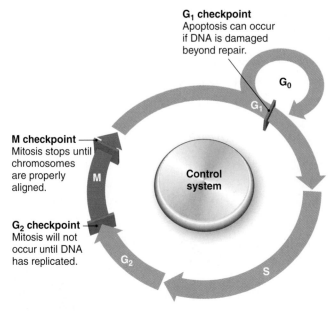

Figure 8.9 Cell cycle checkpoints.

Internal and external signals determine whether the cell is ready to proceed past cell cycle checkpoints. Three important checkpoints are designated.

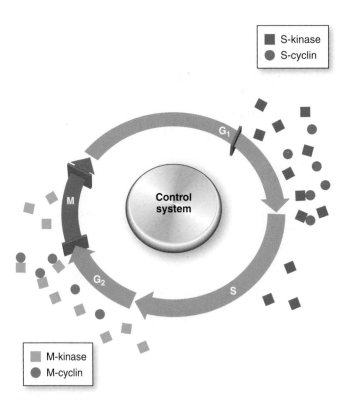

Figure 8.10 Internal signals of the cell cycle.

S-cyclin must combine with S-kinase for the cell cycle to begin DNA replication. M-cyclin must combine with M-kinase for the cell cycle to start mitosis.

normal cells

Cell rounds up, and
nucleus collapses.

Chromatin condenses,
and DNA fragments. —— DNA

Plasma membrane
blisters, blebs form. —— blebs

Cell fragments.

cell fragment ——

—— DNA fragment

apoptotic cell

Figure 8.11 **Apoptosis.**

Apoptosis is a sequence of events that results in a fragmented cell.
The fragments are phagocytized by white blood cells and neighboring
tissue cells.

Even cells that are arrested in G_0 will finish the cell cycle if stimulated to do so by growth factors. For example, epidermal growth factor (EGF) stimulates skin in the vicinity of an injury to finish the cell cycle, thereby repairing the damage. Hormones act on tissues at a distance, and some signal cells to divide. For example, at a certain time in the menstrual cycle of females, the hormone estrogen stimulates cells lining the uterus to divide and prepares the lining for implantation of a fertilized egg.

The cell cycle can be inhibited by cells coming into close contact with other cells. In the laboratory, eukaryotic cells will divide until they line a container in a one-cell-thick sheet. Then they stop dividing, due to a phenomenon termed **contact inhibition.** Some researchers suggest that contact inhibition prevents cells from overgrowing within the body. When tissues are damaged, cells divide to repair the damage. But once the tissue has been repaired, contact inhibition prevents cell overgrowth by halting the cell cycle.

Some years ago, it was noted that mammalian cells in cell culture divide about 70 times, and then they die. Cells seem to "remember" the number of times they divided before, and stop dividing when the usual number of cell divisions is reached. It's as if *senescence,* the aging of cells, is dependent on an internal battery-operated clock that runs down and then stops. We now know that senescence is due to the shortening of telomeres. A **telomere** is a repeating DNA base sequence (TTAGGG) at the end of the chromosomes that can be as long as 15,000 base pairs. Telomeres have been likened to the protective caps on the ends of shoelaces because they ensure chromosomal stability. Each time a cell divides, some portion of a telomere is lost. When telomeres become too short, the cell is "old" and dies by a process called apoptosis.

Apoptosis

Apoptosis is often defined as programmed cell death because the cell progresses through a typical series of events that bring about its destruction (**Fig. 8.11**). The cell rounds up and loses contact with its neighbors. The nucleus fragments, and the plasma membrane develops blisters. Finally, the cell breaks into fragments that are engulfed by white blood cells and/or neighboring cells. Oddly enough, cells routinely harbor the enzymes, called caspases, that bring about apoptosis. These enzymes are ordinarily held in check by inhibitors, but are unleashed by either internal or external signals.

Cell division and apoptosis are two opposing processes that keep the number of cells in the body at an appropriate level. In other words, cell division increases, and apoptosis decreases, the number of **somatic cells** (body cells). Both the cell cycle and apoptosis are normal parts of growth and development. An organism begins as a single cell that repeatedly undergoes the cell cycle to produce many cells, but eventually some cells must die in order for the organism to take shape. For example, when a tadpole becomes a frog, the tail disappears as apoptosis occurs. In humans, the fingers and toes of an embryo are at first webbed, but later they are freed from one another as a result of apoptosis.

Apoptosis occurs all the time, particularly if an abnormal cell that could become cancerous appears. Death through apoptosis prevents a tumor from developing.

8.5 | The Cell Cycle and Cancer

The cell cycle is regulated by signals that either inhibit or promote the cell cycle. When these signals become unbalanced, **cancer** may result. Cancer is a disease of the cell cycle in which cellular reproduction occurs repeatedly without end. Cancers are classified according to their location. *Carcinomas* are cancers of the epithelial tissue that lines organs; *sarcomas* are cancers arising in muscle or connective tissue (especially bone or cartilage); and *leukemias* are cancers of the blood. In this chapter, we consider some general characteristics of cancer cells. Chapter 12 explores specific genetic changes that lead to cancerous growth.

Characteristics of Cancer Cells

Carcinogenesis, the development of cancer, is gradual; it may be decades before a tumor is visible. Cancer cells then have these characteristics:

Cancer cells lack differentiation. Cancer cells lose their specialization and do not contribute to the functioning of a body part. A cancer cell does not look like a differentiated epithelial, muscle, nervous, or connective tissue cell; instead, it looks distinctly abnormal. As mentioned, normal cells can enter the cell cycle about 70 times, and then they die. Cancer cells can enter the cell cycle repeatedly, and in this way they are immortal.

Cancer cells have abnormal nuclei. The nuclei of cancer cells are enlarged and may contain an abnormal number of chromosomes. The chromosomes are also abnormal; some parts may be duplicated, or some may be deleted. In addition, gene amplification (extra copies of specific genes) is seen much more frequently than in normal cells.

Cancer cells do not undergo apoptosis. Ordinarily, cells with damaged DNA undergo apoptosis preventing tumors from developing. Cancer cells, however, do not respond to the normal signals from other cells that trigger apoptosis, and continue to divide.

Cancer cells form tumors. Normal cells anchor themselves to a substratum and/or adhere to their neighbors. Then, they exhibit contact inhibition and stop dividing. Cancer cells, on the other hand, have lost all restraint; they pile on top of one another and grow in multiple layers, forming a **tumor.** They have a reduced need for growth factors, and they no longer respond to inhibitory signals. As cancer develops, the most aggressive cell becomes the dominant cell of the tumor.

Cancer cells undergo metastasis and promote angiogenesis. A **benign tumor** is usually contained within a capsule, and therefore cannot invade adjacent tissue. Additional changes may cause cancer cells to produce enzymes that allow tumor cells to escape the capsule and invade nearby tissues. The tumor has become **malignant,** meaning that it is invasive and may spread. Cells from a malignant tumor may travel through the blood or lymph to start new tumors elsewhere within the body. This process is known as **metastasis (Fig. 8.12).**

An actively growing tumor may only grow so large before it becomes unable to obtain sufficient nutrients to support further growth. Additional mutations in cancer cells allows them to secrete factors that promote **angiogenesis,** the formation of new blood vessels. Additional nutrients and oxygen reach the tumor, allowing it to grow larger. Some modes of cancer treatment are aimed at preventing angiogenesis from occurring. The patient's prognosis (probable outcome) depends on (1) whether the tumor has invaded surrounding tissues, and (2) whether there are metastatic tumors in distant parts of the body.

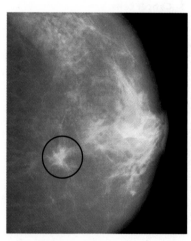

A single cancer cell grows into a tumor.

The tumor becomes malignant and invades nearby tissue.

Cancer cells travel (dotted arrows) through lymphatic (green) and blood (red) vessels, and metastatic tumors form.

a. Development of breast cancer and metastatic tumors

b. Mammogram showing tumor

Figure 8.12 Development of breast cancer.

a. Breast cancer begins as a single abnormal cell grows into a benign tumor. Additional changes allow the tumor to become invasive, or malignant, and invade nearby tissue. Metastasis occurs when the tumor invades blood and lymph vessels, and new tumors develop some distance from the original tumor. **b.** A mammogram (X-ray image of the breast) can find a tumor (circled) too small to be felt.

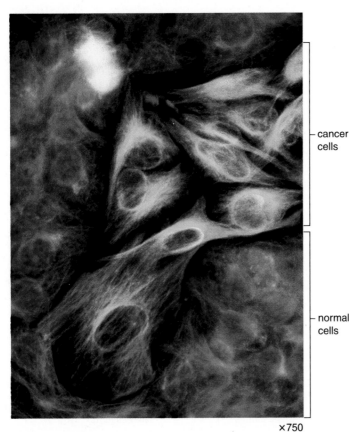

×750

Figure 8.13 Cancer cells.

Compared to the other cells, the brightly colored cancer cells appear distinctly abnormal.

cancer cells

normal cells

Cancer Treatment

Cancer treatments either remove the tumor or interfere with the ability of cancer cells to reproduce. For many solid tumors, removal by surgery is often the first line of treatment. When the cancer is detected at an early stage, surgery may be sufficient to cure the patient by removing all cancerous cells.

Because cancer cells are rapidly dividing, they are susceptible to radiation therapy and chemotherapy (**Fig. 8.13**). The goal of radiation is to kill cancer cells within a specific tumor by directing high-energy beams at the tumor. DNA is damaged to the point that replication can no longer occur, and the cell undergoes apoptosis. Chemotherapy is a way to kill cancer cells that have spread throughout the body. Like radiation, most chemotherapeutic drugs lead to the death of cells by damaging their DNA or interfering with DNA synthesis. Others interfere with the functioning of the mitotic spindle. The drug vinblastine, first obtained from a flowering plant called the periwinkle, prevents the spindle from forming. Taxol, extracted from the bark of the Pacific yew tree, prevents the spindle from functioning as it should. Unfortunately, radiation and chemotherapy often damage cells other than cancer cells, leading to side effects such as nausea and hair loss.

Hormonal therapy also prevents cell division but in a different way. These drugs are designed to prevent cancer cell growth by preventing the cells from receiving signals necessary for their continued growth and division. For example, the drug tamoxifen modifies the receptor for estrogen so that this hormone cannot bind to it and promote the cell cycle. Other drug therapies are also being investigated. One proposed therapy that is now well under way uses drugs designed to inhibit angiogenesis, limiting the ability of a tumor to grow and spread.

Prevention of Cancer

Evidence is clear that the risk of certain types of cancer can be reduced by adopting protective behaviors and following recommended dietary guidelines.

Protective Behaviors

To lower the risk of developing certain cancers, people are advised to avoid smoking, sunbathing, and excessive alcohol consumption.

Cigarette smoking accounts for about 30% of all cancer deaths. Smoking is responsible for 90% of lung cancer cases among men and 79% among women—about 87% altogether. People who smoke two or more packs of cigarettes a day have lung cancer mortality rates 15 to 25 times greater than those of nonsmokers. Smokeless tobacco (chewing tobacco or snuff) increases the risk of cancers of the mouth, larynx, throat, and esophagus.

Skin cancers are often sun-related. Sun exposure is a major factor in the development of the most dangerous type of skin cancer, melanoma, and the incidence of this cancer increases in people living near the equator. Similarly, excessive radon gas[1] exposure in homes increases the risk of lung cancer, especially in cigarette smokers. It is best to test your home and, if necessary, take the proper remedial actions.

Check Your Progress

1. Describe how cancer relates to the cell cycle.
2. List the characteristic features of cancer cells.
3. Contrast the modes of action of radiation and chemotherapy with the mode of action of hormonal therapy.

Answers: 1. Cancer occurs when the cell cycle is not regulated properly.   2. Cancer cells lose their normal functions, have abnormal nuclei, form tumors, do not undergo apoptosis, undergo metastasis, and promote angiogenesis.   3. Radiation and chemotherapy damage DNA or otherwise interfere with the completion of mitosis. Hormonal therapy interferes with cell's reception of an external signal to divide.

[1] Radon gas results from natural radioactive breakdown of uranium in soil, rocks, and water.

Cancers of the mouth, throat, esophagus, larynx, and liver occur more frequently among heavy drinkers, especially when accompanied by tobacco use (cigarettes, cigars, or chewing tobacco).

Protective Diet

The risk of cancer (especially colon, breast, and uterine cancers) is 55% greater among obese women. The risk of cancer is 33% greater among obese men, compared to people of normal weight. Weight loss in these groups can therefore reduce cancer risk. In addition, the following dietary guidelines are recommended (**Fig. 8.14**):

- Increase consumption of foods that are rich in vitamins A and C. Beta-carotene, a precursor of vitamin A, is found in dark green leafy vegetables, carrots, and various fruits. Vitamin C is present in citrus fruits. These vitamins are called antioxidants because in cells they prevent the formation of free radicals (organic ions having an unpaired electron) that can possibly damage DNA. Vitamin C also prevents the conversion of nitrates and nitrites into carcinogenic nitrosamines in the digestive tract.
- Avoid salt-cured or pickled foods because they may increase the risk of stomach and esophageal cancers. Smoked foods, such as ham and sausage, contain chemical carcinogens similar to those in tobacco smoke. Nitrites are sometimes added to processed meats (e.g., hot dogs and cold cuts) and other foods to protect them from spoilage; as mentioned, nitrites are sometimes converted to cancer-causing nitrosamines in the digestive tract.
- Include in the diet vegetables from the cabbage family, which includes cabbage, broccoli, brussels sprouts, kohlrabi, and cauliflower. Consuming these vegetables may reduce the risk of gastrointestinal and respiratory tract cancers.

Figure 8.14   The right diet helps prevent cancer.

Foods that help protect against cancer include fruits and dark green leafy vegetables, including swiss chard (*left*), which are rich in vitamins A and C, and vegetables from the cabbage family, including broccoli.

The Chapter in Review

Summary

8.1 | The Basics of Cellular Reproduction

Cellular reproduction occurs when growth and repair of tissues take place. Cellular reproduction is also necessary to both asexual and sexual reproduction. Following fertilization, a single cell becomes a multicellular organism by cellular reproduction.

Cellular reproduction involves two processes: growth, during which a cell grows and doubles its contents, and cell division, during which the parent cell's contents are split among two daughter cells. During cell division, one-half of the cytoplasm and a copy of the cell's DNA are passed on to each of two daughter cells.

When a cell is not dividing, the DNA is wound around histones, then arranged into a zigzag configuration to form chromatin. Just before cell division, duplicated chromatin condenses into large loops to form duplicated chromosomes consisting of two chromatids held together at a centromere.

8.2 | The Cell Cycle

The cell cycle consists of interphase and the M (mitotic) stage.

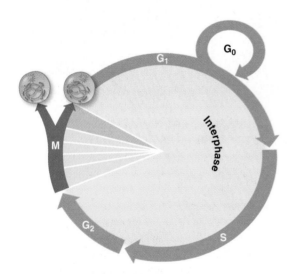

Interphase has the following stages:
- **G_1**: The cell doubles its organelles and accumulates materials that will be used for DNA replication.
- **S**: The cell replicates its DNA.
- **G_2**: The cell synthesizes proteins that will be needed for cell division.

The M (mitotic) stage consists of mitosis and cytokinesis. As a result of mitosis, the daughter nuclei are genetically identical to the parent nucleus and to each other. If the parent nucleus has four chromosomes, the daughter nuclei each have four chromosomes.

8.3 | Mitosis and Cytokinesis

Cell division consists of mitosis and cytokinesis.
Mitosis has four phases:
- During **prophase,** the chromosomes are condensing. Outside the nucleus, the spindle begins to assemble between the separating centrosomes. Prophase continues with the disappearance of the nucleolus and the breakdown of the nuclear envelope. Spindle microtubules from each pole attach to the chromosomes in the region of a centromere.
- During **metaphase,** the chromosomes are aligned at the spindle equator midway between the spindle poles.
- During **anaphase,** the sister chromatids separate and become daughter chromosomes. As the microtubules attached to the chromosomes disassemble, each pole receives a set of daughter chromosomes.
- During **telophase,** the spindle disappears as new nuclear envelopes form around the daughter chromosomes. Each nucleus contains the same number and kinds of chromosomes as the original parent nucleus. Division of the cytoplasm begins.

Cytokinesis differs for animal and plant cells:
- In animal cells, a furrowing process involving actin filaments divides the cytoplasm.
- In plant cells, a cell plate forms from which the plasma membrane and cell wall develop.

8.4 | The Cell Cycle Control System

Checkpoints are interacting signals that promote or inhibit the continuance of the cell cycle. Important checkpoints include these three:
- At G_1, the cell can enter G_0 or undergo apoptosis if DNA is damaged beyond repair. If the cell cycle passes this checkpoint, the cell is committed to complete the cell cycle.
- At G_2, the cell checks to make sure DNA has been replicated properly.
- At the mitotic checkpoint, the cell makes sure the chromosomes are properly aligned and ready to be partitioned to the daughter cells.

Both internal and external signals converge on checkpoints to control the cell cycle. Cyclin-kinase complexes are internal signals that promote either DNA replication or mitosis. Growth factors and hormones are external signals that promote the cell cycle in connection with growth. Inhibitory external signals are responsible for contact inhibition. Internally, when telomeres are too short, the cell cycle stops because the chromosomes become unstable.

When DNA cannot be repaired, apoptosis occurs as enzymes called caspases bring about destruction of the nucleus and the rest of the cell.

8.5 │ The Cell Cycle and Cancer

Cancer results when control of the cell cycle is lost, and cells divide uncontrollably. Cancer cells are undifferentiated, have abnormal nuclei, do not undergo apoptosis, and form tumors. A tumor may stimulate angiogenesis, allowing it to obtain more nutrients and grow larger. Tumors may also undergo metastasis, which allows them to spread throughout the body. Certain behaviors, such as avoiding smoking and sunbathing and adopting a diet rich in fruits and vegetables, are protective against cancer.

Key Terms

anaphase 117	DNA replication 115
angiogenesis 123	histone 115
apoptosis 122	interphase 116
aster 117	kinase 121
benign tumor 123	malignant 123
cancer 123	metaphase 117
carcinogenesis 123	metastasis 123
cell cycle 116	mitosis 116
cell plate 120	nucleosome 115
centromere 115	prophase 117
centrosome 117	signal 121
checkpoint 121	sister chromatid 115
chromatid 116	somatic cell 122
chromatin 115	spindle 117
chromosome 115	spindle equator 117
contact inhibition 122	telomere 122
cyclin 121	telophase 117
cytokinesis 116	tumor 123

Testing Yourself

Choose the best answer for each question.

1. _____ is nuclear division, whereas _____ is division of the cytoplasm.
   a. Cytokinesis, mitosis
   b. Apoptosis, mitosis
   c. Mitosis, apoptosis
   d. Mitosis, cytokinesis

2. The two identical halves of a duplicated chromosome are called
   a. chromosome arms.   c. chromatids.
   b. nucleosomes.   d. homologues.

For questions 3–6, match the items to those in the key. Answers can be used more than once.

Key:

   a. G_1 stage of interphase
   b. G_2 stage of interphase
   c. S stage of interphase

3. This stage follows mitosis.
4. DNA is replicated during this stage.
5. Organelles are doubled in number during this stage.
6. During this stage, the cell produces proteins that will be needed for cell division.

7. Label the stages and phases of the cell cycle on the following diagram. Include anaphase, cytokinesis, G_1 stage, G_2 stage, metaphase, prophase, S stage, and telophase.

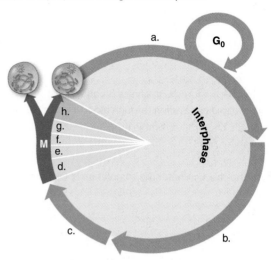

For questions 8–13, match the items to those in the key. Answers can be used more than once, and each question can have more than one answer.

Key:

   a. prophase
   b. metaphase
   c. anaphase
   d. telophase

8. The nucleolus disappears, and the nuclear membrane breaks down.
9. The spindle disappears, and the nuclear envelope forms.
10. Sister chromatids separate.
11. Spindle poles move apart as spindle fibers slide past each other.
12. Chromosomes are aligned on the spindle equator.
13. Each chromosome is composed of two sister chromatids.
14. Plants cannot use a cleavage furrow to undergo cytokinesis because they
    a. lack a plasma membrane.
    b. have a cell wall.
    c. have too many chromosomes.
    d. are too small.
15. Which cell cycle checkpoint allows damaged DNA to be repaired before it is passed on to daughter cells?
    a. G_1 c. S
    b. G_2 d. mitotic
16. Which of the following is not an event of apoptosis?
    a. loss of contact with neighboring cells
    b. blistering of the plasma membrane
    c. increase in the number of mitochondria
    d. fragmentation of the nucleus
17. Rats have 42 chromosomes. How many total chromatids does a rat cell have during G_1 stage? Explain your answer.
    a. 21
    b. 42
    c. 63
    d. 84
    e. 126

18. Rats have 42 chromosomes. How many total chromatids does a rat cell have during G_2 stage? Explain your answer.
    - a. 21
    - b. 42
    - c. 63
    - d. 84
    - e. 126

19. Which of the following is not a feature of cancer cells?
    - a. exhibit contact inhibition
    - b. have enlarged nuclei
    - c. stimulate the formation of new blood vessels
    - d. are capable of traveling through blood and lymph

20. The spindle begins to assemble during
    - a. prophase.
    - b. metaphase.
    - c. anaphase.
    - d. interphase.

21. What is the checkpoint for the completion of DNA synthesis?
    - a. G_1
    - b. S
    - c. G_2
    - d. M

22. Programmed cell death is called
    - a. mitosis.
    - b. meiosis.
    - c. cytokinesis.
    - d. apoptosis.

23. In human beings, mitosis is necessary for
    - a. growth and repair of tissues.
    - b. formation of the gametes.
    - c. maintaining the chromosome number in all body cells.
    - d. the death of unnecessary cells.
    - e. Both a and c are correct.

24. In which phase of mitosis is evidence of cytokinesis present?
    - a. prophase
    - b. metaphase
    - c. anaphase
    - d. telophase
    - e. Both c and d are correct.

25. Which of these is not a behavior that could help prevent cancer?
    - a. maintaining a healthy weight
    - b. eating more dark green leafy vegetables, carrots, and various fruits
    - c. not smoking
    - d. maintaining estrogen levels through hormone replacement therapy
    - e. consuming alcohol only in moderation

26. Cytokinesis in plant cells differs from cytokinesis in animal cells because
    - a. plant cells form a cleavage furrow that constricts the daughter cells.
    - b. in plant cells, many small vesicles form a cell plate that eventually becomes a new cell wall.
    - c. animal cells seldom undergo cytokinesis.
    - d. Both b and c are correct.

Thinking Scientifically

1. The survivors of the atomic bombs that were dropped on Hiroshima and Nagasaki have been the subjects of long-term studies of the effects of ionizing radiation on cancer incidence. The frequencies of different types of cancer in these individuals varied across the decades. In the 1950s, high levels of leukemia and cancers of the lung and thyroid gland were observed. The 1960s and 1970s brought high levels of breast and salivary gland cancers. In the 1980s, rates of colon cancer were especially high. Why do you suppose the rates of different types of cancer varied across time?

2. During prophase of mitosis, the nuclear membrane breaks down. This is undoubtedly a complex, energy-consuming process that has to be carried out again, in reverse, at the end of mitosis. Why do you suppose the nuclear membrane must break down at the beginning of mitosis? What mechanism can you envision for the dismantling of the nuclear membrane?

3. Why is it advantageous that checkpoints evolved during the cell cycle?

Bioethical Issue

Paying for Cancer Treatment

As you have read, new drugs and treatment options are now available for cancers once thought incurable. However, many of these new cures are very expensive. Taxol treatment, for example, can cost in excess of $4,000 per dose. Most of these costs are passed on to the public, largely through higher health insurance rates and higher taxes. Since many of the risk factors for cancer, such as smoking and obesity, are controllable, this has led to an obvious question—should some cancer patients be held responsible for the cost of their treatment?

Many individuals feel that people who develop cancer because of poor lifestyle choices should be held fully responsible for paying for treatment. By doing so, they argue, people have a direct financial incentive to avoid risky behaviors, and are no longer a financial burden on society. But while an individual's poor lifestyle can often be implicated, cancer sometimes strikes otherwise healthy people without any known risk factors. And it can sometimes be difficult to determine whether a person's poor behavior is to blame, even when certain risk factors are present. In such cases, is it really fair to incriminate the patient for their disease when there may be other causes?

Many people also disagree on who should bear the high costs of developing new drugs and treatments. Many, including insurance companies, feel that cancer patients who benefit from new or novel cancer treatments should bear at least some of those costs. Others, however, argue that we as a society owe it to our fellow citizens to fund the research that can often save many innocent lives. There are no easy answers for any of these questions, but as cancer continues to exact a high toll in both human life and financial resources, future generations may soon face some difficult choices.

Essentials of Biology Website

The companion website for *Essentials of Biology* provides a wealth of information organized and integrated by chapter. You will find practice tests, animations, videos, and much more that will complement your learning and understanding of general biology.

http://www.mhhe.com/maderessentials2

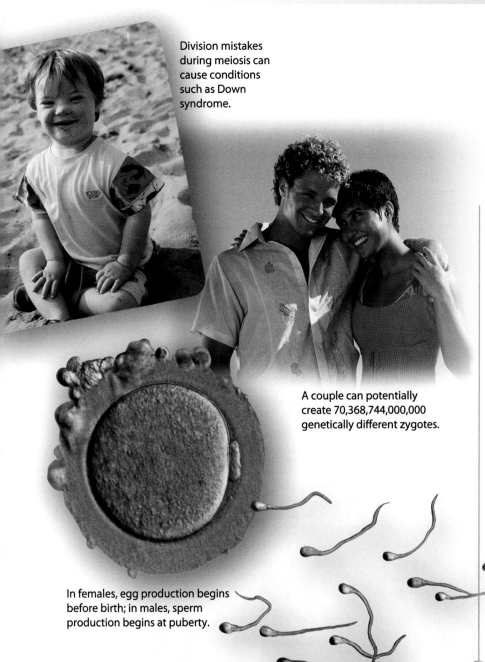

Division mistakes during meiosis can cause conditions such as Down syndrome.

A couple can potentially create 70,368,744,000,000 genetically different zygotes.

In females, egg production begins before birth; in males, sperm production begins at puberty.

9
Sexual Reproduction

Outline

Sexual Reproduction Involves Meiosis

BECAUSE OF A PROCESS called meiosis, two individuals can create offspring that are genetically different from themselves and from each other. This process creates an enormous amount of diversity; in humans, more than 70 trillion different genetic combinations are possible from the mating of just two individuals!

Meiosis is the type of cell division that occurs during sexual reproduction. The process plus maturation result in four cells, called gametes. In humans, the male gametes are sperm, and the female

—Continued next page

Continued—

gametes are eggs. Meiosis that helps produce sperm is called spermatogenesis, and meiosis that helps produce eggs is called oogenesis.

Although spermatogenesis and oogenesis follow the same steps overall, they differ in some ways. One major difference pertains to the age at which the process begins and ends. In males, sperm production does not begin until puberty, but then continues throughout a male's lifetime. In females, the process of producing eggs has started before the female is even born and ends around the age of 50, a time called menopause.

Another difference concerns the number of gametes that can be produced. In males, sperm production is unlimited, whereas females produce only one egg a month. Many more eggs start the process than ever finish it. Several of the birth control methods available to females, such as the pill or "the patch," work by manipulating hormones so that cells called oocytes do not complete meiosis and an egg is never released for fertilization.

In this chapter, you will see how meiosis is involved in the production of gametes. You will also learn how meiosis compares with mitosis, the cell division process discussed in Chapter 8.

9.1 | The Basics of Meiosis

Human beings, like most other animals, practice sexual reproduction in which two parents pass chromosomes to their offspring. But why are the genotypes and phenotypes of siblings different from each other and from either parent? The answer is **meiosis,** a special type of nuclear division that is important in sexual reproduction.

Meiosis serves two major functions: reducing the chromosome number, and also shuffling the chromosomes and genes to produce genetically different gametes, called sperm in males and eggs in females. With **fertilization,** the fusion of sperm and egg, even more shuffling occurs. In the end, each child is unique and has a different combination of chromosomes and genes than either parent. But how does meiosis bring about the distribution of chromosomes to offspring in a way that ensures not only the correct number of chromosomes but also a unique combination of chromosomes and genes? For the answer, let's start by examining the chromosomes.

Homologous Chromosomes

Chromosomes occur in pairs. Both males and females normally have 23 pairs of chromosomes, but one of these pairs is of unequal length in males. The larger chromosome of this pair is the X chromosome, and the smaller is the Y chromosome. In contrast, females have two X chromosomes. The X and Y chromosomes are called the **sex chromosomes** because they contain the genes that determine gender. The other chromosomes, known as **autosomes,** include all the pairs of

Figure 9.1   Homologous chromosomes.

In body cells, the chromosomes occur in pairs called homologous chromosomes. In this micrograph of stained chromosomes from a human cell, the pairs have been numbered. These chromosomes are duplicated, and each one is composed of two sister chromatids.

The 46 chromosomes of a male

chromosomes except the X and Y chromosomes. Twenty-three pairs of chromosomes, or 46 altogether, is called the **diploid (2n) number** of chromosomes in humans. Half this number is the **haploid (n) number** of chromosomes.

The members of a chromosome pair are called **homologous chromosomes,** or **homologues,** because they have the same size, shape, and constriction (location of the centromere) (**Fig 9.1**). When viewed under a microscope, homologous chromosomes have the same characteristic banding pattern upon staining because they contain copies of the same genes arranged in the same order. One homologue of each pair was contributed by each parent. Just as your mother may have long fingers and your father may have short fingers, each homologue may have different versions of a gene. Alternate versions of a gene for a particular trait, (e.g., finger length) are called **alleles.** Alleles occur at the same location on each homologue. Your mother's allele for long fingers will be on one homologue, and your father's allele for short fingers will be on the other homologue.

The Human Life Cycle

The term **life cycle** in sexually reproducing organisms refers to all the reproductive events that occur from one generation to the next. The human life cycle involves two types of nuclear division: mitosis and meiosis (**Fig 9.2**).

During development and after birth, mitosis is involved in the continued growth of the child and the repair of tissues at any time. As a result of mitosis, each somatic (body) cell has the diploid number of chromosomes.

During sexual reproduction, meiosis reduces the chromosome number from the diploid to the haploid number in such a way that the gametes (sperm and egg) have one chromosome derived from each homologous pair of chromosomes. In males, meiosis is a part of **spermatogenesis,** which occurs in the testes and produces sperm. In females, meiosis is a part of **oogenesis,** which occurs in the ovaries and produces eggs. After the sperm and egg join during fertilization, the resulting cell, called the **zygote,** again has homologous pairs of chromosomes. The zygote then undergoes mitosis and differentiation of cells to become a fetus, and eventually a new human being.

Meiosis is important because, if it did not halve the chromosome number, the gametes would contain the same number of chromosomes as the body cells, and the number of chromosomes would double with each new generation. Within a few generations, the cells of sexually reproducing organisms would be nothing but chromosomes! But meiosis, followed by fertilization, keeps the chromosome number constant in each generation.

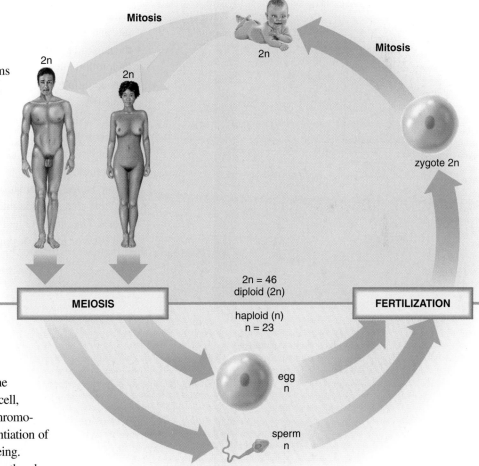

Figure 9.2   Life cycle of humans.

Meiosis in males is a part of sperm production, and meiosis in females is a part of egg production. When a haploid sperm fertilizes a haploid egg, the zygote is diploid. The zygote undergoes mitosis as it eventually develops into a newborn child. Mitosis continues throughout life during growth and repair.

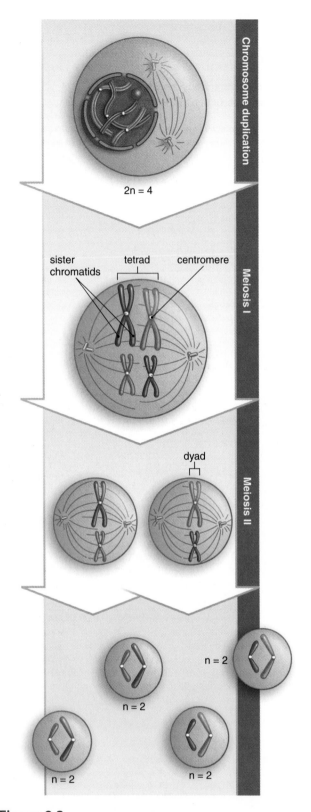

Figure 9.3 **Overview of meiosis.**

Following duplication of chromosomes, the parent cell undergoes two divisions, meiosis I and meiosis II. During meiosis I, homologous chromosomes separate, and during meiosis II, sister chromatids separate. The final daughter cells are haploid. (The blue chromosomes were originally inherited from one parent, and the red chromosomes were originally inherited from the other parent.)

Overview of Meiosis

Meiosis results in four daughter cells because it consists of two divisions, called **meiosis I** and **meiosis II** (**Fig 9.3**). Before meiosis I begins, each chromosome has duplicated and is composed of two sister chromatids. During meiosis I, the homologous chromosomes of each pair come together and line up side-by-side in an event called **synapsis** (**Fig 9.4**). What draws the chromosomes together is still unknown. Synapsis results in a **tetrad,** an association of four chromatids (two homologous chromosomes consisting of two chromatids each). The chromosomes of a tetrad stay in close proximity until they separate. Later in meiosis I, where the homologous chromosomes of each pair separate, one chromosome from each homologous pair goes to each daughter nucleus. No rules restrict which chromosome goes to which daughter nucleus. Therefore, all possible combinations of chromosomes may occur within the gametes.

No duplication of chromosomes is needed between meiosis I and meiosis II. Following meiosis I, the daughter nuclei each have half the number of chromosomes, but the chromosomes are still duplicated (consisting of two chromatids each). The chromosomes are **dyads** because each one is composed of two sister chromatids. During meiosis II, the sister chromatids of each dyad separate and are daughter chromosomes. The resulting four new daughter cells have the haploid number of chromosomes. If the parent cell had four chromosomes, then following meiosis, each daughter cell has two chromosomes. (Remember that counting the centromeres tells you the number of chromosomes in a nucleus.)

Because of meiosis, the gametes have all possible combinations of chromosomes. Notice in Figure 9.3 (bottom) that the daughter cells to the left do not have the same chromosomes as those to the right. Why not? Because the homologous chromosomes of each pair separated during meiosis I. Other chromosome combinations are possible in addition to those depicted. It's possible that the daughter cells could have only red or only blue chromosomes. The daughter cells from meiosis will complete either spermatogenesis and become sperm or oogenesis and become eggs.

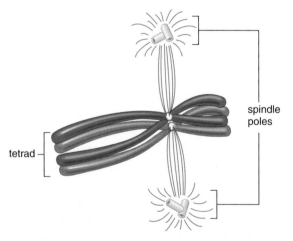

Figure 9.4 **Synapsis.**

During prophase I, homologous chromosomes line up side by side and form a tetrad, consisting of two chromosomes each, containing two chromatids, for a total of four chromatids. The complex is held together by a protein scaffold.

Crossing-Over

When a tetrad forms during synapsis, the nonsister chromatids may exchange genetic material, an event called **crossing-over** (**Fig. 9.5**). A tetrad consists of four chromatids, but only the chromatids belonging to a homologue are sister chromatids, and the ones belonging to the other homologue are **nonsister chromatids.** Crossing-over between the nonsister chromatids shuffles the alleles and serves as the way that meiosis brings about genetic recombination in the gametes.

Recall that the homologous chromosomes carry genes for traits, such as finger length, but that one allele might call for short fingers, for example, while the other allele calls for long fingers. Therefore, when the nonsister chromatids exchange genetic material, the sister chromatids then have a different combination of alleles, and the resulting gametes will be genetically different. So in Figure 9.5, even though two of the gametes will have the same chromosomes, these chromosomes may not have the same combination of alleles as before because of crossing-over. Crossing-over increases the diversity of the gametes and, therefore, of the offspring.

The Importance of Meiosis

Meiosis is important first of all because:

- It assists in keeping the chromosome number constant by producing haploid daughter cells that become the gametes. When a haploid sperm fertilizes a haploid egg, the new individual has the diploid number of chromosomes.
- It introduces genetic variations because (1) crossing-over can result in different types of alleles on the sister chromatids of a homologue, and (2) every possible combination of chromosomes can occur in the daughter cells.

At fertilization, new combinations of chromosomes can occur, and a zygote can have any one of a vast number of combinations of chromosomes. In humans, $(2^{23})^2$ or 70,368,744,000,000 chromosomally different zygotes are possible, even assuming no crossing-over.

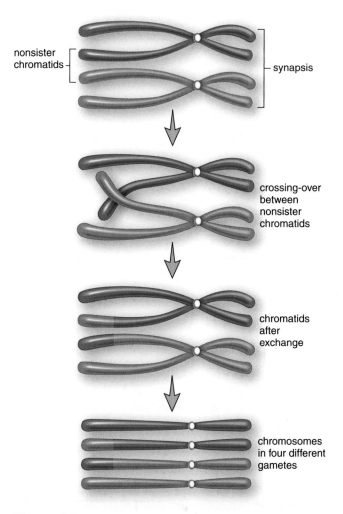

Figure 9.5 Crossing-over.

When homologous chromosomes are in synapsis, the nonsister chromatids exchange genetic material. This illustration shows only one crossover per chromosome pair, but the average is slightly more than two per homologous pair in humans. Following crossing-over, the sister chromatids of a dyad may no longer be identical and instead may have different combinations of alleles.

Check Your Progress

1. List the major differences between meiosis I and meiosis II.
2. Describe what is meant by a tetrad.
3. What are two ways in which meiosis results in gametes that are genetically different?

Answers: 1. During meiosis I, the parent cell is diploid, synapsis and crossing-over occur, and the homologous chromosomes separate. Therefore the daughter cells are haploid. During meiosis II, the parent cell is haploid, the sister chromatids separate, and the daughter cells have the same number of chromosomes as the parent cell.   2. A tetrad contains two duplicated chromosomes, for a total of four chromatids.   3. Crossing-over alters the types of alleles on the chromosomes and the daughter cells have different combinations of the haploid number of chromosomes.

9.2 | The Phases of Meiosis

The same four stages of mitosis—prophase, metaphase, anaphase, and telophase—occur during both meiosis I (**Fig 9.6**) and meiosis II (**Fig 9.7**). However, meiosis I and meiosis II differ from each other and from mitosis in the way chromosomes align during metaphase.

The First Division—Meiosis I

To help you recall the events of meiosis, keep in mind what meiosis accomplishes—namely, the production of gametes that have a reduced chromosome number and are genetically different from each other and from the parent cell. During prophase I, the nuclear envelope fragments and the nucleolus disappears as the spindle appears. As the chromosomes condense, homologues undergo synapsis to produce tetrads. Crossing-over between nonsister chromatids occurs during synapsis and this "shuffles" the alleles on chromosomes.

During metaphase I, the tetrads attach to the spindle and align at the spindle equator, with each homologue facing opposite spindle poles. It does not matter which homologous chromosome faces which pole; therefore, all possible combinations of chromosomes will occur in the gametes. Homologous chromosomes separate during anaphase I. Following re-formation of the nuclear envelopes during telophase and cytokinesis, the daughter nuclei are haploid: Each daughter cell contains only one chromosome from each homologous pair. The chromosomes are

Meiosis I: Homologous chromosomes separate

Prophase I

nuclear envelope fragment

crossing-over

spindle forming

sister chromatids   2n = 4   tetrad

Metaphase I

chromosome attached to a spindle fiber

crossing over

centromere

Anaphase I

chromosomes still duplicated

Telophase I

cleavage furrow

n = 2

Prophase I
Tetrads form, and crossing-over occurs as chromosomes condense; the nuclear envelope fragments.

Metaphase I
Tetrads align at the spindle equator. Either homologue can face either pole.

Anaphase I
Homologues separate, and dyads move to poles.

Telophase I
Daughter nuclei are haploid, having received one duplicated chromosome from each homologous pair.

Figure 9.6   Meiosis I.

As a result of meiosis I, the daughter nuclei are haploid and are genetically dissimilar. They contain a different combination of chromosomes and may have a different combination of alleles on the sister chromatids. The number of centromeres in a cell equals the number of chromosomes.

now dyads, and each still has two sister chromatids. No replication of DNA occurs between meiosis I and II, a period of time called **interkinesis.**

The Second Division—Meiosis II

When you think about it, the events of meiosis II are the same as those for mitosis, except the cells are haploid. At the beginning of prophase II, a spindle appears while the nuclear envelope fragments and the nucleolus disappears. Dyads are present, and each attaches to the spindle. During metaphase II, the dyads are lined up at the spindle equator, with sister chromatids facing opposite spindle poles. During anaphase II, the sister chromatids of each dyad separate and move toward the poles. Each pole receives the same number and kinds of chromosomes. In telophase II, the spindle disappears as nuclear envelopes form.

During cytokinesis, the plasma membrane pinches off to form two complete cells, each of which has the haploid, or n, number of chromosomes. The gametes are genetically dissimilar because they can contain different combinations of chromosomes and because crossing-over changed which alleles are together on a chromosome. Because both cells from meiosis I undergo meiosis II, four daughter cells are produced from the original diploid parent cell. Each daughter cell contains a unique combination of genes.

Check Your Progress

1. Describe the significance of tetrad formation during meiosis I.
2. Contrast the chromosome composition and number in a parent cell before meiosis begins with the chromosome composition and number in the daughter cells when meiosis is complete.

Answers: 1. Tetrad formation allows crossing-over to occur and prepares homologues for separation during anaphase I. 2. The parent cell is diploid and each chromosome has two chromatids; a daughter cell is haploid and each chromosome has one chromatid.

Meiosis II: Sister chromatids separate

Prophase II	Metaphase II	Anaphase II	Telophase II

sister chromatids separate

haploid daughter cells forming

n = 2

Prophase II
Chromosomes condense, and the nuclear envelope fragments.

Metaphase II
The dyads align at the spindle equator.

Anaphase II
Sister chromatids separate, becoming daughter chromosomes that move to the poles.

Telophase II
Four haploid daughter cells are genetically different from each other and from the parent cell.

Figure 9.7 Meiosis II.

During anaphase II, daughter chromosomes consisting of one chromatid each move to the poles. Following meiosis II, there are four haploid daughter cells. Comparing the number of centromeres in the daughter cells with the number in the parent cell at the start of meiosis I verifies that the daughter cells are haploid. Notice that the daughter cells are genetically different from each other and from the original parent cell because they have a different combination of chromosomes and crossing-over occurred during meiosis I.

Table 9.1 | Meiosis I Compared to Mitosis

Meiosis I	Mitosis
Prophase I	*Prophase*
Pairing of homologous chromosomes; crossing-over	No pairing of chromosomes; no crossing-over
Metaphase I	*Metaphase*
Tetrads at spindle equator	Dyads at spindle equator
Anaphase I	*Anaphase*
Homologues of each tetrad separate, and dyads move to poles	Sister chromatids separate, becoming daughter chromosomes that move to the poles
Telophase I	*Telophase*
Two haploid daughter cells, not identical to parent cell	Two diploid daughter cells, identical to the parent cell

Table 9.2 | Meiosis II Compared to Mitosis

Meiosis II	Mitosis
Prophase II	*Prophase*
No pairing of chromosomes	No pairing of chromosomes
Metaphase II	*Metaphase*
Haploid number of dyads at spindle equator	Diploid number of dyads at spindle equator
Anaphase II	*Anaphase*
Sister chromatids separate, becoming daughter chromosomes that move to the poles	Sister chromatids separate, becoming daughter chromosomes that move to the poles
Telophase II	*Telophase*
Four haploid daughter cells, different from each other genetically and from the parent cell	Two diploid daughter cells, genetically identical to the parent cell

9.3 | Meiosis Compared to Mitosis

Figure 9.8 compares meiosis to mitosis. Notice that:

- Meiosis requires two nuclear divisions, but mitosis requires only one nuclear division.
- Meiosis produces four daughter nuclei, and there are four daughter cells following cytokinesis. Mitosis followed by cytokinesis results in two daughter cells.
- Following meiosis, the four daughter cells are haploid and have half the chromosome number as the parent cell. Following mitosis, the daughter cells have the same chromosome number as the parent cell.
- Following meiosis, the daughter cells are genetically dissimilar to each other and to the parent cell. Following mitosis, the daughter cells are genetically identical to each other and to the parent cell.

The specific differences between these nuclear divisions can be categorized according to process and occurrence.

Process

To summarize the processes, **Table 9.1** compares meiosis I to mitosis and **Table 9.2** compares meiosis II to mitosis.

Meiosis I Compared to Mitosis

The following events distinguish meiosis I from mitosis:

- During prophase I of meiosis, synapsis occurs. During synapsis, tetrads form and crossing-over occurs. These events do not occur during mitosis.
- During metaphase I of meiosis, tetrads align at the spindle equator, with homologous chromosomes facing opposite spindle poles. The paired chromosomes have a total of four chromatids each. During metaphase in mitosis, dyads align separately at the spindle equator.
- During anaphase I of meiosis, the homologous chromosomes of each tetrad separate, and dyads (with centromeres intact) move to opposite poles. Sister chromatids do not separate during anaphase I. During anaphase of mitosis, sister chromatids separate, becoming daughter chromosomes that move to opposite poles.

Meiosis II Compared to Mitosis

The events of meiosis II are just like those of mitosis except that in meiosis II, the cells have the haploid number of chromosomes.

Occurrence

Meiosis occurs only at certain times in the life cycle of sexually reproducing organisms, and only in specialized tissues. In humans, meiosis occurs only in the testes and ovaries, where it is involved in the production of gametes. The function of meiosis is to provide gamete variation and to keep the chromosome number constant generation after generation. With fertilization, the full chromosome number is restored. Because unlike gametes fuse, fertilization introduces a great genetic diversity into the offspring.

Mitosis is much more common because it occurs in all tissues during embryonic development and also during growth and repair. The function of mitosis is to keep the chromosome number constant in all the cells of the body so that every cell has the same genetic material. Consider that reproductive cloning results in an individual with the same genes as the parent because a single diploid nucleus gives genes to the new individual (see Fig. 12.2).

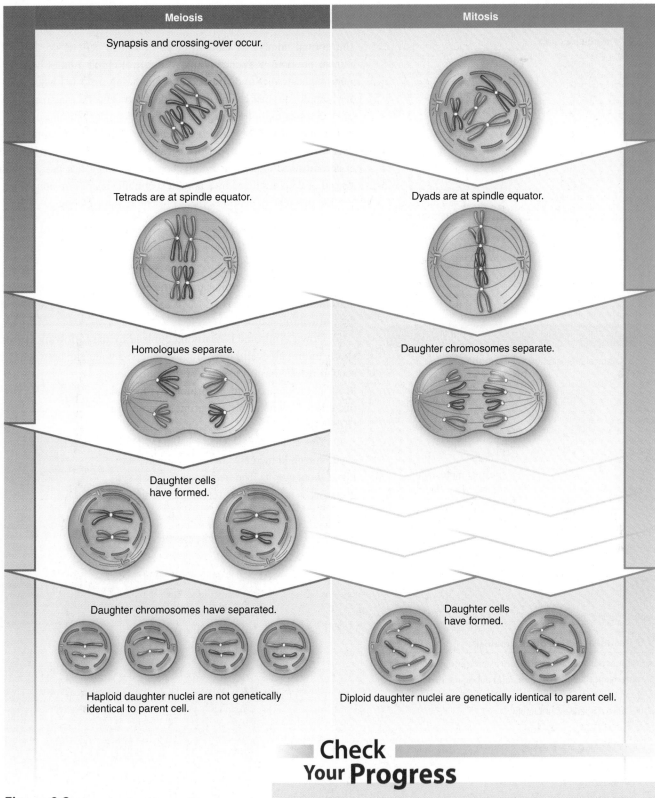

Meiosis	Mitosis
Synapsis and crossing-over occur.	
Tetrads are at spindle equator.	Dyads are at spindle equator.
Homologues separate.	Daughter chromosomes separate.
Daughter cells have formed.	
Daughter chromosomes have separated.	Daughter cells have formed.
Haploid daughter nuclei are not genetically identical to parent cell.	Diploid daughter nuclei are genetically identical to parent cell.

Figure 9.8 **Meiosis compared to mitosis.**

Check Your Progress

1. Compare the attachment of chromosomes to the spindle during metaphase I and metaphase II (or mitosis).

Answer: 1. During metaphase I, the pairs of homologues are attached to the spindle with either homologue facing either spindle pole. During metaphase II (and mitosis), dyads are attached to the spindle with sister chromatids facing opposite poles.

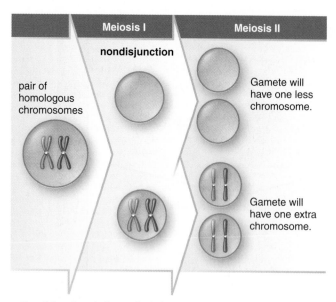

a. Nondisjunction during meiosis I

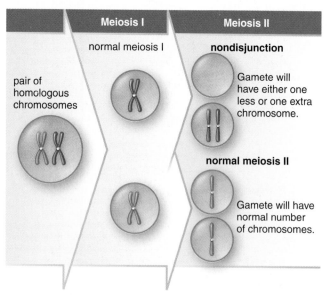

b. Nondisjunction during meiosis II

Figure 9.9   Nondisjunction during meiosis.

Because of nondisjunction, gametes either lack a chromosome or have an extra chromosome. Nondisjunction can occur **(a)** during meiosis I if homologous chromosomes fail to separate, and **(b)** during meiosis II if the sister chromatids fail to separate completely.

9.4 | Abnormal Chromosome Inheritance

The normal number of chromosomes in human cells is 46, but occasionally humans are born with an abnormal number of chromosomes because improper meiosis occurred. **Nondisjunction** occurs during meiosis I when both members of a homologous pair go into the same daughter cell, or during meiosis II when the sister chromatids fail to separate and both daughter chromosomes go into the same gamete (**Fig 9.9**). If an egg that ends up with 24 chromosomes instead of 23 is fertilized with a normal sperm, the result is a *trisomy*, so called because one type of chromosome is present in three copies. If an egg that has 22 chromosomes instead of 23 is fertilized by a normal sperm, the result is a *monosomy*, so called because one type of chromosome is present in a single copy.

Down Syndrome

Down syndrome is *trisomy 21*, in which an individual has three copies of chromosome 21. In most instances, the egg contained two copies of this chromosome instead of one. (In 23% of the cases studied, however, the sperm had the extra chromosome 21.)

Down syndrome is easily recognized by the following characteristics: short stature; an eyelid fold; stubby fingers; a wide gap between the first and second toes; large, fissured tongue; round head; palm crease; and unfortunately, mental disabilities, which can sometimes be severe (**Fig 9.10**).

The chance of a woman having a Down syndrome child increases rapidly with age, starting at about age 40. The frequency of Down syndrome is 1 in 800 births for mothers under 40

Figure 9.10   Down syndrome.

Down syndrome occurs when the egg or the sperm has an extra chromosome 21 due to nondisjunction in either meiosis I or meiosis II. Characteristics include a wide, rounded face and narrow, slanting eyelids. Mental disabilities are often present, but may vary greatly.

years of age and 1 in 80 for mothers over 40 years of age. However, most Down syndrome babies are born to women younger than age 40, because this is the age group having the most babies.

Abnormal Sex Chromosome Number

Nondisjunction during oogenesis or spermatogenesis can result in gametes that have too few or too many X or Y chromosomes. Figure 9.9 can be used to illustrate nondisjunction of the sex chromosomes during oogenesis if we assume that the chromosomes shown represent X chromosomes.

Just as an extra copy of chromosome 21 causes Down syndrome, having additional or missing X or Y chromosomes causes certain syndromes. Newborns with an abnormal sex chromosome number are more likely to survive than those with an abnormal autosome number, and the explanation is quite surprising. Normal females, like normal males, have only one functioning X chromosome. The other X chromosome (or additional X chromosomes) becomes an inactive mass called a **Barr body** (after the person who discovered it).

In humans, the presence of a Y chromosome, not the number of X chromosomes, almost always determines maleness. The *SRY* (*sex-determining region Y*) gene located on the short arm of the Y chromosome produces a hormone called testis-determining factor, which plays a critical role in the development of male genitals. No matter how many X chromosomes are involved, an individual with a Y chromosome is a male, assuming a functional *SRY* is on the Y chromosome. Individuals lacking a functional *SRY* on their Y chromosome have Swyer syndrome, also known as an "XY female."

A person with **Turner syndrome** (45, XO) is a female. The number indicates the total number of chromosomes the individual has, and the O signifies the absence of a second sex chromosome. Turner syndrome females are short, with a broad chest and webbed neck. The ovaries, oviducts, and uterus are very small and underdeveloped. Turner females do not undergo puberty or menstruate, and their breasts do not develop (**Fig 9.11**a). However, some have given birth following in vitro fertilization using donor eggs. These women usually have normal intelligence and can lead fairly normal lives if they receive hormone supplements.

A person with **Klinefelter syndrome** (47, XXY) is a male. A Klinefelter male has two or more X chromosomes in addition to a Y chromosome. The extra X chromosomes become Barr bodies.

In Klinefelter males, the testes and prostate gland are underdeveloped. There is no facial hair, but some breast development may occur (Fig. 9.11b). Affected individuals generally have large hands and feet and very long arms and legs. They are usually slow to learn but not mentally handicapped unless they inherit more than two X chromosomes.

As with Turner syndrome, it is best for parents to know as soon as possible that their child has Klinefelter syndrome because much can be done to help the child lead a normal life.

Check Your **Progress**

1. Describe how a zygote could receive an abnormal chromosome number.
2. Contrast monosomy with trisomy.

Answers: 1. If nondisjunction occurs during meiosis I or II, a gamete, and therefore a zygote, could have an abnormal chromosome number.   2. Monosomy results when an individual is missing one chromosome. Trisomy results when an individual has one extra chromosome.

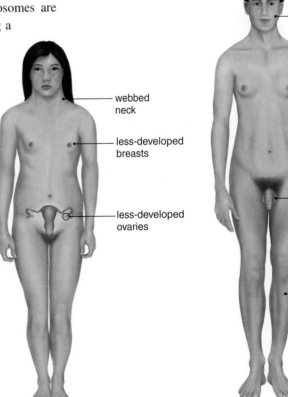

webbed neck

less-developed breasts

less-developed ovaries

no facial hair

some breast development

very long arms

less-developed testes

very long legs

a. A female with Turner (XO) syndrome

b. A male with Klinefelter (XXY) syndrome

Figure 9.11   Abnormal sex chromosome number.

a. A female with Turner syndrome (XO) has a short thick neck, short stature, and lack of breast development. **b.** A male with Klinefelter syndrome (XXY) has immature sex organs and some development of the breasts.

The Chapter in Review

Summary

9.1 | The Basics of Meiosis

Meiosis is a special type of nuclear division that reduces the chromosome number and shuffles the genes between chromosomes. It produces haploid daughter cells with a unique combination of chromosomes and alleles.

Homologous Chromosomes

Homologous chromosomes are the same size with the same centromere position, and contain the same genes. The alleles of a gene may differ between homologues. Diploid cells in humans have 22 homologous pairs of autosomes and 1 pair of sex chromosomes for a total of 46 chromosomes. Males are XY and females are XX. Haploid cells have 22 autosomes and one sex chromosome, either an X or a Y, for a total of 23 chromosomes.

The Human Life Cycle

The human life cycle has two types of nuclear division:
- **Mitosis** ensures that every body cell has 23 pairs of chromosomes. It occurs during growth and repair.
- **Meiosis** occurs during formation of gametes. It ensures that the gametes are haploid and have 23 chromosomes, one from each of the pairs of chromosomes.

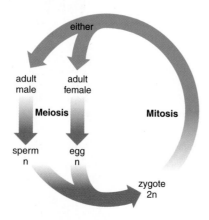

Overview of Meiosis

Meiosis has two nuclear divisions called meiosis I and meiosis II. Therefore, meiosis results in four daughter cells.
- **Meiosis I:** Synapsis and formation of tetrads leads to separation of homologous chromosomes. Because no restrictions govern which member goes to which pole, the gametes will contain all possible combinations of chromosomes.
- **Meiosis II:** The chromosomes are still duplicated at the beginning of meiosis II. Sister chromatids later separate, forming daughter chromosomes. The daughter cells are haploid. If the parent cell has four chromosomes, each of the daughter cells has two chromosomes. All possible combinations of chromosomes occur among the daughter cells.

Crossing-Over Crossing-over between nonsister chromatids during meiosis I "shuffles" the alleles on the chromosomes so that each sister chromatid of a homologue has a different mix of alleles.

The Importance of Meiosis Meiosis is important because the chromosome number stays constant between the generations of individuals and because the daughter cells, and therefore the

gametes, are genetically different. When fertilization occurs, the offspring are genetically unique.

9.2 | The Phases of Meiosis

The First Division—Meiosis I

- **Prophase I:** Tetrads form, and crossing-over occurs as chromosomes condense; the nuclear envelope fragments.
- **Metaphase I:** Tetrads align at the spindle equator. Either homologue can face either pole.
- **Anaphase I:** Homologues of each tetrad separate, and dyads move to the poles.
- **Telophase I:** Daughter nuclei are haploid, having received one duplicated chromosome from each homologous pair.

The Second Division—Meiosis II

- **Prophase II:** Chromosomes condense, and the nuclear envelope fragments.
- **Metaphase II:** The haploid number of dyads align at the spindle equator.
- **Anaphase II:** Sister chromatids separate, becoming daughter chromosomes that move to the poles.
- **Telophase II:** Four haploid daughter cells are genetically different from each other and from the parent cell.

9.3 | Meiosis Compared to Mitosis

The following diagrams show the differences between meiosis and mitosis. Duplication of chromosomes occurs before each begins.

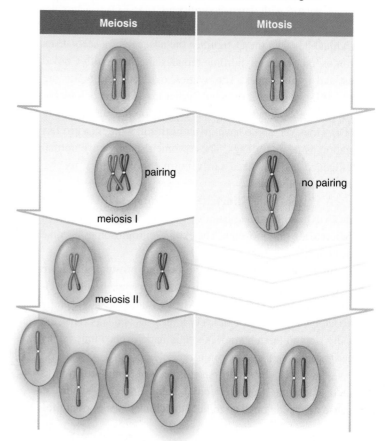

Meiosis

- In humans, meiosis occurs in the testes and ovaries where it produces the gametes.
- Because meiosis has two nuclear divisions, it produces four daughter cells. The daughter cells are haploid and genetically different from each other and from the parent cell.

Mitosis

- Mitosis occurs in the body cells and accounts for growth and repair. Because mitosis has one nuclear division, it produces two daughter cells. The daughter cells are diploid and are genetically identical to each other and to the parent cell.

Meiosis I Compared to Mitosis

Meiosis I differs from mitosis in the following ways:
- During prophase I, tetrads form and crossing-over occurs. No such events occur in mitosis.
- During metaphase I, tetrads are at the spindle equator. During mitosis, dyads are at the spindle equator.
- During anaphase I, homologous chromosomes separate. During mitosis, sister chromatids separate.

Meiosis II Compared to Mitosis

- The events of meiosis II are the same as those of mitosis except that the cells undergoing meiosis II are haploid, whereas those undergoing mitosis are diploid.

9.4 | Abnormal Chromosome Inheritance

Nondisjunction accounts for the inheritance of an abnormal chromosome number. Nondisjunction can occur during meiosis I if homologous chromosomes fail to separate, and both go into one daughter cell and the other daughter cell receives neither. It can also occur during meiosis II if chromatids fail to separate, and both go into one daughter cell and the other daughter cell receives neither.

Down Syndrome

In Down syndrome, an example of an autosomal syndrome, the individual inherits three copies of chromosome 21.

Abnormal Sex Chromosome Number

Examples of syndromes caused by inheritance of abnormal sex chromosomes are Turner syndrome (XO) and Klinefelter syndrome (XXY).
- **Turner syndrome:** An XO individual inherits only one X chromosome. This individual survives because even in an XX individual, one X becomes a Barr body.
- **Klinefelter syndrome:** An XXY individual inherits two X (or more) chromosomes and also a Y chromosome. This individual survives because one (or more) X chromosome becomes a Barr body.

Key Terms

allele   131
autosome   130
Barr body   139
crossing-over   133
diploid (2n) number   131
Down syndrome   138
dyad   132
fertilization   130
haploid (n) number   131

homologous chromosome   131
homologue   131
interkinesis   135
Klinefelter syndrome   139
life cycle   131
meiosis   130
meiosis I   132
meiosis II   132
nondisjunction   138

nonsister chromatid   133
oogenesis   131
sex chromosome   130
spermatogenesis   131

synapsis   132
tetrad   132
Turner syndrome   139
zygote   131

Testing Yourself

Choose the best answer for each question.

1. A human cell contains _____ pair(s) of sex chromosomes.
   - a. 1
   - b. 2
   - c. 22
   - d. 23

2. Mitosis _____ chromosome number, whereas meiosis _____ the chromosome number of the daughter cells.
   - a. maintains, increases
   - b. increases, maintains
   - c. increases, decreases
   - d. maintains, decreases

For questions 3–6, match the items to those in the key. Answers can be used more than once.

Key:
- a. haploid number
- b. three times the haploid number
- c. diploid number
- d. two times the diploid number

3. How many chromatids are in a parent cell entering meiosis?
4. How many chromosomes are in each daughter cell following meiosis I?
5. How many chromatids are in each daughter cell following meiosis I?
6. How many chromosomes are in a parent cell entering meiosis?
7. Which of the following best describes the attachment of chromosomes to the spindle apparatus during meiosis I?
   - a. The homologous pair lines up together, with homologues facing opposite spindle pole.
   - b. The homologous pair lines up together, with homologues facing the same spindle pole.
   - c. Homologous pairs do not line up together, but homologues face opposite spindle poles.
   - d. Homologous pairs do not line up together, with homologues facing the same spindle pole.

For questions 8–15, match the items to those in the key.

Key:
- a. prophase I
- b. metaphase I
- c. anaphase I
- d. telophase I
- e. prophase II
- f. metaphase II
- g. anaphase II
- h. telophase II

8. A cleavage furrow forms, resulting in haploid nuclei. Each chromosome contains two chromatids.
9. Tetrads form, and crossing-over occurs.
10. Dyads align at the spindle equator.
11. Four haploid daughter cells are created.
12. Homologous chromosomes separate and move to opposite poles.
13. Sister chromatids separate.

14. Tetrads align on the spindle equator.

15. Chromosomes in haploid nuclei condense.

16. At each letter in the following diagram, indicate how mitosis differs from meiosis.

Meiosis	Mitosis
a.	
b.	
c.	
d.	

17. The condition that results when an egg carrying 23 chromosomes unites with a sperm carrying 22 chromosomes is called
    a. monosomy.   c. trisomy.
    b. isomy.   d. tetrasomy.

18. An individual with Turner syndrome is conceived when a normal gamete unites with
    a. an egg that was produced by nondisjunction during oogenesis.
    b. a sperm that was produced by nondisjunction during spermatogenesis.
    c. Either a or b is correct.
    d. Neither a nor b is correct.

19. An individual with triple X karyotype (47, XXX) would have how many Barr bodies?
    a. 0   c. 2
    b. 1   d. 3

20. Crossing-over
    a. involves recombination between nonhomologous chromosomes.
    b. involves recombination between sister chromatids.
    c. occurs only between members of a tetrad.
    d. occurs between adjacent tetrads.

21. The separation of sister chromatids of a dyad in anaphase II results in
    a. dyads.
    b. daughter chromosomes.
    c. tetrads.
    d. sister chromatids.

Thinking Scientifically

1. Although most men with Klinefelter syndrome are infertile, some Klinefelter individuals are able to father children. It was found that most fertile individuals with Klinefelter syndrome exhibit mosaicism, in which some cells are normal (46, XY), whereas others contain the extra chromosome (47, XXY). How might this mosaicism come about?

2. In the nineteenth century, physicians noticed that people with Down syndrome were often the youngest children in large families. Some physicians suggested that the disorder is due to "maternal reproductive exhaustion." What might be a reasonable explanation for the relationship between advanced maternal age and incidence of Down syndrome?

 3 As you can reason, meiosis creates a very large amount of geneticvariability within a population. Why might this be advantageous?

Bioethical Issue

The Risks and Benefits of Delaying Childbirth

Many women in today's society are delaying childbirth until their thirties and forties. Between 1991 and 2001, the birthrate among women aged 35 to 39 increased over 30%, while the birthrate among women aged 40 to 44 jumped by almost 70%. Fewer social stigmas, better prenatal care, and improved medical technologies have all enabled this trend. However, there are many who feel that this should be discouraged due to the perceived costs to society.

The decision to delay childbirth does carry risks. The risk of many disorders associated with nondisjunction, such as Down syndrome, increase greatly with age, rising from nearly 1 in 800 at age 30 to 1 in 80 by age 40. The risk of health complications to the mother also increase greatly with age, necessitating more thorough prenatal care. Many of these costs are handed down in the form of higher taxes and health insurance premiums.

Despite the risks, there are some benefits to delaying childbirth. Women over age 35 are usually more affluent, do not often divorce, and are less likely to give birth out of wedlock, lessening the need for social welfare programs. Furthermore, women at more advanced stages of their careers are often able to devote more time to their children than younger women who are just embarking on their careers. Advocates point out that the children of older parents are usually more well cared for. Thus, any extra health-care costs are offset by savings on social programs.

Should we as a society discourage delayed childbirth because of the high costs incurred? Or do the benefits of advanced maternal age make up for the costs incurred?

Essentials of Biology Website

The companion website for *Essentials of Biology* provides a wealth of information organized and integrated by chapter. You will find practice tests, animations, videos, and much more that will complement your learning and understanding of general biology.

http://www.mhhe.com/maderessentials2

Sex chromosomes contain genes for traits unrelated to gender.

Having six fingers is dominant; having five fingers is recessive.

Environmental factors can influence how your genes are expressed.

10

Patterns of Inheritance

Outline

The Environment Influences Our Genetic Traits

JUST HOW MUCH do genes control the development of particular genetic diseases? In cases such as sickle cell disease and cystic fibrosis, the genes control whether a disease develops. Therefore, the probable occurrence of the trait is predictable. In other cases, the influence of the genes alone does not forecast a particular trait. Rather, environmental factors influence the activity of genes and whether certain specific traits develop. Determining the pattern of inheritance for traits that are influenced by the environment is a difficult task.

—Continued next page

Continued—

Tests are available for some alleles associated with genetic diseases. For example, a woman can be tested to see if she has the *BRCA-1* and *BRCA-2* alleles, both of which are associated with breast and ovarian cancer. However, the allele alone is not sufficient to cause the development of cancer—environmental factors are also involved. Therefore, even if a woman tests positive for these alleles, she may not develop cancer; it requires additional undetermined environmental factors for cancer to develop. Unfortunately, it is impossible to predict for sure whether or not cancer will occur. Still, some women have resorted to extraordinary methods, such as radical mastectomy to remove all breast tissue, to prevent the possibility of cancer.

In this chapter, you will learn about the basic principles of inheritance. Stress will be placed on the rules of genetics that allow us to predict the chances of a trait being passed on from one generation to the next. However, we have to bear in mind that sometimes environmental influences affect the passage of traits.

10.1 | Mendel's Laws

Today, most people know that DNA is the genetic material, and they may have heard that scientists have determined the DNA base sequence of human chromosomes. In contrast, they may not know about Gregor Mendel, an Austrian monk who developed certain laws of heredity after doing crosses between garden pea plants in 1860 (**Fig. 10.1**). Gregor Mendel investigated genetics at the organismal level, and this is still the level that intrigues most of us on a daily basis. We observe, for example, that facial and other features run in families, and we would like some way of explaining this observation. And so, it is appropriate to begin our study of genetics at the organismal level and to learn Mendel's laws of heredity.

When Mendel began his work, most plant and animal breeders acknowledged that both sexes contributed equally to a new individual. However, they were unable to account for the presence of definite variations (differences) among the members of a family, generation after generation. Mendel's model of heredity does account for such variations. Therefore, Mendel's model is compatible with the theory of evolution, which states that various combinations of traits are tested by the environment, and those combinations that lead to reproductive success are the ones that are passed on.

a.

Figure 10.1 **Mendel working in his garden.**

a. Mendel grew and tended the garden peas, *Pisum sativum,* he used for his experiments. **b.** Mendel selected these seven traits for study. Before he did any crosses, he made sure the parents bred true. All the offspring either had the dominant trait or the recessive trait.

		Trait	
		Stem length	Pod shape
Characteristics	Dominant	Tall	Inflated
	Recessive	Short	Constricted

b.

Mendel's Experimental Procedure

Mendel's parents were farmers, so he no doubt acquired the practical experience he needed to grow pea plants during childhood. Mendel was also a mathematician; most likely, his background in mathematics prompted him to use a statistical basis for his breeding experiments. He prepared for his experiments carefully and conducted preliminary studies with various animals and plants. He then chose to work with the garden pea, *Pisum sativum*.

The garden pea was a good choice. The plants were easy to cultivate and had a short generation time. And although peas normally self-pollinate (pollen goes only to the same flower), they could be cross-pollinated by hand. Many varieties of peas were available, and Mendel chose 22 of them for his experiments. When these varieties self-pollinated, they were *true-breeding*—meaning that the offspring were like the parent plants and like each other. In contrast to his predecessors, Mendel studied the inheritance of relatively simple and easily detected traits, such as seed shape, seed color, and flower color, and he observed no intermediate characteristics among the offspring. **Figure 10.2** shows Mendel's procedure.

As Mendel followed the inheritance of individual traits, he kept careful records. He used his understanding of the mathematical laws of probability to interpret his results and to arrive at a theory that has since been supported by innumerable experiments. This theory is the *particulate theory of inheritance*, so-called because it is based on the existence of minute particles we now call genes. Inheritance involves the reshuffling of the same genes from generation to generation.

Figure 10.2 Garden pea anatomy and traits.

a. In the garden pea, pollen grains produced in the anther contain sperm, and ovules in the ovary contain eggs. **b.** When Mendel performed crosses, he brushed pollen from one type pea plant onto the stigma of another type pea plant. After sperm fertilized the eggs, the ovules developed into seeds (peas). For a cross involving a parent with round yellow seeds and a parent with wrinkled yellow seeds, he observed and counted a total of 7,324 peas.

1 Cut away anthers.

2 Brush on pollen from another plant.

stigma
anther
ovary

Pollen grains containing sperm are produced in the anther. When pollen grains are brushed onto the stigma, sperm fertilizes eggs in the ovary. Fertilized eggs are located in ovules, which develop into seeds.

a. Flower structure

3 The results of cross from a parent that produces round, yellow seeds × parent that produces wrinkled yellow seeds.

b. Cross pollination

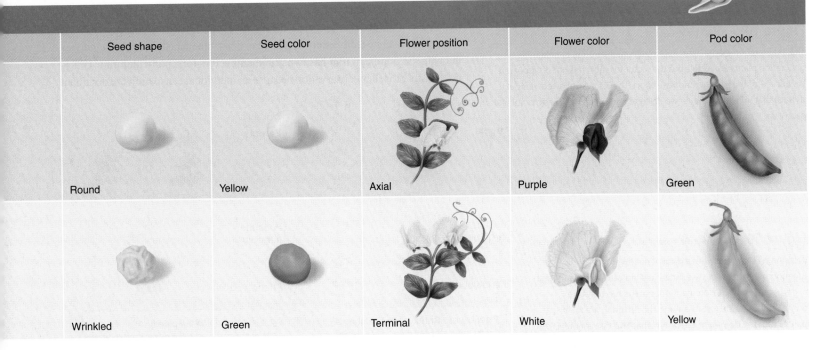

Seed shape	Seed color	Flower position	Flower color	Pod color
Round	Yellow	Axial	Purple	Green
Wrinkled	Green	Terminal	White	Yellow

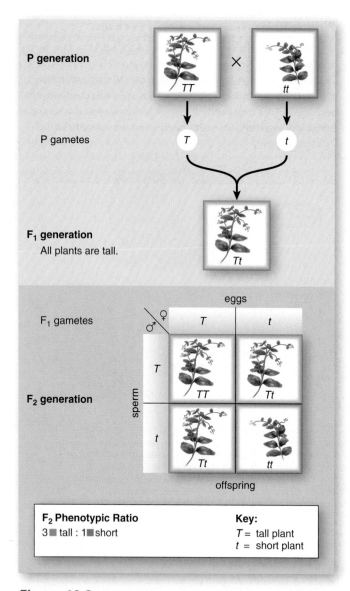

Figure 10.3 **One-trait cross.**

The P generation plants differ in one regard—length of the stem. The F₁ generation plants are all tall, but the factor for short has not disappeared because ¼ of the F₂ generation are short. The 3:1 ratio allowed Mendel to deduce that individuals have two discrete and separate genetic factors for each trait.

One-Trait Inheritance

After ensuring that his pea plants were true-breeding—for example, that his tall plants always had tall offspring and his short plants always had short offspring—Mendel was ready to perform a cross between these two strains. Mendel called the original parents the *P generation*, the first-generation offspring the *F₁* (for filial) *generation*, and the second-generation offspring the *F₂ generation*. The diagram in **Figure 10.3** representing Mendel's F₁ cross is called a Punnett square. In a **Punnett square,** all possible types of sperm are lined up vertically, and all possible types of eggs are lined up horizontally, or vice versa, so that every possible combination of gametes the offspring may inherit occurs within the square.

As Figure 10.3 shows, when Mendel crossed tall pea plants with short pea plants, all the F₁ offspring resembled the tall parent. Did this mean that the other characteristic, shortness, had disappeared permanently? No, because when Mendel allowed the F₁ plants to self-pollinate, ¾ of the F₂ generation were tall and ¼ were short, a 3:1 ratio. Therefore, the F₁ plants had been able to pass on a factor for shortness—it didn't just disappear. Perhaps the F₁ plants were tall because tallness was dominant to shortness?

Mendel's mathematical approach led him to interpret his results differently than previous breeders had done. He reasoned that a 3:1 ratio among the F₂ offspring was possible only if (1) the F₁ parents contained two separate copies of each hereditary factor, one dominant and the other recessive; (2) the factors separated when the gametes were formed, and each gamete carried only one copy of each factor; and (3) random fusion of all possible gametes occurred upon fertilization. Only in this way would shortness reoccur in the F₂ generation.

One-Trait Testcross

Mendel's experimental use of simple dominant and recessive characteristics allowed him to test his interpretation of his crosses. To see if the F₁ carries a recessive factor, Mendel crossed his F₁ generation tall plants with true-breeding, short plants. He reasoned that half the offspring would be tall and half would be short (**Fig. 10.4***a*). And, indeed, those were the results he obtained; therefore, his hypothesis that factors segregate when gametes are formed was supported.

Today, a one-trait **testcross** is used to determine whether or not an individual with the dominant trait has two dominant factors for a particular trait. This is not possible to tell by observation because an individual can exhibit the dominant

Figure 10.4 **One-trait testcross.**

Crossing an individual with a dominant appearance (phenotype) with a recessive individual indicates the genotype. **a.** If a parent with the dominant phenotype has only one dominant factor, the results among the offspring are 1:1. **b.** If a parent with the dominant phenotype has two dominant factors, all offspring have the dominant phenotype.

a.

b.

appearance while having only one dominant factor. Figure 10.4*b* shows that if the individual has two dominant factors, all the offspring will be tall.

After doing one-trait crosses, Mendel arrived at his first law of inheritance—the **law of segregation,** which is a cornerstone of his particulate theory of inheritance. The law of segregation states the following:

- Each individual has two factors for each trait.
- The factors segregate (separate) during the formation of the gametes.
- Each gamete contains only one factor from each pair of factors.
- Fertilization gives each new individual two factors for each trait.

The Modern Genetics View

In the early twentieth century, scientists noted the parallel behavior of Mendel's particulate factors and chromosomes and proposed the *chromosomal theory of inheritance*, which states that chromosomes are carriers of genetic information. Today, we recognize that traits are controlled by alleles (alternate forms of a gene) that occur on the chromosomes at particular locations called the **gene locus.** The **dominant allele** is so named because of its ability to mask the expression of the other allele, called the **recessive allele.** The dominant allele is identified by an uppercase (capital) letter, and the recessive allele by the same but lowercase (small) letter. In humans, for example, alleles for finger length might be *S* for short fingers and *s* for long fingers. While the alleles on one homologue can be the alternates of the alleles on the other homologue, the sister chromatids have the same types of alleles (**Fig. 10.5**). Excluding environmental influences, an individual's traits are determined by the alleles that he or she inherits.

As you learned in Chapter 9, meiosis is the type of cell division that reduces the chromosome number. During meiosis I, the homologues separate. Therefore, the process of meiosis explains Mendel's law of segregation and why only one allele for each trait is in a gamete (Fig. 10.8). When fertilization occurs, the resulting offspring again have two alleles for each trait, one from each parent.

Genotype Versus Phenotype   Different combinations of alleles for a trait can give an organism the same outward appearance. For instance, *SS* and *Ss* individuals both have short fingers. For this reason, it is necessary to distinguish between the alleles present in an organism and the appearance of that organism.

The word **genotype** refers to the alleles an individual receives at fertilization. Genotype may be indicated by letters or by short, descriptive phrases. When an organism has two identical alleles, it is termed **homozygous.** A person that is homozygous dominant for short fingers possesses two dominant alleles (*SS*). All gametes from this individual will contain an allele for short fingers (*S*). Likewise, all gametes produced by a homozygous recessive parent (*ss*) contain an allele for long fingers (*s*). Therefore, all the offspring from this couple will have one allele for short fingers and another for long fingers, as in *Ss*. This individual is **heterozygous,** meaning that they have two different alleles for the trait.

The word **phenotype** refers to the physical appearance of the individual. An organism's phenotype is mostly determined by its genotype. The homozygous dominant (*SS*) individual and the heterozygous (*Ss*) individual both show the dominant phenotype and have short fingers, while the homozygous recessive individual shows the recessive phenotype and has long fingers (**Table 10.1**).

a. Various alleles are located at specific loci.

b. Duplicated chromosomes show that sister chromatids have identical alleles.

Figure 10.5   Homologous chromosomes.

a. The letters represent alleles—that is, alternate forms of a gene. Each allelic pair, such as *Gg* or *Tt*, is located on homologous chromosomes at a particular gene locus. **b.** Sister chromatids carry the same alleles in the same order.

Table 10.1 | Genotype Versus Phenotype

Genotype	Genotype	Phenotype
SS	Homozygous dominant	Short fingers
Ss	Heterozygous	Short fingers
ss	Homozygous recessive	Long fingers

Check Your Progress

1. How is Mendel's law of segregation consistent with the behavior of homologous chromosomes during meiosis?
2. Contrast genotype with phenotype.
3. Explain why animal breeders would do a testcross.

Answers: 1. Mendel observed that the two factors for a trait separate during gamete formation; as we saw in Chapter 9, the members of a pair of homologous chromosomes separate during meiosis.   2. Genotype is the genetic makeup of an individual, while phenotype is the physical appearance of the individual.   3. A testcross can indicate the genotype of an animal exhibiting a dominant trait.

Figure 10.6 **Two-trait cross done by Mendel.**

P generation plants differ in two regards—length of the stem and color of the pod. The F₁ generation shows only the dominant phenotypes, but all possible phenotypes appear among the F₂ generation. The 9:3:3:1 ratio allowed Mendel to deduce that factors segregate into gametes independently of other factors. (Arrow indicates yellow pods).

Two-Trait Inheritance

Mendel performed a second series of crosses in which he crossed true-breeding plants that differed in two traits. For example, he crossed tall plants having green pods (*TTGG*) with short plants having yellow pods (*ttgg*) (**Fig. 10.6**). The F₁ plants showed both dominant characteristics (tall with green pods). As before, Mendel then allowed the F₁ plants to self-pollinate. Two possible results could occur in the F₂ generation:

1. If the dominant factors (*TG*) always go together into the F₁ gametes, and the recessive factors (*tg*) always stay together, then two phenotypes would result among the F₂ plants—tall plants with green pods and short plants with yellow pods.
2. If the four factors segregate into the F₁ gametes independently, then four phenotypes would result among the F₂ plants—tall plants with green pods, tall plants with yellow pods, short plants with green pods, and short plants with yellow pods.

Figure 10.6 shows that Mendel observed four phenotypes among the F₂ plants, supporting the second hypothesis. Therefore, Mendel formulated his second law of heredity—the **law of independent assortment**—which states:

• Each pair of factors segregates (assorts) independently of the other pairs.
• All possible combinations of factors can occur in the gametes.

Note that when all possible sperm have an opportunity to fertilize all possible eggs, such as the example in Figure 10.6, that *the expected phenotypic results of a two-trait cross are always 9:3:3:1 when both parents are heterozygous for the two traits.*

Two-Trait Testcross

The fruit fly *Drosophila melanogaster*, less than one-fifth the size of a housefly, is a favorite subject for genetic research because it has several mutant characteristics that are easily determined. The **wild-type** fly—the type you are most likely to find in nature—has long wings and a gray body, while some mutant flies have short (vestigial) wings and black (ebony) bodies. The key for a cross involving these traits is *L* = long wings, *l* = short wings, *G* = gray body, and *g* = black body.

A two-trait testcross can be used to determine whether an individual is homozygous dominant or heterozygous for either of the two traits. Because it is not possible to determine the genotype of a long-winged, gray-bodied fly by inspection, the genotype may be represented as *L__ G__*.

In a two-trait testcross, an individual with the dominant phenotype for both traits is crossed with an individual with the recessive phenotype for both traits because this fly has a *known* genotype. For example, a long-winged, gray-bodied fly is crossed with a short-winged, black-bodied fly. The heterozygous parent fly (*LlGg*) forms four different types of gametes. The homozygous parent fly (*llgg*) forms only one type of gamete:

P:	*LlGg*	×	*llgg*
Gametes:	*LG, Lg, lG, lg*		*lg*

As **Figure 10.7** shows, all possible combinations of phenotypes occur among the offspring. This 1:1:1:1 phenotypic ratio shows that the *L__G__* fly is heterozygous for both traits and has the genotype *LlGg*. Such an individual is called a **dihybrid.** A Punnett square can also be used to predict the chances of an offspring having a particular phenotype. What are the chances of an offspring with long wings and a gray body? The chances are ¼, or 25%. What are the chances of an offspring with short wings and gray body? The chances are ¼, or 25%, and so forth.

If the *L__G__* fly had been homozygous for both traits, then no offspring would have short wings or a black body when it was crossed with one having the recessive phenotype. If the *L__G__* fly is heterozygous for one trait but not the other (either *LlGG* or *LLGg*), can you predict the expected phenotypic ratio, assuming that the other fly is recessive for both traits?

Mendel's Laws and Probability

When we use a Punnett square to calculate the results of genetic crosses, we assume that each gamete contains one allele for each trait (law of segregation) and that collectively the gametes have all possible combinations of alleles (law of independent assortment). Further, we assume that the male and female gametes combine at random—that is, all possible sperm have an equal chance to fertilize all possible eggs. Under these circumstances, it is possible to use the rules of probability to calculate the expected phenotypic ratio. The **rule of multiplication** says that the chance of two (or more) independent events occurring together is the product of their chances of occurring separately. For example, the chance of getting tails when you toss a coin is ½. The chance of getting two tails when you toss two coins at once is ½ × ½ = ¼.

Let's use the rule of multiplication to calculate the expected results in Figure 10.7. Because each allele pair separates independently, we can treat the cross as two separate one-trait crosses:

Ll	×	*ll*:	Probability of *ll* offspring	=	½
Gg	×	*gg*:	Probability of *gg* offspring	=	½

The probability of the offspring's genotype being *llgg*:

$$½ \, ll \times ½ \, gg = ¼ \, llgg$$

And the same results are obtained for the other possible genotypes among the offspring in Figure 10.7.

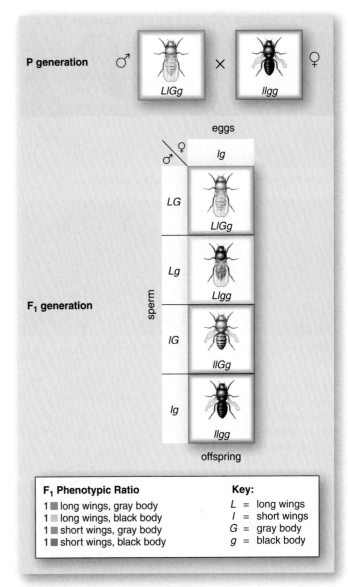

Figure 10.7 Two-trait testcross.

If a fly heterozygous for both traits is crossed with a fly that is recessive for both traits, the expected ratio of phenotypes is 1:1:1:1.

F₁ Phenotypic Ratio

1 ■ long wings, gray body
1 □ long wings, black body
1 ▨ short wings, gray body
1 ■ short wings, black body

Key:

L = long wings
l = short wings
G = gray body
g = black body

■ Check
Your Progress

1. For an individual heterozgous for two traits (*EeFf*), what are the possible combinations of alleles present in the gametes? If two of these individuals reproduce, what are the phenotypic results?
2. A black mouse with a straight tail reproduces with a yellow mouse with a bent tail. The phenotypic ratio among the offspring is 1:1:1:1. What is the genotype of the first mouse?
3. What would the result be if the test individual in Figure 10.7 were
   a. homozygous dominant for both traits? b. Homozygous dominant for one trait but heterozygous for the other?

Answers: 1. *EF, Ef, eF,* and *ef*; 9:3:3:1.   2. Heterozygous for both traits.   3. a. All long-winged, gray-bodied flies. b. Either 1 long wings, gray body : 1 long wings, black body or 1 long wings, gray body : 1 short wings, gray body.

Dominant

Recessive

Unattached earlobes: *EE* or *Ee*

Attached earlobes: *ee*

Widow's peak: *WW* or *Ww*

Straight hairline: *ww*

Mendel's Laws and Meiosis

Today, we are aware that Mendel's laws relate to the process of meiosis. In **Figure 10.8,** a human cell has two pairs of homologous chromosomes, recognized by length—one pair of homologues is short and the other is long. (The color signifies that we inherit chromosomes from our parents; one homologue of each pair is the "paternal" chromosome, and the other is the "maternal" chromosome.) When the homologues separate (segregate), each gamete receives one member from each pair. The homologues separate (assort) independently; it does not matter which member of each pair goes into which gamete. In the simplest of terms, a gamete in Figure 10.8 can receive one short and one long chromosome of either color. Therefore, all possible combinations of chromosomes are in the gametes.

The alleles *E* for unattached earlobes and *e* for attached earlobes are on one pair of homologues, and the alleles *W* for widow's peak and *w* for straight hairline are on the other pair of homologues. Because there are no restrictions as to which homologue goes into which gamete, a gamete can receive either an *E* or an *e* and either a *W* or a *w* in any combination because the chromosomes they are located on assort independently. In the end, collectively, the gametes will have all possible combinations of alleles.

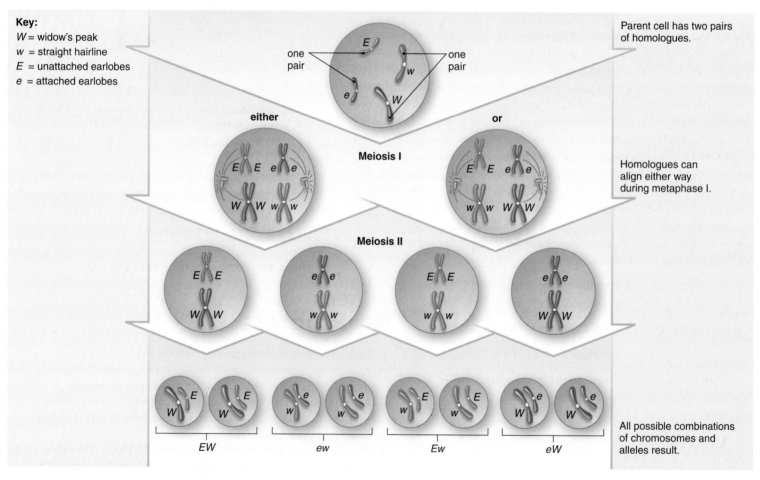

Key:
W = widow's peak
w = straight hairline
E = unattached earlobes
e = attached earlobes

one pair

one pair

Parent cell has two pairs of homologues.

either

or

Meiosis I

Homologues can align either way during metaphase I.

Meiosis II

EW   *ew*   *Ew*   *eW*

All possible combinations of chromosomes and alleles result.

Figure 10.8   **Mendel's laws and meiosis.**

A human cell has 23 pairs of homologous chromosomes (homologues), of which two pairs are represented here. The homologues, and the alleles they carry, segregate independently during gamete formation. Therefore, all possible combinations of chromosomes and alleles occur in the gametes.

10.2 | Beyond Mendel's Laws

Since Mendel's time, variations in the dominant/recessive relationship he described have been discovered. Some alleles are neither dominant nor recessive, and some genes have more than two alleles. Furthermore, some traits are affected by more than one pair of genes, and by the environment. Together, these variations make it clear that the concept of the genotype alone cannot account for all the observable traits of an organism.

C^RC^R C^RC^W C^WC^W

Figure 10.9 Incomplete dominance in four-o'clocks.

In incomplete dominance, the heterozygote is intermediate between the homozygotes. In this case, the heterozygote is pink, whereas the homozygotes are red or white.

Incomplete Dominance

Incomplete dominance is exhibited when the heterozygote has an intermediate phenotype between that of either homozygote. In the flowering plant known as the four-o'clock, a cross between red-flowered four-o'clocks and white-flowered four-o'clocks produces offspring with pink flowers (**Fig. 10.9**). But this is not an example of blending inheritance, because when the pink-flowered plants self-pollinate, ¼ of the offspring have red flowers, ¼ have white flowers, and the rest have pink flowers (a 1:2:1 ratio). The reappearance of the original phenotypes makes it clear that we are still dealing with particulate inheritance of the type described by Mendel.

In humans, people with curly hair have the homozygous recessive genotype, while those with straight hair have the homozygous dominant condition. The heterozygote has wavy hair. We can sometimes explain incomplete dominance by assuming that only the dominant allele codes for a gene product and that the single dose of the product gives the intermediate result.

Multiple-Allele Traits

In ABO blood group inheritance, there are three alleles that determine the presence or absence of antigens on red blood cells and therefore blood type:

I^A = A antigen on red blood cells
I^B = B antigen on red blood cells
i = Neither A nor B antigen on red blood cells

Each person has only two of the three possible alleles, and both I^A and I^B are dominant over i. Therefore, there are two possible genotypes for type A blood (I^AI^A, I^Ai) and two possible genotypes for type B blood (I^BI^B, I^Bi). Type O blood can only result from one genotype (ii) because the i allele is recessive. But, I^A and I^B are fully expressed in the presence of each other. Therefore, if a person inherits one of each of these alleles, that person will have a fourth blood type, AB. **Figure 10.10** shows that matings between certain genotypes can have surprising results in terms of blood type.

Notice that human blood type inheritance is also an example of **codominance**, another type of inheritance that differs from Mendel's findings because more than one allele is fully expressed. When an individual has blood type AB, both A and B antigens appear on the red blood cells. The two different capital letters signify that both alleles are coding for an antigen.

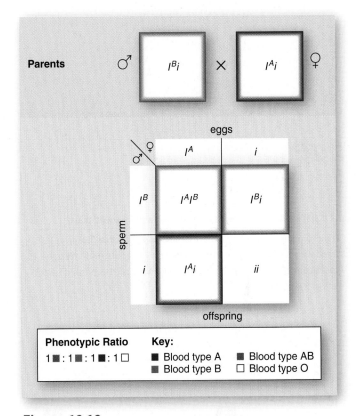

Figure 10.10 Inheritance of ABO blood type.

A mating between individuals with type A blood and type B blood can result in any one of the four blood types. Why? Because the parents are I^Bi and I^Ai. The i allele is recessive; both I^A and I^B are dominant.

Check Your Progress

1. A pink-flowered four-o'clock plant is crossed with a white-flowered four-o'clock plant. What phenotypes appear in the offspring, and in what ratio?
2. What blood types immediately tell you the genotype?

Answers: 1. Pink and white in a 1:1 ratio 2. Type O (ii) and type AB (I^AI^B).

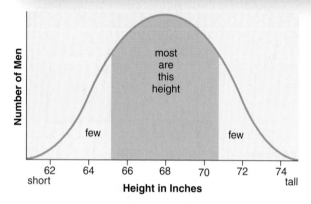

Figure 10.11 Height in humans, a polygenic trait.

When you record the heights of a large group of people chosen at random, the values follow a bell-shaped curve because multiple gene pairs control the trait. Environmental influences, such as nutrition, also affect the phenotype.

Figure 10.12 Coat color in Himalayan rabbits.

Hair growing under an ice pack in these rabbits is black. The dark color on the ears, nose, and feet of these rabbits is believed to be due to a lower body temperature in these areas.

Polygenic Inheritance

Polygenic inheritance occurs when a trait is governed by two or more sets of alleles. The individual has a copy of all allelic pairs, possibly located on many different pairs of chromosomes. Each dominant allele has a quantitative effect on the phenotype, and these effects are additive. The result is a continuous variation of phenotypes, resulting in a distribution that resembles a bell-shaped curve. The more genes involved, the more continuous are the variations and distribution of the phenotypes. Also, environmental effects are involved; in the case of human height, differences in childhood nutrition may modify the phenotype to bring about a smooth, bell-shaped curve (**Fig. 10.11**).

Multifactorial traits are ones that are controlled by polygenes subject to environmental influences. Many genetic disorders, such as cleft lip and/or palate, clubfoot, congenital dislocations of the hip, hypertension, diabetes, schizophrenia, and even allergies and cancers, are most likely multifactorial because they are likely due to the combined action of many genes plus environmental influences. In recent years, reports have surfaced that all sorts of behaviors, including alcoholism, phobias, and even suicide, can be associated with particular genes. No doubt, behavioral traits are somewhat controlled by genes, but again, it is impossible at this time to determine to what degree. And very few scientists would support the idea that these behavioral traits are predetermined by our genes.

Environment and the Phenotype

The relative importance of genetic and environmental influences on the phenotype can vary, but in some instances the environment seems to have an extreme effect. In the water buttercup, the submerged part of the plant has a different appearance from the part above water. This difference is thought to be related to a difference in water intake by the cells.

Temperature can also have a dramatic effect on the phenotypes of plants and animals. Primroses have white flowers when grown above 32°C and red flowers when grown at 24°C. The coats of Siamese cats and Himalayan rabbits are darker in color at the ears, nose, paws, and tail. Himalayan rabbits are known to be homozygous for the allele *ch*, which is involved in the production of melanin. Experimental evidence suggests that the enzyme encoded by this gene is active only at a low temperature and that, therefore, black fur only occurs at the extremities where body heat is lost to the environment (**Fig. 10.12**). When the animal is placed in a warmer environment, new fur on these body parts is light in color.

These examples lend additional support to the observation that human traits controlled by polygenes are also subject to environmental influences. Therefore, many investigators are trying to determine what percentage of various traits is due to nature (inheritance) and what percentage is due to nurture (the environment). Some studies use twins separated from birth, because if identical twins in different environments share the same trait, that trait is most likely inherited. Identical twins are more similar in their intellectual talents, personality traits, and levels of lifelong happiness than are fraternal twins separated from birth. Biologists conclude that all behavioral traits are partly determined by inheritance, and that the genes for these traits act together in complex combinations to produce a phenotype that is modified by environmental influences.

Pleiotropy

Pleiotropy occurs when a single gene has more than one effect. Often, this leads to a *syndrome*, a group of symptoms that appear together and indicate the presence of a particular genetic mutation. For example, persons with Marfan syndrome have disproportionately long arms, legs, hands, and feet; a weakened aorta; and poor eyesight (**Fig. 10.13**). All of these characteristics are due to the production of abnormal connective tissue. Marfan syndrome has been linked to a mutated gene (*FBN1*) on chromosome 15 that ordinarily specifies a functional protein called fibrillin. This protein is essential for the formation of elastic fibers in connective tissue. Without the structural support of normal connective tissue, the aorta can burst, particularly if the person is engaged in a strenuous sport, such as volleyball or basketball. Flo Hyman may have been the best American woman volleyball player ever, but she fell to the floor and died when only 31 years old because her aorta gave way during a game. Now that coaches are aware of Marfan syndrome, they are on the lookout for it among very tall basketball players. Chris Weisheit, whose career was cut short after he was diagnosed with Marfan syndrome, said, "I don't want to die playing basketball."

Many other disorders, including sickle cell disease and porphyria, are also examples of pleiotropic traits. Porphyria is caused by a chemical insufficiency in the production of hemoglobin, the pigment that makes red blood cells red. The symptoms of porphyria are photosensitivity, strong abdominal pain, port-wine-colored urine, and paralysis in the arms and legs. Many members of the British royal family in the late 1700s and early 1800s suffered from this disorder, which can lead to epileptic convulsions, bizarre behavior, and coma.

Check Your **Progress**

1. Polygenic inheritance is controlled by three pairs of alleles. What are the two extreme genotypes?
2. What's the difference in the pattern of inheritance for polygenic inheritance and pleiotropy?

Answers: 1. *AABBCC* and *aabbcc*.   2. Several genes affect one trait in polygenic inheritance, while one gene has several phenotypic effects in pleiotropy.

Figure 10.13   Marfan syndrome, multiple effects of a single human gene.

Individuals with Marfan syndrome exhibit defects in the connective tissue throughout the body. Important changes occur in the skeleton, heart, blood vessels, eyes, lungs, and skin. All these conditions are due to mutation of the gene *FBN₁*, which codes for a constituent of connective tissue. *Most life-threatening.

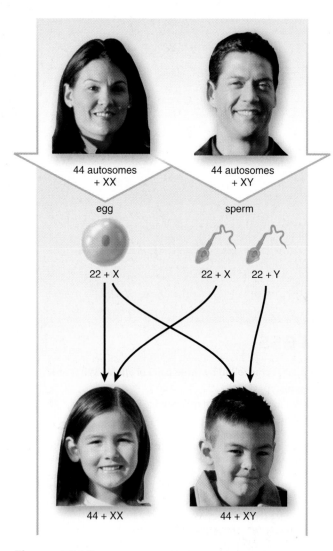

Figure 10.14   Sex determination in human beings.

Sperm determine the sex of an offspring because sperm can carry an X or a Y chomosome. Males inherit a Y chromosome. Genes on the Y chromosome also affect the gender of an individual.

10.3 | Sex-Linked Inheritance

Geneticists of the early twentieth century were convinced that the genes are on the chromosomes because the genes and chromosomes behave similarly during gamete formation (see Fig. 10.8). They also knew that the chromosomes differ between the sexes. As you learned in Chapter 9, the sex chromosomes in females are XX, and those in males are XY (**Fig. 10.14**). Notice that males produce two different types of gametes—those that contain an X and those that contain a Y. Therefore, the chromosomal contribution of the male determines the sex of the new individual.

The much shorter Y chromosome contains only about 80 genes, and most of these genes are concerned with sex differences between men and women. One of the genes on the Y chromosome, *SRY*, does not have a copy on the X chromosome. If the functional *SRY* gene is present, the individual becomes a male, and if it is absent, the individual becomes a female. Thus, female sex development is the "default setting."

In contrast, the X chromosome is quite large and contains nearly 2,000 genes, most of which have nothing to do with the gender of the individual. By tradition, the term **X-linked** refers to such genes carried on the X chromosome. Examples of X-linked traits in humans include hemophilia and red-green color-blindness. The Y chromosome does not carry these genes, and that makes for an interesting inheritance pattern.

X-Linked Alleles

We have already mentioned that the fruit fly is a favorite subject for genetic studies. Flies can be easily and inexpensively raised in simple laboratory glassware; females mate only once and then lay hundreds of eggs during their lifetimes; and the generation time is short, taking only about ten days when conditions are favorable.

Drosophila flies have the same sex chromosome pattern as humans. This facilitates our understanding of a cross performed by early geneticists. When a mutant male with white eyes was crossed with a red-eyed female, the F_1 all had red eyes:

$$\begin{array}{ccc} & ♀ & ♂ \\ \text{P} & \text{red-eyed} \quad \times & \text{white-eyed} \\ F_1 & \text{red-eyed} & \text{red-eyed} \end{array}$$

From these results, researchers knew that red eyes are the dominant characteristic and white eyes are the recessive characteristic. However, the F_2 generation gave an unusual result: The expected 3:1 ratio resulted, but all of the white-eyed flies were males:

$$\begin{array}{ccc} & ♀ & ♂ \\ F_1 \times F_1 & \text{red-eyed} \quad \times & \text{red-eyed} \\ F_2 & \text{red-eyed} & \text{1 red-eyed : 1 white-eyed} \end{array}$$

Obviously, a major difference between the male flies and the female flies was their sex chromosomes. Could it be possible that an allele for eye color was on the Y chromosome but not on the X? This idea was quickly discarded because females usually have red eyes, and they have no Y chromosome. Next, investigators hypothesized that perhaps an allele for eye color was on the X chromosome, but not on the Y chromosome. This explanation turned out to match the results obtained in the experiment (**Fig. 10.15**). These results also support the chromosome theory of inheritance by showing that the behavior of a specific allele corresponds exactly with that of a specific chromosome—the X chromosome in *Drosophila*.

Notice that X-linked alleles have a different pattern of inheritance than alleles on the autosomes. The Y chromosome is blank for these alleles, and so the inheritance of a Y chromosome cannot offset the inheritance of an X-linked recessive allele. For the same reason, *males always receive an X-linked recessive mutant allele from their female parent*; they receive the Y chromosome, which does not have an allele for the trait, from the male parent.

An X-Linked Problem

When solving autosomal genetics problems involving fruit flies, the alleles and genotypes are represented as follows:

Key: L = long wings   Genotypes: = LL, Ll, ll
       l = short wings

As noted in Figure 10.15, however, an X-linked gene is represented by attaching the allele to an X:

Key: X^R = red eyes
       X^r = white eyes

The possible genotypes in both males and females are as follows:

Genotypes: $X^R X^R$ = red-eyed female
             $X^R X^r$ = red-eyed female
             $X^r X^r$ = white-eyed female
             $X^R Y$ = red-eyed male
             $X^r Y$ = white-eyed male

Notice that three genotypes are possible for females, but only two are possible for males. Females can be heterozygous $X^R X^r$, in which case they are carriers. **Carriers** usually do not exhibit a recessive trait, but they are capable of passing on a recessive allele for a trait. Males cannot be carriers; if the dominant allele is on the single X chromosome, they show the dominant phenotype, and if the recessive allele is on the single X chromosome, they show the recessive phenotype.

Males have white eyes when they receive the mutant recessive allele from the female parent. Females can only have white eyes when they receive a recessive allele from both parents.

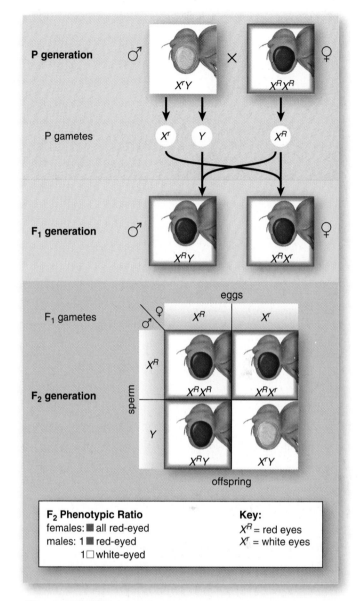

Figure 10.15 **X-linked inheritance.**

Once researchers deduced that the allele for red/white eye color is only on the X chromosome in *Drosophila*, they were able to explain their experimental results. Males with white eyes in the F_2 inherit the recessive allele only from the female parent; they receive a blank Y chromosome from the male parent.

Check Your Progress

1. Why do males produce two different gametes with respect to sex chromosomes?
2. A son inherits sex-linked recessive traits only from his mother and not his father. Explain.

Answers: 1. Being XY, ½ of the gametes carry an X and ½ carry a Y.   2. Males inherit an X and Y chromosome. A father gives his son a Y chromosome, which is blank for sex-linked alleles. A mother gives her son an X chromosome and if it carries a recessive allele, it will be expressed.

10.4 | Inheritance of Linked Genes

Consider that an organism usually has far fewer chromosomes than genes. For example, *Drosophila* has 14,000 genes, but only 4 chromosomes. Therefore, many alleles must be on each chromosome. The alleles that occur on the same chromosome form a *linkage group* because these alleles tend to be inherited together. **Figure 10.16***a* considers a heterozygous individual in which the dominant alleles (*GR*) occur on one homologue, and the recessive alleles (*gr*) occur on the other homologue. It shows you that if linkage is complete, the gametes have to contain either a daughter chromosome with the dominant alleles or a daughter chromosome with the recessive alleles. Figure 10.16*b* shows that if crossing-over occurs between the two alleles, the daughter chromosomes would be recombinant and have the alleles shown. **Recombinant gametes** contain recombined alleles, but occur in reduced numbers because crossing-over is infrequent.

Gene linkage was discoverd by scientists performing *Drosophila* crosses. For example, when scientists performed a two-trait testcross between a gray-bodied, red-eyed fly heterozygous for both traits and a black-bodied, purple-eyed recessive fly, as in **Figure 10.17,** most of the flies resembled one or the other parent. This was unexpected because if the two pairs of alleles are on different chromosomes, independent assortment always occurs, and the heterozygote produces four different types of gametes in equal numbers. In this instance, if the alleles for body color assorted independently of the alleles for eye color, then a 1:1:1:1 ratio would occur among the offspring and 25% of the offspring would be one of four possible genotypes. Instead, only 3% of the offspring showed the recombinant phenotypes. Why? Because the linked alleles usually remained together in the gametes made by the heterozygote, as shown in Figure 10.16*a*, and generally did not resort, as in Figure 10.16*b*. Recombination did occur sometimes, however, and that is why a total of 6% of the offspring had a recombinant phenotype. Their results allowed scientists to conclude that gene linkage was almost complete for these two sets of alleles.

Figure 10.16 **Linked alleles and crossing-over.**

a. Alleles on the same chromosome form a linkage group, and linked alleles tend to be found together in the same daughter cells and, therefore, gametes. **b.** Crossing-over can cause linked alleles to separate and go into different daughter cells, and these become recombinant gametes.

a. Linked alleles usually stay together

b. Crossing-over results in recombination of alleles

Distance Between Gene Loci

It stands to reason that the closer together two genes are, the less likely they are to cross over. Numerous experiments have repeatedly shown that this is the case. Therefore, you can use the percentage of recombinant phenotypes to determine the distance between genes, because there is a direct relationship between the frequency of crossing-over (the percentage of recombinant phenotypes) and the distance between alleles. In the previous example (see Fig. 10.17), a total of 6% of the offspring are recombinants, and this means the rate of crossing-over is 6%. For the sake of mapping the chromosomes, it is assumed that 1% of crossing-over equals one map unit. Therefore, the allele for black body and the allele for purple eyes are six map units apart.

Mapping the Chromosome

Suppose you want to determine the order of three gene loci on the chromosomes. To do so, you perform crosses that involve the three genes. The results will tell you the map distance between the three gene loci. Assume, for instance, that:

1. the distance between the black-body and purple-eye alleles = 6 map units;
2. the distance between the purple-eye and vestigial-wing alleles = 12.5 map units; and
3. the distance between the black-body and vestigial-wing alleles = 18.5 map units.

Therefore, the order of the alleles must be as shown in **Figure 10.18**. Black body must be 6 map units away from purple eyes, and purple eyes must be 12.5 map units away from vestigial wings.

Scientists used linkage data in *Drosophila* to construct a **chromosome map,** which shows the relative distance between the gene loci on a chromosome. However, the possibility of using linkage data to map human chromosomes is limited because we can work only with matings that occur by chance. This, coupled with the fact that humans tend not to have numerous offspring, means that additional methods are needed to sequence the genes on human chromosomes. Today, it is customary also to rely on biochemical methods to map the human chromosomes.

Offspring	Predicted	Observed
GgRr	25%	47%
ggrr	25%	47%
Ggrr	25%	3%
ggRr	25%	3%

F₁ Phenotypic Ratio
1 ■ gray body, red eyes
1 ■ black body, purple eyes
1 ■ gray body, purple eyes
1 ■ black body, red eyes

Key:
G = gray body
g = black body
R = red eyes
r = purple eyes

Figure 10.17 Linked alleles do not assort independently.

When gametes are formed in the heterozygote shown in Figure10.16, the two alleles that occur together on each chromosome segregate together.

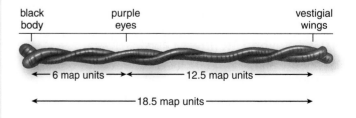

black body   purple eyes   vestigial wings

← 6 map units → ← 12.5 map units →

← 18.5 map units →

Figure 10.18 Mapping chromosomes.

Genes are arranged linearly on the chromosome at specific gene loci. Using the crossing-over data supplied by the text, these genes are sequences as shown.

The Chapter in Review

Summary

10.1 | Mendel's Laws

In 1860, Gregor Mendel, an Austrian monk, developed two laws of heredity based on crosses utilizing the garden pea.

Law of Segregation

Mendel's law of segregation states the following:

- Each individual has two factors for each trait.
- The factors segregate (separate) during the formation of the gametes.
- Each gamete contains only one factor from each pair of factors.
- Fertilization gives each new individual two factors for each trait.

In the context of genetics today,

- Gene pairs are on the chromosomes, one allele on each homologue.
- Alternative forms of a gene are called alleles. An individual may have different alleles of a gene on each homologue.
- Alleles are assigned uppercase letters if they are dominant, or lowercase letters if they are recessive. An individual's genotype may be homozygous dominant (AA), heterozygous (Aa), or homozygous recessive (aa).
- Homologues separate during meiosis, and the gametes have only one allele for each trait—either an A or an a.
- Fertilization gives each new individual two alleles for each trait.

Law of Independent Assortment

Mendel's law of independent assortment states the following:

- Each pair of factors segregates (assorts) independently of the other pairs.
- All possible combinations of factors can occur in the gametes.

In the context of genetics today,

- Each pair of homologues separate independently of the other pairs.
- All possible combinations of chromosomes and their alleles occur in the gametes.
- Mendel's laws are consistent with the manner in which homologues and their alleles separate during meiosis.

Common Autosomal Genetic Crosses

A Punnett square allows all types of sperm to fertilize all types of eggs and gives these results:

Aa	× Aa	3:1 phenotypic ratio
Aa	× aa	1:1 phenotypic ratio
$AaBb$	× $AaBb$	9:3:3:1 phenotypic ratio
$AaBb$	× $aabb$	1:1:1:1 phenotypic ratio

10.2 | Beyond Mendel's Laws

In some patterns of inheritance, the alleles are not just dominant or recessive.

Incomplete Dominance

In incomplete dominance, the heterozygote is intermediate between the two homozygotes. For example, the offspring of red and white four-o'clocks produce pink flowers. The red and white phenotypes reappear when pink four-o'clocks are crossed.

Multiple-Allele Traits

The multiple-allele inheritance pattern is exemplified in humans by blood type inheritance. Every individual has two out of three possible alleles: I^A, I^B, i. Both I^A and I^B are expressed; therefore, this is also a case of codominance.

Polygenic Inheritance

In polygenic inheritance, a trait is controlled by more than one set of alleles. The dominant alleles have an additive effect on the phenotype.

Environment and the Phenotype

Many phenotypes, especially those governed by polygenes, are modified by the environment.

Pleiotropy

In pleiotropy, one gene (consisting of two alleles) has multiple effects on the body. For example, all the disorders common to Marfan syndrome are due to a mutation that leads to a defect in the composition of connective tissue.

10.3 | Sex-Linked Inheritance

Males produce two different types of gametes—those that contain an X and those that contain a Y. The contribution of the male determines the sex and gender of the new individual. XX = female, XY = male.

Gender determination by male

Certain alleles are carried on the X chromosome, but the Y has very few genes. Therefore, a male only needs to inherit one recessive allele on the X chromosome to have a recessive genetic disorder.

Common X-Linked Genetic Crosses

$X^B X^b \times X^B Y$ All daughters will be normal, even though they have 50% chance of being carriers, but sons have a 50% chance of being color blind.

$X^B X^B \times X^b Y$ All children are normal (daughters will be carriers).

10.4 | Inheritance of Linked Genes

Alleles on the same chromosome are linked, and they are generally inherited together, but crossing-over can shuffle them. Crossing-over data can be used to determine the distance between gene loci and to construct a chromosome map, which shows the sequence of gene loci along the chromosome.

Key Terms

Testing Yourself

Choose the best answer for each question.

1. In Mendel's particulate theory of inheritance, the "particles" are now known to be
   a. chromosomes.
   b. genes.
   c. plants.
   d. pollen grains.

2. The offspring ratio from a testcross ($F_1 \times$ homozygous recessive) should be
   a. all dominant.
   b. ¾ dominant, ¼ recessive.
   c. ½ dominant, ½ recessive.
   d. all recessive.

3. Which of the following is not a component of the law of segregation?
   a. Each gamete contains one factor from each pair of factors in the parent.
   b. Factors segregate during gamete formation.
   c. Following fertilization, the new individual carries two factors for each trait.
   d. Each individual has one factor for each trait.

4. When using a Punnett square to predict offspring ratios, we assume that
   a. each gamete contains one allele of each gene.
   b. the gametes have all possible combinations of alleles.
   c. male and female gametes combine at random.
   d. All of these are correct.

5. If you cross a black spaniel with a red spaniel and get a litter of 8 black and 1 red, what is the genotype of the black parent?
   a. *BB*
   b. *Bb*
   c. *bb*
   d. impossible to determine

6. Cystic fibrosis (CF) is an autosomal recessive disorder in humans. If two unaffected individuals have a child with CF, what is the chance that their second child will have CF?
   a. 25%
   b. 50%
   c. 75%
   d. 100%
   e. impossible to determine

7. When one physical trait is affected by two or more pairs of alleles, the condition is called
   a. incomplete dominance.
   b. codominance.
   c. homozygous dominant.
   d. multiple allele.
   e. polygenic inheritance.

8. Fill in the blank spaces in the following Punnett square.

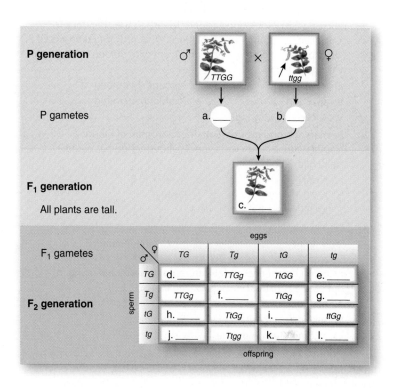

For questions 9–11, consider that coat color and spotting pattern in cocker spaniels depend on two genes. Black (*B*) is dominant to red (*b*), and solid color (*S*) is dominant to spotted (*s*).

9. What phenotypic ratio do you expect for a dihybrid cross?
   a. 1 black solid : 1 red solid : 1 black spotted : 1 red spotted
   b. 9 black solid : 3 red solid : 3 black spotted : 1 red spotted
   c. 1 black solid : 3 red solid : 3 black spotted : 9 red spotted
   d. all black solid

10. If you cross a black spotted dog with a black solid dog and get a ratio of 3 black solid : 3 black spotted : 1 red solid : 1 red spotted, what is the genotype of black spotted parent?
    a. *BBss*
    b. *BbSs*
    c. *Bbss*
    d. *bbSs*
    e. *bbss*

11. What is the genotype of the black solid parent in question 10?
    a. *BBSS*
    b. *BbSS*
    c. *BbSs*
    d. *Bbss*
    e. *bbSs*

For questions 12–15, match the cross with the predicted phenotypic ratios in the key.

Key:

    a. 3:1
    b. 9:3:3:1
    c. 1:1
    d. 1:1:1:1
    e. none of these

12. *AaBb* × *aabb*
13. *Aa* × *Aa*
14. *AAbb* × *aaBB*
15. *Aa* × *aa*
16. *AaBb* × *aabb*
17. *AABB* × *AaBb*

18. *AaBb* × *AaBb*

19. When two monohybrid round squashes are crossed, the offspring ratio is ¼ flat: ½ oblong: ¼ round. Squash shape, therefore, is controlled by incomplete dominance. What offspring ratio would you expect from a cross between a plant with oblong fruit and one with round fruit?
    - a. all oblong
    - b. all round
    - c. ¾ oblong: ¼ round
    - d. ¾ round: ¼ oblong
    - e. ½ oblong: ½ round

20. If a man of blood group AB marries a woman of blood group B whose father was type O, what phenotypes could their children be?
    - a. A only
    - b. A, AB, B, and O
    - c. AB only
    - d. A, AB, and B
    - e. O only

21. Anemia sometimes results from a mutation in a single gene, causing the blood's oxygen supply to be inadequate. A homozygous recessive person has a number of problems, including lack of energy, fatigue, rapid pulse, pounding heart, and swollen ankles. This is an example of
    - a. pleiotropy.
    - b. sex-linked inheritance.
    - c. incomplete dominance.
    - d. polygenic inheritance.
    - e. codominance.

For questions 22–25, list the progeny phenotypes from the following key that would result from each of the crosses in *Drosophila*. Red eye color is dominant over white. The gene for eye color is on the X chromosome. Answers can be used more than once.

Key:

- a. red-eyed female
- b. red-eyed male
- c. white-eyed female
- d. white-eyed male

22. homozygous red-eyed female × white-eyed male

23. heterozygous female × white-eyed male

24. white-eyed female × white-eyed male

25. heterozygous female × red-eyed male

26. Alice and Henry are at the opposite extremes for height, a polygenic trait. Their children will
    - a. exhibit the middle height between their two parents.
    - b. exhibit a 9:3:3:1 ratio of heights between that of their parents.
    - c. exhibit a 3:1 ratio of tall to short heights.
    - d. be the same height as Alice or Henry.

27. Mary's son is color-blind, but Mary is not color-blind. Which of these is a more likely genotype for a relative?
    - a. mother X^cX^c
    - b. father X^cY
    - c. brother X^cX^c
    - d. Both b and c are likely.

Thinking Scientifically

1. Multiple gene pairs may also interact to produce a single phenotype. In peas, genes *C* and *P* are required for pigment production in flowers. Gene *C* codes for an enzyme that converts a compound into a colorless intermediate product. Gene *P* codes for an enzyme that converts the colorless intermediate product into anthocyanin, a purple pigment. A flower, therefore, will be purple only if it contains at least one dominant allele for each of the two genes (C__ P__). What phenotypic ratio would you expect in the F_2 generation following a cross between two double heterozygotes (*CcPp*)?

2. Geneticists often look for unusual events to provide insight into genetic mechanisms. In one such instance, researchers studied XX men and XY women. They found that the XX men contained a chromosomal segment normally found in men but not in women, while XY women were missing that region. What gene do you suppose is on that chromosome piece?

3. What would cause a particular polygenic trait to become more frequent in a population over time?

Bioethical Issue

Selecting the Sex of Your Child

As you know, the sex of a child depends upon whether an X-bearing sperm or a Y-bearing sperm enters the egg. But a new technology that can separate X-bearing and Y-bearing sperm offers prospective parents the opportunity to choose the sex of their child. First, the sperm are dosed with a DNA-staining chemical. Because the X chromosome has slightly more DNA than the Y chromosome, it takes up more dye. When a laser beam shines on the sperm, the X-bearing sperm shine more brightly than the Y-bearing sperm. A machine sorts the sperm into two groups on this basis. The results are not perfect. Following artificial insemination, there's about an 85% success rate for a girl and about a 65% success rate for a boy. But is it morally acceptable to select the sex of your child?

Those who find this practice unethical contend that using such technology is akin to "playing God." They are greatly concerned that this new ability may lead to a society with far more members of one sex than another, which could lead to serious problems. Furthermore, they contend that allowing parents to select the sex of their children could lead to other ethical concerns, such as selecting for specific traits in children.

Proponents of sex-selection technology argue that there are many instances in which the ability to choose the sex of the child may benefit society. For instance, if a mother is a carrier of an X-linked genetic disorder, such as hemophilia or Duchenne muscular dystrophy, this would be the simplest way to ensure a healthy child. Previously, a pregnant woman with these concerns had to wait for the results of an amniocentesis test and then decide whether to abort the pregnancy if it were a boy. In such cases, is it better for all involved to ensure that a child does not have a specific genetic disorder than to take the risk?

Regardless of which side of this issue you fall on, most agree that once you separate reproduction from the sex act, many ethical issues soon follow.

Essentials of Biology Website

The companion website for *Essentials of Biology* provides a wealth of information organized and integrated by chapter. You will find practice tests, animations, videos, and much more that will complement your learning and understanding of general biology.

http://www.mhhe.com/maderessentials2

Every type of tissue cell in your body contains a complete copy of your DNA.

DNA can be transferred from one kind of of species to another. These pigs glow from having received jellyfish bioluminescence genes as embryos.

While keeping track of its progress (screen), a thermal cycler allows scientists to copy just a segment of your DNA over and over again until there are millions of copies.

11

DNA Biology and Technology

Outline

Pig Organs for Transplantation

ONE RECENT ADVANCE in DNA technology is the production of transgenic animals—animals that contain genes from another species. For example, researchers have created a variety of miniature pigs for the purpose of providing organs for human transplants. A unique feature of these pigs is a yellow snout and hooves due to a gene introduced from jellyfish! This makes the pigs easy to identify.

—Continued next page

Continued—

An alternative source of transplant organs is needed because human organs are in limited supply, and many people die each year waiting for transplants. If organs from a normal pig are transplanted into a human, two major problems occur: The organs from pigs are too large for humans and also the organs are quickly rejected. However, transplants from miniature genetically altered pigs avoid both of these complications. First, the miniature size of the pigs makes their organs more appropriate for humans. Second, rejection is avoided because the genes coding for plasma membrane proteins that normally trigger rejection have been "knocked out"—that is, they are not expressed. Pig organs may eventually become a life-saving alternative for people who are unable to acquire human organ transplants.

In this chapter, you will learn about the structure and function of DNA. This knowledge will allow you to appreciate the incredible advances in the field of DNA technology.

11.1 | DNA and RNA Structure and Function

Mendel knew nothing about DNA. It took many years for investigators to come to the conclusion that Mendel's factors, now called genes, are on the chromosomes. Then, researchers wanted to show that DNA was responsible for heredity, not proteins. One experiment involved the use of a virus that attacks bacteria such as *E. coli* (**Fig. 11.1**). A virus is composed of an outer capsid made of protein and an inner core of DNA. The use of radioactive tracers showed that DNA, but not protein, enters bacteria and directs the formation of new viruses. By the early 1950s, investigators not only knew that genes are composed of DNA, they knew that mutated genes result in errors of metabolism. Therefore, DNA in some way must control the cell.

Even though DNA took its name—deoxyribonucleic acid—from the chemical components of its nucleotides, its detailed structure was still to be determined. Finding the structure of DNA was the first step toward understanding how DNA is able to do the following:

- Be variable in order to account for species differences.
- Replicate so that every cell gets a copy during cell division.
- Store information needed to control the cell.
- Undergo mutations, accounting for evolution of new species.

Structure of DNA

Once researchers knew that the genes are composed of DNA, they were racing against time and each other to determine the structure of DNA. They believed that whoever discovered it first would get a Nobel Prize. How James Watson and Francis Crick determined the structure of DNA (and got the Nobel Prize) resembles a mystery, in which each clue was added to the total picture until the breathtaking design of DNA—a double helix—was finally revealed. To achieve this success, Watson and Crick particularly relied on studies done by Rosalind Franklin and Erwin Chargaff.

Figure 11.1 The genes are composed of DNA.

An early experiment by Alfred Hershey and Martha Chase helped determine that DNA is the genetic material. Their experiment involved a virus which infects bacteria such as *E. coli*. They wanted to know which part of the virus—the capsid made of protein or the DNA inside the capsid— entered the bacterium. Radioactive tracers showed that DNA, not protein, enters the bacterium and guides the formation of new viruses. Therefore, DNA must be the genetic material.

Chargaff's Rules

Before Erwin Chargaff began his work, it was known that DNA contains four different types of nucleotides based on their nitrogen-containing bases (**Fig. 11.2**). The bases **adenine (A)** and **guanine (G)** are purines with a double ring, and the bases **thymine (T)** and **cytosine (C)** are pyrimidines with a single ring. With the development of new chemical techniques in the 1940s, Chargaff decided to analyze in detail the base content of DNA in various species.

In contrast to accepted beliefs, Chargaff found that each species has its own percentages of each type of nucleotide. For example, in a human cell, 31% of bases are adenine; 31% are thymine; 19% are guanine; and 19% are cytosine. In all the species Chargaff studied, the amount of A always equaled the amount of T, and the amount of G always equaled the amount of C. These relationships are called *Chargaff's rules*:

1. The amount of A, T, G, and C in DNA varies from species to species.
2. In each species, the amount of A = T and the amount of G = C.

Chargaff's data suggest that DNA has a means to be stable in that A can only pair with T and G can only pair with C. His data also show that DNA can be variable as required for the genetic material. Today, we know that the paired bases may occur in any order, and the amount of variability in their sequences is overwhelming. For example, suppose a chromosome contains 140 million base pairs. Since any of the four possible nucleotide pairs can be present at each pair location, the total number of possible nucleotide pair sequences is $4^{140 \times 10^6}$ or $4^{140,000,000}$.

a.

Phosphate

b.

Sugars

deoxyribose
(DNA only)

ribose
(RNA only)

c.

Nitrogen-containing bases

Adenine (A)

Guanine (G)

Cytosine (C)

Thymine (T)
(DNA only)

Uracil (U)
(RNA only)

d.

Figure 11.2   Nucleotide composition of DNA and RNA.

a. All nucleotides contain phosphate, a 5–carbon sugar, and a nitrogen-containing base, such as cytosine (C). **b.** Structure of phosphate. **c.** In DNA, the sugar is deoxyribose; in RNA the sugar is ribose. **d.** In DNA the nitrogen-containing bases are adenine, guanine, cytosine, and thymine; in RNA the bases are adenine, guanine, cytosine, and uracil.

Check Your Progress

1. Briefly explain an experiment that shows DNA is the genetic material.
2. How did Chargaff's data show that DNA is variable enough to be the genetic material?

Answers: 1. Labeled viral DNA, but not labeled viral protein, enters a host bacterium and brings about formation of new viruses.   2. The amount of A, T, G, and C varies from species to species.

a.

b.

Figure 11.3 **X-ray diffraction pattern of DNA.**

a. Rosalind Franklin X-rayed DNA using crystallography techniques. **b.** The pattern that resulted indicated that DNA is a double helix (see X pattern in the center) and that some part of the molecule is repeated over and over again (see the dark portions at top and bottom). Watson and Crick said that this repeating feature was the paired bases.

Figure 11.4 **Watson and Crick model of DNA.**

James Watson (left) and Francis Crick with their model of DNA.

Franklin's X-Ray Diffraction Data

Rosalind Franklin was a researcher at King's College in London in the early 1950s (**Fig. 11.3***a*). She was studying the structure of DNA using X-ray crystallography. When a crystal (a solid substance whose atoms are arranged in a definite manner) is X-rayed, the X-ray beam is diffracted (deflected), and the pattern that results shows how the atoms are arranged in the crystal.

First, Franklin made a concentrated, viscous solution of DNA and then saw that it could be separated into fibers. Under the right conditions, the fibers were enough like a crystal that when X-rayed, a diffraction pattern resulted. The X-ray diffraction pattern of DNA shows that DNA is a double helix. The helical shape is indicated by the crossed (X) pattern in the center of the photograph in Figure 11.3*b*. The dark portions at the top and bottom of the photograph indicate that some portion of the helix is repeated over and over.

The Watson and Crick Model

In 1951, James Watson, having just earned a Ph.D., began an internship at the University of Cambridge, England. There he met Francis Crick, a British physicist, who was interested in molecular structures. Together, they set out to determine the structure of DNA and to build a model that would explain how DNA varies from species to species, replicates, stores information, and undergoes mutations.

Based on available data, they knew that:

1. DNA is a polymer of four types of nucleotides with the bases adenine (A) and guanine (G), cytosine (C) and thymine (T).
2. Based on Chargaff's rules, the amount of A = T, and the amount of G = C.
3. Based on Rosalind Franklin's X-ray diffraction photograph, DNA is a double helix with a repeating pattern.

Using these data, Watson and Crick built an actual model of DNA out of wire and tin (**Fig. 11.4**). The model showed that the deoxyribose sugar–phosphate molecules are bonded to one another to make up the sides of a twisted ladder. The nitrogenous bases make up the rungs of the ladder—they project into the middle and hydrogen bond with bases on the other strand. Indeed, the pairing of A with T and G with C—now called complementary base pairing—results in rungs of a consistent width as required by the X-ray diffraction data.

Figure 11.5 shows two ways to represent the structure of DNA. Notice that DNA is a double helix because it is double stranded. The two strands are antiparallel and run in opposite directions, as best seen in Figure 11.5*b* where the carbon atoms in deoxyribose are numbered. The 5′ carbon has an attached ℗ group and the 3′ carbon has an attached —OH group, which is circled and colored pink for easy recognition. Further, in the double helix (Fig. 11.5*c*) each strand has a 5′ end where a free ℗ appears and a 3′ end where a free —OH group appears.

The double-helix model of DNA permits the base pairs to be in any order, a necessity for genetic variability between species. Also, the model suggests that complementary base pairing may play a role in the replication of DNA. As Watson and Crick pointed out in their original paper, "It has not escaped our notice that the specific pairing we have postulated immediately suggests a possible copying mechanism for the genetic material."

Figure 11.5 DNA structure.

a. A space-filling model of DNA shows the close stacking of the paired bases, as determined by DNA's X-ray diffraction pattern. **b.** Complementary base pairing dictates that A is hydrogen-bonded to T (as shown), and G is hydrogen-bonded to C in the same manner. The strands of DNA run counter to one another, with the 3′ end of one strand opposite the 5′ end of the other strand. **c.** Diagram of the DNA double helix shows that the molecule resembles a twisted ladder. The paired bases can be in any order.

a. Space-filling model

b. Nucleotide pair

c. Structure of DNA

Check
Your Progress

1. **What feature of DNA's X-ray diffraction pattern indicated that it is a double helix?**
2. **What did Watson and Crick decide about the bases that made their model consistent with Chargaff's rules?**

Answers: 1. The central X suggested that DNA is a double helix. 2. Because A is hydrogen-bonded to T, and G is hydrogen-bonded to C, the amount of A = T and the amount of G = C, as dictated by Chargaff's rules.

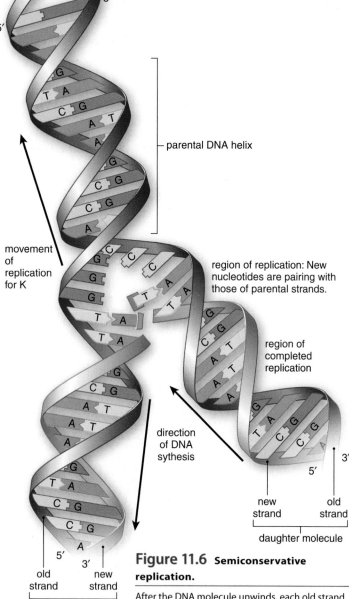

Figure 11.6 **Semiconservative replication.**

After the DNA molecule unwinds, each old strand serves as a template for the formation of a new strand. After replication is complete, there are two daughter DNA molecules identical to each other and to the original double helix.

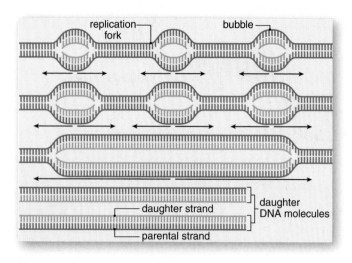

Figure 11.7 **Eukaryotic replication.**

In eukaryotes, replication occurs at numerous replication forks. The replication bubbles created spread out until they meet.

Replication of DNA

When organisms grow or heal any injuries, cells divide. Before division occurs, each new cell requires an exact copy of the parent cell's DNA or it will not be able to function. **DNA replication** refers to the process of copying a DNA molecule.

During DNA replication, the two DNA strands are separated, and each old strand of the parent molecule serves as a template for a new strand in a daughter molecule (**Fig. 11.6**). A **template** is most often a mold used to produce a shape complementary to itself. DNA replication is also **semiconservative** because one of the two old strands is conserved, or present, in each daughter molecule.

Before replication begins, the two strands that make up parental DNA are hydrogen-bonded to one another. Replication requires the following steps:

1. **Unwinding.** The old strands that make up the parent DNA molecule are unwound and "unzipped" (i.e., the hydrogen bonds between the paired bases are broken). A special enzyme called *helicase* unwinds the molecule.
2. **Complementary base pairing.** New complementary nucleotides, always present in the nucleus, are positioned by the process of complementary base pairing.
3. **Joining.** The complementary nucleotides join to form new strands. Each daughter DNA molecule contains an old strand and a new strand.

Step 3 is carried out by an enzyme complex called **DNA polymerase.** Also, we should note that DNA polymerase only joins nucleotides in a 5′ to 3′ direction, as shown in figure 11.6. Then, the enzyme **DNA ligase** seals any breaks in the sugar-phosphate backbone.

In Figure 11.6, the backbones of the parent molecule (original double strand) are bluish. Following replication, the daughter molecules each have an aqua backbone (new strand) and a bluish backbone (old strand). A daughter DNA double helix has the same sequence of base pairs as the parent DNA double helix had originally. Complementary base pairing has allowed this sequence to be maintained.

In eukaryotes, DNA replication begins at numerous origins of replication along the length of the chromosome. Replication proceeds in both directions, forming so-called "replication bubbles." The replication bubbles spread in both directions until they meet (**Fig. 11.7**). Although eukaryotes replicate their DNA at a fairly slow rate—500 to 5,000 base pairs per minute—there are many individual origins of replication. Therefore, eukaryotic cells complete the replication of the diploid amount of DNA (in humans, over 3 billion base pairs) in a matter of hours!

RNA Structure and Function

RNA (ribonucleic acid) is made up of nucleotides containing the sugar ribose, thus accounting for its name. The four nucleotides that make up an RNA molecule have the following bases: adenine (A), **uracil (U)**, cytosine (C), and guanine (G). Notice that in RNA, the base uracil replaces the base thymine (**Fig. 11.8**).

RNA, unlike DNA, is single-stranded, but the single RNA strand sometimes doubles back on itself, allowing complementary base pairing to occur. Similarities and differences between these two nucleic acid molecules are listed in **Table 11.1.**

In general, RNA is a helper to DNA, allowing protein synthesis to occur according to the genetic information that DNA provides. There are three major types of RNA, each with a specific function in protein synthesis.

Messenger RNA

Messenger RNA (mRNA) is produced in the nucleus, or in the nucleoid of prokaryotes. DNA serves as a template for its formation during a process called transcription. Which genes are transcribed is tightly controlled in each type of cell and accounts for why some cells are nerve cells and others are muscle cells, for example. Once formed, mRNA carries genetic information from DNA to the ribosomes in the cytoplasm, where protein synthesis occurs.

Transfer RNA

Transfer RNA (tRNA) is also produced in the nucleus, and a portion of DNA also serves as a template for its production. Because DNA serves as a template for both rRNA and tRNA, it is obvious that not all DNA specifies protein synthesis.

Appropriate to its name, tRNA transfers amino acids to the ribosomes, where the amino acids are joined, forming a protein. There are 20 different types of amino acids in proteins; therefore, at least 20 different tRNAs must be functioning in the cell. Each type of tRNA carries only one type of amino acid.

Ribosomal RNA

In eukaryotic cells, **ribosomal RNA (rRNA)** is produced in the nucleolus of a nucleus, where a portion of DNA also serves as a template for its formation. Ribosomal RNA joins with proteins made in the cytoplasm to form the subunits of **ribosomes,** one large and one small. Each subunit has its own mix of proteins and rRNA. The subunits leave the nucleus and come together in the cytoplasm when protein synthesis is about to begin.

Proteins are synthesized at the ribosomes, which look like small granules in low-power electron micrographs. Ribosomes in the cytoplasm may be free or in clusters called polyribosomes. Often they are found attached to the edge of the *endoplasmic reticulum (ER),* a system of tubules and saccules within the cytoplasm. Proteins synthesized by free ribosomes or polyribosomes are used in the cytoplasm, and those synthesized by attached ribosomes end up in the ER. From there, the protein is carried in a transport vesicle to the Golgi apparatus for modification and transport to the plasma membrane, where it may leave the cell.

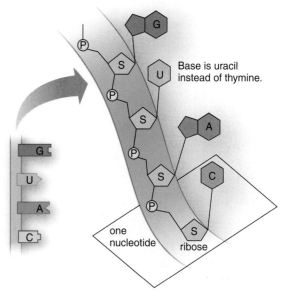

Figure 11.8 Structure of RNA.

Like DNA, RNA is a polymer of nucleotides. In an RNA nucleotide, the sugar ribose is attached to a phosphate molecule and to a base, either G, U, A, or C. Notice that in RNA, the base uracil (U) replaces thymine as one of the bases. RNA is usually single-stranded, whereas DNA is double-stranded.

Table 11.1 | Comparison of DNA and RNA

DNA-RNA SIMILARITIES	
Both are nucleic acids.	
Both are composed of nucleotides.	
Both have a sugar-phosphate backbone.	
Both have four different types of bases.	

DNA-RNA DIFFERENCES	
DNA	**RNA**
Found in nucleus	Found in nucleus and cytoplasm
Genetic material	Helper to DNA
Sugar is deoxyribose	Sugar is ribose
Bases are A, T, C, G	Bases are A, U, C, G
Double-stranded	Single-stranded
Is transcribed (to give mRNA, tRNA, and rRNA)	mRNA is translated (to give proteins)

Check Your Progress

1. Why is DNA replication said to be semiconservative?
2. Contrast the structure of RNA with that of DNA.
3. Contrast the function of messenger RNA with that of transfer RNA.

Answers: 1. Each new DNA molecule contains one new strand and one old strand.   2. RNA contains ribose, while DNA contains deoxyribose. RNA contains uracil instead of thymine. RNA is usually single-stranded, while DNA is double-stranded.   3. Messenger RNA carries information from the nucleus to the ribosomes in the cytoplasm. Transfer RNA carries amino acids to ribosomes.

1. DNA: sequence of bases is geneic information.

Nucleus

template DNA strand

2. Transcription: genetic information is passed from DNA to mRNA.

mRNA

Cytoplasm

mRNA and its codons

ribosome

tRNAs and their anticodons

polypeptide — Gly   Arg   Thr

trait determined by protein

3. Translation: Amino acids in a polypeptide are sequenced as specified by the template DNA strand.

Figure 11.9   Flow of genetic information.

1. Genetic information (genes) consists of a particular sequence of bases. **2.** The process of transcription in the nucleus passes this genetic information to an mRNA molecule. The mRNA moves to a ribosome in the cytoplasm. **3.** During the process of translation at a ribosome in the cytoplasm, the genetic information results in a particular sequence of amio acids in a protein. The activity or inactivity of this protein determines a trait and often whether the phenotype is normal or not normal.

11.2 | Gene Expression

In the early 1900s, the English physician Sir Archibald Garrod observed that family members often have the same metabolic disorder, and he said most likely they all lacked the same functioning enzyme in a metabolic pathway. He introduced the phrase *inborn error of metabolism* to describe this relationship. Garrod's findings were generally overlooked until George Beadle and Edward Tatum devised a way in 1940 to confirm his hypothesis. These investigators performed a series of experiments utilizing red bread mold and found that each of their mutant molds was indeed unable to produce a particular enzyme. This led them to propose that one gene directs the synthesis of one enzyme. Today we know that genes are also responsible for specifying structural proteins in cells.

From DNA to RNA to Protein

It's one thing to know that genes specify proteins and another to explain how they do it. Molecular genetics, which largely began when Watson and Crick discovered the structure of DNA in the 1950s, is able to explain exactly how genes control the building of a specific type protein. Consider that in eukaryotes, DNA resides in the nucleus but RNA is found both in the nucleus and in the cytoplasm where protein synthesis occurs. This means that DNA must pass its genetic information to mRNA, which then actively participates in protein synthesis. The *central dogma of molecular biology* states that genetic information flows from DNA to RNA to protein (**Fig. 11.9**).

Gene expression has occurred when a gene's product—the protein it specifies—is functioning in a cell. Specifically, gene expression requires two processes called transcription and translation. In eukaryotes, transcription takes place in the nucleus and translation takes place in the cytoplasm. During **transcription,** a portion of DNA serves as a template for mRNA formation. During **translation,** the sequence of mRNA bases (which are complementary to those in template DNA) determines the sequence of amino acids in a polypeptide. So, in effect, molecular genetics tells us that genetic information lies in the sequence of the bases in DNA, which through mRNA determines the sequence of amino acids in a protein. tRNA assists mRNA during protein synthesis by bringing amino acids to the ribosomes. Proteins differ from one another by the sequence of their amino acids, and proteins determine the structure and function of cells and indeed the phenotype of the organism.

The Genetic Code

The information contained in DNA and RNA is written in a different chemical language from that in the protein that is specified by the DNA and RNA. The cell needs a way to translate one language into the other, and it uses the genetic code.

But how can the four bases of RNA (A, C, G, U) provide enough combinations to code for the 20 amino acids found in proteins? If the code were a singlet code (one base stands for an amino acid), only four amino acids could be encoded. If the code were a doublet (any two bases stand for one amino acid), it would still not be possible to code for 20 amino acids. But if the code were a triplet, the four bases could supply 64 different triplets, far more than needed to code for 20 different amino acids. It should come as no surprise, then, to learn that the code is a **triplet code.**

Each three-letter (nucleotide) unit of an mRNA molecule is called a **codon,** which codes for a single amino acid (**Fig. 11.10**). Sixty-one triplets correspond to a particular amino acid; the remaining three are stop codons that signal the end of a polypeptide. The codon that stands for the amino acid methionine is also a start codon that signals initiation of translation. Most amino acids have more than one codon, which offers some protection against possibly harmful mutations that could otherwise change the sequence of the bases.

Cracking the genetic code was no simple matter. Researchers performed a series of experiments in which they added artificial mRNA to a medium containing bacterial ribosomes and a mixture of amino acids. By comparing the bases in the mRNA with the resulting polypeptide, they were able to learn the code. For example, an mRNA

Second base

		U	C	A	G	
First base	**U**	UUU ⎤ UUC ⎦ phenylalanine (Phe) UUA ⎤ UUG ⎦ leucine (Leu)	UCU ⎤ UCC ⎥ UCA ⎥ UCG ⎦ serine (Ser)	UAU ⎤ UAC ⎦ tyrosine (Tyr) **UAA** stop **UAG** stop	UGU ⎤ UGC ⎦ cysteine (Cys) **UGA** stop UGG tryptophan (Trp)	U C A G
	C	CUU ⎤ CUC ⎥ CUA ⎥ CUG ⎦ leucine (Leu)	CCU ⎤ CCC ⎥ CCA ⎥ CCG ⎦ proline (Pro)	CAU ⎤ CAC ⎦ histidine (His) CAA ⎤ CAG ⎦ glutamine (Gln)	CGU ⎤ CGC ⎥ CGA ⎥ CGG ⎦ arginine (Arg)	U C A G
	A	AUU ⎤ AUC ⎥ isoleucine (Ile) AUA ⎦ **AUG** methionine (Met) **(start)**	ACU ⎤ ACC ⎥ ACA ⎥ ACG ⎦ threonine (Thr)	AAU ⎤ AAC ⎦ asparagine (Asn) AAA ⎤ AAG ⎦ lysine (Lys)	AGU ⎤ AGC ⎦ serine (Ser) AGA ⎤ AGG ⎦ arginine (Arg)	U C A G
	G	GUU ⎤ GUC ⎥ GUA ⎥ GUG ⎦ valine (Val)	GCU ⎤ GCC ⎥ GCA ⎥ GCG ⎦ alanine (Ala)	GAU ⎤ GAC ⎦ aspartic acid (Asp) GAA ⎤ GAG ⎦ glutamic acid (Glu)	GGU ⎤ GGC ⎥ GGA ⎥ GGG ⎦ glycine (Gly)	U C A G

Third base

Figure 11.10 Messenger RNA codons.

Notice that in this chart, each of the codons are composed of three letters. As an example, find the rectangle where C is the first base and A is the second base. U, C, A, or G can be the third base. CAU and CAC are codons for histidine; CAA and CAG are codons for glutamine.

with a sequence of repeating guanines (GGG′GGG′...) would encode a string of glycine amino acids, so they concluded that the mRNA codon GGG specified the amino acid glycine in a protein.

The genetic code is just about universal in living things. This suggests that the code dates back to the very first organisms on Earth and that all living things are related.

Transcription

During transcription of DNA, a strand of RNA forms that is complementary to a portion of DNA. While all three classes of RNA are formed by transcription, we will focus on transcription to create mRNA.

mRNA Is Formed Transcription begins when the enzyme **RNA polymerase** binds tightly to a **promoter,** a region of DNA with a special nucleotide sequence that marks the beginning of a gene. RNA polymerase opens up the DNA helix just in front of it so that complementary base pairing can occur. Then the enzyme adds new RNA nucleotides that are complementary to the template DNA strand, and an mRNA molecule results (**Fig. 11.11**). The resulting **mRNA transcript** is a complementary copy of the sequence of bases in the template DNA strand. Once transcription is completed, the mRNA is now ready to be processed before it leaves the nucleus for the cytoplasm.

Check Your Progress

1. State the central dogma of molecular biology.
2. Why would you expect the genetic code to be a triplet code?
3. Describe how transcription of DNA results in an mRNA.

Answers: 1. The central dogma states that the information in genes flows from DNA to RNA to protein, and that proteins determine traits. 2. With four bases, a singlet code or doublet code would not give enough combinations, but a triplet code would give enough combinations to code for the 20 amino acids found in cellular proteins. 3. Using a DNA strand as a template, RNA polymerase adds complementary RNA nucleotides to a growing RNA strand.

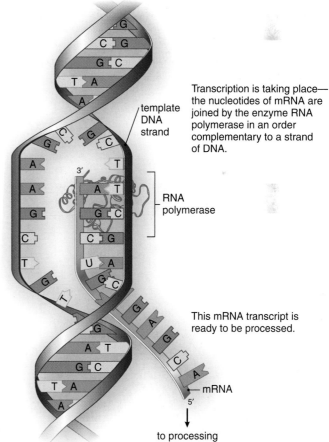

Transcription is taking place—the nucleotides of mRNA are joined by the enzyme RNA polymerase in an order complementary to a strand of DNA.

template DNA strand

RNA polymerase

This mRNA transcript is ready to be processed.

mRNA

to processing

Figure 11.11 Transcription to form mRNA.

During transcription, complementary RNA is made from a DNA template. A portion of DNA unwinds and unzips at the point of attachment of RNA polymerase. A strand of RNA, such as mRNA, is produced when complementary bases join in the order dictated by the sequence of bases in the template DNA strand.

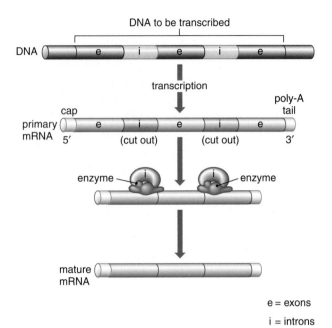

DNA to be transcribed

DNA

transcription

cap   poly-A tail

primary mRNA

5′   (cut out)   (cut out)   3′

enzyme   enzyme

mature mRNA

e = exons
i = introns

Figure 11.12 mRNA processing.

Primary mRNA results when both exons and introns are transcribed from DNA. A "cap" and a poly-A "tail" are attached to the ends of the primary RNA transcript, and the introns are removed so that only the exons remain. This mature mRNA molecule moves into the cytoplasm of the cell where translation occurs.

mRNA Is Processed   The newly synthesized *primary-mRNA* must be processed in order for it to be used properly. Processing occurs in the nucleus of eukaryotic cells. Three steps are required: capping, the addition of a poly-A tail, and splicing (**Fig. 11.12**). After processing, the mRNA is called a *mature mRNA* molecule.

The first nucleotide of the primary-mRNA is modified by the addition of a cap that is composed of an altered guanine nucleotide. On the 3′ end, enzymes add a poly-A tail, a series of adenosine nucleotides. These modifications provide stability to the mRNA; only those that have a cap and tail remain active in the cell.

Most genes in humans are interrupted by segments of DNA that do not code for protein. These portions are called *introns* because they are intervening segments. The other portions of the gene, called *exons,* contain the protein-coding portion of the gene. In *primary-mRNA splicing,* the introns are removed, and the exons joined together. The end result is a mature mRNA molecule consisting of continuous exons. A surprising finding of late has been that introns may play a regulatory function in gene expression.

Ordinarily, processing brings together all the exons of a gene. In some instances, however, cells use only certain exons rather than all of them to form the mature RNA transcript. The result is a different protein product in each cell. In other words, this so-called *alternative mRNA splicing* can potentially increase the possible number of protein products that can be made from a single gene.

After the mRNA strand is processed, it passes from the cell nucleus into the cytoplasm. There it becomes associated with the ribosomes.

Translation: An Overview

Translation is the second step by which gene expression leads to protein (polypeptide) synthesis. Translation requires several enzymes, mRNA, and the other two types of RNA: transfer RNA and ribosomal RNA.

Transfer RNA Brings Amino Acids to the Ribosomes   Each tRNA is a single-stranded nucleic acid that doubles back on itself such that complementary base pairing results in the cloverleaf-like shape shown in **Figure 11.13**. There is at least one tRNA molecule for each of the 20 amino acids found in proteins. The amino acid binds to one end of the molecule. The opposite end of the molecule contains an **anticodon,** a group of three bases that is complementary to a specific codon of mRNA.

During translation, the order of codons in mRNA determines the order in which tRNAs bind at the ribosomes. When a tRNA–amino acid complex comes to the ribosome, its anticodon pairs with an mRNA codon. For example, if the codon is ACC, what is the anticodon, and what amino acid will be attached to the tRNA molecule? From Figure 11.10, we can determine this:

Codon (mRNA)	Anticodon (tRNA)	Amino Acid (protein)
ACC	UGG	Threonine

After translation is complete, a protein contains the sequence of amino acids originally specified by DNA. This is the genetic information that DNA stores and passes on to each cell during the cell cycle, and passes on to the next generation of individuals. DNA's sequence of bases determines the proteins in a cell, and the proteins in turn determine whether an organism is a human being or a giraffe, as an example.

Figure 11.13 **tRNA structure and function.**

a. A tRNA, which is single-stranded but folded, has an amino acid attached to one end and an anticodon at the other end. The anticodon is complementary to a codon. **b.** A ribosome has two binding sites for tRNA, called the P site and the A site. A peptide attached to tRNA at the P site will be passed to an amino acid attached to a tRNA as soon as it arrives at the A site. **c.** The pairing between codon and anticodon at a ribosome ensures that the sequence of amino acids in a polypeptide is the same sequence directed originally by DNA.

Ribosomal RNA Is in Ribosomes Ribosomes are the small structural bodies where translation occurs. Ribosomes are composed of many proteins and several ribosomal RNAs (rRNAs). In eukaryotic cells, rRNA is produced in a nucleolus within the nucleus. There it joins with proteins manufactured in the cytoplasm to form two ribosomal subunits, one large and one small. The subunits leave the nucleus and join together in the cytoplasm to form a ribosome just as protein synthesis begins.

A ribosome has a binding site for mRNA as well as binding sites for two tRNA molecules at a time (Fig. 11.13). These binding sites facilitate complementary base pairing between tRNA anticodons and mRNA codons. The P binding site is for a tRNA attached to a *p*eptide, and the A binding site is for a newly arrived tRNA attached to an *a*mino acid.

As soon as the initial portion of mRNA has been translated by one ribosome, and the ribosome has begun to move down the mRNA, another ribosome attaches to the same mRNA. Therefore, several ribosomes are often attached to and translating the same mRNA, allowing the cell to produce many copies of the same protein at a time. The entire complex is called a **polyribosome** (**Fig. 11.14**).

Figure 11.14 **Polyribosome structure and function.**

a. Several ribosomes, collectively called a polyribosome, move along the same mRNA. Therefore, several proteins can be made at the same time. **b.** Electron micrograph of a polyribosome.

Figure 11.15   Initiation.

A small ribosomal subunit binds to mRNA; an initiator tRNA's anticodon binds to its codon, and the large ribosomal subunit completes the ribosome.

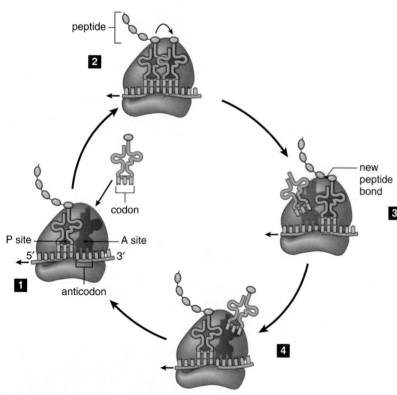

Figure 11.16   Elongation cycle.

1 Two tRNAs can be at a ribosome at one time. **2** The tRNA at the P site passes its peptide to the tRNA at the A site. **3** The tRNA at the P sites leaves. **4** The ribosome moves forward (translocation), and the tRNA-peptide complex is now at the P site. A new tRNA–amino acid complex comes to the A site.

Translation Has Three Phases

Polypeptide synthesis has three phases: initiation, an elongation cycle, and termination. Enzymes are required for each of the steps to occur, and energy is needed for the first two steps.

1. During *initiation,* mRNA binds to the smaller of the two ribosomal subunits; then the larger subunit associates with the smaller one.
2. During an *elongation cycle,* a peptide lengthens one amino acid at a time. The growing peptide is transferred from the outgoing tRNA to the incoming tRNA–amino acid complex, and then the outgoing tRNA leaves. The ribosome then translates the next mRNA codon as it receives a new incoming tRNA–amino acid complex.
3. *Termination* occurs at any one of three special codons that mean stop. The ribosomal subunits and mRNA dissociate, and the completed polypeptide is released.

Initiation   During initiation, a small ribosomal subunit, the mRNA, an *initiator tRNA* bound to the amino acid methionine, and a large ribosomal subunit all come together (**Fig. 11.15**):

- The small ribosomal subunit attaches to the mRNA in the vicinity of the start codon (AUG).
- The anticodon of the initiator tRNA–methionine complex pairs with this codon.
- The large ribosomal subunit joins to the small subunit.

Elongation Cycle   As discussed, a ribosome has two binding sites for tRNA where the tRNA's anticodon binds to a codon of mRNA. During the elongation cycle (**Fig. 11.16**):

- A tRNA at the P site usually bears a peptide built up previously. (See **1**.)
- This tRNA passes its peptide to tRNA–amino acid at the A site. The tRNA at the P site leaves. (See **2** and **3**.)
- During **translocation**, the tRNA-peptide moves to the P site, and the codon at the A site is ready for the next tRNA–amino acid. (See **4**.)

The complete cycle—complementary base pairing of new tRNA, transfer of peptide chain, and translocation—is repeated at a rapid rate (about 15 times each second in the bacterium *Escherichia coli*). The outgoing tRNA picks up another amino acid and is ready to return to the ribosome.

Termination   When a stop codon (see Fig. 11.10) appears at the A site of a ribosome, termination occurs. A protein called a **release factor** binds to the stop codon and also cleaves the polypeptide from the last tRNA. Then, the polypeptide and the assembled components that carried out protein synthesis are separated from one another.

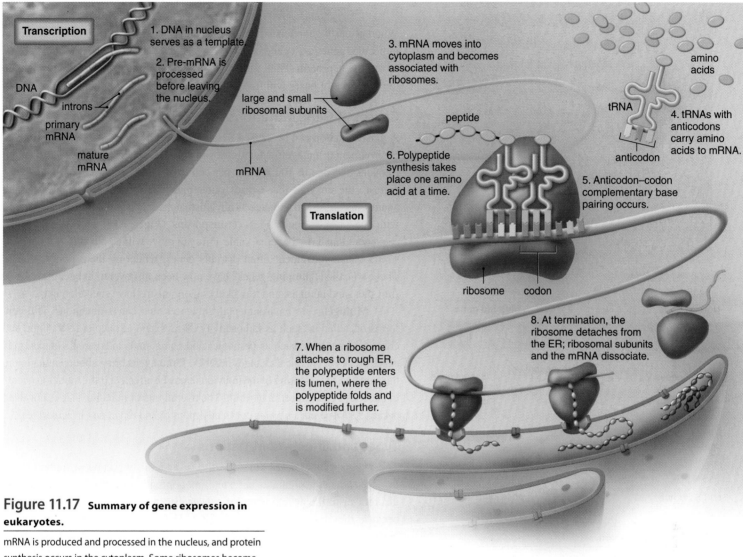

Figure 11.17   Summary of gene expression in eukaryotes.

mRNA is produced and processed in the nucleus, and protein synthesis occurs in the cytoplasm. Some ribosomes become attached to rough ER; the polypeptide then enters the ER, where it folds and may be further modified.

Review of Gene Expression

Genes are segments of DNA that code for all the proteins in a cell. A gene is expressed when its protein product has been made. **Figure 11.17** reviews transcription and mRNA processing in the nucleus, as well as translation during protein synthesis in the cytoplasm of a eukaryotic cell.

Some ribosomes remain free in the cytoplasm, and others become attached to rough ER. In the latter case, the polypeptide enters the lumen of the ER by way of a channel, where it can be further processed by the addition of sugars. Transport vesicles carry the protein to other locations in the cell, including the Golgi apparatus, which may modify it further and package it in a vesicle for transport out of the cell, or cause it to become embedded in the plasma membrane. Proteins have innumerable functions in cells, from enzymatic to structural. Proteins also have functions outside the cell as well. Together they account for the structure and function of cells, tissues, organs, and the organism.

Check Your Progress

1. **Describe the events of translation.**
2. **Explain how termination occurs.**

Answers: 1. During translation, mRNA directs the synthesis of a polypeptide. Messenger RNA has a sequence of codons that dictate the order in which tRNA–amino acid complexes come to a ribosome. Here, during elongation, the peptide is passed from a departing tRNA to an arriving tRNA. In the end, the polypeptide has a sequence of amino acids as dictated by DNA.   2. Termination occurs when the ribosome encounters a special "stop" codon in the mRNA. A release factor binds, the mRNA and ribosomal subunits dissociate, and the polypeptide is released.

a.

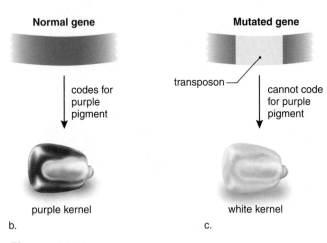

b.   c.

Figure 11.18 Transposon.

a. Barbara McClintock, shown here instructing a student, won the 1983 Nobel Prize in Physiology or Medicine for being the first to discover transposons. She worked with maize (corn). **b.** In corn, a purple-coding gene ordinarily codes for a purple pigment. **c.** A transposon "jumps" into the purple-coding gene. This mutated gene is unable to code for purple pigment, and a white kernel results.

Check Your Progress

1. Define a gene mutation, and describe the two major types of mutations.
2. Explain how a transposon could cause a mutation.

Answers: 1. A gene mutation is a change in the sequence of bases in a gene. Frameshift mutations change the reading frame; point mutations change one base. 2. The movement of transposons can interrupt a gene, rendering it non-functional.

Genes and Gene Mutations

Whereas early geneticists thought of genes as particles on a chromosome, in molecular genetics, a gene is a sequence of DNA bases that code for a product, most often a protein. Therefore, it follows that a **gene mutation** is a change in the sequence of bases within a gene.

Causes of Gene Mutations

A gene mutation can be caused by an error in replication, a transposon, or an environmental mutagen. Mutations due to DNA replication errors are rare; a frequency of one in 100 million per cell division is often quoted. DNA polymerase, the enzyme that carries out replication, proofreads the new strand against the old strand and detects any mismatched pairs, which are then replaced with the correct nucleotides. **Transposons (Fig. 11.18)** are specific DNA sequences that have the remarkable ability to move within and between chromosomes. Their movement often disrupts genes, rendering them nonfunctional. These so-called "jumping genes" have now been discovered in bacteria, plants, fruit flies, and humans, and it is likely all organisms have such elements.

Mutagens are environmental influences that cause mutations. Different forms of radiation, such as radioactivity, X-rays, and ultraviolet (UV) light, and chemical mutagens, such as pesticides and compounds in cigarette smoke, may cause breaks or chemical changes in DNA. If mutagens bring about a mutation in the gametes, the offspring of the individual may be affected. If the mutation occurs in the body cells, cancer may result. The overall rate of mutation is low, however, because *DNA repair enzymes* constantly monitor and repair any irregularities.

Types and Effects of Mutations

The effects of mutations vary greatly. In general, we know a mutation has occurred when the organism has a malfunctioning protein that leads to a genetic disorder or to the development of cancer. But many mutations go undetected because they have no observable effect. The severity of a mutation usually depends on whether it affects one or more codons of a gene. Many mutations have no detectable effect on the protein's function. For example, a base change may not alter the amino acid specified by the gene if it occurs at the third position of a codon (see Fig. 11.10). Such cases are often called *silent mutations*.

Point mutations involve a change in a single DNA nucleotide, and the results can be more or less severe depending on the particular base change that occurs. For example, if valine instead of glutamate occurs in the β chain of hemoglobin at one particular location, sickle cell disease results. The abnormal hemoglobin stacks up inside cells causing them to become sickle-shaped.

A **frameshift mutation** is caused by extra or missing nucleotides in a DNA sequence. They are usually much more severe than point mutations because codons are read from a specific starting point. Therefore, all downstream codons are affected. For instance, if the letter C is deleted from the sentence THE CAT ATE THE RAT, the "reading frame" is shifted. The sentence now reads THE ATA TET HER AT—something that doesn't make sense. Likewise, a frameshift mutation in a gene often renders the protein nonfunctional because it no longer makes sense. The movement of a transposon can cause a frameshift mutation.

11.3 | DNA Technology

An understanding of how DNA functions in cells allowed scientists to manipulate the genes of organisms. We can clone genes (make identical copies) and then use the genes for various purposes. During **genetic engineering,** a cloned gene can be inserted into the genome of an organism, which is then called a **transgenic organism.** Because the genetic code is almost universal, it's possible to transfer cloned genes from virtually any organism into bacteria, plants, and animals.

Recombinant DNA Technology

Recombinant DNA (rDNA) contains DNA from two or more different sources (**Fig. 11.19**). To make rDNA, a researcher needs a **vector,** a piece of DNA that acts as a carrier for the foreign DNA. One common vector is a *plasmid*, which is a small accessory ring of DNA found in bacteria.

Two enzymes are needed to introduce foreign DNA into plasmid DNA: (1) **restriction enzymes** that can cleave DNA at specific places; and (2) DNA ligase, which can seal the foreign DNA into an opening in a plasmid. They are called restriction enzymes because they *restrict* the growth of viruses by cutting their DNA, but they can also be used as molecular scissors to cut DNA at a specific site. For example, the restriction enzyme *Eco*RI always cuts in this manner when DNA has the base sequence GAATTC:

If a plasmid is cut with *Eco*RI, a gap now exists into which a piece of foreign DNA can be placed if it ends in bases complementary to those exposed by the restriction enzyme. To ensure this, it is only necessary to cleave the foreign DNA with the same restriction enzyme. The overhanging bases at the ends of the two DNA molecules are called "sticky ends" because they can bind a piece of foreign DNA by complementary base pairing. Sticky ends facilitate the insertion of foreign DNA into vector DNA.

DNA ligase, the enzyme that functions in DNA replication to seal breaks in a double-stranded helix, seals the foreign piece of DNA into the plasmid. Molecular biologists often give the rDNA to bacterial cells, which readily take up recombinant plasmids when the cells are treated to make them more permeable. Thereafter, as the bacteria replicate the plasmid, the gene is cloned. Cloned genes have many uses. A scientist may allow the **genetically modified** bacterial cells to express the cloned gene and retrieve the protein. Or, in some cases, copies of the cloned gene may be removed from the bacterial cells and then introduced into another organism, such as a corn plant, to produce a transgenic organism.

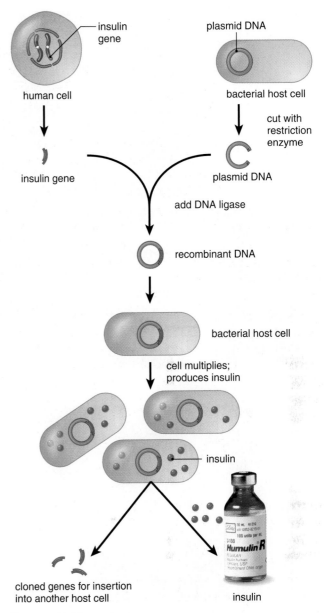

Figure 11.19   Recombinant DNA technology.

Production of insulin is one example of how recombinant DNA technology can benefit humans. Human DNA and plasmid DNA are spliced together. Gene cloning is achieved when a host cell takes up the recombinant plasmid, and as the cell reproduces, the plasmid is replicated. Multiple copies of the gene are now available. If the insulin gene functions normally as expected, insulin may also be retrieved.

Transgenic Organisms

The term **biotechnology** refers to the use of natural biological systems to create a product or achieve some other end desired by human beings. Today, bacteria, plants, and animals can be genetically engineered to make biotechnology products. Various uses of transgenic organisms are shown in **Figure 11.20.**

Transgenic Bacteria

Recombinant DNA technology is used to produce transgenic bacteria, which are grown in huge vats called bioreactors. The gene product is usually collected from the growth medium. Products now on the market that are produced by bacteria include insulin, human growth hormone, t-PA (tissue plasminogen activator), and hepatitis B vaccine. Bacteria can be selected for their ability to degrade a particular substance, such as oil, and this ability can then be enhanced by genetic engineering. Similarly, some mining companies are testing genetically engineered organisms that have improved bioleaching capabilities. Bacteria can also be engineered to produce organic chemicals used in industry.

Transgenic Plants and Animals

Foreign genes have been transferred to cotton, corn, and potato strains to make these plants resistant to pests by causing their cells to produce an insect toxin. Similarly, soybeans have been made resistant to a common herbicide. Some corn and cotton plants are both pest- and herbicide-resistant. These and other genetically engineered crops have increased yields. Genetically engineered plants are revolutionizing both agriculture and medicine—there are now plants that can produce human proteins, such as hormones, clotting factors, and antibodies!

Foreign genes have also been inserted into the eggs of animals, often to give them the gene for bovine growth hormone (bGH). The procedure has resulted in larger fishes, cows, pigs, rabbits, and sheep. Gene "pharming," the use of transgenic farm animals to produce pharmaceuticals, is being pursued by a number of firms. Genes that code for therapeutic and diagnostic proteins are incorporated into an animal's DNA, and the proteins appear in the animal's milk.

To achieve a sufficient number of animals that can produce the product, transgenic animals are sometimes cloned. For many years, it was believed that adult vertebrate animals could not be cloned. Although each cell contains a copy of all the genes, certain genes are turned off in mature, specialized cells, and cloning an adult vertebrate would require that all the genes of an adult cell be turned on again, which had long been thought impossible. In 1997, however, Scottish scientists produced a cloned sheep, which they named Dolly. Since then, calves, goats, and many other animals, including some endangered species, have also been cloned. After enucleated eggs from a donor are microinjected with 2n nuclei from a transgenic animal, they are coaxed to begin development in vitro. Development continues in host females until the clones are born. The offspring are clones because all have the identical genotype of the adult that donated the 2n nuclei. Now that scientists have a way to clone animals, this procedure will undoubtedly be used routinely to procure biotechnology products.

Figure 11.20 Uses of transgenic organisms.

a. Transgenic bacteria, grown industrially, are used to produce medicines. Transgenic animals and plants increase the food supply. **b.** These salmon (*right*) are much heavier and reproduce much sooner because of the growth hormone genes they received as embryos. **c.** Pests didn't consume unblemished peas because the plants that produced them received genes for pest inhibitors. **d.** Transgenic bacteria can also be used for environmental cleanup. These bacteria are able to break down oil. **e.** Transgenic corn can resist tough herbicides used to control weeds, greatly increasing crop yields.

Polymerase Chain Reaction

The **polymerase chain reaction (PCR)** can create millions of copies of a segment of DNA very quickly in a test tube without the use of a vector or a host cell. PCR is very sensitive—it *amplifies* (makes copies of) a targeted DNA sequence that can be less than one part in a million of the total DNA sample!

PCR requires the following: DNA polymerase, the enzyme that carries out DNA replication; a set of *primers* that specify the segment of DNA to be amplified; and a supply of nucleotides for the new DNA strands. Nowadays, PCR is automated, thanks to the discovery of a heat-tolerant DNA polymerase from the bacterium *Thermus aquaticus,* which lives in hot springs. The process is shown in **Figure 11.21.** During a PCR cycle, the mixture of DNA and primers is heated to 95°C to separate the two strands of the double helix. Next, the temperature is lowered so that the primers can bind to the single strands of DNA and DNA polymerase can copy the DNA. Then, the procedure is repeated an additional twenty to thirty times. The amount of DNA doubles with each replication cycle; after one cycle, there are two copies of the targeted DNA sequence; after two cycles, there are four copies, and so forth. Twenty to thirty PCR cycles can yield trillions of copies of the specified DNA segment!

DNA amplified by PCR has many uses. PCR is often used to copy genes for use in creating recombinant DNA. PCR has also been used to amplify mitochondrial DNA sequences in order to decipher the evolutionary history of human populations. Because PCR is so efficient, it has even been possible to sequence DNA taken from a 76,000-year-old mummified human brain.

DNA Fingerprinting

DNA fingerprinting makes use of noncoding sections of DNA that consist of two to five bases repeated over and over again, as in ATCATCATCATC. People differ by how many times such a sequence is repeated, but how can this difference be detected? PCR is used to amplify the entire repeated region. The greater the number of repeats, the longer the segment of DNA that will result after PCR amplification is done. People can be heterozygous for the number of repeats as in **Figure 11.22,** so the DNA from both homologues has to be amplified separately. Following PCR, the DNA is subjected to gel electrophoresis, a process that separates DNA molecules through a jellylike material according to size. The shorter the segment of DNA, the further it migrates.

DNA fingerprinting has many uses. Medically, it can identify the presence of a viral infection or a mutated gene that could predispose someone to cancer. In forensics, DNA fingerprinting from a single sperm is enough to identify a suspected rapist because the DNA is amplified by PCR. It can also be used to identify the parents of a child or identify the remains of someone who died, such as a victim of the September 11, 2001, terrorist attacks. In the future, we will undoubtedly see more applications of recombinant DNA technology that will greatly enrich our lives and improve our health.

▪ Check Your Progress

1. **What two enzymes are necessary to produce recombinant DNA?**
2. **Define transgenic organisms.**

Answers: 1. A restriction enzyme and DNA ligase are needed to produce recombinant DNA.   2. A transgenic organism is one that has been given recombinant DNA (rDNA) so that it has a new or modified trait.

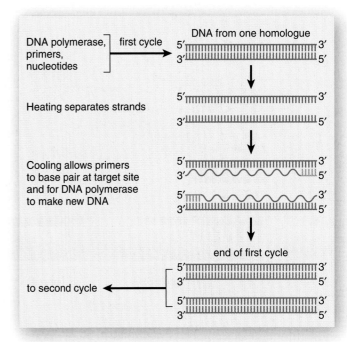

Figure 11.21   Polymerase chain reaction.

At the start of PCR, double-stranded DNA is mixed with DNA polymerase, two primers (gold and pink) that will flank the target DNA sequence, and a supply of the four nucleotides in DNA. At the end of the first cycle, there are two copies of the target DNA sequence. After 25–30 cycles, millions of copies of the target DNA sequence are present.

Figure 11.22   DNA fingerprinting.

Following PCR, gel electrophoresis reveals the comparative pattern of repeats in the DNA. Here, DNA fingerprinting is being used to identify a criminal.

11.4 | Genomics and Proteomics

In the previous century, researchers discovered the structure of DNA, how DNA replicates, and how protein synthesis occurs. Genetics in the twenty-first century concerns **genomics,** the study of genomes—our genes and **intergenic DNA** sequences, and those of other organisms. Researchers now know the sequence of all the base pairs along the length of the human chromosomes. The enormity of the task can be appreciated by knowing that, at the very least, our DNA contains 6 billion base pairs and approximately 20,500 genes. Many organisms have even larger genomes. Genes make up only 5% of the human genome, so genomics may help reveal the function of many intergenic DNA sequences.

Sequencing the Bases of the Human Genome

We now have a working draft of the base-pair sequence in the DNA of all our chromosomes. This feat was accomplished by the Human Genome Project, a 13-year effort that involved both university and private laboratories around the world. How did they do it? First, investigators developed a laboratory procedure that would allow them to decipher a short sequence of base pairs, and then they devised an instrument that would carry out this procedure automatically. Over the project's 13-year span, DNA sequencers were constantly improved to the point where modern instruments can automatically analyze up to 2 million base pairs of DNA in a 24-hour period.

Variations in Base Sequence

Scientists studying the human genome found that many small regions of DNA vary among individuals. For instance, there may be a difference in a single base within a gene (**Fig. 11.23***a, b*), or within an uncharacterized sequence between genes (Fig. 11.23*c*). One surprising finding is that some individuals even have additional copies of some genes (Fig. 11.23*d*)! Many of these differences have no ill effects, but some of them may increase or decrease an individual's susceptibility to disease.

Genome Comparisons

Researchers are comparing the human genome to the genomes of other species for clues to our evolutionary origins. One surprising discovery is that the genomes of all vertebrates are similar. Researchers were not surprised to find that the genes of chimpanzees and humans are 98% alike, but they did not expect to find that our sequence is also 85% like that of a mouse. Scientists also discovered that we share

a. Normal chromosomal DNA

b. Variation in the order of the bases within a gene due to a mutation

c. Variation in the order of the bases within an intergenic sequence due to a mutation

d. Variation in the gene copy number

Figure 11.23   Variations in DNA sequence.

a. A section of DNA containing three genes (A, B, and C) and two intergenic sequences may vary between individuals. **b.** Variation in the sequence of bases within a gene due to a mutation may alter the sequence of amino acids of the protein specified by the gene. **c.** An individual may possess a variation in the order of bases within an intergenic sequence due to a mutation. This mutation would not change the proteins encoded by genes A, B, or C. **d.** In addition to changes in base sequence, DNA is also subject to changes in gene copy number, that is, the number of copies of a particular gene. Note that now there are two copies of gene B.

a number of genes with much simpler organisms, including bacteria! As the genomes of more organisms are sequenced, genome comparisons will likely reveal evolutionary relationships between organisms never previously considered.

One study compared the human genome to that of chromosome 22 in chimpanzees. Among the many genes that differed in sequence were several of particular interest: a gene for proper speech development, several for hearing, and several for smell. The gene necessary for proper speech development is thought to have played an important role in human evolution. You can suppose that changes in hearing may have facilitated using language for communication between people. Changes in smell genes are a little more problematic. The researchers speculated that the olfaction genes may have affected dietary changes or sexual selection. Or, they may have been involved in traits other than smell (**Fig. 11.24**). It was a surprise to find that many of the other genes they located and studied are known to cause human diseases if abnormal. Perhaps comparing genomes would be a way of finding genes associated with human diseases.

Proteomics and Bioinformatics

The known sequence of bases in the human genome predicts that about 20,500 genes are translated into proteins; collectively, all of these proteins are referred to as the human **proteome.** The field of **proteomics** explores the structure and function of these cellular proteins and examines how they interact to produce traits. Because drugs tend to be proteins or small molecules that affect the behavior of proteins, proteomics is crucially important in the development of new drugs for the treatment of disease.

Computer modeling of the three-dimensional shape of these proteins is an important part of proteomics. Researchers may then use the models to predict which molecules may be effective drugs. For example, a compound that fits the active site of an enzyme without being converted into a product may be an effective drug to inhibit that enzyme.

Bioinformatics is the application of computer technologies to the study of the genome and proteome (**Fig. 11.25**). Genomics and proteomics produce raw data, and these fields depend upon computer analysis to find significant patterns in the data. As a result of bioinformatics, scientists hope to find cause-and-effect relationships between an individual's overall genetic makeup and genetic disorders caused by polygenes.

More than half of the human genome consists of uncharacterized sequences that contain no genes and have no known function. Bioinformatics might find that these regions have functions by correlating any sequence changes with resulting phenotypes. New computational tools will most likely be needed in order to accomplish these goals.

■ Check
Your Progress

1. Describe the goals of genomics research.
2. Describe how bioinformatics is being applied to genomics and proteomics.

Answers: 1. Genomics research attempts to understand how genes direct growth and development in order to control the activities of cells and organisms. Genomics also strives to understand the function of the intergenic sequences within the genome. 2. Bioinformatics utilizes computer technology to sift through the enormous amount of data generated from genomics and proteomics to find patterns.

a.

b.

c.

Figure 11.24   Studying genomic differences between chimpanzees and humans.

Researchers found that by comparing the human genome to that of a chimpanzee they were able to discover genes that cause human diseases when abnormal. They also concluded that the genes for (a) speech, (b) hearing, and (c) smell may have influenced the evolution of humans.

Figure 11.25   Bioinformatics.

New computer programs are being developed to make sense out of the raw data generated by genomics and proteomics. Bioinformatics allows researchers to correlate gene activity and protein function, so that the phenotype can be better understood in molecular terms.

The Chapter in Review

Summary

11.1 | DNA and RNA Structure and Function

Structure of DNA

DNA, the genetic material, has the following structure:
- DNA is a polymer in which four nucleotides differ by their bases. The bases are symbolized by the letters A, G, C, and T.
- A is always paired with T, and G is always paired with C, a pattern called complementary base pairing:

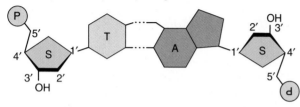

- DNA is a double helix. The deoxyribose sugar–phosphate molecules make up the sides of a twisted ladder and run counter to one another. The paired bases are the rungs of the ladder. Base pairs can be in any order in each type of species.

Replication of DNA

DNA replicates and every new cell gets a copy. During DNA replication, each old strand gives rise to a new strand. Therefore, the process is called semiconservative. Because of complementary base pairing, all these double helical molecules are identical:

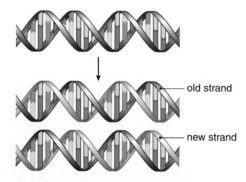

old strand

new strand

RNA Structure and Function

RNA has the following characteristics:
- It is found in the nucleus and cytoplasm of eukaryotic cells.
- It contains the sugar ribose and the bases A, U, C, G.
- It is single-stranded.
- The three forms are mRNA, which carries the DNA message to the ribosomes; tRNA, which transfers amino acids to the ribosomes where protein synthesis occurs; and rRNA, which is found in the ribosomes.

11.2 | Gene Expression

DNA specifies the sequence of amino acids in a protein; therefore, gene expression occurs once the protein product is present. Gene expression requires two steps: transcription and translation.

From DNA to RNA to Protein

Transcription: During transcription, a primary-mRNA forms and is then processed. Processing involves (1) the addition of a cap to one end, (2) the addition of a poly-A tail to the other end, and (3) removal of introns so that only exons remain. The mature mRNA now contains instructions for assembling a protein.

Translation: Translation requires three types of RNA:
- **mRNA** contains codons in which every three bases code for a particular amino acid except when the codon means stop.
- **tRNA** brings amino acids to the ribosomes. One end of the molecule binds to the amino acid, and the other end is the anticodon, three bases that pair with a codon. Each tRNA binds with only one of the 20 types of amino acids in a protein.
- **rRNA** is located in the ribosomes. The P site of a ribosome contains a tRNA attached to a peptide, and the A site contains a tRNA attached to an amino acid.

Three Phases of Translation

The three phases of translation are initiation, the elongation cycle, and termination.

Initiation: During chain initiation at the start codon, the ribosomal subunits, the mRNA, and the tRNA-methionine complex come together.

Elongation cycle: The chain elongation cycle consists of these events:
- A tRNA at the P site passes its peptide to tRNA–amino acid at the A site. The tRNA at the P site leaves.
- The ribosome moves forward one codon (called translocation), and the codon at the A site is ready for the next tRNA–amino acid.

Termination: During chain termination, a stop codon is reached on the mRNA. A release factor binds, the ribosome dissociates, the mRNA departs, and the polypeptide is released.

Genes and Gene Mutations

A gene mutation is a change in the sequence of bases. Mutations can be due to errors in replication, transposons, or environmental mutagens. The results of mutations can vary from no effect to a nonfunctional protein.

11.3 | DNA Technology

Recombinant DNA Technology

Recombinant DNA technology uses restriction enzymes to cleave DNA so that a foreign gene can be inserted into a vector, such as a plasmid:

plasmid DNA

foreign gene

Transgenic Organisms

Biotechnology uses natural biological systems to create a product or to achieve an end desired by human beings. Foreign genes have been transferred to bacteria, to crops, and to farm animals to improve their characteristics and produce commercial products.

Polymerase Chain Reaction

The polymerase chain reaction occurs in a test tube and is often used in conjunction with recombinant DNA technology. Heat separates double-stranded DNA. Primers flank the target DNA, and heat-insensitive DNA polymerase copies the target DNA. The cycle is repeated until millions of copies are produced.

DNA Fingerprinting

DNA fingerprinting helps identify relatives, remains, and criminals. Following PCR if needed, restriction enzymes are used to fragment samples of DNA. When separated by gel electrophoresis, a pattern results that is unique to the individual.

11.4 | Genomics and Proteomics

Sequencing the Bases of the Human Genome

Scientists have sequenced all 3.5 trillion bases of the human genome, and are now employing genomics to study the information it contains. Differences in base sequence between individuals, including single base changes within genes, single base changes in unknown sequences, and differences in copy number of certain genes, together create variation within the population and affect disease susceptibility.

Genome Comparisons The genomes of many organisms, including humans, have been compared to study the evolutionary relationships between them.

Proteomics and Bioinformatics

Proteomics, the study of all the proteins active in a cell, is being employed to predict the structures of proteins and to design drugs that may affect them. Bioinformatics applies computer technology to the study of the genome and proteome. Through the use of these tools, scientists hope to sift through the massive amounts of data generated through genomics and proteomics to discover underlying patterns.

Key Terms

adenine (A) 163	polyribosome 171
anticodon 170	promoter 169
bioinformatics 179	proteome 179
biotechnology 176	proteomics 179
codon 168	recombinant DNA (rDNA) 175
complementary base pairing 166	release factor 172
cytosine (C) 163	restriction enzyme 175
DNA fingerprinting 177	ribosomal RNA (rRNA) 167
DNA ligase 166	ribosome 167
DNA polymerase 166	RNA (ribonucleic acid) 166
DNA replication 166	RNA polymerase 169
frameshift mutation 174	semiconservative 166
gene mutation 174	template 166
genetically modified 175	thymine (T) 163
genetic engineering 175	transcription 168
genomics 178	transfer RNA (tRNA) 167
guanine (G) 163	transgenic organism 175
intergenic DNA 178	translation 168
messenger RNA (mRNA) 167	translocation 172
mRNA transcript 169	transposon 174
mutagen 174	triplet code 168
point mutation 174	uracil (U) 166
polymerase chain reaction (PCR) 177	vector 175

Testing Yourself

1. In the following diagram, label all parts of the DNA molecule.

Choose the best answer for each question.

2. Chargaff's rules state that the amount of A, T, G, and C in DNA
   a. varies from species to species, and the amount of A = T and G = C.
   b. varies from species to species, and the amount of A = G and T = C.
   c. is the same from species to species, and the amount of A = T and G = C.
   d. is the same from species to species, and the amount of A = G and T = C.

3. DNA replication is said to be semiconservative because
   a. each new DNA molecule contains one new strand and one old one.
   b. every other new DNA molecule contains two old strands.
   c. each new DNA strand contains two new strands.
   d. every other new DNA molecule contains one new strand and one old one.

4. Which of the following is not a feature of eukaryotic DNA replication?
   a. Replication bubbles spread in two directions.
   b. A new strand is synthesized using an old one as a template.
   c. Complementary base pairing determines which nucleotides should be added to the new strand.
   d. Each chromosome has one origin of replication.

For questions 5–8, match the items to those in the key. Answers can be used more than once.

Key:

   a. messenger RNA
   b. transfer RNA
   c. ribosomal RNA

5. contains an anticodon
6. carries amino acids to the ribosome
7. carries genetic information from DNA to ribosomes
8. produced in the nucleolus

9. Which of the following statements about the genetic code is not true?
   a. The genetic code is almost universal.
   b. The genetic code is a doublet code.
   c. Multiple codons may encode the same amino acid.
   d. Some special codons mean "stop."

10. Transcription produces _____, while translation produces _____.
   a. DNA, RNA
   b. RNA, polypeptides
   c. polypeptides, RNA
   d. RNA, DNA

11. The application of computer techniques to analyze the genome is called
   a. proteomics.
   b. genomics.
   c. bioinformatics.
   d. genetic code.

For questions 12–16, match the items to the best answer from the key.

Key:

   a. not harmful
   b. sometimes harmful
   c. usually harmful

12. point mutation that changes one codon of a gene
13. frameshift mutation that changes all following codons
14. point mutation that changes a codon into a "stop" codon
15. movement of a transposon into the middle of a gene
16. point mutation that does not change the amino acid specified by the codon

17. Which of the following is not useful in creating recombinant DNA?
   a. restriction endonuclease
   b. DNA ligase
   c. RNA polymerase
   d. plasmid

18. Genome comparisons may yield information about which of the following?
   a. evolutionary relationships between organisms
   b. changes in genes over evolutionary history
   c. links between genes and disease
   d. All of these are correct.

Thinking Scientifically

1. Experiments with bacteria in the 1930s showed that exposing a nonvirulent (noninfectious) strain of bacteria to a heat-killed virulent strain could convert the nonvirulent strain into a virulent one. What could possibly explain this phenomenon?

2. As you have just read, humans and other mammals have a large amount of intergenic DNA. It requires much energy to maintain this "extra" DNA; therefore scientists reason it must have some function. Explain.

Bioethical Issue

Mandatory DNA Fingerprinting

Arthur Lee Whitfield had served 22 years of a 63-year sentence for rape when he was notified that he had gained the right to use DNA fingerprinting to establish his innocence. Luckily, biological evidence had been saved from his jury trial, and DNA fingerprinting showed that his DNA did not match that of the rapist. However, this DNA did match that of another inmate who was serving a life sentence for an unrelated rape conviction.

In response to Whitfield's and other similar cases, law enforcement agencies and criminal defense attorneys alike are pushing to require all individuals who are arrested to submit DNA samples for DNA fingerprinting. Some are even pushing to require DNA fingerprinting of all citizens, citing the potential to resolve many cases and exonerate people wrongfully convicted of crimes. However, many civil libertarians insist that this constitutes search without reasonable cause, which is illegal in the United States. Furthermore, others insist that the government simply cannot be trusted with such a wealth of information, because it might be accessed by hackers or used for purposes other than that for which it was intended.

Should the government require all citizens, or those who are arrested for any reason, to submit DNA samples for fingerprinting? Or do you believe, as some, that it would be a violation of the U.S. constitution?

Essentials of Biology Website

The companion website for *Essentials of Biology* provides a wealth of information organized and integrated by chapter. You will find practice tests, animations, videos, and much more that will complement your learning and understanding of general biology.

http://www.mhhe.com/maderessentials2

lymphoma cancer cell

×7,000

Several mutations are usually necessary for cancer cells to develop.

human papillomavirus

×200,000

A papillomaviral infection can lead to cervical cancer.

blood cancer cells

×700

cervical smear ×200

cervical cancer cell

×5,000

colon cancer cell ×5,000

Chromosomal abnormalities are rampant in cancer cells.

12

Gene Regulation and Cancer

Outline

Faulty Gene Regulation and Cancer

THE TERM CANCER describes a family of diseases having the common characteristic of uncontrolled cell division. You are probably aware that lifestyle choices such as smoking or tanning might increase your odds of developing certain types of cancer. However, have you ever considered that some viral infections can also increase your chances of cancer? A group of viruses termed oncogenic viruses can contribute to the possibility of certain cancers. These viruses are characterized by their ability to insert viral DNA into human chromosomes. Humans have

—Continued next page

Continued—

several genes, called proto-oncogenes, that help regulate cell division. When oncogenic viral DNA inserts a mutated proto-oncogene, called an oncogene, into a human chromosome, the cell cycle may occur repeatedly, and a tumor may result.

Human papillomavirus (HPV) is a well-known oncogenic virus. This virus comes in many forms and often causes benign warts. However, several strains of this virus are known to be oncogenic, and are the primary cause of cervical cancer. Women infected with oncogenic strains of this HPV typically have no symptoms, but an annual Pap smear can detect changes in cervical cells that are consistent with HPV infection. When detected early, infected cells can be removed.

Although many individuals equate mutations with cancer, many more of these genetic changes are usually required in order for cancer to develop. These changes cause a breakdown in gene regulation, which can lead to a loss of control of the cell cycle. In this chapter, you will learn about how genes are regulated and some of the genetic changes that contribute to the development of cancer.

12.1 | Control of Gene Expression

The events of the cell cycle and DNA replication ensure that every cell of the body receives a copy of all the genes. This means that every cell in your body has the potential to become a complete organism. To demonstrate this potential, let's take a look at the processes of reproductive and therapeutic cloning.

Reproductive and Therapeutic Cloning

In **reproductive cloning**, the desired end is an individual that is exactly like the original individual. The cloning of plants has been routine for some time (**Fig. 12.1**). However, at one time it was thought that the cloning of adult animals would be impossible because investigators found it difficult to have the nucleus of an adult cell "start over," even when it was placed in an enucleated egg cell. (An enucleated egg cell has had its own nucleus removed.)

In March 1997, Scottish investigators announced they had cloned a Dorset sheep, which they named Dolly. How was their procedure different from all the others that had been attempted? Again, an adult nucleus was placed in an enucleated egg cell; however, the donor cells had been starved. Starving the donor cells caused them to stop dividing and go into a resting stage (the G_0 stage of the cell cycle). This was the change needed because G_0 stage nuclei are amenable to cytoplasmic signals for initiation of development (**Fig. 12.2**). Now it is common practice to clone farm animals that have desirable genetic traits (**Fig. 12.3**), and even to clone rare animals that might otherwise become extinct.

In the United States, no federal funds can be used for experiments to clone human beings. And at this point, even cloning farm animals is wasteful—in the case of Dolly, out of 29 clones only one was successful. Also, cloned animals may not be healthy. Dolly was put down by lethal injection in 2003 because she was suffering from lung cancer and crippling arthritis. She had lived only half the normal life span for a Dorset sheep.

1 Tiny disks are obtained from carrot root.

2 Each disk produces an undifferentiated tissue mass.

3 Many like carrot plantlets are cloned from each tissue mass.

Figure 12.1 Cloning carrots.
Carrot cells from a root will produce cloned copies of carrot plants in cell culture. This demonstrates that all the nuclei in a carrot root contain all the genes of a carrot plant.

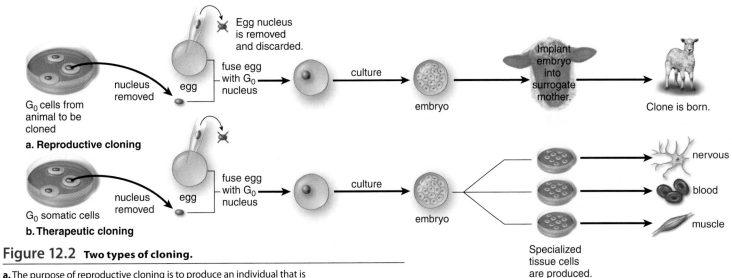

a. Reproductive cloning

Egg nucleus is removed and discarded.

G_0 cells from animal to be cloned

nucleus removed

egg

fuse egg with G_0 nucleus

culture

embryo

Implant embryo into surrogate mother.

Clone is born.

b. Therapeutic cloning

G_0 somatic cells

nucleus removed

egg

fuse egg with G_0 nucleus

culture

embryo

nervous

blood

muscle

Specialized tissue cells are produced.

Figure 12.2 **Two types of cloning.**

a. The purpose of reproductive cloning is to produce an individual that is genetically identical to the one that donated a 2n nucleus. The 2n nucleus is placed in an enucleated egg, and after several divisions, the embryo comes to term in a surrogate mother. **b.** The purpose of therapeutic cloning is to produce specialized tissue cells. A 2n nucleus is placed in an enucleated egg, and after several divisions, the embryonic cells (called embryonic stem cells) are separated and treated to become specialized cells.

In **therapeutic cloning,** the desired end is not an individual organism, but mature cells of various cell types. The purposes of therapeutic cloning are (1) to learn more about how specialization of cells occurs; and (2) to provide cells and tissues that could be used to treat human illnesses such as diabetes, spinal cord injuries, and Parkinson disease.

Therapeutic cloning can be carried out in several ways. The most common method is to isolate **embryonic stem cells** and subject them to treatments that cause them to become particular cell types, such as red blood cells, muscle cells, or nerve cells (see Fig. 12.2*b*). Eventually, it may be possible to produce entire tissues and organs from stem cells. Ethical concerns exist about this type of therapeutic cloning because if the embryo had been allowed to continue development, it would have become an individual.

Another way to carry out therapeutic cloning is to use **adult stem cells.** Stem cells are found in many organs of an adult's body. For example, the skin has stem cells that constantly divide and produce new skin cells, while the bone marrow has stem cells that produce new blood cells. Currently, adult stem cells are limited in the number of cell types that they may become. However, scientists were able to coax adult stem cells from skin into becoming more like embryonic stem cells by adding only four genes. Once researchers learn more about controlling gene expression, they may be able to overcome the current limitations.

Figure 12.3 **Cloned farm animals.**

These pigs were cloned using 2n nuclei from an adult swine. All nuclei contain all the genes of an organism or else cloning wouldn't be possible.

Check Your **Progress**

1. Contrast the goal of reproductive cloning with that of therapeutic cloning.

Answer: 1. The goal of reproductive cloning is to create an individual that is genetically identical to the 2n nucleus donor. The goal of therapeutic cloning is to study how cells become specialized and to provide specialized cells for medical procedures.

Figure 12.4   Gene expression in specialized cells.

All the cells in your body contain all the genes, but gene expression is controlled, and only certain genes are expressed in each type of cell. Housekeeping genes are expressed in all cells; in addition, cells express those genes that account for their specialization.

Levels of Gene Expression Control

The human body contains many types of cells that differ in structure and function. Each cell type must contain its own mix of proteins that make it different from all other cell types. Therefore, only certain genes are active in cells that perform specialized functions, such as nerve, muscle, gland, and blood cells.

Some of these active genes are called housekeeping genes because they govern functions that are common to many types of cells, such as glucose metabolism. But otherwise, the activity of selected genes accounts for the specialization of cells. In other words, gene expression is controlled in a cell, and this control accounts for its specialization (**Fig. 12.4**). Let's begin by examining a simpler system—the control of transcription in prokaryotes.

Gene Expression in Prokaryotes

The bacterium *Escherichia coli* lives in the intestine and can quickly adjust its production of enzymes according to what we eat. If someone drinks a glass of milk, *E. coli* immediately begins to make three enzymes needed to metabolize lactose. The transcription of all three enzymes is under the control of one **promoter,** a short DNA sequence where RNA polymerase attaches. The French microbiologists François Jacob and Jacques Monod called such a cluster of bacterial genes, along with the DNA sequences that control their transcription, an **operon.** They received a Nobel prize in 1961 for their investigations because they were the first to show how gene expression was controlled—specifically, how the *lac* operon was controlled (**Fig. 12.5**).

Jacob and Monod proposed that a **regulatory gene** located outside the operon codes for a **repressor.** The repressor is a protein that, in the *lac* operon, normally binds to the *operator,* which lies next to the promoter. When the repressor is attached to the operator, transcription of the lactose metabolizing genes does not take place because RNA polymerase is unable to bind to the promoter. The lac operon is normally turned off in this way because lactose is usually absent (Fig. 12.5*a*).

a. **Lactose is absent—operon is turned off.**
   Enzymes needed to metabolize lactose are not produced.

Figure 12.5   The *lac* operon.

a. The regulatory gene codes for a repressor. The repressor binds to the operator and prevents RNA polymerase from attaching to the promoter. Therefore, transcription of the lactose metabolizing genes does not occur.
b. When lactose is present, it binds to the repressor, changing its shape so that it becomes inactive and can no longer bind to the operator. Now RNA polymerase binds to the promoter, and the lactose metabolizing genes are expressed.

b. **Lactose is present—operon is turned on.**
   Enzymes needed to metabolize lactose are produced.

What turns the operon on when lactose is present? Lactose binds to the repressor and it changes shape. Now the repressor is unable to bind to the operator and RNA polymerase is able to bind to the promoter (Fig. 12.5*b*). Transcription of the genes needed for lactose metabolism occurs. When lactose is present and glucose, the preferred sugar, is absent, a protein called CAP (not shown) assists the binding of RNA polymerase to the promoter. This further ensures that the lactose-metabolizing enzymes are transcribed when they are needed.

Other bacterial operons, such as those that control amino acid synthesis, are usually turned on. For example, in the *trp* operon, the regulatory gene codes for a repressor that ordinarily is unable to attach to the operator. Therefore, the genes needed to make the amino acid tryptophan are ordinarily expressed. When tryptophan is present, it binds to the repressor. A change in shape activates the repressor and allows it to bind to the operator. Now the operon is turned off.

Gene Expression in Eukaryotes

In bacteria, a single promoter serves several genes that make up a transcription unit, while in eukaryotes, each gene has its own promoter where RNA polymerase binds. Bacteria rely mostly on transcriptional control, but eukaryotes employ a variety of mechanisms to regulate gene expression. These mechanisms affect whether the gene is expressed, the speed with which it is expressed, and how long it is expressed.

Some mechanisms of gene expression occur in the nucleus, and others occur in the cytoplasm (**Fig. 12.6**). In the nucleus, chromatin condensation, DNA transcription, and mRNA processing all play a role in determining which genes are expressed in a particular cell type.

In the cytoplasm, mRNA translation into a polypeptide at the ribosomes can occur right away or be delayed. The mRNA can last a long time or be destroyed immediately, and the same holds true for a protein. These mechanisms control the quantity of gene product and/or how long it is active.

Chromatin Condensation   Eukaryotes utilize chromatin condensation as a way to keep genes turned on or off. The more tightly chromatin is compacted, the less often genes within it are expressed. Darkly staining portions of chromatin, called **heterochromatin,** represent tightly compacted, inactive chromatin.

Check Your Progress

1. Contrast the activity of the *lac* operon when lactose is absent to its activity when lactose is present.
2. List the various levels at which eukaryotes regulate gene expression.

Answers: 1. When lactose is absent, the repressor is active. The active repressor binds the operator and inhibits transcription of the lactose-metabolizing genes. When lactose is present, lactose binds and inactivates the repressor, allowing transcription of the lactose-metabolizing genes.   2. Eukaryotes may regulate transcription at the level of chromatin condensation, DNA transcription, mRNA processing, mRNA translation, and protein activity.

Figure 12.6   **Control of gene expression in eukaryotic cells.**

Gene expression is controlled at various levels in eukaryotic cells. Also, an external signal (red) may alter gene expression.

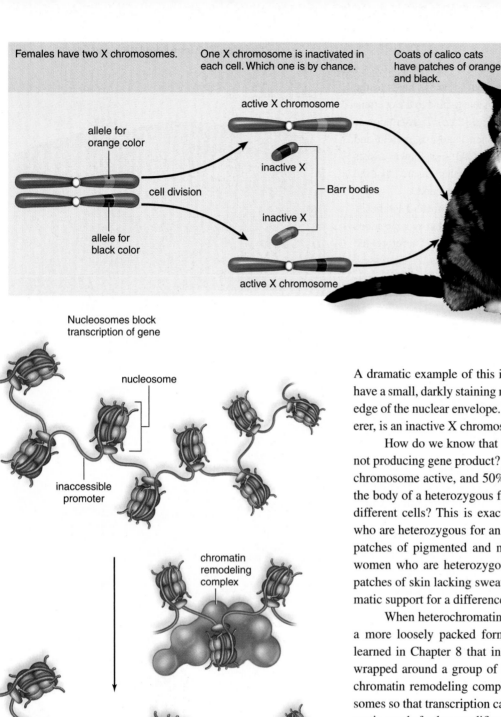

Females have two X chromosomes.

One X chromosome is inactivated in each cell. Which one is by chance.

Coats of calico cats have patches of orange and black.

active X chromosome

allele for orange color

inactive X

Barr bodies

cell division

inactive X

allele for black color

active X chromosome

Figure 12.7 X-inactivation in mammalian females.

In cats, the alleles for black or orange coat color are carried on the X chromosome. Random X-inactivation occurs in females. Therefore, in heterozygous females, 50% of the cells have an allele for black coat color, and 50% of cells have an allele for orange coat color. The result is calico or tortoiseshell cats that have coats with patches of both black and orange. A calico cat, but not a tortoiseshell cat, has patches of white due to the activity of another gene.

Nucleosomes block transcription of gene

nucleosome

inaccessible promoter

chromatin remodeling complex

accessible promoter

Exposed DNA allows gene transcription

Figure 12.8 Histone displacement.

In euchromatin, a chromatin remodeling complex pushes aside the histone portions of nucleosomes so that transcription may begin.

A dramatic example of this is the Barr body in mammalian females. Females have a small, darkly staining mass of condensed chromatin adhering to the inner edge of the nuclear envelope. This structure, called a *Barr body* after its discoverer, is an inactive X chromosome.

How do we know that Barr bodies are inactive X chromosomes that are not producing gene product? Suppose 50% of the cells in a female have one X chromosome active, and 50% have the other X chromosome active. Wouldn't the body of a heterozygous female be a mosaic, with "patches" of genetically different cells? This is exactly what happens. For example, human females who are heterozygous for an X-linked recessive form of ocular albinism have patches of pigmented and nonpigmented cells at the back of the eye. And women who are heterozygous for hereditary absence of sweat glands have patches of skin lacking sweat glands. The female calico cat also provides dramatic support for a difference in X-inactivation in its cells (**Fig. 12.7**).

When heterochromatin undergoes unpacking, it becomes **euchromatin,** a more loosely packed form of chromatin that contains active genes. You learned in Chapter 8 that in eukaryotes, a *nucleosome* is a portion of DNA wrapped around a group of histone molecules. When DNA is transcribed, a chromatin remodeling complex pushes aside the histone portions of nucleosomes so that transcription can begin (**Fig. 12.8**). In other words, even euchromatin needs further modification before transcription can begin. The presence of histones limits access to DNA, and euchromatin becomes genetically active when histones no longer bar access to DNA. Only then is it possible for a gene to be turned on and expressed in a eukaryotic cell.

DNA Transcription Transcription in eukaryotes follows the same principles as in bacteria, except that many more regulatory proteins per gene are involved. The occurrence of so many regulatory proteins allows for greater control, but it also allows for a greater chance of malfunction.

In eukaryotes, **transcription factors** are DNA-binding proteins that help RNA polymerase bind to a promoter. Several transcription factors are needed in each case, and if one is missing, transcription cannot take place. All the transcription factors form a *complex* that also helps pull double-stranded DNA apart and even acts to position RNA polymerase so that transcription can begin. The same transcription factors, in different combinations, are used over again at other promoters, so it is easy to imagine that if one malfunctions, the result

Figure 12.9 Transcription factors and transcription activators.

Transcription factors form a complex at a promoter and a transcription activator binds to an enhancer. RNA polymerase binds to the promoter but transcription does not begin until a hairpin DNA loop brings all regulatory proteins together.

could be disastrous to the cell. The genetic disorder Huntington disease is a devastating psychomotor ailment caused by a defect in a transcription factor.

In eukaryotes, **transcription activators** are DNA-binding proteins that speed transcription dramatically. They bind to a DNA region called an **enhancer** that can be quite a distance from the promoter. A hairpin loop in the DNA can bring the transcription activators attached to enhancers into contact with the transcription factor complex (**Fig. 12.9**).

It's possible for a single transcription activator to have a dramatic effect on gene expression. As an example, investigators have found one DNA-binding protein, MyoD, that alone can activate the genes necessary for fibroblasts to become muscle cells in various vertebrates. Another DNA-binding protein, called Ey, can bring about the formation of not just a single cell type but a complete eye in flies (**Fig. 12.10**).

Check Your **Progress**

1. Contrast heterochromatin with euchromatin, and describe the effect of each chromatin condensation state on gene expression.
2. Describe the composition, location, and function of transcription activators.

Answers: 1. Heterochromatin is tightly packed chromatin, whereas euchromatin is loosely packed. Genes within heterochromatin are seldom expressed, but genes within euchromatin are often transcribed.   2. Transcription activators are proteins that bind to DNA regions called enhancers. They speed transcription by interacting with transcription factors.

Figure 12.10 *Ey* gene.

In fruit flies, the expression of the *EY* gene, in the precursor cells of the leg, triggers the development of an eye on the leg.

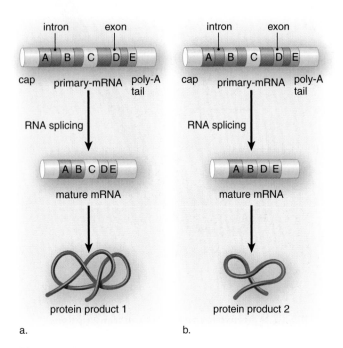

a.   b.

Figure 12.11 **Processing of mRNA transcripts.**

Because the primary-mRNAs are processed differently in these two cells (**a**) and (**b**), distinct proteins result.

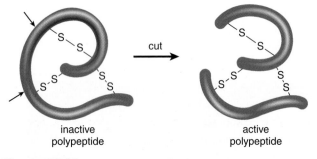

Figure 12.12 **Protein activity.**

The protein insulin is not active until an enzyme removes a portion of the initial polypeptide. The resulting two polypeptide chains are held together by two disulfide (S—S) bonds between like amino acids.

mRNA Processing After transcription, the removal of introns and splicing of exons occurs before mature mRNA leaves the nucleus and passes into the cytoplasm. **Alternative mRNA processing** is a mechanism by which the same primary mRNA can produce different protein products according to which exons are spliced together to form mature mRNAs (**Fig. 12.11**).

The fruit fly gene *DScam* offers a dramatic example of alternative mRNA processing. As many as 38,000 different proteins can be produced according to which of the gene's exons are combined in mature mRNAs. Evidence suggests that these proteins provide each nerve cell with a unique identity as it communicates with other nerve cells in the brain.

mRNA Translation The cytoplasm contains proteins that can control whether translation of mRNA takes place. For example, an initiation factor, known as IF-2, inhibits the start of protein synthesis when it is phosphorylated by a specific protein kinase. Environmental conditions can also delay translation. Red blood cells do not produce hemoglobin unless heme, an iron-containing group, is available.

The longer an mRNA remains in the cytoplasm before it is broken down, the more gene product is produced. During maturation, mammalian red blood cells eject their nuclei, and yet they continue to synthesize hemoglobin for several months thereafter. The necessary mRNAs must be able to persist this entire time. Differences in the length of the poly-A tail can determine how long a particular transcript remains active. Hormones may also affect the stability of certain mRNA transcripts. An mRNA called vitellin persists for three weeks instead of 15 hours if it is exposed to estrogen.

Protein Activity Some proteins are not active immediately after synthesis. For example, insulin is a single long polypeptide that folds into a three-dimensional structure. Only then is a sequence of about 30 amino acids enzymatically removed from the middle of the molecule. This leaves two polypeptide chains bonded together by disulfide (S—S) bonds, and an active insulin molecule results (**Fig. 12.12**). This mechanism allows a protein's activity to be delayed until it is needed.

Many proteins are short-lived in cells because they are degraded or destroyed. Cyclins, which are proteins involved in regulating the cell cycle, are destroyed by giant enzyme complexes called *proteasomes* when they are no longer needed. Proteasomes break down other proteins as well.

Check
Your **Progress**

1. Explain how alternative mRNA processing can alter gene expression.
2. Explain how a mammalian red blood cell can synthesize hemoglobin after the cell ejects its nucleus.

Answers: 1. The exons that are joined together may differ in different cell types, resulting in different mature mRNAs and proteins.   2. The mRNA for hemoglobin survives for several months afer the nucleus is ejected.

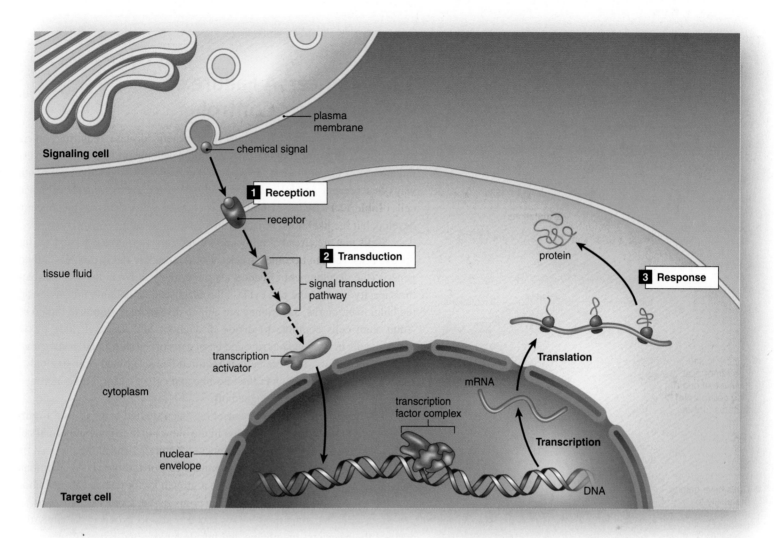

Figure 12.13 Cell-signaling.

1 Reception: A chemical signal is received at a specific receptor protein.
2 Transduction: A signal transduction pathway terminates when a transcription activator is stimulated. It enhances transcription of a specific gene. **3** Response: Translation of the mRNA transcript follows and a protein product results. The protein product is the response to the signal.

Signaling Between Cells in Eukaryotes

In multicellular organisms, cells are constantly sending out chemical signals that influence the behavior of other cells. During animal development, these signals determine the specialized role a cell will play in the organism. Later, the signals help coordinate growth and day-to-day functions. Plant cells also signal each other so that their responses to environmental stimuli, such as direct sunlight, are coordinated.

Typically, **cell-signaling** occurs because a chemical signal binds to a receptor protein in a target cell's plasma membrane. The signal causes the receptor protein to initiate a series of reactions within a **signal transduction pathway.** The end product of the pathway (not the signal) directly affects the metabolism of the cell. For example, growth is only possible if certain genes have been turned on by regulatory proteins.

In **Figure 12.13,** a signaling cell secretes a chemical signal that binds to a specific receptor located in the receiving cell's plasma membrane. The binding activates a series of reactions within a signal transduction pathway. The last reaction activates a transcription activator that enhances the transcription of a specific gene. Transcription leads to the translation of mRNA and a protein product that, in this case, stimulates the cell cycle so that growth occurs.

A protein called Ras functions in signal transduction pathways that lead to the transcription of many genes, several of which promote the cell cycle. Ras is normally inactive, but the reception of a growth factor leads to its activation. If Ras is continually activated, cancer will develop because cell division will occur continuously.

Check Your **Progress**

1. Describe the relationship between a signal transduction pathway and gene expression.

Answer: 1. A signal transduction pathway is a series of reactions that changes the signal to one that affects the activity of a cell, such as increased expression of genes.

12.2 | Cancer: A Failure of Genetic Control

As described in Chapter 8, cancer is a genetic disease requiring several mutations, each propelling cells toward the development of a tumor. Several mutations are usually needed to disrupt the many redundant regulatory pathways that prevent normal cells from taking on the characteristics of cancer cells (**Table 12.1**). Because of this sequence, it takes several years for cancer to develop, but the likelihood of cancer increases as we age.

Cells that are already highly specialized, such as nerve cells and cardiac muscle cells, seldom become cancer cells because they rarely divide. Carcinogenesis (the development of cancer) is more likely to begin in cells that have the capacity to enter the cell cycle. Fibroblasts and cells lining the cavities of the lung, liver, uterus, and kidney are able to divide when stimulated to do so. Adult stem cells continue to divide throughout life. These include blood-forming cells in the bone marrow and basal cells of the skin and digestive tract. Continuous division of these cells is required because blood cells live only a short while, and cells that line the intestine and cells that form the outer layer of the skin are continually sloughed off.

The development of colon cancer has been studied in detail, and **Figure 12.14** is based on those studies. This figure shows that a single abnormal cell begins the process toward the development of a tumor. And along the way, the most aggressive cell becomes the dominant cell of the tumor. As additional mutations occur, the tumor cells release a growth factor that causes neighboring blood vessels to branch into the cancerous tissue, a process called *angiogenesis*. Additional mutations allow cancer cells to produce enzymes that degrade the basement membrane and invade underlying tissues. Cancer cells are motile and able to travel through the blood or lymph vessels to other parts of the body, where they start distant tumors, a process called *metastasis*.

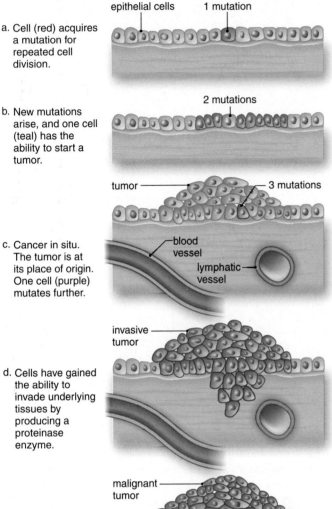

a. Cell (red) acquires a mutation for repeated cell division.

b. New mutations arise, and one cell (teal) has the ability to start a tumor.

c. Cancer in situ. The tumor is at its place of origin. One cell (purple) mutates further.

d. Cells have gained the ability to invade underlying tissues by producing a proteinase enzyme.

e. Cancer cells now have the ability to invade lymphatic and blood vessels.

f. New metastatic tumors are found some distance from the original tumor.

Figure 12.14 Development of cancer.

Development of cancer requires several mutations, each contributing to the development of a tumor. Normally cells exhibit contact inhibition and controlled growth. They stop dividing when they are bordered by other cells. Tumor cells lack contact inhibition and pile up forming a tumor. Cancer cells do not fulfill the functions of a specialized tissue and they have abnormal chromosomes. As more mutations accumulate, the cancer cells may spread and form tumors in other parts of the body.

Table 12.1 | Normal Cells Compared to Cancer Cells

Normal Cells	Cancer Cells
Contact inhibition	No contact inhibition
Controlled growth	Uncontrolled growth (tumor)
Specialized cells	Nonspecialized cells
Normal chromosomes	Abnormal chromosomes
Undergo apoptosis	No apoptosis

Proto-Oncogenes and Tumor Suppressor Genes

Recall that the cell cycle, as discussed on page 118, consists of interphase, followed by mitosis. When cancer develops, the cell cycle occurs repeatedly, in large part, due to mutations in two types of genes. **Figure 12.15** shows that:

1. **Proto-oncogenes:** code for proteins that promote the cell cycle and inhibit apoptosis. They are often likened to the gas pedal of a car because they promote the cell cycle.
2. **Tumor suppressor genes:** code for proteins that inhibit the cell cycle and promote apoptosis. They are often likened to the brakes of a car because they inhibit acceleration of the cell cycle and stop cells from dividing inappropriately. They are called tumor suppressors because tumors may occur when mutations cause these genes to become non-functional.

a.

Growth factor binds to receptor and activates proto-oncogenes through cell-signaling pathway.

Proto-oncogene codes for a protein that promotes the cell cycle. If a proto-oncogene mutates, the resulting oncogenes may lead to uncontrolled cell division.

b.

Tumor suppressor gene codes for a protein that inhibits the cell cycle. Mutant tumor suppressor genes can lose this function.

Proto-Oncogenes Become Oncogenes

When proto-oncogenes mutate, they become cancer-causing genes called **oncogenes**. Proto-oncogenes promote the cell cycle, and oncogenes accelerate the cell cycle. Therefore, a mutation that causes a proto-oncogene to become an oncogene is a "gain of function" mutation.

A **growth factor** is a signal that activates a cell-signaling pathway by bringing about phosphorylation of a signaling protein (Fig. 12.15a). The cell-signaling pathway then activates numerous proteins, many of which promote the cell cycle (Fig. 12.15b). Some proto-oncogenes code for a growth factor or for a receptor protein that receives a growth factor. When these proto-oncogenes become oncogenes, receptor proteins are easy to activate and may even be stimulated by a growth factor produced by the receiving cell. For example, the *Ras* proto-oncogenes promote mitosis when a growth factor binds to a receptor. When *Ras* proto-oncogenes become oncogenes, they promote mitosis even when growth factors are not present. *Ras* oncogenes are found in about 20 to 30 percent of human cancers.

Tumor Suppressor Genes Become Inactive

When tumor suppressor genes mutate, their products no longer inhibit the cell cycle nor promote apoptosis. Therefore, these mutations can be called "loss of function" mutations (Fig. 12.15).

Figure 12.15   Regulation of the cell cycle through control of gene expression.

a. A growth factor binds to a receptor, stimulating a signal transduction pathway (green arrow). **b.** The signal transduction pathway activates expression of proto-oncogenes that promote the cell cycle. Tumor suppressor genes limit cell division by inhibiting the cell cycle (red arrow).

The retinoblastoma protein (RB) controls the activity of a transcription activator called E2F. In the absence of growth factors, RB binds E2F and inhibits entry into the S stage of the cell cycle, but mutations in RB may cause it to release E2F and promote the cell cycle when it is not appropriate.

Cancer develops gradually; multiple mutations usually occur before a cell becomes cancerous. As you can imagine, the conversion of proto-oncogenes into oncogenes coupled with the inactivation of tumor suppressor genes causes the cell cycle to continue unabated, just as a car with a sticking gas pedal and faulty brakes soon careens out of control.

Other Genetic Changes and Cancer

Other genetic influences can contribute to cancer development.

Absence of Telomere Shortening

A **telomere** is a repeating DNA sequence (TTAGGG) at the end of the chromosomes that can be as long as 15,000 base pairs. Just as the caps at the ends of shoelaces protect them from unraveling, so telomeres promote chromosomal stability so that replication can occur. Each time a cell divides, some portion of a telomere is lost; when telomeres become too short, the chromosomes cannot be replicated properly and the cell cycle is stopped.

Embryonic cells and also certain adult cells, such as stem cells and germ cells, have an enzyme called *telomerase* that can rebuild the telomeres. The gene that codes for telomerase is turned on in cancer cells. Now telomeres do not shorten, and cells divide over and over again. Telomerase is believed to become active only after a cell has already started proliferating wildly.

Chromosomal Rearrangements

When the chromosomes of cancer cells become unstable, portions of the DNA double helix may be lost, duplicated, or scrambled. For example, a portion of a chromosome may break off and reattach to another chromosome. Such a **translocation** may lead to cancer, especially if it disrupts genes that regulate the cell cycle.

The so-called *Philadelphia chromosome* is the result of a translocation between chromosomes 9 and 22 (**Fig. 12.16**). Now, the *BCR* (*b*reakpoint *c*luster *r*egion) and *ABL* genes are fused, creating an oncogene called *BCR-ABL* that promotes cell division. This translocation causes nearly 95% of cases of chronic myelogenous leukemia (CML), a cancer of the bone marrow. Recently, researchers have used a new drug called Gleevec to successfully treat CML. Gleevec inhibits the activity of the protein coded for by *BCR-ABL*. Encouraged by this success, scientists are now seeking to develop similar drugs that can inhibit the products of other oncogenes. Although cancer is usually a somatic disease, meaning that it develops only in body cells, some instances of an inherited predisposition to cancer are known.

Figure 12.16   A translocation can create the Philadelphia chromosome.

The Philadelphia chromosome results when a portion of chromosome 9 is swapped for a portion of chromosome 22. This creates the *BCR-ABL* oncogene on chromosome 22. The Philadelphia chromosome is responsible for 95% of cases of chronic myelogenous leukemia.

Inheritance of Cancer-Causing Alleles

BRCA1 and BRCA2   In 1990, DNA linkage studies on large families in which females tended to develop breast cancer identified the first gene allele associated with breast cancer. Scientists named this gene *breast cancer 1* or *BRCA1* (pronounced brak-uh). Later, they found that breast cancer in other families was due to a faulty allele of another gene they called *BRCA2*. Both alleles are mutant tumor suppressor genes that are inherited in an autosomal recessive manner. If one mutated allele is inherited from either parent, a mutation in the other allele is required before the predisposition to cancer is increased. Because the first mutated gene is inherited, it is present in all cells of the body, and then cancer is more likely wherever the second mutation occurs. If the second mutation occurs in the breast, breast cancer may develop (**Fig. 12.17**). If the second mutation is in the ovary, ovarian cancer may develop if additional cancer-causing mutations occur.

RB Gene   The *RB* gene is also a tumor suppressor gene. It takes its name from its association with an eye tumor called a *retino*blastoma, which first appears as a white mass in the retina. A tumor in one eye is most common because it takes mutations in both alleles before cancer can develop (**Fig. 12.18**). Children who inherit a mutated allele are more likely to have tumors in both eyes.

RET Gene   An abnormal allele of the *RET* gene, which predisposes a person to thyroid cancer, can be passed from parent to child. *RET* is a proto-oncogene known to be inherited in an autosomal dominant manner—only one mutated allele is needed to increase a predisposition to cancer. The remaining mutations necessary for thyroid cancer to develop are acquired (not inherited).

Testing for These and Other Genes

Genetic tests are available for the presence of the *BRCA* genes, the *RET* gene, and the *RB* gene. Persons who have inherited these genes may decide to have surgery by choice rather than necessity or be examined frequently for signs of cancer.

Genetic tests are also available for other types of mutated genes that help a physician diagnose cancer. For example, a *ras* oncogene can be detected in stool and urine samples. If the test for *ras* oncogene in the stool is positive, the physician suspects colon cancer, and if the test for *ras* oncogene in the urine is positive, the physician suspects bladder cancer. Tests are even being developed that can screen for many cancers using only saliva!

SEM of breast cancer cell

Figure 12.17   Breast cancer can run in families.

These three sisters have all had breast cancer. Genetic tests can identify women at risk for breast cancer so that they can choose to have frequent examinations to allow for early detection.

Figure 12.18   Inherited retinoblastoma.

A child is at risk for an eye tumor when a mutated *RB* allele is inherited, even though a second mutation in the normal allele is required before the tumor develops.

Check Your Progress

1. Explain how telomeres influence a cell's life span.
2. Explain how chromosomal rearrangements may cause cancer
3. Compare and contrast the *BRCA* gene with the *RB* gene.

Answers: 1. Telomeres become shorter after each cell division. When the telomere becomes too short, chromosomes become unstable and replication cannot occur.   2. When rearrangements occur, genes may be disrupted. When a gene needed for cell cycle control is disrupted, cancer can occur.   3. A mutated *BRCA* is gene is associated with breast cancer, while a mutated RB gene causes retinoblastoma.

The Chapter in Review

Summary

12.1 | Control of Gene Expression

Every cell in the body has a complete set of genes as is proven by the process of cloning, when a 2n nucleus is able to bring about the development of an organism exactly like the 2n nucleus donor or any specialized tissue that matches those of the donor.

Reproductive and Therapeutic Cloning

In reproductive cloning, the desired end is an individual genetically identical to the original individual. To achieve this type of cloning, the early embryo is placed in a surrogate mother until the individual comes to term.

In therapeutic cloning, the cells of the embryo are separated and treated so that they develop into specialized tissues that can be used to treat human disorders. Alternately, adult stem cells can be treated to become specialized tissues.

Levels of Gene Expression Control

The specialization of cells is not due to the presence/absence of genes; it is due to the activity/inactivity of genes.

Gene Expression in Prokaryotes In prokaryotes, gene expression is controlled at the level of transcription.
- In the *lac* operon, the repressor usually binds to the operator. Then RNA polymerase cannot bind, and the operon is turned off. When lactose is present, it binds to the repressor, and now RNA polymerase binds to the promoter, and the genes for lactose metabolism are transcribed.
- In the *trp* operon, the genes are usually transcribed because the repressor is inactive. When tryptophan is present, the repressor becomes active, and RNA polymerase cannot bind to the promoter.

Gene Expression in Eukaryotes Eukaryotes regulate gene expression at various levels:
- Chromatin condensation—conversion of heterochromatin to euchromatin so that genes can be expressed.
- DNA transcription—transcription factors present at the promoter, and transcription activators (bound to enhancers) are brought into contact and promote transcription.
- mRNA processing—alternative mRNA processing leads to a sequence of different exons in mature RNA and, therefore, different proteins.
- mRNA translation—various factors affect how long an mRNA stays active and how many times it is translated.
- Protein activity—activation of a protein may require further steps, and degradation of the protein can occur at once or after a while.

Signaling Between Cells in Eukaryotes Eukaryotic cells constantly communicate with their neighbors by secreting signals.
- Signals are received at receptor proteins located in the plasma membrane.
- A signal transduction pathway (enzymatic reactions) leads to a change in the behavior of the cell. For example, when stimulated via a transduction pathway, a transcription activator stimulates transcription of a particular gene.

12.2 | Cancer: A Failure of Genetic Control

Cancer Is a Genetic Disease

Evidence is strong that cancer is a genetic disease. A series of mutations, particularly in cells that can divide, lead to cancer cells with these characteristics:
- No contact inhibition
- Uncontrolled growth
- Are nonspecialized
- Have abnormal chromosomes
- Do not undergo apoptosis

Proto-Oncogenes and Tumor Suppressor Genes

- Proto-oncogenes ordinarily promote the cell cycle and inhibit apoptosis. Mutation results in a gain of function for these activities.
- Tumor suppressor genes ordinarily suppress the cell cycle and promote apoptosis. Mutation results in loss of function for these activities.

The development of cancer is a gradual process. As control of gene expression is further compromised, cancer cells develop additional characteristics that lead to angiogenesis and metastasis.

Other Genetic Changes and Cancer

Other genetic influences contribute to cancer development:
- When telomerase is present telomeres don't shorten and cells can continue to divide.
- Chromosomal rearrangements, such as translocations, may disrupt genes that regulate the cell cycle.
- Certain inherited genes are known to cause cancer: *BRCA1* and *BRCA2* are mutant genes associated with breast cancer; a mutant *RB* gene is associated with the development of eye tumors; a mutant *RET* gene predisposes a person to thyroid cancer.

Testing for These and Other Genes

Genetic tests are available to help a physician diagnose cancer. These genes are turned off in normal cells, but are active in cancer cells. For example, if the test for the presence of telomerase is positive, the cell is cancerous.

Key Terms

adult stem cell 185	proto-oncogene 193
alternative mRNA processing 190	regulatory gene 186
cell-signaling 191	repressor 186
embryonic stem cell 185	reproductive cloning 184
enhancer 189	signal transduction pathway 191
euchromatin 188	telomere 194
growth factor 193	therapeutic cloning 185
heterochromatin 187	transcription activator 189
oncogene 193	transcription factor 188
operon 186	translocation 194
promoter 186	tumor suppressor gene 193

Testing Yourself

Choose the best answer for each question.

1. Reproductive cloning differs from therapeutic cloning because the goal of reproductive cloning is
   a. the production of specific cell types for medical purposes.
   b. an individual that is genetically identical to the original individual.
   c. an individual that is genetically different from the original individual.
   d. the production of specific tissues and organs for medical transplantation.

2. The major challenge to therapeutic cloning using adult stem cells is
   a. finding appropriate cell types.
   b. obtaining enough tissue.
   c. controlling gene expression.
   d. keeping cells alive in culture.

3. Cancer cells exhibit abnormal characteristics because
   a. they do not produce telomerase.
   b. their tumor suppressor genes stimulate apoptosis.
   c. they have an unlimited food supply.
   d. control of gene expression has been lost.

4. Label this operon with these terms: active repressor, lactose metabolizing genes, operator, RNA polymerase, promoter, regulatory gene, mRNA.

5. Which of the following is not a gene expression control mechanism in eukaryotes?
   a. alternative mRNA processing
   b. rate of ribosome synthesis
   c. longevity of mRNA
   d. chromatin condensation

6. What types of offspring would you expect from a female tortoiseshell cat and a black male? Remember that this coat color gene is due to heterozygous alleles on the X chromosomes.
   a. tortoiseshell females and black males
   b. all black cats
   c. tortoiseshell females, black females, orange males, black males
   d. orange females and black males

For questions 7–11, match the examples to the gene expression control mechanisms in the key. Each answer can be used more than once.

Key:

   a. chromatin condensation
   b. alternative mRNA processing
   c. translation of mRNA
   d. protein activity

7. Insulin does not become active until 30 amino acids are cleaved from the middle of the molecule.

8. The mRNA for vitellin is longer-lived if it is exposed to estrogen.

9. Genes in Barr bodies are inactivated.

10. Each neuron combines the exons of the *DScam* primary-mRNA in different ways to produce a unique combination in mature mRNA.

11. A translation initiation factor inhibits translation when it is phosphorylated by kinase.

12. Suppose a woman is heterozygous for the *RB* gene that causes retinoblastoma. Her husband's mother was homozygous for the mutant *RB* allele, but his father's side of the family has no history of retinoblastoma. What is the probability that this couple will have a child who does not carry the mutant *RB* allele?
    a. ¼   c. ¾
    b. ½   d. 0

13. A cell may be cancerous if it tests positive for a (an)
    a. elongation factor.
    b. transcription factor.
    c. enhancer.
    d. chromosomal translocation.

14. Which of these refers to a protein complex that binds to the promoter of a gene and promotes transcription?
    a. elongation factor
    b. transcription factor
    c. transcriptional activator
    d. chromatin remodeling complex

15. Which of the following statements is true?
    a. Euchromatin contains inactive genes.
    b. Heterochromatin contains both active and inactive genes.
    c. The rearrangement of histones in euchromatin is necessary for transcription to occur.
    d. Barr bodies are composed mainly of genetically active euchromatin.
    e. All of these are correct.

16. How is transcription controlled in eukaryotic cells?
    a. through the use of proto-oncogenes
    b. by means of alternative mRNA processing
    c. using transcription factors and activators
    d. when chromatin is packed to keep genes turned on
    e. None of these are correct.

17. An operon is a short sequence of DNA containing genes regulated together
    a. that prevent RNA polymerase from binding to the promoter.
    b. that prevent transcription from occurring.
    c. and the sequences that control its transcription.
    d. that code for the repressor protein.
    e. that function to prevent the repressor from binding to the operator.

18. The regulatory gene for the *lac* operon
    a. promotes transcription of the operon.
    b. is a sequence of DNA that codes for the repressor.
    c. prevents the repressor from binding to the operator.
    d. keeps the operon off until lactose is present.
    e. Both b and d are correct.

19. What is the purpose of cell signaling in eukaryotes?
    a. to change the behavior of the receiving cell
    b. to initiate a set of reactions through the receptor protein
    c. to turn on a gene
    d. to transmit a nerve impulse, for example
    e. All of these are correct.

20. A transposon may cause cancer if
    a. it stops the cell from producing telomerase.
    b. it no longer stimulates the production of enzymes that degrade the basement membrane.
    c. cell division stops.
    d. a gene needed for cell cycle control is disrupted.

21. Which of the following statements is true?
    a. The gene product of a proto-oncogene inhibits the cell cycle.
    b. Proto-oncogenes are mutant versions of oncogenes that cause cancer.
    c. When mutated tumor suppressor genes lose their function, cancer may result.
    d. When tumor suppressor genes mutate, they become oncogenes.
    e. All of these are true.

22. Which of the following is not an example of why cancer cells are a failure of genetic control?
    a. They exhibit contact inhibition and do not grow in multiple layers.
    b. They lack specialization and do not contribute to the function of a body part.
    c. They have enlarged nuclei and may have an abnormal number of chromosomes.
    d. They fail to undergo apoptosis even though they are abnormal cells.
    e. They release a growth factor that causes neighboring blood vessels to branch into the tumor and therefore feed the cancerous tissue.

Thinking Scientifically

1. Cyclin D is a proto-oncogene that is necessary for entry into the S stage of the cell cycle. When bound to the correct kinase, RB is modified, and cell division is promoted. Then, cyclin D is destroyed. How might a mutation in cyclin D lead to uncontrolled cell growth?

2. New findings indicate that mutations outside of genes may cause disease, such as in some cases of Hirschsprung disease and multiple endocrine neoplasia. Explain how such a mutation might disrupt the regulation of gene expression.

3. Researchers have observed more alternative mRNA splicing in complex animals, such as humans, than in less complex animals, such as fruit flies. Why might alternative mRNA splicing be advantageous in complex animals?

Bioethical Issue

Should BPA be Banned?

Environmental estrogens are hormonelike chemicals that can compete with sex hormones for receptors, altering gene expression and promoting cell division. Recently, a link has been established between the estrogen-like chemical bisphenol A (BPA) and an increased risk of certain cancers, leading to calls from many to regulate or ban the compound.

BPA is used in the manufacture of polycarbonate plastics, found in many consumer products such as food containers and baby bottles. Studies demonstrate that cleaning polycarbonate with alkali or bleach may cause leaching of BPA, potentially exposing consumers to the chemical. Studies found BPA responsible for enlarged female reproductive organs in mice kept in polycarbonate cages; follow-up studies indicated that mice exposed to BPA had a three- to four-fold elevated risk of breast cancer. Additional studies suggest that exposure to BPA during fetal development may have similar effects. Citing these and other government-funded studies, many have called for government regulation of BPA, with many more demanding a ban on the use of polycarbonate baby bottles. In 2008, Health Canada weighed in on the issue, stating that BPA "is toxic to human health."

However, not all scientists agree. Industry scientists and other critics point to industry-funded studies that show little to no risk to humans, especially to adults. Furthermore, they insist that diligence by consumers when cleaning polycarbonate plastics can greatly decrease whatever risk is posed by the plastic, since avoiding bleach or alkali almost totally eliminates BPA leaching.

Do you believe that polycarbonate plastic food containers should be banned? If not, would a ban on polycarbonate baby bottles be reasonable? Or, as some contend, does polycarbonate pose no more risk to humans than other common household materials?

Essentials of Biology Website

The companion website for *Essentials of Biology* provides a wealth of information organized and integrated by chapter. You will find practice tests, animations, videos, and much more that will complement your learning and understanding of general biology.

http://www.mhhe.com/maderessentials2

Genetic Counseling

This machine sequences DNA at the rate of 350,000 nucleotides a day.

A fluorescent probe (complementary DNA) shows the location of a gene.

×150

One cell of an embryo can be removed and tested for genetic mutations.

Outline

Ensuring a Healthy Baby

TODAY IT IS possible to screen embryos for genetic abnormalities before they are implanted in the uterus. This process is called preimplantation genetic diagnosis (PGD). Couples who are at high risk of having a child with a genetic disorder can use PGD to increase their chances of having a normal child. The couple donates a number of egg cells and sperm cells, and the embryos are created in a lab by a process called in vitro fertilization. Once the embryos are large enough, one cell from each embryo can be removed and tested for genetic mutations.

—Continued next page

Continued—

If an embryo is found to have a mutation for a genetic disorder of concern, it can be discarded. Embryos that appear free of genetic flaws can be implanted into the mother. If all goes well, a healthy baby will be born.

Some people have ethical concerns about using PGD to select genetically healthy embryos. Even more controversial, however, is the use of this technology purely to choose the sex of an embryo. Some couples so desperately want a boy or a girl that they use PGD to guarantee the gender of their child, and some physicians justify this use of the technology for the achievement of "gender-balance" within a family. This use of PGD is legal in the United States, but has been banned in many countries.

In this chapter, you will learn about some of the techniques currently being used to counsel couples about their chances of passing on a genetic disorder, as well as technologies that may be used to test and treat parents and children for genetic disorders in the future.

13.1 | Counseling for Chromosomal Disorders

Potential parents are becoming aware that many illnesses are caused by abnormal chromosomal inheritance or by gene mutations. Therefore, more couples are seeking **genetic counseling,** which is available at many major hospitals as a means to determine the risk of inherited disorders in a family. For example, a couple might be prompted to seek counseling after several miscarriages, when several relatives have a particular medical condition, or if they already have a child with a genetic defect. The counselor helps the couple understand the mode of inheritance, the medical consequences of a particular genetic disorder, and the decisions they might wish to make (**Fig. 13.1**).

Various human disorders may result from abnormal chromosome number or structure. When a pregnant woman is concerned that her unborn child might have a chromosomal defect, the counselor may recommend karyotyping the fetus's chromosomes.

Karyotyping

A **karyotype** is a visual display of the chromosomes arranged by size, shape, and banding pattern. Any cell in the body except red blood cells, which lack a nucleus, can be a source of chromosomes for karyotyping. In adults, it is easiest to use white blood cells separated from a blood sample for this purpose. In fetuses, whose chromosomes are often examined to detect a syndrome, cells can be obtained by either amniocentesis or chorionic villus sampling.

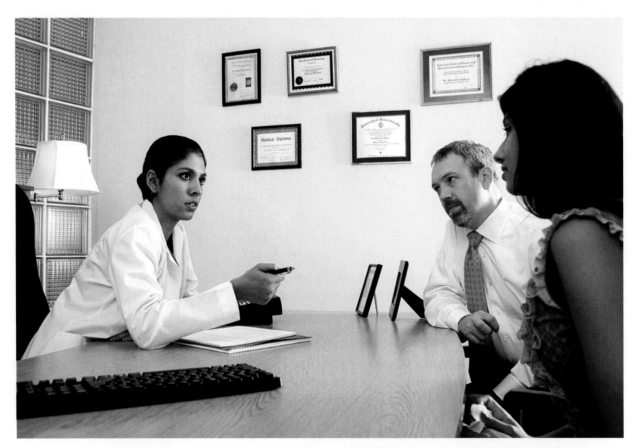

Figure 13.1 Genetic counseling.

A genetic counselor uses genetic information about the family to predict the chances a couple will have a child affected by a genetic disorder. If the woman is pregnant, tests can determine whether the child will be born free of genetic disorders.

Figure 13.2 Testing for chromosomal mutations.

To test the fetus for an alteration in the chromosome number or structure, fetal cells can be acquired by (**a**) amniocentesis or (**b**) chorionic villus sampling. **c.** Karyotyping then will reveal chromosomal mutations. In this case, the karyotype shows that the newborn will have Down syndrome (see page 140).

a. During amniocentesis, a long needle is used to withdraw amniotic fluid containing fetal cells.

b. During chorionic villi sampling, a suction tube is used to remove cells from the chorion, where the placenta will develop.

Amniocentesis is a procedure for obtaining a sample of amniotic fluid from the uterus of a pregnant woman. Blood tests and the age of the mother are considered when determining whether the procedure should be done. The risk of spontaneous abortion increases by about 0.3% due to amniocentesis, and doctors use the procedure only if it is medically warranted. Amniocentesis is not usually performed until the fourteenth to the seventeenth week of pregnancy.

A long needle is passed through the abdominal and uterine walls to withdraw a small amount of fluid, which also contains fetal cells (**Fig. 13.2***a*). Tests are done on the amniotic fluid, and the cells are cultured for karyotyping. Karyotyping the chromosomes may be delayed as long as four weeks so that the cells can be cultured to increase their number.

Chorionic villus sampling (CVS) is a procedure for obtaining chorionic cells in the region where the placenta will develop. This procedure can be done as early as the fifth week of pregnancy. A long, thin suction tube is inserted through the vagina into the uterus (Fig. 13.2*b*). Ultrasound, which gives a picture of the uterine contents, is used to place the tube between the uterine lining and the chorionic villi. Then, a sampling of chorionic cells is obtained by suction. The cells do not have to be cultured, and karyotyping can be done immediately. But testing amniotic fluid is not possible because no amniotic fluid is collected. Also, CVS carries a greater risk of miscarriage than amniocentesis—0.8% compared to 0.3%. The advantage of CVS is getting the results of karyotyping at an earlier date.

After a cell sample has been obtained, the cells are stimulated to divide in a culture medium. A chemical is used to stop mitosis during metaphase when chromosomes are the most highly compacted and condensed. The cells are then killed, spread on a microscope slide, and dried. Stains are applied to the slides, and the cells are photographed. Staining causes the chromosome to have dark and light crossbands of varying widths, and these can be used, in addition to size and shape, to help pair up the chromosomes. Today, technicians use a computer to arrange the chromosomes in pairs. As described in Chapter 9, the karyotype of a person who has Down syndrome usually has three copies of chromosome number 21 instead of two (Fig. 13.2*c*).

c. Karyotype of a person with Down syndrome. Note the three number 21 chromosomes.

Check Your Progress

1. Describe the relationship between amniocentesis and karyotyping.
2. List an advantage and a disadvantage of chorionic villus sampling compared to amniocentesis.
3. Explain why karyotypes utilize metaphase chromosomes of mitosis.

Answers: 1. During amniocentesis, a sample of amniotic fluid is collected. Cells from this fluid are cultured and used for karyotyping.   2. Chorionic villus sampling can be performed earlier in pregnancy and does not require a cell culture period; however, CVS carries a higher risk of miscarriage than amniocentesis.   3. The chromosomes are the most condensed during metaphase.

a.

b.

Figure 13.3 Deletion.

a. When chromosome 7 loses an end piece, the result is Williams syndrome.
b. These children, although unrelated, have the same appearance, health, and behavioral problems that are characteristic of Williams syndrome.

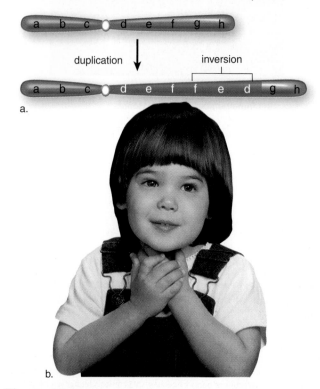

a.

b.

Figure 13.4 Duplication.

a. When a piece of chromosome 15 is duplicated and inverted, inv dup 15 syndrome results. **b.** Children with this syndrome have poor muscle tone and autistic characteristics.

Chromosomal Mutations

Most chromosomal mutations can be detected microscopically. A karyotype reveals changes in chromosome number, and a skilled technician is able to detect differences in chromosome structure based on a change in the normal banding patterns of the chromosomes (see Fig. 13.2c).

In humans, only a few variations in chromosome number, such as Down syndrome, Turner syndrome, and Klinefelter syndrome, are typically seen. Changes in chromosome structure, however, are much more common in the population. Syndromes that result from changes in chromosome structure are due to the breakage of chromosomes and their failure to reunite properly. Various environmental agents—radiation, certain organic chemicals, or even viruses—can cause chromosomes to break apart. Ordinarily, when breaks occur in chromosomes, the segments reunite to give the same sequence of genes. But their failure to do so results in one of several types of mutations: deletion, duplication, translocation, or inversion. Chromosomal mutations can occur during meiosis, and if the offspring inherits the abnormal chromosome, a syndrome may result.

Deletions and Duplications

A **deletion** occurs when a single break causes a chromosome to lose an end piece, or when two simultaneous breaks lead to the loss of an internal chromosome segment. An individual who inherits a normal chromosome from one parent and a chromosome with a deletion from the other parent no longer has a pair of alleles for each trait, and a syndrome can result.

Williams syndrome occurs when chromosome 7 loses a tiny end piece (**Fig. 13.3**). Children who have this syndrome look like pixies, with turned-up noses, wide mouths, a small chin, and large ears. Although their academic skills are poor, they exhibit excellent verbal and musical abilities. The gene that governs the production of the protein elastin is missing, and this affects the health of the cardiovascular system and causes their skin to age prematurely. Such individuals are very friendly but need an ordered life, perhaps because of the loss of a gene for a protein that is normally active in the brain.

Cri du chat (cat's cry) syndrome occurs when chromosome 5 is missing an end piece. The affected individual has a small head, is mentally disabled, and has facial abnormalities. Abnormal development of the glottis and larynx results in the most characteristic symptom—the infant's cry resembles that of a cat.

In a **duplication,** a chromosome segment is repeated so that the individual has more than two alleles for certain traits. An inverted duplication is known to occur in chromosome 15. (**Inversion** means that a segment joins in the direction opposite from normal.) Children with this syndrome, called *inv dup 15 syndrome*, have poor muscle tone, mental disabilities, seizures, a curved spine, and autistic characteristics that include poor speech, hand flapping, and lack of eye contact (**Fig. 13.4**).

Translocation

A **translocation** is the exchange of chromosome segments between two non-homologous chromosomes. A person who has both of the involved chromosomes has the normal amount of genetic material and a normal phenotype, unless the chromosome exchange breaks an allele into two pieces or fuses two genes together. The person who inherits only one of the translocated chromosomes will no doubt have only one copy of certain alleles and three copies of other alleles. A genetic counselor begins to suspect a translocation has occurred when spontaneous abortions are commonplace and family members suffer from various syndromes. A special microscopic technique allows a technician to determine that a translocation has occurred.

In 5% of Down syndrome cases, a translocation that occurred in a previous generation between chromosomes 21 and 14 is the cause. As long as the two chromosomes are inherited together, the individual is normal. But in future generations a person may inherit two normal copies of chromosomes 21 and the abnormal chromosome 14 that contains a segment of chromosome 21. In these cases, Down syndrome is not related to parental age but instead tends to run in the family of either the father or the mother.

Figure 13.5 shows a father and daughter who have a translocation between chromosomes 2 and 20. Although they have the normal amount of genetic material, they have the distinctive face, abnormalities of the eyes and internal organs, and severe itching characteristic of Alagille syndrome. People with this syndrome ordinarily have a deletion on chromosome 20; therefore, it can be deduced that the translocation disrupted an allele on chromosome 20 in the father. The symptoms of Alagille syndrome range from mild to severe, so some people may not be aware they have the syndrome. This father did not realize it until he had a child with the syndrome.

Inversion

An inversion occurs when a segment of a chromosome is turned 180°. You might think this is not a problem because the same genes are present, but the reverse sequence of alleles can lead to altered gene activity if it disrupts control of gene expression.

Inversions usually do not cause problems, but may lead to an increased occurrence of abnormal chromosomes during sexual reproduction. Crossing-over between an inverted chromosome and the non-inverted homologue can lead to recombinant chromosomes that have both duplicated and deleted segments. This happens because alignment between the two homologues is only possible when the inverted chromosome forms a loop (**Fig. 13.6**).

a.

b.

Figure 13.5 Translocation.

a. When chromosomes 2 and 20 exchange segments, Alagille syndrome results because the translocation disrupts as allele on chromosome 20. **b.** Distinctive facial features are one of the characteristics of Alagille syndrome.

Check Your **Progress**

1. **Contrast a duplication chromosome with a deletion chromosome.**
2. **Contrast a translocation chromosome with an inversion chromosome.**
3. **Explain why an individual with a translocation might appear normal.**

Answers: 1. A duplication chromosome contains a repeated segment, while a deletion chromosome is missing a segment.   2. A translocation chromosome contains a segment from another, nonhomologous chromosome. An inversion chromosome contains a segment (from the same chromosome) that has been flipped 180°.   3. As long as the translocated chromosomes are inherited together and control of gene expression is not disrupted, the individual still has the normal genotype and would appear normal.

homologous chromosomes

nonsister chromatids

duplication and deletion in both

Figure 13.6 Inversion.

(Left) A segment is inverted in the sister chromatids of one homologue. Notice that, in the red chromosome, *edc* occurs instead of *cde*. *(Middle)* The nonsister chromatids can align only when the inverted sequence forms an internal loop. After crossing-over, a duplication and a deletion can occur. *(Right)* The nonsister chromatid at the left arrow has *AB* and *ab* alleles and neither *fg* nor *FG* alleles. The nonsister chromatid at the right arrow has *gf* and *GF* alleles and neither *AB* nor *ab* alleles.

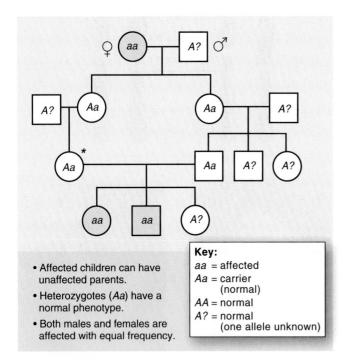

- Affected children can have unaffected parents.
- Heterozygotes (*Aa*) have a normal phenotype.
- Both males and females are affected with equal frequency.

Key:
aa = affected
Aa = carrier (normal)
AA = normal
A? = normal (one allele unknown)

Figure 13.7   Autosomal recessive pedigree.

The list gives ways to recognize an autosomal recessive disorder. How would you know that the individual at the * is heterozygous?

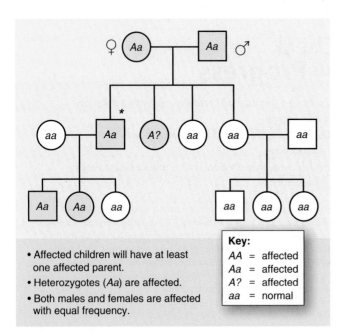

- Affected children will have at least one affected parent.
- Heterozygotes (*Aa*) are affected.
- Both males and females are affected with equal frequency.

Key:
AA = affected
Aa = affected
A? = affected
aa = normal

Figure 13.8   Autosomal dominant pedigree.

The list gives ways to recognize an autosomal dominant disorder. How would you know that the individual at the * is heterozygous?

* The designated individual in Figure 13.7 does not have the disorder but she has children with the disorder. The designated individual in Figure 13.8 does have the disorder but he has children without the disorder.

13.2 | Counseling for Genetic Disorders

Even if no chromosomal abnormality is likely, amniocentesis still might be done because it is now possible to perform biochemical tests on amniotic fluid to detect over 400 different disorders caused by specific genes in a fetus. The genetic counselor determines ahead of time what tests might be warranted. To do this, the counselor needs to know the medical history of the family in order to construct a pedigree.

Family Pedigrees

A **pedigree** is a chart of a family's history with regard to a particular genetic trait. Some traits are carried on the autosomes (nonsex chromosomes) and are called **autosomal traits;** other traits are carried on the sex chromosomes and are, therefore, called **sex-linked traits.** In a pedigree, males are designated by squares and females by circles. Shaded circles and squares represent individuals expressing the genetic disorder. A line between a square and a circle represents a union. A vertical line going downward leads directly to a single child; if there are more children, they are placed off a horizontal line.

From the counselor's knowledge of genetic disorders, he or she might already know the *pattern of inheritance* of a trait—that is, whether it is autosomal dominant, autosomal recessive, or X-linked recessive. The counselor can then determine the chances that any child born to the couple will have the abnormal phenotype.

Pedigrees for Autosomal Disorders

A family pedigree for an *autosomal recessive disorder* is shown in **Figure 13.7.** In this pattern, a child can be affected when neither parent is affected. Such heterozygous parents are *carriers* because, although they are unaffected, they are capable of having a child with the genetic disorder. If the family pedigree suggests that the parents are carriers for an autosomal recessive disorder, the counselor might suggest confirming this by doing the appropriate genetic test. Then, if the parents so desire, it would be possible to do prenatal testing of the fetus for the genetic disorder.

Figure 13.7 lists other ways that a counselor may recognize an autosomal recessive pattern of inheritance. Notice that in this pedigree, cousins are the parents of three children, two of whom have the disorder. Aside from illustrating that reproduction between related individuals is more likely to bring out recessive traits, this pedigree also shows that "chance has no memory"; therefore, each child born to heterozygous parents has a 25% chance of having the disorder. In other words, if a heterozygous couple has four children, each child might have the condition.

Figure 13.8 shows an *autosomal dominant* pattern of inheritance. In this pattern, a child can be unaffected even when the parents are heterozygous and therefore affected. Figure 13.8 lists other ways to recognize an autosomal dominant pattern of inheritance. This pedigree illustrates that when both parents are unaffected, all their children are unaffected. Why? Because neither parent has a dominant gene that causes the condition to be passed on.

Pedigrees for Sex-Linked Disorders

Figure 13.9 gives a pedigree for an *X-linked recessive disorder*. Recall that sons inherit X-linked recessive traits from their mothers because their only X chromosome comes from their mothers. More males than females have the disorder because recessive alleles on the X chromosome are always expressed in males—the Y chromosome lacks an allele. Females who have the condition inherited the allele from both their mother and their father, and all the sons of such a female will have the condition.

If a male has an X-linked recessive condition, his daughters are all carriers even if his partner is normal. Therefore, X-linked recessive conditions often appear to pass from grandfather to grandson. Figure 13.9 lists other ways to recognize an X-linked recessive disorder.

Still fewer traits are known to be *X-linked dominant*. If a condition is X-linked dominant, daughters of affected males have a 100% chance of having the condition. Females can pass an X-linked dominant allele to both sons and daughters. If a female is heterozygous and her partner is normal, each child has a 50% chance of escaping an X-linked dominant disorder, depending on which of the mother's X chromosomes is inherited.

Still fewer genetic disorders involve genes carried on the Y chromosome. For example, the *SRY* gene (sex determining region Y) is involved in determining gender during development. A counselor would recognize a *Y-linked pattern of inheritance* because Y-linked disorders are present only in males and are passed directly from father to *all* sons but not to daughters. (Can you explain why?)

Check Your Progress

1. Explain what a genetic counselor can learn from a pedigree.
2. Contrast the pedigree for an autosomal recessive disorder with that of an autosomal dominant disorder when the two parents are unaffected.
3. Explain why males cannot be carriers for X-linked recessive disorders.

Answers: 1. The counselor might be able to determine the inheritance pattern of a disorder and the chances that a child born to the couple will have the disorder.   2. An autosomal recessive disorder may show up in the children of two unaffected parents, whereas none of the children will have an autosomal dominant disorder if the parents are unaffected.   3. Males only possess one X chromosome. Therefore, males with a recessive allele on the X chromosome have no other allele and always exhibit recessive X-linked disorders.

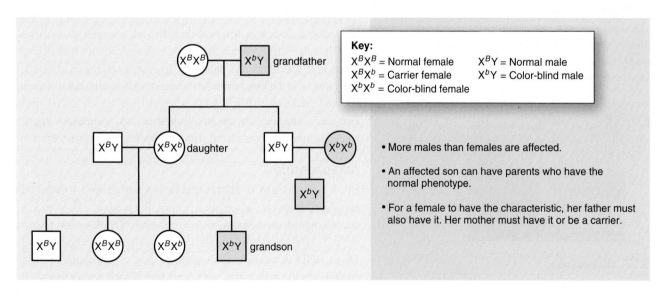

Figure 13.9   X-linked recessive pedigree.

The list gives ways of recognizing an X-linked recessive disorder—in this case, color blindness.

Figure 13.10 **Methemoglobinemia.**

The hand of the woman on the right appears blue due to chemically induced methemoglobinemia. Individuals with the disorder lack the enzyme diaphorase and are unable to clear methemoglobin from the blood.

— nebulizer

— percussion vest

Figure 13.11 **Cystic fibrosis.**

This person is undergoing antibiotic and percussion therapy for cystic fibrosis. Antibiotic therapy is used to control lung infections of cystic fibrosis patients. The antibiotic can be aerosolized and administered using a nebulizer. A percussion vest loosens mucus in the lungs.

Figure 13.12 **Alkaptonuria.**

Alkaptonuria is caused by the inability to metabolize a compound called homogentisic acid. Excess homogentisic acid accumulates in the urine, and its presence causes the urine to turn black when the urine is exposed to air.

Genetic Disorders of Interest

Medical genetics has traditionally focused on disorders caused by single gene mutations, and we will discuss a few examples.

Autosomal Disorders

Autosomal disorders are caused by mutated alleles on the autosomal chromosomes (all the chromosomes except the sex chromosomes). Some of these disorders are recessive, and therefore an individual must inherit two affected alleles before having the disorder. Others are dominant, meaning that it takes only one affected allele to cause the disorder. Dominant and recessive inheritance was discussed in Chapter 10.

Methemoglobinemia

Methemoglobinemia is a relatively harmless disorder that results from an accumulation of methemoglobin, an alternative form of hemoglobin, in the blood. Since methemoglobin is blue instead of red, the skin of people with the disorder appears bluish-purple in color (**Fig. 13.10**).

A persistent and determined physician finally solved the age-old mystery of what causes methemoglobinemia through blood tests and pedigree analysis of a family with the disorder, the blue Fugates of Troublesome Creek in Kentucky. On a hunch, the physician tested the Fugates for the enzyme diaphorase, which normally converts methemoglobin back to hemoglobin, and found that they indeed lacked the enzyme. Next, he treated a patient with the disorder in a simple but rather unconventional manner—by injecting a dye called methylene blue! The dye is a strong reducing agent capable of donating electrons to methemoglobin, converting it back into hemoglobin. The results were striking but immediate—the patient's skin quickly turned pink again.

Cystic Fibrosis

Cystic fibrosis is an autosomal recessive disorder that occurs among all ethnic groups, but it is the most common lethal genetic disorder among Caucasians in the United States. Research has demonstrated that chloride ions (Cl^-) fail to pass through a plasma membrane channel protein in the cells of these patients. Ordinarily, after chloride ions have passed through the membrane, sodium ions (Na^+) and water follow. It is believed that the lack of water then causes abnormally thick mucus in the bronchial tubes and pancreatic ducts, thus interfering with the function of the lungs and pancreas. To ease breathing in affected children, the thick mucus in the lungs must be loosened periodically, but still the lungs become infected frequently (**Fig. 13.11**). Clogged pancreatic ducts prevent digestive enzymes from reaching the small intestine, and to improve digestion, patients take digestive enzymes mixed with applesauce before every meal.

Alkaptonuria

Black urine disease, or **alkaptonuria,** is a rare genetic disorder that follows an autosomal recessive inheritance pattern. People with alkaptonuria lack a functional copy of the homogentisate oxygenase (*HGD*) gene found on chromosome 3. HGD normally breaks down a compound called homogentisic acid. When HGD is missing, homogentisic acid accumulates in the blood and is passed into the urine. The compound turns black upon exposure to air, giving the urine a characteristic color and odor (**Fig. 13.12**). Homogentisic acid also accumulates in joint spaces and connective tissues, leading to darkening of the tissues and eventual arthritis by adulthood.

Sickle Cell Disease

Sickle cell disease is an autosomal recessive disorder in which the red blood cells are not biconcave disks like normal red blood cells, but are irregular in shape (**Fig. 13.13**). In fact, many are sickle-shaped. A single base change in the globin gene causes hemoglobin to differ from normal hemoglobin by a single amino acid. Now, the abnormal hemoglobin molecules stack up and form insoluble rods, and the red blood cells become sickle-shaped.

Because sickle-shaped cells can't pass along narrow capillary passageways as well as disk-shaped cells can, they clog the vessels and break down. This is why persons with sickle cell disease suffer from poor circulation, anemia, and low resistance to infection. Internal hemorrhaging leads to further complications, such as jaundice, episodic pain in the abdomen and joints, and damage to internal organs.

Sickle cell heterozygotes have the sickle cell trait, in which the blood cells are normal, unless they experience dehydration or mild oxygen deprivation. Still, at present, most experts believe that persons with the sickle cell trait are generally healthy and do not need to restrict their physical activity.

Marfan Syndrome

Marfan syndrome (see Chapter 10, page 153), an autosomal dominant disorder, is caused by a defect in an elastic connective tissue protein called fibrillin. This protein is normally abundant in the lens of the eye; the bones of limbs, fingers, and ribs; and the wall of the aorta. Thus, the affected person often has a dislocated lens, tall stature, long limbs and fingers, and a caved-in chest. The aorta wall is weak and can possibly burst without warning. A tissue graft can strengthen the aorta, but affected individuals still should not overexert themselves.

Huntington Disease

Huntington disease is a dominant neurological disorder that leads to progressive degeneration of neurons in the brain (**Fig. 13.14**). The disease is caused by a single mutated copy of the gene for a protein called huntingtin. Most patients appear normal until they are middle-aged and have already had children, who may also have the inherited disorder. There is no effective treatment, and death comes 10 to 15 years after the onset of symptoms.

Several years ago, researchers found that the gene for Huntington disease was located on chromosome 4. A test was developed for the presence of the gene, but few people want to know if they have inherited the gene because there is no cure. But now we know that the disease stems from an unusual mutation. Extra codons cause the huntingtin protein to have a series of extra glutamines. Whereas the normal version of huntingtin has stretches of between 10 and 25 glutamines, mutant huntingtin may contain 36 or more. Because of the extra glutamines, the huntingtin protein changes shape and forms large clumps inside neurons. Even worse, it attracts and causes other proteins to clump with it. One of these proteins, called CBP, helps nerve cells survive. Researchers hope to combat the disease by boosting CBP levels.

Figure 13.13
Sickle cell disease.

Persons with sickle cell disease have sickle-shaped red blood cells because of an abnormal hemoglobin molecule.

×2500

Many neurons in normal brain

Loss of neurons in Huntington brain

Figure 13.14 **Huntington disease.**

Huntington disease is characterized by increasingly serious psychomotor and mental disturbances because of a loss of neurons in the brain.

Check Your Progress

1. Compare and contrast the physiological bases of methemoglobinemia and alkaptonuria.
2. Explain how a single base mutation in the globin gene may cause sickle cell disease.
3. Explain why a heterozygote for Huntington disease will exhibit the disorder.

Answers: 1. Both result from the buildup of a substrate that is not broken down properly. The substance is methemoglobin for methemoglobinemia and homogentisic acid for alkaptonuria.   2. The single base change in the gene alters a codon, changing the amino acid specified. The abnormal globin protein is then prone to forming rods that alter the shape of the red blood cells.   3. Huntington is a dominant disorder. The huntingtin protein changes shape and clumps, and other proteins nearby aggregate too.

X-Linked Recessive Disorders

As you know, X-linked recessive disorders are caused by mutated alleles on the X chromosome. A son inherits an X-linked recessive condition from his mother.

Color Blindness   Color blindness is a common X-linked recessive disorder. About 8% of Caucasian men have red-green color blindness. Most of them see brighter greens as tans, olive greens as browns, and reds as reddish-browns. A few cannot tell reds from greens at all; they see only yellows, blues, blacks, whites, and grays.

Duchenne Muscular Dystrophy   Duchenne muscular dystrophy is an X-linked recessive disorder characterized by wasting away of the muscles. The absence of a protein, now called dystrophin, is the cause of the disorder. Much investigative work determined that dystrophin is involved in the release of calcium from the sarcoplasmic reticulum in muscle fibers. The lack of dystrophin causes calcium to leak into the cell, which promotes the action of an enzyme that dissolves muscle fibers. When the body attempts to repair the tissue, fibrous tissue forms (**Fig. 13.15**), and this cuts off the blood supply so that more and more cells die.

Symptoms such as waddling gait, toe walking, frequent falls, and difficulty in rising may appear as soon as the child starts to walk. Muscle weakness intensifies until the individual is confined to a wheelchair. Death usually occurs by age 20; therefore, affected males are rarely fathers. The recessive allele remains in the population through passage from carrier mother to carrier daughter.

As therapy, immature muscle cells can be injected into muscles, and for every 100,000 cells injected, dystrophin production occurs in 30–40% of muscle fibers.

fibrous tissue

normal tissue

Figure 13.15   Muscular dystrophy.

In muscular dystrophy, the calves enlarge because fibrous tissue develops as muscles waste away due to a lack of the protein dystrophin.

Testing for Genetic Disorders

Following genetic testing, a genetic counselor can explain to prospective parents the chances a child of theirs will have a disorder that runs in their family. If a woman is already pregnant, the parents may want to know whether the unborn child has the disorder. If the woman is not pregnant, the parents may opt for testing of the embryo or egg before she does become pregnant, as described shortly.

Testing depends on the genetic disorder of interest. In some instances, it is appropriate to test for a particular protein, and in others to test for the mutated gene.

Testing for a Protein

Some genetic mutations lead to disorders caused by a lack of enzyme activity. For example, in the case of methemoglobinemia, it is possible to test for the quantity of the enzyme diaphorase in a blood sample and from that, determine whether the individual is likely homozygous normal, a carrier, or has methemoglobinemia. If the parents are carriers, each child has a 25% chance of having methemoglobinemia. This knowledge may lead prospective parents to opt for testing of the embryo or egg, as described in Figures 13.19 and 13.20.

Testing the DNA

Two types of DNA testing are possible: testing for a genetic marker and using a DNA probe.

Genetic Markers   Testing for a genetic marker relies on a difference in the DNA due to the presence of the abnormal allele. As an example, consider that individuals with sickle cell trait or Huntington disease have an abnormality in a gene's base sequence. This abnormality in sequence is a **genetic marker.** As you know, restriction enzymes cleave DNA at particular base sequences. Therefore, the fragments that result from the use of a restriction enzyme may be different for people who are normal than for those who are heterozygous or homozygous for a mutation (**Fig. 13.16**).

Genetic Profiling   New technologies have made DNA testing easy and inexpensive. For example, it is now possible to place thousands of known disease-associated mutant alleles onto a **DNA microarray**—a small silicon chip containing many DNA samples, in this case the mutant alleles (**Fig. 13.17**). Genomic DNA from the subject to be tested is labeled with a fluorescent dye, and then added to the microarray. The spots on the microarray fluoresce if the DNA binds to the mutant alleles on the chip, indicating that the subject may have a particular disorder or is at risk for developing it later in life. An individual's complete genotype, including all the various mutations, is called the **genetic profile.**

With the help of a genetic counselor, individuals can be educated about their genetic profile. It's possible that a person has or will have a genetic disorder caused by a single pair of alleles. However, polygenic traits are more common, and in these instances, the genetic profile can indicate an increased or decreased risk for a disorder. Risk information can be used to design a program of medical surveillance and to foster a lifestyle aimed at reducing the risk. For example, suppose an individual has mutations common to people with colon cancer. It would be helpful to have an annual colonoscopy so that any abnormal growths can be detected and removed before they become invasive.

a. Normal fragmentation pattern   b. Genetic disorder fragmentation pattern

Figure 13.16   Use of a genetic marker to test for a genetic mutation.

a. In this example, DNA from a normal individual has certain restriction enzyme cleavage sites. **b.** DNA from another individual lacks one of the cleavage sites, and this loss indicates that the person has a mutated gene. In heterozygotes, half of their DNA would have the cleavage site and half would not have it. (In other instances, the gain in a cleavage site could be an indication of a mutation.)

Figure 13.17   Use of a DNA microarray to test for mutated genes.

This DNA microarray contains many disease-associated mutant alleles. Fluorescently labeled genomic DNA from an individual has been added to the microarray. Any fluorescent spots indicate that binding has occurred and that the individual may have the genetic disorder or is at risk for developing it later in life.

Check
Your Progress

1. The allele for Duchenne muscular dystrophy is seldom inherited from a father. Explain.
2. Describe how changes in base sequence may be used to test for genetic disorders.
3. Describe the benefits of knowing your genetic profile.

Answers: 1. Affected males seldom survive long enough to reproduce.   2. The change in base sequence of a mutant allele may cause restriction enzymes to cleave DNA at unexpected places, altering the normal fragmentation pattern.   3. A genetic profile will tell you whether or not you are at risk for a genetic disorder. This information can be used to tailor your lifestyle to reduce the risk of developing disease.

Figure 13.18 Ultrasound.

Today, an ultrasound exam can produce three-dimensional images, such as the normal one shown here. Ultrasound exams can allow physicians to detect serious abnormalities, such as neural tube abnormalities.

Testing the Fetus

If a woman is already pregnant, ultrasound can detect serious fetal abnormalities, and it is also possible to obtain and test the DNA of fetal cells for genetic defects.

Ultrasound Ultrasound images help doctors evaluate fetal anatomy. An ultrasound probe scans the mother's abdomen, and a transducer transmits high-frequency sound waves that are transformed into a picture on a video screen. This picture shows the fetus inside the uterus (**Fig. 13.18**). Ultrasound can be used to determine the baby's age and size, and whether there is more than one baby. Also, some chromosomal abnormalities, such as Down syndrome, Edwards syndrome (three copies of chromosome 18), and Patua syndrome (3 copies of chromosome 13), cause anatomical abnormalities during fetal development that may be detected by ultrasound by the twentieth week of pregnancy. For this reason, a routine ultrasound at this time is considered essential to good prenatal care.

Many other conditions such as spina bifida, can be diagnosed by an ultrasound. Spina bifida results when the spine fails to close properly during the first month of pregnancy. Surgery to close a newborn's spine in such a case is generally performed within 24 hours after birth.

Testing Fetal Cells Fetal cells can be tested for various genetic disorders. If the fetus has an incurable disorder, the parents may wish to consider an abortion.

For testing purposes, fetal cells may be acquired through amniocentesis or chorionic villus sampling, as described earlier in this chapter. In addition, fetal cells may be collected from the mother's blood. As early as nine weeks into the pregnancy, a small number of fetal cells can be isolated from the mother's blood using a cell sorter. While mature red blood cells lack a nucleus, immature red blood cells do have a nucleus, and they also have a shorter life span than mature red blood cells. Therefore, if nucleated fetal red blood cells are collected from the mother's blood, they are known to be from this pregnancy.

Only about 1/70,000 blood cells in a mother's blood are fetal cells, and therefore PCR is used to amplify the DNA from the few cells collected. The procedure poses no risk whatsoever to the fetus.

8-celled embryo

Embryonic cell is removed.

Cell is genetically healthy.

Embryo develops normally in uterus.

Figure 13.19 Testing the embryo.

Genetic diagnosis is performed on one cell removed from an 8-celled embryo. If this cell is found to be free of the genetic defect of concern, and the 7-celled embryo is implanted in the uterus, it develops into a newborn with a normal phenotype.

Testing the Embryo and Egg

As discussed in Chapter 29, in vitro fertilization (IVF) is carried out in laboratory glassware. The physician obtains eggs from the prospective mother and sperm from the prospective father, and places them in the same receptacle, where fertilization occurs. Following IVF, now a routine procedure, it is possible to test the embryo. Prior to IVF, it is possible to test the egg for any genetic defect. In any case, only normal embryos are transferred to the uterus for further development.

Testing the Embryo   If prospective parents are carriers for one of the genetic disorders discussed earlier, they may want assurance that their offspring will be free of the disorder. Genetic diagnosis of the embryo will provide this assurance.

Following IVF, the zygote (fertilized egg) divides. When the embryo has six to eight cells, one of these cells can be removed for diagnosis, with no effect on normal development (**Fig. 13.19**). Only embryos that test negative for the genetic disorders of interest are placed in the uterus to continue developing.

So far, about 1,000 children worldwide have been born free of alleles for genetic disorders that run in their families following embryo testing. In the future, it's possible that embryos that test positive for a disorder could be treated by gene therapy, so that those embryos, too, would be allowed to continue to term.

Testing the Egg   Recall that meiosis in females results in a single egg and at least two polar bodies. Polar bodies, which later disintegrate, receive very little cytoplasm, but they do receive a haploid number of chromosomes, and thus can be useful for genetic diagnosis. When a woman is heterozygous for a recessive genetic disorder, about half the polar bodies receive the mutated allele, and in these instances the egg receives the normal allele. Therefore, if a polar body tests positive for a mutated allele, the egg probably received the normal allele (**Fig. 13.20**). Only normal eggs are then used for IVF. Even if the sperm should happen to carry the mutation, the zygote will, at worst, be heterozygous. But the phenotype will appear normal.

If gene therapy becomes routine in the future, it's possible that an egg could be given genes that control traits desired by the parents, such as musical or athletic ability, prior to IVF. Such genetic manipulation, called **eugenics,** carries many ethical concerns.

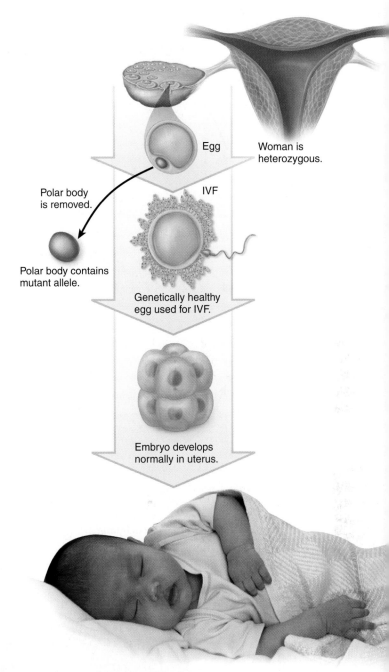

Figure 13.20   Testing the egg.

Genetic diagnosis is performed on a polar body removed from an egg. If the egg is free of a genetic defect, it is used for IVF, and the embryo is transferred to the uterus for further development.

Check Your Progress

1. Describe how fetal blood cells can be separated from maternal blood cells.
2. Explain how it is possible to test an embryo for a genetic disorder.
3. Explain the rationale for testing a polar body to determine if an egg contains a genetic mutation.

Answers: 1. Fetal blood cells still have nuclei, allowing them to be separated from maternal blood cells by using a cell sorter.   2. When the embryo contains six to eight cells, one cell is removed and used for testing.   3. If a woman is heterozygous for a genetic disorder, the first polar body receives one allele while the egg receives the other. If the polar body contains the mutant allele, the egg will probably be genetically normal.

1 Remove bone marrow stem cells.

defective gene

bone marrow

2 Use a virus to carry the normal gene into bone marrow stem cells.

recombinant DNA

normal gene

virus

3 Recombinant DNA molecules carry the normal gene into the genome of stem cells.

4 Return genetically engineered stem cells to the patient.

recombinant DNA

normal gene

Figure 13.21   Ex vivo gene therapy in humans.

Bone marrow stem cells are withdrawn from the body. A virus is used to insert a normal gene into the host genome, and then the cells are returned to the body.

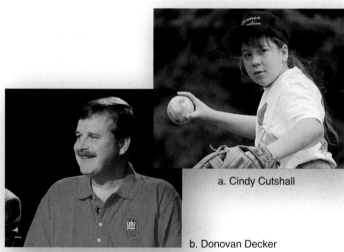

a. Cindy Cutshall

b. Donovan Decker

Figure 13.22   Gene therapy patients.

a. This patient was treated with the *ADA* gene to cure severe combined immunodeficiency. **b.** This patient was treated with the *VEGF* gene to alleviate poor coronary circulation.

13.3 | Gene Therapy

Gene therapy is the insertion of genetic material into human cells for the treatment of a disorder. It includes procedures that give a patient healthy genes to make up for faulty genes, as well as the use of genes to treat various other human illnesses, such as cardiovascular disease and cancer. Gene therapy includes both ex vivo (outside the body) and in vivo (inside the body) methods. Viruses genetically modified to be safe can be used to ferry a normal gene into cells (**Fig. 13.21**) and so can liposomes, which are microscopic globules of lipids specially prepared to enclose the normal gene. On the other hand, sometimes the gene is injected directly into a particular region of the body. Below, we give examples of **ex vivo gene therapy** (the gene is inserted into cells that have been removed and then returned to the body) and **in vivo gene therapy** (the gene is delivered directly into the body).

Ex Vivo Gene Therapy

Children who have SCID (severe combined immunodeficiency) lack the enzyme ADA (adenosine deaminase), which is involved in the maturation of cells that produce antibodies. In order to carry out gene therapy, bone marrow stem cells are removed from the blood and infected with a virus that carries a normal gene for the enzyme. Then the cells are returned to the patient. Bone marrow stem cells are preferred for this procedure because they divide to produce more cells with the same genes. Patients who have undergone this procedure show significantly improved immune function associated with a sustained rise in the level of ADA enzyme activity in the blood (**Fig. 13.22***a*).

Ex vivo gene therapy is also used in the treatment of familial hypercholesterolemia, a genetic disorder in which high levels of plasma cholesterol make the patient subject to fatal heart attacks at a young age. A small portion of the liver is surgically excised and then infected with a retrovirus containing a normal gene for a cholesterol receptor before the tissue is returned to the patient. Several patients have experienced lowered plasma cholesterol levels following this procedure.

Some cancers are being treated with ex vivo gene therapy procedures. In one procedure, immune system cells are removed from a cancer patient and genetically engineered to display tumor antigens. After these cells are returned to the patient, they stimulate the immune system to kill tumor cells.

In Vivo Gene Therapy

Most cystic fibrosis patients lack a gene that codes for a regulator of a transmembrane carrier for the chloride ion. In gene therapy trials, the gene needed to cure cystic fibrosis is sprayed into the nose or delivered to the lower respiratory tract by adenoviruses or by the use of liposomes (microscopic vesicles that spontaneously form when lipoproteins are put into a solution). Investigators are trying to improve uptake of the genes and are also hypothesizing that a combination of all three vectors might be more successful.

Genes are also being used to treat medical conditions such as poor coronary circulation. It has been known for some time that VEGF (vascular endothelial growth factor) can cause the growth of new blood vessels. The gene that codes for this growth factor can be injected alone, or within a virus, into the heart to stimulate branching of coronary blood vessels. Patients who have received this treatment report that they have less chest pain and can run longer on a treadmill (Fig. 13.22b).

Rheumatoid arthritis, a crippling disorder in which the immune system turns against a person's own body and destroys joint tissue, has recently been treated with in vivo gene therapy methods. Clinicians inject adenoviruses that contain an anti-inflammatory gene into the affected joint. The added gene reduces inflammation within the joint space and lessens the patient's pain and suffering. Clinical trials have been promising, and animal studies have even shown that gene therapy may stave off arthritis in at-risk individuals.

Figure 13.23 summarizes how gene therapy is being used to treat various illnesses.

Brain
(gene transfer by injection)*
• Huntington disease
• Alzheimer disease
• Parkinson disease
• brain tumors

Skin
(gene transfer by modified blood cells)*
• skin cancer

Lungs
(gene transfer by aerosol spray)*
• cystic fibrosis
• hereditary emphysema

Liver
(gene transfer by retroviral implants)**
• familial hypercholesterolemia

Blood
(gene transfer by bone marrow transplant)**
• sickle cell disease

Endothelium (blood vessel lining)
(gene transfer by implantation of genetically altered endothelium)**
• hemophilia
• diabetes mellitus

Muscle
(gene transfer by injection)*
• Duchenne muscular dystrophy

Joint
(gene transfer by injection)
• rheumatoid arthritis

Bone marrow
(gene transfer by implantation of genetically altered stem cells)**
• ADA deficient SCID
• sickle cell disease

| * in vivo |
| ** ex vivo |

Figure 13.23 Sites of ex vivo and in vivo somatic gene therapy.

Ex vivo: Tissue is removed, genetically modified, and returned to patient.

In vivo: Targeted areas (e.g., brain, lungs) receive copies of normal genes by various methods of gene transfer.

The Chapter in Review

Summary

13.1 | Counseling for Chromosomal Disorders

A counselor can detect chromosomal mutations by studying a karyotype of the individual.

Karyotyping

A karyotype is a display of the chromosomes arranged by pairs; the autosomes are numbered from 1 to 22. The sex chromosomes are not numbered.

Chromosomal Mutations

Chromosomal mutations involve changes in chromosome number or structure. Abnormal chromosome number results from nondisjunction during meiosis. Changes in chromosome structure include deletions, duplications, translocations, and inversions:

In Williams syndrome, one copy of chromosome 7 has a deletion; in cri du chat syndrome, one copy of chromosome 5 has a deletion; and in inv dup 15 syndrome, chromosome 15 has an inverted duplication. Down syndrome can be due to a translocation between chromosomes 14 and 21 in a previous generation.

An inversion can lead to chromosomes that have a deletion and a duplication. During synapsis, the homologue with the inverted sequence must loop back to align with the normal homologue. Following crossing-over, one nonsister chromatid has a deletion and the other has a duplication.

13.2 | Counseling for Genetic Disorders

To assist a couple in determining whether they have or their prospective children will have a genetic disorder, a genetic counselor can (1) construct family pedigrees for genetic disorders of interest, (2) order genetic testing of prospective parents, the fetus, the embryo, or the egg, and even (3) suggest gene therapy.

Family Pedigrees

A family pedigree is a visual representation of the history of a genetic disorder in a family. Constructing pedigrees helps a genetic counselor decide whether a genetic disorder that runs in a family is autosomal recessive (see Fig. 13.7) or dominant (see Fig. 13.8); X-linked (see Fig. 13.9); or some other pattern of inheritance.

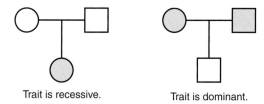

Trait is recessive.   Trait is dominant.

Genetic Disorders of Interest

Autosomal Disorders

- Methemoglobinemia (inability to clear methemoglobin from the blood)
- Cystic fibrosis (faulty regulator of chloride channel)
- Alkaptonuria (inability to metabolize homogentisic acid)
- Sickle cell disease (sickle-shaped red blood cells)
- Marfan syndrome (defective elastic connective tissue)
- Huntington disease (abnormal huntingtin protein)

X-Linked Disorders

- Color blindness (inability to detect colors normally)
- Duchenne muscular dystrophy (absence of dystrophin leads to muscle weakness)

Testing for Genetic Disorders

A counselor can order the appropriate test to detect a disorder.

Testing for Proteins Blood or tissue samples can be tested for enzyme activity. Lack of enzyme activity can sometimes indicate that a genetic disorder exists.

Testing DNA Cut DNA with restriction enzymes and then compare the fragment pattern to the normal pattern. Label DNA with fluorescent tags, and see if fragments bind to DNA probes on a DNA chip that contain the mutation.

Testing the Fetus An ultrasound can detect disorders due to certain chromosomal abnormalities and other conditions, such as spina bifida, which are due to the inheritance of mutant alleles. Fetal cells can be obtained by amniocentesis or chorionic villus sampling, or by sorting out fetal cells from the mother's blood.

Testing the Embryo Following in vitro fertilization (IVF), it is possible to test the embryo. A cell is removed from an eight-celled embryo, and if it is found to be genetically healthy, the embryo is implanted in the uterus, where it develops to term.

Testing the Egg Before IVF, a polar body can be tested. If the woman is heterozygous, and the polar body has the genetic defect, the egg does not have it. Following fertilization, the embryo is implanted in the uterus.

13.3 | Gene Therapy

During gene therapy, a genetic defect is treated by giving the patient a foreign gene.

- **Ex vivo therapy:** Cells are removed from the patient, treated, and returned to the patient.
- **In vivo therapy:** A foreign gene is given directly to the patient via nasal spray, liposomes, or adenovirus.

Key Terms

Testing Yourself

Choose the best answer for each question.

1. The major advantage of chorionic villus sampling over amniocentesis is that it
   a. allows karyotyping to be done earlier.
   b. produces karyotypes with better images of chromosomes.
   c. carries a lower risk of spontaneous abortion.
   d. produces a sample that is not contaminated by cells from the mother.

2. Which of the following disorders is not caused by a change in chromosome structure?
   a. cri du chat syndrome
   c. Klinefelter syndrome
   b. inv dup 15 syndrome
   d. Alagille syndrome

3. Which of the following chromosomal mutations would not result in an individual possessing three alleles of a gene?
   a. an extra chromosome
   c. a duplication
   b. a translocation
   d. a deletion

4. Fill in the genotypes a.–i. of the family members in the following pedigree for an autosomal recessive trait. Shaded individuals are affected.

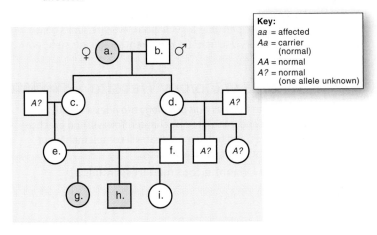

Key:
aa = affected
Aa = carrier (normal)
AA = normal
A? = normal (one allele unknown)

5. Fill in the genotypes a.–h. of the family members in the following pedigree for an autosomal dominant trait. Shaded individuals are affected.

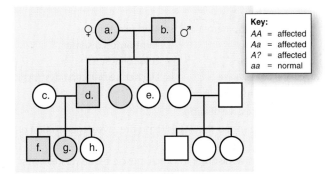

Key:
AA = affected
Aa = affected
A? = affected
aa = normal

6. Fill in the genotypes a.–e. of the family members in the following pedigree for an X-linked recessive trait. Shaded individuals are affected.

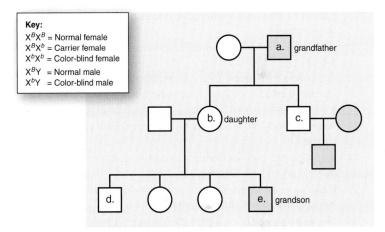

Key:
$X^B X^B$ = Normal female
$X^B X^b$ = Carrier female
$X^b X^b$ = Color-blind female
$X^B Y$ = Normal male
$X^b Y$ = Color-blind male

For questions 7–12, match the pedigree characteristics to those in the key. Answers can be used more than once. Some questions may have more than one answer.

Key:
   a. X-linked recessive
   b. X-linked dominant
   c. autosomal recessive
   d. autosomal dominant
   e. Y-linked

7. All daughters and no sons from affected males express the trait.
8. Males express the trait more frequently than females.
9. Affected children always have at least one affected parent.
10. Affected children can have unaffected parents.
11. Two affected parents may have unaffected children.
12. The trait is passed directly from father to son, but never from mother to son.
13. Disorders that result from the inability to break down a substance are
   a. cystic fibrosis and color blindness.
   b. methemoglobinemia and alkaptonuria.
   c. color blindness and methemoglobinemia.
   d. muscular dystrophy and cystic fibrosis.

For questions 14–21, match the descriptions to the conditions in the key. Answers can be used more than once. Some questions may have more than one answer.

Key:

a. methemoglobinemia
b. cystic fibrosis
c. alkaptonuria
d. sickle cell disease
e. Marfan syndrome
f. Huntington disease
g. color blindness
h. Duchenne muscular dystrophy

14. autosomal dominant disorder
15. autosomal recessive disorder
16. neurological disorder that leads to progressive degeneration of brain cells
17. X-linked recessive disorder in which muscle tissue wastes away
18. the most common lethal genetic disorder among U.S. Caucasians
19. X-linked recessive disorder
20. results from the lack of the enzyme diaphorase, resulting in the inability to clear homogentisic acid
21. caused by abnormal hemoglobin, resulting in red blood cells that are not round
22. Which of the following methods may be used in genetic testing of a fetus?
   a. in vitro fertilization
   b. ultrasound
   c. testing a polar body
   d. genetic markers
23. Parents who do not have methemoglobinemia produce a child who has methemoglobinemia. What is the genotype of all persons involved?
24. A 25-year-old man has Huntington disease. Is it possible for him to father a normal son? Explain.
25. A girl has color blindness. What is her genotype? What are the possible genotypes of the parents?
26. The advantage of using a DNA microarray over testing for genetic markers is that a DNA microarray
   a. tests for additional or missing restriction enzyme cleavage sites.
   b. can test for many different mutations at once.
   c. is the only method that can distinguish between a heterozygous and homozygous individual.
   d. does not require the use of PCR to amplify DNA.
27. An ultrasound can be used to
   a. determine if a fetus is heterozygous or homozygous for a trait.
   b. identify anatomical abnormalities that may suggest a genetic defect.
   c. amplify fetal blood cells for DNA extraction.
   d. perform a karyotype.
28. Testing a polar body may be done if
   a. both parents are homozygous for a genetic disorder.
   b. the male parent has a translocation.
   c. the male parent is a carrier for a genetic disorder.
   d. the female parent is a carrier for a genetic disorder.
29. Ex vivo gene therapy involves
   a. delivery of genes to cells within the body.
   b. only the use of liposomes to deliver genes.
   c. removal of cells from the body before treatment.
   d. All of the above is correct.

Thinking Scientifically

1. As you learned in Chapter 11, some individuals possess extra copies of certain genes. Recently, a few genetic diseases were linked to having extra copies of specific genes. How might extra copies of a gene be harmful in some cases?
2. The use of adenoviruses to deliver genes used in gene therapy has sometimes proven problematic. Recently, adenoviruses were used in gene therapy trials performed on ten infants with X-linked severe combined immunodeficiency syndrome (XSCID), also known as "bubble boy disease." Although the trial was considered a success because the gene was expressed and the children's immune systems were restored, researchers were shocked and disappointed when two children developed leukemia. How might you explain the development of leukemia in these gene therapy patients?

3. Genetic counselors note that certain debilitating diseases are maintained in a population, even when they are disadvantageous. Why wouldn't they be removed by the evolutionary process?

Bioethical Issue

Gene Therapy Trials

Despite past successes and lofty promises, some gene therapy clinical trials have had disappointing or even tragic results. Eighteen-year-old Jesse Gelsinger died in a 1999 gene therapy trial because of a massive immune response to the adenovirus used to deliver the gene. In 2007, 36-year-old Jolee Mohr died from a massive fungal infection while undergoing gene therapy treatment for rheumatoid arthritis.

These highly publicized cases raised concern among scientists, government regulators, and patient advocates. Many of them insist that increased scrutiny of gene therapy clinical trials is needed. Indeed, a subsequent review of gene therapy clinical trials revealed that 37 of 970 adverse effects were reported to the NIH, constituting serious legal and ethical violations. However, many researchers contend that these and other cases were isolated events, and insist that current safeguards are sufficient to protect test subjects. For example, they point out that none of the other 125 patients participating in Mohr's clinical trial suffered adverse effects from the treatment.

Do you believe that the NIH should increase supervision of gene therapy trials and increase penalties for violations, even if it limits gene therapy trials and hampers research? Or are current safeguards sufficient?

Essentials of Biology Website

The companion website for *Essentials of Biology* provides a wealth of information organized and integrated by chapter. You will find practice tests, animations, videos, and much more that will complement your learning and understanding of general biology.

http://www.mhhe.com/maderessentials2

14

Darwin and Evolution

Only in Hawaii did honeycreepers evolve into over 50 species.

Akiapolaau

Laysan Finch

Due to artificial selection, you can choose from more than 150 breeds of dogs.

Like dogs and honeycreepers, bacteria evolve. A growing problem in hospitals is the number of bacteria that have become resistant to antibiotics.

Outline

Evolution Accounts for Diversity

WHAT DO THE many breeds of dogs, the honeycreepers of Hawaii, and a child's antibiotic-resistant ear infection have in common? Evolution! Without **evolution**—change in a line of descent over time—we wouldn't see such a great variety of living things about us.

—Continued next page

Continued—

But aside from its many benefits, evolution also sometimes causes problems for humans.

Some bacteria have evolved to the point that they are resistant to the antibiotics once successfully used to cure the diseases they cause. For example, antibiotics originally cured bacterial ear infections within a few days. Unseen, however, were the one or two bacteria with just the right mutation to resist a particular drug. All the descendants of these bacteria were also resistant, causing the antibiotic to be useless as a cure for this type of ear infection. The antibiotic is considered the *selective agent* because it allowed the resistant bacteria to flourish while killing their relatives.

What was the selective agent for the many breeds of dogs available as pets today? Humans were, of course. Over the years, humans selected which dogs to mate to produce today's varieties. This process is called *artificial selection*.

In nature, the selective agent isn't usually a human or some foreign substance like an antibiotic. It is some aspect of the natural environment, and the process is called *natural selection*. An environment may offer a variety of food sources to any given species. Different individuals may utilize some resources but not others. Over time, specializations to acquire certain foods can occur. At least 50 species of birds called honeycreepers were once found on the Hawaiian Islands and nowhere else on Earth. Of the few species that remain today, some species feed on insects, others on flower nectar, and still others on seeds. Through analysis, all honeycreepers have been traced back to one unspecialized ancestor arriving on the islands an estimated 4–7 million years ago.

This chapter describes how the work of Charles Darwin contributes to our current knowledge of the evolutionary process and how other studies give evidence that all life has a common source.

14.1 | Darwin's Theory of Evolution

Charles Darwin was only 22 in 1831 when he accepted the position of naturalist aboard the HMS *Beagle*, a British naval ship about to sail around the world (**Fig. 14.1**). Darwin had a suitable background for this position. He was a dedicated student of nature and had long been a collector of insects. His sensitive nature had prevented him from studying medicine, and he went to divinity school at Cambridge instead. Even so, he attended many lectures in both biology and geology, and he was also tutored in these subjects by the Reverend John Henslow. Darwin spent the summer of 1831 doing fieldwork with Cambridge geologist Adam Sedgwick, before Henslow recommended him to the captain of the *Beagle* as the ship's naturalist. The voyage was to take two years, but ended up taking five, traversing the Southern Hemisphere, where we now know that life is most abundant and varied. Along the way, Darwin encountered forms of life very different from those in his native England.

Darwin's major mission was to expand the navy's knowledge of natural resources (e.g., water and food) in foreign lands. By the time the trip was over, Darwin had made observations that led him to conclude that biological evolution occurs, contrary to most beliefs at the time.

Figure 14.1 **Voyage of the HMS *Beagle*.**

The map shows the journey of the HMS *Beagle* around the world. As Darwin traveled along the east coast of South America, he noted that a bird called a rhea looked like the African ostrich. On the Galápagos Islands, marine iguanas, found no other place on Earth, use their large claws to cling to rocks and their blunt snouts for eating marine algae.

Charles Darwin

HMS *Beagle*

marine iguanas

Galápagos Islands

rhea

Before Darwin

Prior to Darwin, most people had an entirely different way of viewing the world. They believed that the Earth was only a few thousand years old—and that since the time of creation, species had remained exactly the same. Even so, studying the anatomy of organisms and then classifying them interested many investigators who wished to show that species were created to be suitable to their environment. Explorers and collectors traveled the world and brought back currently existing species and fossils to be classified. **Fossils** are the remains of once-living species often found in strata. **Strata** are layers of rock formed from sedimentary material **(Fig. 14.2*a*)**. A newfound fossil shows that snakes had hip bones and hind limbs some 90 MYA (million years ago) and that they evolved on land **(Fig. 14.3*b*)**.

A noted zoologist of the early 19th century, Georges Cuvier, founded the science of *paleontology*, the study of fossils. Cuvier was faced with a problem. He believed in the fixity of species, yet Earth's strata clearly showed a succession of different life-forms over time. To explain these observations, he hypothesized that a local catastrophe had caused a mass extinction whenever a new stratum of that region showed a new mix of fossils. After each catastrophe, the region was repopulated by species from surrounding areas, which accounted for the appearance of new fossils in the new stratum. The result of all these catastrophes was change appearing over time. Some of Cuvier's followers, who came to be called *catastrophists*, even suggested that worldwide catastrophes had occurred, and that after each of these events, new sets of species had been created.

In contrast to Cuvier, Jean-Baptiste de Lamarck, another biologist, hypothesized that evolution occurs and that adaptation to the environment was the cause of diversity. Therefore, after studying the succession of life-forms in strata, Lamarck concluded that more complex organisms are descended from less complex organisms. To explain the process of adaptation to the environment, Lamarck proposed the idea of *inheritance of acquired characteristics*, in which the use and disuse of a structure can bring about inherited change. One example Lamarck gave—and the one for which he is most famous—is that the long neck of giraffes developed over time because giraffes stretched their necks to reach food high in trees and then passed on a long neck to their offspring (see Fig.14.8). However, this hypothesis of the inheritance of acquired characteristics has never been substantiated by experimentation. For example, if acquired characteristics were inherited, people who become blind by accident would have blind children, and circumcised males would have boys that lack a foreskin from birth. Modern genetics explains why the idea of acquired characteristics cannot be substantiated. Phenotypic changes acquired during an organism's lifetime do not result in genetic changes that can be passed to subsequent generations.

a. Visible strata

Figure 14.2   Fossils and strata.

a. Due to erosion, it is often possible to see a number of strata, layers of rock or sedimentary material that contain fossils. The oldest fossils are in the lowest stratum. **b.** A fossil snake dated 90 MYA.
Reprinted by permission from Macmillan Publishers Ltd: Nature (A Cretaceous Terrestrial Snake With Robust Hindlimbs and a Sacrum), copyright 2006

ribs

hip bone

b. A fossil snake with hip bones

Check Your Progress

1. Describe the pre-Darwinian view of the world.
2. How did catastrophists explain change within the fossil record?
3. Describe Cuvier's contribution to evolutionary theory.

Answers: 1. The Earth is only thousands of years old. Species do not change over time. A creator is responsible for the diversity of life. Observations should support the current view of the world. 2. Catastrophists proposed that previous species were eliminated by a catastrophe and new ones replaced them from surrounding areas or they were created anew.   3. Cuvier founded the science of paleontology. He said that fossils are found in strata, and a change in species can be observed over time.

Darwin's ideas were close to those of Lamarck. For example, Darwin said that living things share common characteristics because they have a common ancestry. One of the most unfortunate misinterpretations of this statement was that humans are descendants of apes. For Darwin, however, humans and apes share a common ancestor just as, say, you and your cousins can trace your ancestry back to the same grandparents. In contrast to Lamarck, Darwin's observations led him to conclude that species are suited to the environment through no will of their own but by natural selection. He saw the process of natural selection as the means by which different species come about (see Fig. 14.8).

Darwin's Conclusions

Darwin's conclusions that organisms are related through common descent and that adaptation to various environments results in diversity were based on several types of data, including his study of geology, fossils, and biogeography. **Biogeography** is the study of the distribution of life-forms on Earth.

Darwin's Study of Geology and Fossils

Darwin took Charles Lyell's book *Principles of Geology* on the *Beagle* voyage. In contrast to former beliefs, this book gave evidence that Earth was subject to slow but continuous cycles of erosion and uplift. Weathering causes erosion; thereafter, dirt and rock debris are washed into the rivers and transported to oceans. When these loose sediments are deposited, strata result **(Fig. 14.3a)**. Then the strata, which often contain fossils, are uplifted from below sea level to form land. Given enough time, slow natural processes can account for extreme geological changes. Lyell went on to propose the theory of **uniformitarianism,** which stated that these slow changes occurred at a uniform rate. Even though uniformitarianism has

Figure 14.3 **Formation of strata.**

a. This diagram shows how water brings sediments into the sea; the sediments then become compacted to form a stratum (sing). Fossils are often trapped in strata, and as a result of a later geological upheaval, the strata may be located on land. **b.** Fossil remains of freshwater snails, *Turritella*, in a stratum.

a.

b.

been rejected, modern geology certainly substantiates a hypothesis of slow and continual geological change. Darwin, too, was convinced that Earth's massive geological changes are the result of slow processes, and that Earth, therefore, must be very old.

On his trip, Darwin observed massive geological changes firsthand. When he explored what is now Argentina, he saw raised beaches for great distances along the coast. In Chile, he witnessed the effects of an earthquake that caused the land to rise several feet and left marine shells inland, well above sea level. When Darwin also found marine shells high in the cliffs of the impressive Andes Mountains, he became even more convinced that Earth was subject to slow geological changes. While Darwin was making geological observations, he also collected fossil specimens that differed somewhat from modern species (**Fig. 14.4**). Once Darwin accepted the supposition that Earth must be very old, he began to think that there would have been enough time for descent with modification—that is, evolution—to occur. Living forms could be descended from extinct forms known only from the fossil record. It would seem that species were not fixed; instead, they change over time.

Darwin's Study of Biogeography

Darwin could not help but compare the animals of South America to those with which he was familiar in England. For example, instead of rabbits, he found the Patagonian hare in the grasslands of South America. The Patagonian hare, a rodent also native to South America (**Fig. 14.5**), has long legs and ears but the face of a guinea pig. Did the Patagonian hare resemble a rabbit because the two types of animals were adapted to the same type of environment—an outcome we call convergent evolution today? Did the two distantly related species converge on the same overall body form because both live in similar habitats and have similar behaviors? Did the Patagonian hare have the face of a guinea pig because of common ancestry with guinea pigs?

a.

b.

Figure 14.4   A glyptodont compared to an armadillo.

a. The giant armadillo-like glyptodont is known only by the study of its fossil remains. Darwin found such fossils and came to the conclusion that this extinct animal must be related to living armadillos. The glyptodont weighed 2,000 kilograms. **b.** A modern armadillo weighs about 4.5 kilograms.

Check Your **Progress**

1. Describe Lamarck's idea of inheritance of acquired characteristics.
2. Why did Lyell think the Earth was very old?

Answers: 1. Use and disuse of structures can produce heritable changes.   2. Lyell found evidence that massive geological changes have occurred but at a slow rate.

a. Patagonian hare   b. European rabbit

Figure 14.5   The Patagonian hare and European rabbit.

a. The Patagonian hare, native to South America, has long legs and other adaptations similar to those of a rabbit but has the face of a guinea pig.
b. The characteristics of the Patagonian hare resemble those of the European rabbit, which does not occur naturally in South America.

As Darwin sailed southward along the eastern coast of South America, he saw how similar species replaced one another. For example, the greater rhea (an ostrichlike bird) found in the north was replaced by the lesser rhea in the south. Therefore, Darwin reasoned that related species could be modified according to environmental differences caused by change in latitude. When he reached the Galápagos Islands, he found further evidence of this. The Galápagos Islands are a small group of volcanic islands located 965 kilometers off the western coast of South America. These islands are too far from the mainland for most terrestrial animals and plants to colonize, yet life was present. The types of plants and animals found there were slightly different from species Darwin had observed on the mainland; even more important, they also varied from island to island according to their own unique environment. Where did the species inhabiting these islands come from, and what caused the islands to have different species?

Finches Although some of the finches on the Galápagos Islands seemed like mainland finches, others were quite different (**Fig. 14.6**). Today, there are ground-dwelling finches with beak sizes dependent on the sizes of the seeds they eat. Tree-dwelling finches have beak sizes and shapes dependent on the sizes of their insect prey. A cactus-eating finch possesses a more pointed beak, enabling access to nectar within cactus flowers. The most unusual of the finches is a woodpecker-type finch. A woodpecker normally has a sharp beak to chisel through tree bark and a long tongue to probe for insects. The Galápagos woodpecker-type finch has the sharp beak but lacks the long tongue. To make up for this, the bird carries a twig or cactus spine in its beak and uses it to poke into crevices. Once an insect emerges, the finch drops this tool and seizes the insect with its beak.

Later, Darwin speculated as to whether all the different species of finches he had seen could have descended from a mainland finch. In other words, he wondered if a mainland finch was the common ancestor to all the types on the Galápagos Islands. Had **speciation,** the formation of a new species, occurred because the islands allowed isolated populations of birds to evolve independently? Could the present-day species have resulted from accumulated changes occurring within each of these isolated populations?

Tortoises Each of the Galápagos Islands also seemed to have its own type of tortoise, and Darwin began to wonder if this could be correlated with a difference in vegetation among the islands. Long-necked tortoises seemed to inhabit only dry areas, where food was scarce, and most likely the longer neck was helpful in reaching tall-growing cacti. In moist regions with relatively abundant ground foliage, short-necked tortoises were found. Had an ancestral tortoise from the mainland of South America given rise to these different types, each adapted to a different environment? An **adaptation** is any characteristic that makes an organism more suited to its environment. It often takes many generations for an adaptation to develop.

a. Ground-dwelling finch

b. Woodpecker finch

c. Cactus finch

Figure 14.6 Galápagos finches.

Each of the present-day 13 species of finches has a beak adapted to a particular way of life. **a.** The heavy beak of the large ground-dwelling finch is suited to a diet of seeds. **b.** The beak of the woodpecker finch is suited for using tools to probe for insects found in holes or crevices. **c.** The long, somewhat decurved beak and split tongue of the cactus finch is suited to probing cactus flowers for nectar.

Natural Selection and Adaptation

Once Darwin realized that adaptations develop over time, he began to think about a mechanism by which adaptations might arise. Eventually, he proposed **natural selection** as the mechanism. Natural selection is a process that results in the evolution of organisms well adapted to their environment. Natural selection requires these steps:

1. The members of a population have heritable variations (**Fig. 14.7**).
2. The population produces more offspring than the resources of an environment can support.
3. The individuals who have favorable traits survive and reproduce to a greater extent than those who lack these traits.
4. Across generations, a larger proportion of the population possesses the favorable traits and the population becomes adapted to the environment.

Notice that because natural selection utilizes only variations that happen to be provided by genetic changes, it lacks any directedness or anticipation of future needs. Natural selection is an ongoing process because the environment of living things is constantly changing. Extinction (loss of a species) can occur when previous adaptations are no longer suitable to a changed environment.

Organisms Have Variations

Darwin emphasized that the members of a population vary in their functional, physical, and behavioral characteristics. Prior to Darwin, variations were considered imperfections that should be ignored since they were not important to the description of a species. Darwin, on the other hand, realized that variations were essential to the natural selection process. Darwin suspected—but did not have the evidence we have today—that the occurrence of variations is completely random; they arise by accident and for no particular purpose. Also, new variations are more likely to be harmful than beneficial to an organism.

The variations that make adaptation to the environment possible are those that are passed on from generation to generation. The science of genetics was not yet well established, so Darwin was never able to determine the cause of variations or how they are passed on. Today, we realize that genes, together with the environment, determine the phenotypes of an organism. Mutations, along with chromosomal rearrangements and assortment of chromosomes during meiosis and fertilization, can cause new variations to arise.

Organisms Struggle to Exist

In Darwin's time a socioeconomist named Thomas Malthus stressed the reproductive potential of human beings. He proposed that death and famine are inevitable because the human population tends to increase faster than the supply of food. Darwin applied this concept to all organisms and saw that the available resources were not sufficient to allow all members of a population to survive. He calculated the reproductive potential of elephants, and concluded that, after only 750 years, the descendants of a single pair of elephants would number about 19 million! Obviously, no environment has the resources to support an elephant population of this magnitude. Because each generation has the same reproductive potential as the previous generation, there is a constant struggle for existence, and only certain members of a population survive and reproduce each generation.

Figure 14.7 **Variations in shells of a marine snail,** *Liguus fascitus.*

For Darwin, variations such as these in a species of snails were highly significant and were required in order for natural selection to result in adaptation to the environment.

Check Your Progress

1. Explain why Darwin's observations about Galápagos finches were significant to him.
2. Outline the process of natural selection, as described by Darwin.

Answers: 1. The finches were all descended from a mainland finch, and yet they were adapted differently according to the environment of the various islands. 2. Members of a population have heritable variations. More offspring are produced than resources can support. The most adapted individuals survive and reproduce. Across generations, a larger proportion of a species becomes more adapted to the environment.

Lamarck's proposal	Darwin's proposal
Originally, giraffes had short necks.	Originally, giraffe neck length varied.
Giraffes stretched their necks in order to reach food.	Struggle to exist causes long-necked giraffes to have the most offspring.
With continual stretching, most giraffes now have long necks.	Due to natural selection, most giraffes now have long necks.

Figure 14.8   Mechanism of evolution.

This diagram contrasts Jean-Baptiste Lamarck's proposal of acquired characteristics with Charles Darwin's proposal of natural selection. Only natural selection is supported by data.

Organisms Differ in Fitness

Fitness is the reproductive success of an individual relative to other members of the population. Fitness is determined by comparing the number of surviving fertile offspring that are produced for each member of the population. The most-fit individuals are the ones that capture a disproportionate amount of resources, and convert these resources into a larger number of viable offspring. Because organisms vary anatomically and physiologically, and because the challenges of local environments vary as well, what determines fitness varies for different populations. For example, among western diamondback rattlesnakes living on lava flows, the most fit are those that are black. Among those living on desert soil, the most fit are typically light colored with brown blotching. Background matching helps an animal both capture prey and avoid being captured; therefore, it is expected to lead to survival and increased reproduction.

Darwin noted that when humans help carry out **artificial selection,** they breed selected animals with particular traits to reproduce. For example, prehistoric humans probably noted desirable variations among wolves and selected particular individuals for breeding. Therefore, the desired traits increased in frequency in the next generation. This same process was repeated many times, resulting in today's numerous varieties of dogs, all descended from the wolf. In a similar way, several varieties of vegetables can be traced to a single ancestor. Chinese cabbage, brussels sprouts, and kohlrabi are all derived from a single species, *Brassica oleracea*.

In nature, interactions with the environment determine which members of a population reproduce to a greater degree than other members. In contrast to artificial selection, the result of natural selection is not predesired. Natural selection occurs because certain members of a population happen to have a variation that allows them to survive and reproduce to a greater extent than other members. For example, any variation that increases the speed of a hoofed animal helps it escape predators and live longer; a variation that reduces water loss is beneficial to a desert plant; and one that increases the sense of smell helps a wild dog find its prey. Therefore, we expect the organisms with these characteristics to have increased fitness. **Figure 14.8** contrasts Lamarck's ideas with those of Darwin.

Organisms Become Adapted

An adaptation (see page 223) may take many generations to evolve. We can especially recognize an adaptation when unrelated organisms living in a particular environment display similar characteristics. For example, manatees, penguins, and sea turtles all have flippers, which help them move through the water—also an example of convergent evolution. Adaptations also account for why organisms are able to escape their predators **(Fig. 14.9)** or why they are suited to their way of life **(Fig. 14.10)**. Natural selection causes adaptive traits to be increasingly represented in each succeeding generation. There are other processes of evolution aside from natural selection (see Chapter 15), but natural selection is the only process that results in adaptation to the environment.

Check Your Progress

1. What is the main difference between artificial selection and natural selection?

Answer: 1. Artificial selection is carried out by humans, and natural selection is carried out by environmental forces, other than humans.

Darwin and Wallace

After the HMS *Beagle* returned to England in 1836, Darwin waited more than 20 years to publish his book *On the Origin of Species*. During the intervening years, he used the scientific process to support his hypothesis that today's diverse life-forms arose by descent from a common ancestor and that natural selection is a mechanism by which species can change and new species can arise. Darwin was prompted to publish his book after reading a similar hypothesis formulated by Alfred Russel Wallace.

Wallace (1823–1913) was an English naturalist who, like Darwin, was a collector at home and abroad. He went on collecting trips, each of which lasted several years, to the Amazon and the Malay Archipelago. After studying the animals of every island within the Malay Archipelago, he divided the islands into a western group, with organisms like those found in Asia, and an eastern group, with organisms like those of Australia. The sharp line dividing these two island groups within the archipelago is now known as Wallace's Line. A narrow but deep strait occurs at Wallace's Line. At times during the past 50 million years, this strait persisted even when sea levels were low and land bridges appeared between the other islands. Therefore, the strait would have always served as a barrier to the dispersal of organisms between the two groups of islands.

While traveling, Wallace also wrote an essay called "On the Law Which Has Regulated the Introduction of New Species." In this essay he said that "every species has come into existence coincident both in time and space with a preexisting closely allied species." A year later, after reading Malthus's treatise on human population increase, Wallace conceived the idea of "survival of the fittest." He quickly completed an essay proposing natural selection as an agent for evolutionary change and sent it to Darwin for comment. Darwin was stunned. Here was the hypothesis he had formulated, but had never dared to publish. He told his friend and colleague, Charles Lyell, that Wallace's ideas were so similar to his own that even Wallace's "terms now stand as heads of my chapters."

Darwin suggested that Wallace's paper be published immediately, even though Darwin himself as yet had nothing in print. However, Lyell and others who knew of Darwin's detailed work substantiating the process of natural selection suggested that a joint paper be read to the renowned Linnean Society. On July 1, 1858, Darwin presented an abstract of *On the Origin of Species*, his famous book, which was published in 1859. The title of Wallace's section was "On the Tendency of Varieties to Depart Indefinitely from the Original Type." This professional presentation served as the announcement to the world that species share a common descent and have diverged through natural selection.

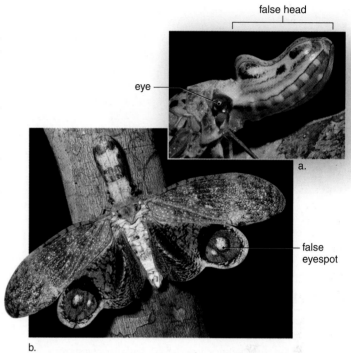

Figure 14.9 Adaptations of the alligator bug.

The alligator bug of the Brazilian rain forest has antipredator adaptations. **a.** The insect blends into its background, but if discovered, the false head, which resembles a miniature alligator, may frighten a predator. **b.** If the predator is not frightened, the insect suddenly reveals huge false eyespots on its hindwings.

Figure 14.10 Adaptations of the vampire bat.

Vampire bats of the rain forests of Central and South America have the adaptations of a nocturnal, winged predator. The bat uses its enlarged ears and echolocation to locate prey in the dark. It bites its prey with two pointed incisors. Saliva containing an anticoagulant called draculin runs into the bite; the bat then licks the flowing blood. The vampire bat's forelimbs are modified to form wings, and it roosts by using its grasping hindlimbs.

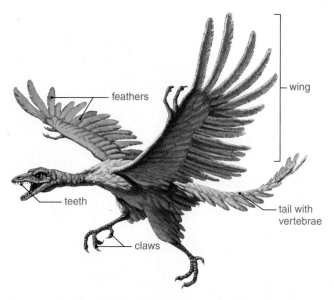

Figure 14.11 **Recreation of *Archaeopteryx*.**

The fossil record suggests that *Archaeopteryx* had a feather-covered, reptilian-type tail, which shows up well in this artist's representation. (Red labels = reptilian characteristics; green labels = bird characteristics.)

Modern toothed whales

The reduced hind limbs of *Rodhocetus kasrani* could not have aided it in walking or swimming. *Rodhocetus* swam with an up-and-down motion, as do modern whales.

Ambulocetus natans probably walked on land (as do modern sea lions) and swam by flexing its backbone and paddling with its hind limbs (as do modern otters).

Hypothetical mesonychid skeleton

Figure 14.12 **Evolution of whales.**

The discoveries of *Ambulocetus* and *Rodhocetus* have filled in the gaps in the evolution of whales from extinct hoofed mammals that lived on land to the ocean-dwelling mammals we know today. MYA = million years ago

14.2 | Evidence for Evolution

The theory of evolution states that all living things have a common ancestor, but each is adapted to a particular way of life. Many lines of evidence consistently support this hypothesis that organisms are related through common descent. A hypothesis becomes a scientific theory only when a variety of evidence from independent investigators supports the hypothesis. The theory of evolution is a unifying principle in biology because it can explain so many different observations in various fields of biology. The theory of evolution has the same status in biology that the germ theory of disease has in medicine.

Darwin cited much of the evidence for evolution we will discuss, except that he had no knowledge of the biochemical data that became available after his time.

Fossil Evidence

The fossils trapped in rock strata are the **fossil record** that tell us about the history of life. One of the most striking patterns in the fossil record is a succession of life-forms from the simple to the more complex. Occasionally this pattern is reversed, showing that evolution is not unidirectional. Particularly interesting are the fossils that serve as transitional links between groups. Even in Darwin's day, scientists knew of the *Archaeopteryx* fossils, which show that birds have reptilian features, including jaws with teeth, and long jointed tails. *Archaeopteryx* also had feathers and wings (**Fig. 14.11**).

Other transitional links between fossil vertebrates have more recently been found, including the lobe-finned fish *Tiktaalik roseae*, the reptilelike amphibian *Seymouria*, and the mammal-like synapsids. These fossils allow us to deduce that fishes evolved before amphibians, which evolved before reptiles, including birds. The synapsids show that reptiles and mammals share an amniote ancestry.

Recently, four-legged aquatic mammals, the mesonychids, were discovered that provide important insights concerning the evolution of whales from land-living, hoofed ancestors (**Fig. 14.12**). The fossilized whale *Ambulocetus* may have been amphibious, walking on land and still swimming in the sea. *Rodhocetus* swam with an up-and-down tail motion like modern whales do; its reduced hindlimbs could not have helped in swimming.

In 2006, a snake fossil dated 90 MYA was discovered with hip bones and hind limbs—a trait absent in all living snakes (see Fig. 14.2*b*). Some snakes, such as pythons, have vestigial hind limbs, but these snakes lack the hip bones present in this fossil. Since all lizards have hip bones and most have limbs, this fossil is now serving as a transitional fossil between lizards and snakes.

Biogeographical Evidence

Biogeography is the study of the distribution of organisms throughout the world. Such distributions are consistent with the hypothesis that, when forms are related, they evolved in one locale and then spread to accessible regions. Therefore, you would expect a different mix of plants and animals whenever geography separates continents, islands, or seas. As previously mentioned, Darwin noted that South America lacked rabbits, even though the environment was quite suitable for them. He concluded that rabbits evolved elsewhere and had no means of reaching South America.

To take another example, both cacti and spurges (*Euphorbia*) are plants adapted to a hot, dry environment—both are succulent, spiny, flowering plants. Why do cacti grow in American deserts and most *Euphorbia* grow in African deserts, when each would do well on the other continent? It seems obvious that they evolved similar adaptations because their respective continents had similar environments.

The islands of the world are home to many unique species of animals and plants found no place else, even when the soil and climate are the same. Why do so many species of finches live on the Galápagos Islands and so many species of honeycreepers, a type of finch, live in the Hawaiian Islands when these species are not on the mainland? The reasonable explanation is that an ancestral finch migrated to all the different islands. Then geographic isolation allowed the ancestral finch to evolve into a different species on each island.

Also, long ago South America, Antarctica, and Australia were connected (see Fig. 16.13*a*). Marsupials (pouched mammals) and placental mammals arose at this time, but today marsupials are plentiful only in Australia and placental mammals are plentiful in South America. Why are marsupials only plentiful in Australia (**Fig. 14.13**)? After marsupials arose, Australia separated and drifted away, and marsupials were free to evolve into many different forms because they had no competition from placental mammals. In the Americas, the placental mammals were able to successfully compete against the marsupials, and the opossum is the only marsupial in the Americas. In some cases, marsupial and placental mammals physically resemble one another— for example, the marsupial wombat and the marmot, and the marsupial Tasmanian wolf and the wolf. This supports the hypothesis that evolution is influenced by the environment and by the mix of plants and animals in a particular continent—that is, by biogeography, not by design.

Sugar glider, a tree dweller

Kangaroo, a herbivore of plains and forests

Koala, a tree dweller

Coarse-haired wombat, nocturnal and living in burrows

Figure 14.13 Marsupials of Australia.

Each type of marsupial in Australia is adapted to a different way of life. All of them presumably evolved from a common ancestor that entered Australia some 60 million years ago.

Check Your Progress

1. Explain why evolution is no longer considered a hypothesis.
2. Explain how biogeographical information about Galápagos finches supports the theory of evolution.

Answers: 1. Evolution is supported by many diverse and independent lines of evidence, such as the fossil record, anatomical similarities, and developmental stages. 2. The finches show that a common ancestor can give rise to other species through adaptation to unique environments—in this case, the environment on different islands.

Figure 14.14   Significance of structural similarities.

Although the specific details of vertebrate forelimbs are different, the same basic bone stucture and position are present (color-coded here). This unity of anatomy is evidence of a common ancestor.

Anatomical Evidence

Darwin was able to show that a common descent hypothesis offers a plausible explanation for vestigial structures and anatomical similarities among organisms.

Vestigial structures are anatomical features that are fully developed in one group of organisms but reduced and nonfunctional in other similar groups. Most birds, for example, have well-developed wings used for flight. However, some bird species (e.g., ostrich) have greatly reduced wings and do not fly. Similarly, whales (see Fig. 14.12) and snakes have no use for hindlimbs, and yet extinct whales and snakes have remnants of hip bones and legs. Humans have a tailbone but no tail. The presence of vestigial structures can be explained by the common descent hypothesis. Vestigial structures occur because organisms inherit their anatomy from their ancestors; they are traces of an organism's evolutionary history.

Vertebrate forelimbs are used for flight (birds and bats), orientation during swimming (whales and seals), running (horses), climbing (arboreal lizards), or swinging from tree branches (monkeys). Yet all vertebrate forelimbs contain the same sets of bones organized in similar ways, despite their dissimilar functions (**Fig. 14.14**). The most plausible explanation for this unity is that the basic forelimb plan belonged to a common ancestor, and then the plan became modified in the succeeding groups as each continued along its own evolutionary pathway. Anatomically similar structures explainable by inheritance from a common ancestor are called **homologous structures.** In contrast, **analogous structures** serve the same function, but are not constructed similarly, and therefore could not have a common ancestry. The wings of birds and insects are analogous structures.

The homology shared by vertebrates extends to their embryological development (**Fig. 14.15**). At some time during development, all vertebrates have a postanal tail and exhibit paired pharyngeal pouches supported by cartilaginous bars. In fishes and amphibian larvae, these pouches develop into functioning gills. In humans, the first pair of pouches becomes the cavity of the middle ear, and the auditory tube. The second pair becomes the tonsils; the third and fourth pairs become the thymus and parathyroid glands. Why should pharyngeal pouches, which have lost their original function, develop and then become modified in terrestrial vertebrates? The most likely explanation is that new structures (or structures with novel functions) originate by "modifying" the preexisting structures of one's ancestors.

Figure 14.15   Significance of developmental similarities.

At this comparable developmental stage, a chick embryo and a pig embryo have many features in common, which suggests they evolved from a common ancestor.

Types of Organisms

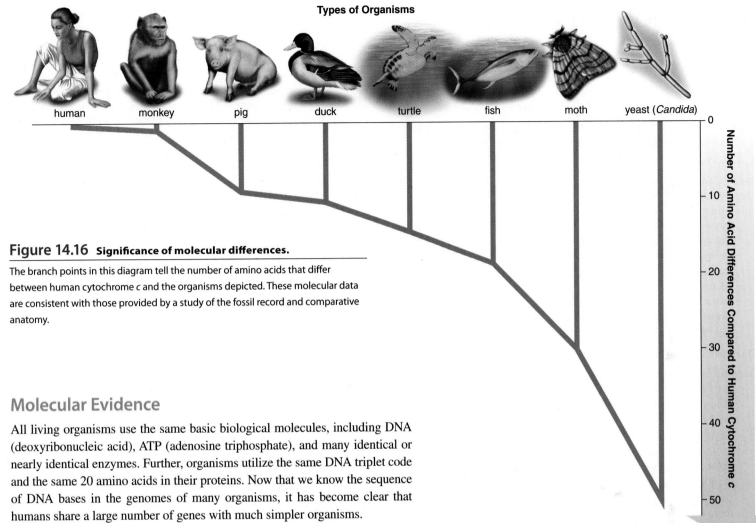

human   monkey   pig   duck   turtle   fish   moth   yeast (*Candida*)

Figure 14.16 Significance of molecular differences.

The branch points in this diagram tell the number of amino acids that differ between human cytochrome *c* and the organisms depicted. These molecular data are consistent with those provided by a study of the fossil record and comparative anatomy.

Number of Amino Acid Differences Compared to Human Cytochrome c

0 — 10 — 20 — 30 — 40 — 50

Molecular Evidence

All living organisms use the same basic biological molecules, including DNA (deoxyribonucleic acid), ATP (adenosine triphosphate), and many identical or nearly identical enzymes. Further, organisms utilize the same DNA triplet code and the same 20 amino acids in their proteins. Now that we know the sequence of DNA bases in the genomes of many organisms, it has become clear that humans share a large number of genes with much simpler organisms.

Also of interest, evolutionary developmental biologists have found that many developmental genes are shared in animals ranging from worms to humans. It appears that life's vast diversity has come about by only slight differences in the same genes. The result has been widely divergent body plans. For example, a similar gene in arthropods and vertebrates determines the back to front axis. But, although the base sequences are similar, the genes have opposite effects. In arthropods, such as fruit flies and crayfish, the nerve cord is toward the front, whereas in vertebrates, such as chickens and humans, the nerve cord is toward the back.

When the degree of similarity in DNA base sequences of genes, or the degree of similarity in amino acid sequences of proteins is examined, the data are as expected, assuming common descent. Cytochrome *c* is a molecule used in the electron transport chain of all the organisms shown in **Figure 14.16.** Data regarding differences in the amino acid sequence of cytochrome *c* show that the sequence in a human differs from that in a monkey by only one amino acid, from that in a duck by 11 amino acids, and from that in *Candida*, a yeast, by 51 amino acids. These data are consistent with other data regarding the anatomical similarities of these organisms, and therefore their relatedness.

Check Your Progress

1. Explain how vestigial structures support the theory of evolution.
2. Contrast homologous structures with analogous structures.
3. With reference to Figure 14.16, why would it make sense that the most distantly related species have greater structural differences and greater DNA differences?

Answers: 1. These structures show that present-day organisms have an ancestor from the past. Vestigial structures are useless today but they were functional in the ancestor. 2. Homologous structures share an anatomical similarity that reflects a common ancestry. Analogous structures have a common function but do not serve as evidence of a common ancestry. 3. DNA determines the structure and function of organisms. Organisms whose ancestry separated sometime in the past will have experienced different DNA changes.

The Chapter in Review

Summary

14.1 | Darwin's Theory of Evolution

Charles Darwin took a position as a naturalist aboard the HMS *Beagle* and made a trip around the world, largely in the Southern Hemisphere.

Before Darwin

Before Darwin, people believed that Earth was young, species did not change, variations were imperfections, and observations could substantiate these views.

- Cuvier was an early paleontologist who believed that species do not change. He observed species come and go in the fossil record, and he said these changes were due to catastrophies.
- Lamarck was a zoologist who hypothesized that evolution and adaptation to the environment do occur. He suggested that acquired characteristics could be inherited. For example, he said giraffes stretched their necks to reach food in trees, and then this change was inherited by the next generation.

Darwin's Conclusions

Darwin's conclusions based on geology and fossils are:
- The Earth is very old, giving time for evolution to occur.
- Living things are descended from extinct forms known only from the fossil record.

Darwin's conclusions based on biogeography are:
- Living things evolve where they are. This explains, for example, why South America has the Patagonian hare, whereas England has the rabbit.
- Living things are adapted to local environments. This explains why there are many types of finches and tortoises in the Galápagos Islands.

Natural Selection and Adaptation

According to Darwin, the result of natural selection is a population better adapted to its local environment than previous generations:

	Observation	Result	Conclusion
1	a. Organisms have variations.	b. New adaptations to the environment arise.	Organisms become more adapted with each generation.
2	a. Organisms struggle to exist.	b. More organisms are present than can survive.	
3	a. Organisms differ in fitness.	b. Organisms best suited to the environment survive and reproduce.	

Darwin and Wallace

Alfred Russel Wallace was a naturalist who, like Darwin, traveled to other continents in the Southern Hemisphere. He also collected evidence of common descent, and his reading of Malthus caused him to develop the same mechanism for adaptation (natural selection) as Darwin. Darwin's work was more thorough, as evidenced by his book *On the Origin of Species*.

14.2 | Evidence for Evolution

A theory in science is a concept supported by much evidence, and the theory of evolution is supported by several types of evidence:

- The fossil record indicates the history of life in general and allows us to trace the descent of a particular group.
- Biogeography shows that the distribution of organisms on Earth is explainable by assuming organisms evolved in one locale.
- Anatomy and development of organisms reveals homologies explainable only by common ancestry.
- Biochemical molecules of the same type occur in all organisms. Differences indicate the degree of relatedness.

Key Terms

adaptation 222	fossils 219
analogous structure 228	homologous structure 228
artificial selection 224	natural selection 223
biogeography 220	speciation 222
evolution 217	strata 219
fitness 224	uniformitarianism 220
fossil record 226	vestigial structure 228

Testing Yourself

Choose the best answer for each question.

1. _____ developed the idea that acquired characteristics can be inherited.
   - a. Darwin
   - b. Lamarck
   - c. Wallace
   - d. Sedgwick
   - e. Cuvier

2. The major feature Darwin noticed that differed among species of Galápagos finches was
   - a. coloration.
   - b. beak size.
   - c. beak color.
   - d. length of tail feathers.

3. Which of the following is not an example of natural selection?
   - a. Insect populations exposed to pesticides become resistant to the chemicals.
   - b. Plant species that produce fragrances to attract pollinators produce more offspring.
   - c. Rabbits that sprint quickly are more likely to escape predation.
   - d. On a tree, leaves that grow in the shade are larger than those that grow in the sun.

4. The variations necessary for natural selection
   - a. occur randomly.
   - b. are influenced by environment.
   - c. can be caused by mutation.
   - d. can be caused by recombination during meiosis.
   - e. All of these are correct.

5. Which is most likely to be favored during natural selection, but not artificial selection?
   - a. fast seed germination rate
   - b. short generation time
   - c. efficient seed dispersal
   - d. lean pork meat production

6. Natural selection is the only process that results in
   a. genetic variation.
   b. adaptation to the environment.
   c. phenotypic change.
   d. competition among individuals in a population.
7. Why was it helpful to Darwin to learn that Lyell concluded the Earth was very old?
   a. An old Earth has more fossils than a new Earth.
   b. It meant there was enough time for evolution to have occurred slowly.
   c. There was enough time for the same species to spread out into all continents.
   d. Darwin said artificial selection occurs slowly.
   e. All of these are correct.
8. In the following diagram, contrast Lamarck's proposal with Darwin's proposal by matching the phrases in the key to the letters in the diagram.

Key:

Originally, giraffe neck length varied.

Giraffes stretched their necks in order to reach food.

Struggle to exist causes long-necked giraffes to have the most offspring.

Originally, giraffes had short necks.

Today most giraffes have long necks. (used twice)

9. All the finches on the Galápagos Islands
   a. are unrelated but descended from a common ancestor.
   b. are descended from a common ancestor, and therefore related.
   c. rarely compete for the same food source.
   d. Both a and c are correct.
   e. Both b and c are correct.
10. Evolution is considered a
    a. hypothesis because it is supported by data from the fossil record.
    b. hypothesis because it is supported by multiple types of data.
    c. theory because it is supported by data from the fossil record.
    d. theory because it is supported by multiple types of data.
11. Catastrophists were not able to explain
    a. multiple extinctions.
    b. the replacement of one group of organisms by another.
    c. successive changes that link groups of organisms in the fossil record.
    d. More than one of these are correct.

For questions 12–15, match the description with the type of evidence for evolution it supports, as listed in the key. Answers can be used more than once.
   a. biogeographical
   b. anatomical
   c. biochemical
12. The genetic code is the same for all organisms.
13. The human knee bone and spine were derived from ancestral structures that supported four-legged animals.
14. The South American continent lacks rabbits even though the environment is quite suitable.
15. The amino acid sequence of human hemoglobin is more similar to that of rhesus monkeys than to that of mice.
16. Fossils that serve as transitional links allow scientists to
    a. determine how prehistoric animals interacted with each other.
    b. deduce the order in which various groups of animals arose.
    c. relate climate change to evolutionary trends.
    d. determine why evolutionary changes occur.
17. Among vertebrates, the flipper of a dolphin and the fin of a tuna are
    a. homologous structures.
    b. homogeneous structures.
    c. analogous structures.
    d. reciprocal structures.
18. Which of these pairs is mismatched?
    a. Charles Darwin—natural selection
    b. Cuvier—series of catastrophes explains the fossil record
    c. Lamarck—uniformitarianism
    d. All of these are correct.
19. According to the inheritance of acquired characteristics hypothesis,
    a. if a man loses his hand, then his children will also be missing a hand.
    b. changes in phenotype are passed on by way of the genotype to the next generation.
    c. organisms are able to bring about a change in their phenotype.
    d. evolution is striving toward particular traits.
    e. All of these are correct.

20. Organisms
    a. compete with other members of their species.
    b. differ in fitness.
    c. are adapted to their environment.
    d. are related by descent from common ancestors.
    e. All of these are correct.
21. DNA nucleotide differences between organisms
    a. indicate how closely related organisms are.
    b. indicate that evolution occurs.
    c. explain why there are phenotypic differences.
    d. are to be expected.
    e. All of these are correct.
22. The fossil record offers direct evidence for common descent because you can
    a. see that the types of fossils change over time.
    b. sometimes find common ancestors.
    c. trace the ancestry of a particular group.
    d. trace the biological history of living things.
    e. All of these are correct.
23. Molecular evidence is increasingly used today to establish relationships because
    a. genes are composed of DNA.
    b. the more similar the genes, the more likely two organisms are related.
    c. this evidence is objective and not subject to opinion.
    d. All of these are correct.
24. For there to be homologous structures,
    a. a common ancestor had to have existed.
    b. analogous structures also have to exist.
    c. the bones have to be used similarly.
    d. All of these are correct.

Thinking Scientifically

1. The human appendix, a vestigial extension off the large intestine, is homologous to a structure called a caecum in other mammals. A caecum, generally larger than our appendix, houses bacteria that aid in digesting cellulose, the main component of plants. How might the presence of the appendix be used to show our common ancestry with other mammals, and what might it tell us about the dietary history of humans?

2. Geneticists compare DNA base sequences among organisms and from this data determine a gene's rate of evolution. Different genes have been found to evolve at different rates. Explain why some genes might have faster rates of evolution than other genes as populations adapt to their environments.

3. Both Darwin and Wallace concluded that natural selection is the mechanism for biological evolution while observing life on islands. The Hawaiian and nearby islands once had at least 50 species of honeycreepers and these occurred nowhere else on Earth. Natural selection occurs everywhere and in all species. What characteristics of islands allow the outcome of natural selection to be so obvious?

Essentials of Biology Website

The companion website for *Essentials of Biology* provides a wealth of information organized and integrated by chapter. You will find practice tests, animations, videos, and much more that will complement your learning and understanding of general biology.

http://www.mhhe.com/maderessentials2

15

Evolution on a Small Scale

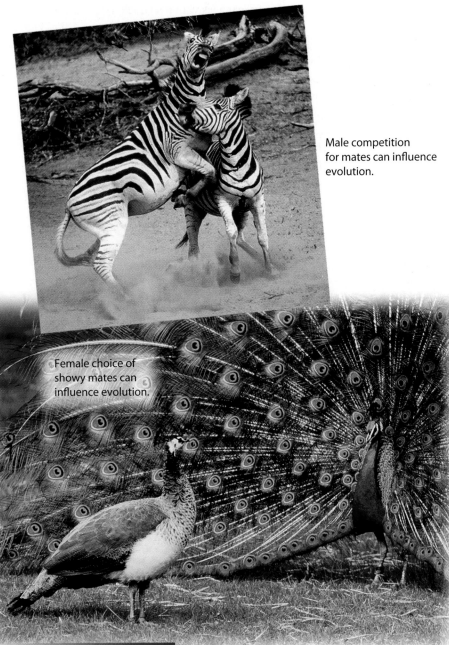

Male competition for mates can influence evolution.

Female choice of showy mates can influence evolution.

Outline

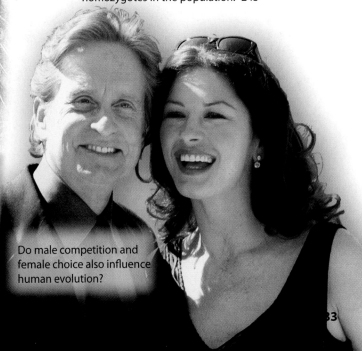

Sexual Selection

YOU MIGHT THINK that fitness means keeping in shape, but to an evolutionary biologist it means having more fertile offspring than other individuals. Think about it—only if an animal reproduces can that animal's genes be passed on and become prevalent in the next generation. Adaptation to the environment increases the chance of reproducing,

—Continued next page

Do male competition and female choice also influence human evolution?

Continued—

but so does *sexual selection,* which occurs because of an advantage that helps an animal acquire a mate.

Such advantages as increased size, brilliant feathers, evolution of horns, or enlarged canine teeth help males fight for and attract females. Females, in turn, must choose carefully. Perhaps to a female, the showier male is healthier or more appealing. If so, the same characteristics will be advantageous for her sons! Sexual selection increases the chances of reproducing, but can shorten the life span of males possessing such advantageous traits. A large, showy male that is consistently fighting most likely doesn't live as long as a small, inconspicuous male.

Do male competition and female choice occur among humans? Some think so. They point out that human males tend to be larger and more aggressive than females, and that statistically males have a shorter life span. Also, wealthy and successful males are more apt to be attractive to women, and sometimes older men marry younger women who are still fertile, thereby increasing their own fitness.

Sexual selection is one of the factors that influences evolution on a small scale (microevolution), which is the subject of this chapter.

15.1 | Microevolution

The idea that individuals evolve is a common misconception. Instead, as evolution occurs, genetic and therefore phenotypic changes occur within a population. A **population** is all the members of a single species occupying a particular area at the same time and reproducing with one another. **Microevolution** pertains to small measurable evolutionary changes within a population from generation to generation.

Darwin stressed that the members of a population vary, as in **Figure 15.1,** but he did not know how variations were generated or how they were passed down from parents to offspring. Today, DNA sequencing in a number of plants and animals shows that each gene in sexually reproducing organisms has many alleles. The reshuffling of alleles during sexual reproduction can result in a range of phenotypes among the members of a population (see Chapter 10). Even though sexual reproduction can cause genotypes and phenotypes to vary, it cannot in and of itself bring about microevolution. Microevolution is influenced by several other circumstantial factors.

Figure 15.1 Population of perennial flowers.

Among the members of a flowering plant population, petal color can be a continuous variation, so that there is a range of phenotypes. Few individuals have the extreme phenotypes, white petals and yellow petals, and most individuals have the intermediate phenotype, pink petals. A graph of the phenotypes results in the bell-shaped curve shown above a photo of the population.

Evolution in a Genetic Context

It was not until the 1930s that biologists were able to apply the principles of genetics to populations and thereafter to develop a way to recognize when evolution has occurred and measure how much population has changed.

In **population genetics,** the various alleles at all the gene loci in all individuals make up the **gene pool** of the population. It is customary to describe the gene pool of a population in terms of genotype and allele frequencies. Let's take an example based on the peppered moths, which can be light colored or dark colored (see Fig. 15.3). Suppose you research the literature and find that the color of peppered moths is controlled by a single set of alleles and you decide to use the following key:

$$D = \text{dark color}$$

$$d = \text{light color}$$

Furthermore, you find that in one Great Britain population before pollution fully darkened the trees, only 4% (0.04) of the moths were homozygous dominant; 32% (0.32) were heterozygous, and 64% (0.64) were homozygous recessive. From these genotype frequencies you can calculate the allele frequencies in the population:

The frequency of the gametes (sperm and egg) produced by this population will necessarily be the same as the allele frequencies. Assuming random mating (all possible gametes have an equal chance to combine with any other), we can use these gamete frequencies to calculate the ratio of genotypes in the next generation by using a Punnett square (**Fig. 15.2**).

There is an important difference between a Punnett square that represents a cross between individuals and the one shown in Figure 15.2. In Figure 15.2, we are using the gamete frequencies in the population to determine the genotype frequencies in the next generation. As you can see, the results show that the genotype frequencies (and therefore the allele frequencies) in the next generation are the same as they were in the previous generation. In other words, the homozygous dominant moths are still 0.04; the heterozygous moths are still 0.32, and the homozygous recessive moths are still 0.64 of the population. This remarkable finding tells us that *sexual reproduction alone cannot bring about a change in genotype and allele frequencies.* Also, the dominant allele need not increase from one generation to the next. Dominance does not cause an allele to become a common allele. The potential constancy, or equilibrium state, of gene pool frequencies was independently recognized in 1908 by G. H. Hardy, an English mathematician, and W. Weinberg, a German physician. They used the binomial equation to calculate the genotype and allele frequencies of a population as is illustrated in Figure 15.2.

$$p = \text{frequency of the dominant allele}$$

$$q = \text{frequency of the recessive allele}$$

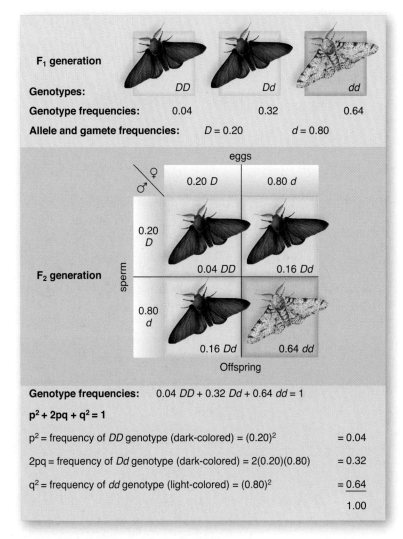

Figure 15.2   Hardy-Weinberg equilibrium.

Using the gamete frequencies in a population, it is possible to employ a Punnett square to calculate the genotype frequencies of the next generation. When this is done, it can be shown that sexual reproduction alone does not alter a Hardy-Weinberg equilibrium; the genotype and therefore allele frequencies remain the same. Hardy-Weinberg used the binomial expression to calculate the genotype frequencies of a population.

a.

b.

Figure 15.3 Industrial melanism and microevolution.

Microevolution has occurred when there is a change in gene pool frequencies—in this case, due to natural selection. **a.** When birds cannot see light-colored moths on light tree trunks, the light-colored phenotype is more frequent in the population. **b.** When birds cannot see dark-colored moths on dark tree trunks, the dark-colored phenotype is more frequent in the population. Therefore, when trees become sooty due to pollution, the percentage of the dark-colored phenotype increases.

The *Hardy-Weinberg principle* states that an equilibrium of genotype frequencies in a gene pool, calculated by using the expression $p^2 + 2pq + q^2 = 1$, will remain in effect in each succeeding generation of a sexually reproducing population as long as five conditions are met:

1. No mutations: Allelic changes do not occur, or changes in one direction are balanced by changes in the opposite direction.
2. No gene flow: Migration of alleles into or out of the population does not occur.
3. Random mating: Individuals pair by chance and not according to their genotypes or phenotypes.
4. No genetic drift: The population is very large and changes in allele frequencies due to chance alone are insignificant.
5. No selection: No selective agent favors one genotype over another.

These conditions are rarely, if ever, met, and genotype and allele frequencies in the gene pool of a population do change from one generation to the next. Therefore, microevolution does occur and the extent of change can be measured. The significance of the Hardy-Weinberg principle is that it tells us what factors cause evolution—those that violate the conditions listed. Microevolution can be detected and measured by noting the amount of deviation from a Hardy-Weinberg equilibrium of genotype frequencies in the gene pool of a population

For genotype frequencies to be subject to natural selection, they must result in a change of phenotype frequencies. **Industrial melanism,** an increase in the frequency of a dark phenotype due to pollution, provides us with an example. We supposed that only 36% of our moth population was dark colored (homozygous dominant plus heterozygous). Why might that be? Previous to the rise of industry, dark-colored moths rested on light tree trunks where they were seen and eaten by birds. However, with the advent of industry, the trunks of trees darkened, and the light-colored moths became visible and were eaten **(Fig. 15.3)**. Predatory birds acted as a selective agent, and microevolution occurred—in the mid-1950s the number of dark-colored moths in some Great Britain populations exceeded 80%. Aside from showing that natural selection can occur within a short period of time, our example illustrates that a change in gene pool frequencies does occur as microevolution occurs.

Check Your Progress

1. How does a population geneticist know that microevolution has occurred?
2. There being only two alleles, then $p + q = 1$. If $p = 0.62$ in a population, what is q?

Answers: 1. Evolution has occurred if allele frequencies within a population have changed from one generation to the next. 2. $q = 1 - 0.62 = 0.38$

Causes of Microevolution

Any conditions that deviate from the list of conditions for allelic equilibrium cause evolutionary change. Thus, these five conditions can cause a divergence from the Hardy-Weinberg equilibrium: mutation, gene flow, nonrandom mating, genetic drift, and natural selection.

Genetic Mutations

Mutations are the ultimate source for allele differences (**Fig. 15.4**). Without mutations, there could be no new variations among members of a population on which natural selection can act.

Prokaryotes do not reproduce sexually and therefore are more dependent on mutations to introduce variations. All mutations that occur and result in phenotypic differences can be tested by the environment. However, in sexually reproducing organisms, mutations, if recessive, do not immediately affect the phenotype.

In a changing environment, even a seemingly harmful mutation that results in a phenotypic difference can be the source of an adaptive variation. For example, the water flea *Daphnia* ordinarily thrives at temperatures around 20°C, but a mutation exists that requires *Daphnia* to live at temperatures between 25°C and 30°C. The adaptive value of this mutation is entirely dependent on environmental conditions.

Genetic diversity is promoted when a trait is polygenic, with many alleles for each gene locus. Certain combinations of these alleles might be more adaptive than others in a particular environment. The most favorable genotype may not occur until just the right combination of alleles are grouped in a single individual. For example, if alleles *K* and *T* are present in one parent, and alleles *Q* and *R* are present in the other parent, an offspring could inherit all of these alleles, and the result might be a protein with a different structure that enables the offspring to prefer different environmental conditions than its parents did.

Increasingly, investigators are sequencing DNA to discover the amount of hidden genetic diversity in a population. This has allowed them to find various genomic loci that exhibit *single nucleotide polymorphisms,* or *SNPs* (pronounced snips). In a SNP, there is only a single base difference between two loci. Due to the redundancy of the genetic code (see page 169), a SNP need not result in different amino acid sequences but may still cause regulatory differences. Therefore, researchers are in the process of trying to discover if ethnic groups share the same patterns of SNPs and if this could affect what diseases they are subject to.

Figure 15.4 Freckles.

A dominant allele causes freckles—so why doesn't everyone have freckles? The Hardy-Weinberg principle, which states that sexual reproduction in and of itself doesn't change allele frequencies, explains why dominant alleles don't become more prevalent with each generation.

Check Your **Progress**

1. List the five conditions that can cause microevolution to occur.
2. What is a Hardy-Weinberg equilibrium? What happens to this equilibrium when microevolution occurs?
3. Describe the significance of mutations in terms of evolution.

Answers: 1. Mutations, gene flow, nonrandom mating, genetic drift, and natural selection.   2. A Hardy-Weinberg equilibrium is the constancy of gene pool frequencies from generation to generation. Microevolution alters the equilibrium, and gene pool frequencies do change.   3. Mutations are ultimately responsible for all the genetic variability upon which evolution depends.

Gene Flow

Gene flow, also called *gene migration*, is the movement of alleles among populations by migration of breeding individuals. Gene flow can increase the variation within a population by introducing novel alleles that were produced by mutation in another population. Continued gene flow due to migration of individuals makes gene pools similar and reduces the possibility of allele frequency differences among populations now and in the future. Indeed, gene flow among populations can prevent speciation from occurring. Due to gene flow, the snake populations featured in **Figure 15.5** are **subspecies**—different populations within the same species. Despite somewhat distinctive characteristics, these subspecies of the species *Pantherophis obsoleta* can readily interbreed when they come into contact with one another.

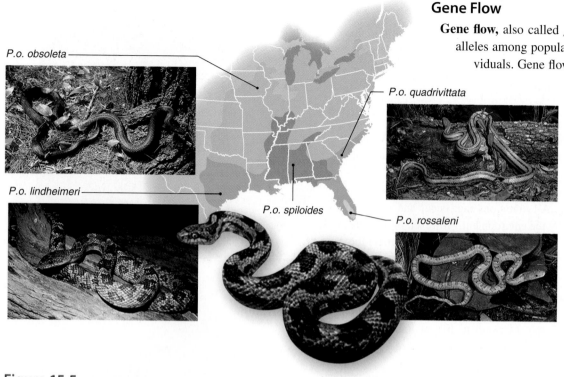

Figure 15.5 Gene flow.

Each rat snake represents a separate population of snakes. Because the populations are adjacent to one another, interbreeding occurs, and so does gene flow between the populations. This keeps their gene pools somewhat similar, and each of these populations is considered a subspecies of the species *Pantherophis obsoleta*. Therefore, each has a three-part name.

Nonrandom Mating

Random mating occurs when individuals select mates and pair by chance and not according to their genotypes or phenotypes. Inbreeding, or mating between relatives, is also an example of **nonrandom mating.** Inbreeding does not change allele frequencies, but it does gradually increase the proportion of homozygotes, because the homozygotes that result must necessarily produce only homozygotes.

Assortative mating occurs when individuals tend to mate with those that have the same phenotype with respect to a certain characteristic. For example, in humans, tall people tend to mate with each other. Assortative mating causes the population to subdivide into two phenotypic classes, between which gene exchange is reduced. Homozygotes for the gene loci that control the trait in question increase in frequency, and heterozygotes for these loci decrease in frequency.

Sexual selection favors characteristics that increase the likelihood of obtaining mates, and in this way it promotes nonrandom mating. In most species, males that compete best for access to females and/or have a phenotype that attracts females are more apt to mate and have increased fitness (see the introduction to this chapter.)

Genetic Drift

Genetic drift refers to changes in the allele frequencies of a gene pool due to chance. This pattern of evolution is called genetic drift because allele frequencies "drift" over time. They can increase or decrease due to which members of a population die, survive, or reproduce with one another. Although genetic drift occurs in both large and small populations, a larger population is expected to suffer less of a sampling error than a smaller population. Suppose you had a large bag containing 1,000 green balls and 1,000 blue balls, and you randomly drew 10%, or 200, of the balls. Because there is a large number of balls of each color in the bag, you can reasonably expect to draw 100 green balls and 100 blue balls, or at least a ratio close to this. It is extremely unlikely that you would draw 200 green or 200 blue balls. But suppose you had a bag containing only 10 green balls and 10 blue balls, and you drew 10%, or only 2 balls. You could easily draw two green balls or two blue balls, or one of each color.

When a population is small, there is a greater chance that some rare genotype might not participate at all in the production of the next generation. Suppose in a small population of frogs, certain frogs by chance do not pass on their traits. Certainly, the next generation will have a change in allele frequencies (**Fig. 15.6**). When genetic drift leads to a loss of one or more alleles, other alleles over time become *fixed* in the population.

In an experiment involving brown eye color, each of 107 *Drosophila* populations was kept in its own culture bottle. Every bottle contained eight heterozygous flies of each sex. There were no homozygous recessive or homozygous dominant flies. From the many offspring, the experimenter chose at random eight males and eight females. This action, which represented genetic drift, continued for 19 generations. By the nineteenth generation, 25% of the populations contained only homozygous recessive flies, and 25% contained only homozygous dominant flies having the allele for brown eye color.

Genetic drift is a random process, and therefore it is not likely to produce the same results in several populations. In California, there are a number of cypress groves, each a separate population. The phenotypes within each grove are more similar to one another than they are to the phenotypes in the other groves. Some groves have longitudinally shaped trees, and others have pyramidally shaped trees. The bark is rough in some colonies and smooth in others. The leaves are gray to bright green or bluish, and the cones are small or large. Because the environmental conditions are similar for all the groves, and no correlation has been found between phenotype and environment across groves, scientists hypothesize that these variations among populations are due to genetic drift.

Bottleneck Effect   Sometimes a species is subjected to near extinction because of a natural disaster (e.g., earthquake or fire) or because of overharvesting and habitat loss. It is as though most of the population has stayed behind and only a few survivors have passed through the neck of a bottle (**Fig. 15.7**). This so-called **bottleneck effect** prevents the majority of genotypes from participating in the production of the next generation.

The extreme genetic similarity found in cheetahs is believed to be due to a bottleneck. In a study of 47 different enzymes, each of which can occur in several different forms in other types of cats, all the cheetahs studied had exactly the same form. This demonstrates that genetic drift can cause certain alleles to be lost from a population. Exactly what caused the cheetah bottleneck is not known. It is speculated that perhaps cheetahs were slaughtered by nineteenth-century cattle farmers protecting their herds, or were captured by Egyptians as pets 4,000 years ago, or were decimated by a mass extinction tens of thousands of years ago. Today, cheetahs suffer from relative infertility because of the intense inbreeding that occurred after the bottleneck.

Figure 15.6   Genetic drift.

Genetic drift occurs when by chance only certain members of a population (in this case, green frogs) reproduce and pass on their genes to the next generation. The allele frequencies of the next generation's gene pool may be markedly different from those of the previous generation.

Figure 15.7   Bottleneck effect.

A bottleneck effect may occur when a catastrophic reduction in a population occurs, such as after a major epidemic or a hurricane. For example, a parent population contains roughly equal numbers of genotypes represented by blue, yellow, purple, green, and red marbles. The chance survivors of the catastrophe have genotypes represented by blue and yellow, resulting in a new population with altered gene pool frequencies.

Check
Your Progress

1. **Explain how gene flow can alter allele frequencies.**
2. **Explain why assortative mating is a type of nonrandom mating.**
3. **Explain why genetic drift is more likely to happen in a small population.**

Answers: 1. Individuals can add and remove alleles when they move into and out of populations.   2. When individuals choose a phenotype like their own, mating is not random.   3. The chance that a rare genotype might not or might participate in passing alleles to the next generation is greater. The result would be a decrease or increase in the frequency of this genotype.

Table 15.1 | Natural Selection

Evolution by natural selection requires:
1. Variation. The members of a population differ from one another.
2. Inheritance. Many of these differences are heritable genetic differences.
3. Degrees of successful reproduction. Individuals that are better adapted to their environment are more likely to reproduce, and their fertile offspring will make up a greater proportion of the next generation.

Founder Effect The **founder effect** is an example of genetic drift in which rare alleles, or combinations of alleles, occur at a higher frequency in a population isolated from the general population. After all, founding individuals contain only a fraction of the total genetic diversity of the original gene pool. The alleles carried by their founder or founders are dictated by chance alone. The Amish of Lancaster County, Pennsylvania, are an isolated group that was founded by German settlers. Today, as many as one in 14 individuals carries a recessive allele that causes an unusual form of dwarfism (affecting only the lower arms and legs) and polydactylism (extra fingers) **(Fig. 15.8)**. In most populations, only one in 1,000 individuals has this allele.

15.2 | Natural Selection

Natural selection is the process that results in adaptation of a population to the biotic (living) and abiotic (nonliving) environments. In the biotic environment, organisms acquire resources through competition, predation, and parasitism. The abiotic environment includes weather conditions, dependent chiefly upon temperatures and precipitation. Charles Darwin became convinced that species evolve with time and suggested natural selection as the mechanism for adaptation to the environment (see Chapter 14). In **Table 15.1,** Darwin's hypothesis of natural selection is stated in a way that is consistent with modern genetics.

As a result of natural selection, the most *fit* individuals become more prevalent in a population, and in this way, a population changes over time. The most fit individuals are those that reproduce more than others. The more fit individuals are likely those that are better adapted to the environment.

Types of Selection

Most of the traits on which natural selection acts are polygenic and controlled by more than one pair of alleles located at different gene loci. Such traits have a range of phenotypes, the frequency distribution of which usually resembles a bell-shaped curve, as shown in Figures 15.9–15.11.

Three types of natural selection are possible for any particular trait: directional selection, stabilizing selection, and disruptive selection.

Figure 15.8 Founder effect.

A member of the founding population of Amish in Pennsylvania had a recessive allele for a rare kind of dwarfism linked with polydactylism. The percentage of the Amish population now carrying this allele is much higher compared to that of the general population.

Check
Your **Progress**

1. How is the founder effect an example of genetic drift?

Answer: 1. The alleles carried by a few founding individuals could by chance represent only a small nonrandom sample of the larger population's various genes. The frequency of these genes could now increase in future generations by chance.

Directional Selection

Directional selection occurs when an extreme phenotype is favored and the distribution curve shifts in that direction. Such a shift can occur when a population is adapting to a changing environment.

Industrial melanism, discussed earlier and depicted in Figure 15.3, is an example of directional selection. Drug resistance is another. As you may know, widespread use of antibiotics and pesticides results in populations of bacteria and insects that are resistant to these chemicals. When an antibiotic is administered, some bacteria may survive because they are genetically resistant to the antibiotic. These are the bacteria that are likely to pass on their genes to the next generation. As a result, the number of resistant bacteria keeps increasing. Drug-resistant strains of bacteria that cause tuberculosis have now become a serious threat to the health of people worldwide.

Another example of directional selection is the human struggle against malaria, a disease caused by an infection of the liver and the red blood cells. The *Anopheles* mosquito transmits the disease-causing protozoan *Plasmodium* from person to person. In the early 1960s, international health authorities thought malaria would soon be eradicated. A new drug, chloroquine, seemed effective against *Plasmodium*, and spraying of DDT (an insecticide) had reduced the mosquito population. But by the mid-1960s, *Plasmodium* was showing signs of chloroquine resistance, and worse yet, mosquitoes were becoming resistant to DDT. A few drug-resistant parasites and a few DDT-resistant mosquitoes had survived and multiplied, shifting the distribution curve toward the resistant type of parasite.

The gradual increase in the size of the modern horse, *Equus*, is an example of directional selection that can also be correlated with a change in the environment—in this case, from forest conditions to grassland conditions **(Fig. 15.9)**. Even so, as discussed previously, the evolution of the horse should not be viewed as a straight line of descent because we know of many side branches that became extinct.

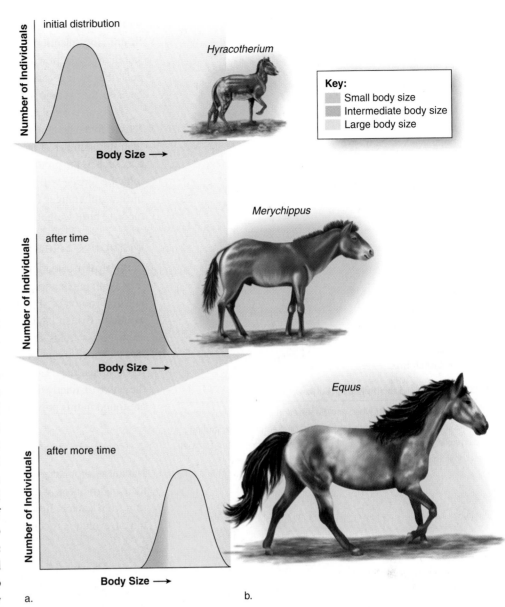

Figure 15.9   Directional selection.

a. Natural selection favors one extreme phenotype, and the distribution curve shifts. **b.** *Equus*, the modern-day horse, evolved from *Hyracotherium*, which was about the size of a dog. This small animal could have hidden among trees, and it had low-crowned teeth for browsing. When grasslands began to replace forests, the ancestors of *Equus* may have been subject to selection pressure for the development of strength, intelligence, speed, and durable grinding teeth. Larger animals were stronger and more successful in combat, those with larger skulls and brains had better sensory processing, those with longer legs and better-developed hooves could escape enemies, and animals with durable teeth could feed more efficiently on grass. Animals with these characteristics tended to produce more offspring.

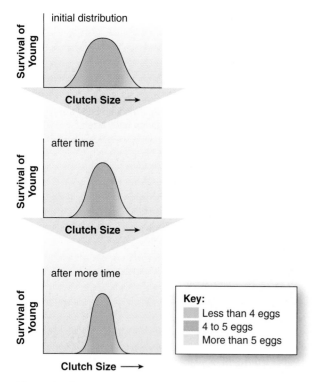

Figure 15.10 Stabilizing selection.

Stabilizing selection occurs when natural selection favors the intermediate phenotype over the extremes. For example, Swiss starling birds that lay four to five eggs (usual clutch size) have more young survive than those that lay less than four eggs or more than five eggs.

Stabilizing Selection

Stabilizing selection occurs when an intermediate phenotype is favored. With stabilizing selection, extreme phenotypes are selected against, and individuals near the average are favored. This is the most common form of selection because the average individual is well adapted to its environment.

As an example, consider that when Swiss starlings (*Sturnus vulgaris*) lay four to five eggs, more young survive than when the female lays more or less than this number (**Fig. 15.10**). Genes determining physiological characteristics, such as the production of yolk, and behavioral characteristics, such as how long the female will mate, are involved in determining clutch size. In humans, birth weight ranges from 0.89 to 4.9 kilograms. But most babies have an intermediate birth weight—the weight that favors survival.

Disruptive Selection

In **disruptive selection,** two or more extreme phenotypes are favored over any intermediate phenotype. Therefore, disruptive selection favors *polymorphism,* the occurrence of different forms in a population of the same species. For example, British land snails (*Cepaea nemoralis*) are found in low-vegetation areas (grass fields and hedgerows) and in forests. In low-vegetation areas, thrushes feed mainly on snails with dark shells that lack light bands, and in forest areas, they feed mainly on snails with light-banded shells. Therefore, these two distinctly different phenotypes, each adapted to its own environment, are found in this population (**Fig. 15.11**).

Adaptations Are Not Perfect

Natural selection doesn't always produce organisms that are perfectly adapted to their environment. Why not? First, it is important to realize that evolution doesn't start from scratch. Just as you can only bake a cake with the ingredients available to you, evolution is constrained by the available variations. Light-weight titanium

Figure 15.11 Disruptive selection.

a. Disruptive selection favors two extreme phenotypes, no banding and banding. **b.** Today, British land snails mainly comprise these two different phenotypes, each adapted to a different habitat.

a.   b.

bones might benefit birds but their bones contain calcium and other minerals the same as the other reptiles. When you mix the ingredients for a cake, you probably follow the same steps taught to you by your elders. Similarly, the processes of development prevent the emergence of novel features and therefore the wing of a bird has the same bones as those of other vertebrate forelimbs.

Imperfections are common because of necessary compromises. The success of humans is attributable to their dexterous hands but the spine is subject to injury because the vertebrate spine did not originally evolve to stand erect. A feature that evolves has a benefit that is worth the cost. The benefit of freeing the hands must have been worth the cost of spinal injuries from assuming an erect posture, for example. We should also consider that sexual selection, discussed in the opening story, has a reproductive benefit but not necessarily an adaptive benefit.

Maintenance of Variations

A population always shows some genotypic variation. The maintenance of variation is beneficial because populations with limited variation may not be able to adapt to new conditions should the environment change and may become extinct. How can variation be maintained in spite of selection constantly working to reduce it?

First, we must remember that the forces that promote variation are still at work: Mutation still generates new alleles, recombination and independent assortment still shuffles the alleles during gametogenesis, and fertilization still creates new combinations of alleles from those present in the gene pool. Second, gene flow might still occur. If the receiving population is small and is mostly homozygous, gene flow can be a significant source of new alleles. Finally, natural selection favors certain phenotypes, but the other types may still remain in reduced frequency. Disruptive selection even promotes polymorphism in a population. Diploidy and the heterozygote also help because they maintain recessive alleles in the gene pool.

Diploidy and the Heterozygote

Only alleles that are expressed (cause a phenotypic difference) are subject to natural selection. In diploid organisms, this fact makes the heterozygote a potential protector of recessive alleles that might otherwise be weeded out of the gene pool. Because the heterozygote remains in a population, so does the possibility of the recessive phenotype, which might have greater fitness in a changing environment. When natural selection favors the ratio of two or more phenotypes in generation after generation, it is called *balanced polymorphism*. Sickle cell disease offers an example of balanced polymorphism.

Sickle Cell Disease   Individuals with sickle cell disease have the genotype Hb^SHb^S (Hb = hemoglobin, the oxygen-carrying protein in red blood cells; s = sickle cell) and tend to die at an early age due to hemorrhaging and organ destruction. Those who are heterozygous and have sickle cell trait (Hb^AHb^S; A = normal) are better off because their red blood cells usually become sickle-shaped only when the oxygen content of the environment is low. Ordinarily, those with a normal genotype (Hb^AHb^A) are the most fit.

Geneticists studying the distribution of sickle cell disease in Africa have found that the recessive allele (Hb^S) has a higher frequency (0.2 to as high as 0.4 in a few areas) in regions where malaria is also prevalent (**Fig. 15.12**). What

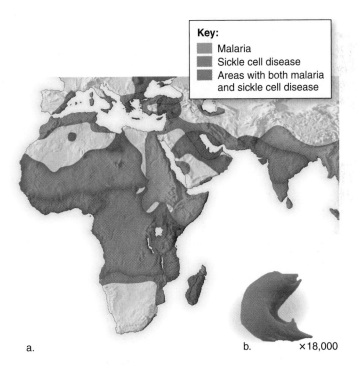

Key:
- Malaria
- Sickle cell disease
- Areas with both malaria and sickle cell disease

a.   b.   ×18,000

Figure 15.12   Sickle cell disease.

a. Red shows the areas where malaria was prevalent in Africa, the Middle East, southern Europe, and southern Asia in 1920, before eradication programs began; shown in blue are the areas where sickle cell disease most often occurred. The overlap of these two distributions (purple) suggested a causal connection. **b.** Scanning electron micrograph of a sickled red blood cell.

Table 15.2 | Sickle Cell Disease

Genotype	Phenotype	Result
Hb^AHb^A	Normal	Dies due to malarial infection
Hb^AHb^S	Sickle cell trait	Lives due to protection from both
Hb^SHb^S	Sickle cell disease	Dies due to sickle cell disease

is the connection between higher frequency of the recessive allele and malaria? Malaria is caused by a parasite that lives in and destroys the red blood cells of the normal homozygote (Hb^AHb^A). However, the parasite is unable to live in the red blood cells of the heterozygote (Hb^AHb^S) because the infection causes the red blood cells to become sickle-shaped. Sickle-shaped red blood cells lose potassium, and this causes the parasite to die. In an environment where malaria is prevalent, the heterozygote is favored. Each of the homozygotes is selected against but is maintained because the heterozygote is favored in parts of Africa subject to malaria. **Table 15.2** summarizes the effects of the three possible genotypes.

Check Your Progress

1. List forces that help maintain genetic variability in a population.
2. Explain the higher incidence of sickle cell disease in populations subject to malaria.

Answers: 1. Mutation, recombination, gene flow, disruptive selection, and diploidy.
2. Sickle cell disease occurs at a higher incidence in populations subject to malaria because the heterozyote ($Hb^A Hb^S$) is favored by natural selection in these regions.

The Chapter in Review

Summary

15.1 | Microevolution

Evolution in a Genetic Context

Microevolution involves several elements:
- All the various genes of a population make up its gene pool.
- Hardy-Weinberg equilibrium is present when gene pool allele frequencies remain the same from generation to generation. Certain conditions have to be met to achieve an equilibrium.
- The conditions are (1) no mutations, (2) no gene flow, (3) random mating, (4) no genetic drift, and (5) no selection. Since these conditions are rarely met, a change in gene pool frequencies is likely.
- When gene pool frequencies change, microevolution has occurred. Deviations from a Hardy-Weinberg equilibrium allow us to determine when evolution has taken place and to measure the extent of change.

Causes of Microevolution

Microevolution is caused by five conditions:
- Mutations are the ultimate source for variation. Certain genotypic variations may be of evolutionary significance, only if the environment changes. Genetic diversity is promoted when there are several alleles for each loci and when single nucleotide polymorphisms (SNPs) exist.
- Gene flow occurs when a breeding individual (in animals) migrates to another population or when gametes and seeds (in plants) are

carried into another population. Constant gene flow between two populations causes their gene pools to become similar.
- Nonrandom mating occurs when relatives mate (inbreeding) or when assortative mating takes place. Both of these cause an increase in homozygotes. Sexual selection, which occurs when a characteristic that increases the chances of mating is favored, favors random mating.
- Genetic drift occurs when allele frequencies are altered by chance— that is, by random sampling errors. Genetic drift is particularly evident after a bottleneck, when severe inbreeding occurs, or when founders start a new population.
- Natural selection (see Section 15.2).

15.2 | Natural Selection

Adaptation occurs when the more fit individuals reproduce more than others. These individuals usually possess traits better suited for survival in the environment, and over generations, the frequency of this adaptive trait increases within the population.

Types of Selection

Most of the traits of evolutionary significance are polygenic, and a range of phenotypes in a population result in a bell-shaped curve. Three types of selection occur:
- **Directional selection:** The curve shifts in one direction, as when dark-colored peppered moths become prevalent in polluted areas.

- **Stabilizing selection:** The peak of the curve increases, as when most human babies have the intermediate birth weight. Babies that are very small or very large are less fit than those of intermediate weight.
- **Disruptive selection:** The curve has two peaks, as when British land snails vary because a wide geographic range causes selection to vary.

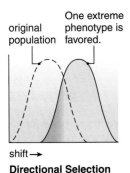

original population — One extreme phenotype is favored.

Directional Selection

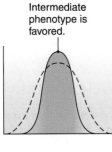

Intermediate phenotype is favored.

Stabilizing Selection

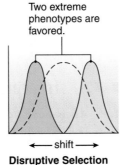

Two extreme phenotypes are favored.

Disruptive Selection

Adaptations Are Not Perfect

Adaptations may not be perfect because evolution builds on what variation exists. Only certain types of variations are available, and developmental procedures tend toward the same types of results. The result is often a compromise between a benefit and a cost. Sexual selection has a reproductive benefit that may not pertain to adaptation.

Maintenance of Variations

Despite constant natural selection, variation is maintained because:

- Mutations and recombination still occur; gene flow among populations can introduce new alleles; and natural selection may not eliminate less favored phenotypes.
- In sexually reproducing diploid organisms, the heterozygote acts as a repository for recessive alleles whose frequency is low. In sickle cell disease, the heterozygote is more fit in areas where malaria occurs, and therefore both homozygotes are maintained in the population.

▮▮ Key Terms

assortative mating   238
bottleneck effect   239
directional selection   241
disruptive selection   242
founder effect   240
gene flow   238
gene pool   235
genetic drift   238
industrial melanism   236

microevolution   234
mutation   237
natural selection   240
nonrandom mating   238
population   234
population genetics   235
sexual selection   238
stabilizing selection   242
subspecies   238

▮▮ Testing Yourself

Choose the best answer for each question.

1. A population consists of 48 *AA*, 54 *Aa*, and 22 *aa* individuals. What is the frequency of the *A* allele?
   a. 0.60
   b. 0.40
   c. 0.62
   d. 0.42
   e. 0.58

2. Which of the following is the binomial equation?
   a. $2p^2 + 2pq + 2q^2$
   b. $p^2 + pq + q^2$
   c. $2p^2 + pq + 2q^2$
   d. $p^2 + 2pq + q^2$

For questions 3 and 4, consider that about 70% of white North Americans can taste the chemical phenylthiocarbamide. The ability to taste is due to the dominant allele *T*. Nontasters are *tt*. Assume this population is in Hardy-Weinberg equilibrium.

3. What is the frequency of *t*?
   a. 0.30
   b. 0.70
   c. 0.55
   d. 0.09
   e. 0.60

4. What is the frequency of heterozygous tasters?
   a. 0.495
   b. 0.21
   c. 0.42
   d. 0.2475
   e. 0.45

5. Typically, mutations are immediately expressed and tested by the environment in
   a. prokaryotes.
   b. eukaryotes.
   c. prokaryotes and eukaryotes.
   d. neither prokaryotes nor eukaryotes.

6. The offspring of better-adapted individuals are expected to make up a larger proportion of the next generation. The most likely explanation is
   a. mutations and nonrandom mating.
   b. gene flow and genetic drift.
   c. mutations and natural selection.
   d. mutations and genetic drift.

7. The northern elephant seal went through a severe population decline as a result of hunting in the late 1800s. The population has rebounded but is now homozygous for nearly every gene studied. This is an example of
   a. negative assortative mating.
   b. migration.
   c. mutation.
   d. a bottleneck.
   e. disruptive selection.

8. Which of the following generally results in a net gain in genetic variability?
   a. genetic drift
   b. mutation
   c. directional selection
   d. bottleneck
   e. stabilizing selection

For questions 9–15, indicate the effect of each of the conditions of the Hardy-Weinberg principle on genotype and allele frequencies. Each answer may be used more than once.

Key:
   a. alters genotype and allele frequencies
   b. alters genotype frequency only
   c. alters allele frequency only
   d. does not alter genotype or allele frequency

9. mutation
10. gene flow
11. inbreeding
12. assortative mating
13. genetic drift
14. bottleneck
15. natural selection

16. A small, reproductively isolated religious sect called the Dunkers was established by 27 families that came to the United States from Germany 200 years ago. The frequencies for blood group alleles in this population differ significantly from those in the general U.S. population. This is an example of
   a. negative assortative mating.   d. bottleneck effect.
   b. natural selection.   e. gene flow.
   c. founder effect.

17. Assuming a Hardy-Weinberg equilibrium, 21% of a population is homozygous dominant, 50% is heterozygous, and 29% is homozygous recessive. What percentage of the next generation is predicted to be homozygous recessive?
   a. 21%   d. 42%
   b. 50%   e. 58%
   c. 29%

18. When a population is small, there is a greater chance of
   a. gene flow.   d. mutations occurring.
   b. genetic drift.   e. sexual selection.
   c. natural selection.

19. Which of the following is not expected to help maintain genetic variability?
   a. gene flow   d. disruptive selection
   b. mutation   e. genetic drift
   c. recombination

20. The sickle cell allele is maintained in regions where malaria is prevalent because
   a. the allele confers resistance to the parasite.
   b. gene flow is high in those regions.
   c. disruptive selection is occurring.
   d. genetic drift randomly selects for the allele.

21. Complete the following by drawing a curve in the other two graphs to show the effect of directional selection.

22. Complete the following by drawing a curve in the other two graphs to show the effect of stabilizing selection.

23. Complete the following by drawing curves in the other two graphs to show the effect of disruptive selection.

Thinking Scientifically

1. A behaviorist studying mountain bluebird reproductive behavior found that male aggression toward another male was highest during nest construction, and then it lessened after the first egg is laid, and again after hatching occurred. **a.** Knowing that natural selection favors behaviors that save energy, explain the reproductive behavior of the male mountain bluebirds. **b.** What might make you rule out genetic drift as an explanation?

Bioethical Issue

Who Should Reproduce?

The highly regarded population geneticist Sir Ronald Aylmer Fisher published a book in 1930 entitled *The Genetical Theory of Natural Selection*. In the book, he claims that civilizations fail because individuals with the highest level of fitness (those at the top of the societal ladder) do not reproduce as often as less affluent individuals. He suggests that high-income couples be paid to have children in order to improve the population. In other words, Fischer is suggesting that we carry out selection within the human population.

Can you envision any scenario in which our society should encourage reproduction by the most-fit and discourage it by the least-fit? If so, how should it be done? What characteristics would be appropriate to select for or against?

Essentials of Biology Website

The companion website for *Essentials of Biology* provides a wealth of information organized and integrated by chapter. You will find practice tests, animations, videos, and much more that will complement your learning and understanding of general biology.

http://www.mhhe.com/maderessentials2

The common green iguana, a strong swimmer native to South America, could be ancestral to other types of iguanas.

16

Evolution on a Large Scale

The rhinoceros iguana, with bumps on its snout, lives to the northeast of South America, only on Hispaniola.

The unique marine iguana feeds at sea and lives to the west of South America, only in the Galápagos Islands.

Outline

Green Iguanas in a Family Tree

IGUANAS ARE A type of lizard known for a very long tail, which can be up to three times their body length. The green iguana is common throughout the South American continent. Its claws enable it to climb trees, where it feeds on succulent leaves and soft fruits along riverbanks. It is an excellent swimmer and uses rivers to travel to new feeding areas.

Green iguanas do not inhabit islands to the west and northeast of South America. Instead, the Galápagos Islands to the west have one type of land iguana and one type of marine iguana. Marine iguanas, black in color, are

—Continued next page

Continued—

unique to the Galápagos Islands—they occur no place else on Earth. These air-breathing reptiles can dive to a depth exceeding 10 meters (33 feet) and remain submerged for more than 30 minutes. They use their claws to cling to the bottom rocks while feeding on algae.

Hispaniola,[1] an island to the northeast of South America, is inhabited by the rhinoceros iguana, which has three horny bumps on its snout and is dark brown to black in color. This iguana lives mainly in dry forests with rocky, limestone habitats and feeds on a wide variety of plants.

The theory of evolution explains the occurrence of the unique iguanas on the Galápagos Islands and Hispaniola. It is hypothesized and supported by various data that ancestral iguanas swam or hitchhiked on floating driftwood from South America to the Galápagos to the west and to Hispaniola to the northeast. After arrival, these iguana populations were cut off from other iguana populations, and this allowed them to evolve into these new and different species. The origin of species and how they are classified and studied is the topic of this chapter.

16.1 | Macroevolution

Chapter 15 considered evolution on a small scale—that is, microevolution, small changes over a small period of time. In this chapter, we turn our attention to evolution on a large scale—that is, **macroevolution,** large changes over a very long period of time. The history of life on Earth is a part of macroevolution. Macroevolution requires **speciation,** the splitting of one species into two or more new species. Speciation involves the gene pool changes that we studied in Chapter 15.

Species originate, adapt to their environments, and then may become extinct (**Fig. 16.1**). Without the origin and extinction of species, life on Earth would not have the ever-changing history that we find in the fossil record.

Defining Species

Before we consider the origin of species, we first need to define a species. Appearance is not always a good criterion. The members of different species can look quite similar, while the members of a single species can be diverse in appearance. Although many definitions have been proposed, the *biological species concept* offers a testable way to define a species that does not depend on appearance: The members of a species interbreed and have a shared gene pool, and each species is reproductively isolated from every other species. For example, the flycatchers in **Figure 16.2** are members of separate species because they do not interbreed in nature.

According to the biological species concept, gene flow occurs between the populations of a species but not between populations of different species. The red maple and the sugar maple are found over a wide geographical range

[1] Hispaniola is divided into Haiti and the Dominican Republic.

Figure 16.1 **Dinosaurs.**

Triceratops (left) and *Tyrannosaurus rex (right)* were dinosaurs that lived about 65 million years ago when flowering plants were already flourishing. The prevailing forms of life on Earth change as new species come into existence, flourish, and then decline or become extinct.

in the eastern half of the United States, and each species is made up of many populations. However, the members of each species' population rarely hybridize in nature. Therefore, these two types of plants are separate species. In contrast, the human species has many populations that certainly differ in physical appearance (**Fig. 16.3**). We know, however, that all humans belong to one species because the members of these populations can produce fertile offspring.

The biological species concept is useful, as we shall see, but even so, it has its limitations. For example, it applies only to sexually reproducing organisms and cannot apply to asexually reproducing organisms. Then, too, sexually reproducing organisms are not always as reproductively isolated as we would expect. Some North American orioles live in the western half of the continent and some in the eastern half of the continent. Yet even the two most genetically distant oriole species, as recognized by analyzing their mitochondrial DNA, will hybridize where they meet in the middle of the continent.

There are other definitions of species aside from the biological definition. Later in this chapter, we will define species as a category of classification below the rank of genus. Species in the same genus share a recent common ancestor. A **common ancestor** is a single ancestor for two or more different groups. Your grandmother is the common ancestor for you and your cousins. Similarly, there is a common ancestor for all species of roses.

Acadian flycatcher,
Empidonax virescens

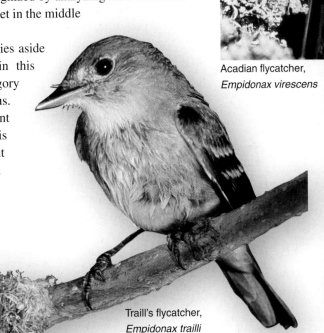

Traill's flycatcher,
Empidonax trailli

least flycatcher,
Empidonax minimus

Figure 16.2   Three species of flycatchers.

Although these flycatcher species are nearly identical in appearance, we know they are separate species because they are reproductively isolated—the members of each species reproduce only with one another. Each species has a characteristic song and its own particular habitat during the mating season as well.

a.   b.

Figure 16.3   Human populations.

The Massai of East Africa (**a**) and the Eskimos that live near the Arctic Circle (**b**) are both members of the species *Homo sapiens* because Massai can reproduce with Eskimos and produce fertile offspring.

Reproductive Barriers

As mentioned, for two species to be separate, they must be reproductively isolated—that is, gene flow must not occur between them. Reproductive barriers are isolating mechanisms that prevent successful reproduction from occurring (**Fig. 16.4**). In evolution, reproduction is successful only when it produces fertile offspring.

Prezygotic (before the formation of a zygote) **isolating mechanisms** are those that prevent reproductive attempts and make it unlikely that fertilization will be successful if mating is attempted. Habitat isolation, temporal isolation, behavioral isolation, mechanical isolation, and gamete isolation make it highly unlikely that particular genotypes will contribute to the gene pool of a population.

Habitat isolation When two species occupy different habitats, even within the same geographic range, they are less likely to meet and attempt to reproduce. This is one of the reasons that the flycatchers in Figure 16.2 do not mate, and that the red maple and sugar maple do not exchange pollen. In tropical rain forests, many animal species are restricted to a particular level of the forest canopy, and in this way they are isolated from similar species.

Temporal isolation Two species can live in the same locale, but if each reproduces at a different time of year, they do not attempt to mate. For example, *Reticulitermes hageni* and *R. virginicus* are two species of termites. The former has mating flights in March through May, whereas the latter mates in the fall and winter months.

Behavioral isolation Many animal species have courtship patterns that allow males and females to recognize one another (**Fig. 16.5**). Female fireflies recognize males of their species by the pattern of the male's flashings; similarly, female crickets recognize males of their species by the male's chirping. Many males recognize females of their species by sensing

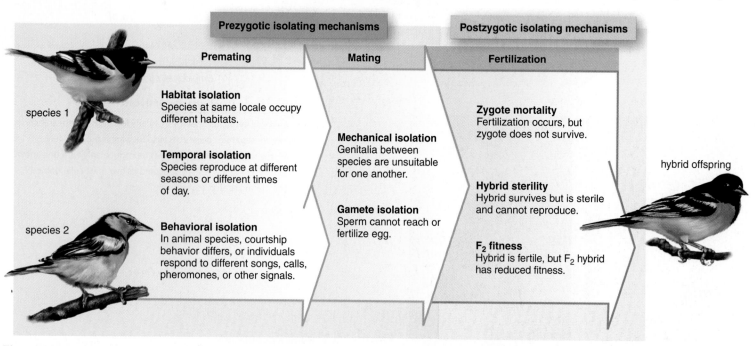

Figure 16.4 **Reproductive barriers.**

Prezygotic isolating mechanisms prevent mating attempts or a successful outcome should mating take place; for example, between two species of orioles. No zygote is ever formed if these mechanisms are successful. Postzygotic isolating mechanisms prevent offspring from reproducing—that is, if a hybrid oriole should result, it would be unable to breed successfully.

chemical signals called pheromones. For example, female gypsy moths secrete chemicals from special abdominal glands. These chemicals are detected downwind by receptors on the antennae of males.

Mechanical isolation When animal genitalia or plant floral structures are incompatible, reproduction cannot occur. Inaccessibility of pollen to certain pollinators can prevent cross-fertilization in plants, and the sexes of many insect species have genitalia that do not match, or other characteristics that make mating impossible. For example, male dragonflies have claspers that are suitable for holding only the females of their own species.

Gamete isolation Even if the gametes of two different species meet, they may not fuse to become a zygote. In animals, the sperm of one species may not be able to survive in the reproductive tract of another species, or the egg may have receptors only for sperm of its species. Also, in each type flower, only certain pollen grains can germinate so that sperm successfully reach the egg.

Postzygotic (after the formation of a zygote) **isolating mechanisms** prevent hybrid offspring (reproductive product of two different species) from developing or breeding, even if reproduction attempts have been successful.

Zygote mortality The hybrid zygote may not be viable, and so it dies. A zygote with two different chromosome sets may fail to go through mitosis properly, or the developing embryo may receive incompatible instructions from the maternal and paternal genes so that it cannot continue to exist.

Hybrid sterility The hybrid zygote may develop into a sterile adult. As is well known, a cross between a horse and a donkey produces a mule, which is usually sterile—it cannot reproduce (**Fig. 16.6**). Sterility of hybrids generally results from complications in meiosis that lead to an inability to produce viable gametes. A cross between a cabbage and a radish produces offspring that cannot form gametes, even though the diploid number is 18, an even number, most likely because the cabbage chromosomes and the radish chromosomes cannot align during meiosis.

F_2 *fitness* If hybrids can reproduce, their offspring are unable to reproduce. In some cases, mules are fertile, but their offspring (the F_2 generation) are not fertile.

Check Your **Progress**

1. Define a species according to the biological species concept.
2. Describe the limitations of the biological species concept.
3. Contrast prezygotic isolating mechanisms with postzygotic isolating mechanisms.

Answers: 1. A biological species is a group of organisms that can interbreed. The members of this species are reproductively isolated from members of other species. 2. The biological species concept cannot be applied to asexually reproducing organisms. In addition, reproductive isolation may not be complete, so some breeding between species may occur. 3. Prezygotic isolating mechanisms do not allow zygotes to be formed. Postzygotic mechanisms allow zygotes to be formed, but the zygotes either die or become offspring that are infertile.

Figure 16.5 Prezygotic isolating mechanism.

An elaborate courtship display allows the blue-footed boobies of the Galápagos Islands to select a mate. The male lifts his feet in a ritualized manner that shows off their bright blue color.

horse   donkey

mating

fertilization

mule (F_1 hybrid)

Usually mules cannot reproduce. If an F_2 offspring does result, it cannot reproduce.

Figure 16.6 Postzygotic isolating mechanism.

Mules are horse-donkey hybrids. Mules are infertile due to a difference in the chromosomes inherited from their parents.

Northern ancestral population

Eastern migration southward establishes these populations.

Western migration southward establishes these populations.

Central Valley

Genetic differences build up between populations.

In the end, two reproductively isolated populations cannot mate in Southern California.

Figure 16.7 **Allopatric speciation.**

In this example of allopatric speciation, the Central Valley of California is separating a range of populations descended from the same northern ancestral species. The limited contact between the populations on the west and those on the east allow genetic changes to build up to such an extent that members of the two southern populations rarely reproduce with each other and are designated as subspecies.

Models of Speciation

The introduction to this chapter suggests that iguanas of South America may be the common ancestor for both the marine iguana on the Galápagos Islands (to the west of South America) and the rhinoceros iguana on Hispaniola (an island to the northeast of South America). If so, how could it have happened? Green iguanas are strong swimmers, so by chance a few could have migrated to these islands, where they formed populations separate from each other and from the parent population back in South America. Each population continued on its own evolutionary path as new mutations, genetic drift, and natural selection occurred. Eventually, reproductive isolation developed and there were three species of iguanas. A speciation model based on geographic isolation of populations is called **allopatric speciation** because *allo* means different and *patria* loosely means homeland.

Figure 16.7 features an example of allopatric speciation, which has been extensively studied in California. Members of an ancestral population of *Ensatina* salamanders existing in the Pacific Northwest migrated southward, establishing a range of populations. Each population was exposed to its own selective pressures along the coastal mountains and along the Sierra Nevada Mountains. Due to the barrier created by the Central Valley of California, limited gene flow occurred between the eastern populations and the western populations. Genetic differences increased from north to south, resulting in two distinct forms of *Ensatina* salamanders in southern California that differ dramatically in color and interbreed only rarely.

With **sympatric speciation,** a population develops into two or more reproductively isolated groups without prior geographic isolation. One of the best examples to easily illustrate this type of speciation is found among plants, where it can occur by means of *polyploidy,* an increase in the number of sets of chromosomes to $3n$ or above. The presence of sex chromosomes makes it difficult for polyploidy speciation to occur in animals. In plants, hybridization between two species can be followed by a doubling of the chromosome number. Such polyploid plants are reproductively isolated by a postzygotic mechanism; they can reproduce successfully only with other similar polyploids, and backcrosses with their parents are sterile. Therefore, three species instead of two species result. **Figure 16.8** shows that the

Wild wheat
$2n = 28$

× Wild wheat
$2n = 14$

→ Sterile hybrid
$2n = 21$

Doubling of chromosome number →

Bread wheat
$2n = 42$

Figure 16.8 **Sympatric speciation.**

In this example of sympatric speciation, two populations of wild wheat hybridized many years ago. The hybrid is sterile, but chromosome doubling allowed some plants to reproduce. These plants became today's bread wheat.

parents of our present-day bread wheat had 28 and 14 chromosomes, respectively. The hybrid with 21 chromosomes is sterile, but polyploid bread wheat with 42 chromosomes is fertile because the chromosomes can align during meiosis.

Adaptive Radiation

A clear example of speciation through adaptive radiation is provided by the finches on the Galápagos Islands, which are often called Darwin's finches because Darwin first realized their significance as an example of how evolution works. During **adaptive radiation,** many new species evolve from a single ancestral species. The many species of finches that live on the Galápagos Islands are hypothesized to be descendants from a single type of ancestral finch from the mainland (**Fig. 16.9**). The populations on the various islands were subjected to the founder effect involving genetic drift, genetic mutations, and the process of natural selection. Because of natural selection, each population became adapted to a particular habitat on its island. In time, the various populations became so genotypically different that now, when by chance they reside on the same island, they do not interbreed and are therefore separate species. There is evidence that the finches use beak shape to recognize members of the same species during courtship. Rejection of suitors with the wrong type of beak is a behavioral prezygotic isolating mechanism.

Similarly, inhabiting the Hawaiian Islands is a wide variety of honeycreepers, all descended from a common goldfinch-like ancestor that arrived from Asia or North America about 5 million years ago. Today, honeycreepers have a range of beak sizes and shapes for feeding on various food sources, including seeds, fruits, flowers, and insects (as discussed in Chapter 1, Fig. 1.5).

Check Your Progress

1. **Contrast allopatric speciation with sympatric speciation.**
2. **Explain why Darwin's finches are an example of adaptive radiation.**

Answers: 1. Allopatric speciation occurs when a geographic barrier stops gene flow, while sympatric speciation produces reproductively isolated groups without geographic isolation. 2. One mainland finch species was probably ancestral to all existing species on the Galápagos Islands. Each new species evolved in response to the unique environment on the island it inhabits.

Figure 16.9 **Darwin's finches.**

Each of Darwin's finches is adapted to gathering and eating a different type of food. Tree finches have beaks largely adapted to eating insects and, at times, plants. Ground finches have beaks adapted to eating the flesh of the prickly–pear cactus or different–sized seeds.

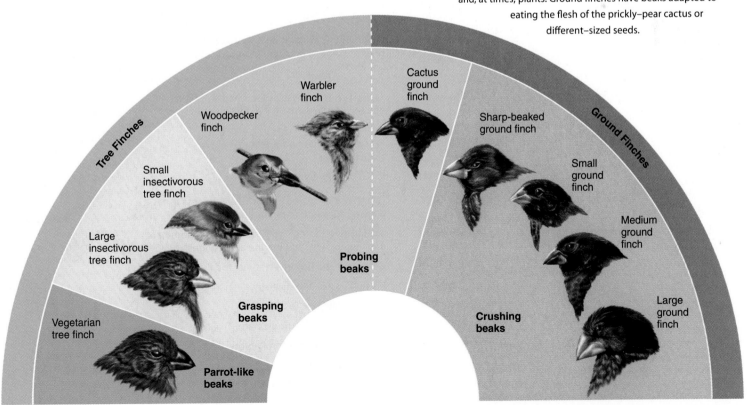

Table 16.1 | The Geological Timescale: Major Divisions of Geological Time and Some of the Major Evolutionary Events That Occurred

Era	Period	Epoch	MYA	Plant and Animal Life
Cenozoic	Quaternary	Holocene	0–0.01	AGE OF HUMAN CIVILIZATION; Destruction of ecosystems accelerates extinctions.
				SIGNIFICANT MAMMALIAN EXTINCTION
		Pleistocene	0.01–2	Modern humans appear; modern plants spread and diversify.
		Pliocene	2–6	First hominids appear; modern angiosperms flourish.
		Miocene	6–24	Apelike mammals, grazing mammals, and insects flourish; grasslands spread; and forests contract.
	Tertiary	Oligocene	24–37	Monkeylike primates appear; modern angiosperms appear.
		Eocene	37–58	All modern orders of mammals are present; subtropical forests flourish.
		Paleocene	58–65	Primates, herbivores, carnivores, insectivores are present; angiosperms diversify.
				MASS EXTINCTION: DINOSAURS AND MOST REPTILES
Mesozoic	Cretaceous		65–144	Placental mammals and modern insects appear; angiosperms spread and conifers persist.
	Jurassic		144–208	Dinosaurs flourish; birds and angiosperms appear.
				MASS EXTINCTION
	Triassic		208–250	First mammals and dinosaurs appear; forests of conifers and cycads dominate land; corals and molluscs dominate seas.
				MASS EXTINCTION
Paleozoic	Permian		250–286	Reptiles diversify; amphibians decline; and gymnosperms diversify.
	Carboniferous		286–360	Amphibians diversify; reptiles appear; and insects diversify. Age of great coal-forming forests.
				MASS EXTINCTION
	Devonian		360–408	Jawed fishes diversify; insects and amphibians appear; seedless vascular plants diversify and seed plants appear.
	Silurian		408–438	First jawed fishes and seedless vascular plants appear.
				MASS EXTINCTION
	Ordovician		438–510	Invertebrates spread and diversify; jawless fishes appear; nonvascular plants appear on land.
	Cambrian		510–543	Marine invertebrates with skeletons are dominant and invade land, and marine algae flourish.
Precambrian time			600	Oldest soft-bodied invertebrate fossils.
			1,400–700	Protists evolve and diversify.
			2,000	Oldest eukaryotic fossils.
			2,500	O_2 accumulates in atmosphere.
			3,500	Oldest known fossils (prokaryotes).
			4,500	Earth forms.

16.2 | The Fossil Record

The history of the origin and extinction of species on Earth is best discovered by studying the fossil record (**Fig. 16.10**). **Fossils** are the traces and remains of past life or any other direct evidence of past life. **Paleontology** is the science of discovering and studying the fossil record and, from it, making decisions about the history of species.

The Geological Timescale

Because life-forms have evolved over time, the strata (layers of sedimentary rock, see Fig. 14.2a) of the Earth's crust contain different fossils. By studying the strata and the fossils they contain, geologists have been able to construct a geological timescale (**Table 16.1**). It divides the history of life on Earth into eras, then periods, and then epochs, and describes the types of fossils common to each of these divisions of time. Notice that in the geological time-scale, only the periods of the Cenozoic era are divided into epochs, meaning that more attention is given to the evolution of primates and flowering plants than to the earlier evolving organisms. Despite an epoch assigned to modern civilization, humans have only been around about .04% of the history of life.

In contrast, prokaryotes existed alone for some 2 billion years before the eukaryotic cell and multicellularity arose during Precambrian time. Some prokaryotes became the first photosynthesizers to add oxygen to the atmosphere. The presence of oxygen may have spurred the evolution of the eukaryotic cell and multicellularity during the Precambrian. All major groups of animals evolved during what is sometimes called the Cambrian explosion. The fossil record for Precambrian time is meager, but the fossil record for the Cambrian period is rich. The evolution of the invertebrate external skeleton accounts for this increase in the number of fossils. Perhaps this skeleton, which impedes the uptake of oxygen, couldn't evolve until oxygen was plentiful. Or perhaps the external skeleton was merely a defense against predation.

The origin of life on land is another interesting topic. During the Paleozoic era, plants are present on land before animals. Non-vascular plants preceded vascular plants, and among these, cone-bearing plants (gymnosperms) preceded flowering plants (angiosperms). Among vertebrates, the fishes are aquatic, and the amphibians invaded land. The reptiles, which include dinosaurs and birds, share an amniote ancestor with the mammals. The number of species in the world has continued to increase until the present time, despite the occurrence of five mass extinctions, including one significant mammalian extinction during the history of life on Earth.

The Pace of Speciation

Darwin theorized that evolutionary changes occur gradually. In other words, he supported a *gradualistic model* to explain the pace of evolution. Speciation probably occurs after populations become isolated, with each group continuing slowly on its own evolutionary pathway. The gradualism model often shows the evolutionary history of groups of organisms by drawing the type of **evolutionary tree**

Figure 16.10 Fossils.

a. A fern leaf from 245 million years ago (MYA) remains because it was buried in sediment that hardened to rock. **b.** This midge (40 MYA) became embedded in amber (hardened resin from a tree). **c.** The shells of ammonites (135 MYA) left a mold filled in by minerals. **d.** Most fossils, such as this early insectivore mammal (47 MYA) are remains of hard parts because they do not decay as the soft parts do. **e.** This dinosaur footprint (135 MYA) is a sign of ancient life. **f.** This tree trunk (190 MYA) is petrified because minerals have replaced the original tissues.

Time

Change

a. Gradualistic model

Time

Change

b. Punctuated equilibrium model

Figure 16.11   **Pace of evolution.**

a. According to the gradualistic model, new species evolve slowly from an ancestral species. **b.** According to the punctuated equilibrium model, new species appear suddenly and then remain largely unchanged until they become extinct.

shown in **Figure 16.11***a*. Note that in this diagram, an ancestral species has given rise to two separate species, represented by a slow change in plumage color. The gradualistic model suggests that it is difficult to indicate when speciation has occurred because there would be so many transitional links. In some cases, it has been possible to trace the evolution of a group of organisms by finding transitional links.

More often, however, species appear quite suddenly in the fossil record, and then they remain essentially unchanged phenotypically until they undergo extinction. Some paleontologists have therefore developed a *punctuated equilibrium model* to explain the pace of evolution. The model says that a period of equilibrium (no change) is punctuated (interrupted) by speciation. Figure 16.11*b* shows the type of diagram paleontologists prefer to use when representing the history of evolution over time. This model suggests that transitional links are less likely to become fossils and less likely to be found. Speciation most likely involves only an isolated subpopulation at one locale. Only when this new subpopulation expands and replaces existing species is it apt to show up in the fossil record.

The differences between these two models are subtle, especially when we consider that the "sudden" appearance of a new species in the fossil record could represent many thousands of years because geological time is measured in millions of years.

Mass Extinctions of Species

As researchers have noted, most species exist for only a limited amount of time (measured in millions of years), and then they die out (become extinct). **Mass extinctions** are the disappearance of a large number of species or a higher taxonomic group within a relatively short period of time. The geological timescale in Table 16.1 shows the occurrence of five mass extinctions: at the ends of the Ordovician, Devonian, Permian, Triassic, and Cretaceous periods. Also, there was a significant mammalian extinction at the end of the Pleistocene epoch. While many factors contribute to mass extinctions, two possible types of events (continental drift and meteorite impacts) have particular significance.

Continental drift—the movement of continents—has contributed to several extinctions. You may have noticed that the coastlines of several continents are mirror images of each other. For example, the outline of the west coast of Africa matches that of the east coast of South America. Also, the same geological structures are found in many of the areas where the continents at some time in the past touched. A single mountain range runs through South America, Antarctica, and Australia, for example. But the mountain range is no longer continuous because the continents have drifted apart. The reason the continents drift is explained by a branch of geology known as *plate tectonics*, which is based on the fact that the Earth's crust is fragmented into slablike plates that float on a lower, hot mantle layer (**Fig. 16.12**). The continents and the ocean basins are a part of these rigid plates, which move like conveyor belts.

The loss of habitats is a significant cause of extinctions, and continental drift can lead to massive habitat changes. We know that 250 million years ago (MYA), at the time of the Permian mass extinction, all the Earth's continents came together to form one supercontinent called Pangaea (**Fig. 16.13***a*). The result was dramatic environmental change through the shifting of wind patterns, ocean currents, and most importantly, the amount of available shallow marine habitat. Marine organisms suffered as the oceans merged, and the amount of shoreline, where many marine organisms live, was drastically reduced. Species diversity did not recover until some continents drifted away from the poles, shorelines increased,

Figure 16.12 **Plate tectonics.**

The Earth's surface is divided into several solid tectonic plates floating on the fluid magma beneath them. At rifts in the ocean floor, two plates gradually separate as fresh magma wells up and cools, enlarging the plates. Mountains, including volcanoes, are raised where one plate pushes beneath another at subduction zones. Where two plates slowly grind past each other at a fault line, tension builds up, which is released occasionally in the form of eathquakes.

and warmth returned (Fig. 16.13*b*). Terrestrial organisms were affected as well because the amount of interior land, which tends to have a drier and more erratic climate, increased. Immense glaciers developing at the poles withdrew water from the oceans and chilled even once-tropical land.

The other event that is known to have contributed to mass extinctions is the impact of a meteorite as it crashed into the Earth. The result of a large meteorite striking Earth could have been similar to that of a worldwide atomic bomb explosion: A cloud of dust would have mushroomed into the atmosphere, blocking out the sun and causing plants to freeze and die. This type of event has been proposed as a primary cause of the Cretaceous extinction that saw the demise of the dinosaurs. Cretaceous clay contains an abnormally high level of iridium, an element that is rare in the Earth's crust but more common in meteorites. A layer of soot has been identified in the strata alongside the iridium, and a huge crater that could have been caused by a meteorite was found in the Caribbean–Gulf of Mexico region on the Yucatán peninsula.

Check Your **Progress**

1. **How is the fossil record the best evidence for macroevolution?**
2. **Explain the phrase "punctuated equilibrium model."**
3. **How do you explain that the outline of the west coast of Africa matches that of the east coast of South America?**

Answers: 1. Fossils allow you to see changes occurring over time.   2. The fossil record shows times of equilibrium (no change) being interrupted by relatively sudden, rapid changes.   3. At one time, these continents were joined before they drifted apart.

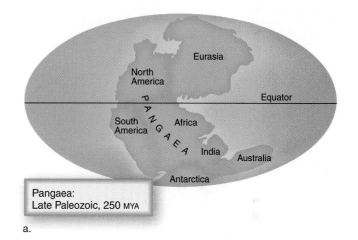

Pangaea:
Late Paleozoic, 250 MYA

a.

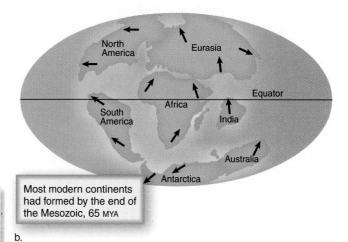

Most modern continents had formed by the end of the Mesozoic, 65 MYA

b.

Figure 16.13 **Continental drift.**

a. About 250 MYA, all the continents were joined into a supercontinent called Pangaea. **b.** By 65 MYA, all the continents had begun to separate. This process is continuing today. North America and Europe are separating at a rate of about 2 centimeters per year.

Domain Bacteria	Domain Archaea	Domain Eukarya

Kingdom — Plantae

Phylum — Anthophyta

Class — Monocotyledones

Order — Asparagales

Family — Orchidaceae

Genus — *Cypripedium*

Species — *Cypripedium acaule*

Figure 16.14 Taxonomy hierarchy.

A domain is the most inclusive of the classification categories. The plant kingdom is in the domain Eukarya. In the plant kingdom are several phyla, each represented here by lavender circles. The phylum Anthophyta has only two classes (the monocots and eudicots). The class Monocotyledones encompasses many orders. In the order Orchidales are many families; in the family Orchidaceae are many genera; and in the genus *Cypripedium* are many species—for example, *Cypripedium acaule*. (This illustration is diagrammatic and doesn't necessarily show the correct number of subcategories.)

16.3 | Systematics

All fields of biology, but especially **systematics,** are dedicated to understanding the evolutionary history of life on Earth. Systematics is very analytical and relies on a combination of data from the fossil record and comparative anatomy and development, with an emphasis today on molecular data, to determine **phylogeny,** the evolutionary history of a group of organisms. Classification is a part of systematics because ideally organisms are classified according to our present understanding of evolutionary relationships.

Linnaean Classification

Taxonomy is the branch of biology concerned with identifying, naming, and classifying organisms. A **taxon** (pl., taxa) is an organism or a group of organisms at a particular level in a classification system. The binomial system of nomenclature assigns a two-part name to each type organism. For example, the plant in **Figure 16.14** has been named *Cypripedium acaule.* This name means that the plant is in the genus *Cypripedium* and that the specific epithet is *acaule.* Notice that the scientific name is in italic and only the genus is capitalized. The genus can be abbreviated to a single letter if the full name has been given previously and if it is used with a specific epithet. For example, *C. acaule* is an acceptable way to designate this plant. The name of an organism usually tells you something about the organism. In this instance, the genus name, *Cypripedium* refers to the slipper shape of the flower, and the specific epithet *acaule* says that the flower has no independent stem.

Why do organisms need scientific names? And why do scientists use Latin, rather than common names, to describe organisms? There are several reasons. First, a common name will vary from country to country because different countries use different languages. Second, even people who speak the same language sometimes use different common names to describe the same organism, for example, bowfin, grindle, choupique, and cypress trout describe the same common fish, *Amia calva.* Furthermore, between countries the same common name is sometimes given to different organisms. A "robin" in England is very different from a "robin" in the United States, for example. Latin, on the other hand, is a universal language that not too long ago was well known by most scholars, many of whom were physicians or clerics. When scientists throughout the world use the same scientific binomial name, they know they are speaking of the same organism.

Today, taxonomists use several categories of classification created by Swedish biologist Carl Linnaeus in the 1700s to show varying levels of similarity: **species, genus, family, order, class, phylum,** and **kingdom.** Recently, a higher taxonomic category, the **domain,** has been added to this list. There can be several species within a genus, several genera within a family, and so forth. In this hierarchy, the higher the category, the more inclusive it is (Fig. 16.14). Therefore species in the same domain have general traits in common, while those in the same genus have quite specific traits in common.

In most cases, each category of classification can be subdivided into three additional categories, as in superorder, order, suborder, and infraorder. Considering these, there are more than 30 categories of classification.

Phylogenetic Trees

Figure 16.15 shows how the classification of groups of organisms allows us to construct a **phylogenetic tree,** a diagram that indicates common ancestors and lines of descent (lineages). The common ancestor at the base of the tree has traits that are shared by all the other groups in the tree. For example, the Artiodactyla are characterized by having hoofs with an even number of toes. On the other hand notice that the Cervidae have antlers but the Bovidae have no antlers. Finally, among the Cervidae, the antlers are highly branched in the red deer but palmate (having the shaped of a hand) in reindeer. Notice that as the lineage moves from common ancestor to common ancestor the traits become more specific to just particular groups of animals. It is this progression in specificity that allows classification categories to serve as a basis for depicting a phylogenetic tree.

Tracing Phylogeny

While Figure 16.15 makes use of morphological data, systematists particularly today use a multitude of data in order to discover the evolutionary relationships between species. They rely heavily on a combination of fossil record data, morphological data, and molecular data to determine the correct sequence of common ancestors in any particular group of organisms.

Morphological data include homologies which are similarities among organisms that stem from having a common ancestor. Comparative anatomy, including embryological evidence and fossil data, provides information regarding homology. **Homologous structures** are related to each other through common descent. The forelimbs of vertebrates are homologous because they contain the same bones organized in the same general way as in a common ancestor (see Fig. 14.14). This is the case, even though a horse has but a single digit and toe (the hoof), while a bat has four lengthened digits that support its membranous wings.

Deciphering homology is sometimes difficult because of convergent evolution. **Convergent evolution** is the acquisition of the same or similar traits in distantly related lines of descent. Similarity due to convergence is termed **analogy.** The wings of an insect and the wings of a bat are analogous. You may recall from Chapter 14 that **analogous structures** have the same function in different groups, but organisms with these structures do not have a recent common ancestor. *Analogous structures arise because of adaptations to the same type of environment.* Both cacti and spurges are adapted similarly to a hot, dry environment, and both are succulent, spiny, flowering plants. However, the details of their flower structure indicate that these plants are not closely related.

Figure 16.15 **Classification and phylogeny.**

The classification and phylogenetic tree for a group of organisms is ideally constructed to reflect their phylogenetic history. A species is most closely related to other species in the same genus, more distantly related to species in other genera of the same family, and so forth, on through order, class, phylum, and kingdom.

Check Your Progress

1. List the categories of classification, in order, from the largest category to the smallest.
2. Contrast homologous structures with analogous structures.
3. What is a phylogenetic tree?

Answers: 1. Domain, kingdom, phylum, class, order, family, genus, species. 2. Homologous structures are similar in anatomy because they are derived from a common ancestor. Analogous structures have the same function in different groups and do not share a common ancestry. 3. A phylogenetic tree is a diagram that indicates common ancestors and lines of descent.

Check Your Progress

1. What types of data are used to determine evolutionary relationships?
2. Explain why the sequencing of ribosomal RNA (rRNA) is done for evolutionary studies.

Answers: 1. The fossil records, anatomical homologies, and molecular data are used.
2. Because rRNA molecules are found in ribosomes that carry out protein synthesis, their nucleotide sequence changes very little over time. Therefore, similarities in rRNA sequence can be reliably used to determine who is related to whom.

Speciation occurs when mutations bring about changes in the base-pair sequences of DNA. Systematists, therefore, assume that the more closely species are related, the fewer changes will be found in DNA base-pair sequences. Because molecular data are straightforward and numerical, they can sometimes sort out relationships obscured by inconsequential anatomical variations or convergence. Computer software breakthroughs have made it possible to analyze nucleotide sequences quickly and accurately. Also, these analyses are available to anyone doing comparative studies through the Internet, so each investigator doesn't have to start from scratch. The combination of accuracy and availability of vast amounts of data, even entire genomes, has made molecular systematics a standard way to study the relatedness of organisms today.

All cells have ribosomes essential for protein synthesis, and the genes that code for ribosomal RNA (rRNA) have changed very slowly during evolution because drastic changes lead to malfunctioning cells. Therefore, comparative rRNA sequencing provides a reliable indicator of the similarity between organisms. Ribosomal RNA sequencing helped investigators conclude that all living things can be divided into the three domains.

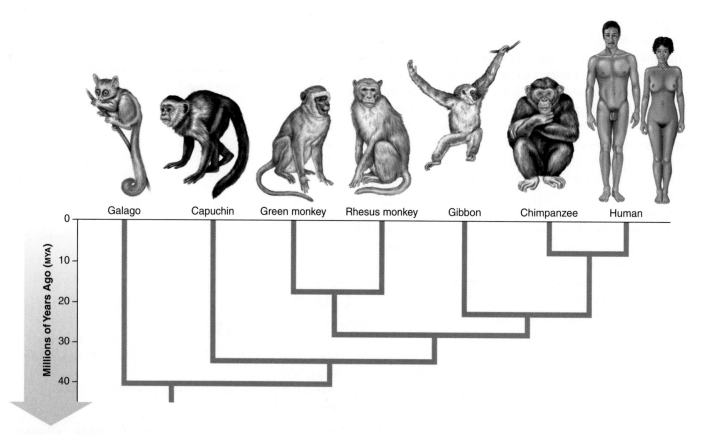

Figure 16.16 **Molecular data.**

The relationships of certain primate species are based on a study of their genomes. The length of the branches indicates the relative number of DNA base-pair differences between the groups. These data, along with knowledge of the fossil record for one divergence, make it possible to suggest a date for the other divergences in the tree.

One study involving DNA differences produced the data shown in **Figure 16.16.** Notice the close relationship between chimpanzees and humans. This relationship has recently been recognized by the designation of a new subfamily called Homininae that includes not only chimpanzees and humans but also gorillas. Molecular data indicates that gorillas and chimpanzees are more closely related to humans than they are to orangutans. Below the taxon subfamily, humans and chimpanzees are placed together in their own tribe, a rarely used classification category.

Cladistics and Cladograms

Cladistics is a way to trace the evolutionary history of a group by using shared traits derived from a common ancestor to determine relationships. These traits are then used to construct phylogenetic trees called cladograms. A **cladogram** depicts the evolutionary history of a group based on the available data.

The first step when constructing a cladogram is to draw up a table that summarizes the traits of the species being compared. At least one but preferably several species is/are considered an *outgroup*. The outgroup is not part of the study group also called the *ingroup*. In **Figure 16.17a**, lancelets are the outgroup because unlike the species in the ingroup they are not vertebrates. Any trait, such as a notochord, found in both the outgroup and the ingroup is a shared *ancestral trait* presumed to have been present in a common ancestor to both the outgroup and ingroup. Ancestral traits are not shared derived traits and therefore are not used to construct the cladogram. They merely help us determine which traits will be used to construct the cladogram.

A rule that many cladists follow is the principle of *parsimony*, which states that the least number of assumptions is the most probable. Thus, they construct a cladogram that minimizes the number of assumed evolutionary changes, or leaves the fewest number of derived traits unexplained. Therefore, any trait in the table found in scattered species (in this case, four bony limbs and a long cylindrical body) is not used to construct the cladogram because we would have to assume that these traits evolved more than once among the species of the study group. The other differences are designated as *shared derived traits*—that is they are homologies shared by only certain species of the study group. Combining the data regarding shared derived traits will tell us how the members of the ingroup are related to one another.

The Cladogram

A cladogram contains several clades; each **clade** includes a common ancestor (at the circles) and all of its descendant species. The cladogram in Figure 16.17b has three clades (1–3) that differ in size because the first includes the other two, and so forth. All the species in the study group belong to a clade that has vertebrae; only newts, snakes, and lizards are in a clade that has lungs and a three-chambered heart; and only snakes and lizards are in a clade that has an amniotic egg and internal fertilization. (An amniotic egg has a sac that surrounds and protects the embryo—fish and amphibian eggs do not have this sac.) Following the principle of parsimony, this is the sequence in which these traits must have evolved during the evolutionary history of vertebrates. Any other arrangement of species would produce a less parsimonious evolutionary sequence—that is, a tree that would be more complicated.

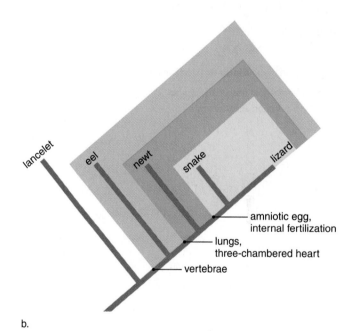

a.

b.

Figure 16.17 Constructing a cladogram.

a. First, a table is drawn up that lists characters for all the taxa. An examination of the table shows which characters are ancestral (notochord, aqua) and which are derived (lavender, orange, and yellow). The shared derived characters distinguish the taxa. **b.** In a cladogram, the shared derived characters are sequenced in the order they evolved and are used to define clades. A clade contains a common ancestor and all the species that share the same derived characters (homologies). Four bony limbs and a long cylindrical body were not used in constructing the cladogram because they are in scattered taxa.

A cladogram is objective—it lists the traits used to construct the cladogram. Cladists typically use much morphological, fossil, and molecular data to construct a cladogram. Still, cladists regard a cladogram as a hypothesis. Whether our tree is consistent with the one true evolutionary history of life can be tested, and modifications can be made on the basis of additional data.

Linnaean Classification Versus Cladistics

Figure 16.18 illustrates the types of problems that arise when trying to reconcile Linnaean classification with the principles of cladistics. Figure 16.18, which is based on cladistics, shows that birds are in a clade with crocodiles with whom they share a recent common ancestor. This ancestor had a gizzard. An examination of the skulls of crocodiles and birds would show other derived traits that they share. Birds have scaly skin and share this ancestor with other reptiles as well. Yet Linnaean classification places birds in their own group separate from crocodiles and separate from reptiles in general. In many other instances, also, Linnaean classification is not consistent with new understandings about phylogenetic relationships. Therefore some cladists have proposed a different system of classification called the International Code of Phylogenetic Nomenclature, or PhyloCode, which sets forth rules to follow in naming of clades. Other biologists are hoping to modify Linnaean classification to be consistent with the principles of cladistics.

Two major problems may be unsolvable: (1) Clades are hierarchically as are Linnaean categories. However, more clades may exist than do Linnaean taxonomic categories and it is therefore difficult to equate clades with taxons. (2) The taxons are not necessarily equivalent in the Linnaean system. To take an example, the family taxon within Kingdom Plantae may not be equivalent to the family taxon in Kingdom Animalia. Because of such problems, some cladists recommend abandoning Linnaeus altogether.

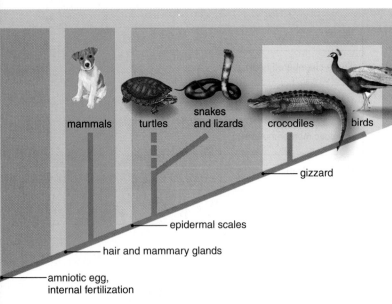

Figure 16.18 **Cladistic classification.**

Taxonomic designations are based upon evolutionary history. Each taxon includes a common ancestor and all of its descendants.

Check
Your **Progress**

1. Describe what cladists mean by a "clade."
2. Explain how cladists employ the principle of parsimony.
3. Why is Linnaean classification difficult to reconcile with the principles of cladistics?
4. If B and D are fishes, what other animal (designated by a letter) in this cladogram is also a fish and explain?

Answers: 1. A clade is a common ancestor and all its descendant groups which share the same derived traits.   2. Cladists use the minimum number of assumptions to construct the least complicated cladogram possible.   3. Linnaean classification sometimes fails to place common ancestors and all their descendants in the same group even though they all share the same derived traits.   4. C is a fish because it shares a common ancestor with B and D.

The Three-Domain System

Classification systems change over time. Historically, most biologists utilized the **five-kingdom system** of classification, which contains kingdoms for the plants, animals, fungi, protists, and monerans. Organisms were placed into these kingdoms based on type of cell (prokaryotic or eukaryotic), level of organization (unicellular or multicellular), and type of nutrition. In the five-kingdom system, the monerans were distinguished by their structure—they were prokaryotic (lack a membrane-bounded nucleus)—whereas the organisms in the other kingdoms were eukaryotic (have a membrane-bound nucleus). The kingdom Monera contained all prokaryotes, which according to the fossil record evolved first.

Sequencing the genes for rRNA provided new information that called into question the five-kingdom system of classification. Aside from molecular data, cellular data also suggests that there are two groups of prokaryotes, named the Bacteria and the Archaea. These two groups are so fundamentally different from each other they have been assigned to separate domains, a category of classification that is higher than the kingdom category. The Bacteria arose first, followed by the Archaea and then the Eukarya (**Fig. 16.19**). The Archaea and Eukarya are more closely related to each other than either is to the Bacteria. Systematists, using the **three-domain system** of classification, are in the process of sorting out what kingdoms belong within **domain Bacteria** and **domain Archaea. Domain Eukarya** contains kingdoms for protists, animals, fungi, and plants. Later in this text, we will study the individual kingdoms that occur within the domain Eukarya. The protists do not share one common ancestor, and some suggest that the kingdom should be divided into many different kingdoms. The number of kingdoms is still being determined among systematists, illustrating that classification can be changed as new data becomes available.

eukaryotes

plants
fungi
animals

Eukarya
• Eukaryotic, unicellular to multicellular organisms
• Membrane-bounded nucleus
• Sexual reproduction
• Phenotypes and nutrition are diverse
• Each kingdom has specializations
• Flagella, if present, have a 9 + 2 organization

protists

cyanobacteria

Bacteria
• Prokaryotic, unicellular organisms
• Lack a membrane-bounded nucleus
• Reproduce asexually
• Heterotrophic by absorption
• Autotrophic by chemosynthesis or by photosynthesis
• Move by flagella

heterotrophic bacteria

Archaea
• Prokaryotic, unicellular organisms
• Lack a membrane-bounded nucleus
• Reproduce asexually
• Many are autotrophic by chemosynthesis; some are heterotrophic by absorption
• Unique rRNA base sequence
• Distinctive plasma membrane and cell wall chemistry

prokaryotes

common ancestor

Figure 16.19   Three-domain system.

In this system, the prokaryotes are in the domains Bacteria and Archaea. The eukaryotes are in the domain Eukarya, which contains four kingdoms for the protists, animals, fungi, and plants.

The Chapter in Review

Summary

16.1 | Macroevolution

Macroevolution is evolution on a large scale because it considers the history of life on Earth. Macroevolution begins with speciation, the origin of new species. Without speciation and the extinction of species, life on Earth would not have a history.

Defining Species

The biological concept of species

- recognizes a species by its inability to produce viable fertile offspring with members of another group.
- is useful because species can look similar, and members of the same species can have different appearances.
- has its limitations. Hybridization does occur between some species, and it is only relevant to extant* sexually reproducing organisms.

Reproductive Barriers

Prezygotic and postzygotic barriers keep species from reproducing with one another.

- Prezygotic isolating mechanisms involve habitat isolation, temporal isolation, and behavioral isolation.
- Postzygotic isolating mechanisms prevent hybrid offspring from developing or breeding if reproduction has been successful.

Models of Speciation

Allopatric speciation and sympatric speciation are two models of speciation.

- The allopatric speciation model proposes that a geographic barrier keeps groups of populations apart. Meanwhile, prezygotic and postzygotic isolating mechanisms develop, and these prevent successful reproduction if these two groups come into future contact.
- The sympatric speciation model proposes that a geographic barrier is not required for speciation to occur.

16.2 | The Fossil Record

The fossil record, as outlined by the geological timescale, traces the history of life in broad terms. It has been possible to absolutely date fossils by using radioactive dating techniques.

The fossil record

- can be used to support a gradualistic model: Two groups of organisms arise from an ancestral species and gradually become two different species.
- also supports the punctuated equilibrium model: A period of equilibrium (no change) is interrupted by speciation within a relatively short period of time.
- shows that at least five mass extinctions, including one significant mammalian extinction, have occurred during the history of life on Earth. Two major contributors to mass extinctions are the loss of habitats due to continental drift and the disastrous results from a meteorite impacting Earth.

---

* extant = living organisms, not extinct.

16.3 | Systematics

- Systematics is the study of the evolutionary relationships among all organisms, past and present. Systematics includes classification. In the Linnaean system of classification, every organism is assigned a scientific name, which immediately indicates its genus and specific epithet. Species are also assigned to a family, order, class, phylum, kingdom, and domain according to their molecular and structural similarities as well as evolutionary relationships to other species.
- Phylogeny depicts the evolutionary history of a group of organisms. Systematics relies on the fossil record, homology, and molecular data to determine relationships among organisms.
- Cladists use shared derived characters to construct cladograms. In a cladogram, a clade consisting of a common ancestor and all the species derived from that ancestor, the species have shared derived characteristics.
- Linnaean classification has come under severe criticism because it does not always follow the principles of cladistics in the grouping of organisms.

Classification Systems

- The **three-domain system** uses molecular data to designate three evolutionary domains: Bacteria, Archaea, and Eukarya:

Three-Domain System

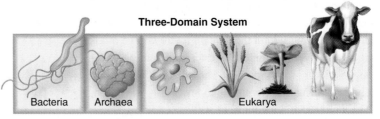

Bacteria   Archaea   Eukarya

- Domains Bacteria and Archaea contain prokaryotes.
- Domain Eukarya contains kingdoms for the protists, animals, fungi, and plants.

Key Terms

adaptive radiation   253
allopatric speciation   252
analogous structure   259
analogy   259
clade   261
cladistics   261
cladogram   261
class   258
common ancestor   249
convergent evolution   259
domain   258
domain Archaea   263
domain Bacteria   263
domain Eukarya   263
evolutionary tree   255
family   258
five-kingdom system   263
fossil   255
genus   258

homologous structure   259
kingdom   258
macroevolution   248
mass extinction   256
order   258
paleontology   255
phylogenetic tree   259
phylogeny   258
phylum   258
postzygotic isolating
   mechanism   251
prezygotic isolating mechanism   250
speciation   248
species   258
sympatric speciation   252
systematics   258
taxon (pl., taxa)   258
taxonomy   258
three-domain system   263

Testing Yourself

Choose the best answer for each question.

1. A biological species
   a. always looks different from other species.
   b. always has a different chromosome number from that of other species.
   c. is reproductively isolated from other species.
   d. never occupies the same niche as other species.

For questions 2–9, indicate the type of isolating mechanism described in each scenario.

Key:

   a. habitat isolation
   b. temporal isolation
   c. behavioral isolation
   d. mechanical isolation
   e. gamete isolation
   f. zygote mortality
   g. hybrid sterility
   h. low F_2 fitness

2. Males of one species do not recognize the courtship behaviors of females of another species.
3. One species reproduces at a different time than another species.
4. A cross between two species produces a zygote that always dies.
5. Two species do not interbreed because they occupy different areas.
6. A hybrid between two species produces gametes that are not viable.
7. Two species of plants do not hybridize because they are visited by different pollinators.
8. The sperm of one species cannot survive in the reproductive tract of another species.
9. The offspring of two hybrid individuals exhibit poor vigor.
10. Complete the following diagram illustrating allopatric speciation by using these phrases: genetic changes (used twice), geographic barrier, species 1, species 2, species 3.

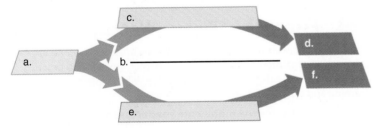

11. Transitional links are least likely to be found if evolution proceeds according to the
    a. gradualistic model.
    b. punctuated equilibrium model.
12. Which is the scientific name of an organism?
    a. *Rosa rugosa*
    b. *Rosa*
    c. *rugosa*
    d. *rugosa rugosa*
    e. Both a and d are correct.
13. Which of these statements best pertains to taxonomy? Species
    a. always have three-part names, such as *Homo sapiens sapiens*.
    b. are always reproductively isolated from other species.
    c. share ancestral traits, but may have their own unique derived traits.
    d. always look exactly alike.
    e. Both c and d are correct.

14. Which of the following groups are domains? Choose more than one answer if correct.
    a. bacteria
    b. archaea
    c. eukarya
    d. animals
    e. plants
15. Which pair is mismatched?
    a. homology—character similarity due to a common ancestor
    b. molecular data—DNA strands match
    c. fossil record—bones and teeth
    d. homology—functions always differ
    e. molecular data—DNA and RNA data
16. One benefit of the fossil record is
    a. that hard parts are more likely to fossilize.
    b. that fossils can be dated.
    c. its completeness.
    d. that fossils congregate in one place.
    e. All of these are correct.
17. The discovery of common ancestors in the fossil record, the presence of homologies, and molecular data similarities help scientists determine
    a. how to classify organisms.
    b. the proper cladogram.
    c. how to construct phylogenetic trees.
    d. how evolution occurred.
    e. All of these are correct.
18. In cladistics,
    a. a clade must contain the common ancestor plus all its descendants.
    b. shared derived characters help construct cladograms.
    c. the principle of parsimony states that the simpler hypothesis is preferred.
    d. the species in a clade share homologous structures.
    e. All of these are correct.
19. Answer these questions about the following cladogram.
    a. This cladogram contains how many clades? How are they designated in the diagram?
    b. What character is shared by all animals in the study group? What characters are shared by only snakes and lizards?
    c. Which animals share the most recent common ancestor? How do you know?

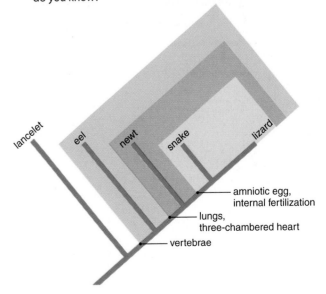

20. Allopatric, but not sympatric, speciation requires
    a. reproductive isolation.
    b. geographic isolation.
    c. spontaneous differences in males and females.
    d. prior hybridization.
    e. rapid rate of mutation.
21. Which kingdom is mismatched?
    a. Protista—domain Bacteria
    b. Protista—single-celled algae
    c. Plantae—flowers and mosses
    d. Animalia—arthropods and humans
    e. Fungi—molds and mushrooms
22. Many new species evolving in various environments from a common ancestor is called
    a. cladistics.
    b. the gradualistic model of evolution.
    c. adaptive radiation.
    d. convergent evolution.
    e. the PhyloCode.

Thinking Scientifically

1. Using as many terms as necessary (from both X and Y axes), fill in the proposed phylogenetic tree for vascular plants.

	Ferns	Conifers	Ginkgos	Monocots	Eudicots
vascular tissue	X	X	X	X	X
produce seeds		X	X	X	X
naked seeds		X	X		
needle-like leaves		X			
fan-shaped leaves			X		
enclosed seeds				X	X
one embryonic leaf				X	
two embryonic leaves					X

vascular tissue

2. The Hawaiian Islands are located thousands of kilometers from any mainland. Each island arose from the sea bottom and was colonized by plants and animals that drifted in on ocean currents or winds. Each island has a unique environment in which its inhabitants have evolved. Consequently, most of the plant and animal species on the islands do not exist anywhere else in the world.

   In contrast, on the islands of the Florida Keys, there are no unique or indigenous species. All of the species on those islands also exist on the mainland. Suggest an explanation for these two different patterns of speciation.

Bioethical Issue

Funding for Phylogenies

Reconstructing evolutionary relationships can have its benefits. For example, an emerging virus is apt to jump to a related species rather than to an unrelated species. We now know, for example, that the HIV virus jumped from the chimpanzee—with whom we share 95% of our DNA sequence—to humans. Chimpanzees don't become ill from the virus; therefore, studying their immune system might help us develop strategies to combat AIDS in humans.

   Cladists want to acquire more information about how genetic and environmental factors influence bone development. Recently, cladograms based on bone shape and arrangement do not match up well with those based on DNA base-pair sequencing. If we knew more about bone development under different environmental conditions, we might be able to discover the reason why they don't match and develop evolutionary (phylogenetic) trees in which all have confidence. To acquire the necessary data could cost millions of dollars of research funding.

   Should the public be willing to fund all types of research or only research that has immediate medical benefits, as was described for HIV research? Would you be willing to fund research that helps us understand our evolutionary past, even if the medical benefit is not immediately known?

Essentials of Biology Website

The companion website for *Essentials of Biology* provides a wealth of information organized and integrated by chapter. You will find practice tests, animations, videos, and much more that will complement your learning and understanding of general biology.

http://www.mhhe.com/maderessentials2

17

The First Forms of Life

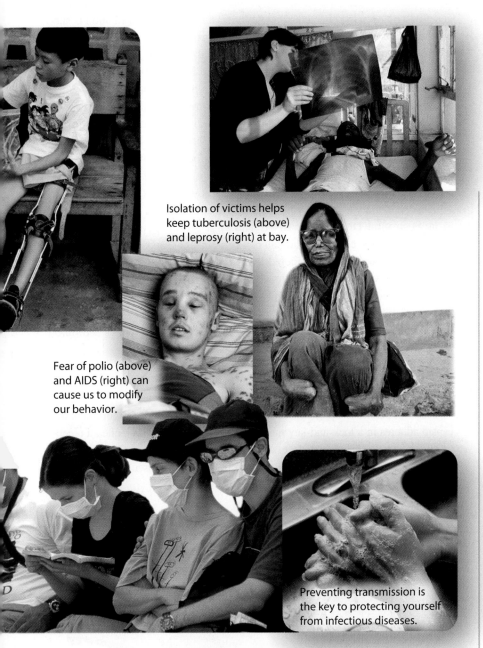

Isolation of victims helps keep tuberculosis (above) and leprosy (right) at bay.

Fear of polio (above) and AIDS (right) can cause us to modify our behavior.

Preventing transmission is the key to protecting yourself from infectious diseases.

Outline

17.1 | The Viruses
- All viruses have an outer capsid composed of protein and an inner core of nucleic acid. Some animal viruses have an outer membranous envelope. 268
- Viruses reproduce in plant and animal cells much the same way as they do in bacteria. 268
- Emerging viral diseases, including those caused by retroviruses, are of special concern today. 270

17.2 | Viroids and Prions
- Viroids that attack crops, and prions that attack animals, are less complex than viruses. 272

17.3 | The Prokaryotes
- Prokaryotes—the bacteria and archaea—lack a nucleus and most of the other cytoplasmic organelles found in eukaryotic cells. 272
- The first cells were prokaryotes. 273
- Prokaryotes reproduce asexually by binary fission. 274
- Most bacteria are free-living heterotrophs that play important roles in the environment, food science, and biotechnology. Some bacteria are disease-causing parasites. 276
- Biochemical characteristics distinguish archaea from bacteria and eukaryotes. 278
- Many archaea are quite specialized and are well known for living in extreme habitats. 278

17.4 | The Protists
- Endosymbiosis played a role in the origin of the eukaryotic cell. 279
- The protists are largely unicellular, but are quite varied in structure and life cycle. 279
- This text groups the protists according to their modes of nutrition. 280

Disease-Causing Microbes

AIDS, POLIO, TUBERCULOSIS, gonorrhea, SARS, leprosy, malaria. *Microbes,* named for their microscopic size, cause these diseases. Disease-causing microbes, discussed in this chapter, include tiny viruses, some bacteria, and certain protists. Although microscopic size unifies the microbes, their structures differ. Viruses are noncellular, bacteria are prokaryotes, and protists are eukaryotes. Viruses always cause an infection, but only some bacteria and a few protists do. Among the protists, algae rarely cause disease because they usually make their

—Continued next page

Continued—

own food. The parasitic way of life allows the disease-causing bacteria and protozoans (protists) to get their food and reproduce. Viruses don't need food—after entering a cell, they cause the cell to make hundreds of viruses at one time.

We didn't always know that microbes cause disease. The germ theory of disease didn't take hold until the twentieth century. After that, rapid progress was made in identifying which microbes cause which diseases and how they are transmitted. For example, we now know that AIDS and gonorrhea are sexually transmitted, tuberculosis and SARS are contracted by way of the respiratory tract, and the *Anopheles* mosquito spreads malaria. Each microbe lives in a different part of the body: HIV replicates in just certain types of white blood cells, tuberculosis usually centers in the lungs, and the protozoan that causes malaria reproduces in red blood cells.

This chapter provides basic information about microbes and the various ways they reproduce. Such knowledge enables us not only to fight the diseases microbes sometimes cause, but also to take advantage of the many benefits they provide.

17.1 | The Viruses

The study of **viruses** has contributed much to our understanding of disease, genetics, and even the characteristics of living things. Their contribution is surprising because viruses are not included in the classification of organisms. They are noncellular, while organisms are always cellular. Viruses can be amazingly small; when 0.2 microns, a virus is only about one-fifth the size of a bacterium.

Each type of virus always has at least two parts: an outer **capsid,** composed of protein subunits, and an inner core of nucleic acid—either DNA or RNA. The adenovirus shown in **Figure 17.1** also has spikes (formed from a glycoprotein) that are involved in attaching the virus to the host cell. In some animal viruses, the capsid is surrounded by an outer membranous envelope with glycoprotein spikes. The envelope is actually a piece of the host's plasma membrane that also contains proteins produced by the virus. The interior of a virus can contain various enzymes such as the polymerases, which are needed to produce viral DNA and/or RNA. The viral genome has at most several hundred genes; by contrast, a human cell contains thousands of genes.

Should viruses be considered living? Both scientific and philosophical debates have raged with regard to this question. Viruses are *obligate intracellular parasites* because they can only reproduce inside a living cell. Outside a living cell, viruses can be stored as chemicals or even synthesized in the laboratory from chemicals. Still, viruses do have a genome that mutates and functions to direct their reproduction when inside a cell.

Viral Reproduction

Viruses are specific to a particular host cell because a spike or some portion of the capsid adheres in a lock-and-key manner to a specific molecule (called a receptor) on the host cell's outer surface. A virus cannot infect a host cell to which it is unable to attach. For example, the tobacco mosaic virus (discussed on p. 270)

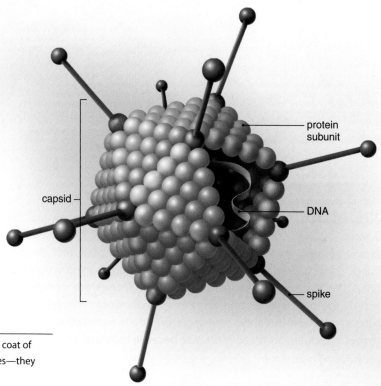

Figure 17.1 **Adenovirus anatomy.**

Typical of viruses, adenoviruses have a nucleic acid core and a coat of protein, the so-called capsid. Note the projections called spikes—they help a virus enter a cell.

cannot infect an exposed human because its capsid cannot attach to the receptors on the surfaces of human cells. Once inside a host cell, the viral genome *takes over the metabolic machinery of the host cell.* In large measure, the virus uses the host's enzymes, ribosomes, transfer RNA (tRNA), and ATP for its reproduction.

Reproduction of Bacteriophages

A **bacteriophage,** or simply *phage,* is a virus that reproduces in a bacterium. Bacteriophages are named in different ways; the one shown in **Figure 17.2** is called lambda (λ). When phage λ reproduces, it can undergo the lytic cycle or the lysogenic cycle. The **lytic cycle** (**fig. 17.3***a*) may be divided into five stages: attachment, penetration, biosynthesis, maturation, and release. **1** During *attachment,* the capsid combines with a receptor in the bacterial cell wall. **2** During *penetration,* a viral enzyme digests away part of the cell wall, and viral DNA is injected into the bacterial cell. **3** *Biosynthesis* of viral components begins after the virus inactivates host genes not necessary to viral replication. The machinery of the host cell then carries out viral DNA replication and production of multiple copies of the capsid protein subunits. **4** During *maturation,* viral DNA and capsids assemble to produce several hundred viral particles. Lysozyme, an enzyme coded for by a viral gene, disrupts the cell wall, and **5** *release* of phage particles occurs. The bacterial cell dies as a result.

In the **lysogenic cycle** (fig. 17.3*b*), the infected bacterium does not immediately produce phage, but may do so sometime in the future. In the meantime, the phage is *latent*—not actively reproducing. Following attachment and penetration, *integration* occurs: Viral DNA becomes incorporated into bacterial DNA with no destruction of host DNA. While latent, the viral DNA is called a *prophage.* The prophage is replicated along with the host DNA, and all subsequent cells, called lysogenic cells, carry a copy of the prophage. Certain environmental factors, such as ultraviolet radiation, can induce the prophage to enter the lytic stage of biosynthesis, followed by maturation and release.

Figure 17.2 Bacteriophage lambda(λ).

The micrograph shows many viral particles attached to a bacterium and the blow-up shows how DNA from a virus enters the bacterium.

5 **Release:** New viruses leave host cell.

a. Lytic Cycle

4 **Maturation:** Assembly of viral components.

3 **Biosynthesis:** Viral components are synthesized.

1 **Attachment:** Capsid combines with receptor.

2 **Penetration:** Viral DNA enters host.

Integration

b. Lysogenic Cycle

prophage

Cloning of viral DNA

Figure 17.3 Lytic and lysogenic cycles of a virus called lambda.

a. In the lytic cycle, viral particles escape when the cell is lysed (broken open). **b.** In the lysogenic cycle, viral DNA is integrated into host DNA. At some time in the future, the lysogenic cycle can be followed by the last three steps of the lytic cycle.

Figure 17.4 Infected tobacco plant.

Mottling and a distorted leaf shape are typical of a viral infection in plants.

Plant Viruses

Crops and garden plants are also subject to viral infections. Plant viruses tend to enter through damaged tissues and then move about the plant through *plasmodesmata,* cytoplasmic strands that extend between plant cell walls. The best-studied plant virus is tobacco mosaic virus, a long, rod-shaped virus with only one type of protein subunit in its capsid (**Fig. 17.4**). Not all viruses are deadly, but over time, they often debilitate a plant.

Viruses are passed from one plant to another by insects, and by gardening tools, which move sap from one plant to another. Viral particles are also transmitted by way of seeds and pollen. Unfortunately, no chemical can control viral diseases. Until recently, the only way to control viral diseases was to destroy symptomatic plants and to control the insect vector, if there is one. Now that bioengineering is routine, it is possible to transfer genes conferring disease resistance between plants. One of the most successful examples to date is the creation of papaya plants resistant to papaya ring spot virus (PRSV) in Hawaii. One transgenic line is now completely resistant to PRSV.

Animal Viruses

Viruses that cause diseases in animals, including humans, reproduce in a manner similar to that of bacteriophages (**Fig. 17.5**). However, there are modifications. In particular, some, but not all, animal viruses have an outer membranous envelope beyond their capsid. After attachment to a receptor in the plasma membrane, viruses with an envelope either fuse with the plasma membrane or enter by endocytosis. After an enveloped virus enters, uncoating follows—that is, the capsid is removed by enzymes within the host cell (see **1** to **3**). Once the viral genome, either DNA or RNA, is free of its covering, biosynthesis plus the other steps then proceed (see **7** to **9**). While a naked animal virus exits the host cell in the same manner as does a bacteriophage, those with an envelope bud from the cell. During budding, the virus picks up its envelope, consisting mainly of lipids and proteins, from the host plasma membrane. Spikes, those portions of the envelope that allow the virus to enter a host cell, are coded for by viral genes.

The herpesviruses, which cause cold sores, genital herpes, and chickenpox in humans, are examples of infections that remain latent much of the time. Herpesviruses linger in spinal ganglia until stress, excessive sunlight, or some other stimulus causes them to undergo the lytic cycle. HIV (human immunodeficiency virus), the cause of AIDS, remains relatively latent in lymphocytes, slowly releasing new viruses.

Retroviruses

Retroviruses are RNA animal viruses that have a DNA stage. Figure 17.5 illustrates the reproduction of HIV, the retrovirus that causes AIDS. A retrovirus contains a special enzyme called reverse transcriptase, which carries out transcription of RNA to

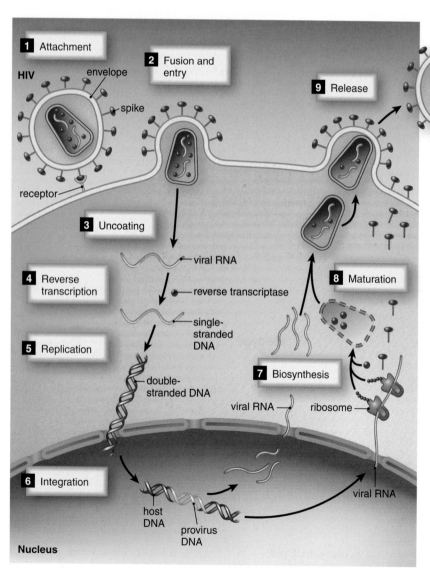

Figure 17.5 Reproduction of HIV.

HIV, the virus that causes AIDS, goes through the steps noted in the boxes. Because HIV is a retrovirus with an RNA genome, the enzyme reverse transcriptase is utilized to produce a single-stranded DNA copy of the genome. After synthesis of a complementary strand, the double-stranded viral DNA integrates into the host DNA.

DNA. The enzyme synthesizes one strand of DNA using the viral RNA as a template, and another DNA strand that is complementary to the first one. Using host enzymes, the resulting double-stranded DNA is integrated into the host genome. The viral DNA (or *provirus*) remains in the host genome and is replicated when host DNA is replicated (see **4** to **6**). When and if this DNA is transcribed, new viruses are produced by the steps we have already cited: biosynthesis, maturation, and release. Being an animal virus with an envelope, HIV buds from the cell.

Figure 17.6   Emerging diseases.

Emerging diseases, such as those noted here according to their country of origin, are new or demonstrate increased prevalence. These disease-causing agents may have acquired new virulence factors, or environmental factors may have encouraged their spread to an increased number of hosts. Avian influenza (bird flu) is a new emerging disease.

Emerging Viruses

HIV is an **emerging virus,** the causative agent of a disease that only recently has infected large numbers of people. Other examples of emerging viruses are West Nile virus, SARS virus, hantavirus, Ebola virus, and avian influenza (bird flu) virus (**Fig. 17.6**).

Infectious diseases emerge in several different ways. In some cases, the virus is simply transported from one location to another. The West Nile virus is making headlines because it changed its range: It was transported into the United States and is taking hold in bird and mosquito populations. Severe acute respiratory syndrome (SARS) was clearly transported from Asia to Toronto, Canada. A world in which you can begin your day in Bangkok and end it in Los Angeles is a world in which disease can spread at an unprecedented rate.

Other factors can also cause infectious viruses to emerge. Viruses are well known for their high mutation rates. Some of these mutations affect the structure of the spikes or capsids, so a virus that previously could only infect a particular animal species can now also infect the human species. For example, the diseases AIDS and Ebola fever are caused by viruses that at one time infected only monkeys and apes. SARS is another example of a mutant virus that most likely jumped species. A related virus was isolated from the palm civet, a catlike carnivore sold for food in China. Wild ducks are resistant to avian influenza viruses that can spread from them to chickens, which increases the likelihood the disease, often called bird flu, will spread to humans. Another possible way a virus could emerge is by a change in the mode of transmission. What would happen if a little-known virus were suddenly able to be transmitted like the common cold?

Drug Control of Human Viral Diseases

Because viruses reproduce using the metabolic machinery of the cell, it has been difficult to develop antiviral drugs. If a patient has not been vaccinated against the agent, he or she is usually told to simply let the virus run its course. However, some antiviral drugs are available. Most antiviral compounds, such as ribavirin and acyclovir, are structurally similar to nucleotides, and therefore they interfere with viral genome synthesis. Compounds related to acyclovir are commonly used to suppress herpes outbreaks. HIV is treated with antiviral compounds specific to a retrovirus. The well-publicized drug AZT and others block reverse transcriptase. And HIV protease inhibitors block enzymes required for the maturation of viral proteins.

Check Your Progress

1. **All viruses contain what two components?**
2. **Contrast a lytic with a lysogenic viral life cycle.**
3. **What happens when animal viruses bud from the host cell?**
4. **What is a retrovirus?**

Answers: 1. All viruses are composed of an outer capsid made of protein and an inner core of DNA or RNA.   2. In the lytic cycle, the bacterium is used to immediately produce more virus particles. In the lysogenic cycle, the phage is latent, but it may at some point be stimulated to enter the lytic cycle.   3. They acquire an outer envelope, which in part consists of the host plasma membrane.   4. An animal virus that uses the enzyme reverse transcriptase to make a DNA copy of its RNA genome.

Figure 17.7 Kuru.

Kuru is a fatal prion disease that infected the Fore (pronounced foray), a tribe of the remote highlands of Papua New Guinea, prior to the 1950s. Following their death, family members were ritualistically cooked and eaten, with the closest female relatives and children usually consuming the brain, the organ most likely to pass on kuru. A misshapen protein is the causative agent of all prion diseases, whether they occur in animals (e.g., mad cow disease) or in humans (e.g., kuru, CJD). The inset shows a diseased monkey brain.

17.2 | Viroids and Prions

About a dozen crop diseases have been attributed not to viruses but to **viroids,** which are naked strands of RNA (not covered by a capsid). Like viruses, though, viroids direct the cell to produce more viroids.

Some diseases in humans have been attributed to **prions,** a term coined for *proteinaceous infectious* particles. The discovery of prions began when it was observed that members of a primitive tribe in the highlands of Papua New Guinea died from a disease called kuru (meaning trembling with fear) after participating in the cannibalistic practice of eating a deceased person's brain (**Fig. 17.7**). The causative agent was smaller than a virus—it was a misshapen protein. It appears that a normal protein changes shape so that its polypeptide chain is in a different configuration. The result is a fatal prion infection and a neurodegenerative disorder.

It is believed that a prion can interact with normal proteins to turn them over to the "dark side," but the mechanism is unclear. The process has been best studied in a disease called scrapie, which attacks sheep. Other prion diseases include the widely publicized mad cow disease, human maladies such as Creutzfeldt-Jakob syndrome (CJD), and a variety of chronic wasting syndromes in several other animal species.

Check Your **Progress**

1. Contrast a viroid with a prion.
2. Prions cause disease. Why then can proteins that become prions be found in healthy brains?

Answers: 1. A viroid is composed of naked RNA, while a prion is composed of protein. 2. Prions are derived from normal proteins that have become prions due to a change in shape.

17.3 | The Prokaryotes

What was the first cellular life on Earth like? What features in modern organisms are the most primitive? The answers to these questions will probably come from a study of prokaryotes. These cells harbor no nucleus to contain their genome. They have no sealed compartments and no membrane-bounded organelles to perform specific functions. Yet these structurally simple, unicellular organisms are endowed with far greater metabolic capabilities than more structurally complex organisms; some of them need neither air nor organic matter to survive. The two types of **prokaryotes** are the bacteria and the archaea. First, we discuss the origin of cells, then examine the bacteria, the best-known prokaryotes, followed by the archaea.

Until the 19th century, many thought that prokaryotes could arise spontaneously. But in 1850, Louis Pasteur showed that previously sterilized broth cannot become cloudy with growth unless it is exposed directly to the air, where bacteria are abundant. The cell theory formulated about this time states that all organisms are composed of cells and cells only come from preexisting cells. How, then, did the first cells arise? Much work has gone into studying this question, which pertains to the origin of life, because all living organisms are composed of cells.

The Origin of Cells

The first living cells were prokaryotes, possessing DNA but lacking a nucleus. Fossilized prokaryotes have been found in rocks that are 3.5 billion years old; scientists think that prokaryotes are likely to have existed for millions of years prior to these most ancient fossils. The first cells were preceded by biological macromolecules, such as proteins and nucleic acids. Today, amino acids, nucleotides, and other building blocks for biological macromolecules are routinely produced by living cells; this is known as *biotic synthesis*. Prior to cellular life, macromolecules must have formed by *abiotic synthesis*.

Conditions on the early Earth were very different from today. Initially, temperatures were very high. Although there was little free oxygen (O_2) in the atmosphere, there would have been an abundance of other gases such as water vapor (H_2O), carbon dioxide (CO_2), and nitrogen (N_2), along with lesser amounts of hydrogen (H_2), methane (CH_4), ammonia (NH_3), hydrogen sulfide (H_2S), and carbon monoxide (CO). As the early Earth cooled, water vapor condensed to liquid water, and rain began to fall, producing the oceans. The abiotic synthesis of organic monomers under these special conditions may have occurred with the input of energy from a variety of possible sources, including lightning, sunlight, meteorite impacts, volcanic activity, or radioactive decay (**Fig. 17.8**). The monomers polymerized to form macromolecules, perhaps enabled by inorganic catalysts such as iron and zinc.

Protocells, cell-like structures complete with an outer membrane, may have resulted from the self-assembly of macromolecules, and eventually given rise to cellular life. In fact, researchers have studied cell-like structures that arise from collections of biological macromolecules under laboratory conditions. Some are water-filled spheres with an outer layer similar to that of a cell. If enzymes are trapped inside a sphere, they can catalyze chemical reactions like those of cellular metabolism.

An important development in the origin of life would have been a molecule capable of passing information about metabolism and structure from one generation to the next, the role fulfilled by DNA in today's cells. Some researchers think the first hereditary molecule may have been RNA. It is hypothesized that, over time, the more stable DNA replaced RNA as a long-term repository for genetic information, leading to the self-replicating system seen in all living cells today.

Bacteria

Bacteria are the most diverse and prevalent organisms on Earth. Billions of bacteria exist in nearly every square meter of soil, water, and air. They also make a home on your skin and in your intestines. Although tens of thousands of different bacteria have been identified, this is likely only a very small fraction of living bacteria. Less than 1% of bacteria in the soil can be grown in the laboratory. Molecular genetic techniques are now being used to discover the extent of bacterial diversity.

Figure 17.8   Origin of the first cell(s).

There was an increase in the complexity of macromolecules, leading to a self-replicating system (DNA → RNA → protein) enclosed by a plasma membrane. The protocell underwent biological evolution becoming a true cell.

Check Your Progress

1. Since the first cells arose in a world without free oxygen, what could you conclude about their oxygen-using capability?
2. What is a protocell? Would the components of protocells have formed by abiotic or biotic synthesis?

Answers: 1. They were anaerobes and did not use oxygen.   2. A cell-like structure with an outer membrane, most likely formed through the abiotic synthesis of component macromolecules.

a. Sphere-shaped streptococci   ×8,000

b. Rod-shaped
E. coli
×10,500

c. Spirochete, T. pallidum

Figure 17.9   Shapes of bacteria.

a. Streptococci, which exist as chains of cocci, cause a number of illnesses, including strep throat.
b. *Escherichia coli*, which lives in your intestine, is a rod-shaped bacillus. **c.** *Treponema pallidum*, the cause of syphilis , is a spirochete.

plasma membrane

cell wall

capsule

hook

filament

basal body

Figure 17.10   Flagella.

Each flagellum of a bacterium contains a basal body, a hook, and a filament. The red-dashed arrows indicate that the hook (and filament) rotates 360°. Flagella rotating counterclockwise result in forward motion of the bacterium; a clockwise rotation causes the bacterium to move backward.

General Biology of Bacteria

Bacterial Structure   Bacteria have a number of different shapes; however, most bacteria are spheres, called cocci, or rods, called bacilli, or spirals, called spirilla (sing., spirillum) if they are rigid or spirochetes if they are flexible (**Fig. 17.9**). Intermediate forms between a sphere and rod are coccobacilli; a slightly curved rod is called a vibrio. Many bacteria grow as single cells, but some form doublets (the diplococci and diplobacilli). Others form chains as do the streptococci, the cause of strep throat. A third growth habit resembles a bunch of grapes, as in the staphylococci, a cause of food poisoning.

Figure 4.4 on page 55 reviews the structure of a bacterium. A bacterium, being a prokaryote, has no nucleus. A single, closed circle of double-stranded DNA constitutes the chromosome, which occurs in a limited area of the cell called the **nucleoid.** In some cases, extrachromosomal DNA molecules called **plasmids** are also found in bacteria.

Bacteria have ribosomes but not membrane-bounded organelles such as mitochondria and chloroplasts. Those that are photosynthetic have thylakoid membranes, but these are not enclosed by another membrane. Motile bacteria generally use **flagella** for locomotion, but never cilia. The bacterial flagellum is not structured like a eukaryotic flagellum (**Fig. 17.10**). It has a filament composed of three strands of the protein flagellin wound in a helix. The filament is inserted into a hook that is anchored by a basal body. The fully reversible 360° rotation causes the bacterium to spin as it moves forward and backward.

Bacteria have an outer cell wall strengthened not by cellulose, but by **peptidoglycan,** a complex of polysaccharides linked by amino acids. The cell wall prevents bacteria from bursting or collapsing due to osmotic changes. Parasitic bacteria are further protected from host defenses by a polysaccharide capsule that surrounds the cell wall.

Bacterial Reproduction   Bacteria (and archaea) reproduce asexually by means of **binary fission.** The single, circular chromosome replicates, and then the two copies separate as the cell enlarges. The newly formed plasma membrane and cell wall partition the two new cells, with a chromosome in each one (**Fig. 17.11***a, b*). Mitosis, which requires the formation of a spindle apparatus, does not occur in prokaryotes. Binary fission turns one cell into two cells, two cells into four cells, four cells into eight cells, and so on until billions of cells have been produced.

In eukaryotes, genetic recombination occurs as a result of sexual reproduction. Sexual reproduction does not occur among prokaryotes, but three means of genetic recombination have been observed in bacteria. **Conjugation** occurs when a donor cell passes DNA directly to a recipient cell. During conjugation, donor and recipient are temporarily linked together, often by means of a *conjugation pilus*. While they are linked, the donor cell passes DNA to the recipient cell. **Transformation** occurs when a bacterium picks up (from the surroundings) free pieces of DNA secreted by live prokaryotes or released by dead prokaryotes. During **transduction,** bacteriophages carry portions of bacterial DNA from one cell to another. Plasmids, which sometimes carry genes for resistance to antibiotics, can be transferred between infectious bacteria by any of these means.

When faced with unfavorable environmental conditions, some bacteria form **endospores** (Fig. 17.11*c*). A portion of the cytoplasm and a copy of the chromosome dehydrate and are then encased by three heavy, protective spore coats. The rest of the bacterial cell deteriorates, and the endospore is released. Spores survive in the harshest of environments—desert heat and desiccation, boiling temperatures, polar ice, and extreme ultraviolet radiation. They also survive for very long periods. When anthrax spores 1,300 years old germinate, they can still cause a severe infection (usually seen in cattle and sheep). Humans also fear a deadly but uncommon type of food poisoning called botulism, which is caused by the germination of endospores inside cans of food. Spore formation is not a means of reproduction, but it does allow survival and dispersal of bacteria to new places.

Bacterial Nutrition   Prokaryotes are more metabolically diverse than eukaryotes. For example, plants are **photoautotrophs** (many times called autotrophs or photosynthesizers) that can only perform oxygenic photosynthesis: They depend on solar energy to split water and energize electrons for the reduction of carbon dioxide. Among bacteria, the **cyanobacteria** are also photoautotrophs and do the same. The cyanobacteria may well represent the oldest lineage of oxygenic organisms (**Fig. 17.12***a*). Some fossil cyanobacteria have been dated at 3.5 billion years old. The evolution of cyanobacteria drastically altered the atmosphere of the early Earth by adding vast amounts of oxygen. Many cyanobacteria are capable of fixing atmospheric nitrogen and reducing it to an organic form. Therefore, they need only minerals, air, sunlight, and water for growth.

Other bacterial photosynthesizers don't release oxygen because they take electrons from a source other than water; some split hydrogen sulfide (H_2S) and release sulfur (S) in marshes where they live anaerobically.

The **chemoautotrophs** (many times called chemosynthesizers) don't use solar energy at all. They reduce carbon dioxide using energetic electrons derived from inorganic molecules such as ammonia or hydrogen gas. Electrons can also be extracted from certain minerals, such as iron. Some chemoautotrophs oxidize sulfur compounds spewing from deep-sea vents 2.5 kilometers below sea level. The organic compounds they produce support the growth of communities of organisms found at vents, where only darkness prevails (Fig. 17.12*b*).

Most bacteria are **chemoheterotrophs** (often referred to as simply heterotrophs) and, like animals, take in organic nutrients that they use as a source of energy and building blocks to synthesize macromolecules. The first cells were most likely chemoheterotrophs that fed upon the abundant organic molecules in their environment. (Autotrophs would have appeared later as the nutrient supply

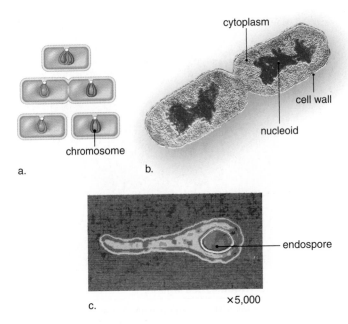

a.   b.

Figure 17.11   Bacterial reproduction and survival.

a, b. When conditions are favorable to growth, prokaryotes divide to multiply. **c.** Formation of endospores allows bacteria to survive unfavorable environmental conditions.

a. Cyanobacterium

b. Hosts for chemoautotrophic bacteria

Figure 17.12   Bacterial autotrophs.

a. Cyanobacteria are photoautotrophs that photosynthesize in the same manner as plants—they split water and release oxygen. **b.** Certain chemoautotrophic bacteria live inside tubeworms, where they produce organic compounds without the need of sunlight. In this way they help support ecosystems at hydrothermal vents deep in the ocean.

Figure 17.13 **Nodules of a legume.**

Although some free-living bacteria carry on nitrogen fixation, those of the genus *Rhizobium* invade the roots of legumes, with the resultant formation of nodules. Here the bacteria convert atmospheric nitrogen to an organic nitrogen that the plant can use. These are nodules on the roots of a soybean plant.

a.

b.

after   before

Figure 17.14 **Bioremediation.**

a. Bacteria have been used for many years in sewage treatment plants to break down human wastes. **b.** Increasingly, the ability of bacteria to break down pollutants is being researched and enhanced. Here technicians are relying on bacteria to clean up an oil spill.

was depleted and the ability to make one's own food became advantageous.) Unlike animals, chemoheterotrophic bacteria are **saprotrophs** that send enzymes into the environment and decompose almost any large organic molecule to smaller ones that are absorbable. There is probably no natural organic molecule that cannot be digested by at least one bacterial species. Bacteria play a critical role in recycling matter and making inorganic molecules available to photosynthesizers.

Heterotrophic bacteria may be either free-living or **symbiotic,** meaning that they form (1) mutualistic (both partners benefit), (2) commensalistic (one partner benefits, the other is not harmed), or (3) parasitic (one partner benefits, the other is harmed) relationships. Mutualistic bacteria that live in human intestines release vitamins K and B_{12}, which we can use to help produce blood components. In the stomachs of cows and goats, special mutualistic prokaryotes digest cellulose, enabling these animals to feed on grass.

Commensalism often occurs when one population modifies the environment in such a way that a second population benefits. Obligate anaerobes can live in our intestines only because the bacterium *Escherichia coli* uses up the available oxygen. The parasitic bacteria cause diseases, as discussed on page 277.

Environmental and Medical Importance of Bacteria

Bacteria in the Environment   For an ecosystem to sustain its populations, the chemical elements available to living things must eventually be recycled. A fixed and limited amount of elements are available to living things—the rest are either buried too deep in the Earth's crust or present in forms that are not usable. All living organisms, including producers, consumers, and decomposers, are involved in the important process of cycling elements to sustain life (see Fig. 1.4). *Bacteria are decomposers* that digest dead organic remains and return inorganic nutrients to producers. Without the work of decomposers, life would soon come to a halt.

While decomposing, bacteria perform reactions needed for biogeochemical cycling, such as for the carbon and nitrogen cycles. Let's examine how bacteria participate in the nitrogen cycle. Plants are unable to fix atmospheric nitrogen (N_2), but they need a source of ammonia or nitrate in order to produce proteins. Bacteria in the soil can fix atmospheric nitrogen and/or change nitrogen compounds into forms that plants can use. In addition, mutualistic bacteria live in the root nodules of soybean, clover, and alfalfa plants where they reduce atmospheric nitrogen to ammonia, which is used by plants (**Fig. 17.13**). Without the work of bacteria, nitrogen would not be available for plants to produce proteins, nor available to animals that feed on plants or other animals.

Bioremediation is the biological cleanup of an environment that contains harmful chemicals called *pollutants.* To help correct the situation, the vast ability of bacteria to break down almost any substance, including sewage, is being exploited (**Fig. 17.14***a*). People have added thousands of tons of slowly degradable pesticides and herbicides, nonbiodegradable detergents, and plastics to the environment. Strains of bacteria are being developed specifically for cleaning up these particular types of pollutants. Some strains have been used to remove Agent

Orange, a potent herbicide, from soil samples, and dual cultures of two types of bacteria have been shown to degrade PCBs, chemicals formerly used as coolants and industrial lubricants. The occasional oil spill spoils beaches and kills wildlife. The ability of bacteria to degrade petroleum has been improved by biotechnology and the addition of a growth-promoting fertilizer (Fig. 17.13*b*). Without the fertilizer, the lack of nitrogen and phosphate in seawater limits their growth.

Bacteria in Food Science and Biotechnology   A wide variety of food products are created through the action of bacteria (**Fig. 17.15**). Under anaerobic conditions, bacteria carry out fermentation, which results in a variety of acids. One of these acids is lactate, a product that pickles cucumbers, curdles milk into cheese, and gives these foods their characteristic tangy flavor. Other bacterial fermentations can produce flavor compounds, such as the propionic acid in Swiss cheese. Bacterial fermentation is also useful in the manufacture of such products as vitamins and antibiotics—in fact, most antibiotics known today were discovered in soil bacteria.

As you know, biotechnology can be used to alter the genome and the products generated by bacterial cultures. Bacteria can be genetically engineered to produce medically important products such as insulin, human growth hormone, and vaccines against a number of human diseases. The natural ability of bacteria to perform all manner of reactions is also enhanced through biotechnology.

Bacterial Diseases in Humans   Microbes that can cause disease are called **pathogens.** Pathogens are able to (1) produce a toxin, and/or (2) adhere to surfaces, and sometimes (3) invade organs or cells.

Toxins are small organic molecules, or small pieces of protein or parts of the bacterial cell wall that are released when bacteria die. Toxins are poisonous, and bacteria that produce a toxin usually cause serious diseases. In almost all cases, the growth of the microbes themselves does not cause disease; the toxins they release cause disease. When someone steps on a rusty nail, bacteria may be introduced deep into damaged tissue. The damaged area does not have good blood flow and can become anaerobic. *Clostridium tetani,* the cause of tetanus (lockjaw), proliferates under these conditions. The bacteria never leave the site of the wound, but the tetanus toxin they produce does move throughout the body. This toxin prevents the relaxation of muscles. In time, the body contorts because all the muscles have contracted. Eventually, the person suffocates.

Adhesion factors allow a pathogen to bind to certain cells, and this determines which organs or cells of the body will be its host. Like many bacteria that cause dysentery (severe diarrhea), *Shigella dysenteriae* is able to stick to the intestinal wall. In addition, *S. dysenteriae* produces a toxin called Shiga toxin that makes it even more life-threatening. Also, invasive mechanisms that give a pathogen the ability to move through tissues and into the bloodstream result in a more medically significant disease than if it were localized. Usually a person can recover from food poisoning caused by *Salmonella*. But some strains of *Salmonella* have virulence factors—including a needle-shaped apparatus that injects toxin into body cells—that allow the bacteria to penetrate the lining of the colon and move beyond this organ. Typhoid fever, a life-threatening disease, can then result.

Because bacteria are cells in their own right, a number of antibiotic compounds are active against bacteria and are widely prescribed. One problem with antibiotic therapy has been increasing bacterial resistance to antibiotics (see p. 286).

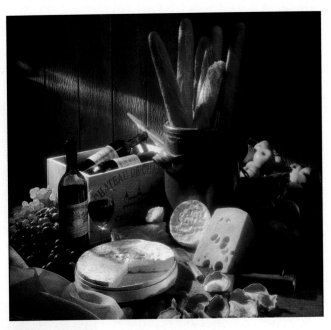

Figure 17.15   Bacteria in food processing.

Bacteria and fungi (see Chapter 18) are used to help produce food products. Fungal fermentation results in carbon dioxide and alcohol, which help produce bread and wine, respectively. Bacterial fermentation results in acids that give some types of cheeses their characteristic taste.

Check
Your Progress

1. List the three most common shapes of bacteria.
2. Describe the chromosome of a bacterium.
3. List the three ways in which genetic recombination can occur in bacteria.
4. In what ways do bacteria help and harm humans?

Answers: 1. Spheres (cocci), rods (bacilli), and spirals (spirilla or spirochetes). 2. A single, closed circle of double-stranded DNA.   3. Conjugation, transformation, and transduction.   4. Bacteria break down dead organic remains and pollutants; they help produce food and products; they also cause diseases.

Figure 17.16   Methanogen habitat and structure.

a. A swamp where methanogens live. **b.** Micrograph of *Methanosarcina mazei*, a methanogen.

Figure 17.17   Halophile habitat and structure.

a. Great Salt Lake, Utah, where halophiles live. **b.** Micrograph of *Halobacterium salinarium*, a halophile.

Figure 17.18   Thermoacidophile habitat and structure.

a. Boiling springs and geysers in Yellowstone National Park where thermoacidophiles live. **b.** Micrograph of *Sulfolobus acidocaldarius*, a thermoacidophile. (The dark central region has yet to be identified and could be an artifact of staining.)

Archaea

As previously discussed in Chapter 16, the tree of life now contains three domains: Archaea, Bacteria, and Eukarya. Because many **archaea** and some bacteria are found in extreme environments (hot springs, thermal vents, salt basins), they may have diverged from a common ancestor relatively soon after life began. Later, the eukarya are believed to have split off from the archaeal line of descent. Archaea and eukarya share some of the same ribosomal proteins (not found in bacteria), initiate transcription in the same manner, and have similar types of tRNA.

Structure

The plasma membranes of archaea contain unusual lipids that allow many of them to function at high temperatures. The archaea have also evolved diverse cell wall types, which facilitate their survival under extreme conditions. The cell walls of archaea do not contain peptidoglycan as do the cell walls of bacteria. In some archaea, the cell wall is largely composed of polysaccharides, and in others, the wall is pure protein. A few have no cell wall.

Types of Archaea

Archaea are often discussed in terms of their unique habitats. The **methanogens** (methane makers) are found in anaerobic environments, such as in swamps, marshes, and the intestinal tracts of animals. Those found in animal intestines exist as mutualists or commensals, not as parasites—that is, archaea are not known to cause infectious diseases. Methanogens are chemoautotrophs that couple the production of methane (CH_4) from hydrogen gas (H_2) and carbon dioxide to the formation of ATP (**Fig. 17.16**). This methane, which is also called *biogas*, is released into the atmosphere where it contributes to the greenhouse effect and global warming. About 65% of the methane in our atmosphere is produced by methanogenic archaea.

The **halophiles** require high salt concentrations for growth (usually 12–15%; by contrast, the ocean is about 3.5% salt). Halophiles have been isolated from highly saline environments such as the Great Salt Lake in Utah, the Dead Sea, solar salt ponds, and hypersaline soils (**Fig. 17.17**). These archaea have evolved a number of mechanisms to survive in high-salt environments. They depend on a pigment related to the rhodopsin in our eyes to absorb light energy for the purpose of pumping chloride, and another similar type pigment for the purpose of synthesizing ATP.

A third major type of archaea are the **thermoacidophiles** (**Fig. 17.18**). These archaea are isolated from extremely hot, acidic environments such as hot springs, geysers, submarine thermal vents, and around volcanoes. They reduce sulfur to sulfides and survive best at temperatures above 80°C; some can even grow at 105°C (remember that water boils at 100°C)! Metabolism of sulfides results in acidic sulfates, and these bacteria grow best at pH 1 to 2.

Check Your **Progress**

1. **What are the two domains of prokaryotes?**
2. **Which group of archaea would be most abundant in sulfur-rich hot springs? In the gastrointestinal tract of a cow?**

Answers: 1. Archaea and Bacteria.   2. Thermoacidophiles, methanogens.

17.4 | The Protists

Like ancient creatures from another planet, the **protists** inhabit the oceans and other watery environments of the world. Their morphological diversity is their most outstanding feature—unicellular diatoms are encrusted in silica "petri dishes"; dinoflagellates have plates of armor; and ciliates shaped like slippers have complex structures.

Many protists are unicellular, but all are eukaryotes with a nucleus and a wide range of organelles. We now have ample evidence that the organelles of eukaryotic cells arose from close symbiotic associations between bacteria and primitive eukaryotes (**Fig. 17.19**). This so-called **endosymbiotic theory** is supported by the presence of double membranes around mitochondria and chloroplasts. Also, these organelles have their own genomes, although incomplete, and their ribosomal genes point to bacterial origins. The mitochondria appear closely related to certain bacteria, and the chloroplasts are most closely related to cyanobacteria.

To explain the diversity of protists, we can well imagine that once the eukaryotic cell arose, it provided the opportunity for many different lineages to begin. Some unicellular protists have organelles not seen in other eukaryotes. For example, food is digested in food vacuoles, and excess water is expelled when contractile vacuoles discharge their contents.

Protists also possibly give us insight into the evolution of a multicellular organism with differentiated tissues. Some protists are a colony of single cells, with certain cells specialized to produce eggs and sperm, and others are multicellular, with tissues specialized for various purposes. Perhaps the first type of organization preceded the second in a progression toward multicellular organisms.

General Biology of Protists

The complexity and diversity of protists make it difficult to classify them. The variety of protists is so great that it's been suggested they could be split into more than a dozen kingdoms. Due to limited space, this text groups the protists according to modes of nutrition.

Traditionally, the term **algae** means aquatic photosynthesizer. At one time, botanists classified algae as plants because they contain chlorophyll *a* and carry on photosynthesis. In aquatic environments, algae are a part of the phytoplankton, photosynthesizers that lie suspended in the water. They are producers, which serve as a source of food for other organisms and pour oxygen into the environment. In terrestrial systems, algae are found in soils, on rocks, and in trees. One type of alga is a symbiote of animals called corals, which depend on them for food as they build the coral reefs of the world. Others partner with fungi in lichens capable of living in harsh terrestrial environments.

The definition of a **protozoan** as a unicellular chemoheterotroph explains why protozoans were originally classified with the animals. Often a protozoan has some form of locomotion, either by flagella, pseudopods, or cilia. In aquatic environments, protozoans are a part of the zooplankton, suspended microscopic heterotrophs that serve as a food source for animals. While most are free-living, some protozoans are human pathogens, often causing diseases of the blood or intestines. In many cases, their complex life cycles inhibit the development of suitable treatments.

There is still one other group of protists: the slime molds and water molds. These microorganisms are chemoheterotrophs, but the slime molds ingest their food in the same manner as the protozoan called an amoeba, while the water molds are saprotrophic, like fungi.

Figure 17.19   Evolution of the eukaryotic cell.

Invagination of the plasma membrane accounts for the formation of the nucleus and certain other organelles. The endosymbiotic theory states that mitochondria and chloroplasts are derived from prokaryotes that were taken up by a much larger eukaryotic cell.

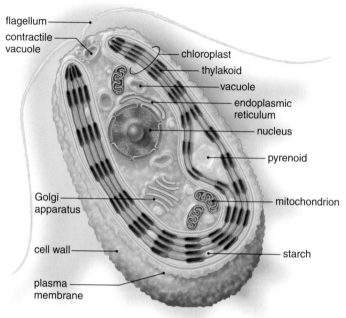

Figure 17.20 *Chlamydomonas.*

Chlamydomonas, a green alga, has the organelles and other structures typical of a motile algal cell.

Algae

The green alga *Chlamydomonas* serves as our model for algal structure (**Fig. 17.20**). The most conspicuous organelle in the algal cell is the chloroplast. Algal chloroplasts share many features with those of plants, and the two groups likely share a common origin; for example, the photosynthetic pigments are housed in thylakoid membranes. Not surprisingly, then, algae perform photosynthesis in the same manner as plants. Pyrenoids are organelles found in algae that are active in starch storage and metabolism. Vacuoles are seen in algae, along with mitochondria. Algae generally have a cell wall, and many produce a slime layer that can be harvested and used for food processing. Some algae are nonmotile, while others possess flagella.

Algae can reproduce asexually or sexually in most cases. Asexual reproduction can occur by binary fission, as in bacteria. Some proliferate by forming flagellated spores called zoospores, while others simply fragment, with each fragment becoming a progeny alga. Sexual reproduction generally requires the formation of gametes that combine to form a zygote.

One traditional way to classify algae is based on the color of the pigments in their chloroplasts: **green algae, red algae, golden-brown algae,** and **brown algae** (**Fig. 17.21**). The green algae are most closely related to plants, and they are commonly represented by three species: *Chlamydomonas*; colonial *Volvox*, a large, hollow sphere with dozens to hundreds of cells; and *Spirogyra*, a filamentous alga

Figure 17.21 **Algal diversity.**

a. *Acetabularia*, a unicellular green alga; **b.** *Bossiella*, a coralline red alga; **c.** *Fucus*, a brown alga; **d.** *Licmorpha*, a stalked diatom; **e.** *Chlamydomonas*, a unicellular green alga; **f.** *Volvox*, a colonial green alga with daughter colonies inside; **g.** *Spirogyra*, a filamentous green alga undergoing conjugation to form zygotes; **h.** *Ceratium*, an armored dinoflagellate; **i.** *Macrocystis*, a brown alga; **j.** *Sargassum*, a brown alga.

in which the chloroplasts form a green spiral ribbon. The coralline red algae deposit calcium carbonate in their cell walls and contribute to the formation of coral reefs. The golden-brown algae are represented by the diatoms, which have a petri dish structure with each half an elaborate shell composed of silica. The green algae, red algae, and brown algae include the multicellular seaweeds.

The Protozoans

Protozoans are unicellular, but their cells are very complex. Many of the functions we normally associate with the organs of multicellular organisms are performed by organelles in protozoans. Some protozoans have more than one nucleus. In some cases, the two nuclei are identical in size and function. Other protozoans, such as *Paramecium*, have a large macronucleus and a small micronucleus (**Fig. 17.22**). The macronucleus produces mRNA and directs metabolic functions. The micronucleus is important for reproduction. Protozoans usually reproduce asexually by binary fission.

Protozoans are heterotrophic, and many feed by engulfing food particles. Phagocytic vacuoles act as their "stomachs" into which digestive enzymes and acid are added. Secretory vacuoles release enzymes that may enhance any pathogenicity. Contractile vacuoles permit osmoregulation, particularly in fresh water.

The various protozoans shown in **Figure 17.23** illustrate that they are usually motile. The **ciliates,** so named because they move by cilia, are repre-

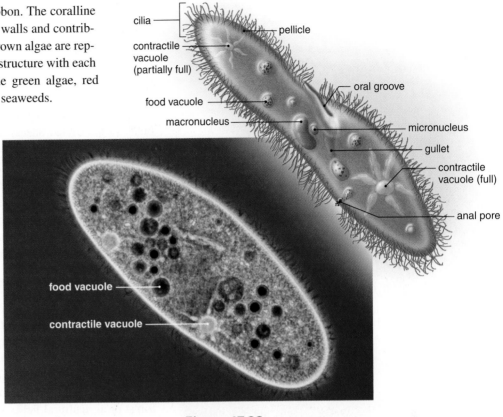

Figure 17.22 A paramecium.

A paramecium is a ciliate, a type of complex protozoan that moves by cilia.

Figure 17.23 Protozoan diversity.

a. *Entamoeba invadens,* a parasitic amoeba; **b.** *Globigerina* sp., a foraminiferan; **c.** *Plasmodium* sp., a sporozoan that causes malaria (several are seen here infecting a red blood cell); **d.** *Trypanosoma brucei,* a zooflagellate that causes African sleeping sickness; **e.** *Giardia lamblia,* a zooflagellate that causes severe diarrhea.

sented by *Paramecium*, and the **amoeboids** by amoebas, which move by cytoplasmic extensions called **pseudopods.** The **radiolarians** and **foraminiferans** are two types of marine amoeboids with calcium carbonate skeletons important in limestone formations—including the White Cliffs of Dover in Britain. The amoeba *Entamoeba histolytica* causes amoebic dysentery. The **zooflagellates,** which move by flagella, also cause diseases. In this group, a **trypanosome** is the cause of African sleeping sickness, a blood disease that cuts off circulation to the brain. In the United States, *Giardia lamblia* contaminates water supplies and causes severe diarrhea. Apicomplexans, commonly called **sporozoans** because they produce spores, are unlike other protozoan groups because they are not motile. One genus, *Plasmodium*, causes malaria, the most widespread and dangerous protozoan disease. According to the CDC (Centers for Disease Control and Prevention), there are approximately 350–500 million cases of malaria worldwide every year; more than a million people die of the infection. Malaria is transmitted by a mosquito. Toxoplasmosis, another protozoan disease, is commonly transmitted by cats.

Slime Molds and Water Molds

In forests and woodlands, **slime molds** feed on, and therefore help dispose of, dead plant material. They also feed on bacteria, keeping their population sizes under control. Usually, plasmodial slime molds exist as a plasmodium—a diploid, multinucleated, cytoplasmic mass enveloped by a slime sheath that creeps along, phagocytizing decaying plant material in a forest or agricultural field. At times unfavorable to growth, such as during a drought, the plasmodium develops many sporangia. A sporangium is a reproductive structure that produces spores resistant to dry conditions. When favorable moist conditions return, the spores germinate, releasing a flagellated cell or an amoeboid cell. Eventually, two of them fuse to form a zygote that feeds and grows, producing a multinucleated plasmodium once again. **Figure 17.24** shows the life cycle of a plasmodial slime mold.

Plasmodium phagocytizes food

Sporangia produce spores

Figure 17.24 Plasmodial slime molds.

As long as conditions are favorable, a plasmodial slime mold exists as a multinucleated diploid plasmodium that creeps along the forest floor phagocytizing organic remains. When conditions become unfavorable, sporangia form where meiosis produces haploid spores, structures that can survive unfavorable times. Spores germinate to release independent haploid cells. Fusion of these cells produces a zygote that becomes a mature plasmodium once again.

The **water molds** decompose remains but are also significant parasites of plants and animals in ecosystems. The potato blight that brought about famine in nineteenth-century Ireland was caused by a water mold. Slime molds and water molds were once classified as fungi, but unlike fungi, all have flagellated cells at some time during their life cycles. Only the water molds have a cell wall, but it contains cellulose, not the chitin of fungal cell walls. Both slime molds and water molds form spores, each a small, single-celled reproductive body capable of becoming a new organism. The spores of water molds are flagellated, but those of slime molds are windblown.

Although grouped together here, slime molds and water molds may not be very closely related to one another. Some experts suggest that water molds may be near relatives of diatoms and brown algae, while slime molds are associated with the amoeboids. Indeed, the vegetative state of the slime molds is mobile and amoeboid, and like the amoeboids, they ingest their food by phagocytosis.

Check Your **Progress**

1. List the features of mitochondria and chloroplasts that support the endosymbiotic theory.
2. Compare and contrast algae and protozoans.
3. What food sources are used by slime molds in terrestrial ecosystems?

Answers: 1. Double membranes and distinct genomes.   2. Both algae and protozoans are protists and belong to domain Eukarya. Algae are photoautotrophs; protozoans are chemoheterotrophs.   3. Decaying plant materials and bacteria.

The Chapter in Review

Summary

17.1 | The Viruses

Viruses are noncellular particles.

Structure

- Viruses have at least two parts: an outer capsid composed of protein subunits and an inner core of nucleic acid, either DNA or RNA.
- Animal viruses are either naked (no outer envelope) or they have an outer membranous envelope.

Reproduction

Viruses are obligate intracellular parasites that can reproduce only inside living cells. Bacteriophages can have a lytic or lysogenic life cycle. The **lytic cycle** consists of these steps:

- **1** Attachment
- **2** Penetration
- **3** Biosynthesis
- **4** Maturation
- **5** Release

In the **lysogenic cycle**, viral DNA is integrated into bacterial DNA for an indefinite period of time, but it can undergo the last three steps of the lytic cycle at any time.

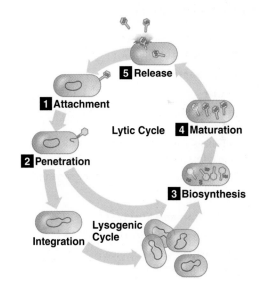

Plant Viruses

Crops and garden plants are subject to viral infections. Not all viruses are deadly, but over time they debilitate a plant.

Animal Viruses

The reproductive cycle in animal viruses has the same steps as in a bacteriophage, with modifications if the virus has an envelope, in which case:

- Fusion or endocytosis brings virus into cell.
- Uncoating is needed to free the genome from the capsid.
- Budding releases the viral particles from the cell.

HIV, the AIDS virus, is an RNA retrovirus. These viruses have an enzyme, reverse transcriptase, which carries out reverse transcription. This produces single-stranded DNA, which replicates, forming a double helix that becomes integrated into host DNA.

Drug Control of Human Viral Diseases

Antiviral drugs are structurally similar to a nucleotide and interfere with viral genome synthesis. HIV protease inhibitors block enzymes required for maturation of viral proteins.

17.2 │ Viroids and prions

Viroids are naked (not covered by a capsid) strands of RNA that can cause disease.

Prions are protein molecules that have a misshapen tertiary structure. Prions cause such diseases as CJD in humans and mad cow disease in cattle when they cause other proteins of their own kind to also become misshapen.

17.3 │ The Prokaryotes

The bacteria and archaea are prokaryotes. Prokaryotes lack a nucleus and most of the other cytoplasmic organelles found in eukaryotic cells.

The Origin of Life

The first cells were prokaryotes. Before the first cells appeared, biological macromolecules formed spontaneously in the unique conditions of the early Earth's atmosphere. Scientists have shown that collections of biological macromolecules assemble to form nonliving, cell-like structures under laboratory conditions.

General Biology of Bacteria

Bacterial Structure

- Structures include rods (bacilli), spheres (cocci), and spirals (spirilla or spirochetes).
- Single, closed circle of double-stranded DNA (chromosome) is in a nucleoid.
- Flagellum is unique and rotates, causing the organism to spin.
- Cell wall contains peptidoglycan.

Bacterial Reproduction and Survival

- Reproduction is asexual by binary fission.
- Genetic recombination occurs by means of conjugation, transformation, and transduction.
- Endospore formation allows bacteria to survive an unfavorable environment.
- Endospores are extremely resistant to destruction; the genetic material can thereby survive unfavorable conditions.

Comparison of Viruses and Prokaryotes

Characteristic of Life	Viruses	Prokaryotes
Consist of cell	No	Yes
Metabolize	No	Yes
Respond to stimuli	No	Yes
Multiply	Yes (always inside living cell)	Yes (usually independently)
Evolve	Yes	Yes

Bacterial Nutrition

- Bacteria can be autotrophic. Cyanobacteria are photoautotrophs—they photosynthesize, as do plants. Chemoautotrophs oxidize inorganic compounds, such as hydrogen gas, hydrogen sulfide, and ammonia, to acquire energy to make their own food. Chemoautotrophs are chemosynthesizers that support communities at deep-sea vents.
- Like animals, most bacteria are chemoheterotrophs (heterotrophs), but they are saprotrophic decomposers. Many heterotrophic prokaryotes are symbiotic. The mutualistic nitrogen-fixing bacteria live in nodules on the roots of legumes.

Environmental and Medical Importance of Bacteria

Bacteria in the Environment

- As decomposers, bacteria keep inorganic nutrients cycling in ecosystems.
- The reactions they perform keep the nitrogen cycle going.
- Bacteria play an important role in bioremediation.

Bacteria in Food Science and Biotechnology

- Bacterial fermentations are important in the production of foods.
- Genetic engineering allows bacteria to produce medically important products.

Bacteria Diseases in Humans

- Bacterial pathogens that can cause diseases are able to (1) produce a toxin, (2) adhere to surfaces, and (3) sometimes invade organs or cells.
- Indiscriminate antibiotic therapy has led to bacterial resistance to some antibiotics.

Archaea

The archaea (domain Archaea) are a second type of prokaryote. Some characteristics of archaea are:

- They appear to be more closely related to the eukarya than to the bacteria.
- They do not have peptidoglycan in their cell walls, as do the bacteria, and they share more biochemical characteristics with the eukarya than do bacteria.
- Some are well known for living under harsh conditions, such as anaerobic marshes (methanogens), salty lakes (halophiles), and hot sulfur springs (thermoacidophiles).

17.4 | The Protists
General Biology of Protists

- Protists are eukaryotes. Endosymbiotic events account for the presence of mitochondria and chloroplasts in eukaryotic cells. Some are multicellular with differentiated tissues.
- Protists have great ecological importance because in largely aquatic environments they are the producers (algae) or sources (algae and protozoans) of food for other organisms.

Algae

- Algae possess chlorophylls. They store reserve food as starch and have cell walls, as do plants.
- Algae are divided according to their pigments, being either green, brown, golden-brown, or red. Many are unicellular, but the green, red, and brown algae include multicellular seaweeds.

Protozoans

- Protozoans are heterotrophic and usually motile by means of cilia (paramecia), pseudopods (amoebas), or flagella (zooflagellates).
- Sporozoans (apicomplexans) cause diseases, including malaria, the most serious protozoan disease.
- A zooflagellate (trypanosome) causes African sleeping sickness; *Giardia* causes severe diarrhea.

Slime Molds and Water Molds

- Slime molds are decomposers in forests and woodlands.
- Water molds have cell walls of cellulose, which distinguish them from fungi.

Key Terms

algae (sing., alga)   279
amoeboid   282
archaea   278
bacteria   273
bacteriophage   269
binary fission   274
bioremediation   276
brown algae   280
capsid   268
chemoautotroph   275
chemoheterotroph   275
ciliate   281
conjugation   275
cyanobacteria   275
emerging virus   271
endospore   275
endosymbiotic theory   279
flagella   274
foraminiferan   282
golden-brown algae   280
green algae   280
halophile   278
lysogenic cycle   269
lytic cycle   269
methanogen   278
nucleoid   274
pathogen   277
peptidoglycan   274
photoautotroph   275
plasmid   274
prion   272
prokaryote   272
protist   279
protocell   273
protozoan   279
pseudopod   282
radiolarian   282
red algae   280
retrovirus   270
saprotroph   276
slime mold   282
sporozoan   282
symbiotic   276
thermoacidophile   278
transduction   275
transformation   275
trypanosome   282
viroid   272
virus   268
water mold   282
zooflagellate   282

Testing Yourself

Choose the best answer for each question.

1. A virus always contains
   a. a cell wall.
   b. a plasma membrane.
   c. nucleic acid.
   d. cytoplasm.
   e. More than one of these are correct.

2. Label the parts of a virus in the following illustration.

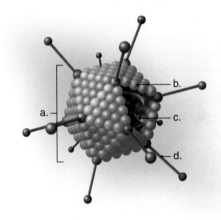

3. The five stages of the bacteriophage lytic cycle occur in this order:
   a. penetration, attachment, release, maturation, biosynthesis
   b. attachment, penetration, release, biosynthesis, maturation
   c. biosynthesis, attachment, penetration, maturation, release
   d. attachment, penetration, biosynthesis, maturation, release
   e. penetration, biosynthesis, attachment, maturation, release

4. Animal viruses
   a. contain both DNA and RNA.
   b. sometimes have an envelope.
   c. sometimes infect bacteria.
   d. do not reproduce inside cells.

5. The enzyme that is unique to retroviruses is
   a. reverse transcriptase.   c. DNA gyrase.
   b. DNA polymerase.   d. RNA polymerase.

6. The _____ cycle is a phage life cycle in which the infected bacterium does not immediately produce phage.
   a. lytic   c. pathogenic
   b. lysogenic   d. transformation

7. Which of these is mismatched?
   a. amoeboids—pseudopods   c. algae—variously colored
   b. sporozoans—disease agents   d. slime molds—trypanosomes

8. Which of the following statements about viroids is false?
   a. They are composed of naked RNA.
   b. They cause plant diseases.
   c. They die once they reproduce.
   d. They cause infected cells to produce more viroids.

9. Prion proteins cause disease when they
   a. enlarge in size.
   b. break into small pieces.
   c. cause normal proteins to change shape.
   d. interact with DNA.

10. Bacterial cells contain
    a. ribosomes.
    b. nuclei.
    c. mitochondria.
    d. vacuoles.
    e. More than one of these are correct.

11. The primary producers at deep-sea vents are
    a. heterotrophs.
    b. symbionts.
    c. chemoautotrophs.
    d. photoautotrophs.

For questions 12–17, determine which type of organism is being described. Each answer in the key may be used more than once.

Key:

   a. bacteria
   b. archaea
   c. both bacteria and archaea
   d. neither bacteria nor archaea

12. peptidoglycan in cell wall
13. methanogens
14. sometimes parasitic
15. contain a nucleus
16. plasma membrane contains lipids
17. reproduction by binary fission

18. Unlike plants, algae contain
    a. chloroplasts.
    b. vacuoles.
    c. mitochondria.
    d. pyrenoids.
    e. cell walls.

19. Which are unicellular, golden-brown algae with a silica shell?
    a. diatoms
    b. radiolarians
    c. trypanosomes
    d. sporozoans
    e. foraminiferans

20. Label the parts of a paramecium in the following illustration.

Thinking Scientifically

1. Based on the evolutionary tree below, are the protists more closely related to bacteria, or to archaea? What evidence supports this relationship? Why are viruses not shown on the tree?

2. In the former Soviet Union and Eastern Europe, bacteriophage therapy has long been used as an alternative to antibiotic drugs for treating bacterial infections. How could introducing bacteriophages into a patient's body help fight a bacterial infection? What are some potential benefits and shortcomings of bacteriophage therapy?

Bioethical Issue

Misuse of Antibiotics

The Food and Drug Administration (FDA) estimates that physicians annually write 50 million unnecessary prescriptions for antibiotics to treat viral infections. Of course, these antibiotics are ineffective against viruses. Furthermore, the frequent exposure of bacteria to these drugs has resulted in the development of numerous strains with antibiotic resistance. Doctors may prescribe antibiotics because patients demand them. In addition, if the cause of the illness is not known, it may be safer to prescribe an antibiotic that turns out to be ineffective than to withhold the antibiotic when it would have helped. The FDA has, therefore, initiated a new policy requiring manufacturers to label antibiotics with precautions about their misuse.

Do you think this type of labeling information will educate consumers enough to significantly reduce the inappropriate use of antibiotics? Physicians already know about the dangers of antibiotic resistance, but may prescribe them anyway. This is probably due, in large part, to the fact that they are liable for erring on the side of withholding an antibiotic but not for inappropriately prescribing one. Should physicians be held more accountable for prescribing antibiotics? If so, how?

Essentials of Biology Website

The companion website for *Essentials of Biology* provides a wealth of information organized and integrated by chapter. You will find practice tests, animations, videos, and much more that will complement your learning and understanding of general biology.

http://www.mhhe.com/maderessentials2

Without plants, there would be no animals.

Twelve plants, wheat among them, stand between humans and starvation.

Cotton is a textile fiber that accounts for about half of all the fabric sold in the world.

18

Land Environment: Plants and Fungi

Outline

Thanking Plants and Fungi Too

HAVE YOU THANKED a green plant today? Without plants, we and our civilized way of life could not exist. Land plants provide us with our food, either directly or indirectly, and with the oxygen we breathe. Oxygen also rises into the stratosphere and becomes the ozone shield that absorbs ultraviolet rays and makes life on land possible.

The bulk of the human diet comes from just 12 plants. Wheat is associated with Europe, corn with the Americas, and rice with the Far East.

—Continued next page

Continued—

Other significant food crops include white and sweet potatoes, cassava, soybeans, common beans, sugarcane, sugar beets, coconuts, and bananas.

Eons ago, the bodies of land plants became the coal we now burn to produce much of our electricity. The wood of trees is also commonly used as fuel. Then, too, fermentation of plant materials produces alcohol, which can be used directly to fuel automobiles or as a gasoline additive.

Much of our clothing comes from land plants. People still like the feel of cotton next to their skin, while linen from flax makes a stronger cloth. Cotton is used in all sorts of other cloth products, such as towels, sheets, and upholstered furniture. Even the manufacture of rayon depends on cellulose from plant cell walls.

Thousands of other household products come from land plants. Wood is used to make furniture as well as the houses that contain the furniture. Jute is used for rope, carpet, insulation, and burlap bags. Coconut oil is found in soaps, shampoos, and suntan lotions. Even toothpastes contain such plant flavorings as peppermint and spearmint. Suffice it to say, it would be impossible to mention all the products from plants that we depend on for our daily needs.

Fungi also have great economic importance, from the mushrooms topping your pizza to the yeast that makes the dough rise. More importantly, the roles of fungi in ecosystems are indispensable to life on Earth. The majority of fungi are decomposers that help break down dead organisms and recycle their nutrients. Others have symbiotic relationships that enhance the growth of land plants.

In this chapter, you will explore the evolution and diversity of two kingdoms: Plantae and Fungi.

18.1 | Onto Land

Plants (kingdom Plantae) are multicellular, photosynthetic eukaryotes which have become adapted to a land existence. Although a land environment does offer advantages such as plentiful light, it also has certain challenges such as the constant threat of desiccation (drying out). Most important all stages of reproduction—gametes, zygote, and embryo—must be protected from the drying effects of air. To keep the internal environment of cells moist, a land plant must acquire water, and transport it to all parts of the body, while keeping the body in an erect position. We will see how plants have adapted to these problems by evolving an internal vascular system.

The evolutionary history of plants begins in the water. Most likely land plants evolved from a form of freshwater green algae some 500 MYA (million years ago): Both green algae and plants (1) contain chlorophylls *a* and *b* and various accessory pigments, (2) store excess carbohydrates as starch, and (3) have cellulose in their cell walls. A comparison of DNA and RNA base sequences suggest that land plants are most closely related to a group of freshwater green algae known as **charophytes.** Although *Spirogyra* (see Fig. 17.21) is a charophyte, molecular scientists tell us that the ancestors of land plants were more closely related to the charophytes shown in **Figure 18.1.** Although the common ancestor of modern charophytes and land plants no longer exists, if it did, it would have features resembling those of the *Chara* and *Coleochaete*.

Let's take a look at these filamentous green algae. *Chara* are commonly known as stoneworts because they are encrusted with calcium carbonate deposits. The body consists of a single file of very long cells anchored in mud by thin filaments. Whorls of branches occur at regions called nodes, located between the cells of the main axis. Male and female reproductive structures grow at the nodes. A *Coleochaete* looks flat like a pancake, but the body is actually composed of long, branched filaments of cells. Most important to the evolution of plants, charophytes protect the zygote. Land plants not only protect the zygote, they also protect and nourish the resulting embryo—this may be the first derived feature that separates land plants from green algae.

To illustrate the evolution of land plants, we will be discussing five groups. It is possible to associate each group with one of five evolutionary events, each one an adaptation to a land existence.

Mosses are low-lying plants that lack vascular tissue and therefore have no means of transporting water, but they do protect the body of the plant from drying out and protect the embryo within a special structure.

The lycophytes, which evolved around 420 MYA, are among the first plants to have a vascular system that transports water and solutes from the roots to the leaves of land plants. Plants with vascular tissue have true roots, stems, and leaves. The leaves of lycophytes are very narrow. Ferns are well-known plants with large leaves called **megaphylls.** The evolution of branching and megaphylls allows a plant to increase the amount of photosynthesis and carbohydrate produced. Without an adequate production of food, a plant can't increase in size. The next evolutionary event was the evolution of **seeds.** A seed contains an embryo and stored organic nutrients within a protective coat (see Fig. 18.10). Seeds are highly

a. *Chara*　　　　b. *Coleochaete*

Figure 18.1 Close algal relatives of plants.

The closest living relatives of land plants are the filamentous green algae known as charophytes. **a.** *Chara*, commonly called stoneworts, favor an environment rich in the mineral calcium carbonate, which becomes deposited on the surface of the alga. **b.** The body of this *Coleochaete* is also composed of filaments of cells; even so, it is only the size of a pinhead, and one cell layer thick.

resistant structures well suited to protect a plant embryo from drying out until conditions are favorable for germination. The gymnosperms were the first seed plants to appear about 360 MYA. The final evolutionary event of interest to us is the evolution of the **flower,** a reproductive structure found in angiosperms. Flowers attract pollinators such as insects, and they also give rise to fruits that cover seeds. Plants with flowers may have evolved about 160 MYA. **Figure 18.2** traces the evolutionary history of land plants and will serve as a backdrop as we discuss major groups of plants in the pages that follow.

Check Your **Progress**

1. **List the similarities between charophytes and land plants.**
2. **List the five evolutionary events that allowed plants to successfully inhabit land.**

Answers: 1. Both charophytes and plants contain chlorophylls *a* and *b* plus accessory pigments, store carbohydrates as starch, have cellulose in their cell walls, and form structures that protect and nourish the zygote.   2. Protection of the embryo, the presence of vascular tissue, presence of megaphylls, seed production, and the presence of flowers.

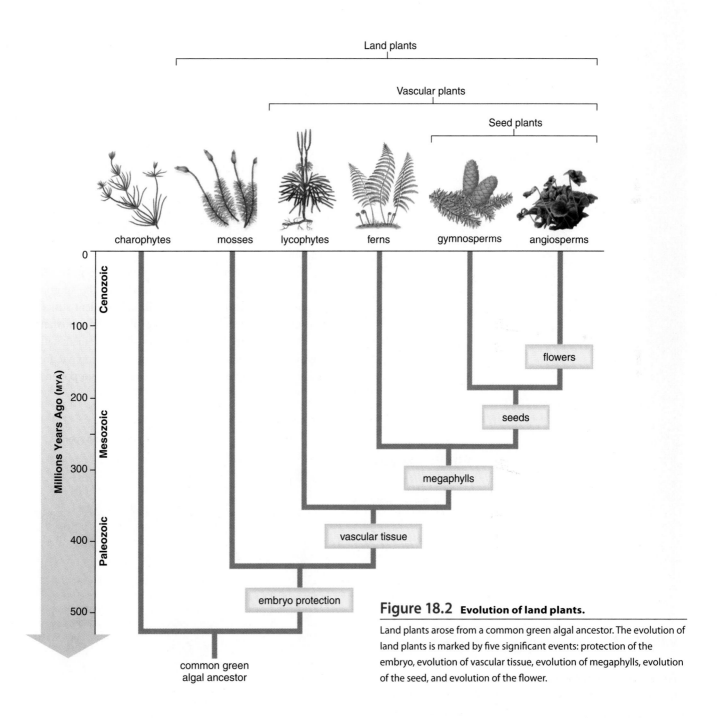

Figure 18.2   Evolution of land plants.

Land plants arose from a common green algal ancestor. The evolution of land plants is marked by five significant events: protection of the embryo, evolution of vascular tissue, evolution of megaphylls, evolution of the seed, and evolution of the flower.

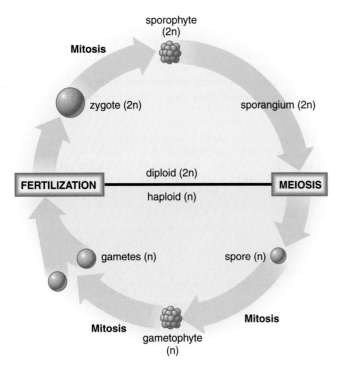

Figure 18.3 **Alternation of generations life cycle.**

Alternation of Generations

The life cycle of land plants is quite different from that of animals. All land plants undergo an **alternation of generations,** which means that each type of plant exists in two forms (**Fig. 18.3**). One form is the diploid sporophyte, and the other is the haploid gametophyte. The **sporophyte** (2n) is named for its production of haploid spores by meiosis. A **spore** is a reproductive cell that develops into a new organism without the need to fuse with another reproductive cell. In the plant life cycle, a spore undergoes mitosis and becomes a gametophyte. The **gametophyte** (n) is named for its production of gametes. A sperm and egg fuse, forming a zygote that undergoes mitosis and becomes the sporophyte.

You'll want to make two observations in Figure 18.3. First, note that, in plants, meiosis produces haploid spores. This is consistent with the realization that the sporophyte is the diploid generation, and spores are haploid. Second, note that mitosis occurs as a spore becomes a gametophyte, and it also occurs as a zygote becomes a sporophyte. Mitosis must happen in both places in order to have two generations.

The Dominant Generation

When you bring a land plant to mind, you're probably thinking of the dominant generation—the generation that is larger, lasts longer, and is the generation we recognize as the plant. In nonvascular plants, the gametophyte is the dominant generation, and in vascular plants, the sporophyte is the dominant generation. In the history of plants, *only the sporophyte evolves vascular tissue*; therefore, the trend toward sporophyte dominance is an adaptation to life on land. As the sporophyte gains in dominance, the gametophyte becomes microscopic. It also becomes dependent upon the sporophyte, the generation best adapted to a dry land environment.

Note the appearance of the generations in **Figure 18.4.** In mosses (**bryophytes**), the gametophyte is much larger than the sporophyte. In lycophytes and ferns, the gametophyte is a small, independent structure. In contrast, the female gametophyte of cone-bearing plants (**gymnosperms**) and flowering plants (**angiosperms**) is microscopic—it is retained within the body of the sporophyte plant. This protects the female gametophyte from drying out. Also, the male gametophyte of seed plants lies within a pollen grain. Pollen grains

Figure 18.4 **Reduction in the size of the gametophyte.**

Notice the reduction in the size of the gametophyte among these representatives of today's plants. This trend occurred as plants became adapted for life on land.

have strong protective walls and are transported by wind, insects, or birds to reach the egg. In the life cycle of seed plants, the spores, the gametes, and the zygote are protected from drying out in the land environment.

18.2 | Diversity of Land Plants

In the evolution of land plants, the nonvascular plants evolved before the vascular plants.

Nonvascular Plants

The **nonvascular plants,** often called the *bryophytes*, do not have true roots, stems, and leaves—which, by definition, must contain well-developed vascular tissue. In bryophytes, the gametophyte is the dominant generation.

The most familiar bryophytes are the liverworts and mosses, which are low-lying plants. In bryophytes, the gametophyte consists of leafy shoots, which produce the gametes (**Fig. 18.5**). Flagellated sperm swim in a film of water to reach an egg. After a sperm fertilizes an egg, the resulting zygote becomes an embryo that develops into a sporophyte. The sporophyte is attached to, and derives its nourishment from, the photosynthetic gametophyte. The sporophyte produces spores in a structure called a **sporangium.** The spores are released into the air where they can be dispersed by the wind, an adaptation to life on land. The spores will germinate if they land in moist surroundings. Upon germination, male and female gametophytes develop.

The common name of several organisms implies that they are mosses when they are not. Irish moss is an edible red alga that grows in leathery tufts along northern seacoasts. Reindeer moss, a lichen, is the dietary mainstay of reindeer and caribou in northern lands. Club mosses, discussed later in this chapter, are vascular plants, and Spanish moss, which hangs in grayish clusters from trees in the southeastern United States, is a flowering plant of the pineapple family.

Adaptations and Uses of Bryophytes

The lack of well-developed vascular tissue and the presence of swimming sperm largely account for the short height of bryophytes such as mosses. Still, mosses can be found from the Antarctic through the tropics to parts of the Arctic. Although most mosses prefer damp, shaded locations in the temperate zone, some survive in deserts, and others inhabit bogs and streams. In forests, they frequently form a mat that covers the ground and rotting logs. In dry environments, they may become shriveled, turn brown, and look completely dead. As soon as it rains, the plant becomes green and resumes metabolic activity. Mosses are much better than flowering plants at living on stone walls, on fences, and in shady cracks of hot, exposed rocks. When bryophytes colonize bare rock, the rock degrades to soil that can be used for their own growth and the growth of other organisms.

In areas such as bogs, where the ground is wet and acidic, dead mosses, especially *Sphagnum*, do not decay. The accumulated moss, called peat or bog moss, can be used as fuel. Peat moss is also commercially important in another way. Because it has special nonliving cells that can absorb moisture, peat moss is often used in gardens to improve the water-holding capacity of the soil. The so-called copper mosses live only in the vicinity of copper, and can serve as an indicator of copper deposits.

Figure 18.5   Mosses.

In bryophytes, exemplified by mosses, the gametophyte is the dominant generation. Sperm swim from male shoots to female shoots, and the zygote develops into an attached sporophyte. Within structures called sporangia, the sporophyte produces haploid spores that develop into gametophyte shoots.

Check Your Progress

1. **Is a gametophyte haploid (n) or diploid (2n)? A sporophyte?**
2. **Is a spore haploid (n) or diploid (2n)? By what process does a sporophyte produce spores?**

Answers: 1. A gametophyte is haploid (n); a sporophyte is diploid (2n).
2. A spore is haploid (n). Sporophytes produce spores by meiosis.

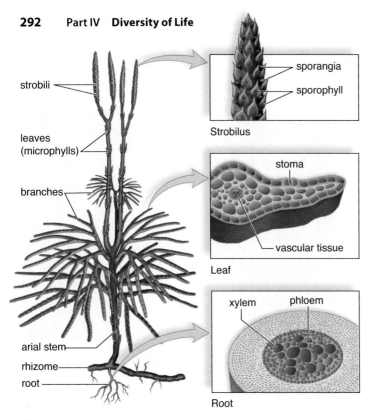

Figure 18.6 **Ground pine,** *Lycopodium*.

Lycophytes such as *Lycopodium* have vascular tissue and thus true roots, stems, and leaves. The *Lycopodium* sporophyte develops an underground rhizome system. A rhizome is an underground stem; this rhizome produces roots along its length.

Figure 18.7 **Diversity of ferns.**

Vascular Plants

All the other land plants we will study are **vascular plants.** The vascular plants usually have *true* roots, stems, and leaves. The roots absorb water from the soil, and the stem conducts water to the leaves. The leaves are fully covered by a waxy cuticle except where it is interrupted by stomata, little pores for gas exchange, the opening and closing of which can be regulated to control water loss.

Vascular tissue consists of **xylem,** which conducts water and minerals up from the roots, and **phloem,** which conducts organic nutrients from one part of a plant to another. The walls of conducting cells in xylem are strengthened by **lignin,** an organic compound that makes them stronger, more waterproof, and resistant to attack by parasites and predators. Only because of strong cell walls can plants reach great heights.

Seedless Vascular Plants

Certain vascular plants (e.g., lycophytes and ferns) are seedless; the other two groups of vascular plants (gymnosperms and angiosperms) are seed plants. In seedless vascular plants, the dominant sporophyte produces windblown spores, and the independent gametophyte produces flagellated sperm that require outside moisture to swim to an egg (see Fig. 18.8).

Lycophytes

Lycophytes, also called **club mosses,** were among the first land plants to have vascular tissue. Unlike true mosses (bryophytes), the lycophytes have well-developed vascular tissue in roots, stems, and leaves (**Fig. 18.6**). Typically, a fleshy underground and horizontal stem, called a rhizome, sends up upright arial stems. Tightly packed, scalelike leaves cover the stems and branches, giving the plant a mossy look. The small leaves, termed **microphylls,** each have a single vein composed of xylem and phloem. The sporangia are borne on terminal clusters of leaves, called strobili, which are club-shaped. The spores are sometimes harvested and sold as lycopodium powder, or vegetable sulfur, for use in pharmaceuticals and in fireworks because it is highly flammable. *Lycopodium* featured in Figure 18.6 are common in moist woodlands in temperate climates, where they are called ground pines; they are also abundant in the tropics and subtropics.

Ferns

Ferns are a widespread group of plants that are well known for their attractiveness. Unlike lycophytes, ferns have **megaphylls,** or large leaves with branched veins. Megaphylls provide a large surface area for capturing the sunlight needed for photosynthesis, and the veins conduct water and minerals throughout the leaf tissue. Ferns and other land plants with megaphylls are better able to produce food, and thus can grow and reproduce more efficiently than plants with microphylls. Fern megaphylls are called **fronds.** The maidenhair fern has fronds that are broad, with subdivided leaflets; those of the royal fern stand about 1.8 meters tall; and those of the hart's tongue fern are straplike and leathery (**Fig. 18.7**). Sporangia are often located in clusters called **sori** (sing., sorus), on the undersides of the fronds, where they may be shielded by thin protective structures called *indusia* (**Fig. 18.8**).

Adaptations and Uses of Ferns   Ferns are most often found in a moist environment because the small, water-dependent gametophyte, which lacks vascular tissue, is separate from the sporophyte. Also, flagellated sperm require an outside source of moisture in which to swim to the eggs (see Fig. 18.8). Once established, some ferns, such as the bracken fern, can spread into drier areas because their rhizomes, which grow horizontally in the soil, produce new plants.

At first it may seem that ferns do not have much economic value, but they are frequently used by florists in decorative bouquets and as ornamental plants in the home and garden. Wood from tropical tree ferns is often used as a building material because it resists decay, particularly by termites. Ferns, especially the ostrich fern, are used as food—in the northeastern United States, many restaurants feature fiddleheads (that season's first growth) as a special treat. Ferns also have medicinal value; many Native Americans use ferns as an astringent during childbirth to stop bleeding, and the maidenhair fern is the source of a cold medicine.

Coal Age Plants

Ferns and the other seedless vascular plants we have been discussing were as large as trees and more abundant during the Carboniferous period when a great swamp forest encompassed what is now northern Europe, the Ukraine, and the Appalachian Mountains in the United States (**Fig. 18.9**). A large number of these plants died but did not decompose completely. Instead, they were compressed to form the coal that we still mine and burn today; therefore, seedless vascular plants are sometimes called the *Coal Age* plants. (Oil has a similar origin, but it most likely formed in marine sedimentary rocks and included animal remains.)

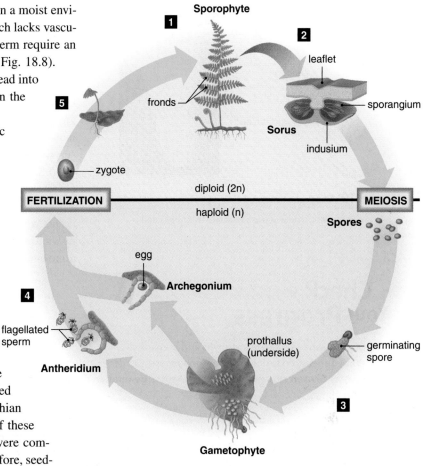

Figure 18.8   Fern life cycle.

1 The sporophyte is dominant in ferns. **2** In the fern shown here, sori are on the underside of a fern. Each sorus, protected by an indusium, contains sporangia, in which meiosis occurs and spores are produced and then released. **3** A spore germinates into a prothallus (the gametophyte), which has sperm-bearing and egg-bearing structures on its underside. **4** Fertilization takes place when moisture is present, because the flagellated sperm must swim in a film of water to the egg. **5** The resulting zygote begins its development inside an archegonium, and eventually a young sporophyte becomes visible. The young sporophyte develops, and fronds appear.

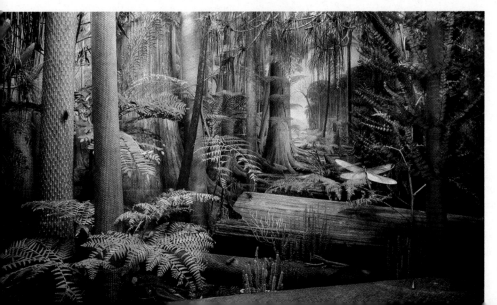

Figure 18.9   The Carboniferous period.

Growing in the swamp forests of the Carboniferous period were treelike club mosses (*left*), treelike horsetails (*right*), and lower, fernlike foliage (*left*). When the trees fell, they were covered by water and did not decompose completely. Sediment built up and turned to rock, the pressure of which caused the organic material to become coal, a fossil fuel that helps run our industrialized society today.

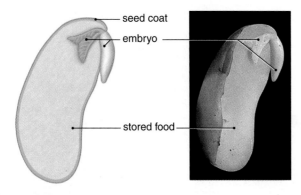

Figure 18.10   Seed anatomy.

A split bean seed showing the seed coat, sporophyte embryo, and stored food.

Seed Plants

Seed plants are the most plentiful land plants in the biosphere today. Most trees are seed plants, and so are almost all your garden plants. The major parts of a seed are shown in **Figure 18.10.** The seed coat and stored food protect the sporophyte embryo and allow it to survive harsh conditions during a period of dormancy (arrested state), until environmental conditions become favorable for growth. Seeds can even remain dormant for hundreds of years. When a seed germinates, the stored food is a source of nutrients for the growing seedling. The survival value of seeds largely accounts for the dominance of seed plants today.

Seed plants have two types of spores and produce two kinds of gametophytes—male and female. The gametophytes are microscopic and consist of just a few cells. **Pollen grains** are drought resistant male gametophytes. **Pollination** occurs when a pollen grain is brought to the vicinity of the female gametophyte by wind or a pollinator. Later, sperm move toward the female gametophyte through a growing pollen tube, and fertilization occurs. *Note that no external water is needed to accomplish fertilization.* The whole male gametophyte, rather than just the sperm as in seedless plants, moves to the female gametophyte.

A female gametophyte develops within an **ovule,** which eventually becomes a seed (**Fig. 18.11**). The embryo in a seed will be the sporophyte plant. In gymnosperms, the ovules are not completely enclosed by sporophyte tissue at the time of pollination. In angiosperms, the ovules are completely enclosed within diploid sporophyte tissue (an ovary), which becomes a fruit. We shall see that the flower has many advantages, one of which is the production of fruit.

Check Your Progress

1. Distinguish between microphylls and megaphylls.
2. What generation is dominant among mosses? In other groups of land plants? Give an explanation for this trend of generation dominance among land plants.

Answers: 1. Microphylls are small leaves, each having a single vein. Megaphylls are large leaves with branched veins.   2. The gametophyte is dominant among mosses, and the sporophyte is dominant among the other groups of land plants. The sporophyte evolved vascular tissue, and vascular tissue is a distinct advantage in a land environment.

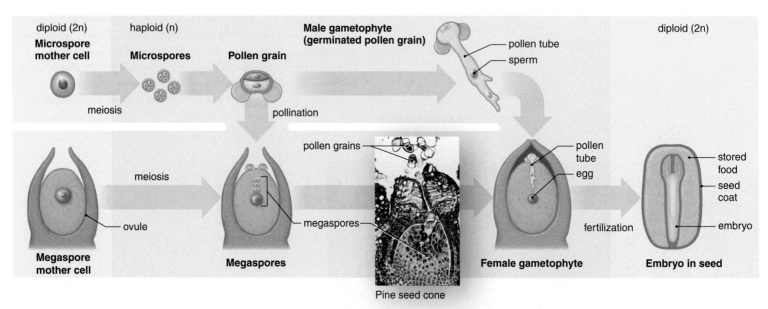

Figure 18.11   Production of a seed.

Development of the male gametophyte (*above*) begins when a microspore mother cell undergoes meiosis to produce microspores, each of which becomes a pollen grain. Development of the female gametophyte (*below*) begins in an ovule, where a megaspore mother cell undergoes meiosis to produce megaspores, only one of which will undergo mitosis to become the female gametophyte. During pollination, the pollen grain is carried to the vicinity of the ovule. The pollen grain germinates, and a nonflagellated sperm travels in a pollen tube to the egg produced by the female gametophyte. Following fertilization, the zygote becomes the sporophyte embryo; tissue within the ovule becomes the stored food, and the ovule wall becomes the seed coat.

Gymnosperms

In **gymnosperms,** ovules and seeds are exposed on the surface of a cone scale (modified leaf). The term gymnosperm means "naked seed." Ancient gymnosperms, including cycads, were present in the swamp forests of the Carboniferous period (**Fig. 18.12**). Of these, the conifers have become a dominant plant group today.

Conifers

Pines, spruces, firs, cedars, hemlocks, redwoods, cypresses, and others are all **conifers** (**Fig. 18.13**). The name *conifer* signifies plants that bear **cones** containing the reproductive structures of the plant, but other types of gymnosperms are also cone-bearing. The coastal redwood, a conifer native to northwestern California and southwestern Oregon, is the tallest living vascular plant; it may attain nearly 100 meters in height. Another conifer, the bristlecone pine of the White Mountains of California, is the oldest living tree—one specimen is 4,900 years of age.

Adaptations and Uses of Pine Trees Pine trees are well adapted for dry conditions. For instance, vast areas of northern temperate regions are covered in evergreen coniferous forests. The tough, needlelike leaves of pines conserve water because they have a thick cuticle and recessed stomata. This type of leaf helps them live in areas where frozen topsoil makes it difficult for the roots to obtain plentiful water.

A substance called resin protects leaves and other parts of the tree from insect and fungal attacks. The resin of certain pines is harvested; the liquid portion, called turpentine, is a paint thinner, while the solid portion is used on stringed instruments. The wood of pines is used extensively in construction, and vast forests of pines are planted for this purpose. The wood consists primarily of xylem tissue that lacks some of the more rigid cell types found in flowering trees. Therefore, it is considered a "soft" rather than a "hard" wood.

a. Male cycad with a pollen-bearing cone.

b. Female cycad with a seed-bearing cone.

Figure 18.12 Cycads.

Cycads are an ancient group of gymnosperms that are threatened today because they are slow growing. Herbivorous dinosaurs of the Mesozoic era most likely fed on cycad seeds.

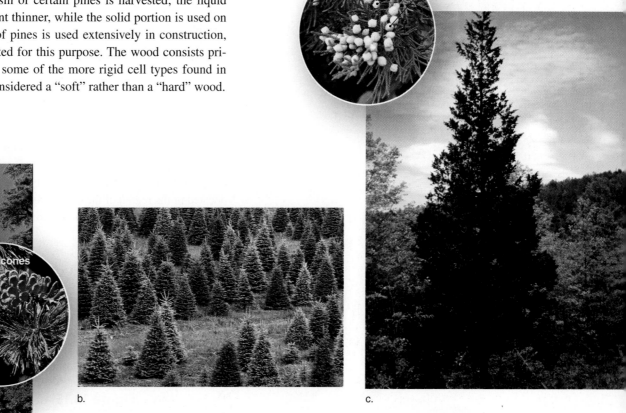

fleshy seed cones (juniper berries)

pollen cones

seed cones

a.

b.

c.

Figure 18.13 Conifers.

a. Pine trees are the most common of the conifers. The pollen cones (male) are smaller than the seed cones (female) and produce plentiful pollen. A cluster of pollen cones may produce more than one million pollen grains. Other conifers include (**b**) the spruces, which make beautiful Christmas trees, and (**c**) junipers, which have fleshy seed cones.

Figure 18.14 *Amborella trichopoda.*

Molecular data suggest this plant is most closely related to the first flowering plants.

Angiosperms

The **angiosperms** (meaning covered seeds) are an exceptionally large and successful group of land plants, with 240,000 known species—six times the number of species of all other plant groups combined. Angiosperms, also called the flowering plants, live in all sorts of habitats, from freshwater to desert, and from the frigid north to the torrid tropics. They range in size from the tiny, almost microscopic duckweed to *Eucalyptus* trees over 100 meters tall. Most plants in your garden produce flowers, and therefore are angiosperms. In northern climates, the trees that lose their leaves are flowering plants. In subtropical and tropical climates, flowering trees as well as gymnosperms tend to keep their leaves all year.

Although the first fossils of angiosperms are no older than about 135 million years, the angiosperms probably arose much earlier. Indirect evidence suggests the possible ancestors of angiosperms may have originated as long ago as 160 MYA. To help solve the mystery of their origin, botanists have turned to DNA comparisons to find a living plant that is most closely related to the first angiosperms. Their data point to *Amborella trichopoda* as having the oldest lineage among today's angiosperms (**Fig. 18.14**). This small shrub, which has small, cream-colored flowers, lives only on the island of New Caledonia in the South Pacific.

The Flower

Most flowers have certain parts in common despite their dissimilar appearances. The flower parts, called **sepals, petals, stamens,** and **carpels,** occur in whorls (circles) (**Fig. 18.15**). The sepals, collectively called the **calyx,** protect the flower bud before it opens. The sepals may drop off or may be colored like the petals. Usually, however, sepals are green and remain in place. The petals, collectively called the **corolla,** are quite diverse in size, shape, and color. The petals often attract a particular pollinator. Each stamen consists of a stalk called a **filament** and an **anther,** where pollen is produced in pollen sacs. In most flowers, the anther is positioned where the pollen can be carried away by wind or a pollinator. One or more carpels is at the center of a flower. A carpel has three major regions: ovary, style, and stigma. The swollen base is the **ovary,** which contains from one to hundreds of ovules. The **style** elevates the **stigma,** which is sticky or otherwise adapted for the reception of pollen grains. Glands located in the region of the ovary produce nectar, a nutrient that is gathered by pollinators as they go from flower to flower.

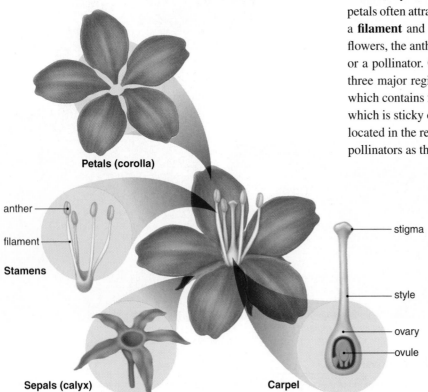

Petals (corolla)

anther

filament

Stamens

stigma

style

ovary

ovule

Sepals (calyx)

Carpel

Figure 18.15 **Generalized flower.**

A flower has four main parts: sepals, petals, stamens, and carpels. A stamen has a filament and an anther. A carpel has an ovary, a style, and a stigma. The ovary contains ovules.

Flowering Plant Life Cycle

In angiosperms, the flower produces seeds enclosed by fruit. The ovary of a carpel contains several ovules, and each of these eventually holds an egg-bearing female gametophyte called the embryo sac. During pollination, a pollen grain is transported by various means from the anther of a stamen to the stigma of a carpel, where it germinates. The **pollen tube** carries the two sperm into a small opening of an ovule. During **double fertilization,** one sperm unites with an egg nucleus, forming a diploid zygote, and the other sperm unites with two other nuclei, forming a triploid (3n) **endosperm (Fig. 18.16).** In angiosperms, the endosperm is the stored food.

Ultimately, the ovule becomes a seed that contains a sporophyte embryo. In some seeds, the endosperm is absorbed by the seed leaves, called **cotyledons,** whereas in other seeds, endosperm is digested as the seed germinates. When you open a peanut, the two halves are the cotyledons. If you look closely, you will see the embryo between the cotyledons. A **fruit** is derived from an ovary and possibly accessory parts of the flower. Some fruits (e.g., apple) provide a fleshy covering for their seeds, and other fruits provide a dry covering (e.g., pea pod, peanut shell).

Figure 18.16   Life cycle of a flowering eudicot.

Eudicots have seeds with two cotyledons.

Figure 18.17 Pollinators.

a. A bee-pollinated flower is typically a color other than red (bees cannot see red). **b.** Butterfly-pollinated flowers are wide, allowing the butterfly to land. **c.** Hummingbird-pollinated flowers are curved back, allowing the bird's beak to reach the nectar. **d.** Bat-pollinated flowers are large and sturdy and able to withstand rough treatment.

Adaptations and Uses of Angiosperms

Successful completion of sexual reproduction in angiosperms requires the effective dispersal of pollen and then seeds. Adaptations have resulted in various means of dispersal of pollen and seeds. Wind-pollinated flowers are usually not showy, whereas insect-pollinated flowers and bird-pollinated flowers are often colorful (**Fig. 18.17***a–c*). Night-blooming flowers attract nocturnal mammals or insects; these flowers are usually aromatic and white or cream-colored (Fig. 18.17*d*). Although some flowers disperse their pollen by wind, many are adapted to attract specific pollinators, such as bees, wasps, flies, butterflies, moths, and even bats, which carry only particular pollen from flower to flower. For example, bee-pollinated flowers are usually blue or yellow and have ultraviolet shadings that lead the pollinator to the location of nectar at the base of the flower. In turn, the mouthparts of bees are fused into a long tube that is able to obtain nectar from this location. Today, there are some 240,000 species of flowering plants and over 900,000 species of insects. This diversity suggests that the success of angiosperms has contributed to the success of insects and vice versa.

The fruits of flowers protect and aid in the dispersal of seeds. Dispersal occurs when seeds are transported by wind, gravity, water, and animals to another location. Fleshy fruits may be eaten by animals, which transport the seeds to a new location and then deposit them when they defecate. Because animals live in particular habitats and/or have particular migration patterns, they are apt to deliver the fruit-enclosed seeds to a suitable location for seed germination (initiation of growth) and development of the plant.

Economic Benefits of Plants

One of the primary economic benefits of land plants is the use of their fruits as food. Botanists use the term "fruit" in a much broader way than do laypeople. Among the foods mentioned in the introduction, you would have no trouble recognizing a banana as a fruit. A coconut is also a fruit, as are the grains (corn, wheat, rice) and the pods that contain beans or peas (**Fig. 18.18**). Cotton is derived from the cotton boll, a fruit containing seeds with seed hairs that become textile fibers used to make cloth.

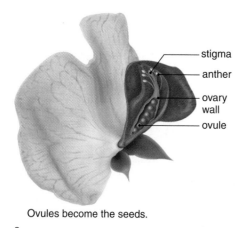

stigma
anther
ovary wall
ovule

Ovules become the seeds.

a.

b.
c.

Ovary becomes the fruit.

d.

Figure 18.18 Pea flower and the development of a pea pod.

a. Pea flower. **b.** Pea flower still has its petals soon after fertilization. **c.** Petals fall and the pea pod is quite noticeable. **d.** Pea pod is clearly visible developing from an ovary.

Other economic benefits of land plants include foods and commercial products made from roots, stems, and leaves. Cassava and sweet potatoes are edible roots; white potatoes are the tubers of underground stems. Most furniture and paper are made from the wood of a tree trunk (**Fig. 18.19**). Also, the many chemicals produced by plants make up 50% of all pharmaceuticals and various other types of products we can use. The cancer drug taxol originally came from the bark of the Pacific yew tree. Today, plants are even bioengineered to produce certain substances of interest.

Indirectly, the economic benefits of land plants are often dependent on pollinators. Only if pollination occurs can these plants produce a fruit and propagate themselves. In recent years, the populations of honeybees and other pollinators have been declining worldwide, principally due to a parasitic mite but also due to the widespread use of pesticides. Consequently, some plants are endangered because they have lost their normal pollinators. Because of our dependence on flowering plants, we should protect pollinators!

Figure 18.19   **Humans use wood and wood products in daily life.**

Ecological Benefits of Plants

The ecological benefits of flowering plants are so important that we could not exist without them. Land plants produce food for themselves and directly or indirectly for all other organisms in the biosphere. The oxygen they produce is also used by themselves and all organisms that carry out cellular respiration.

Forests are an important part of the water cycle and the carbon cycle. The roots of trees in particular hold soil in place and absorb water that returns to the atmosphere. Without these functions of trees and other plants, rainwater runs off and contributes to flooding. The absorption of carbon dioxide by plants lessens the amount in the atmosphere. CO_2 in the atmosphere contributes to global warming because it and other gases trap heat near the surface of the Earth. The burning of tropical rain forests is double trouble for global warming because it adds CO_2 to the atmosphere and removes trees that otherwise would absorb CO_2. Some plants can also be used to clean up toxic messes. For example, poplar, mustard, and mulberry species take up lead, uranium, and other pollutants from the soil.

In addition to all the other uses of land plants, we should not forget their aesthetic value. Almost everyone prefers to vacation in a natural setting and enjoys the sight of trees and flowers in the yard and in the house.

Check Your Progress

1. What structure in seed plants becomes a seed? What are the components of a seed?
2. List the components of the stamen. Where is pollen formed? List the components of the carpel. Which part becomes a seed? The fruit?
3. Contrast the adaptations of a moss and a flowering plant to the land environment.

Answers: 1. The ovule becomes a seed. A seed contains an embryo, stored food, and a seed coat.   2. The stamen contains the anther and the filament. Pollen forms in the pollen sacs of the anther. The carpel contains the stigma, style, and ovary. An ovule in the ovary becomes a seed, and the ovary becomes the fruit.   3. A moss plant protects the sporophyte embryo from drying out and produces windblown spores for dispersal of offspring. A moss is not fully adapted to land because it lacks vascular tissue, and the gametophyte produces flagellated sperm that must swim to the egg. A flowering plant is fully adapted to land. The dominant sporophyte, which possesses vascular tissue, produces pollen. Pollen is windblown or carried by animals. The sporophyte also retains the embryo in the ovule, which becomes a seed. Animals help disperse seeds

Figure 18.20 Diversity of fungi.

a. Scarlet hood, an inedible mushroom.
b. Spores exploding from a puffball.
c. Common bread mold. **d.** A morel, an edible fungus.

Figure 18.21 Body of a fungus.

a. The body of a fungus is called a mycelium. **b.** A mycelium contains many individual chains of cells, and each chain is called a hypha.

18.3 | The Fungi

Asked whether **fungi** are animal or vegetable, most would choose vegetable. But this would be wrong because fungi do not have chloroplasts, and they can't photosynthesize. Then, too, fungi are not animals, even though they are chemoheterotrophs like animals. Animals ingest their food, but fungi must grow into their food. Fungi release digestive enzymes into their immediate environment and then absorb the products of digestion. Also, animals are motile, but the majority of fungi are nonmotile and do not have flagella at any stage in their life cycle. The fungal life cycle differs from that of both animals and plants because fungi produce windblown spores during both an asexual and a sexual life cycle.

Table 18.1 contrasts fungi with land plants and animals. The many unique features of fungi indicate that although fungi are multicellular eukaryotes (except for the unicellular yeasts and chytrids), they are not closely related to any other group of organisms. DNA sequence data suggest fungi are distantly related to animals rather than plants, and may be descendants of a flagellated protist.

General Biology of a Fungus

Representative members of kingdom Fungi are shown in **Figure 18.20**. All parts of a typical fungus are composed of **hyphae** (sing., hypha), which are thin filaments of cells. The hyphae are packed closely together to form a complex structure, such as a mushroom. However, the main body of a fungus is not the mushroom or the puffball or the morel; these are just temporary reproductive structures. Mold shows best that the main body of a fungus is a mass of hyphae called a **mycelium** (**Fig. 18.21**). The mycelium penetrates the soil, the wood, or the bread in which the fungus is growing. Mycelia in the soil can become large enough to cover acres, making them the largest organisms on Earth and earning them the label "the humongous fungus among us."

Fungal cells typically have thick cell walls, but unlike plants, the fungal cell walls do not contain cellulose. They are made of another polysaccharide in which the glucose monomers contain amino groups (amino sugars) and form a polymer called chitin. This polymer is also the major structural component of the exoskeleton of insects and arthropods, such as lobsters and crabs. Walls or *septa* (sing., septum) divide the cells of a hypha in many types of fungi. Septa have pores that allow the cytoplasm to pass from one cell to the other along the length of the hypha. The hyphae give the mycelium quite a large surface area, which facilitates the ability of the mycelium to absorb nutrients. Hyphae extend toward a food source by growing at their tips, and the hyphae of a mycelium absorb and then pass nutrients on to the growing tips.

Table 18.1 | How Fungi Differ from Land Plants and Animals

Feature	Fungi	Land Plants	Animals
Nutrition	Chemoheterotrophic by absorption	Photosynthetic	Chemoheterotrophic by ingestion
Movement	Most nonmotile	Nonmotile	Motile
Body	Mycelium of hyphae	Specialized tissues/organs	Specialized tissues/organs
Adult chromosome number	Haploid	Haploid/diploid	Diploid
Cell wall	Composed of chitin	Composed of cellulose	No cell wall
Reproduction	Most have spores/mating hyphae	Spores/gametes	Gametes

Black Bread Mold

Multicellular organisms are characterized by specialized cells. Black bread mold demonstrates that the hyphae of a fungus may be specialized for various purposes (**Fig. 18.22**). In this fungus, horizontal hyphae exist on the surface of the bread; other hyphae grow into the bread, anchoring the mycelium and carrying out digestion; and some form stalks that bear sporangia.

The mycelia of two different mating types are featured in the center and bottom of Figure 18.22. During asexual reproduction, each mycelium produces sporangia, where spore formation occurs. Spores are resistant to environmental damage, and they are often made in large numbers. Fungal spores are windblown, a distinct advantage for a nonmotile organism living on land. When spores encounter a moist environment, they germinate into new mycelia without going through any developmental stages, another feature that distinguishes fungi from animals.

Sexual reproduction in fungi involves the conjugation of hyphae from different mating types (usually designated + and −). In black bread mold, the tips of + and − hyphae join, the nuclei fuse, and a thick-walled zygospore results. The zygospore undergoes a period of dormancy before it germinates, producing sporangia. Meiosis occurs within the sporangia, producing spores of both mating types. The spores, which are dispersed by air currents, give rise to new mycelia. In fungi, only the zygote is diploid, and all other stages of the life cycle are haploid, a distinct difference from animals.

Figure 18.22 Life cycle of black bread mold.

During asexual reproduction, sporangia produce asexual spores (purple and blue arrows). During sexual reproduction, two hyphae tips fuse, and then two nuclei fuse, forming a zygote that develops a thick, resistant wall (zygospore). When conditions are favorable, the zygospore germinates, and meiosis within a sporangium produces windblown spores (yellow arrows).

a. Sexual reproduction

Figure 18.23 **Sexual reproduction produces mushrooms.**

a. Fusion of + and − hyphae tips results in hyphae that form the mushroom (a fruiting body). The nuclei fuse in clublike structures (basidia) attached to the gills of a mushroom, and meiosis produces spores. **b.** A circle of mushrooms is called a fairy ring.

Mushroom

When you eat a mushroom, you are eating a *fruiting body*, whose function is to produce spores. In mushrooms, when the tips of + and − hyphae fuse, the haploid nuclei do not fuse immediately. Instead, so-called dikaryotic (two nuclei) hyphae form a mushroom consisting of a stalk and a cap. Club-shaped structures called basidia (sing., basidium) project from the gills located on the underside of the cap (**Fig. 18.23***a*). Fusion of nuclei inside these structures is followed by meiosis and production of windblown spores. Each cap of a mushroom produces tens of thousands of spores.

In the soil, a mycelium is absorbing nutrients and may form a ring. Therefore, when the weather turns warm and rain is plentiful, the mycelium may put forth a circle of mushrooms on your lawn. At one time, people believed this ring indicated where fairies had joined hands and danced the night before (Fig. 18.23*b*).

Chytrids

Chytrids are a unique group of fungi characterized by their motility (**Fig. 18.24**). Both spores and gametes of chytrids have flagella, a feature which was lost at some point in the evolution of the other fungi. Some of the chytrids are unicellular, while others form hyphae without septa. Another oddity distinguishes the chytrids: Some have an alternation of generations life cycle, much like that of green plants and certain algae, and very uncommon among fungi. Most inhabit water and soil, although some live as parasites of plants and animals.

Ecological Benefits of Fungi

Most fungi are **saprotrophs** that decompose the remains of plants, animals, and microbes in the soil. Fungal enzymes can degrade cellulose and even lignin in the woody parts of plants. That is why fungi so often grow on dead trees. This also means that fungi can be used to remove excess lignin from paper pulp. Ordinarily, lignin is difficult to extract from pulp and ends up being a pollutant once it is removed.

Along with the bacteria, which are also decomposers, fungi play an indispensable role in the environment by returning inorganic nutrients to photosynthesizers. Many people take advantage of the activities of bacteria and fungi by composting their food scraps or yard waste. When a gardener makes a compost pile and provides good conditions for decomposition to occur, the result is a dark, crumbly material that serves as an excellent fertilizer. And while the material may smell bad as decomposition is occurring, the finished compost looks and smells like rich, moist earth.

Some fungi eat animals that they encounter as they feed on their usual meals of dead organic remains. For example, the oyster fungus secretes a substance that anesthetizes any nematodes (roundworms) feeding on it (**Fig. 18.25**). After the worms become inactive, the fungal hyphae penetrate and digest their bodies, absorbing the nutrients. Other fungi snare, trap, or fire projectiles into nematodes and other small animals before digesting them. The animals serve as a source of nitrogen for the fungus.

20 μm

Figure 18.24 **Chytrids.**

Chytriomyces hyalinus, a chytrid, attacking a dinoflagellate, a unicellular alga.

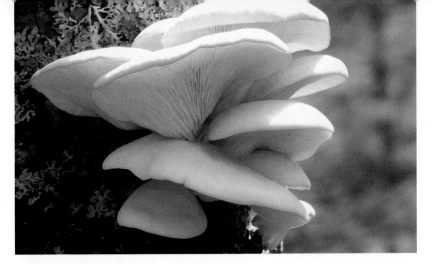

Figure 18.25 Carnivorous fungus.

Fungi like this oyster fungus, a type of bracket fungus, grow on trees because they can digest cellulose and lignin. If this fungus meets a roundworm in the process, it immobilizes the worm and digests it also. The worm is a source of nitrogen for the fungus.

Mutualistic Relationships

In a mutualistic relationship, two different species live together and help each other out. **Lichens** are a mutualistic association between a particular fungus and cyanobacteria or green algae (**Fig. 18.26**). The fungal partner is efficient at acquiring nutrients and moisture, and therefore lichens can survive in poor soils, as well as on rocks with no soil. The organic acids given off by fungi release from rocks the minerals that can be used by the photosynthetic partner. Lichens are ecologically important because they produce organic matter and create new soil, allowing plants to invade the area.

Lichens occur in three varieties: compact crustose lichens, often seen on bare rocks or tree bark; shrublike fruticose lichens; and leaflike foliose lichens. Regardless, the body of a lichen has three layers. The fungal hyphae form a thin, tough upper layer and a loosely packed lower layer. These layers shield the photosynthetic cells in the middle layer. Specialized fungal hyphae that penetrate or envelop the photosynthetic cells transfer organic nutrients to the rest of the mycelium. The fungus not only provides minerals and water to the photosynthesizer, but also offers protection from predation and desiccation. Lichens can reproduce asexually by releasing fragments that contain hyphae and an algal cell. At first, the relationship between the fungi and algae was likely a parasite-and-host interaction. Over evolutionary time, the relationship apparently became more mutually beneficial, although how to test this hypothesis is a matter of debate at the present time.

Mycorrhizal fungi form mutualistic relationships with the roots of most plants, helping the plants grow more successfully in dry or poor soils, particularly those deficient in inorganic nutrients (**Fig. 18.27**). The fungal hyphae greatly increase the surface area from which the plant can absorb water and nutrients. It has been found beneficial to encourage the growth of mycorrhizal fungi when restoring lands damaged by strip mining or chemical pollution.

Mycorrhizal fungi may live on the outside of roots, enter between root cells, or penetrate root cells. The fungus and plant cells exchange nutrients, with the fungus bringing water and minerals to the plant, and the plant providing organic carbon to the fungus. Early plant fossils indicate that the relationship between fungi and plant roots is an ancient one, and therefore it may have helped plants adapt to life on dry land. The general public is not familiar with mycorrhizal fungi, but a few people relish truffles, the fruiting bodies of a mycorrhizal fungus that grows in oak and beech forests. Truffles are considered a gourmet delicacy.

Crustose lichen

reproductive unit

fungal hyphae   algal cells

fungal hyphae

Fruticose lichen

Foliose lichen

a.   b.

Figure 18.26 Lichens.

a. Morphology of a lichen. **b.** Examples of lichens.

Figure 18.27 Plant growth experiment.

A soybean plant without mycorrhizal fungi grows poorly compared to two others infected with different strains of mycorrhizal fungi.

Figure 18.28 **Commercial importance of fungi.**

Delicacies such as various types of mushrooms are fungi, and some of our other favorite foods require the participation of fungi to produce them. Significant medicines have also come from fungi.

a. asexually produced spore

upper surface of leaf

fungal spores

fungal hypha

plant cell

b.

Figure 18.29 **Plant fungal disease.**

Peach leaf curl causes lesions on the leaf by invading the leaf, as shown in (**a**) the photo and (**b**) the diergram.

Economic Benefits of Fungi

Fungi help us produce medicines and many types of foods (**Fig. 18.28**). The mold *Penicillium* was the original source of penicillin, a breakthrough antibiotic that led to an important class of cillin antibiotics. Cillin antibiotics have saved millions of lives.

Yeast fermentation is utilized to make bread, beer, wine, and distilled spirits. During fermentation, yeast cells produce the carbon dioxide that causes dough to rise when gas pockets are preserved as the bread bakes. Ethanol is also a product of yeast fermentation, and is used for the making of alcoholic beverages. Other types of fungal fermentations contribute to the manufacture of various cheeses and soy sauce from soybeans. Another commercial application of interest is the use of fungi to soften the centers of certain candies.

Fungi as Food

In the United States, the consumption of mushrooms has been steadily increasing. In 2001, total consumption of all mushrooms totaled 1.13 billion pounds—21% greater than in 1991. In addition to adding taste and texture to soups, salads, and omelets, and being used in stir-fries, mushrooms are an excellent low-calorie meat substitute with great nutritional value and lots of vitamins. Although there are thousands of mushroom varieties in the world, the white button mushroom, *Agaricus bisporus*, dominates the U.S. market. However, in recent years, sales of brown-colored variants have surged in popularity and have been one of the fastest-growing segments of the mushroom industry. Portabella is a marketing name used by the mushroom industry for the more flavorful brown strains of *A. bisporus*. The mushroom is brown because it is allowed to open, exposing the mature gills with their brown spores; crimini is the same brown strain, but it is not allowed to open before it is harvested. Non-*Agaricus* varieties, especially shiitake and oyster, have slowly gained in popularity over the past decade. Shiitake is touted for lowering cholesterol levels and having antitumor and antiviral properties.

Fungi as Disease-Causing Organisms

Fungi cause diseases in both plants and animals, including humans.

Fungi and Plant Diseases

Fungal pathogens, which usually gain access to plants by way of the stomata or a wound, are a major concern for farmers. Serious crop losses occur each year due to fungal disease. As much as a third of the world's rice crop is destroyed each year by rice blast disease. Corn smut is a major problem in the midwestern United States. Various rusts attack grains, and leaf curl is a disease of fruit trees (**Fig. 18.29**).

The life cycle of rusts may be particularly complex, since it requires two different host species to complete the cycle. Black stem rust of wheat uses barberry bushes as an alternate host. Eradication of barberry bushes in areas where wheat is grown helps control this rust. Fungicides are regularly applied to crops to limit the negative effects of fungal pathogens. Wheat rust can also be controlled by producing new and resistant strains of wheat.

Fungi and Animal Diseases

As is well known, certain mushrooms are poisonous. The ergot fungus that grows on grain can result in ergotism when a person eats contaminated bread. Ergotism is characterized by hysteria, convulsions, and sometimes death.

Mycoses are diseases caused by fungi. Mycoses have three possible levels of invasion: Cutaneous mycoses only affect the epidermis; subcutaneous mycoses affect deeper skin layers; and systemic mycoses spread their effects throughout the body by traveling in the bloodstream. Fungal diseases that can be contracted from the environment include ringworm from soil, rose gardener's disease from thorns, Chicago disease from old buildings, and basketweaver's disease from grass cuttings. Opportunistic fungal infections now seen in AIDS patients stem from fungi that are always present in the body but take the opportunity to cause disease when the immune system becomes weakened.

Candida albicans causes the widest variety of fungal infections. Disease occurs when antibacterial treatments kill off the microflora community, allowing *Candida* to proliferate. Vaginal *Candida* infections are commonly called "yeast infections" in women. Oral thrush is a *Candida* infection of the mouth common in newborns and AIDS patients (**Fig. 18.30***a*). In individuals with inadequate immune systems, *Candida* can move throughout the body, causing a systemic infection that can damage the heart, brain, and other organs.

Ringworm is a group of related diseases caused, for the most part, by fungi in the genus *Tinea*. Ringworm is a cutaneous infection that does not penetrate the skin. The fungal colony grows outward, forming a ring of inflammation. The center of the lesion begins to heal, giving the lesion its characteristic appearance, a red ring surrounding an area of healed skin (**Fig. 18.31***b*). Athlete's foot is a form of tinea that affects the foot, mainly causing itching and peeling of the skin between the toes (Fig. 18.31*c*).

Batrachochytrium dendrobatidis is a parasitic chytrid that causes a cutaneous infection called chytridiomycosis in frogs around the world. The disease is thought to have originated in South Africa. It began to spread in the 1930s after African clawed frogs were captured and sold as pets and for use in medicine and research. Chytridiomycosis has recently decimated frog populations in Australia and Central and South America.

The vast majority of people living in the midwestern United States have been infected with *Histoplasma capsulatum*. This common soil fungus, often associated with bird droppings, leads in most cases to a mild "fungal flu." However, about 3,000 of these cases develop into a severe disease. About 50 persons die each year from histoplasmosis when the fungus grows within cells of the immune system. Lesions are formed in the lungs, leaving calcifications that are visible in X-ray images and resemble those of tuberculosis.

Because fungi are eukaryotes and more closely related to animals than to bacteria, it is hard to design an antibiotic against fungi that does not also harm animals. Thus, researchers exploit any biochemical differences they can discover between animals and fungi.

c.

Figure 18.30   Human fungal diseases.

a. Thrush, or oral candidiasis, is characterized by the formation of white patches on the tongue. **b.** Ringworm and (**c**) athlete's foot are caused by *Tinea* spp.

Check Your Progress

1. Contrast sexual reproduction in black bread mold with sexual reproduction in a mushroom.
2. How do fungi contribute to ecological cycling?
3. What organisms make up a lichen? Why is the relationship mutualistic?

Answers: 1. In black bread mold, meiosis produces spores in sporangia. In a mushroom, meiosis produces spores in clublike structures attached to the gills.   2. When fungi secrete enzymes that digest the remains of plants and animals in the environment, inorganic nutrients are returned to plants.   3. A lichen is composed of a fungus and a cyanobacterium or a green alga. Each partner gives to the other. The fungus provides protection, water, and minerals, and the photosynthesizer provides organic nutrients.

The Chapter in Review

Summary

18.1 | Onto Land

The Ancestry of Plants

Plants (kingdom Plantae) probably evolved from a multicellular, freshwater green alga about 500 MYA. The freshwater green algae known as charophytes appear to be the closest living relatives of land plants. Whereas algae are adapted to life in the water, plants are adapted to living on land. The charophytes have some characteristics that could be helpful on land. During the evolution of plants, five significant events are associated with adaptation to a land existence: Evolution of the (1) embryo protection, (2) vascular tissue, (3) megaphylls, (4) seeds, and (5) the flower.

Alternation of Generations

Land plants have an alternation of generations life cycle, in which each type of plant exists in two forms, the sporophyte (2n) and the gametophyte (n):

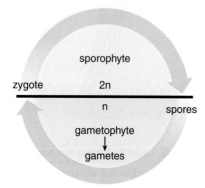

18.2 | Diversity of Land Plants

Nonvascular Plants

Bryophytes, represented by the mosses, are plants with the following characteristics:

- There is no well-developed vascular tissue.
- The gametophyte is dominant, and flagellated sperm swim in external moisture to the egg.
- The sporophyte is dependent on the gametophyte.
- Windblown spores disperse the gametophyte.

sporophyte

gametophyte

Vascular Plants

In vascular plants, the dominant sporophyte has two kinds of well-defined conducting tissues. Xylem is specialized to conduct water and dissolved minerals, and phloem is specialized to conduct organic nutrients and hormones. Certain vascular plants are seedless (lycophytes and ferns). They were large and abundant during the Carboniferous period.

Lycophytes have true roots, stems, and leaves because they contain vascular tissue. Vascular tissue allows lycophytes and other vascular plants to attain greater size due to more efficient transport of water, minerals, nutrients, and hormones. Lycophytes have narrow leaves called microphylls.

Ferns have large leaves with branching veins called megaphylls. Megaphylls enhance food production by photosynthesis. Ferns also have the following life cycle which is typical of seedless vascular plants:

- The sporophyte generation is dominant and produces windblown spores.
- The gametophyte is separate and independent. Flagellated sperm swim in external moisture to the egg.

sporophyte

Seed Plants have reproductive structures that are protected from drying out. Seed plants have male and female gametophytes. Gametophytes are reduced in size. The female gametophyte is retained within an ovule, and the male gametophyte is the mature pollen grain. The fertilized ovule becomes the seed, which contains a sporophyte embryo, food, and a seed coat.

- **Gymnosperms** are cone-bearing plants, represented by the pine tree. They have "naked seeds" because the seeds are not enclosed by fruit, as are those of flowering plants.
- **Angiosperms** are the flowering plants. Angiosperm reproductive organs are in the flower. Pollen is produced in pollen sacs inside an anther. Pollen is transported by wind, or from flower to flower by birds, insects, or bats. Fertilized ovules in the ovary become seeds, and the ovary becomes the fruit. Thus, angiosperms have "covered seeds."

18.3 | The Fungi

Kingdom Fungi includes unicellular yeasts, multicellular mushrooms and molds, and flagellated chytrids.

General Biology of a Fungus

Fungi are neither animals nor plants.

- The body of a typical fungus is composed of thin filaments of cells, called hyphae, that form a mass called a mycelium.
- The cell wall contains chitin.
- Most fungi produce windblown spores during both asexual and sexual reproduction.

Ecological Benefits of Fungi

- Fungi are saprotrophs that carry on external digestion. As decomposers, fungi keep ecological cycles going in the biosphere.
- Lichens (fungi plus cyanobacteria or green algae) are primary colonizers in poor soils or on rocks.
- Mycorrhizal fungi grow on or in plant roots and help the plant absorb minerals and water.

Economic Benefits of Fungi

- Fungi help produce foods and medicines, as well as serving as a source of food themselves.

Fungi as Disease-Causing Organisms

- Fungal pathogens of plants include blasts, smuts, and rusts that attack crops of great economic importance, such as rice and wheat.
- Animal diseases caused by fungi include thrush, ringworm, chytridiomycosis, and histoplasmosis.

Key Terms

Testing Yourself

Choose the best answer for each question.

1. Which of the following is not a plant adaptation to land?
   a. recirculation of water
   b. protection of embryo in maternal tissue
   c. development of flowers
   d. presence of vascular tissue
   e. seed production

2. Plant spores are
   a. haploid (n).
   b. produced by gametophytes.
   c. produced by sporophytes.
   d. diploid (2n).
   e. Both a and c are correct.

3. Charophytes
   a. are freshwater green algae.
   b. lack vascular tissue.
   c. are the closest living relatives of land plants.
   d. enclose their zygotes within protective structures.
   e. All of these are correct.

4. Which of the following is a true statement?
   a. People don't eat mushrooms because they might be poisonous.
   b. Penicillin is derived from a fungus.
   c. The alcohol from yeast fermentation makes bread rise.
   d. Fungi are prokaryotes like bacteria.
   e. Fungi ingest their food in the same way animals do.

5. Label the parts of the generalized plant life cycle in the following illustration.

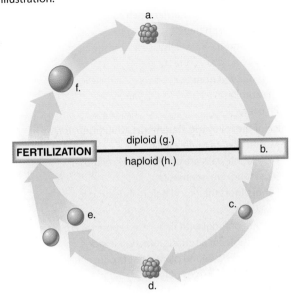

For questions 6–13, identify the group(s) to which each feature belongs. Each answer in the key may be used more than once. Each question may have more than one answer.

Key:

a. mosses
b. lycophytes
c. ferns
d. gymnosperms
e. angiosperms

6. have megaphylls
7. exhibit alternation of generations
8. produce seeds
9. lack true roots, stems, and leaves
10. produce ovules that are not completely surrounded by sporophyte tissue
11. produce swimming sperm
12. produce flowers
13. protect the embryo as well as the zygote
14. Label the parts of the flower in the following illustration.

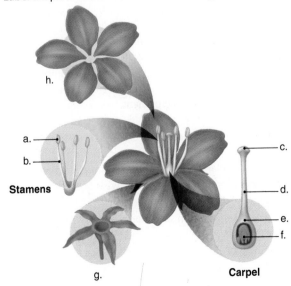

15. Lichens
   a. can live on bare rocks.
   b. can only survive in nutrient-rich soils.
   c. are parasitic on trees.
   d. can reproduce asexually.
   e. Both a and d are correct.

16. A fruit is derived from
   a. the corolla.   c. an ovule.
   b. an ovary.   d. the calyx.

17. Which of the following statements about fungi is false?
   a. Most fungi are multicellular.
   b. Fungal cell walls are composed of cellulose.
   c. Most fungi are nonmotile.
   d. Fungi digest their food before ingesting it.

18. Chytrids are
   a. a group of fungi with flagellated spores and gametes.
   b. responsible for a serious infection in frogs.
   c. a group of flowering plants that produce "naked seeds."
   d. the immature leaves of ferns.
   e. Both a and b are correct.

19. A fungal spore
   a. contains an embryonic organism.
   b. germinates directly into an organism.
   c. is often windblown.
   d. is most often diploid.
   e. Both b and c are correct.

20. Mycorrhizal fungi
   a. are a type of lichen.
   b. help plants gather solar energy.
   c. help plants gather inorganic nutrients.
   d. All of these are correct.

21. Which is the largest group of land plants?
   a. angiosperms   d. gymnosperms
   b. lycophytes   e. ferns
   c. bryophytes

22. The gametophyte is the dominant generation in
   a. ferns.   d. angiosperms.
   b. mosses.   e. More than one of these are correct.
   c. gymnosperms.

23. A seed is a mature
   a. embryo.   c. ovary.
   b. ovule.   d. pollen grain.

24. A mushroom is like a plant because it
   a. is a multicellular eukaryote.   d. is photosynthetic.
   b. produces spores.   e. All but d are correct.
   c. is adapted to a land environment.

25. Xylem is
   a. a polysaccharide found in the cell walls of all fungi.
   b. a plant vascular tissue that transports water and minerals.
   c. a plant vascular tissue that transports organic compounds.
   d. a polysaccharide found in the cell walls of all plants.
   e. the part of a flower that produces pollen.

26. The male gametophyte in seed plants is the
   a. pollen grain.   d. megaspore mother cell.
   b. endosperm.   e. seed.
   c. cone.

27. Which of the following is a saprotroph?
   a. photosynthetic organism   d. decomposer
   b. pine tree   e. fern leaf
   c. nonvascular plant

28. A mycelium is
   a. a mass of fungal filaments.
   b. a type of fungus with flagellated spores and gametes.
   c. the main body of a typical fungus.
   d. a mutualistic association between a fungus and a green alga or cyanobacterium.
   e. Both a and c are correct.

29. Which of these were among the first land plants with well-developed vascular tissue?
   a. angiosperms   d. gymnosperms
   b. ferns   e. bryophytes
   c. lycophytes

Thinking Scientifically

1. Bare-root pine tree seedlings transplanted into open fields often grow very slowly. However, pine seedlings grow much more vigorously if they are dug from their native environment and then transplanted into a field, as long as some of the original soil is retained on the seedlings. Why is it so important to retain some native soil on the seedlings?

2. Evolutionary trees like the one here indicate that members of kingdoms Plantae and Fungi both had protist ancestors. All three groups of organisms belong to which domain? What characteristics would distinguish the protist ancestors of plants from those of fungi?

Bioethical Issue

Scientific Limitations

Bristlecone pines are among the oldest trees on Earth. In 1964, a graduate student working in the southwestern United States took core samples from several trees to determine their age. One tree was found to be over 4,000 years old. When the student's coring tool broke, the U.S. Forest Service gave him permission to cut down the tree in order to accurately determine its age. The tree was found to be 4,862 years old—the oldest known living creature on Earth.

   Did the student do anything wrong, scientifically or morally, considering that he was given permission to cut down the tree? Who should be responsible for protecting unique trees like the bristlecone pine?

Essentials of Biology Website

The companion website for *Essentials of Biology* provides a wealth of information organized and integrated by chapter. You will find practice tests, animations, videos, and much more that will complement your learning and understanding of general biology.

http://www.mhhe.com/maderessentials2

Both a cat and a duckbill platypus, which has a bill like a duck, are mammals.

Both a gull, which flies in the air, and a penguin, which doesn't, are birds.

Outline

Who's Related to Who?

FINDING RELATIONSHIPS ISN'T ALL that easy. A duckbill platypus has webbed feet and a paddlelike tail for swimming. It uses its bill to find and collect food. Millions of tiny, jelly-filled pits lining the toothless bill are electroreceptors that detect tiny electrical charges given off by the platypus's prey. A platypus eats worms, crustaceans, fish, tadpoles, and adult frogs, which it grinds with two hornlike plates at the back of its bill. To reproduce, the duckbill platypus lays hard-shelled eggs.

—*Continued next page*

Continued—

Somewhat surprisingly, the duckbill platypus is a mammal, the same as a dog or cat. The two characteristics of a mammal are hair and mammary glands, and the duckbill platypus has both. Thick hair covers its body, except for its bill and feet. Female platypuses produce milk from mammary glands, but because they have no nipples, the young lick the milk from their fur.

It's clear to us that gulls are birds. After all, they fly in the air. But penguins don't fly. When on land, they walk around on webbed hind feet with a flipper hanging at each side. Penguins are excellent divers and spend most of their time at sea feeding on fish, squid, and small crustaceans called krill. The Emperor penguin has been known to dive to a depth of more than 1,500 feet and stay down for over 18 minutes. Penguins' solid bones—not hollow like those of other birds—help them stay underwater.

So what makes a penguin a bird? Like the gull, the attractive coat of a penguin is made of feathers, and its flipper, built like a bird's wing, allows it to "fly" through the water. Penguins lay hard-shelled eggs and raise their chicks on land. A penguin might look like a mammal from a distance, but it is clearly a bird!

The content of this chapter will help you get to know some of the other major groups of animals.

19.1 | Evolution of Animals

Animals can be contrasted with plants and fungi. Like both of these, animals are multicellular eukaryotes, but unlike plants, which make their food through photosynthesis, they are chemoheterotrophs and must acquire nutrients from an external source. So must fungi, but fungi digest their food externally and absorb the breakdown products. Animals ingest (eat) their food and digest it internally.

Animals usually carry on sexual reproduction and begin life only as a fertilized diploid egg. From this starting point, they undergo a series of developmental stages to produce an organism that has specialized tissues, usually within organs that carry on specific functions. Two types of tissues in particular—muscles and nerves—characterize animals. The presence of these tissues allows an animal to exhibit motility and a variety of flexible movements. The evolution of these tissues enables many types of animals to search actively for their food and to prey on other organisms. Coordinated movements also allow animals to seek mates, shelter, and a suitable climate—behaviors that have resulted in the vast diversity of animals. The more than 30 animal phyla we recognize today have all evolved from a single ancestor.

Figure 19.1 illustrates the characteristics of a complex animal using the frog as an example. A frog goes through a number of embryonic stages to become

Adult frog

Figure 19.1 Animals.

Animals begin life as a fertilized egg. The egg undergoes development to produce a multicellular organism that has specialized tissues. Animals depend on a source of external food to carry on life's processes. This series of images shows the development and metamorphosis of the frog, a complex animal.

Stages in development, from zygote to embryo.

Stages in metamorphosis, from hatching to tadpole.

1 Flagellates form an aggregate.	**2** Colony of cells forms a hollow sphere.	**3** Specialization of cells for reproduction.	**4** Infolding creates tissues.

single flagellate

reproductive cells

Figure 19.2   The colonial flagellate hypothesis.

The numbered statements explain the hypothesis.

a larval form (the tadpole) with specialized organs, including muscular and nervous systems that enable it to swim. A larva is an immature stage that typically lives in a different habitat and feeds on different foods than the adult. By means of a change in body form called *metamorphosis*, the larva, which only swims, turns into a sexually mature adult frog that swims and hops. The aquatic tadpole lives on plankton, and the terrestrial adult typically feeds on insects and worms. A large African bullfrog will try to eat just about anything, including other frogs, as well as small fish, reptiles, and mammals.

Ancestry of Animals

In Chapter 18, we discussed evidence that plants most likely share a green algal ancestor with the charophytes. What about animals? Did they also evolve from a protist, most likely a particular protozoan? The *colonial flagellate hypothesis* states that animals are descended from an ancestor that resembled a hollow spherical colony of flagellated cells. **Figure 19.2** shows how the process would have begun with an aggregate of a few flagellated cells. From there a larger number of cells could have formed a hollow sphere. Individual cells within the colony would have become specialized for particular functions, such as reproduction. Two tissue layers could have arisen by an infolding of certain cells into a hollow sphere. Tissue layers do arise in this manner during the development of animals today. The colonial flagellate hypothesis is also attractive because it implies that **radial symmetry** preceded **bilateral symmetry** as the evolutionary tree of animals predicts (see Fig. 19.4).

Among the protists, choanoflagellates (collared flagellates) most likely resemble the last unicellular ancestor of animals, and molecular data tell us that they are the closest living protist relative of animals! A choanoflagellate is a single cell, 3–10 μm in diameter, with a flagellum surrounded by a collar of 30–40 microvilli (**Fig. 19.3**).

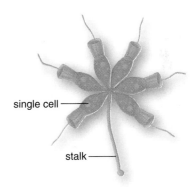

single cell

stalk

Figure 19.3   Choanoflagellates.

The choanoflagellates are the living protozoans most closely related to animals and may resemble their last unicellular ancestor. Some choanoflagellates live in colonies like these, in which a group of cells are attached to a surface by means of a stalk.

Check Your **Progress**

1.   **Compare and contrast nutrition in animals with nutrition in fungi.**
2.   **Compare and contrast nutrition in animals with nutrition in plants.**

Answers: 1. Both animals and fungi are chemoheterotrophs. Fungi digest their food externally, while animals digest their food internally.   2. Animals are chemoheterotrophs, while plants are photosynthesizers.

The Evolutionary Tree of Animals

Because so many types of animals arose during the Cambrian period in such a short time (see p. 308) it has been impossible to trace the evolutionary history of animals with certainty. However, molecular data and the developmental stages of today's living animals in particular have produced the evolutionary tree shown in **Figure 19.4.**

Evolutionary Trends

Animals differ in biological organization. Sponges, like all animals, are *multicellular*, but they have no true tissues and therefore have the cellular level of organization. Cnidarians, such as *Hydra*, with two tissue layers in their body wall, have true tissues, which can be associated with two germ layers when they are embryos. The animals in all the other phyla have three germ layers called ectoderm, endoderm, and mesoderm as embryos, and these shape their organs as they develop. **Germ layers** are so-called because they give rise to all other tissues and organs in an animal's body.

Animals differ in symmetry. Many sponges have no particular symmetry, and are therefore asymmetrical. *Radial symmetry*, as seen in cnidarians, means that the animal is organized circularly, similar to a wheel. No matter where the animal is sliced longitudinally, two mirror images are obtained (**Fig. 19.5**a). *Bilateral symmetry*, as seen in flatworms, means that the animal has

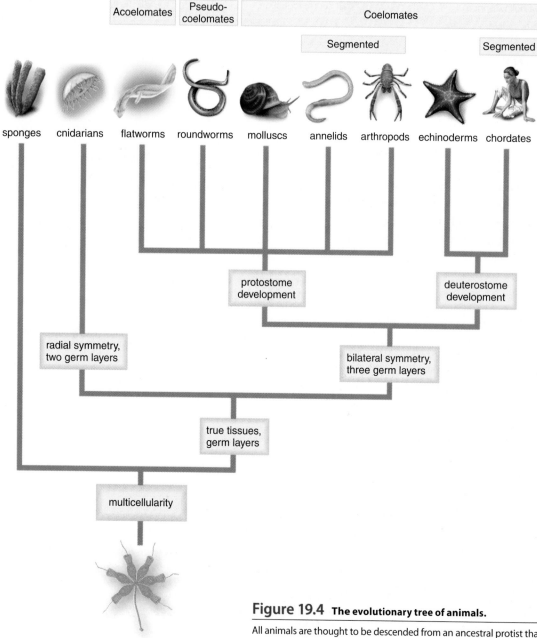

Figure 19.4 **The evolutionary tree of animals.**

All animals are thought to be descended from an ancestral protist that may have resembled a living Choanoflagellate.

definite right and left halves; only a longitudinal cut down the center of the animal will produce mirror images (Fig. 19.5b). During the evolution of animals, the trend toward bilateral symmetry is accompanied by **cephalization,** localization of a brain and specialized sensory organs at the anterior end of an animal. Therefore, bilateral symmetry was an important precursor for the evolutionary trend of increasing complexity of the nervous system and sensory organs. The appearance of bilateral symmetry marked the evolution of animals that were able to exploit environmental resources in new ways through their more active lifestyles. When using molecular data, it is assumed that the more closely related two organisms are, the more nucleotide sequences they will have in common. Molecular data and developmental data suggest that the groups designated as **protostomes** are more closely related to each other than they are to the **deuterostomes.**[1] The protostomes differ with regard to the presence of a **coelom,** or body cavity. Some protostomes have no body cavity and are acoelomates as are flatworms (**Fig. 19.6a**). **Acoelomates** are packed solid with mesoderm. In contrast, a body cavity provides a space for the various internal organs. Roundworms are **pseudocoelomates,** and their body cavity is incompletely lined by mesoderm—that is, a layer of mesoderm exists beneath the body wall but not around the gut (Fig. 19.6b). The other protostomes and the deuterostomes are coelomates in which the body cavity is completely lined with mesoderm (Fig. 19.6c). In animals with a coelom, mesentery, which is composed of strings of mesoderm, supports the internal organs. In coelomate animals, such as earthworms, lobsters, and humans, the mesoderm can interact not only with the ectoderm but also with the endoderm. Therefore, body movements are freer because the outer wall can move independently of the organs, and the organs have the space to become more complex. In animals without a skeleton, a coelom even acts as a so-called hydrostatic skeleton.

Animals can be nonsegmented or *segmented*. **Segmentation** is the repetition of body parts along the length of the body. Annelids, such as earthworms; arthropods, such as lobsters; and chordates, such as ourselves, are segmented. To illustrate your segmentation, run your hand along your backbone, which is composed of a series of vertebrae. Segmentation leads to specialization of parts because the various segments can become differentiated for specific purposes. In the case of your backbone, the two small vertebrae just beneath your skull are specialized to permit you to move your head up and down and side to side. The vertebrae of your lower back are much larger and sturdier, as they support the weight of your upper body.

[1] Molecular data suggest that the protostomes can be divided into two subgroups called the lophotrochozoa and the ecdysozoa, but the text does not consider this because, at present, a consensus has not been reached by the scientific community.

Check Your Progress

1. What type of symmetry does the human body have? Does the human body have cephalization?
2. Are humans protostomes or deuterostomes?
3. Which would you expect to have a more complex nervous system, cnidarians or flatworms?

Answers: 1. Humans have bilateral symmetry and cephalization.   2. Humans are deuterostomes.   3. For the most part, flatworms, with bilateral symmetry and cephalization, have more complex nervous systems than the radially symmetrical cnidarians.

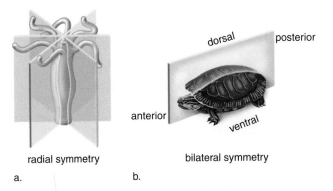

radial symmetry
a.

bilateral symmetry
b.

Figure 19.5   Radial versus bilateral symmetry.

a. With radial symmetry, two mirror images are obtained no matter how the animal is sliced longitudinally. Radially symmetrical animals tend to stay in one place and reach out in all directions to get their food. **b.** With bilateral symmetry, mirror images are obtained only if the animal is sliced down the middle. Bilaterally symmetrical animals tend to actively go after their food.

a. Acoelomate (flatworms)

b. Pseudocoelomate (roundworms)

c. Coelomate (molluscs, annelids, arthropods, echinoderms, chordates)

Figure 19.6   Type of body cavity.

a. Flatworms don't have a body cavity; they are acoelomates, and mesoderm fills the space between ectoderm and endoderm. **b.** Roundworms are pseudocoelomates; they have a body cavity, and mesoderm lies next to the ectoderm but not the endoderm. **c.** Other animals are coelomates, and mesoderm lines the entire body cavity.

a. Yellow tube sponge   b. Collar tube

Figure 19.7 **Sponge anatomy.**

Water enters a sponge through pores and circulates past collar cells before exiting at the mouth, or osculum. Collar cells digest small particles that become trapped by their appendages, flagella in the microvilli of their collar. Amoeboid cells transport nutrients from cell to cell.

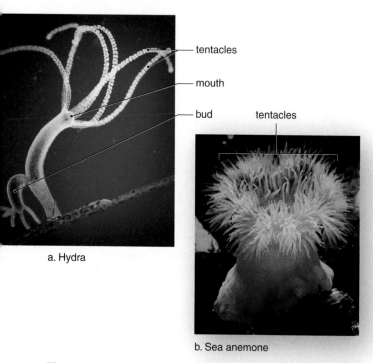

a. Hydra

b. Sea anemone

Figure 19.8 **Cnidarians.**

Hydras (**a**) and sea anemones (**b**) are solitary polyps that use tentacles laden with stinging cells to capture their food.

19.2 | Introducing the Invertebrates

This section describes the invertebrate animals—the sponges, cnidarians, flatworms, and roundworms. Several key evolutionary trends are seen among these phyla (see Fig. 19.4).

Sponges: Multicellularity

Sponges (phylum Porifera) have saclike bodies perforated by many pores (**Fig. 19.7***a*). Sponges are aquatic, largely marine animals that vary greatly in size, shape, and color. Sponges are multicellular but lack organized tissues. Therefore, *sponges have the cellular level of organization*. Molecular data place sponges at the base of the evolutionary tree of animals.

The body wall of a sponge is lined internally with flagellated cells called *collar cells*, or choanocytes (Fig. 19.7*b*). The beating of the flagella produces water currents that flow through the pores into the central cavity and out through the osculum, the upper opening of the body. Even a simple sponge only 10 centimeters tall is estimated to filter as much as 100 liters of water each day. It takes this much water to supply the needs of the sponge. A sponge is a sedentary **filter feeder,** an organism that filters its food from the water by means of a straining device—in this case, the pores of the walls and the microvilli making up the collar of collar cells. Microscopic food particles that pass between the microvilli are engulfed by the collar cells and digested by them in food vacuoles.

Sponges can reproduce both asexually and sexually. They reproduce asexually by fragmentation or by *budding*. During budding, a small protuberance appears and gradually increases in size until a complete organism forms. Budding produces colonies of sponges that can become quite large. During sexual reproduction, eggs and sperm are released into the central cavity, and the zygote develops into a flagellated larva that may swim to a new location. If the cells of a sponge are mechanically separated, they will reassemble into a complete and functioning organism! Like many less specialized organisms, sponges are also capable of regeneration, or growth of a whole from a small part.

Some sponges have an endoskeleton composed of *spicules*, small needle-shaped structures with one to six rays. Most sponges have fibers of spongin, a modified form of collagen; a bath sponge is the dried spongin skeleton from which all living tissue has been removed. Today, however, commercial "sponges" are usually synthetic.

Cnidarians: True Tissues

Cnidarians (phylum Cnidaria) are an ancient group of invertebrates with a rich fossil record. Cnidarians are radially symmetrical and capture their prey with a ring of tentacles that bear specialized stinging cells, called cnidocytes (**Fig. 19.8**). Each cnidocyte has a capsule called a **nematocyst,** containing a long, spirally coiled, hollow thread. When the trigger of the cnidocyte is touched, the nematocyst is discharged. Some nematocysts merely trap a prey or predator; others have spines that penetrate and inject paralyzing toxins before the prey is captured and drawn into the gastrovascular cavity. Most cnidarians live in the sea, though there are a few freshwater species.

During development, cnidarians have two germ layers (ectoderm and endoderm), and as adults *cnidarians have the tissue level of organization.* Two basic body forms are seen among cnidarians—the polyp and the medusa. The mouth of a polyp is directed upward from the substrate, while the mouth of the medusa is directed downward. A medusa has much jellylike packing material and is commonly called a "jellyfish." Polyps are tubular and generally attached to a rock (**Fig. 19.9**).

Cnidarians, as well as other marine animals, have been the source of medicines, particularly drugs that counter inflammation.

Flatworms: Bilateral Symmetry

Flatworms (phylum Platyhelminthes) have *bilateral symmetry.* Like all the other animal phyla we will study, they also have three germ layers. However, the flatworms have no coelom—they are acoelomates. The presence of mesoderm in addition to ectoderm and endoderm gives bulk to the animal and leads to greater complexity.

Free-living flatworms, called **planarians,** have several body systems (**Fig. 19.10**), including a ladderlike nervous system. A small anterior brain and two lateral nerve cords are joined by cross-branches called transverse nerves. Planarians exhibit cephalization; aside from a brain, the "head" end has light-sensitive organs (the eyespots) and chemosensitive organs located on the auricles. Their three muscle layers—an outer circular layer, an inner longitudinal layer, and a diagonal layer—allow for varied movement. Their ciliated lower epidermis allows planarians to glide along a film of mucus.

Figure 19.9 More cnidarians.

a. The Portuguese man-of-war is a colony of polyp and medusa types of individuals. One polyp becomes a gas-filled float, and the other polyps are specialized for feeding. **b.** The calcium carbonate skeletons of corals form the coral reefs, areas of biological diversity. **c.** In jellyfishes, the medusa is the primary stage of the life cycle, and a polyp form remains quite small and inconspicuous.

a. Portuguese man-of-war

b. Cup coral

c. Jellyfish

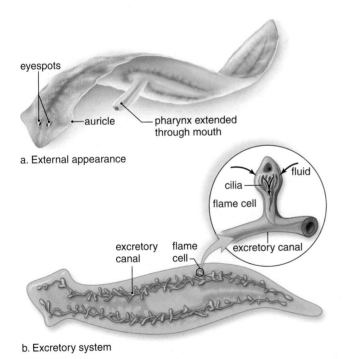

a. External appearance

b. Excretory system

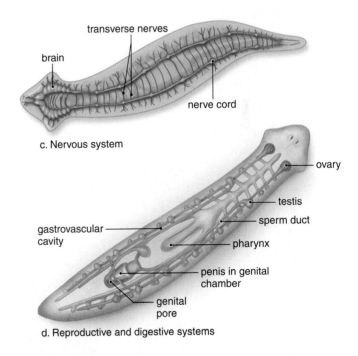

c. Nervous system

d. Reproductive and digestive systems

Figure 19.10 Anatomy of a planarian.

a. This drawing shows that flatworms are bilaterally symmetrical and have a head region with eyespots. **b.** The excretory system with flame cells is shown in detail. **c.** The nervous system has a ladderlike appearance. **d.** The reproductive system (shown in brown) has both male and female organs, and the digestive system (shown in pink) has a single opening. When the pharynx is extended as shown in (**a**), a planarian sucks food up into a gastrovascular cavity that branches throughout its body.

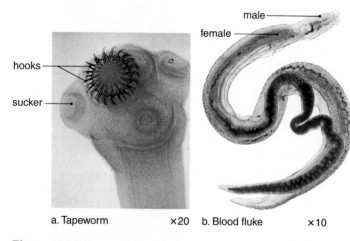

a. Tapeworm ×20 b. Blood fluke ×10

Figure 19.11 Parasitic flatworms.

a. The anterior end of a tapeworm has hooks and suckers for attachment to the intestinal wall. **b.** Sexes are separate in blood flukes, which cause schistosomiasis.

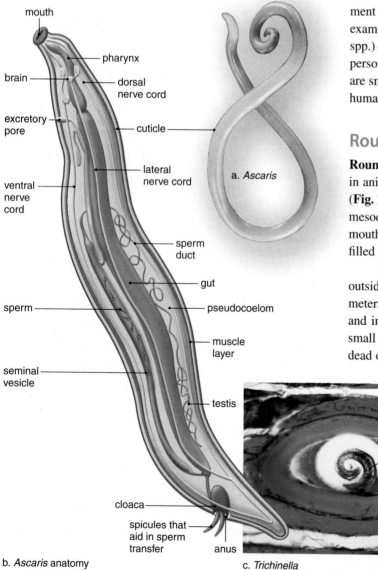

a. *Ascaris*

b. *Ascaris* anatomy

c. *Trichinella* ×124

The animal captures food by wrapping itself around the prey, entangling it in slime, and pinning it down. Then the planarian extends a muscular pharynx, and by a sucking motion tears up and swallows its food. The pharynx leads into a three-branched gastrovascular cavity where digestion occurs. The digestive tract is incomplete because it has only one opening.

Planarians are **hermaphrodites,** meaning that they possess both male and female sex organs. The worms practice cross-fertilization: The penis of one is inserted into the genital pore of the other, and a reciprocal transfer of sperm takes place. The fertilized eggs hatch in 2–3 weeks as tiny worms.

The parasitic flatworms belong to two classes: the tapeworms and the flukes. As adults, tapeworms are endoparasites (internal parasites) of various vertebrates, including humans (**Fig. 19.11**a). They vary in size from a few millimeters to nearly 20 meters. Tapeworms have a well-developed anterior region called the scolex, which bears hooks and suckers for attachment to the intestinal wall of the host. Behind the scolex, a series of reproductive units called proglottids contain a full set of female and male sex organs. After fertilization, the organs within a proglottid disintegrate, and it becomes filled with mature eggs. The eggs, or else the mature proglottids, are eliminated in the feces of the host.

Flukes are all endoparasites of various vertebrates. The anterior end of these animals has an oral sucker and at least one other sucker used for attachment to the host. Flukes are usually named for the organ they inhabit; for example, there are blood, liver, and lung flukes. Blood flukes (*Schistosoma* spp.) occur predominantly in the Middle East, Asia, and Africa. Nearly 800,000 persons die each year from blood fluke infection, called schistosomiasis. Adults are small (approximately 2.5 centimeters long) and may live for years in their human hosts (Fig. 19.11b).

Roundworms: Pseudocoelomates

Roundworms (phylum Nematoda) possess two anatomical features not seen in animals discussed previously: a *body cavity* and a *complete digestive tract* (**Fig. 19.12**). The body cavity is a pseudocoelom and is incompletely lined with mesoderm (see Fig. 19.6). The digestive tract is complete because it has both a mouth and an anus. Worms in general do not have a skeleton, but the fluid-filled pseudocoelom supports muscle contraction and enhances flexibility.

The roundworms are nonsegmented, meaning that they have a smooth outside body wall. Roundworms are generally colorless and less than 5 centimeters in length, and they occur almost everywhere—in the sea, in fresh water, and in the soil—in such numbers that thousands of them can be found in a small area. Many are free-living and feed on algae, fungi, microscopic animals, dead organisms, and plant juices, causing great agricultural damage. Parasitic

Figure 19.12 Roundworm anatomy.

a. The roundworm *Ascaris*. **b.** Roundworms such as *Ascaris* have a pseudocoelom and a complete digestive tract with a mouth and an anus. The sexes are separate; this is a male roundworm. **c.** The larvae of the roundworm *Trichinella* penetrate striated muscle fibers, where they coil in a sheath formed from the muscle fiber.

roundworms live anaerobically in every type of animal and many plants. Several parasitic roundworms infect humans.

A female *Ascaris lumbricoides,* a human parasite, is very prolific, producing over 200,000 eggs daily. The eggs are eliminated with host feces, and under the right conditions they can develop into a worm within two weeks. The eggs enter the body via uncooked vegetables, soiled fingers, or ingested fecal material, and hatch in the intestines. The juveniles make their way into the cardiovascular system and are carried to the heart and lungs. From the lungs, the larvae travel up the trachea, where they are swallowed and eventually reach the intestines. There the larvae mature and begin feeding on intestinal contents.

Trichinosis is a fairly serious human infection rarely seen in the United States. The female trichina worm burrows into the wall of the host's small intestine; here she deposits live larvae that are carried by the bloodstream to the skeletal muscles, where they encyst. Once the adults are in the small intestine, digestive disorders, fatigue, and fever occur. After the larvae encyst, the symptoms include aching joints, muscle pain, and itchy skin. Humans catch the disease when they eat infected meat.

Elephantiasis is caused by a roundworm called a filarial worm, which utilizes mosquitoes as a secondary host. Because the adult worms reside in lymphatic vessels, fluid return is impeded, and the limbs of an infected human can swell to an enormous size, even resembling those of an elephant. When a mosquito bites an infected person, it can transport larvae to a new host.

Other roundworm infections are more common in the United States. Children frequently acquire pinworm infection, and hookworm is seen in the southern states, as well as worldwide. A hookworm infection can be very debilitating because the worms attach to the intestinal wall and feed on blood. Good hygiene, proper disposal of sewage, thorough cooking of meat, and regular deworming of pets usually protect people from parasitic roundworms.

19.3 | Protostomes and Deuterostomes Compared

Protostomes and deuterostomes can be distinguished on the basis of embryological development. Notice that cell division to form an embryo is different in protostomes (flatworms, roundworms, molluscs, annelids, and arthropods) compared to that in deuterostomes (echinoderms and chordates). In protostomes and in deuterostomes, the first embryonic opening is called the blastopore. In protostomes the blastopore becomes the mouth, and in deuterostomes, it becomes the anus. Actually, this is the feature that explains their names; the mouth is the first opening in protostomes (*proto*, before) and the second opening in deuterostomes (*deutero*, second). The coelom, when present, develops by a splitting of the mesoderm in protostomes and develops as outpockets of the primitive gut in deuterostomes (**Fig. 19.13**).

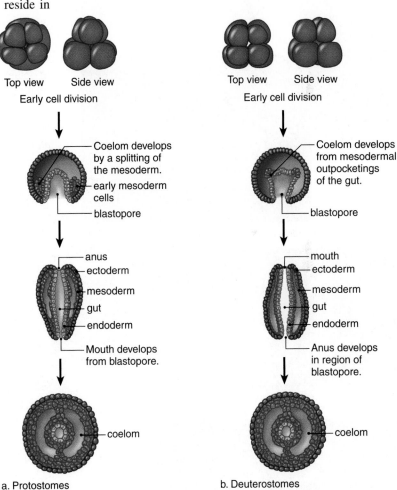

a. Protostomes

b. Deuterostomes

Figure 19.13 Development in protostomes and deuterostomes.
a. Protostomes are characterized by spiral cell division when the embryo first forms and a blastopore that becomes the mouth. Further, if a coelom is present, it forms by a splitting of the mesoderm. **b.** Deuterostomes are characterized by radial cell division when the embryo first forms and a blastopore that becomes the anus. Also, the primitive gut outpockets to form the coelom.

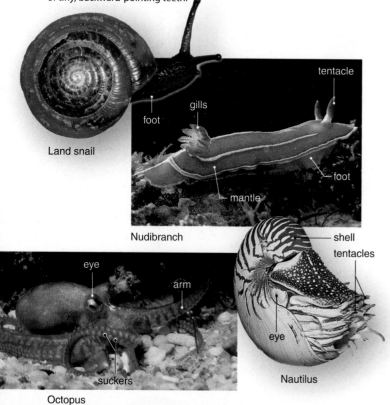

heart · shell

coelom

mantle

visceral mass

mantle cavity

anus

gill

radula

mouth

foot

a.

radula teeth

mouth

b.

Figure 19.14 Body plan of molluscs.

Molluscs have a three-part body: **a.** a muscular foot, a visceral mass, and a mantle. **b.** In the mouth, the radula is a tonguelike organ that bears rows of tiny, backward-pointing teeth.

Land snail

tentacle

gills

foot

foot

mantle

Nudibranch

shell

tentacles

eye

arm

eye

suckers

Nautilus

Octopus

eye

tentacle

valve

Scallop

valve

Mussels

19.4 | Molluscs, Annelids, and Arthropods

Molluscs, annelids, and arthropods are the types of protostomes that have a coelom.

Molluscs

Despite being a very large and diversified group, all molluscs (phylum Mollusca) have a body composed of at least three distinct parts: (1) the *visceral mass* is the soft-bodied portion that contains internal organs; (2) the *foot* is the strong, muscular portion used for locomotion; and (3) the *mantle* is a membranous or sometimes muscular covering that envelops, but does not completely enclose, the visceral mass. In addition, the *mantle cavity* is the space between the two folds of the mantle. The mantle may secrete an exoskeleton called a *shell*. Another feature often present is a rasping, tonguelike *radula*, an organ that bears many rows of teeth and is used to obtain food (**Fig. 19.14**).

In **gastropods** (meaning stomach-footed), including nudibranchs, conchs, and snails, the foot is ventrally flattened, and the animal moves by muscle contractions that pass along the foot (**Fig. 19.15**, *top*). Many gastropods are herbivores that use their radulas to scrape food from surfaces. Others are carnivores, using their radulas to bore through surfaces such as bivalve shells to obtain food. In snails that are terrestrial, the mantle is richly supplied with blood vessels and functions as a lung.

In **cephalopods** (meaning head-footed), including octopuses, squids, and nautiluses, the foot has evolved into tentacles about the head (Fig. 19.15, *middle*). The tentacles seize prey, and then a powerful beak and a radula tear it apart. Cephalopods possess well-developed nervous systems and complex sensory organs. The large brain is formed from a fusion of ganglia, and nerves leaving the brain supply various parts of the body. An especially large pair of nerves controls the rapid contraction of the mantle, allowing these animals to move quickly by jet propulsion of water. Rapid movement and the secretion of a brown or black pigment from an ink gland help cephalopods escape their enemies. In the squid and octopus, a well-developed eye resembles that of vertebrates, having a cornea, lens, retina, and iris. Octopuses have no shell, and squids have only a remnant of one concealed beneath the skin. Scientific experiments reveal that octopuses in particular are highly intelligent.

Clams, oysters, scallops, and mussels are called **bivalves** because of the two parts to their shells (Fig. 19.15, *bottom*). They have a muscular foot that projects ventrally from the shell. In a clam, such as the freshwater clam, the calcium carbonate shell has an inner layer of mother-of-pearl. The clam is a filter feeder. Food particles and water enter the mantle cavity by way of the incurrent siphon, a posterior opening between the two valves. Mucous secretions cause smaller particles to adhere to the gills, and ciliary action sweeps them toward the mouth.

Figure 19.15 Molluscan diversity.

Top row: Snails (*left*) and nudibranchs (*right*) are gastropods. Middle row: Octopuses (*left*) and nautiluses (*right*) are cephalopods. Bottom row: Scallops (*left*) and mussels (*right*) are bivalves.

Molluscs have some economic importance as a source of food and pearls. If a foreign body is placed between the mantle and the shell of a clam, concentric layers of shell are deposited about the particle to form a pearl.

Annelids: Segmented Worms

Annelids (phylum Annelida) are *segmented,* as can be seen externally by the rings that encircle the body of an earthworm. Partitions called septa divide the well-developed, fluid-filled coelom, which is used as a hydrostatic skeleton to facilitate movement. In annelids, the complete digestive tract body plan has led to specialization of parts (**Fig. 19.16**). For example, the digestive system may include a pharynx, esophagus, crop, gizzard, intestine, and accessory glands. Annelids have an extensive closed circulatory system with blood vessels that run the length of the body and branch to every segment. The nervous system consists of a brain connected to a ventral nerve cord, with ganglia in each segment. The excretory system consists of nephridia in most segments. A **nephridium** is a tubule that collects waste material and excretes it through an opening in the body wall.

Most annelids are polychaetes (having many setae per segment) that live in marine environments. Setae are bristles that anchor the worm or help it move. A clam worm is a predator. It preys on crustaceans and other small animals, capturing them with a pair of strong, chitinous jaws that extend with a part of the pharynx (**Fig. 19.17***a*). In support of its predatory way of life, a clam worm has a well-defined head region, with eyes and other sense organs. Other polychaetes are sedentary (sessile) tube worms, with tentacles that form a funnel-shaped fan. Water currents created by the action of cilia trap food particles that are directed toward the mouth (Fig. 19.17*b*).

The oligochaetes (few setae per segment) include the earthworms (see Fig. 19.16). Earthworms do not have a well-developed head, and they reside in soil where there is adequate moisture to keep the body wall moist for gas exchange. They are scavengers that feed on leaves or any other organic matter, living or dead, that can conveniently be taken into the mouth along with dirt.

Leeches have no setae but have the same body plan as other annelids. Most are found in fresh water, but some are marine or even terrestrial. The medicinal leech can be as long as 20 centimeters, but most leeches are much shorter. A leech has two *suckers,* a small one around the mouth and a large posterior one. While some leeches are free-living, most are fluid feeders that penetrate the surface of an animal by using a proboscis or their jaws, and then suck in fluids with their powerful pharynx. Leeches are able to keep blood flowing and prevent clotting by means of a substance in their saliva known as hirudin, a powerful anticoagulant. This secretion has added to their potential usefulness in the field of medicine today (Fig. 19.17*c*).

a.

b.

Figure 19.16   **Earthworm anatomy.**

a. Internal anatomy of the anterior part of an earthworm. **b.** Cross section of an earthworm.

Figure 19.17   **Other annelids.**

A polychaete can be predacious like this marine clam worm (**a**) or live in a tube like this fan worm called the Christmas tree worm (**b**). **c.** The medicinal leech, also an annelid, is sometimes used to remove blood from tissues after surgery.

a. Clam worm

b. Christmas tree worm

c. Medicinal leech

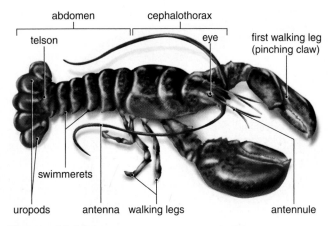

Figure 19.18   Exoskeleton and jointed appendages.

Arthropods, such as a lobster, have various appendages attached to the head region of a cephalothorax and five pairs of walking legs attached to the thorax region. Appendages called swimmerets, used in reproduction and swimming, are attached to the abdomen. The uropods and telson make up a fan-shaped tail.

Figure 19.19   Monarch butterfly metamorphosis.

a. A caterpillar (larva) eats and grows. **b.** After the larva goes through several molts, it builds a chrysalis around itself and becomes a pupa. **c.** Inside the pupa, the larva undergoes changes in organ structure to become an adult. **d, e.** The adult butterfly emerges from the chrysalis and reproduces, and the cycle begins again.

Arthropods: Jointed Appendages

Arthropods (phylum Arthropoda) are extremely diverse. Over one million species have been discovered and described, but some experts suggest that as many as 30 million arthropods may exist—most of them insects. The success of arthropods can be attributed to the following six characteristics:

1. *Jointed appendages.* Basically hollow tubes moved by muscles, jointed appendages have become adapted to different means of locomotion, food gathering, and reproduction (**Fig. 19.18**). These modifications account for much of the diversity of arthropods.

2. *Exoskeleton.* A rigid but jointed exoskeleton is composed primarily of **chitin,** a strong, flexible, nitrogenous polysaccharide. The exoskeleton serves many functions, including protection, prevention of desiccation, attachment for muscles, and locomotion.

   Because an exoskeleton is hard and nonexpandable, arthropods must undergo **molting,** or shedding of the exoskeleton, as they grow larger. During molting, arthropods are vulnerable and are attacked by many predators.

3. *Segmentation.* In many species, the repeating units of the body are called segments. Each has a pair of jointed appendages. In others, the segments are fused into a head, thorax, and abdomen.

4. *Well-developed nervous system.* Arthropods have a brain and a ventral nerve cord. The head bears various types of sense organs, including compound and simple eyes. Many arthropods also have well-developed touch, smell, taste, balance, and hearing capabilities. Arthropods display many complex behaviors and communication skills.

5. *Variety of respiratory organs.* Marine forms utilize gills; terrestrial forms have book lungs (e.g., spiders) or air tubes called tracheae. Tracheae serve as a rapid way to transport oxygen directly to the cells. The circulatory system is open, with the dorsal heart pumping blood into various sinuses throughout the body.

6. *Reduced competition through metamorphosis.* Many arthropods undergo a change in form and physiology as a larva becomes an adult. Metamorphosis allows the larva to have a different lifestyle than the adult (**Fig. 19.19**). For example, larval crabs live among and feed on plankton, while adult crabs are bottom dwellers that catch live prey or scavenge dead organic matter. Among insects such as butterflies, the caterpillar feeds on leafy vegetation, while the adult feeds on nectar.

Crustaceans, a name derived from their hard, crusty exoskeleton, are a group of largely marine arthropods that include barnacles, shrimps, lobsters, and crabs (**Fig. 19.20**). There are also some freshwater crustaceans, including the crayfish, and some terrestrial ones, including the sowbug, or pillbug. Although crustacean anatomy is extremely diverse, the head usually bears a pair of compound eyes and five pairs of appendages. The first two pairs of appendages, called antennae and antennules, respectively, lie in front of the mouth and have sensory functions. The other three pairs are mouthparts used in feeding. In a lobster, the thorax bears five pairs of walking legs. The first walking leg is a pinching claw. The gills are situated above the walking legs. The head and thorax are fused into a cephalothorax, which is covered on the top and sides by a nonsegmented carapace. The abdominal segments, which are largely muscular, are equipped with swimmerets, small paddlelike structures. The last two segments bear the uropods and the telson, which make up a fan-shaped tail to propel the lobster backward (see Fig. 19.18).

Figure 19.20 Crustacean diversity.

a. A copepod uses its long antennae for floating and its feathery maxillae for filter feeding. Shrimp (**b**) and crabs (**c**) are decapods—they have five pairs of walking legs. Shrimp resemble crayfish more closely than they do crabs, which have a reduced abdomen. Marine shrimp feed on copepods. **d.** The gooseneck barnacle is attached to an object by a long stalk. Barnacles have no abdomen and a reduced head; the thoracic legs project through a shell to filter feed. Barnacles often live on man-made objects such as ships, buoys, and cables.

Crustaceans play a vital role in the food chain. Tiny crustaceans known as krill are a major source of food for baleen whales, sea birds, and seals. Countries such as Japan are harvesting krill for human use. Copepods and other small crustaceans are primary consumers in marine and aquatic ecosystems. Many species of lobsters, crabs, and shrimp are important in the seafood industry. Some barnacles are destructive to wharfs, piers, and boats.

Among arthropods, the **arachnids** include spiders, scorpions, ticks, mites, and harvestmen ("daddy longlegs") (**Fig. 19.21***a*). Spiders have a narrow waist that separates the cephalothorax from the abdomen. Most spiders inject venom into their prey and digest their food externally before sucking it into the stomach. Spiders use silk threads for all sorts of purposes, from lining their nests to catching prey. The internal organs of spiders also show how they are adapted to a terrestrial way of life. Malpighian tubules work in conjunction with rectal glands to reabsorb ions and water before a relatively dry nitrogenous waste (uric acid) is excreted. Invaginations of the inner body wall form lamellae ("pages") of spiders' so-called book lungs.

Scorpions are the oldest terrestrial arthropods (Fig. 19.21*b*). Today, they occur in the tropics, subtropics, and temperate regions worldwide. They are nocturnal and spend most of the day hidden under a log or a rock. Ticks and mites are parasites. Ticks suck the blood of vertebrates and sometimes transmit diseases, such as Rocky Mountain spotted fever or Lyme disease. Chiggers, the larvae of certain mites, feed on the skin of vertebrates.

The horseshoe crab is grouped with the arachnids because the first pair of appendages are pincerlike structures used for feeding and defense. Horseshoe crabs have pedipalps, which they use as feeding and sensory structures, and four pairs of walking legs (Fig. 19.21*c*). Horseshoe crabs are of great interest to medical science. An extract from the horseshoe crab's blood cells is used to ensure that vaccines are free of bacterial toxins. Other compounds from the horseshoe crab are being investigated for antibiotic, antiviral, and anticancer properties.

Figure 19.21 More arthropods.

a. The black widow spider is a venomous spider that spins a web. **b.** Scorpions have large pincers in front and a long abdomen, which ends with a stinger containing venom. **c.** Horseshoe crabs are common along the North American east coast. **d.** A millipede has two pairs of legs on most segments. **e.** A centipede has a pair of appendages on almost every segment.

Housefly

Tortoiseshell scale

Snout beetle

Green lacewing

Flea

Walking stick

Scarlet and green leafhopper

Dragonfly

True bug

While millipedes (Fig. 19.21*d*), with two pairs of legs on most segments, are herbivorous, centipedes (Fig. 19.21*e*), with a pair of appendages on every segment, are carnivorous. The head appendages of these animals are similar to those of insects, which are the largest group of arthropods, or indeed animals.

Insects are so numerous (probably over one million species) and so diverse that the study of this one group is a major specialty in biology called entomology (**Fig. 19.22**). Some insects show remarkable behavior adaptations, as exemplified by the social systems of bees, ants, termites, and other colonial insects.

Insects are adapted to an active life on land, although some have secondarily invaded aquatic habitats. The body is divided into a head, a thorax, and an abdomen. The head usually bears a pair of sensory antennae, a pair of compound eyes, and several simple eyes. The mouthparts are adapted to each species' particular way of life: A grasshopper has mouthparts that chew, and a butterfly has a long tube for siphoning the nectar from flowers. The abdomen contains most of the internal organs; the thorax bears three pairs of legs and the wings—either one or two pairs, or none. Wings enhance an insect's ability to survive by providing a way of escaping enemies, finding food, facilitating mating, and dispersing the offspring. The exoskeleton of an insect is lighter and contains less chitin than that of many other arthropods. The male has a penis, which passes sperm to the female. The female, as in the grasshopper, may have an ovipositor for laying the fertilized eggs. Some insects, such as butterflies, undergo complete metamorphosis, involving a drastic change in form (see Fig. 19.19).

Check Your Progress

1. List the three body parts of a mollusc.
2. List the main features of annelids.
3. List the five major features that have helped to make arthropods successful.

Answers: 1. The mollusc's body has a visceral mass, a foot, and a mantle.   2. Annelids are segmented, with a complete digestive tract body plan. They contain a digestive system, closed circulatory system, nervous system, and excretory system.   3. Rigid, jointed exoskeleton; segmented body; well-developed nervous system; variety of respiratory organs; many go through metamorphosis.

Grasshopper Mouthparts Chewing

right mandible

left mandible

right maxilla with maxillary palp

left maxilla with maxillary palp

labium with labial palps

labrum

Honeybee

Luna moth

Lubber grasshopper

Figure 19.22 Insect diversity.

19.5 | Echinoderms and Chordates

The deuterostomes include the echinoderms and the chordates as noted in the evolutionary tree of animals (see Fig. 19.4).

Echinoderms

It may seem surprising that **echinoderms** (phylum Echinodermata) lack features associated with vertebrates such as humans, and yet of all the invertebrates, they are most closely related to chordates (**Fig. 19.23**). For example, the echinoderms are often radially, not bilaterally, symmetrical. Their larva is a free-swimming filter feeder with bilateral symmetry, but it metamorphoses into a radially symmetrical adult. Also, adult echinoderms do not have a head, brain, or segmentation. The nervous system consists of nerves in a ring around the mouth extending outward radially. Nevertheless, they are deuterostomes as are the chordates.

Echinoderm locomotion depends on a water vascular system. In the sea star, water enters this system through a sieve plate, or madreporite. Eventually it is pumped into many tube feet, expanding them. When the foot touches a surface, the center withdraws, producing suction that causes the foot to adhere to the surface. By alternating the expansion and contraction of its many tube feet, a sea star moves slowly along.

Echinoderms don't have a complex respiratory, excretory, or circulatory system. Fluids within the coelomic cavity and a water vascular system carry out many of these functions. For example, gas exchange occurs across the skin gills and the tube feet. Nitrogenous wastes diffuse through the coelomic fluid and the body wall.

In ecosystems, most echinoderms feed variously on organic matter in the sea or substratum, but sea stars prey upon crustaceans, molluscs, and other invertebrates. From the human perspective, sea stars cause extensive economic loss because they consume oysters and clams before they can be harvested. Fishes and sea otters eat echinoderms, and scientists favor echinoderms for embryological research.

Figure 19.23 Echinoderm diversity.

a. Anatomy of a sea star. **b.** Sea lilies are immobile, but feather stars can move about. They usually cling to coral or sponges where they feed on plankton. **c.** Sea cucumbers have a long, leathery body that resembles a cucumber, except for the feeding tentacles about the mouth. **d.** Brittle stars have a central disk from which long, flexible arms radiate. **e.** Sea urchins and sand dollars have spines for locomotion, defense, and burrowing.

a.

b. Sea lily (above), feather star (right)

c. Sea cucumber

d. Brittle star

e. Sea urchins (left), sand dollar (right)

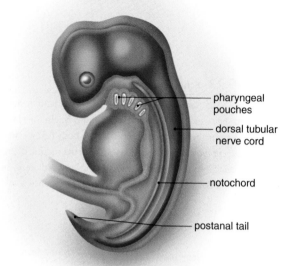

Figure 19.24 **The four chordate characteristics.**

Chordates

At some time in their life history, a **chordate** (phylum Chordata) has the four characteristics depicted in **Figure 19.24** and listed here:

1. A *dorsal supporting rod,* called a **notochord.** The notochord is located just below the nerve cord toward the back (i.e., dorsal). **Vertebrates** have an endoskeleton of cartilage or bone, including a vertebral column that has replaced the notochord during development.

2. A *dorsal tubular nerve cord.* Tubular means that the cord contains a canal filled with fluid. In vertebrates, the nerve cord is protected by the vertebrae. Therefore, it is called the spinal cord because the vertebrae form the spine.

3. *Pharyngeal pouches.* These structures are seen only during embryonic development in most vertebrates. In the invertebrate chordates, the fishes, and some amphibian larvae, the pharyngeal pouches become functioning gills. Water passing into the mouth and the pharynx goes through the gill slits, which are supported by gill arches. In terrestrial vertebrates that breathe with lungs, the pouches are modified for various purposes. In humans, the first pair of pouches become the auditory tubes. The second pair become the tonsils, while the third and fourth pairs become the thymus gland and the parathyroids.

4. A *tail.* Because the tail extends beyond the anus, it is called a postanal tail.

The Invertebrate Chordates

There are a few invertebrate chordates in which the notochord is never replaced by the vertebral column. **Tunicates** live on the ocean floor and take their name from a tunic that makes the adults look like thick-walled, squat sacs. They are also called sea squirts because they squirt water from one of their siphons when disturbed (**Fig. 19.25***a*). The tunicate larva is bilaterally symmetrical and has the four chordate characteristics. Metamorphosis produces the sessile adult in which numerous cilia move water into the pharynx and out numerous gill slits, the only chordate characteristic that remains in the adult.

    Lancelets are marine chordates only a few centimeters long. They have the appearance of a lancet—a small, two-edged surgical knife (Fig. 19.25*b*). Lancelets are found in the shallow water along most coasts, where they usually lie partly buried in sandy or muddy substrates with only their anterior mouth and gill apparatus exposed. They feed on microscopic particles filtered out of the constant stream of water that enters the mouth and exits through the gill slits. Lancelets retain the four chordate characteristics as adults. In addition, segmentation is present, as witnessed by the fact that the muscles are segmentally arranged and the dorsal tubular nerve cord has periodic branches.

Evolutionary Trends Among the Chordates

Figure 19.26 depicts the evolutionary tree of the chordates and previews the animal groups we will discuss in the remainder of this chapter. The figure also lists at least one main evolutionary trend that distinguishes each group of animals from the preceding ones. The tunicates and lancelets are invertebrate chordates; they don't have vertebrae. The vertebrates are the fishes, amphibians, reptiles, including birds, and mammals.

Figure 19.25 **Invertebrate chordates.**

a. Tunicates (sea squirts) have numerous gill slits, the only chordate characteristic that remains in the adult. **b.** Lancelets have all four chordate characteristics as adults.

As embryos, vertebrates have the four chordate characteristics. During embryonic development, the notochord is generally replaced by a vertebral column composed of individual vertebrae. Remnants of the notochord are seen in the intervertebral discs, which are compressible cartilaginous pads found between the vertebrae. The vertebral column, which is a part of the flexible but strong endoskeleton, gives evidence that vertebrates are segmented.

Fishes with an endoskeleton of cartilage and some bone in their scales were the first to have jaws. Early bony fishes had lungs. Amphibians were the first group to clearly have jointed appendages and to invade land. The typical life cycle of amphibians includes a larval stage that lives in the water, as described in Figure 19.1. However, amphibians exploit a wide range of habitats and many reproduce on land. Reptiles, birds, and mammals have a means of reproduction more suited to land. During development, an amnion and other extraembryonic membranes are present. These membranes carry out all the functions needed to support the embryo as it develops into a young offspring, capable of feeding on its own.

Check Your **Progress**

1. What are unique features of echinoderms?
2. List the four distinguishing features of chordates.
3. Name two groups of invertebrate chordates in which the notochord is never replaced by vertebrae.

Answers: 1. Echinoderms have radial symmetry as adults and a water vascular system with tube feet for locomotion.   2. Chordates have a notochord, dorsal tubular nerve cord, pharyngeal pouches, and postanal tail.   3. Tunicates, commonly known as sea squirts, and lancelets are invertebrate chordates.

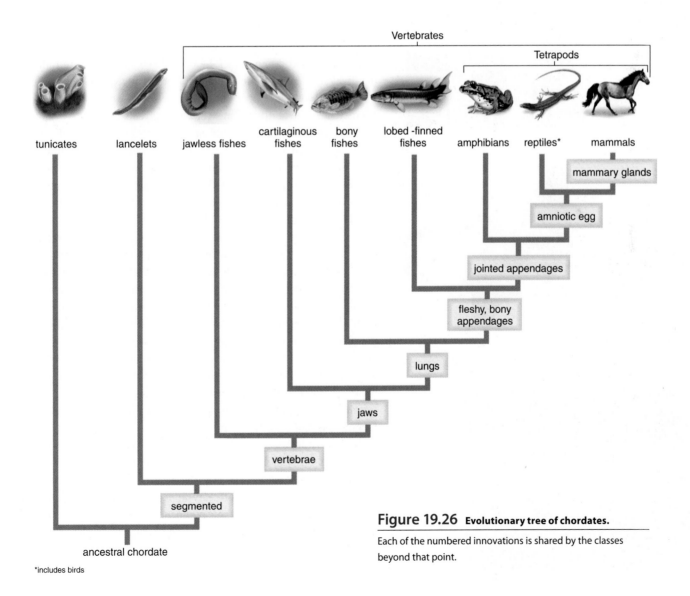

Figure 19.26   Evolutionary tree of chordates.

Each of the numbered innovations is shared by the classes beyond that point.

*includes birds

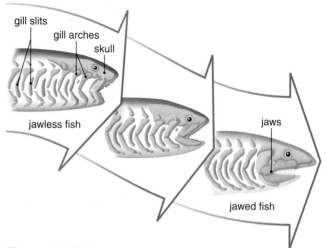

gill slits
gill arches
skull

jawless fish

jaws

jawed fish

Figure 19.27 Evolution of jaws.

Jaws evolved from the anterior gill arches of ancient jawless fishes.

Figure 19.28

Diversity of fishes.

a. The lamprey is a jawless fish. Note the toothed oral disk. **b.** The shark is a cartilaginous fish. **c.** The soldierfish, a bony fish, has the typical appearance of a ray-finned fish.

toothed oral disk

gill slits
(seven pairs)

a. Lamprey, a jawless fish

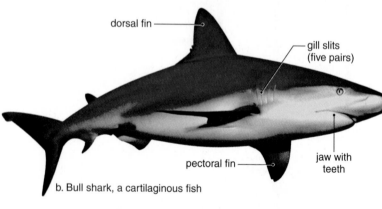

dorsal fin

gill slits
(five pairs)

pectoral fin

jaw with teeth

b. Bull shark, a cartilaginous fish

caudal fin   anal fin   second dorsal fin   first dorsal fin

gill slit
(one pair)

operculum
(covering
for gills)

pectoral fin

pelvic fin

c. Soldierfish, a bony fish

Fishes: First Jaws and Lungs

The first vertebrates were jawless fishes, which wiggled through the water and sucked up food from the ocean floor. Today, there are three living classes of fishes: jawless fishes, cartilaginous fishes, and bony fishes. The two latter groups have jaws, tooth-bearing bones of the head. Jaws are believed to have evolved from the first pair of gill arches, structures that ordinarily support gills (**Fig. 19.27**). The presence of jaws permits a predatory way of life.

Living representatives of the **jawless fishes** are cylindrical and up to a meter long. They have smooth, scaleless skin and no jaws or paired fins. The two groups of living jawless fishes are *hagfishes* and *lampreys* (**Fig. 19.28a**). The hagfishes are scavengers, feeding mainly on dead fishes, while some lampreys are parasitic. When parasitic, the round mouth of the lamprey serves as a sucker. The lamprey attaches itself to another fish and taps into its circulatory system. Water cannot move in through the lamprey's mouth and out over the gills, as is common in all other fishes. Instead, water moves in and out through the gill openings.

Cartilaginous fishes are the sharks (Fig. 19.28b), the rays, and the skates, which have skeletons of cartilage instead of bone. The small dogfish shark is often dissected in biology laboratories. One of the most dangerous sharks inhabiting both tropical and temperate waters is the hammerhead shark. The largest sharks, the whale sharks, feed on small fishes and marine invertebrates and do not attack humans. Skates and rays are rather flat fishes that live partly buried in the sand and feed on mussels and clams.

Three well-developed senses enable sharks and rays to detect their prey: (1) they have the ability to sense electric currents in water—even those generated by the muscle movements of animals; (2) they, and all other types of fishes, have a lateral line system, a series of cells that lie within canals along both sides of the body and can sense pressure caused by a fish or another animal swimming nearby; and (3) they have a keen sense of smell—the part of the brain associated with this sense is twice as large as the other parts. Sharks can detect about one drop of blood in 115 liters (25 gallons) of water.

Bony fishes are by far the most numerous and diverse of all the vertebrates. Most of the bony fishes we eat, such as perch, trout, salmon, and haddock, are **ray-finned fishes** (Fig. 19.28c). Their fins, which are used in balancing and propelling the body, are thin and supported by bony spikes. Ray-finned fishes have various ways of life. Some, such as herring, are filter feeders; others, such as trout, are opportunists; and still others, such as piranhas and barracudas, are predaceous carnivores.

Ray-finned fishes have a swim bladder, which usually serves as a buoyancy organ. By secreting gases into the bladder or by absorbing gases from it, these fishes can change their density, and thus go up or down in the water. The streamlined shape, fins, and muscle action of ray-finned fishes are all suited to locomotion in the water. Their skin is covered by bony scales that protect the body but do not prevent water loss. When ray-finned fishes respire, the gills are kept continuously moist by the passage of water through the mouth and out the gill slits. As the water passes over the gills, oxygen is absorbed by the blood, and carbon dioxide is given off. Ray-finned fishes have a single-circuit circulatory system. The heart is a simple pump, and the blood flows through the chambers, including a nondivided atrium and ventricle, to the gills. Oxygenated blood leaves the gills and goes to the body proper, eventually returning to the heart for recirculation.

Another type of bony fish are called the **lobe-finned fishes.** Ancient lobe-finned fishes gave rise to the amphibians (**Fig. 19.29a**). These fishes not only had fleshy appendages that could be adapted to land locomotion, but most also had a **lung,** which was used for respiration.

Amphibians: Jointed Vertebrate Limbs

Amphibians, whose name means living on both land and in the water, are represented today by frogs, toads, newts, and salamanders. Amphibians are variously colored. Some brightly colored ones have skin toxins that make predators sick or even die. Usually, the pattern of color is protective because it allows them to remain unnoticed.

Aside from *jointed limbs* (Fig. 19.29b), amphibians have other features not seen in bony fishes: eyelids for keeping their eyes moist, a sound-producing larynx, and ears adapted to picking up sound waves. The brain is larger than that of a fish. Adult amphibians usually have small lungs. Air enters the mouth by way of nostrils, and when the floor of the mouth is raised, air is forced into the relatively small lungs. Respiration is supplemented by gas exchange through the smooth, moist, and glandular skin. The amphibian heart has only three chambers, compared to the four of mammals. Mixed blood is sent to all parts of the body; some is sent to the skin, where it is further oxygenated.

Most members of this group lead an amphibious life—that is, the larval stage lives in the water, and the adult stage lives on the land (see Fig. 19.1). While metamorphosis is a distinctive characteristic of amphibians, some do not practice it. Some salamanders and even some frogs are direct developers; the adults do not return to the water to reproduce, and there is no tadpole stage. A great variety of reproductive strategies are seen among amphibians. In the gastric-brooding frogs, the female ingests as many as 20 fertilized eggs, which undergo development in her stomach, and then she vomits them up as tadpoles or froglets depending on the species. The great variety of life histories observed among amphibians shows that they have been extremely successful on land over 360 million years since they evolved.

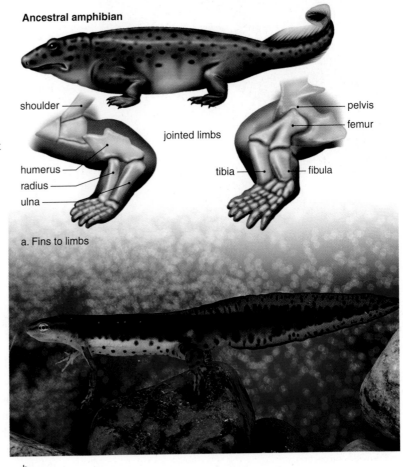

a. Fins to limbs

b.

Figure 19.29   Evolution of amphibians.

a. A lobe-finned fish compared with an amphibian. A shift in the position of the bones in the forelimbs and hind limbs lifted and supported the body. **b.** Newt (shown here) and frogs (see Fig.19.1) are types of living amphibians.

Check Your Progress

1. Explain the evolutionary significance of the development of a jaw.
2. From which group of bony fishes did amphibians evolve?
3. Among amphibians, what features in particular are adaptations to life on land?

Answers: 1. The jaw allowed vertebrates to become predators.   2. Amphibians are descended from ancient lobe-finned fishes.   3. Amphibians have jointed limbs, eyelids for keeping their eyes moist, ears for picking up sound waves, a sound-producing larynx, and lungs in adults.

Figure 19.30   Reptiles.

Snakes wave a forked tongue in the air to collect chemical molecules, which are then brought back into the mouth and delivered to an organ in the roof of the mouth. Analyzed chemicals help the snake trail its prey, recognize a predator, or find a mate.

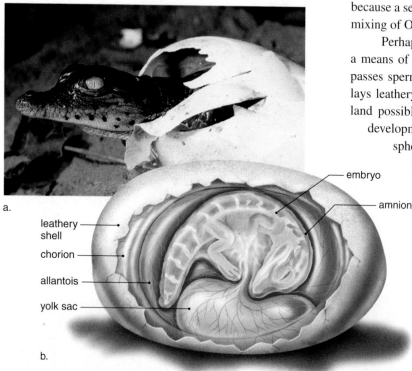

a.

leathery shell

chorion

allantois

yolk sac

embryo

amnion

b.

Figure 19.31   Reproductive strategy.

a. A baby American crocodile hatching out of its shell. Note that the shell is leathery and flexible, not brittle like a bird's egg. **b.** Inside the egg, the embryo is surrounded by extraembryonic membranes. The chorion aids gas exchange, the yolk sac provides nutrients, the allantois stores waste, and the amnion encloses a fluid that prevents drying out and provides protection.

Reptiles: Amniotic Egg

The **reptiles** diversified and were most abundant between 245 and 66 MYA. These animals included the dinosaurs, which are remembered for their great size. *Brachiosaurus*, a herbivore, was about 23 meters (75 feet) long and about 17 meters (56 feet) tall. *Tyrannosaurus rex*, a carnivore, was 5 meters (16 feet) tall when standing on its hind legs. The bipedal stance of some reptiles was preadaptive for the evolution of wings in birds.

The reptiles living today include turtles, crocodiles, snakes, lizards, and birds (discussed next). The typical reptilian body is covered with hard, keratinized scales, which protect the animal from desiccation and from predators. Another adaptation for a land existence is the manner in which snakes use their tongue as a sense organ (**Fig. 19.30**). Reptiles have well-developed lungs enclosed by a protective rib cage. Most reptiles have a three-chambered heart because a septum that divides a third chamber is incomplete. This allows some mixing of O_2-rich and O_2-poor blood in this chamber.

Perhaps the most outstanding adaptation of the reptiles is that they have a means of reproduction suitable to a land existence. The penis of the male passes sperm directly to the female. Fertilization is internal, and the female lays leathery, flexible, shelled eggs. The *amniote egg* made development on land possible and eliminated the need for a swimming larval stage during development. The amniote egg provides the developing embryo with atmospheric oxygen, food, and water; removes nitrogenous wastes; and protects the embryo from drying out and from mechanical injury. This is accomplished by the presence of extraembryonic membranes (**Fig. 19.31**).

Fishes, amphibians, and reptiles (other than birds) are **ectothermic,** meaning that their body temperature matches the temperature of the external environment. If it is cold externally, they are cold internally; if it is hot externally, they are hot internally. Most reptiles regulate their body temperatures by exposing themselves to the sun if they need warmth or by hiding in the shadows if they need cooling off.

Birds

Birds share a common ancestor with crocodiles and have traits, such as a tail with vertebrae, clawed feet, and the presence of scales that show they are in fact reptiles. Perhaps you have noticed the scales on the legs of a chicken; in addition, a bird's *feathers* are actually modified reptilian scales. However, birds lay a hard-shelled amniote egg rather than the leathery egg of other reptiles. The exact history of birds is still in dispute, but gathering evidence indicates that birds are closely related to bipedal dinosaurs and that they should be classified as such.

Nearly every anatomical feature of a bird can be related to its ability to fly (**Fig. 19.32**). The forelimbs are modified as wings. The hollow, very light bones are laced with air cavities. A horny beak has replaced jaws equipped with teeth, and a slender neck connects the head to a rounded, compact torso. Respiration is efficient since the lobular lungs form anterior and posterior air sacs. The presence of these sacs means that the air moves one way through the lungs, and gases are continuously exchanged across respiratory tissues. Another benefit of air sacs is that they lighten the body and aid flying.

Birds have a four-chambered heart that completely separates O$_2$-rich blood from O$_2$-poor blood. Birds are **endotherms** and generate internal heat. Many endotherms can use metabolic heat to maintain a constant internal temperature. This may be associated with their efficient nervous, respiratory, and circulatory systems. Also, their feathers provide insulation. Birds have no bladder and excrete uric acid in a semidry state.

Birds have particularly acute vision and well-developed brains. Their muscle reflexes are excellent. These adaptations are suited to flight. An enlarged portion of the brain seems to be the area responsible for instinctive behavior. A ritualized courtship often precedes mating. Many newly hatched birds require parental care before they are able to fly away and seek food for themselves. A remarkable aspect of bird behavior is the seasonal migration of many species over very long distances. Birds navigate by day and night, whether it's sunny or cloudy, by using the sun and stars and even the Earth's magnetic field to guide them.

Traditionally, the classification of birds was particularly based on beak and foot types (**Fig. 19.33**), and to some extent on habitat and behavior. The various orders include birds of prey with notched beaks and sharp talons; shorebirds with long, slender, probing bills and long, stiltlike legs; woodpeckers with sharp, chisel-like bills and grasping feet; waterfowl with webbed toes and broad bills; penguins with wings modified as flippers; and songbirds with perching feet. Now genetics is used to determine relationships among birds.

upstroke

downstroke

Figure 19.32 Bird flight.

Birds fly by flapping their wings. Bird flight requires an airstream and a powerful wing downstroke for lift, a force at right angles to the airstream.

Check Your **Progress**

1. Describe how reptiles are adapted to reproduce on land.
2. What characteristics do birds share with other reptiles?
3. List the anatomical features of birds that help them fly.

Answers: 1. Fertilization is internal in reptiles and results in the production of an amniote egg. 2. Birds have scales (feathers are modified scales) and clawed feet, have internal fertilization, and produce amniote eggs. 3. Birds' forelimbs are modified as wings; their bones are usually hollow with air cavities; air sacs involved in respiration also lighten the body; and they have a well-developed nervous system and muscles.

Figure 19.33 Bird beaks.

a. A cardinal's beak allows it to crack tough seeds. **b.** A flamingo's beak strains food from the water with bristles that fringe the mandibles. **c.** A bald eagle's beak allows it to tear prey apart.

Mammals: Hair and Mammary Glands

Mammals are amniotes, and as such, they share an amniote ancestor with reptiles. The first **mammals** were small, about the size of mice. During all the time the dinosaurs flourished (165 MYA), mammals were a minor group that changed little. Some of the earliest mammalian groups are still represented today by the monotremes and marsupials, but they are not abundant. The placental mammals that evolved later went on to live in many habitats.

The two chief characteristics of mammals (class Mammalia) are hair and milk-producing mammary glands. Almost all mammals are endotherms and generate internal heat. Many of the adaptations of mammals are related to temperature control. Hair, for example, provides insulation against heat loss and allows mammals to be active, even in cold weather.

Mammary glands enable females to feed (nurse) their young without leaving them to find food. Nursing also creates a bond between mother and offspring that helps ensure parental care while the young are helpless. In most mammals, the young are born alive after a period of development in the uterus, a part of the female reproductive system. Internal development shelters the young and allows the female to move actively about while the young are maturing.

Monotremes are mammals that, like birds, have a *cloaca*, a terminal region of the digestive tract serving as a common chamber for feces, excretory wastes, and sex cells. They also lay hard-shelled amniote eggs. They are represented by the spiny anteater and the duckbill platypus, both of which live in Australia. The female duckbill platypus lays her eggs in a burrow in the ground. She incubates the eggs, and after hatching, the young lick up milk that seeps from mammary glands on her abdomen. The spiny anteater has a pouch on the belly side formed by swollen mammary glands and longitudinal muscle (**Fig. 19.34***a*). Hatching takes place in this pouch, and the young remain there for about 53 days. Then they stay in a burrow, where the mother periodically visits and nurses them.

The young of **marsupials** begin their development inside the female's body, but they are born in a very immature condition. Newborns crawl up into a pouch on their mother's abdomen. Inside the pouch, they attach to the nipples of mammary glands and continue to develop. Frequently, more are born than can be accommodated by the number of nipples, and it's "first come, first served."

a.

b.

c.

Figure 19.34 **Monotremes and marsupials.**

a. The spiny anteater is a monotreme that lives in Australia.
b. The opossum is the only marsupial in the United States. The Virginia opossum is found in a variety of habitats. **c.** The koala is an Australian marsupial that lives in trees.

The Virginia opossum is the only marsupial that occurs north of Mexico (Fig. 19.34*b*). In Australia, however, marsupials underwent adaptive radiation for several million years without competition. Thus, marsupial mammals are now found mainly in Australia, with some in Central and South America as well. Among the herbivorous marsupials, koalas are tree-climbing browsers (Fig. 19.34*c*), and kangaroos are grazers. The Tasmanian wolf or tiger, thought to be extinct, was a carnivorous marsupial about the size of a collie dog.

The vast majority of living mammals are **placental mammals** (**Fig. 19.35**). In these mammals, the extraembryonic membranes of the amniote egg (see Fig. 19.31) have been modified for internal development within the uterus of the female. The chorion contributes to the fetal portion of the placenta, while a part of the uterine wall contributes to the maternal portion. Here, nutrients, oxygen, and waste are exchanged between fetal and maternal blood.

Mammals are adapted to life on land and have limbs that allow them to move rapidly. In fact, an evaluation of mammalian features leads us to the obvious conclusion that they lead active lives. The brain is well developed; the lungs are expanded not only by the action of the rib cage but also by the contraction of the diaphragm, a horizontal muscle that divides the thoracic cavity from the abdominal cavity; and the heart has four chambers. The internal temperature is constant, and hair, when abundant, helps insulate the body.

The mammalian brain is enlarged due to the expansion of the cerebral hemispheres that control the rest of the brain. The brain is not fully developed until after birth, and young learn to take care of themselves during a period of dependency on their parents.

Mammals can be distinguished by their methods of obtaining food and their mode of locomotion. For example, bats have membranous wings supported by digits; horses have long, hoofed legs; and whales have paddlelike forelimbs. The specific shape and size of the teeth may be associated with whether the mammal is a herbivore (eats vegetation), a carnivore (eats meat), or an omnivore (eats both meat and vegetation). For example, mice have continuously growing incisors; horses have large, grinding molars; and dogs have long canine teeth.

Figure 19.35   Placental mammals.

Placental mammals have adapted to various ways of life. **a.** Deer are herbivores that live in forests. **b.** Lions are carnivores on the African plain. **c.** Monkeys typically inhabit tropical forests. **d.** Whales are sea-dwelling placental mammals.

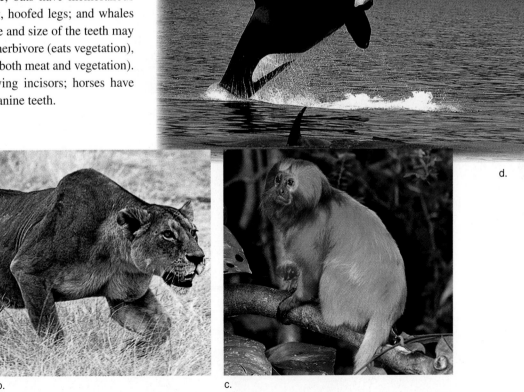

a.

b.

c.

d.

19.6 | Human Evolution

The evolutionary tree in **Figure 19.36** shows that all primates share one common ancestor and that the other types of primates diverged from the human line of descent (called a **lineage**) over time. The **prosimians** include the lemurs, tarsiers, and lorises. The **anthropoids** include the monkeys, apes, and humans. The designation hominid includes only African apes, humans, and the closest extinct relatives of humans.

Primates are adapted to an arboreal life—meaning, for living in trees. Primate limbs are mobile, and the hands and feet have five digits each. Many primates have both an opposable big toe and a thumb—that is, the big toe or thumb can touch each of the other toes or fingers. (Humans don't have an opposable big toe, but the thumb is opposable, resulting in a grip that is both powerful and precise.) The opposable thumb allows a primate to easily reach out and bring food such as fruit to the mouth. When locomoting, primates grasp and release tree limbs freely because nails have replaced claws.

The evolutionary trend among primates is toward a larger and more complex brain—the brain size is smallest in prosimians and largest in modern

Figure 19.36   Evolutionary tree of primates.

Primates are descended from an ancestor that may have resembled a tree shrew. The descendants of this ancestor adapted to the new way of life and developed traits such as a shortened snout and nails instead of claws. The time when each type of primate diverged from the main line of descent is known from the fossil record. A common ancestor was living at each point of divergence. For example, there was a common ancestor for hominids (humans and the African apes) about 7 MYA; and one for anthropoids about 45 MYA.

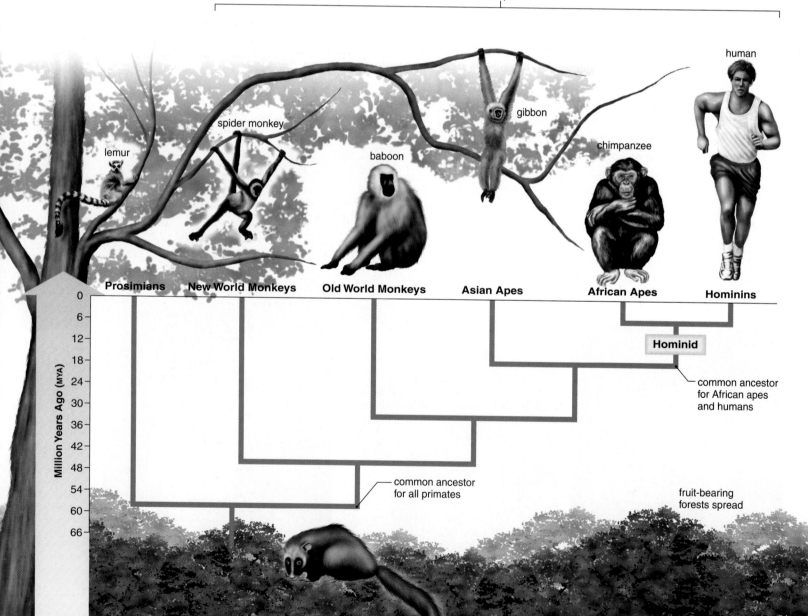

humans. In humans, the cerebral cortex has many association areas and expands so much that it becomes extensively folded. The portion of the brain devoted to smell gets smaller, and the portions devoted to sight increase in size and complexity during primate evolution. Also, more and more of the brain is involved in controlling and processing information received from the hands and the thumb. The result is good hand-eye coordination in humans.

Notice that prosimians were the first type of primate to diverge from the human line of descent, and African apes were the last group to diverge from our line of descent. The evolutionary tree also indicates that humans are most closely related to African apes. One of the most unfortunate misconceptions concerning human evolution is that Darwin and others suggested that humans evolved from apes. On the contrary, humans and apes share a common apelike ancestor. Today's apes are our distant cousins, and we couldn't have evolved from our cousins because we are contemporaries—living on Earth at the same time. Presently, researchers hypothesize that the last common ancestor for African apes and humans lived about 7 MYA.

Evolution of Humanlike Hominins

All of the fossils shown in **Figure 19.37** are all humanlike **hominins.** (The term **hominid** is now a family designation for both humans and African apes.) This illustration tells when these various hominins lived and emphasizes their type of face and brain size. However, to be a humanlike hominin, it is only necessary that the fossil have an anatomy suitable for standing erect and walking on two feet, a characteristic called **bipedalism.** The bars in Figure 19.37 extend from the date of a species' appearance in the fossil record to the date it became extinct.

Check Your **Progress**

1. What groups of animals are primates?
2. What groups of animals are anthropoids?
3. Did humans evolve from apes?

Answers: 1. Prosimians, monkeys, apes, and humans are primates. 2. Monkeys, apes, and humans are anthropoids. 3. No, humans and apes share a common ancestor.

Figure 19.37 Human evolution.

Several groups of extinct humanlike hominins preceded the evolution of modern humans. These groups have been divided into the early hominins (orange), the australopithecines (green), and *Homo* (purple). Only modern humans are defined as *Homo sapiens*.

Figure 19.38 *Australopithecus afarensis.*

a. A reconstruction of Lucy on display at the St. Louis Zoo. **b.** These fossilized footprints occur in ash from a volcanic eruption some 3.7 MYA. The larger footprints are double, and a third, smaller individual was walking to the side. (A female holding the hand of a youngster may have been walking in the footprints of a male.) The footprints suggest that *A. afarensis* walked bipedally.

Early Humanlike Hominins

In Figure 19.37, early hominins are represented by orange-colored bars. Scientists have now found several fossils dated around the time the ape lineage and the human lineage are believed to have split, and one of these is *Sahelanthropus tchadensis.* Only the braincase has been found and dated at 7 MYA. Although the braincase is very apelike, the location of the opening for the spine at the back of the skull suggests *bipedalism.* Also, the canines are smaller and the tooth enamel is thicker than those of an ape.

Another early hominin, *Ardipithecus ramidus,* is representative of the ardipithecines of 4.5 MYA. So far, only skull fragments of *A. ramidus* have been described. Indirect evidence suggests that the species was possibly bipedal, and that some individuals may have been 122 centimeters tall. The teeth seem intermediate between those of earlier apes and later hominids. Recently, fossils dated 4 MYA show a direct link between *A. ramidus* and the australopithecines, discussed next.

Australopithecines

It's possible that one of the **australopithecines,** a group of hominins that evolved and diversified in Africa about 4 MYA, is a direct ancestor of humans. More than 20 years ago, a team led by Donald Johanson unearthed nearly 250 fossils of a hominid called *Australopithecus afarensis.* A now-famous female skeleton dated at 3.18 MYA is known worldwide by its field name, Lucy. (The name derives from the Beatles song "Lucy in the Sky with Diamonds.") Although her brain was quite small (400 cc), the shapes and relative proportions of Lucy's limbs indicate that she stood upright and walked bipedally (**Fig. 19.38***a*). Even better evidence of bipedal locomotion comes from a trail of footprints dated about 3.7 MYA. The larger prints are double, as though a smaller-sized individual was stepping in the footfalls of another—and there are additional small prints off to the side, within hand-holding distance (Fig. 19.38*b*).

Since the australopithecines were apelike above the waist (small brain) and humanlike below the waist (walked erect), it seems that human characteristics did not evolve all at one time. The term **mosaic evolution** is applied when different body parts change at different rates and therefore at different times.

Homo habilis

Homo habilis, dated between 2.0 and 1.9 MYA, may be ancestral to modern humans. Some of these fossils have a brain size as large as 775 cc, which is about 45% larger than the brain of *A. afarensis.* The cheek teeth are smaller than those of any of the australopithecines. Therefore, it is likely that this early *Homo* was omnivorous and ate meat in addition to plant material. Bones at the campsites of *H. habilis* bear cut marks, indicating that they used tools to strip the meat from bones.

The stone tools made by *H. habilis,* whose name means "handyman," are rather crude. These hominins may have used sharp flakes of broken rocks to scrape away hide, cut tendons, and easily remove meat from bones.

The skulls of *H. habilis* suggest that the portions of the brain associated with speech areas were enlarged. We can speculate that the ability to speak may have led to hunting cooperatively. Some members of the group may have remained plant gatherers, and if so, both hunters and gatherers most likely ate together and shared their food. In this way, society and culture could have begun. **Culture,** which encompasses human behavior and products (such as technology and the arts), is dependent upon the capacity to speak and transmit knowledge. We can further speculate that the advantages of a culture to *H. habilis* may have hastened the extinction of the australopithecines.

Homo erectus

Homo erectus and like fossils are found in Africa, Asia, and Europe and dated between 1.9 and 0.3 MYA. Although all fossils assigned the name *H. erectus* are similar in appearance, there is enough discrepancy to suggest that several different species have been included in this group. Compared to *H. habilis*, *H. erectus* had a larger brain (about 1,000 cc) and a flatter face. The recovery of an almost complete skeleton of a 10-year-old boy indicates that *H. erectus* was much taller than the hominins discussed thus far. Males were 1.8 meters tall (about 6 feet), and females were 1.55 meters (approaching 5 feet). Indeed, these hominins were erect and most likely had a striding gait like that of modern humans. The robust and most likely heavily muscled skeleton still retained some australopithecine features. Even so, the size of the birth canal indicates that infants were born in an immature state that required an extended period of care.

It is believed that *H. erectus* first appeared in Africa and then migrated into Asia and Europe. At one time, the migration was thought to have occurred about 1 MYA, but recently *H. erectus* fossil remains in Java and the Republic of Georgia have been dated at 1.9 and 1.6 MYA, respectively. These remains push the evolution of *H. erectus* in Africa to an earlier date than has yet been determined. In any case, such an extensive population movement is a first in the history of humankind and a tribute to the intellectual and physical skills of the species.

H. erectus was the first hominin to use fire, and it also fashioned more advanced tools than earlier *Homo* species. These hominids used heavy, teardrop-shaped axes and cleavers, as well as flakes that were probably used for cutting and scraping. Some investigators believe *H. erectus* was a systematic hunter that brought kills to the same site over and over again. In one location, researchers have found over 40,000 bones and 2,647 stones. These sites could have been "home bases" where social interaction occurred and a prolonged childhood allowed time for learning. Perhaps a language evolved and a culture more like our own developed.

Evolution of Modern Humans

Most researchers believe that *Homo sapiens* (modern humans) evolved from *H. erectus*, but they differ as to the details. The hypothesis that *H. sapiens* evolved separately from *H. erectus* in Asia, Africa, and Europe is called the **multiregional continuity hypothesis** (**Fig. 19.39***a*). This hypothesis proposes that evolution to modern humans was essentially similar in several different places. If so, each region should show a continuity of its own anatomical characteristics from the time when *H. erectus* first arrived.

Opponents argue that it seems highly unlikely that evolution would have produced essentially the same result in these different places. They suggest, instead, the **replacement model,** or **out-of-Africa hypothesis,** which proposes that *H. sapiens* evolved from *H. erectus* only in Africa, and thereafter *H. sapiens* migrated to Europe and Asia about 100,000 years BP (before present) (Fig. 19.39*b*). If so, fossils dated 200,000 BP and 100,000 BP are expected to be markedly different from each other.

According to which hypothesis would modern humans be most genetically alike? The multiregional continuity hypothesis states that human populations have been evolving separately for a long time; therefore, genetic differences would be expected between groups. According to the replacement model, we are all descended from a few individuals from about 100,000 years BP. Therefore, the replacement model suggests that we are more genetically similar.

A few years ago, a study attempted to show that all the people of Europe (and the world, for that matter) have essentially the same mitochondrial DNA.

a. Multiregional continuity

b. Out of Africa

Figure 19.39   **Evolution of modern humans.**

a. The multiregional continuity hypothesis proposes that *Homo sapiens* evolved separately in at least three different places: Asia, Africa, and Europe. Therefore, continuity of genotypes and phenotypes is expected in each region, but not between regions. **b.** The replacement model (out-of-Africa hypothesis) proposes that *Homo sapiens* evolved only in Africa; then this species migrated and supplanted populations of *Homo* in Asia and Europe about 100,000 years ago.

Check Your Progress

1. Contrast the australopithecines with *Homo habilis*.
2. What are some distinctive features of *Homo erectus* compared to *Homo habilis*?
3. What is the replacement model (out-of-Africa hypothesis)?
4. Who were the Neandertals?

Answers : 1. Australopithecines were apelike above the waist and humanlike below the waist. *Homo habilis* had a larger brain and smaller cheek teeth than the australopithecines, and was able to use stone tools and possibly speech.   2. *Homo erectus* had a larger brain and flatter face, most likely a striding gait as we do, used fire, and made better tools.   3. The replacement model proposes that *Homo sapiens* evolved only in Africa and then migrated to other parts of the world.   4. The Neandertals were a stocky and powerfully muscular people capable of surviving harsh Ice Age conditions. They are thought to have been culturally advanced.

Called the "mitochondrial Eve" hypothesis by the press (note that this is a misnomer because no single ancestor is proposed), the statistics that calculated the date of the African migration were found to be flawed. Still, the raw data—which indicate a close genetic relationship among all Europeans—support the replacement model.

These opposing hypotheses have sparked many other innovative studies. The final conclusions are still being determined.

Neandertals

Neandertals (*H. neandertalensis*) take their name from Germany's Neander Valley, where one of the first Neandertal skeletons, dated some 200,000 years ago, was discovered. It's possible that the Neandertals were already present in Eurasia when modern humans represented by Cro-Magnons (discussed next) arrived on the scene. In that case, competition with the Cro-Magnons may have made the Neandertals become extinct. The Neandertals had massive brow ridges, and their nose, jaws, and teeth protruded far forward. The forehead was low and sloping, and the lower jaw lacked a chin. The Neandertals were heavily muscled, especially in the shoulders and neck. The bones of their limbs were shorter and thicker than those of modern humans. The Neandertals lived in Europe and Asia during the last Ice Age, and their sturdy build could have helped conserve heat.

The Neandertals give evidence of being culturally advanced. Most lived in caves, but those living in the open may have built houses. They manufactured a variety of stone tools, including spear points, which could have been used for hunting, and scrapers and knives, which would have helped with food preparation. They most likely successfully hunted bears, woolly mammoths, rhinoceroses, reindeer, and other contemporary animals. They buried their dead with flowers and tools and may have had a religion.

Cro-Magnons

Cro-Magnons are the oldest fossils to be designated *H. sapiens*. Cro-Magnons, who are named after a fossil location in France, had a thoroughly modern appearance. They made advanced tools, including compound tools, as when stone flakes were fitted to a wooden handle. They may have been the first to make knifelike blades and to throw spears, enabling them to kill animals from a distance. They were such accomplished hunters that some researchers believe they were responsible for the extinction of many larger mammals, such as the giant sloth, the mammoth, the saber-toothed tiger, and the giant ox, during the late Pleistocene epoch.

Cro-Magnons hunted cooperatively, and were perhaps the first to have a language. They are believed to have lived in small groups, with the men hunting by day while the women remained at home with the children, gathering and processing food items. Probably, the women also were engaged in maintenance tasks. The Cro-Magnon culture included art. They sculpted small figurines out of reindeer bones and antlers. They also painted beautiful drawings of animals, some of which have survived on cave walls in Spain and France (**Fig. 19.40**).

Figure 19.40   Cro-Magnons.

Cro-Magnon people are the first to be designated *Homo sapiens*. Their tool-making ability and other cultural attributes, including their artistic talents, are legendary.

The Chapter in Review

Summary

19.1 | Evolution of Animals

Animals are motile, multicellular heterotrophs that ingest their food.

Ancestry of Animals

The choanoflagellates are protists that most likely resemble the last unicellular ancestor of animals. A colony of flagellated cells could have led to a multicellular animal that formed tissues by invagination.

The Evolutionary Tree of Animals

The evolutionary tree of animals is chiefly based on molecular and developmental data. During the evolution of animals, multicellularity preceded true tissues and bilateral symmetry. Bilateral symmetrical animals either undergo protostome development or deuterostome development. Certain of the protostomes and all deuterostomes have a internal cavity called a coelom. Some of the coelomate animals are segmented.

19.2 | Introducing the Invertebrates

Sponges are multicellular (lack tissues) and have various symmetries.

Cnidarians have two tissue layers, are radially symmetrical, have a saclike digestive cavity, and possess stinging cells and nematocysts.

Flatworms have ectoderm, endoderm, and mesoderm but no coelom. They are bilaterally symmetrical and have a saclike digestive cavity.

Roundworms have a pseudocoelom and a complete digestive tract.

Incomplete   Complete

19.3 | Protostomes and Deuterostomes Compared

On the basis of molecular and developmental data:

- Flatworms, roundworms, molluscs, annelids, and arthropods are protostomes; the first embryonic opening becomes the mouth. In the coelomate protostomes, the coelom develops by a splitting of mesoderm.

Development in protostomes

- Echinoderms and chordates are deuterostomes; the second embryonic opening becomes the mouth, and the coelom forms by outpocketing of the primitive gut.

Development in deuterostomes

19.4 | Molluscs, Annelids, and Arthropods

These groups of animals are the coelomate protostomes.

Molluscs

The body of a mollusc typically contains a visceral mass, a mantle, and a foot.

Gastropods Snails, representatives of this group, have a flat foot, a one-part shell, and a mantle cavity that carries on gas exchange.

Cephalopods Octopuses and squids display marked cephalization, move rapidly by jet propulsion, and have a closed circulatory system.

Bivalves Bivalves such as clams have a hatchet foot and a two-part shell, and are filter feeders.

Annelids: Segmented Worms

Annelids are segmented worms; segmentation is seen both externally and internally.

Marine Worms Polychaetes are worms that have many setae. A clam worm is a predaceous marine worm with a defined head region.

Earthworms Earthworms are oligochaetes that scavenge for food in the soil and do not have a well-defined head region.

Leeches Leeches are annelids that feed by sucking blood.

Arthropods: Jointed Appendages

Arthropods are the most varied and numerous of animals. Their success is largely attributable to a flexible exoskeleton and specialized body regions.

Crustaceans have a head that bears compound eyes, antennae, and mouthparts. Five pairs of walking legs are present.

Arachnids include spiders, scorpions, ticks, mites, harvestmen and horseshoe crabs. Spiders live on land and spin silk, which they use to capture prey as well as for other purposes.

Insects have three pairs of legs attached to the thorax. Insects have adaptations to a terrestrial life, such as wings for flying.

19.5 | Echinoderms and Chordates

Echinoderms

Echinoderms have radial symmetry as adults (not as larvae) and endoskeletal spines. Typical echinoderms have tiny skin gills, a central nerve ring with branches, and a water vascular system for locomotion, as exemplified by the sea star.

Chordates

Chordates (tunicates, lancelets, and vertebrates) have a notochord, a dorsal tubular nerve cord, pharyngeal pouches, and a postanal tail at some time in their life history.

Invertebrate Chordates Adult tunicates lack chordate characteristics except gill slits, but adult lancelets have the four chordate characteristics and show obvious segmentation.

Vertebrate Chordates Vertebrate chordates include the fishes, amphibians, reptiles, birds, and mammals.

Fishes

- The first vertebrates, represented by hagfishes and lampreys, lacked jaws and fins.
- Cartilaginous fishes, represented by sharks and rays, have jaws and a skeleton made of cartilage.
- Bony fishes have jaws and fins supported by bony spikes; the bony fishes include those that are ray-finned and a few that are lobe-finned. Some of the lobe-finned fishes have lungs.

Amphibians (frogs, toads, newts, and salamanders) evolved from ancient lobe-finned fishes and have two pairs of jointed vertebrate limbs. Frogs usually return to the water to reproduce; and metamorphose into terrestrial adults.

Reptiles (snakes, lizards, turtles, crocodiles, and birds) lay a shelled egg, which contains extraembryonic membranes, including an amnion that allows them to reproduce on land.

Birds are feathered reptiles, which helps them maintain a constant body temperature. They are adapted for flight; their bones are hollow with air cavities; lungs form air sacs that allow one-way ventilation; and they have well-developed sense organs.

Mammals are amniotes that have hair and mammary glands. The former helps them maintain a constant body temperature, and the latter allows them to nurse their young.

- Monotremes lay eggs.
- Marsupials have a pouch in which the newborn matures.
- Placental mammals, which are far more varied and numerous, retain offspring inside the uterus until birth.

19.6 | Human Evolution

Arboreal primates are mammals adapted to living in trees. During the evolution of primates, various groups diverged in a particular sequence. Prosimians (tarsiers and lemurs) diverged first, followed by the monkeys, then the apes, and then humans. Molecular biologists tell us we are most closely related to the African apes, with which we share a common ancestor dated at about 7 MYA.

Evolution of Hominins

Human evolution occurred in Africa, where it has been possible to find the remains of several early hominins that date back to 7 MYA. The evolution of hominins includes these innovations:

Bipedal	• The most famous australopithecine is Lucy (3.18 MYA) whose brain was small but who walked bipedally.
Tool Use	• *Homo habilis,* present about 2 MYA, is certain to have made tools.
Brain Size Increased	• *Homo erectus,* with a brain capacity of 1,000 cc and a striding gait, was the first to migrate out of Africa.

Evolution of Modern Humans

Two contradicting hypotheses have been suggested about the origin of modern humans:

Multiregional Continuity Hypothesis Modern humans originated separately in Asia, in Europe, and in Africa.

Replacement Model (Out-of-Africa Hypothesis) Modern humans originated in Africa and, after migrating into Europe and Asia, replaced the other *Homo* species (including the Neandertals) found there. Cro-Magnon was the first hominid to be considered *Homo sapiens.*

Key Terms

acoelomate	313	*Homo sapiens*	335
amphibian	327	insect	322
annelid	319	jawless fishes	326
anthropoid	332	lancelet	324
arachnid	321	leech	319
arthropod	320	lineage	332
australopithecine	334	lobe-finned fishes	327
bilateral symmetry	311	lung	327
bipedalism	333	mammal	330
bird	328	marsupial	330
bivalve	318	molting	320
bony fishes	326	monotreme	330
cartilaginous fishes	326	mosaic evolution	334
cephalization	313	multiregional continuity	
cephalopod	318	hypothesis	335
chitin	320	Neandertal	336
chordate	324	nematocyst	314
cnidarian	314	nephridium	319
coelom	313	notochord	324
Cro-Magnon	336	out-of-Africa hypothesis	335
crustacean	320	placental mammal	331
culture	334	planarian	315
deuterostome	313	prosimian	332
echinoderm	323	protostome	313
ectothermic	328	pseudocoelomate	313
endotherm	329	radial symmetry	311
filter feeder	314	ray-finned fishes	326
flatworm	315	replacement model	335
gastropod	318	reptile	328
germ layers	312	roundworm	316
hermaphrodite	316	segmentation	313
hominid	333	sponge	314
hominin	333	tunicate	324
Homo erectus	335	vertebrate	324
Homo habilis	334		

Testing Yourself

Choose the best answer for each question.

1. Which of these protists is hypothesized to be ancestral to animals?
   a. A green algal protist produced both the plants and animals.
   b. A choanoflagellate that moves by flagella.
   c. An amoeboid protist that moves by pseudopods.
   d. Animals unlike plants do not have a protist ancestor.

2. In the following diagram, label the major regions of the bodies of acoelomates, pseudocoelomates, and coelomates.

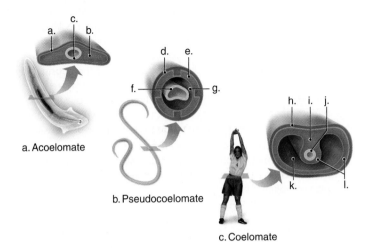

a. Acoelomate

b. Pseudocoelomate

c. Coelomate

3. Sponges are ancestors of
    a. cnidarians.
    b. protostomes.
    c. deuterostomes.
    d. None of these are correct.

For questions 4–10, identify the group(s) to which each feature belongs. Each answer may be used more than once. Each question may have more than one answer.

Key:

    a. flatworms
    b. roundworms
    c. molluscs
    d. annelids
    e. arthropods
    f. echinoderms
    g. chordates

4. most have an endoskeleton
5. typically bilaterally symmetrical as adults
6. nonsegmented
7. lack any kind of coelom
8. chitin exoskeleton
9. contain a mantle
10. move by pumping water
11. Insects have wings and three pairs of legs attached to the
    a. abdomen.
    b. thorax.
    c. head.
    d. midsection.
12. Cartilaginous fishes detect their prey by sensing
    a. electrical currents.
    b. odors.
    c. pressure changes.
    d. More than one of these are correct.
    e. All of these are correct.
13. Which of the following is not a feature of mammals?
    a. hair
    b. milk-producing glands
    c. ectothermic
    d. four-chambered heart
    e. diaphragm to help expand lungs

14. This species was probably the first to have the use of fire.
    a. *Homo habilis*
    b. *Homo erectus*
    c. *Homo sapiens*
    d. *Sahelanthropus tchadensis*
    e. *Australopithecus afarensis*
15. Which of these is matched correctly?
    a. *Australopithecus afarensis*—bipedal but small brain
    b. *Homo habilis*—made tools and migrated often
    c. *Homo erectus*—had fire and migrated out of Europe to Africa
    d. *Homo sapiens*—projecting face and had culture
    e. Both a and c are correct.
16. Cnidarians are considered to be organized at the tissue level because they contain
    a. ectoderm and endoderm.
    b. ectoderm.
    c. ectoderm and mesoderm.
    d. endoderm and mesoderm.
    e. mesoderm.
17. A cnidocyte is
    a. a digestive cell.
    b. a reproductive cell.
    c. a stinging cell.
    d. an excretory cell.
18. Which of the following is not a feature of a coelomate?
    a. radial symmetry
    b. three germ layers
    c. complete digestive tract body plan
    d. organ level of organization
19. A mollusc's shell is secreted by the
    a. foot.
    b. head.
    c. visceral mass.
    d. mantle.
20. The type of mollusc that has tentacles is a
    a. gastropod.
    b. bivalve.
    c. univalve.
    d. cephalopod.
21. A feature of annelids is
    a. a segmented body.
    b. acoelomate.
    c. a sac body plan.
    d. radial symmetry.
22. Which of the following is not a feature of an insect?
    a. compound eyes
    b. eight legs
    c. antennae
    d. an exoskeleton
    e. jointed legs
23. Which of the following undergoes metamorphosis?
    a. insects
    b. mammals
    c. amphibians
    d. reptiles
    e. Both a and c are correct.
24. Which of the following is not an arachnid?
    a. spider
    b. tick
    c. mite
    d. scorpion
    e. beetle
25. Unlike bony fishes, amphibians have
    a. ears.
    b. jaws.
    c. a circulatory system.
    d. a heart.
26. Which of the following is not an adaptation for flight in birds?
    a. air sacs
    b. modified forelimbs
    c. bones with air cavities
    d. acute vision
    e. well-developed bladder

27. Examples of monotremes include the
    a. spiny anteater and duckbill platypus.
    b. opossum and koala.
    c. badger and skunk.
    d. porcupine and armadillo.

28. Mammals are distinguished based on
    a. size and hair type.
    b. mode of reproduction.
    c. number of limbs and method of caring for young.
    d. number of mammary glands and number of offspring.

29. The first humanlike feature to evolve in the hominins was
    a. a large brain.   c. a slender body.
    b. massive jaws.   d. bipedal locomotion.

30. In *H. habilis*, enlargement of the portions of the brain associated with speech probably led to
    a. cooperative hunting.
    b. the sharing of food.
    c. the development of culture.
    d. All of these are correct.

31. Mitochondrial DNA data support which hypothesis for the evolution of humans?
    a. multiregional continuity
    b. replacement model (out-of-Africa)

32. Which of the following is an anthropoid, but not a hominid?
    a. human   d. chimpanzee
    b. gibbon   e. gorilla
    c. orangutan

33. Which of the following does not have cephalization?
    a. sponge   d. snail
    b. planarian   e. scorpion
    c. frog

34. Which of the following is endothermic?
    a. roundworm   d. opossum
    b. octopus   e. Both c and d are correct.
    c. eagle

35. Which of the following is a bilaterally symmetrical, hermaphroditic acoelomate?
    a. planarian   d. earthworm
    b. sponge   e. sea star
    c. jellyfish

Thinking Scientifically

 1. Recently, three fossil skulls of *Homo sapiens,* dating to about 160,000 years BP, were discovered in eastern Africa. These skulls fill a gap between the 100,000-year-old *H. sapiens* skulls found in Africa and Israel and 500,000-year-old skulls of archaic *H. sapiens* found in Ethiopia. Supporters of the replacement model (out-of-Africa hypothesis) argue that this discovery strengthens their position by documenting the succession of human ancestors from 6 MYA through this most recent group. Does this finding prove that the replacement model is correct? If not, what fossil evidence might yet be found to support the multiregional continuity hypothesis?

2. Think of the animals in this chapter that are radially symmetrical (cnidarians, many adult echinoderms). How is their lifestyle different from that of bilaterally symmetrical animals? How does their body plan complement their lifestyle?

Bioethical Issue

Use of Animals in Research

People who approve of laboratory research involving animals point out that even today it would be difficult to develop new vaccines and medicines against infectious diseases, new surgical techniques for saving human lives, or new treatments for spinal cord injuries without the use of animals. Even so, most scientists favor what are now called the "three Rs": (1) replacement of animals by in vitro, or test-tube, methods whenever possible; (2) reduction of the number of animals used in experiments; and (3) refinement of experiments to cause less suffering to animals.

F. Barbara Orlans of the Kennedy Institute of Ethics at Georgetown University says, "It is possible to be both pro research and pro reform." She feels that animal activists need to accept that sometimes animal research is beneficial to humans, and all scientists need to consider the ethical dilemmas that arise when animals are used for laboratory research. Do you approve of this compromise?

Essentials of Biology Website

The companion website for *Essentials of Biology* provides a wealth of information organized and integrated by chapter. You will find practice tests, animations, videos, and much more that will complement your learning and understanding of general biology.

http://www.mhhe.com/maderessentials2

20

Plant Anatomy and Growth

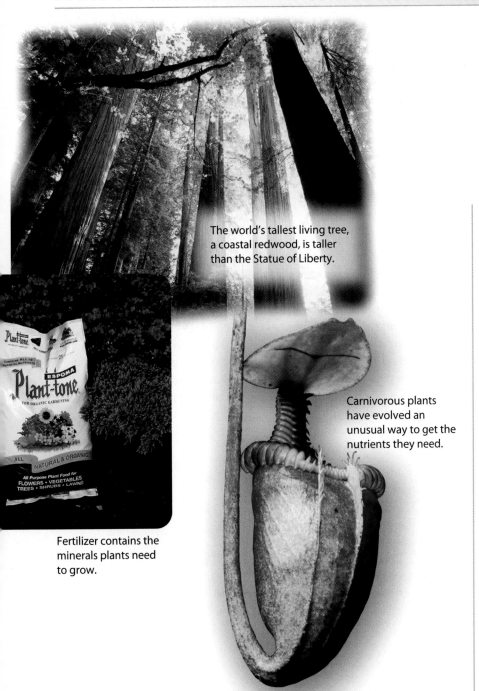

The world's tallest living tree, a coastal redwood, is taller than the Statue of Liberty.

Carnivorous plants have evolved an unusual way to get the nutrients they need.

Fertilizer contains the minerals plants need to grow.

Outline

Carnivorous Plants are Adapted to Harsh Conditions

THE PITCHER PLANT is an unusual plant that likes to grow in areas that lack nitrogen and phosphorus, which are required nutrients for plants. What's their secret? Pitcher plants feed on animals, particularly insects, to get the nutrients they need. That's right—like a few other types of plants, pitcher plants are carnivorous.

A pitcher plant is named for its leaves, which are shaped like a container we call a pitcher. The pitcher attracts flying insects because it has a scent,

—Continued next page

Continued—

is brightly colored, and provides nectar that insects can eat. When an insect lands on the lip of the pitcher, a slippery substance and downward-pointing hairs encourage it to slide into a pit. There, juices secreted by the leaf begin digesting the insect. The plant then absorbs these nutrients into its tissues. Although the pitcher is designed to attract insects, animals as large as rats have been found in pitcher plants! The unique adaptations of pitcher plants allow them to thrive in hostile environments where few plants can survive.

Some animals are able to turn the tables on the pitcher plant by taking advantage of the liquid they provide. Larvae of mosquitoes and flies have been known to develop safely inside a pitcher plant.

Carnivorous plants such as the pitcher plant are fairly unusual in the plant world, but so are many others, such as the coastal redwoods—the tallest trees on planet Earth. In this chapter, you will learn about the organs and structures of flowering plants, the nutrition they need, and how they transport nutrients within plant tissues.

20.1 | Plant Organs

A flowering plant, whether a cactus, a daisy, or an apple tree, has a shoot system and a root system. The **shoot system** consists of the stem, the leaves, and the flowers which carry out sexual reproduction, while the **root system** simply consists of the roots. The three vegetative organs—the leaf, the stem, and the root—perform functions that allow a plant to live and grow (**Fig. 20.1**). The structure and function of the flower are discussed in Chapter 21.

Activity at the **terminal bud,** the bud at the tip of a main shoot or a branch shoot, causes the shoot to increase in length, while activity at the tip of the root causes the root system to increase in length. This is called *primary growth*, and it would be equivalent to you increasing in height by growing from your head and your feet!

Leaves

Leaves are usually the chief organs of photosynthesis, and as such they require a supply of solar energy, carbon dioxide, and water. Broad and thin foliage leaves have a maximum surface area for the collection of solar energy and the absorption of carbon dioxide. Leaves receive water from the root system by way of vascular tissue that terminates in the leaves. **Deciduous** plants lose their leaves, often due to a yearly dry season or due to the onset of winter. Other trees, called **evergreens,** retain their leaves for the entire year.

Leaves may have several other functions aside from photosynthesis (**Fig. 20.2**). Leaves may be modified as tendrils that allow the plant to attach to objects (Fig. 20.2a) or as traps for catching insects (Fig. 20.2b). The leaves of a cactus are spines that reduce water loss and also serve to protect the plant from browsing animals (Fig. 20.2c).

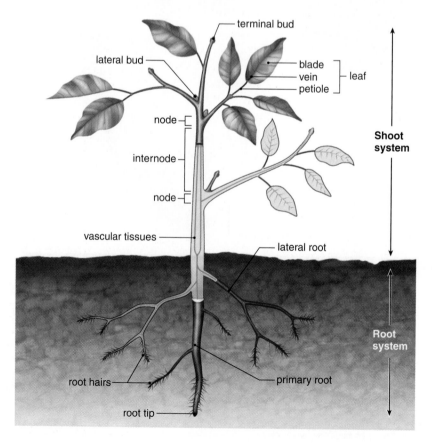

Figure 20.1 The vegetative body of a plant.

A plant has a root system, which extends below ground, and a shoot system, which is composed of the stem and leaves. Activity at the terminal bud and the root tip allows a plant to increase in length, a process called primary growth.

Stems

A **stem** supports the leaves, transports materials between the roots and leaves, and produces new tissue. At the end of a stem, a terminal bud, or apical bud, produces new leaves and other tissues during primary growth. Lateral (side) branches grow from a **lateral bud,** also called an axillary bud, located at the angle where a leaf joins the stem. A **node** occurs where a leaf or leaves are attached to the stem, and an **internode** is the region between the nodes (see Fig. 20.1). Vascular tissue transports water and minerals from the roots through the stem to the leaves and transports the products of photosynthesis, usually in the opposite direction.

Stems may have functions other than those mentioned. In a cactus, the stem is the primary photosynthetic organ and serves as a water reservoir (see Fig. 20.2*c*), and the tuber of a potato plant is a food storage portion of an underground stem (Fig. 20.2*d*). **Perennial plants** are able to regrow each season from varied underground structures, such as tubers, bulbs, and rhizomes, all of which bear nodes that can produce a new shoot system.

Roots

A plant's root system supports the plant by anchoring it in the soil, as well as absorbing water and minerals from the soil for the entire plant. As a rule of thumb, the root system is at least equivalent in size and extent to its shoot system. Therefore, an apple tree has a much larger root system than a corn plant. Also, the extent of a root system depends on the environment. A single corn plant may have roots as deep as 2.5 meters (m), while a mesquite tree that lives in the desert may have roots that penetrate to a depth of 20 m.

Roots have a cylindrical shape and a slimy surface that allows them to penetrate the soil as they grow and permit water to be absorbed from all sides. In a special zone near the root tip, there are many **root hairs** that greatly increase the absorptive capacity of the root. Root hairs are so numerous that they increase the absorptive surface of a root tremendously. It has been estimated that a single rye plant has about 14 billion root-hair cells, and if placed end to end, the root hairs would stretch 10,626 kilometers. Root-hair cells are constantly being replaced, so this same rye plant forms about 100 million new root-hair cells every day. You probably know that a plant yanked out of the soil will not fare well when transplanted. This is because small lateral roots and root hairs are torn off. Transplantation is more apt to be successful if you take a part of the surrounding soil along with the plant, leaving as many of the lateral roots and the root hairs intact as possible.

Just as stems and leaves are diverse, so are roots. A carrot plant has one main *taproot*, which stores the products of photosynthesis (**Fig. 20.3***a*). Grass has fibrous roots that cling to the soil (Fig. 20.3*b*), and corn plants have *prop roots* that grow from the stem for better support (Fig. 20.3*c*).

a. Cucumber

b. Venus's flytrap

c. Cactus

d. Potato

Figure 20.2   Leaf and stem diversity.

a. The tendrils of a cucumber are leaves modified to attach the plant to a support. **b.** Venus's flytrap leaves serve as the trap for unsuspecting insects. After digesting this fly, the plant will absorb the nutrients released. **c.** The spines of a cactus protect the fleshy stem from herbivores. The cactus uses its fleshy stem for both food production and water storage. **d.** The familiar potato is actually an underground stem, or tuber, that is used for food storage.

Figure 20.3   Root diversity.

a. A taproot has one main root. **b.** A fibrous root has many slender roots with no main root. **c.** Prop roots are organized for support.

a. Taproot

b. Fibrous root system

c. Prop roots

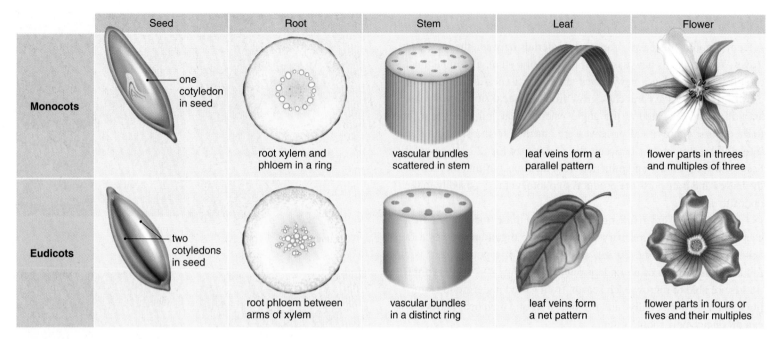

	Seed	Root	Stem	Leaf	Flower
Monocots	one cotyledon in seed	root xylem and phloem in a ring	vascular bundles scattered in stem	leaf veins form a parallel pattern	flower parts in threes and multiples of three
Eudicots	two cotyledons in seed	root phloem between arms of xylem	vascular bundles in a distinct ring	leaf veins form a net pattern	flower parts in fours or fives and their multiples

Figure 20.4 Flowering plants are either monocots or eudicots.

Five features are typically used to distinguish monocots from eudicots: number of cotyledons; arrangement of vascular tissue in roots, stems, and leaves; and number of flower parts.

Monocot Versus Eudicot Plants

Figure 20.4 shows how flowering plants can be divided into two major groups. One of the differences between the two groups concerns the **cotyledons,** which are embryonic leaves present in seeds. (The cotyledons wither after the first true leaves appear.) Plants whose embryo has one cotyledon are known as monocotyledons, or **monocots.** Other embryos have two cotyledons, and these plants are known as eudicotyledons, or **eudicots** (true dicots). The cotyledons of eudicots supply nutrients for seedlings, but the cotyledons of monocots store some nutrients and act as a transfer tissue for nutrients stored elsewhere.

Although the distinction between monocots and eudicots may seem minimal at first glance, there are many differences in their structures. For example, the location and arrangement of vascular tissue differs between monocots and eudicots. Vascular plants contain two main types of transport tissue: the **xylem** for water and minerals, and the **phloem** for organic nutrients. In a sense, xylem and phloem are to plants what veins and arteries are to animals. In the monocot root, vascular tissue occurs in a ring around the center. But in the eudicot, vascular tissue is located in the center. The xylem forms a star shape, and phloem is located between the points of the star. In a stem, vascular tissue occurs in the **vascular bundles.** The vascular bundles are scattered in the monocot stem, and they occur in a ring in the eudicot stems.

In a leaf, vascular tissue forms **leaf veins.** In monocots, the veins are parallel, while in eudicots, the leaf veins form a net pattern. Monocots and eudicots also have different numbers of flower parts, and this difference will be discussed further in Chapter 21.

The eudicots are the larger group and include some of our most familiar flowering plants—from dandelions to oak trees. The monocots include grasses, lilies, orchids, and palm trees, among others. Some of our most significant food sources are monocots, including rice, wheat, and corn.

Check Your Progress

1. List the three vegetative organs in a plant.
2. What are some possible leaf modifications and what functions do they serve?
3. Explain how root hairs increase the absorptive capacity of a root.
4. List the major differences in the organization of vascular tissue between monocots and eudicots.

Answers: 1. Vegetative organs are the leaves, stem, and root. 2. Some leaves (tendrils) attach to objects, some leaves capture insects, and others (spines) protect the plant from being eaten. 3. Root hairs provide greater surface area for absorbing water and minerals. 4. Monocots: xylem and phloem in a ring in the root; scattered vascular bundles in the stem; parallel leaf veins. Eudicots: phloem located between arms of xylem in the root; vascular bundles form a ring in the stem; netted leaf veins.

20.2 | Plant Tissues and Cells

Unlike humans, flowering plants grow their entire lives because they have **meristem** (embryonic) tissue, composed of undifferentiated cells that divide. **Apical meristem** is located in the terminal bud of the shoot system and in the root tip. When apical meristem cells divide, one of the daughter cells remains a meristem cell, and the other differentiates into one of the three types of specialized tissues of a plant:

1. **Epidermal tissue** forms the outer protective covering of a plant.
2. **Ground tissue** fills the interior of a plant and helps carry out the functions of a particular organ.
3. **Vascular tissue** transports water and nutrients in a plant and provides support.

Another type of meristem tissue is associated with vascular tissue and is called **vascular cambium.** When the meristem cells of vascular cambium divide, they give rise to new vascular tissue, which causes a plant to increase in girth (called *secondary growth*).

Epidermis and Ground Tissue

The entire body of a plant is covered by an **epidermis,** a layer of closely packed cells that act as a barrier. The walls of epidermal cells that are exposed to air are covered with a waxy **cuticle** to minimize water loss. The cuticle also protects against bacteria and other organisms that might cause disease. As mentioned in Section 20.1, roots have root hairs, long slender projections of epidermal cells that increase the surface area of the root for absorption of water and minerals (**Fig. 20.5***a*). In leaves, the epidermis often contains **stomata** (sing., stoma). A stoma is a small opening surrounded by two guard cells (Fig. 20.5*b*). When the stomata are open, gas exchange and water loss occur.

In the trunk of a tree, the epidermis is replaced by cork, which is a part of bark (Fig. 20.5*c*). New cork cells are made by a meristem called *cork cambium.* As the new cork cells mature, they increase slightly in volume, and their walls become encrusted with *suberin,* a lipid material, so that they are waterproof and chemically inert. These nonliving cells protect the plant and make it resistant to attack by fungi, bacteria, and animals.

Ground tissue forms the bulk of leaves, stems, and roots. Ground tissue contains three types of cells (**Fig. 20.6**). **Parenchyma** cells are the least specialized of the cell types and are found in all the organs of a plant. They may contain chloroplasts and carry on photosynthesis, or they may contain colorless plastids that store the products of photosynthesis. **Collenchyma** cells are like parenchyma cells except that they have irregularly shaped corners and thickened primary cell walls. Collenchyma cells often form bundles just beneath the epidermis and give flexible support to immature regions of a plant body. The familiar strands in celery stalks are composed mostly of collenchyma cells. **Sclerenchyma** cells have thick secondary cell walls containing *lignin,* which makes plant cell walls tough and hard. If we compare a cell wall to reinforced concrete, cellulose fibrils would play the role of steel rods, and lignin would be analogous to the cement. Most sclerenchyma cells are nonliving; their primary function is to support the mature regions of a plant. The hard outer shells of nuts are made of sclerenchyma cells. The long fibers in plants, composed of strings of sclerenchyma, make them useful for a number of purposes. For example, cotton and flax fibers can be woven into cloth, and hemp fibers can make strong rope.

a. Root hairs

corn seedling
root hairs
elongating tip of root

chloroplasts
nucleus
stoma
epidermal cells
guard cell

b. Stoma of leaf

cork
cork cambium

×500

c. Cork of older stem

Figure 20.5 Modifications of epidermal tissue.

a. Root epidermis has root hairs, which help to absorb water. **b.** Leaf epidermis contains stomata (sing., stoma) for gas exchange. **c.** Cork, which is a part of bark, replaces epidermis in older, woody stems.

×100

a. Parenchyma cells

×340

b. Collenchyma cells

×340

c. Sclerenchyma cells

Figure 20.6 Ground tissue cells.

a. Parenchyma cells are the least specialized of the plant cells. **b.** Collenchyma cells have thicker and more irregular cell walls than parenchyma cells. **c.** Sclerenchyma cells have very thick walls and are nonliving—their only function is to give strong support.

a.

b.

Figure 20.7 Vascular tissue.

a. Photomicrograph of xylem vascular tissue, with (*left*) drawing of a vessel (composed of vessel elements) and (*right*) drawing of tracheids. **b.** Photomicrograph of phloem vascular tissue, with drawing of a sieve tube. Each sieve-tube member has a companion cell.

Vascular Tissue

Vascular tissue extends from the root through the stem to the leaves and vice versa (see Fig. 20.1). In the root, the vascular tissue is located in a central cylinder; in the stem, vascular tissue is in vascular bundles; and in the leaves, it is found in leaf veins. Although both types of vascular tissue are usually found together, they have different functions. Xylem transports water and minerals from the roots to the leaves. Phloem transports sugar, in the form of sucrose, and other organic compounds, such as hormones, often from the leaves to the roots.

Xylem contains two types of conducting cells: vessel elements and tracheids (**Fig. 20.7a**). Both types of conducting cells are hollow and nonliving, but the **vessel elements** are larger, have perforated end walls, and are arranged to form a continuous pipeline for water and mineral transport. The end walls and side walls of **tracheids** have pits that allow water to move from one tracheid to another.

The conducting cells of phloem are **sieve-tube members,** so-called because they contain a cluster of pores in their end-walls. The pores are collectively known as a *sieve plate*. The sieve-tube members are arranged to form a continuous sieve tube (Fig. 20.7b). Sieve-tube members contain cytoplasm but no nuclei. Each sieve-tube member has a companion cell, which does have a nucleus. The companion cells may very well be involved in the transport function of phloem.

20.3 | Organization of Leaves

Figure 20.8 shows the general structure of a leaf. The wide portion of a foliage leaf is called the **blade.** The **petiole** is a stalk that attaches the blade to the stem. The blade of a leaf is often undivided, or *simple*, as in a cottonwood (Fig. 20.8a). But in some leaves, the blade is divided, or *compound*, as in a shagbark hickory or honey locust (Fig. 20.8b,c). Notice that it is possible to tell from the placement of the lateral bud whether you are looking at several individual leaves or one compound leaf.

Figure 20.9 shows a typical eudicot leaf of a temperate-zone plant in cross section. The outer layer of the leaf, or epidermis, has an outer, waxy cuticle that prevents loss of water and also prevents gas exchange because it is not gas permeable. The lower epidermis of eudicot leaves and both surfaces of monocot leaves contain stomata. At the stomata, carbon dioxide for photosynthesis enters a leaf, and oxygen, the by-product of photosynthesis, exits. The epidermis may also bear protective hairs and/or glands that secrete irritating substances. These adaptations discourage insects from eating leaves.

The interior of a leaf is composed of **mesophyll,** the tissue that carries on photosynthesis. Notice how leaf veins terminate at the mesophyll. Leaf veins transport water and minerals to a leaf and transport the product of photosynthesis, a sugar, away from the leaf. Mesophyll has two distinct regions: **palisade mesophyll,** containing elongated cells, and **spongy mesophyll,** containing irregular cells bounded by air spaces. The loosely packed arrangement of the cells in the spongy layer increases the amount of surface area for gas exchange and water loss. Water brought into a leaf by leaf veins evaporates from spongy mesophyll and exits at the stomata.

Check Your **Progress**

1. List the three types of tissue in a plant and the cells that make up these tissues.
2. Contrast the function of xylem with that of phloem.
3. List the features of the epidermis of a leaf.

Answers: 1. Epidermal tissue: epidermal cells; ground tissue: parenchyma, collenchyma, and sclerenchyma cells; vascular tissue: xylem (vessel elements and tracheids) and phloem (sieve-tube members and companion cells). 2. Xylem transports water and minerals from roots to leaves. Phloem transports organic compounds throughout the plant. 3. Most of the epidermis is composed of tightly packed epidermal cells covered by a waxy cuticle. Many stomata and their guard cells are also present. Sometimes hairs or glands play a protective role.

blade

lateral bud

c. Twice compound leaf

a. Simple leaf

b. Compound leaf

blade

leaflet

petiole

lateral bud

Figure 20.8 **Simple versus compound leaves.**

a. The cottonwood tree has a simple leaf. **b.** The shagbark hickory has a compound leaf. **c.** The honey locust has a twice compound leaf. A single lateral bud appears where the leaf attaches to the stem.

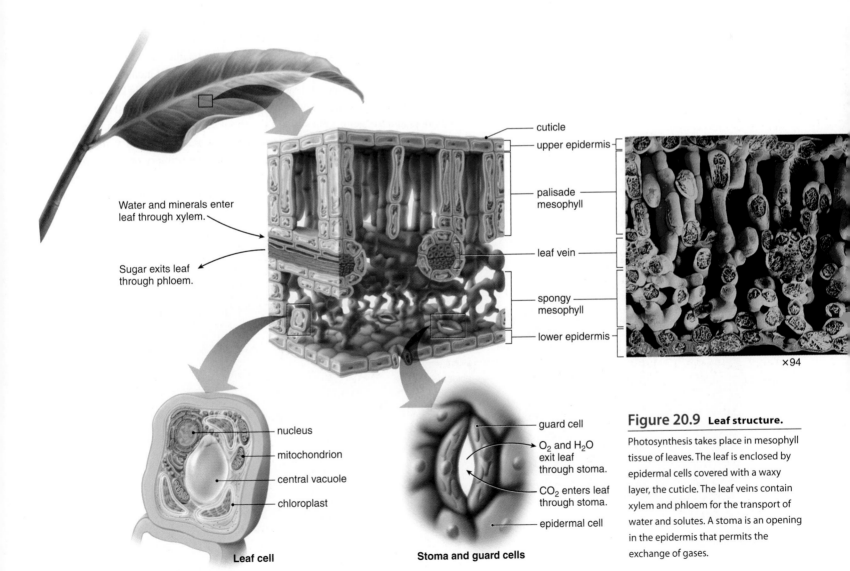

Water and minerals enter leaf through xylem.

Sugar exits leaf through phloem.

cuticle
upper epidermis
palisade mesophyll
leaf vein
spongy mesophyll
lower epidermis

×94

nucleus
mitochondrion
central vacuole
chloroplast

Leaf cell

guard cell

O₂ and H₂O exit leaf through stoma.

CO₂ enters leaf through stoma.

epidermal cell

Stoma and guard cells

Figure 20.9 **Leaf structure.**

Photosynthesis takes place in mesophyll tissue of leaves. The leaf is enclosed by epidermal cells covered with a waxy layer, the cuticle. The leaf veins contain xylem and phloem for the transport of water and solutes. A stoma is an opening in the epidermis that permits the exchange of gases.

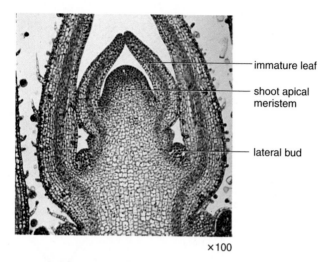

×100

Figure 20.10   Shoot tip.

The terminal bud contains the apical meristem that produces leaves and also cells that become the tissues of stems.

20.4 | Organization of Stems

During the primary growth of stems, the shoot apical meristem in the terminal bud produces new cells that elongate and then differentiate to become the tissues and organs of the shoot system, namely the leaves and the stem. The shoot apical meristem is protected by the immature leaves that envelop it. Also, note the lateral buds from which lateral (side) branches (including leaves) develop (**Fig. 20.10**).

Nonwoody Stems

A stem that experiences only primary growth is nonwoody. Plants that have nonwoody stems, such as zinnias and daisies, are termed **herbaceous** plants. As with leaves, the outermost tissue of a herbaceous stem is the epidermis, which is covered by a waxy cuticle to prevent water loss. Beneath the epidermis of eudicot stems is the **cortex,** a narrow band of parenchyma cells. The cortex is sometimes green and carries on photosynthesis. The so-called **pith** consists of ground tissue and is found in the center. A monocot stem lacks an organized cortex or pith.

Herbaceous stems have distinctive vascular bundles, where xylem and phloem are found. In each bundle, xylem is typically found toward the inside of the stem, and phloem is found toward the outside. In the herbaceous eudicot stem, the vascular bundles are arranged in a distinct ring that separates the cortex from the pith. In a herbaceous monocot stem, the vascular bundles are scattered throughout the stem. **Figure 20.11** contrasts herbaceous eudicot and monocot stems.

The vascular tissue of a stem supports the growth of the shoot system. The sclerenchyma cells of vascular tissue and the strong walls of vessel members and tracheids help support the shoot system as it increases in length. Also, the vascular bundles bring water and minerals to the leaf veins, which distribute them to the mesophyll cells of leaves, the primary organs of photosynthesis in most plants. The vascular bundles also distribute the products of photosynthesis to the root system and to any immature leaves that are not yet able to carry on photosynthesis.

Figure 20.11   Nonwoody (herbaceous) stems.

a. In eudicot stems, the vascular bundles are arranged in a ring around well-defined ground tissue called pith. **b.** In monocot stems, the vascular bundles are scattered within the ground tissue.

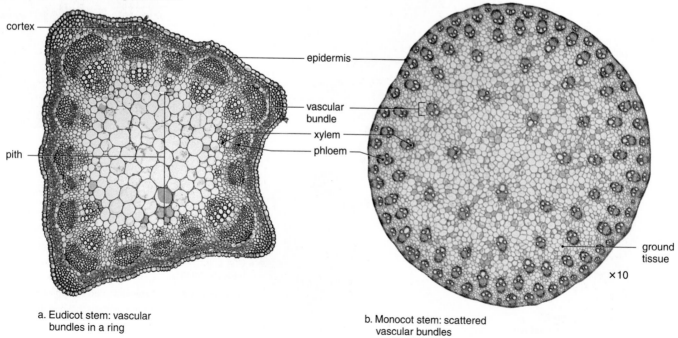

a. Eudicot stem: vascular bundles in a ring

b. Monocot stem: scattered vascular bundles

×10

Woody Stems

Plants with woody stems, such as trees and shrubs, experience both primary and secondary growth. *Secondary growth* increases the girth of stems, branches, and roots. It occurs because of a difference in the location and activity of vascular cambium, which, as you may recall from Section 20.2, is a type of meristem tissue. In herbaceous eudicot plants, vascular cambium is usually present between the xylem and phloem of each vascular bundle. In woody plants, the vascular cambium forms a ring of meristem that divides parallel to the surface of the plant and produces new xylem and phloem each year. Eventually, a woody eudicot stem has three distinct areas: the bark, the wood, and the pith. Vascular cambium occurs between the bark and the wood (**Fig. 20.12**).

Bark

The **bark** of a tree contains cork, cork cambium, cortex, and phloem. It is very harmful to remove the bark of a tree, because without phloem, organic nutrients cannot be transported. Although new phloem tissue is produced each year by vascular cambium, it does not build up in the same manner as xylem (Fig. 20.12).

Cork cambium is located beneath the epidermis. When cork cambium first begins to divide, it produces tissue that disrupts the epidermis and replaces it with cork cells. **Cork cells** are impregnated with suberin, a waxy layer that makes them waterproof but also causes them to die. In a woody stem, gas exchange is impeded, except at *lenticels*, which are pockets of loosely arranged cork cells not impregnated with suberin.

Wood

Wood is *secondary xylem* that builds up year after year, thereby increasing the girth of trees. In trees that have a growing season, vascular cambium is dormant during the winter. In the spring, when moisture is plentiful and leaves require much water for growth, vascular cambium produces secondary xylem tissue that contains wide vessels with thin walls. In this so-called *spring wood*, wide vessels transport sufficient water to the growing leaves. Later in the season, moisture is scarce, and the wood at this time, called *summer wood*, has a lower proportion of vessels. Strength is required because the tree is growing larger, and summer wood contains numerous thick-walled tracheids. At the end of the growing season, just before the cambium becomes dormant again, only heavy fibers with especially thick secondary walls may develop. When the trunk of a tree has spring wood followed by summer wood, the two together make up one year's growth, or an **annual ring** (Fig. 20.12). You can tell the age of a tree by counting the annual rings.

Wood is one of the most useful and versatile materials known. It is used for building structures and for making furniture, and in addition to its functionality, its beauty gives us pleasure. Much wood is also used for heating and cooking, as well as for producing paper, chemicals, and pharmaceuticals. The resins derived from wood also have commercial importance.

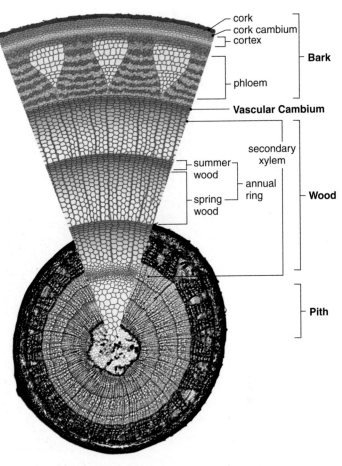

Figure 20.12   Organization of a woody stem.

In a woody stem, vascular cambium produces new secondary phloem and secondary xylem each year. Secondary xylem builds up to form wood consisting of annual rings. Counting the annual rings tells you that this woody stem is three years old.

Check Your Progress

1. Contrast a nonwoody eudicot stem with a monocot stem.
2. List the components of bark.
3. Explain what causes wood to consist of annual rings.
4. Compare and contrast spring wood with summer wood.

Answers: 1. Eudicot stems contain an epidermis, cortex, and pith, with vascular bundles arranged in a ring between the cortex and pith. Monocot stems lack a distinct cortex or pith, and vascular bundles are scattered.   2. Bark is composed of cork, cork cambium, cortex, and phloem.   3. The new secondary xylem produced by vascular cambium every growing season builds up, forming the annual rings of wood.   4. An annual ring often consists of spring wood followed by summer wood. In the spring, when water is plentiful, wood contains wide vessels with thin walls. In the summer, when conditions are dry, wood contains numerous thick-walled tracheids and also fibers.

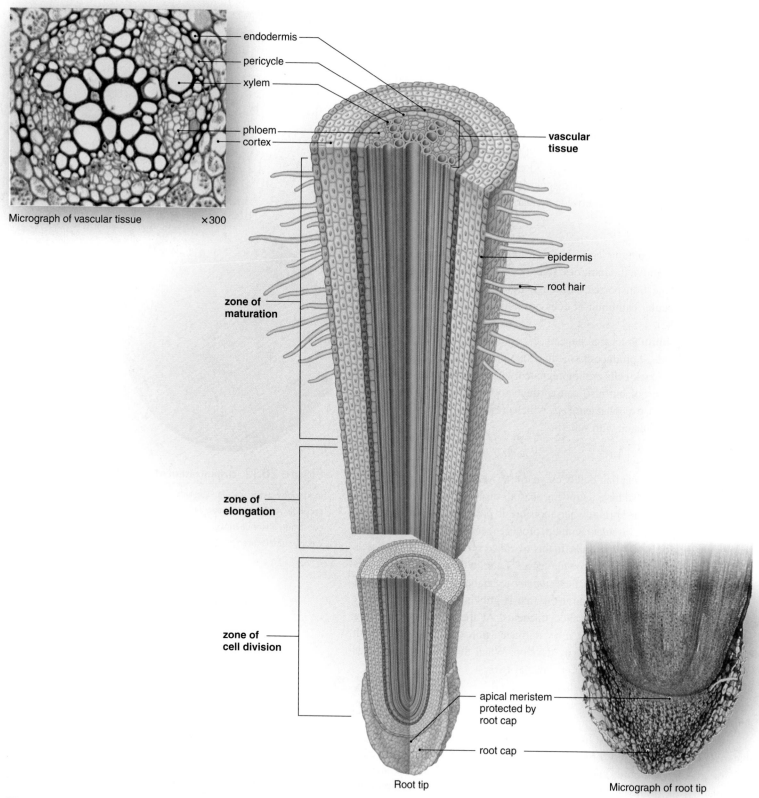

Micrograph of vascular tissue   ×300

endodermis
pericycle
xylem
phloem
cortex

vascular tissue

epidermis
root hair

zone of maturation

zone of elongation

zone of cell division

apical meristem protected by root cap

root cap

Root tip

Micrograph of root tip

Figure 20.13   Eudicot root tip.

The root tip is divided into three zones, best seen in a longitudinal section such as this. In the eudicot, xylem is typically star-shaped, and phloem lies between the points of the star. Water and minerals must eventually pass through the cytoplasm of endodermal cells in order to enter xylem; endodermal cells regulate the passage of minerals into xylem.

20.5 | Organization of Roots

Both monocot and eudicot roots contain the same growth zones. **Figure 20.13,** a longitudinal section of a eudicot root, reveals these zones where cells are in various stages of differentiation as primary growth occurs. The apical meristem of a root contains actively dividing cells and is protected by a **root cap** (Fig. 20.13, *lower right*). As the cells of the apical meristem divide, it provides cells to the zone of elongation. In the zone of elongation, the cells lengthen as they become specialized. In the zone of maturation, which contains fully differentiated cells, many of its epidermal cells bear root hairs.

Tissues of a Root

The structure and function of tissues in a root are as follows:

Vascular tissue

Both monocot and eudicot roots have a **vascular cylinder** that contains xylem and phloem, but they are arranged differently. In a eudicot root (Fig. 20.13, *upper left*), the xylem is star-shaped because several xylem arms radiate from a common center. Phloem is found in separate regions between the points of the star. In a monocot root (**Fig. 20.14**), the vascular cylinder consists of alternating xylem and phloem bundles that surround a pith. Pith can function as a storage site.

Pericycle The **pericycle** is the first layer of cells inside the endodermis. It has retained its capacity to divide and can start the development of branch, or lateral, roots (**Fig. 20.15**).

Endodermis The **endodermis** is a single layer of rectangular cells that fit snugly together. Further, a layer of impermeable material on all but two sides forces water and minerals to pass through endodermal cells. In this way, the endodermis regulates the entrance of minerals into the vascular tissue of the root.

Cortex Large, thin-walled parenchyma cells make up the cortex of the root. The cells contain starch granules, and the cortex may function in food storage.

Epidermis The epidermis, which forms the outer layer of the root, consists of only a single layer of largely thin-walled, rectangular cells. In the zone of maturation, many epidermal cells have root hairs.

Growth of Roots

Both roots and stems have apical meristem regions that account for their primary growth. The branching of a stem occurs at a lateral bud, whereas the branching of a root occurs some distance from apical meristem and is initiated by the pericycle. Most eudicot roots, like woody stems, experience secondary growth. Vascular cambium located between the arms of the xylem and the phloem produces secondary xylem and phloem. As secondary xylem builds up, a root increases in girth.

epidermis
cortex
endodermis
pericycle
xylem
phloem
pith

Figure 20.14   Monocot root.

In a monocot root, the same zones and tissues are present, but xylem and phloem form a ring surrounding the pith in the center.

branch root grows from pericycle

a.   ×40

branch root breaking through cortex and epidermis

b.   ×40

Figure 20.15   Branching of eudicot root.

These cross sections of a willow show the origination (**a**) and growth (**b**) of a branch root from the pericycle.

Check Your Progress

1. Describe the relationship between the root apical meristem and the root cap.
2. List the functions of the tissues of a eudicot root.
3. Compare and contrast roots with stems.

Answers: 1. The root apical meristem is located at the tip of the root and is covered by the root cap.   2. Cortex: food storage; endodermis: control of mineral uptake; pericycle: formation of branch roots; epidermis: forms outer layer, bears root hairs; vascular tissue: transports water, minerals, and organic nutrients.   3. Both roots and stems have apical meristems that allow them to increase in length. They are capable of branching; stems branch at lateral buds, while root branches originate in the pericycle.

solution lacks nitrogen   complete nutrient solution

Figure 20.16   Nitrogen deficiency.

This experiment shows that sunflower plants respond poorly if their growth medium lacks nitrogen.

— root

— root nodule

a.

b.

Figure 20.17   Adaptations of roots for mineral uptake.

a. Nitrogen-fixing bacteria live in nodules on the roots of plants, particularly legumes. **b.** Rough lemon plants with mycorrhizal fungi (*right*) grow much better than plants without mycorrhizal fungi (*left*).

20.6 | Plant Nutrition

Plant nutrition is remarkable to us because plants require only inorganic nutrients, and from these they make all the organic compounds that compose their bodies. Of course, they require carbon, hydrogen, and oxygen, which they can acquire from carbon dioxide and water, but they also need other elements, or minerals, which they obtain from the environment. An element is termed an essential nutrient if a plant cannot live without it. The essential nutrients are divided into *macronutrients* and *micronutrients*, according to their relative concentrations in plant tissue. The following diagram and slogan show which are the macronutrients and which are the micronutrients:

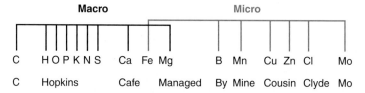

All of these elements play vital roles in plant cells. For example, nitrogen and phosphorus are required to make nucleic acids, and nitrogen and sulfur are important parts of amino acids. **Figure 20.16** shows how an insufficient supply of the macronutrient nitrogen can affect a plant. The micronutrients are often cofactors for enzymes in various metabolic pathways.

It's interesting to observe that humans make use of a plant's ability to take minerals from the soil. For example, we depend on plants for supplies of calcium to build our bones and teeth, and for iron to help carry oxygen to our cells. Minerals such as copper and zinc, which we acquire by eating plants, are also cofactors for our own enzymes.

Adaptations of Roots for Mineral Uptake

Minerals enter a plant at its root system, and two mutualistic relationships assist roots in fulfilling this function. The air all about us contains about 78% nitrogen (N_2), but plants can't make use of it. Most plants depend on bacteria in the soil to *fix* nitrogen—that is, the bacteria change atmospheric nitrogen (N_2) to nitrate (NO_3^-) or ammonium (NH_4^+), both of which plants can take up and use. Some plants, such as legumes, have roots colonized by bacteria that are able to take up atmospheric nitrogen and reduce it to a form suitable for incorporation into organic compounds (**Fig. 20.17***a*). The bacteria live in **root nodules,** and the plant supplies the bacteria with carbohydrates; the bacteria in turn furnish the plant with nitrogen compounds.

Another mutualistic relationship involves fungi and almost all plant roots (Fig. 20.17*b*). This association is known as a **mycorrhizal association,** which literally means "fungal roots." The hyphae of the fungus increase the surface area available for water uptake and break down organic matter, releasing inorganic nutrients that the plant can use. In return, the root furnishes the fungus with sugars and amino acids. Plants are extremely dependent on mycorrhizal fungi. For example, orchid seeds, which are quite small and contain limited nutrients, do not germinate until a mycorrhizal fungus has invaded their cells. The visual appearance of plants that do not have adequate nitrogen convinces us how important these associations are (see Fig. 20.16).

Check Your **Progress**

1. List the macronutrients needed for plant growth.
2. Explain how plants obtain nitrogen from the air.

Answers: 1. Carbon, hydrogen, oxygen, phosphorus, potassium, nitrogen, sulfur, and magnesium. 2. Bacteria in soil or root nodules fix aerial nitrogen. Those in the soil form nitrate or ammonium that plant roots absorb. Those in root nodules form nitrogen compounds that are passed to the plant.

20.7 | Transport of Nutrients

Water and Mineral Transport in Xylem

Water and minerals are taken up by root hairs at the same time, and they are both transported within xylem. The vessel elements, assisted by the tracheids, constitute a continuous system for water and mineral transport in a plant (see Fig. 20.7). In other words, a pipeline for water transport exists from the roots through the stem to the leaves. People have long wondered how plants, especially very tall trees, lift water to the leaves against gravity. Water enters root cells by osmosis, and sure enough, this does create a positive pressure called root pressure. But this pressure is not nearly enough to account for the movement of water and minerals all the way to the leaves.

The mechanism by which water and minerals travel up the xylem is called the **cohesion-tension model.** To understand how it works, we have to turn our attention to the anatomy of leaves and the properties of water. A leaf, as you know, has small openings called stomata, and the stomata open to air spaces in the leaf. Dry air passing across the leaves causes water to evaporate from the surface of spongy mesophyll cells. Water vapor in the air spaces of a leaf exits a leaf at the stomata, a phenomenon called **transpiration.** The evaporation of water creates a *tension* that is sufficient to pull a water column up from the roots to the leaves (**Fig. 20.18**). Why? Because water is both cohesive and adhesive. Water molecules are cohesive—they cling tightly to each other because of hydrogen bonding. As a result, the water molecules tend to stay together in a column as the molecules are pulled upward by transpiration. Also, water molecules are adhesive, because they are polar and are attracted to polar surfaces, such as the sides of the xylem vessels. The adhesion of water molecules to the vessel walls prevents them from slipping back.

The total amount of water a plant loses through transpiration over a long period of time is surprisingly large. At least 90% of the water taken up by roots is eventually lost at the leaves. A single corn plant loses somewhere between 135 and 200 liters of water through transpiration during a growing season. As transpiration occurs, the water column is pulled upward, first within the leaf, then from the stem, and finally from the roots.

Evaporation of water (transpiration) creates tension that pulls the water column from the roots to the leaves.

H_2O

Cohesion and adhesion of water molecules keeps the water column intact within xylem.

Water enters a plant at root cells.

H_2O

Figure 20.18 **Cohesion-tension model of xylem transport.**

How does water rise to the top of tall trees? Xylem vessels are water-filled pipelines from the roots to the leaves. When water evaporates from spongy mesophyll into the air spaces of leaves, this water column is pulled upward due to the cohesion of water molecules with one another and the adhesion of water molecules to the sides of the vessels.

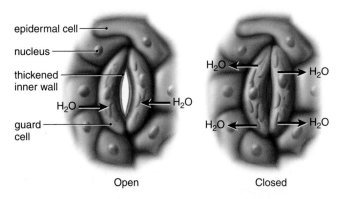

Figure 20.19   Opening and closing of stomata.

Stomata open when water enters guard cells and turgor pressure increases. Stomata close when water exits guard cells and turgor pressure is lost.

Opening and Closing of Stomata

Stomata open and close due to turgor pressure changes within guard cells (**Fig. 20.19**). When water enters the guard cells, turgor pressure is created; when water leaves the guard cells, turgor pressure is lost. For transpiration to occur, the stomata must stay open. But when a plant is under stress and about to wilt from lack of water, the stomata close. Now the plant is unable to take up carbon dioxide from the air, and photosynthesis ceases.

In plants kept in the dark, stomata open and close on a 24-hour basis, just as if they were responding to the presence of sunlight in the daytime and the absence of sunlight at night. Such circadian rhythms (behaviors that occur every 24 hours) are areas of intense investigation at this time.

Organic Nutrient Transport in Phloem

Phloem transports organic nutrients from stored supplies in roots, or from mature leaves that are photosynthesizing, to all other parts of a plant, including young leaves that have not yet reached their full photosynthetic potential and flowers in the process of making seeds and fruits. Nutrients are transported to any part of a plant, including roots, when they lack a ready supply.

Just as xylem is continuous throughout a plant, so is phloem. In phloem, the sieve-tube members align end to end, and plasmodesmata extend through sieve plates from one sieve-tube member to the other, allowing direct cytoplasmic exchange of organic materials. Sieve tubes, therefore, form a continuous pathway for transport of organic nutrients, primarily the sugar sucrose.

The mechanism by which nutrients are transported through phloem is called the **pressure-flow model** (**Fig. 20.20**). During the growing season, photosynthesizing leaves are making sugar and are therefore a *source* of sugar. This sugar is actively transported into sieve tubes. As the concentration of sugar rises inside the sieve tube cells, water follows by osmosis since the concentration of water outside phloem is now lower than inside phloem. The buildup of water within sieve tubes creates a positive pressure that starts a flow of phloem contents. The other

Source:
Sugar is actively transported into sieve tubes, and water follows by osmosis.

This creates a positive pressure that causes a flow within phloem.

Sink:
Sugar is actively transported out of sieve tubes, and water follows by osmosis.

Source
sieve-tube member
companion cell } phloem
mature leaf cells

water molecule

sugar molecule

sieve plate

Sink

Figure 20.20   Pressure-flow model of phloem transport.

Sugar and water enter sieve-tube members at a source. This creates a positive pressure, which causes phloem contents to flow. Sieve-tube members form a continuous pipeline from a source to a sink, where sugar and water exit sieve-tube members.

parts of a plant, such as roots that are not photosynthesizing, are a *sink* for sugar, because they require a supply of sugar. Here, sugar is actively transported out of phloem, and water again follows by osmosis. In this way, phloem contents continue to flow from a source (e.g., leaves) to a sink (e.g., root).

The experiment described in **Figure 20.21** supports the pressure-flow model of phloem transport. Two bulbs connected by a glass tube are placed in a split container of water. Each bulb is bounded by a membrane that is permeable to water, but not to sugar. The first bulb, the source, contains a concentrated sugar solution, while the second bulb, the sink, contains a dilute sugar solution. Analogous to phloem, water flows into the first bulb by osmosis because the first bulb contains a lower water concentration (higher solute concentration) than does the container. The entrance of water creates a *pressure* that causes water to *flow* toward the second bulb. This pressure not only drives the solution to the second bulb, but it is strong enough to overcome the force of osmosis. Water moves out through the membrane of the second bulb even though the second bulb contains a lower water concentration than does the container.

The pressure-flow model can thus account for the flow of sugar in either direction, depending on which is the source and which is the sink. For example, in the spring, recently formed leaves can be a sink, and roots that have held stored sugar over the winter can be a source. Thus, organic nutrients will flow from the roots up to the leaves, instead of the other way around.

Figure 20.21   Experiment supporting the pressure-flow model.

When a concentrated sugar solution is placed in the first bulb, water enters and causes a flow of both water and sugar to the second bulb. The pressure is sufficient to cause water to exit the second bulb, even though the surrounding water contains no sugar. As water enters the first bulb, water level in the container decreases (*on left*); as water exits the second bulb, water level increases (*on right*).

Check Your Progress

1. Explain the role of transpiration in xylem transport.
2. Explain how the cohesive and adhesive properties of water assist water transport in xylem.
3. Explain how sugar moves from a source to a sink in phloem.

Answers: 1. Transpiration, evaporation of water at the leaves, causes water to move upward in xylem.   2. Because of cohesion and adhesion, a continuous water column exists in xylem. Without a continuous sequence of water molecules, transpiration would be unable to pull water from roots to the leaves.   3. Sugar is transported into phloem at a source; when water follows by osmosis, a pressure builds that causes phloem sap to flow toward a sink, where sugar is transported out of phloem.

The Chapter in Review

Summary

20.1 | Plant Organs

A flowering plant has a shoot system and a root system. Both grow from meristems at their tips.

Leaves carry on photosynthesis, but may be modified for other purposes.

Stems support leaves, conduct materials to and from roots and leaves, and produce new tissues.

Roots anchor a plant, absorb water and minerals, and store the products of photosynthesis.

Monocot Versus Eudicot Plants

Flowering plants are divided into eudicots and monocots according to:
- Number of cotyledons in the seed.
- Arrangement of vascular tissue in leaves, stems, and roots.
- Number of flower parts:

monocot flower   eudicot flower

20.2 | Plant Tissues and Cells

Plants have three types of specialized tissue:

Epidermal tissue is composed of only epidermal cells.

Ground tissue contains:

- Parenchyma cells, which are thin-walled and capable of photosynthesis when they contain chloroplasts.
- Collenchyma cells, which have thicker walls for flexible support.
- Sclerenchyma cells, which are hollow, nonliving support cells with secondary walls.

Vascular tissue consists of:

- Xylem, which contains vessels composed of vessel elements and tracheids. Xylem transports water and minerals.
- Phloem, which contains sieve tubes composed of sieve-tube members, each of which has a companion cell. Phloem transports organic nutrients.

xylem   phloem

20.3 | Organization of Leaves

- A leaf has a cuticle-covered epidermis, with stomata mostly in the lower layer.
- Stomata allow water vapor and oxygen to escape and carbon dioxide to enter the leaf.
- Mesophyll (palisade and spongy) forms the body of a leaf and carries on photosynthesis.

20.4 | Organization of Stems

Primary growth of a stem is due to the activity of the shoot apical meristem, which is protected within a terminal bud.

Nonwoody Stems

- Nonwoody (herbaceous) eudicots have epidermis, cortex tissue, vascular bundles in a ring, and an inner pith.
- Monocot stems have scattered vascular bundles and lack a distinct cortex or pith.

eudicot stem   monocot stem

Woody Stems

- Secondary growth of a woody stem is due to vascular cambium, which produces new xylem and phloem every year.
- Cork replaces epidermis in woody plants.
- A woody stem has bark (contains cork, cork cambium, cortex, and phloem). Wood contains annual rings of xylem.

20.5 | Organization of Roots

A monocot and eudicot root tip has three zones:

- Zone of cell division (contains root apical meristem) protected by the root cap.
- Zone of elongation, where cells elongate and differentiate.
- Zone of maturation (has root hairs).

Tissues of a Root

Within the zone of maturation, a monocot and eudicot root contains the following tissues:

- **Vascular tissue:** In eudicot roots, xylem is arranged in a star shape with phloem between the arms of the xylem. In monocot roots, a central pith is surrounded by a ring of vascular tissue containing alternating bundles of xylem and phloem.
- **Pericycle:** Cells that can divide and form lateral roots.
- **Endodermis:** Regulates entrance of minerals into vascular tissue.
- **Cortex:** Parenchyma cells that function in food storage.
- **Epidermis:** A single layer of cells that forms the outer layer and may possess root hairs.

Growth of Roots

Like stems, roots undergo primary growth from an apical meristem. Stems branch from a lateral bud, but roots branch some distance from the apical meristem. Root branching is initiated by the pericycle.

20.6 | Plant Nutrition

Plants need only inorganic nutrients to make all the organic compounds that make up their bodies. Some nutrients are essential, being either macronutrients or micronutrients.

Adaptations of Roots for Mineral Uptake

Some roots have nodules where bacteria fix nitrogen and produce forms that plants can use. Most roots have mycorrhizal fungi.

20.7 | Transport of Nutrients

Water and Mineral Transport in Xylem

The cohesion-tension model of xylem transport states that transpiration (evaporation of water from spongy mesophyll at stomata) creates tension, which pulls water upward in xylem. This method works only because water molecules are cohesive and adhesive and form a water column in xylem.

Organic Nutrient Transport in Phloem

The pressure-flow model of phloem transport states that sugar is actively transported into phloem at a source, and water follows by osmosis. The resulting increase in pressure creates a flow, which moves water and sugar to a sink.

Key Terms

annual ring   349
apical meristem   345
bark   349
blade   346
cohesion-tension model   353
collenchyma   345
cork cambium   349
cork cell   349
cortex   348
cotyledon   344
cuticle   345
deciduous   342
endodermis   351
epidermal tissue   345
epidermis   345
eudicot   344
evergreen   342
ground tissue   345
herbaceous   348
internode   343
lateral bud   343
leaf   342
leaf vein   344
meristem   345
mesophyll   346
monocot   344
mycorrhizal association   352
node   343

palisade mesophyll   346
parenchyma   345
perennial plant   343
pericycle   351
petiole   346
phloem   344
pith   348
pressure-flow model   354
root cap   351
root hair   343
root nodule   352
root system   342
sclerenchyma   345
shoot system   342
sieve-tube member   346
spongy mesophyll   346
stem   343
stomata (sing., stoma)   345
terminal bud   342
tracheid   346
transpiration   353
vascular bundle   344
vascular cambium   345
vascular cylinder   351
vascular tissue   345
vessel element   346
wood   349
xylem   344

Testing Yourself

Choose the best answer for each question.

1. It is possible to distinguish between stems and roots by looking for
   a. lateral branching on stems. c. nodes on stems.
   b. petioles on roots.   d. nodes on roots.

2. A monocot stem differs from a eudicot stem because in the monocot stem,
   a. xylem and phloem are in a ring surrounding the pith.
   b. xylem and phloem are in bundles scattered throughout the stem.
   c. there is an organized cortex and pith present.
   d. xylem forms a star in the center, with phloem between the points of the star.
   e. the epidermis does not have a waxy cuticle surrounding it.

3. Because of transpiration, water
   a. evaporates directly from the leaf surface.
   b. flows from leaf veins through the stem, towards the roots.
   c. exits the leaf through stomata, pulling water into the leaf from leaf veins.
   d. exits xylem in leaf veins and enters phloem by osmosis.

4. Hard materials in plants, such as the husks of nuts, are composed primarily of
   a. vascular tissue.   d. parenchyma cells.
   b. collenchyma cells.   e. epidermal tissue.
   c. sclerenchyma cells.

5. Label the parts of a plant in the following diagram.

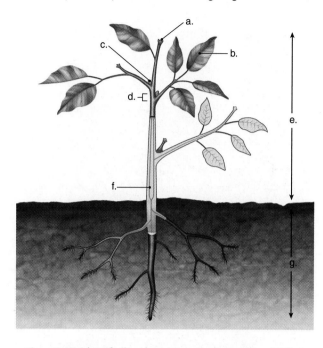

For questions 6–10, identify the plant organ that contains each feature. Each answer may be used more than once. Each question may have more than one answer.

Key:

    a. nonwoody stem
    b. woody stem
    c. root (nonwoody)
    d. leaf

6. cork cambium
7. vascular bundles
8. ground tissue
9. vascular cambium
10. apical meristem

11. During secondary growth, a tree adds more xylem and phloem through the activity of the
    a. apical meristems.   c. pericycle.
    b. vascular cambium.   d. cork cambium.

12. Root nodules are important because they
    a. encourage the growth of mycorrhizal fungi.
    b. represent areas of extensive branch root growth.
    c. contain nitrogen-fixing bacteria.
    d. provide extra oxygen to the plant root system.

13. The closing of stomata and the presence of a waxy cuticle act similarly to
    a. minimize water loss.   c. enhance water absorption.
    b. enhance gas exchange.   d. allow nitrogen fixation.

14. According to the pressure-flow model, sugar is actively transported into phloem, and
    a. enters xylem where it is moved toward the leaves due to transpiration.
    b. creates pressure to move water toward the roots.
    c. is transported out of the leaves through stomata.
    d. water follows by osmosis, providing a pressure that moves sugar along.

15. Label the parts of a leaf in the following illustration.

16. Label the parts of the eudicot stem in the following illustration.

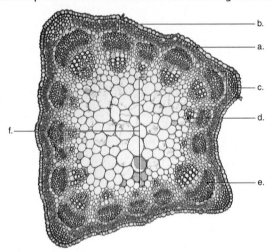

17. Label the parts of the root in the following illustration.

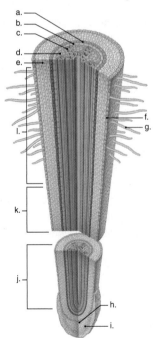

Thinking Scientifically

1. Scientists observe that the roots of legumes grow around nitrogen-fixing bacteria capable of forming nodules. How do the roots recognize these bacteria (see Fig. 4.6)?

2. Plants of the genus *Welwitschia* live in the deserts in Africa. Annual rainfall averages only 2.5 centimeters per year but every night a fog rolls in. Why might these plants be adapted to opening their stomata at night?

Bioethical Issue

Does forest thinning prevent wildfires?

In 2007, forest fires devastated large areas in California and revived the debate over forest fire prevention. Many blamed the numerous wildfires on poor management of national forests, proposing that limited logging of old-growth forests on federal lands occur. This logging, they contend, would clear flammable underbrush and thin out forests so that fires could not spread as easily.

Many environmentalists, however, are skeptical. They point out that thinning the forests would not only disturb critical habitat for many endangered species, but also it could make forests even more combustible by allowing more sunlight to penetrate the canopy and dry out organic matter on the forest floor. Many feel that the proposed logging would merely enrich the timber industry without accomplishing much.

In light of the loss of life and property due to these devastating wildfires, should the federal government allow the thinning of national forests? Or would this only exacerbate the problem?

Essentials of Biology Website

The companion website for *Essentials of Biology* provides a wealth of information organized and integrated by chapter. You will find practice tests, animations, videos, and much more that will complement your learning and understanding of general biology.

http://www.mhhe.com/maderessentials2

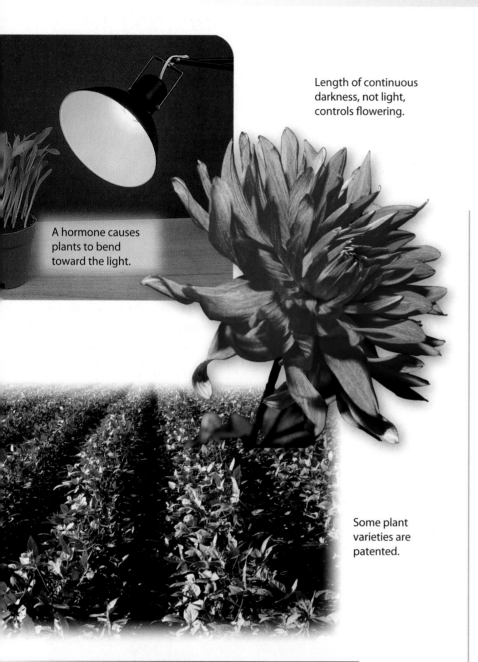

Length of continuous darkness, not light, controls flowering.

A hormone causes plants to bend toward the light.

Some plant varieties are patented.

21

Plant Responses and Reproduction

Outline

Patents for Genetically Engineered Plants

ISSUING PATENTS FOR plants is not a new practice. The first plant to be patented was a peach tree in 1932, and since then numerous plants have been patented. In fact, the United States Patent and Trademark Office has a special patent class just for asexually reproducing plants.

Why might you patent a plant? Plant breeders develop plants that are more beautiful, more useful, or more resistant to disease. A patent gives the plant breeder exclusive rights to sell a particular plant for 20 years. One of the more controversial plant patents is the one for Roundup Ready® plants.

—Continued next page

Continued—

These plants have been genetically modified to be resistant to Roundup,® a herbicide widely used to kill weeds that compete with plants for nutrients and water. The benefit of Roundup Ready wheat, cotton, or soybeans is that after fields are sprayed with Roundup, the crop survives but the weeds are destroyed. This method of fighting weeds saves time and labor compared to traditional means of controlling weeds. Although Roundup Ready crops have these benefits, what about possible risks? The biggest drawback to Roundup Ready crops is that their pollen could possibly transmit the genes for herbicide resistance to other plants, weeds in particular, and then these weeds would not be controlled by Roundup. A catastrophe of great magnitude could be in the making.

In this chapter, you will learn about other ways plants of commercial importance have been genetically engineered. But first, the chapter introduces the plant hormones involved in plant responses and flowering. It also reviews sexual and asexual reproduction in flowering plants in more detail than those topics were covered in Chapter 18.

21.1 | Plant Hormones

Plants usually respond to environmental stimuli such as light, gravity, and seasonal changes by altering their pattern of growth in some way. Most of these responses occur at the cellular level, and are mediated by hormones. Plant hormones are small organic molecules produced by the plant that serve as chemical signals between cells and tissues. Currently, the five commonly recognized groups of plant hormones are auxins, gibberellins, cytokinins, abscisic acid, and ethylene.

Auxins

The most common naturally occurring **auxin** is indoleacetic acid (IAA). It is produced in the shoot apical meristem and is found in young leaves and in flowers and fruits. Therefore, you would expect auxin to affect many aspects of plant growth and development, and it does.

Auxin is involved in phototropism, observable when stems bend toward a light source (**Fig. 21.1**). When a plant is exposed to unidirectional light, auxin moves to the shady side, where it binds to receptors and activates an ATP-driven pump that transports hydrogen ions (H^+) out of the cell into the cell wall. The acidic environment weakens cellulose fibrils, and activated enzymes further degrade the cell wall. Water now enters the cell, and the resulting increase in turgor pressure causes the cells on the shady side to elongate and the stem to bend toward the light.

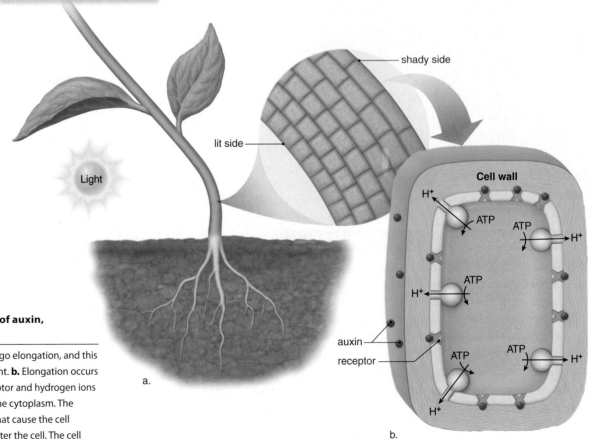

Figure 21.1 **Mode of action of auxin, a plant hormone.**

a. Plant cells on the shady side undergo elongation, and this causes a stem to bend toward the light. **b.** Elongation occurs after auxin (red balls) binds to a receptor and hydrogen ions (H^+) are actively transported out of the cytoplasm. The resulting acidity activates enzymes that cause the cell wall to weaken and allow water to enter the cell. The cell then elongates.

No doubt you have noticed that plants seldom produce branches at or near the shoot tip. This phenomenon is due to **apical dominance** caused by the action of auxin (**Fig. 21.2**). Auxin produced in the apical meristem of the terminal bud is transported downward, inhibiting the growth of lateral buds. Release from apical dominance occurs when pruning removes the shoot tip. Then, the lateral buds grow and the plant takes on a bushier appearance. When auxin is applied to the stump, apical dominance is restored.

Auxin also stimulates root formation. If you apply a paste that contains auxin to a stem cutting, roots begin to grow. Similarly, auxin production by seeds promotes the growth of fruit. And as long as auxin is concentrated in leaves or in fruits they do not fall off the plant. Therefore, trees are often sprayed with auxin to keep mature fruit from falling to the ground.

Gibberellins

Gibberellins were discovered in 1926 when a Japanese scientist was investigating a fungal disease of rice plants called "foolish seedling disease." Rapid stem elongation weakened the plants and caused them to collapse. The fungus infecting the plants produced an excess of a chemical called gibberellin, named after the fungus *Gibberella fujikuroi*. It wasn't until 1956 that a form of gibberellin now known as gibberellic acid was isolated from a flowering plant rather than from a fungus. We now know of about 70 gibberellins. The most common of these is gibberellic acid, GA_3 (the subscript designation distinguishes it from other gibberellins). Sources of gibberellin in flowering plant parts are young leaves, roots, embryos, seeds, and fruits.

Gibberellins are growth promoting hormones that bring about elongation of the cells. When gibberellins are applied externally to plants, the most obvious effect is stem elongation between the nodes (**Fig. 21.3**). Gibberellins can cause dwarf plants to grow, cabbage plants to become as much as 2 meters tall, and bush beans to become pole beans.

Dormancy is a period during which a plant or a seed does not grow, even though conditions may be favorable for growth. The dormancy of seeds and buds can be broken by applying gibberellins. Research with barley seeds has shown how GA_3 is involved in the germination of seeds. Endosperm is the tissue that serves as food for the embryo and seedling as they undergo development. Barley seeds have a large, starchy endosperm that must be broken down into sugars to provide energy for the embryo to grow. After GA_3 attaches to a receptor in the plasma membrane, a cell signaling pathway activates the gene that codes for amylase (see Fig. 12.13). Amylase then acts on starch to release sugars as a source of energy for embryonic and seedling growth.

a.

b.

Figure 21.2 **Apical dominance.**

a. Lateral bud growth is inhibited when a plant retains its terminal bud. **b.** When the terminal bud is removed, lateral branches develop and the plant is bushier.

Figure 21.3 **Effect of gibberellins.**

The plant on the right was treated with gibberellins; the plant on the left was not treated. Gibberellins are often used to promote stem elongation in economically important plants, but the exact mode of action remains unclear.

Check Your Progress

1. Describe the effects of auxin on plant growth.
2. Explain how auxin promotes apical dominance.
3. Describe the effects of gibberellins on plant growth.

Answers: 1. Auxin is involved in phototropism and in establishing apical dominance. Auxin also prevents fruit and leaves from dropping and promotes root growth.   2. Auxin is produced in the apical meristem and transported away from the shoot tip. The large amount of auxin near the shoot tip inhibits the development of nearby lateral buds.   3. Gibberellins promote stem elongation and cause seeds and buds to break dormancy.

Twig during winter

Twig in spring

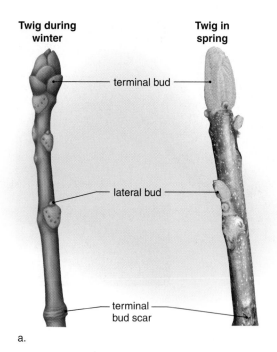

terminal bud

lateral bud

terminal bud scar

a.

H₂O

H₂O

H₂O   K⁺

1 K⁺ exits guard cells, and water follows.

H₂O

2 Stoma closes.   ×370

b.

Figure 21.4 Effects of abscisic acid.

a. Abscisic acid encourages the formation of winter buds (left), and a reduction in the amount of abscisic acid breaks bud dormancy (right). **b.** Abscisic acid also brings about the closing of a stoma by influencing the movement of potassium ions (K⁺) out of guard cells.

Cytokinins

Cytokinins were discovered as a result of attempts to grow plant tissues and organs in culture vessels in the 1940s. It was found that cell division occurs when coconut milk (a liquid endosperm) and yeast extract are added to the culture medium. Although the effective agent or agents could not be isolated at the time, they were collectively called cytokinins because, as you may recall, cytokinesis means division of the cytoplasm. A naturally occurring cytokinin was not isolated until 1967. Because it came from the kernels of maize (*Zea*), it was called zeatin.

Cytokinins influence plant growth by promoting cell division. Cytokinins are found in plant meristems, young leaves, root tips, and also in seeds and fruits. Whenever a plant grows, cytokinins are no doubt involved.

Other Effects of Cytokinins

Cytokinins also prevent **senescence,** or aging, of plant organs. As cytokinin levels drop within a plant organ, such as a leaf, growth slows or even stops. Then, the leaf loses its natural color as large molecules are broken down and transported to other parts of the plant. Senescence is a necessary part of a plant's growth. For example, as some plants grow taller, they naturally lose their lower leaves.

Plant organ formation is influenced by cytokinins. For example, researchers are well aware that the ratio of auxin to cytokinin and the acidity of the culture medium determine whether a plant tissue forms an undifferentiated mass, called a *callus*, or differentiates to form roots, vegetative shoots, leaves, or floral shoots. Some reports suggest that chemicals called oligosaccharins (chemical fragments released from the cell wall) are effective in directing differentiation. Perhaps the reception of auxin and cytokinins, which leads to the activation of enzymes, releases these fragments from the cell wall.

Abscisic Acid

Abscisic acid is sometimes called the stress hormone because it initiates and maintains seed and bud dormancy and brings about the closure of stomata. Dormancy has begun when a plant stops growing and prepares for adverse conditions (even though conditions at the time may be favorable for growth). For example, it is believed that abscisic acid moves from leaves to vegetative buds in the fall, and thereafter these buds are converted to winter buds. A winter bud is covered by thick, hardened scales (**Fig. 21.4***a*). A reduction in the level of abscisic acid and an increase in the level of gibberellins are believed to break seed and bud dormancy. Then seeds germinate, and buds send forth leaves.

Abscisic acid brings about the closing of stomata when a plant is under water stress (Fig. 21.4*b*). In some unknown way, abscisic acid causes potassium ions (K⁺) to leave guard cells. Thereafter, the guard cells lose water, and a stoma closes.

It was once believed that abscisic acid functioned in **abscission,** the dropping of leaves, fruits, and flowers from a plant. Although the external application of abscisic acid promotes abscission, this hormone is no longer believed to function naturally in this process. Instead, the hormone ethylene, discussed next, is thought to bring about abscission.

Ethylene

Ethylene is a gas that can move freely in the air. Like the other hormones studied, ethylene works with other hormones to bring about certain effects. For example, low levels of auxin and perhaps gibberellin, as compared to the levels in the stem, probably initiate abscission. But once the process of abscission has begun, ethylene stimulates certain enzymes, such as cellulase, which cause leaf, fruit, or flower drop (**Fig. 21.5***a,b*). Cellulase hydrolyzes cellulose in plant cell walls.

In the early 1900s, it was common practice to prepare citrus fruits for market by placing them in a room with a kerosene stove. Only later did researchers realize that ethylene, an incomplete combustion product of kerosene, was ripening the fruit (Fig. 21.5*c*). Because it is a gas, ethylene can act from a distance. A barrel of ripening apples can induce ripening in a bunch of bananas, even if they are in different containers. If a plant is wounded due to physical damage or infection, ethylene is released at the wound site. This is why one rotten apple spoils the whole barrel.

Table 21.1 summarizes the effects of the five groups of plant hormones.

a. No abscission   b. Abscission

c.

Check Your **Progress**

1. Describe the effects of cytokinins on plant growth.
2. Describe the effects of abscisic acid on plant growth.
3. Describe the effects of ethylene on plant growth.

Answers: 1. Cytokinins promote cell division in actively growing parts of a plant. They also prevent senescence and initiate leaf growth.   2. Abscisic acid initiates and maintains seed and bud dormancy. It also causes stomata to close when a plant is water-stressed.   3. Ethylene stimulates abscission of leaves, fruits, and flowers. It also enhances fruit ripening.

Figure 21.5 Functions of ethylene.

a. Normally, there is no abscission when a holly twig is placed under a glass jar for a week. **b.** When an ethylene-producing ripe apple is also under the jar, abscission of the holly leaves occurs. **c.** Ethylene causes fruits to ripen, making them luscious to eat.

Table 21.1 | Functions of the Major Plant Hormones

Hormone	Functions	Where Produced or Found in Plant
Auxins	Maintain apical dominance; involved in phototropism and gravitropism; promote growth of roots in tissue culture; prevent leaf and fruit drop.	Apical meristems, immature leaves
Gibberellins	Promote stem elongation between nodes; break seed and bud dormancy and influence germination of seeds.	Apical meristem, immature leaves, seeds
Cytokinins	Promote cell division; prevent senescence; along with auxin, promote differentiation leading to roots, shoots, leaves, or floral shoots.	Root apical meristem
Abscisic acid	Initiates and maintains seed and bud dormancy; promotes formation of winter buds; promotes closure of stomata.	Endosperm, roots, tissues containing chloroplasts
Ethylene	Promotes abscission (leaf, fruit, or flower drop); promotes ripening of fruit.	Aging parts of plant and ripening fruit; apical meristem and nodes of stem

Figure 21.6 **Positive phototropism.**

The stem of a plant curves toward the light. This response is due to the accumulation of auxin on the shady side of the stem.

Figure 21.7 **Negative gravitropism.**

The stem of a plant curves away from the direction of gravity 24 hours after the plant was placed on its side. This response is due to the accumulation of auxin on the lower side of the stem.

21.2 | Plant Responses

Plant responses are strongly influenced by such environmental stimuli as light, day length, gravity, and touch. The ability of a plant to respond to environmental signals fosters the survival of the plant and the species in a particular environment.

Plant responses to environmental signals can be rapid, as when stomata open in the presence of light, or they can take some time, as when a plant flowers in season. Despite their variety, most plant responses to environmental signals are due to growth and sometimes differentiation, brought about at least in part by particular hormones.

Tropisms

Plant growth toward or away from a directional stimulus is called a **tropism.** Differential growth causes one side of an organ to elongate faster than the other, and the result is a curving toward or away from the stimulus. The following two well-known tropisms were each named for the stimulus that causes the response:

> Phototropism: a movement in response to a light stimulus
> Gravitropism: a movement in response to gravity

Growth toward a stimulus is called a positive tropism, and growth away from a stimulus is called a negative tropism. In **Figure 21.6**, a positive **phototropism** is illustrated as when a stem curves toward the light. **Figure 21.7** illustrates negative **gravitropism**—stems curve away from the direction of gravity. Roots, of course, exhibit positive gravitropism.

The role of auxin in the positive phototropism of stems has been studied for quite some time. Because blue light in particular causes phototropism to occur, it is believed that a yellow pigment related to the vitamin riboflavin acts as a photoreceptor for light. Following reception, auxin migrates from the bright side to the shady side of a stem. The cells on that side elongate faster than those on the bright side, causing the stem to curve toward the light (see Fig. 21.1). Negative gravitropism of stems occurs because auxin moves to the lower part of a stem when a plant is placed on its side.

Photoperiodism

Flowering in angiosperms is a striking response to environmental seasonal changes. In some plants, flowering occurs according to the **photoperiod,** which is the ratio of the length of day to the length of night over a 24-hour period. Plants can be divided into three groups:

1. **Short-day plants/long-night plants** flower when the day length is shorter and the night is longer than a definite length of time called the critical length. (Examples are cocklebur, poinsettia, and chrysanthemum.)
2. **Long-day plants/short-night plants** flower when the day length is longer and the night is shorter than a critical length. (Examples are wheat, barley, clover, and spinach.)
3. **Day-neutral plants** do not depend on day/night length for flowering, and instead rely on other environmental stimuli. (Examples are tomato and cucumber.)

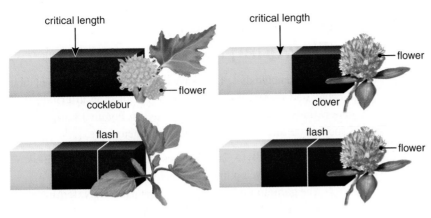

a. Short-day (long-night) plants

b. Long-day (short-night) plant

Further, we should note that both long-day plants and short-day plants can have the same critical length. **Figure 21.8** illustrates that the cocklebur and the clover have the same critical length. The cocklebur flowers when the day is shorter (night is longer) than 8.5 hours, and clover flowers when the day is longer (night is shorter) than 8.5 hours.

Experiments have shown that the *length of continuous darkness, not light, controls flowering in many plants*. For example, the cocklebur will not flower if a suitable length of darkness is interrupted by a flash of light. On the other hand, clover will flower when an unsuitable length of darkness is interrupted by a flash of light. (Interrupting the light period with darkness has no effect on flowering.) Nurseries use these kinds of data to make all kinds of flowers available throughout the year (**Fig. 21.9**).

Phytochrome and Plant Flowering

If flowering is dependent on night length, plants must have some way to detect these periods. In some plants, this appears to be the role of a blue-green leaf pigment called **phytochrome.** The proportion of red light to far-red light determines the particular form of phytochrome:

P_r (phytochrome red) absorbs red light and is converted into P_{fr}.
P_{fr} (phytochrome far-red) absorbs far-red light and is converted into P_r.

During the day, sunlight contains more red light than far-red light, and P_r is converted into P_{fr}. But at dusk, more far-red light is available. Now, P_{fr} is converted to P_r. There is also a slow metabolic replacement of P_{fr} by P_r during the night. Is this the timing device that tells the plant the length of darkness? Researchers have been looking for a flowering hormone for many years, but perhaps phytochrome itself triggers flowering through a signaling pathway.

Other Functions of Phytochrome

Apparently, the presence of P_{fr} indicates to some seeds that sunlight is present and conditions are favorable for germination. Such seeds must be only partly covered with soil when planted. Phytochrome may also affect leaf expansion and stem branching. In the absence of P_{fr}, stems elongate as a way to reach sunlight. Accordingly, seedlings grown in the dark have lengthened stems with smaller than normal leaves. But once the seedling is exposed to sunlight and P_{fr} is present, the seedling begins to grow normally—the leaves expand, and the stem branches.

Figure 21.8   Length of darkness controls flowering.

a. The cocklebur flowers when days are short and nights are long (*top*). If a long night is interrupted by a flash of light, the cocklebur will not flower (*bottom*). **b.** Clover flowers when days are long and nights are short (*top*). If a long night is interrupted by a flash of light, clover will still flower (*bottom*). Therefore, we can conclude that the length of continuous darkness controls flowering.

Figure 21.9   Flowering.

Nurseries know how to regulate the photoperiod so that many types of flowers are available year-round for special occasions.

Check Your Progress

1. Compare and contrast positive and negative tropisms.
2. Contrast short-day, long-day, and day-neutral plants.
3. List the functions of phytochrome in plants.

Answers: 1. Both are differential growth responses to external stimuli. Whereas positive tropisms cause plant growth toward a stimulus, negative tropisms cause plant growth away from a stimulus.   2. Short-day plants flower when the night is longer than a critical length; long-day plants flower when the night is shorter than a critical length; and day-neutral plants do not depend on day/night length to stimulate flowering.   3. Phytochrome functions in detection of photoperiod, stimulation of seed germination, and development of normal stems and leaves.

Figure 21.10 **Alternation of generations in flowering plants.**

In flowering plants, there are two types of spores and two gametophytes, male and female. Flowering plants are adapted to a land existence: The spores, the gametophytes, and the zygote are protected from drying out, in large part by the sporophyte.

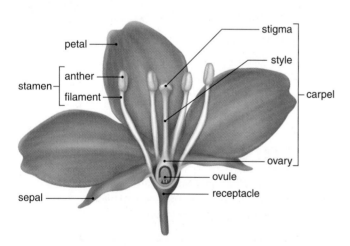

Figure 21.11 **Anatomy of a flower.**

A complete flower has all flower parts: sepals, petals, stamens, and at least one carpel.

21.3 | Sexual Reproduction in Flowering Plants

In Chapter 18, we noted that plants have two multicellular stages in their life cycle, and therefore their life cycle is called an **alternation of generations.** In this life cycle, depicted in Figure 18.3, a diploid sporophyte alternates with a haploid gametophyte:

> The **sporophyte** (2n) produces haploid spores by meiosis. The spores develop into gametophytes.
> The **gametophytes** (n) produce gametes. Upon fertilization, the cycle returns to the 2n sporophyte.

Overview of the Plant Life Cycle

Flowering plants have an alternation of generations life cycle, but with the modifications shown in **Figure 21.10.** First, we will give an overview of the flowering plant life cycle, and then we will discuss the life cycle in more depth. In flowering plants, the sporophyte is dominant, and it is the generation that bears flowers. The flower is the reproductive structure of angiosperms. The flower of the sporophyte produces two types of spores: microspores and megaspores. A **microspore** develops into a male gametophyte, which is a pollen grain. A **megaspore** develops into a female gametophyte, the embryo sac, which is microscopic and retained within the flower.

A pollen grain is either windblown or carried by an animal to the vicinity of the embryo sac. At maturity, a pollen grain contains two nonflagellated sperm. The embryo sac contains an egg.

A pollen grain develops a pollen tube, and the sperm move down the pollen tube to the embryo sac. After a sperm fertilizes an egg, the zygote becomes an embryo, still within the flower. The structure that houses the embryo develops into a seed. The seed also contains stored food and is surrounded by a seed coat. The seeds are often enclosed by a fruit, which aids in dispersing the seeds. When a seed germinates, a new sporophyte emerges and develops.

As you learned in Chapter 18, the life cycle of flowering plants is well adapted to a land existence. No external water is needed to transport the pollen grain to the embryo sac, or to enable the sperm to reach the egg. All stages of the life cycle are protected from drying out.

Flowers

The **flower** is a unique reproductive structure found only in angiosperms (**Fig. 21.11**). Flowers produce the spores and protect the gametophytes. They often attract pollinators that aid in transporting pollen from plant to plant. Flowers also produce the fruits that enclose the seeds. The success of angiosperms, with over 240,000 species, is largely attributable to the evolution of the flower.

In monocots, flower parts occur in threes and multiples of three; in eudicots, flower parts are in fours or fives and multiples of four or five (**Fig. 21.12**).

A typical flower has four whorls of modified leaves attached to a **receptacle** at the end of a flower stalk.

sepal

carpel

stamens

petal

a. Daylily, a monocot

carpel

stamens

petal

b. Wild geranium, a eudicot

Figure 21.12 **Monocot versus eudicot flowers.**

a. Monocots, such as daylilies, have flower parts in threes. In particular, note the three petals. **b.** Geraniums are eudicots. They have flower parts in fours or fives; note the five petals of this flower.

1. The **sepals** are the most leaflike of all the flower parts. They are usually green, but sometimes can resemble petals (Fig. 21.12*a*). Sepals protect the bud as the flower develops.
2. An open flower also has a whorl of **petals,** whose color accounts for the attractiveness of many flowers. The size, the shape, and the color of petals are attractive to a specific pollinator. Wind-pollinated flowers may have no petals at all.
3. **Stamens** are the "male" portion of the flower. Each stamen has two parts: the **anther,** a saclike container, and the **filament,** a slender stalk. Pollen grains develop from the microspores produced in the anther.
4. At the very center of a flower is the **carpel,** a vaselike structure that represents the "female" portion of the flower. A carpel usually has three parts: the **stigma,** an enlarged sticky knob; the **style,** a slender stalk; and the **ovary,** an enlarged base that encloses one or more **ovules.** The ovule becomes the seed, and the ovary becomes the fruit.

A flower can have a single carpel or multiple carpels. Sometimes several carpels are fused into a single structure, in which case the ovary has several chambers, each of which contains ovules. A carpel usually contains many ovules, which increases the number of seeds the plant may produce.

Not all flowers have sepals, petals, stamens, and a carpel. Those that do are said to be *complete*, and those that do not are said to be *incomplete*. Flowers that have both stamens and carpels are called bisexual flowers; those with only stamens are male flowers. Those with only carpels are female flowers. If both male and female flowers are on one plant, the plant is called *monoecious* (**Fig. 21.13**). But if male and female flowers occur on separate plants, the plant is called *dioecious*. Holly trees are dioecious, and if red berries are a priority, it is necessary to acquire a plant with male flowers and another plant with female flowers.

a. Male flowers

b. Female flowers

Figure 21.13 **Corn plants are monoecious.**

A single corn plant has clusters of male flowers (**a**) and female flowers (**b**). Male flowers possess stamens and produce the pollen that is carried by wind to the carpels of the female flowers, where an ear of corn develops.

Check Your **Progress**

1. Briefly describe the alteration of generations in the life cycle of plants.
2. List the four types of modified leaves that make up a typical flower.

Answers: 1. A plant's life cycle alternates between a diploid sporophyte, which produces haploid spores, and a haploid gametophyte, which produces gametes that fuse to form a diploid zygote. 2. A typical flower is composed of sepals, petals, stamens, and carpels.

From Spores to Fertilization

Now that we have some acquaintance with the flowering plant life cycle, we will examine it in more detail. As you may recall, the sporophyte of seed plants produces two types of spores: microspores, which become **male gametophytes** (mature pollen grains); and megaspores, which becomes the **female gametophyte,** or **embryo sac.** Just exactly where does this happen, and how do these events contribute to the life cycle of flowering plants?

Microspores develop into pollen grains in the anthers of stamens (**Fig. 21.14**). A **pollen grain** at first consists of two cells. The larger cell will eventually produce a pollen tube. The smaller cell divides, either now or later, to become two sperm. This is why the stamen is called the "male" portion of the flower.

Figure 21.14 Life cycle of a flowering plant.

Follow the development of a microspore into a germinated pollen grain that contains sperm. Also follow the development of a megaspore into the female gametophyte that contains an egg. Note the occurrence of double fertilization, in which one sperm from the pollen tube fertilizes the egg, producing a zygote, and the other joins with two other female gametophyte cells to become endosperm. Now the ovule becomes a seed.

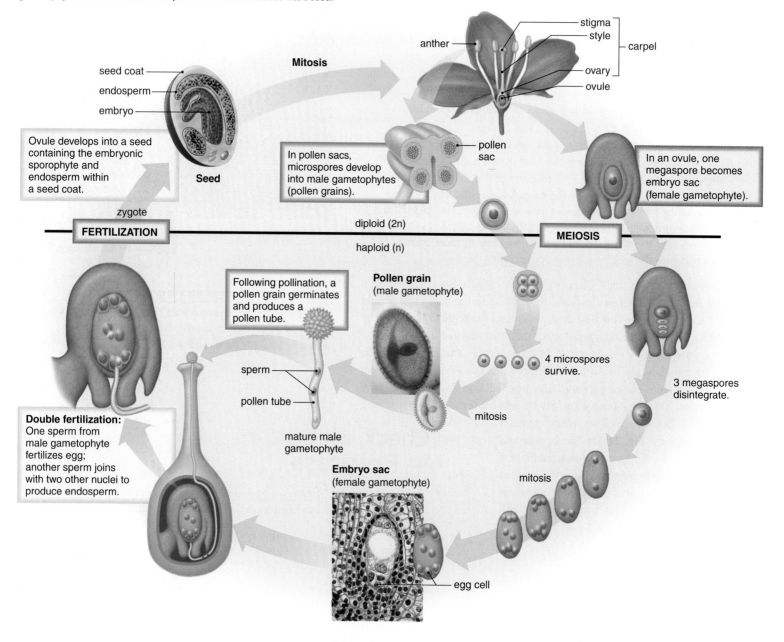

seed coat
endosperm
embryo

Mitosis

stigma
style — carpel
anther
ovary
ovule

Ovule develops into a seed containing the embryonic sporophyte and endosperm within a seed coat.

Seed

In pollen sacs, microspores develop into male gametophytes (pollen grains).

pollen sac

In an ovule, one megaspore becomes embryo sac (female gametophyte).

zygote

FERTILIZATION

diploid (2n)

MEIOSIS

haploid (n)

Following pollination, a pollen grain germinates and produces a pollen tube.

Pollen grain
(male gametophyte)

4 microspores survive.

3 megaspores disintegrate.

sperm

pollen tube

mitosis

Double fertilization:
One sperm from male gametophyte fertilizes egg; another sperm joins with two other nuclei to produce endosperm.

mature male gametophyte

Embryo sac
(female gametophyte)

mitosis

egg cell

Pollen grains are distinctive to the particular plant (**Fig. 21.15**), and **pollination** is simply the transfer of pollen from the anther to the stigma of a carpel. Plants often have adaptations that favor *cross-pollination*, which occurs when the pollen landing on the stigma is from a different plant of the same species. For example, the carpels may mature only after the anthers have released their pollen. Cross-pollination may also be brought about with the assistance of an animal pollinator. If a pollinator such as a bee goes from flower to flower of only one type of plant, cross-pollination is more likely to occur in an efficient manner. The secretion of nectar is one way that insects are attracted to plants, and over time, certain pollinators have become adapted to reach the nectar of only one type of flower. In the process, pollen is inadvertently picked up and taken to another plant of the same type. Plants attract particular pollinators in still other ways. For example, through the evolutionary process, orchids of the genus *Ophrys* have flowers that look like female wasps. Males of that species pick up pollen when they attempt to mate with these flowers!

Figure 21.14 also allows you to follow the development of the megaspore. In an ovule, within the ovary of a carpel, meiosis produces four megaspores. One of the cells develops into the female gametophyte, or so-called embryo sac, which is a seven-celled structure containing a single egg cell. Fertilization of the egg occurs after a pollen grain lands on the stigma of a carpel and develops a pollen tube. A pollen grain that has germinated and produced a pollen tube is the mature male gametophyte (Fig. 21.14, *middle*). A pollen tube contains two sperm. Once it reaches the ovule, **double fertilization** occurs. One sperm unites with the egg, forming a 2n zygote. The other sperm unites with two nuclei centrally placed in the embryo sac, forming a 3n endosperm cell.

Check Your **Progress**

1. How does each of the two cells in a pollen grain contribute to the fertilization process?
2. Briefly describe the development of the megaspore into the female gametophyte.
3. What are the products of double fertilization in angiosperms?

Answers: 1. The larger cell produces a pollen tube that allows the two sperm produced when the smaller cell divides to reach the embryo sac where double fertilization occurs. 2. The megaspore undergoes mitosis to produce the female gametophyte, a seven-celled embryo sac that contains one egg cell. 3. When one sperm fertilizes the egg, a zygote results. When the second sperm joins with the other two nuclei of the embryo sac, the endosperm results.

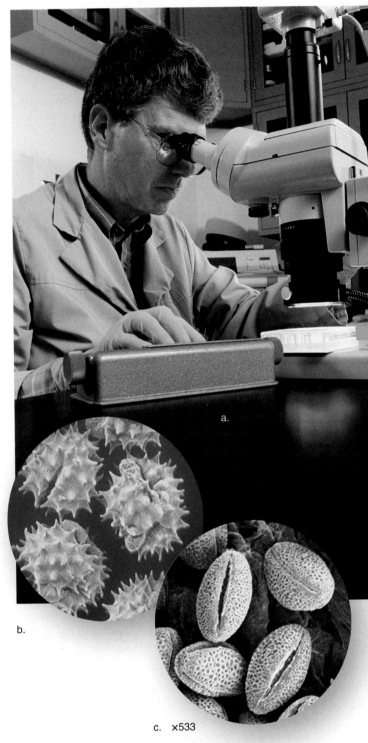

a.

b.

c. ×533

Figure 21.15 Pollen.

a. Pollen grains are so distinctive that a paleontologist can use fossilized pollen to date the appearance of a plant in a particular area. Pollen grains can become fossils because their strong walls are resistant to chemical and mechanical damage. **b.** Pollen grains of Canadian goldenrod. **c.** Pollen grains of a pussy willow.

Figure 21.16   Development of the seed in a eudicot.

Development begins with (**a**) the zygote and ends with (**f**) the seed. As the embryo develops, it progresses through the stages noted from (**b**) to (**e**).

Development of the Seed in a Eudicot

It is now possible to account for the three parts of a seed. The ovule wall will become a protective covering called the *seed coat*. Double fertilization, as discussed earlier, has resulted in an endosperm nucleus and a zygote. Cell division produces a multicellular *embryo* and a multicellular endosperm, which is the *stored food* of a seed.

As the embryo passes through the stages noted in **Figure 21.16**, tissue specialization occurs as evidenced by the eventual appearance of the plant axis (Fig. 21.16*e*), as well as the shoot tip and root tip, each of which contains apical meristem. When the seed germinates, the shoot tip produces a shoot system, and the root tip produces a root system.

During seed development, note that the so-called **cotyledons,** or embryonic leaves, absorb the developing endosperm and become large and fleshy. The food stored by the cotyledons will nourish the embryo when it resumes growth. Cotyledons wither when the first *true leaves* grow and become functional. The common garden bean is a good example of a eudicot seed with large cotyledons and no endosperm (see Fig. 21.19). A **seed** contains the embryo and stored food within a seed coat.

Monocots Versus Eudicots

Whereas eudicot embryos have two cotyledons, monocot embryos have only one cotyledon. In monocots, the cotyledon functions in food storage, and it also absorbs food molecules from the endosperm and passes them to the embryo. In other words, the endosperm is retained in monocot seeds. In eudicots, the cotyledons usually store all the nutrient molecules that the embryo uses. Therefore, the endosperm disappears because it has been taken up by the two cotyledons.

A corn plant is a monocot; consequently, in a corn kernel, there is only one cotyledon and the endosperm is present (see Fig. 21.20).

Fruit Types and Seed Dispersal

Seeds develop from ovules, and fruits develop from ovaries and sometimes other parts of a flower. In other words, flowering plants have seeds enclosed by a **fruit** (**Fig. 21.17**).

Fruits may be quite diverse. *Dry fruits* are generally a dull color with a thin and dry ovary wall so that the potential food for animals is largely confined to the seeds. In grains such as wheat, corn, and rice, the fruit looks like a seed. Nuts (e.g., walnuts, pecans) have a hard outer shell covering a single seed. A legume, such as a pea, has a several-seeded fruit that splits open to release the seeds. A legume illustrates that what we call a vegetable may actually be a fruit.

Figure 21.17 Examples of fruits.

Some fruits are dry, as in walnuts and peas. Some fruits are fleshy, as in peaches and apples. To a botanist, any plant product derived from an ovary plus perhaps other flower parts is a fruit.

In contrast to dry fruits, *fleshy fruits* have an usually juicy portion that is brightly colored to attract animals. A drupe (e.g., peach, cherry, olive) is a "stone fruit"—the outer part of the ovary wall is fleshy, but there is an inner stony layer. Inside the stony layer is the seed. A berry, such as a tomato, contains many seeds. An apple is a *pome*, in which a dry ovary covers the seeds, and the fleshy part is derived from the receptacle of the flower. A strawberry is an interesting fruit because the flesh is derived from the receptacle, and what appear to be the seeds are actually dry fruits!

Dispersal of Seeds

For plants to be widely distributed, their seeds have to be dispersed—that is, distributed preferably long distances from the parent plant. Various means of dispersal are well known. Birds and mammals sometimes eat fruits, including the seeds, which then pass out of the digestive tract with the feces some distance from the parent plant (**Fig. 21.18***a*). Squirrels and other animals gather seeds and fruits, which they bury some distance away. Some plants have evolved unusual ways to ensure dispersal. The hooks and spines of clover, cocklebur, and burdock fruits attach to the fur of animals and the clothing of humans that carry them far away from the parent plant (Fig. 21.18*b)*. Other seeds are dispersed by wind. Woolly hairs, plumes, and wings are all adaptations for this type of dispersal. The dandelion fruit is attached to several hairs that function like a parachute and aid dispersal (Fig. 21.18*c*). The winged fruit of a maple tree, which contains two seeds, has been known to travel up to 10 kilometers from its parent (Fig. 21.18*d*). Different still, a touch-me-not plant has seed pods that swell as they mature. A passing animal may cause the swollen pods to burst, hurling the ripe seeds a great distance away from the plant.

Although we tend to disregard it, when farmers plant their crops, they are dispersing seeds.

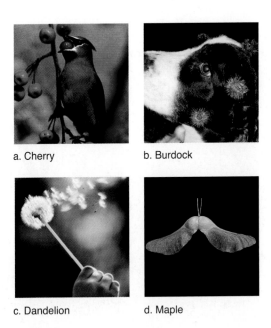

a. Cherry   b. Burdock

c. Dandelion   d. Maple

Figure 21.18 Methods of seed dispersal.

Plants often have adaptations to ensure that their seeds are spread some distance from the parent plant. **a.** The waxwing will carry the seed of a cherry some distance away. **b.** The spines of burdock fruit stick to a passerby. **c.** Dandelion and (**d**) maple fruits have adaptations that allow them to be carried long distances by wind.

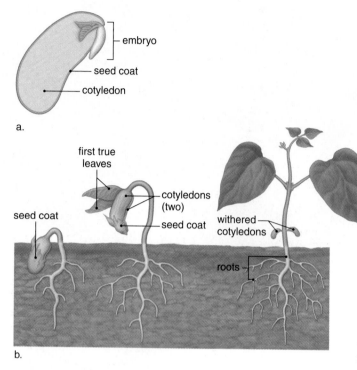

a.

b.

Figure 21.19 Common garden bean, a eudicot.

a. Seed structure. **b.** Germination and development of the seedling. Notice that there are two cotyledons and that the leaves are net-veined.

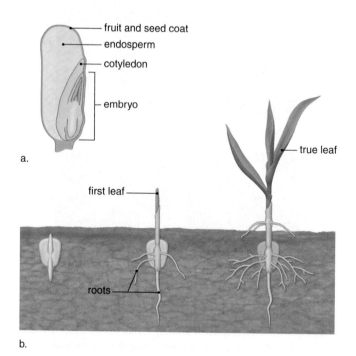

a.

b.

Figure 21.20 Corn, a monocot.

a. Corn kernel structure. **b.** Germination and development of the seedling. Notice that there is one cotyledon and that the leaves are parallel-veined.

Germination of Seeds

Following dispersal, if all goes well, the seeds will **germinate.** As growth occurs, a seedling appears. Germination doesn't usually take place until there is sufficient water, warmth, and oxygen to sustain growth. In deserts, germination does not occur until there is adequate moisture. These requirements help ensure that seeds do not germinate until the most favorable growing season has arrived.

Some seeds do not germinate until they have been dormant for a period of time. For seeds, dormancy is the time during which no growth occurs, even though conditions may be favorable for growth. In the temperate zone, seeds often have to be exposed to a period of cold weather before dormancy is broken. Fleshy fruits (e.g., apples, pears, oranges, and tomatoes) contain inhibitors so that germination does not occur until the seeds are removed and washed. Aside from water, bacterial action and even fire can act on the seed coat, allowing it to become permeable to water. The uptake of water causes the seed coat to burst and germination to occur.

Eudicot Versus Monocot Germination

If the two cotyledons of a bean seed are parted, you can see the cotyledons and a rudimentary plant with immature leaves. As the eudicot seedling emerges from the soil, the shoot is hook-shaped to protect the immature leaves as they start to grow. The cotyledons shrivel up as the true leaves of the plant begin photosynthesizing (**Fig. 21.19**). A corn kernel is actually a fruit, and therefore the outer covering is the fruit and seed coat combined (**Fig. 21.20**). Inside is the single cotyledon. Also, both the immature leaves and the root are covered by sheaths. The sheaths are discarded when the seedling begins growing, and the immature leaves become the first true leaves of the corn plant.

Check Your **Progress**

1. **What are cotyledons, and what is their function?**
2. **Contrast dry fruits with fleshy fruits.**
3. **Describe the role of fruits in seed dispersal.**

Answers: 1. Cotyledons are embryonic leaves that are present in seeds. Cotyledons store nutrients derived from endosperm (in eudicots), or both store and take nutrients from the endosperm (in monocots).   2. Dry fruits have a thin, dry ovary wall, and most of the potential food for animals is confined to the seeds. Fleshy fruits have a juicy portion that is brightly colored.   3. Fruits are designed to ensure the seeds are carried far away from the parent plant. Some fruits are eaten by animals and passed into their feces; others are gathered and buried by animals. Other fruits may have adaptations such as hooks, spines, plumes, and wings to ensure seed dispersal.

21.4 | Asexual Reproduction in Flowering Plants

Many plants can also reproduce asexually because they contain undifferentiated cells in the meristem. In asexual reproduction, there is only one parent, instead of two as in sexual reproduction. For example, complete strawberry plants can grow from the nodes of stolons, and iris plants can grow from the nodes of rhizomes (**Fig. 21.21**). White, or Irish, potatoes are actually portions of underground stems, and each eye is a bud that will produce a new potato plant if it is planted with a portion of the swollen tuber. Sweet potatoes are modified roots; they can be propagated by planting sections of the root. You may have noticed that the roots of some fruit trees, such as cherry and apple trees, produce "suckers," small plants that can be used to grow new trees.

There are several ways that seedless fruits can arise, but usually some sort of stimulation causes the plant to produce fruit, even though fertilization never occurred. In the seedless watermelon, which is triploid and cannot go through meiosis to form sperm and eggs, pollination is used as a stimulus to cause the plants to form fruit. Often, hormones, such as auxins, gibberellins, or cytokinins, are used to stimulate seedless fruit formation. Seedless grapes actually contain seeds, but the hard seed casing fails to develop and this causes the embryo to abort. Then, the grapes are sprayed with gibberellin to increase their size.

Propagation of Plants in Tissue Culture

Tissue culture is the growth of a tissue in an artificial liquid or solid culture medium. Plant cells are **totipotent,** which means that each plant cell can become an entire plant. This has led to the commercial production of embryos from the somatic cells of plants (**Fig. 21.22**). Adult cells are first treated to remove their

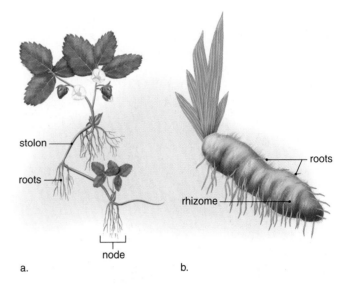

a.   b.

Figure 21.21 Asexual reproduction in plants.

a. A strawberry plant has an aboveground horizontal stem called a stolon. Every other node produces a new plant. **b.** An iris has an underground horizontal stem called a rhizome, whose nodes produce new plants.

a. Protoplasts (naked cells)   b. Cell wall regeneration   c. Aggregates of cells

d. Callus (undifferentiated mass)   e. Somatic embryo   f. Plantlet

Figure 21.22 Tissue culture.

a. When plant cell walls are removed by digestive enzyme action, cells without cell walls, called protoplasts, result. **b.** Cell walls regenerate, and cell division begins. **c.** Cell division produces aggregates of cells. **d.** An undifferentiated mass, called a callus, develops. **e.** From the callus, somatic embryos such as this one appear. **f.** The embryos develop into plantlets that can be transferred to soil for growth into adult plants.

Figure 21.23   **Whole plants from meristem tissue.**

Each of these flasks contains meristem tissue and a growth medium. In the presence of light and with slow rotation of the flasks, a new plant will appear in a few weeks.

cell walls. Then, thousands or even millions of embryos can be produced by following the procedure shown in Figure 21.22. The embryos can be encapsulated in artificial seeds and shipped anywhere. Many vegetable plants found in nurseries, such as tomato, celery, and asparagus, are produced this manner. Occasionally mutations arise during the production process, and this is another way that plant breeders can produce new plants with desirable traits.

Instead of using mature plant tissues as the starting material, you can use meristem tissue as a source of plant cells. In this case, the end result is a population of *clonal plants* that have the same traits as the original plant (**Fig. 21.23**). If the correct proportions of hormones are added to the liquid medium, many new shoots will develop from a single shoot tip. When these are removed, more shoots form. An advantage to using meristem culture is that meristem tissue, unlike other portions of a plant, is virus-free. Therefore, the plants produced will also be virus-free. (The presence of plant viruses weakens plants and makes them less productive.)

Anther culture is a technique in which mature anthers are cultured in a medium containing vitamins and growth factors. The haploid cells within the pollen grains divide, producing *proembryos* consisting of as many as 20 to 40 cells. Finally, the pollen grains rupture, releasing haploid embryos. The experimenter can then generate a haploid plant, or chemical agents can be added that encourage chromosomal doubling. After chromosomal doubling, the resulting plants are diploid, and the homologous chromosomes carry the same genes. Anther culture is another way to produce plants that are certain to have the same traits as the parent plant.

When the desired product is not an entire plant, but merely a substance the plant produces, scientists may use a technique called *cell suspension culture*. Rapidly growing calluses are cut into small pieces and shaken in a liquid nutrient medium so that single cells or small clumps of cells break off and form a suspension. These cells often produce the same chemicals as the entire plant. For example, cell suspension cultures of cells from the quinine tree produce quinine, and those of the wooly foxglove produce digitoxin, both of which are useful drugs for humans.

Genetic Engineering of Plants

Traditionally, **hybridization,** the crossing of different varieties of plants or even species, was used to produce plants with desirable traits. Hybridization, followed by vegetative propagation of the mature plants, generated a large number of identical plants with these traits. Today, it is possible to directly alter the genes of organisms, and in that way produce new varieties with desirable traits.

Tissue Culture and Genetic Engineering

Genetic engineering can be done utilizing plant cells lacking cell walls in tissue culture (see Fig. 21.22). A foreign piece of DNA isolated from any type of organism—plant, animal, or bacteria—is placed in the tissue culture medium. High-voltage electrical pulses can then be used to create pores in the plasma membrane so that the foreign gene enters the cells. In one such procedure, a gene for the production of the jellyfish green fluorescent protein was inserted into tobacco, and the adult plants glowed when exposed to ultraviolet light. Likewise, scientists have even attempted to produce vaccines by inserting nonlethal viral genes into crop plants!

Check Your Progress

1. Explain how plants with seedless fruits can be propagated.
2. Compare and contrast tissue culture plant propagation with meristem culture.

Answers: 1. Plants with seedless fruits are propagated asexually, usually by cuttings.   2. Both are ways of producing many plant embryos from a single plant. Plants propagated in tissue culture can differ when mutations arise, but the plant embryos derived from meristem culture are identical.

Today, it is possible to use a gene gun to bombard a callus with microscopic DNA-coated metal particles. Then, genetically altered embryos develop into genetically altered adult plants. Many plants, including corn and wheat varieties, have been genetically engineered by this method. Such plants are called **genetically modified plants (GMPs),** or **transgenic plants,** because they carry a foreign gene and have new and different traits. **Figure 21.24** shows two types of transgenic plants.

Agricultural Plants with Improved Traits

Corn and cotton plants, in addition to soybean and potato plants, have been genetically engineered to be resistant to either herbicides or insect pests. Some corn and cotton plants have been developed that are resistant to both insects and herbicides. In 2001, transgenic crops were planted on more than 72 million acres worldwide, and the acreage is expected to triple in about five years. If crops are resistant to a broad-spectrum herbicide and weeds are not, then the herbicide can be used to kill the weeds. When herbicide-resistant plants were planted, weeds were easily controlled, less tillage was needed, and soil erosion was minimized. However, wildlife and native plants may decline when increased amounts of herbicides are used.

Some citizens are concerned about GMPs because of possible effects on their health. This concern has been prompted by limited laboratory data suggesting that some people may be allergic to GMPs (**Fig. 21.25**). Some feel that drastic consequences might occur in the future, even though no life-threatening effects of GMPs have so far been recorded. In 2000, such fears even prompted the recall of several brands of taco shells from stores after it was discovered that the shells contained traces of a genetically modified variety of corn not approved for humans. Ecologists also worry that GMPs may harm the environment. It's possible that beneficial insects feeding on the GMPs might be harmed by them, or that GMPs might pass their herbicide resistance genes to certain weeds.

In the meantime, researchers are trying to produce more types of GMPs (**Fig. 21.26**). A salt-tolerant tomato will soon be field-tested. First, scientists identified a gene coding for a channel protein that transports Na^+ into a vacuole, preventing it from interfering with plant metabolism. Then the scientists used the gene to engineer tomato plants that maximally produce the channel protein. The GMPs thrived when watered with a salty solution. This technology may turn once unusable land into arable farmland. Salt-tolerant and also drought- and cold-tolerant cereals, rice, and sugarcane might help provide enough food for the growing world population.

Potato blight is the most serious potato disease in the world. About 150 years ago, it was responsible for the Irish potato famine, which caused the death of millions of people. By placing a gene from a naturally blight-resistant wild potato into a farmed variety, researchers have now created potato plants that are no longer vulnerable to a range of blight strains. A similar technique is

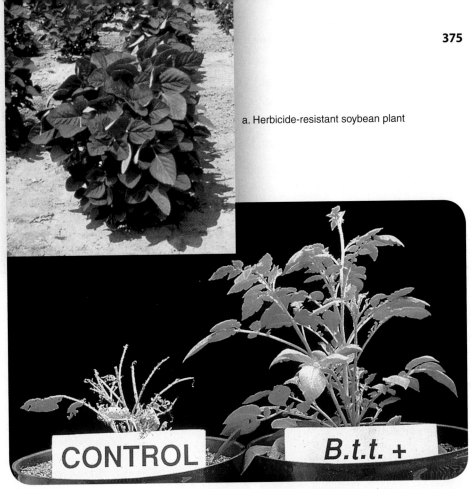

a. Herbicide-resistant soybean plant

b. Nonresistant potato plant

c. Pest-resistant potato plant

Figure 21.24 Transgenic plants.

a. These soybean plants have been genetically modified to be resistant to herbicides. Nonresistant potato plants (**b**). can be genetically modified to be resistant to pests (**c**).

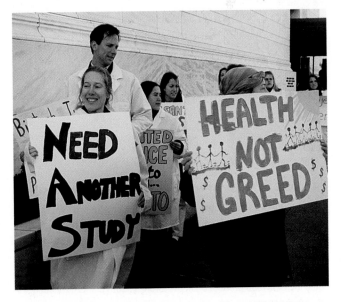

Figure 21.25 Regulation of genetically modified plants.

Activists protest the present regulation of genetically modified plants, which they believe to be too permissive.

Transgenic Crops of the Future

Improved Agricultural Traits

Herbicide resistant	Wheat, rice, sugar beets, canola
Salt tolerant	Cereals, rice, sugarcane, canola
Drought tolerant	Cereals, rice, sugarcane
Cold tolerant	Cereals, rice, sugarcane
Improved yield	Cereals, rice, corn, cotton
Disease protected	Wheat, corn, potatoes

Improved Food-Quality Traits

Fatty acid/oil content	Corn, soybeans
Protein/starch content	Cereals, potatoes, soybeans, rice, corn
Amino acid content	Corn, soybeans

Medicinal Benefits

Vaccine production	Corn, soybeans, tomatoes

a. Desirable traits

Salt intolerant

c. Transgenic corn for antibody production

b. Salt tolerant

Figure 21.26 Genetically modified crops of the future.
a. Genetically modified crops of the future include those with improved agricultural traits, better food quality, or medicinal benefits. **b.** A salt-tolerant tomato plant has been engineered. The plant to the right does poorly when watered with a salty solution, but the engineered plant to the left is tolerant of the solution. The development of salt-tolerant crops could increase food production in the future. **c.** Corn plants have been engineered to produce vaccines to many diseases.

being employed in the fight against the chestnut blight, which nearly wiped out the American chestnut in the early twentieth century.

Some progress has also been made in increasing the food quality of crops. Soybeans have been developed that mainly produce monounsaturated fatty acids, a change that may improve human health. These altered plants also produce acids that can be used as hardeners in paints and plastics. The necessary genes were taken from other plants and transferred into the soybean DNA.

Other types of genetically engineered plants are expected to increase productivity. Stomata might be altered to take in more carbon dioxide or to lose less water. A team of Japanese scientists is working on introducing the C_4 photosynthetic capability into rice. Unlike C_3 plants, C_4 plants do well in hot, dry weather (see page 94). These modifications would require a more complete reengineering of plant cells than the single-gene transfers that have been done so far.

Commercial Products

Single-gene transfers also have allowed plants to produce various products, including human hormones, clotting factors, and antibodies. One type of antibody made by corn can deliver radioisotopes to tumor cells, and another antibody may help prevent HIV infection (Fig. 21.26c). The tobacco mosaic virus has been used as a vector to introduce a human gene into adult tobacco plants in the field. (Note that this technology bypasses the need for tissue culture completely.) Tens of grams of α-galactosidase, an enzyme needed for the treatment of a human lysosome storage disease, were harvested per acre of tobacco plants. And it only took 30 days to get tobacco plants to produce antibodies to treat non-Hodgkin's lymphoma after being sprayed with a genetically engineered virus.

Check Your Progress

1. List the concerns some individuals have regarding genetically modified plants (GMPs).
2. List the potential benefits of GMPs.

Answers: 1. People might be allergic to GMPs, and GMPs could also harm the environment. 2. GMPs may be more tolerant of high salt, drought, and cold; they may be more disease and pest resistant; they may have improved nutritional quality; and they may also be modified in order to make products of medical importance to humans.

The Chapter in Review

Summary

21.1 | Plant Hormones

Plant hormones lead to physiological changes within the cell. The five commonly recognized groups of plant hormones are:

- **Auxins:** Affect growth patterns; cause apical dominance and phototropism.
- **Gibberellins:** Promote stem elongation, break seed dormancy.
- **Cytokinins:** Promote cell division, prevent senescence of leaves, influence differentiation of plant tissues.
- **Abscisic acid:** Initiates and maintains seed and bud dormancy, closing of stomata.
- **Ethylene:** Causes abscission of leaves, fruits, and flowers; ripens fruits.

21.2 | Plant Responses

Environmental signals play a significant role in plant growth and development.

Tropisms

Tropisms are growth responses toward (positive) or away (negative) from unidirectional stimuli.

> Phototropism: Response to a light
> Gravitropism: Response to gravity

- Auxin is responsible for the negative gravitropism exhibited by stems. Stems grow upward opposite the direction of gravity.

Photoperiodism

Flowering is a response to seasonal changes, namely length of the night.

critical length

Short-day (long-night) plants

- Short-day plants flower when nights are longer than a critical length.
- Long-day plants flower when nights are shorter than a critical length.
- Some plants are day/night-neutral.

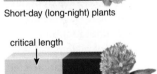

critical length

Long-day (short-night) plants

Phytochrome and Plant Flowering

Phytochrome is a plant pigment that responds to daylight. Phytochrome in plant cells brings about flowering and encourages seed germination. In addition, phytochrome influences leaf expansion and affects stem branching.

21.3 | Sexual Reproduction in Flowering Plants

The life cycle of flowering plants is adapted to a land existence.

- Flowering plants have an alternation of generations life cycle with separate male and female gametophytes.
- Pollen grain is the male gametophyte
- The embryo sac, located within the ovule of a flower, is the female gametophyte
- Double fertilization occurs, and the zygote and endosperm result.

Development of the Seed in a Eudicot

The zygote undergoes a series of stages to become an embryo. In eudicots, the embryo has two cotyledons, which absorb the endosperm. In monocots, the embryo has a single cotyledon and also endosperm. Aside from the embryo and stored food, a seed has a seed coat.

eudicot seed   monocot seed

Fruit Types and Seed Dispersal

A fruit is a mature, ripened ovary and may also be composed of other flower parts. Some fruits are dry (e.g., nuts, legumes), and some are fleshy (e.g., apples, peaches). In general, fruits aid the dispersal of seeds. Following dispersal, a seed germinates.

21.4 | Asexual Reproduction in Flowering Plants

Many flowering plants reproduce asexually.

- Nodes located on stems (either aboveground or underground) give rise to entire plants.
- Roots produce new shoots.

Propagation of Plants in Tissue Culture

The production of clonal plants utilizing tissue culture is now a commercial venture. Plant cells in tissue culture can produce chemicals of medical and commercial importance.

Genetic Engineering of Plants

The practice of plant tissue culture facilitates genetic engineering to produce plants that have improved agricultural or food-quality traits. Plants can also be engineered to produce chemicals of use to humans.

Key Terms

Testing Yourself

Choose the best answer for each question. For questions 1–8, identify the plant hormone in the key that is associated with each phenomenon. Each answer may be used more than once.

Key:

    a. auxin
    b. gibberellin
    c. cytokinin
    d. abscisic acid
    e. ethylene

1. initiates and maintains seed and bud dormancy
2. stimulates root development
3. capable of moving from plant to plant through the air
4. causes phototropism in stems
5. responsible for apical dominance
6. stimulates leaf, fruit, and flower drop
7. needed to break seed and bud dormancy
8. prevents senescence
9. A tropism is defined as
    a. a growth response to a directional stimulus.
    b. a plant response to hormones.
    c. a hormonal response to overcrowding by other plants.
    d. abscission of leaves due to plant hormones.
10. Stigma is to carpel as anther is to
    a. sepal.   c. ovary.
    b. stamen.   d. style.

11. An incomplete flower lacks
    a. sepals.   c. carpels.
    b. petals.   d. stamens.
                        e. any one or more of the above.
12. The embryo of a flowering plant can be found in the
    a. pollen.   c. microspore.
    b. anther.   d. seed.
13. Label the parts of the flowering plant life cycle in the following illustration.

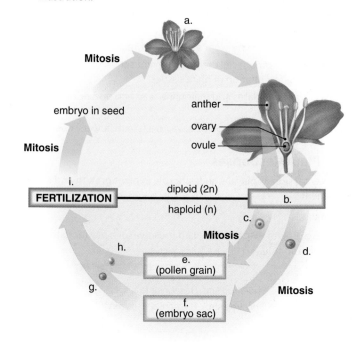

14. Plant tissue culture takes advantage of
    a. phototropism.
    b. gravitropism.
    c. asexual reproduction from stolons.
    d. totipotency.
15. Which of the following is not caused by the action of auxin?
    a. accelerated root development
    b. abscission of leaves
    c. phototropism
    d. apical dominance
16. Double fertilization refers to the formation of a _____ and a ___ _____.
    a. zygote, zygote   c. zygote, megaspore
    b. zygote, pollen grain   d. zygote, endosperm
17. Which is the correct order of the following events:
(1) megaspore becomes embryo sac, (2) embryo formed, (3) double fertilization, (4) meiosis?
    a. 1, 2, 3, 4   c. 4, 3, 2, 1
    b. 4, 1, 3, 2   d. 2, 3, 4, 1
18. In the absence of abscisic acid, plants may have difficulty
    a. forming winter buds.   c. Both a and b are correct.
    b. closing the stomata.   d. Neither a nor b is correct.
19. Phytochrome plays a role in
    a. flowering.   c. leaf growth.
    b. stem growth.   d. All of these are correct.

20. A plant requiring a dark period of at least 14 hours will
    a. flower if a 14-hour night is interrupted by a flash of light.
    b. not flower if a 14-hour night is interrupted by a flash of light.
    c. flower if the days are 14 hours long.
    d. flower when the nights drop below 14 hours long.

21. Short-day plants
    a. are apt to flower in the fall.
    b. are apt to flower in the summer.
    c. do not have a critical photoperiod.
    d. will not flower if a short day is interrupted by bright light.
    e. All of these are correct.

22. Which of the following is a natural method of asexual reproduction in plants?
    a. meristem culture
    b. vegetative propagation from stolons or rhizomes
    c. anther culture
    d. cell suspension culture

23. Which of the following plant hormones causes stem elongation between nodes?
    a. auxin
    b. cytokinins
    c. gibberellins
    d. abscisic acid
    e. ethylene

24. Which of the following types of modified leaves is not a part of a flower?
    a. tendrils
    b. anthers
    c. sepals
    d. carpels

25. The megaspore is similar to the microspore in that both
    a. have the diploid number of chromosomes.
    b. become an embryo sac.
    c. become a gametophyte that produces a gamete.
    d. are necessary for seed production.
    e. Both c and d are correct.

26. Phytochrome is found in which of the following plant organs?
    a. seeds
    b. leaves
    c. flowers
    d. roots

27. Which of the following is not a component of the carpel?
    a. stigma
    b. filament
    c. ovary
    d. ovule
    e. style

28. Label the parts of the flower in the following diagram.

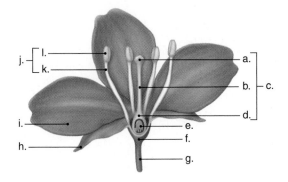

29. The embryo sac is derived from
    a. one of four microspores.
    b. the second sperm nucleus of the pollen grain.
    c. the endosperm.
    d. the cotyledons.

30. In contrast to eudicots, monocot
    a. embryos have two cotyledons.
    b. seeds contain endosperm.
    c. cotyledons store food.
    d. embryos undergo a heart stage of development.

31. Foreign genes may be introduced into plants by using
    a. microinjected seeds.
    b. genetically modified fungi.
    c. electrical pulses.
    d. DNA replication.

32. Plant biotechnology can lead to
    a. increased crop production.
    b. disease-resistant plants.
    c. treatment of human disease.
    d. All of these are correct.

33. Plants may be propagated asexually in the laboratory through
    a. tissue culture.
    b. anther culture.
    c. root or stem cutting.
    d. cell suspension culture.
    e. All of these are correct.

34. Plant cells are totipotent, which means that
    a. they may reproduce asexually at any time.
    b. each plant cell has the capability of becoming an entire plant.
    c. they are only capable of reproducing sexually.
    d. each cell has had its cell wall removed.

35. Plants propagated by tissue culture rather than anther culture differ because plants produced by tissue culture
    a. are subject to mutations and may not be identical.
    b. are haploid rather than diploid.
    c. have exactly the same characteristics as the parent plant.
    d. are produced from calluses.

36. Transgenic plants
    a. are produced by anther culture or tissue culture.
    b. are genetically identical.
    c. are a new plant species.
    d. may possess improved traits.

37. Some individuals are concerned about genetically modified plants because
    a. some people could be allergic to them.
    b. they could spread the foreign genes they contain to other plants.
    c. they might reproduce out of control.
    d. All of these are correct.

38. The calyx is composed of
    a. sepals.
    b. petals.
    c. anthers.
    d. ovaries.
    e. stigmas.

39. A seed is a mature
    a. embryo.
    b. ovule.
    c. ovary.
    d. pollen grain.

40. Label the parts of the eudicot seed and seedling in the following diagram.

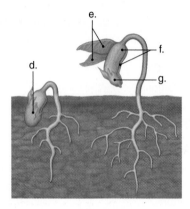

Thinking Scientifically

1. In late November every year, florists ship truckloads of poinsettia plants to stores around the country. Typically, the plants are individually wrapped in plastic sleeves. If the plants remain in the sleeves for too long during shipping and storage, their leaves begin to curl under and eventually fall off. What plant hormone do you suppose causes this response? How do you suppose the plastic sleeves affect the response?

2. Snow buttercups (*Ranunculus adoneus*) live in alpine regions and produce sun-tracking flowers. The flowers face east in the morning to absorb the sun's warmth in order to attract pollinators and speed the growth of fertilized ovules. The flowers track the sun all day, continually bending to face the sun. How might you determine whether the flowers or stems are responsible for sun tracking? Assuming you have determined that the stem is responsible, how would you determine which region of the stem follows the sun?

3. Some plants, such as Hepatica, bloom in the late winter or very early spring when weather conditions are harsh and there are relatively few insects available to pollinate their flowers. What advantage might this give a plant? And what types of modifications would you expect the flowers of these plants to possess?

Bioethical Issue

Unintended Consequences of Genetic Engineering

Witchweed is a serious parasitic weed that destroys 40% of Africa's cereal crop annually. Because it is intimately associated with its host (the cereal crop), witchweed is difficult to selectively destroy with herbicides. One control strategy is to create genetically modified cereals with herbicide resistance. Then, herbicides will kill the weed without harming the crop. Herbicide-resistant sorghum was created to help solve the witchweed problem. However, scientists discovered that the herbicide resistance gene could be carried via pollen into Johnson grass, a relative of sorghum and a serious weed in the United States. Farmers would have a new challenge if Johnson grass could no longer be effectively controlled with herbicides. Consequently, efforts to create herbicide-resistant sorghum were put on hold.

Do you think farmers in Africa should be able to grow herbicide-resistant sorghum, allowing them to control witchweed? Or, should the production of herbicide-resistant sorghum be banned worldwide in order to avoid the risk of introducing the herbicide resistance gene into Johnson grass?

Essentials of Biology Website

The companion website for *Essentials of Biology* provides a wealth of information organized and integrated by chapter. You will find practice tests, animations, videos, and much more that will complement your learning and understanding of general biology.

http://www.mhhe.com/maderessentials2

22

Being Organized and Steady

The internal environment is composed of blood and tissue fluid.

Insulin maintains blood glucose levels.

Blood is a liquid tissue.

How Homeostasis Works

TO KEEP YOU healthy, your body is always maintaining homeostasis, the relative constancy of the internal environment. Your cells are bathed by tissue fluid, which continually exchanges substances with the blood. Therefore, the internal environment consists of blood and tissue fluid.

How might homeostasis be maintained? Let's take an example. Despite extreme fluctuations in our diet, the blood glucose (sugar) concentration usually stays within an acceptable range range, day in and day out. This is primarily due to the work of two hormones, called insulin and glucagon.

—Continued next page

Outline

Continued—

Just after you have eaten, the blood glucose level usually rises. In response, your pancreas releases insulin, which causes the body's cells, including those of the liver, to take up glucose from the blood. Now the blood glucose level lowers back into the normal range. The liver ordinarily stores glucose, so if your blood runs out of glucose before you eat again, the other hormone, glucagon, causes the liver to release sugar into the blood.

If by chance the liver runs out of stored glucose, it will convert other molecules such as amino acids from protein to glucose in order to maintain the blood glucose level. Liver glucose stores might become extremely low if you have diabetes mellitus, and your pancreas does not produce enough insulin or if you are on a low-carbohydrate diet. The inability of the pancreas to produce insulin and the inability of the liver to maintain the blood glucose concentration are serious conditions that require good medical care. If homeostasis cannot be maintained, death can occur due to either an extremely high or extremely low blood glucose level. Of all the organs in the body, it is particularly important that the brain receive a constant supply of glucose.

Understanding the structure and function of the various types of tissues in the body is of critical importance to understanding the functions of various organ systems. In this chapter, you will learn about basic tissue types, and how they come together to form organs and organ systems. In addition, you will learn how all these components of the body maintain homeostasis.

22.1 | The Body's Organization

In Part One of this text, you studied the general structure and function of a plant cell and an animal cell. Now it is time to consider that an animal's body is composed of specialized cells that perform particular functions. Your body contains trillions of cells, of which there are a hundred different types. Cells of the same type occur within a tissue. A **tissue** is a group of similar cells performing a similar function. An **organ** contains different types of tissues arranged in a certain fashion. In other words, the structure and function of an organ are dependent on the tissues it contains. That is why it is sometimes said that tissues, not organs, are the structural and functional units of the body. Several organs are found within an organ system, and the organs of a system work together to perform necessary functions for the organism.

Let's take an example (**Fig. 22.1**). In humans, the urinary system contains these organs: kidneys, ureters, bladder, and urethra. The epithelial cells of the kidneys remove waste molecules from the blood and produce urine. The tubular shape of the ureters is suitable for passing urine to the bladder, which can store urine because it is expandable. Urine passes out of the body by way of the tubular urethra. The functions of the urinary system—to produce, store, and rid the body of metabolic wastes—are dependent on its organs, which in turn are dependent on the tissues making up these organs.

cell

tissue
(cuboidal epithelium)

organ
(kidney)

organ system
(urinary system)

organism

Figure 22.1 Levels of biological organization.

A tissue is composed of specialized cells all having the same structure and performing the same functions. An organ is composed of those types of tissues that help it perform particular functions. An organ system contains several organs and has functions necessary to the continued existence of an organism.

Other examples also show that the work of an organ is dependent on its tissues. In the digestive system, the intestine absorbs nutrients. The cells of the tissue lining the lumen (cavity) of the small intestine have tiny fingerlike projections called microvilli that increase the available surface area for absorption. Muscle cells, within muscle tissue, shorten when they contract because they have intracellular components that move past one another. Nerve cells, within nervous tissue, have long, slender projections that carry impulses to distant body parts. The biological axiom that "structure suits function" and vice versa begins with the specialized cells within a tissue. And thereafter, this truism applies also to organs and organ systems.

From the many different types of animal cells, biologists have been able to categorize tissues into just four major types:

- *Epithelial tissue* (epithelium) covers body surfaces and lines body cavities.
- *Connective tissue* binds and supports body parts.
- *Muscular tissue* moves the body and its parts.
- *Nervous tissue* receives stimuli and conducts nerve impulses.

Except for nervous tissue, each type of tissue is subdivided into even more types (**Fig. 22.2**). We look at each of these in the sections that follow to see how their particular structure allows them to perform their functions.

Check Your **Progress**

1. List the levels of biological organization, and explain how they are related to each other.
2. List the four major types of tissues in animals.

Answers: 1. An organism contains organ systems, which are composed of groups of organs that work together. An organ is composed of tissues, which determine its structure and function. Each tissue is composed of like cells.   2. The four tissue types are epithelial, connective, muscular, and nervous.

Figure 22.2   Types of vertebrate tissues.

The four classes of vertebrate tissues are epithelial (pink), connective (blue), muscular (tan), and nervous (yellow).

Nervous tissue
neuron

Epithelial tissue
cuboidal
columnar
squamous

Connective tissue
bone
loose fibrous connective tissue
blood

Muscular tissue
skeletal muscle
cardiac muscle
smooth muscle

Epithelial Tissue Protects

Epithelial tissue, also called *epithelium,* forms the external and internal linings of many organs and covers the surface of the body. Therefore, for a substance to enter or exit the body at the digestive tract, the lungs, or the genital tract, it must cross an epithelial tissue. Epithelial cells adhere to one another, but an epithelium is generally only one cell thick. This enables an epithelium to serve a protective function and substances have to pass through epithelial cells in order to reach a tissue beneath them.

Epithelial cells differ in shape (**Fig. 22.3**). **Squamous epithelium,** such as that lining the air spaces of the lungs and blood vessels, is composed of flattened cells. **Cuboidal epithelium,** which lines the **lumen** (cavity) of the kidney tubules, contains cube-shaped cells. **Columnar epithelium** has cells resembling rectangular pillars or columns, with nuclei usually located near the bottom of each cell. Columnar epithelium is found lining the lumen of the digestive tract. These epithelia are simple because they are composed of one layer of cells. Pseudostratified epithelium appears to be layered but actually has only one layer of cells.

The outer region of the skin, called the epidermis, is a stratified (layered) squamous epithelium in which the cells have been reinforced by keratin, a protein that provides

×250

Squamous
• lines the lungs
• protects

a.

×250

Cuboidal
• lines the kidney tubules
• absorbs molecules

b.

×250

goblet cell secretes mucus

Columnar
• lines the small intestine
• absorbs nutrients

c.

×250

cilia

Pseudostratified
• lines the trachea
• sweeps impurities toward throat

goblet cell secretes mucus

d.

Figure 22.3 **Epithelial tissue.**

Epithelial tissue covers the surfaces and lines the cavities of internal organs. It also makes up the outer layer of skin, called the epidermis. The functions of epithelial tissue are associated with protection, secretion, and absorption. **a.** Squamous epithelium. **b.** Cuboidal epithelium. **c.** Columnar epithelium. **d.** Pseudostratified (appears to be layered but is not) epithelium.

×250

Stratified
• epidermis of skin
• protects

Figure 22.4 Skin.

The outer portion of skin, called the epidermis, is a stratified epithelium. The many layers of tightly packed cells reinforced by the protein keratin make the skin protective against water loss and pathogen invasion. New epidermal cells arise in the most inner layer and are shed at the outer layer. In reptiles, such as this gila monster, the epidermis forms scales, which are simply projections hardened with keratin. When the epidermis is shed, the scales have to be replaced. Pigmented epidermal cells account for coloration of the skin.

strength and waterproofing (**Fig. 22.4**). A stratified epithelium allows skin to protect the body from injury, drying out, and possible pathogen (virus and bacterium) invasion.

The lining of the urinary bladder is a transitional epithelium, whose structure suits its function. When the walls of the bladder are relaxed, the transitional epithelium consists of several layers of cuboidal cells. When the bladder is distended with urine, the epithelium stretches, and the outer cells take on a squamous appearance. The cells are able to slide in relation to each other, while at the same time forming a barrier that prevents any urine from diffusing into other regions of the body.

One or more epithelial cells are the primary components of glands, which produce products and secrete them by exocytosis. During secretion, the contents of a vesicle are released when the vesicle fuses with the plasma membrane. Some glands have a duct, and others are ductless. Each mucus-secreting goblet cell within the lining of the digestive tract is a single-celled gland that produces a protective mucus (Fig. 22.3c). Pseudostratified epithelium lines the trachea (windpipe) (Fig. 22.3d). Along the trachea, mucus traps foreign particles, and the upward motion of cilia carries the mucus to the back of the throat, where it may be either swallowed or expelled. Smoking can cause a change in mucus secretion and inhibit ciliary action, and the result is an inflammatory condition called chronic bronchitis.

Epithelial tissue is well known for constantly replacing its cells. This feature is particularly useful along the digestive tract, where rough food particles and enzymes can damage the lining. Think about this fact the next time you burn the roof of your mouth with a bite of your "too hot" pizza and notice how quickly it heals. The liver, which is composed of cells of epithelial origin, can regenerate whole portions removed during surgery. But there is a price to pay for the ability of epithelial tissue to constantly divide—it is more likely to become cancerous than are other tissues. Cancers of epithelial tissue within the digestive tract, lungs, and breast are called carcinomas.

Check
Your Progress

1. Describe the composition and activity of a gland.
2. Explain why cancers are more likely to develop in epithelial tissue than in other types of tissue.

Answers: 1. A gland is usually composed primarily of epithelium. The epithelial cells produce and secrete a product, such as mucus.   2. Epithelial tissue is constantly dividing, so errors that might occur during cell division and result in cancer are more likely to occur.

Connective Tissue Connects and Supports

The many different types of **connective tissue** are all involved in binding organs together and providing support and protection. As a rule, connective tissue cells are widely separated by a **matrix,** a noncellular material that varies from solid to semifluid to fluid. The matrix usually has fibers, notably collagen fibers. Collagen is the most common protein in the human body, which gives you some idea of how prevalent connective tissue is.

Loose Fibrous and Related Connective Tissues

Let's consider **loose fibrous connective tissue** first and then compare the other types to it (see Fig. 22.2). This tissue occurs beneath an epithelium and connects it to other tissues within an organ. It also forms a protective covering for many internal organs, such as muscles, blood vessels, and nerves. Its cells are called **fibroblasts** because they produce a matrix that contains fibers, including collagen fibers and elastic fibers. The presence of loose fibrous connective tissue in the walls of lungs and arteries allows these organs to expand and then return to their original shape.

Adipose tissue is a type of loose connective tissue in which the fibroblasts enlarge and store fat, and there is limited matrix (**Fig. 22.5**). Adipose tissue is located beneath the skin and around organs such as the heart and kidneys, where it cushions and protects these organs and serves as long-term energy storage.

Compared to loose fibrous connective tissue, **dense fibrous connective tissue** contains more collagen fibers, and they are packed closely together

Figure 22.5   Types of connective tissue.

The human knee provides examples of most types of connective tissue.

Dense fibrous tissue   ×250

Adipose tissue   ×250

fat

Compact bone   ×320

osteon

central canal

cell in lacuna

matrix

cell within a lacuna

Hyaline cartilage   ×250

(Fig. 22.5). This type of tissue has more specific functions than does loose fibrous connective tissue. For example, dense fibrous connective tissue is found in **tendons,** which connect muscles to bones, and in **ligaments,** which connect bones to other bones at joints.

In **cartilage,** the cells lie in small chambers called **lacunae,** separated by a matrix that is solid yet flexible. **Hyaline cartilage,** the most common type of cartilage, contains only very fine collagen fibers (Fig. 22.5). The matrix has a white, translucent appearance when unstained. Hyaline cartilage is found in the nose and at the ends of the long bones and the ribs, and it forms rings in the walls of respiratory passages. The human fetal skeleton is also made of this type of cartilage, which is later replaced by bone. Cartilaginous fishes, such as sharks, have a cartilaginous skeleton throughout their lives.

Bone is the most rigid connective tissue (Fig. 22.5). It consists of an extremely hard matrix of inorganic salts, notably calcium salts, deposited around collagen fibers. The inorganic salts give bone rigidity, and the protein fibers provide elasticity and strength, much as steel rods do in reinforced concrete. The inorganic salts found in bone also act as a storehouse of calcium and phosphate ions for the body. **Compact bone,** the most common type, consists of cylindrical structural units called osteons. The central canal of each osteon is surrounded by rings of hard matrix. Bone cells are located in spaces called lacunae between the rings of matrix. Blood vessels in the central canal carry nutrients that allow bone to renew itself.

Blood

Blood is composed of several types of cells suspended in a liquid matrix called plasma. Blood is unlike other types of connective tissue in that the matrix (i.e., plasma) is not made by the cells (**Fig. 22.6**). Some people do not classify blood as connective tissue; instead, they suggest a separate tissue category called vascular tissue.

Blood serves the body well. It transports nutrients and oxygen to cells and removes their wastes. It helps distribute heat and also plays a role in fluid, ion, and pH balance. Also, various components of blood help protect us from disease, and blood's ability to clot prevents fluid loss.

Blood contains two types of cells. **Red blood cells** are small, biconcave, disk-shaped cells without nuclei. The presence of the red pigment hemoglobin makes the cells red, and in turn, makes the blood red. Hemoglobin combines with oxygen, and in this way red blood cells transport oxygen.

White blood cells may be distinguished from red blood cells by the fact that they are usually larger, have a nucleus, and without staining would appear translucent. White blood cells fight infection in two primary ways. Some white blood cells are phagocytic and engulf infectious pathogens, while other white blood cells produce antibodies, molecules that combine with foreign substances to inactivate them.

Platelets are another component of blood, but they are not complete cells; rather, they are fragments of giant cells present only in bone marrow. When a blood vessel is damaged, platelets form a plug that seals the vessel, and along with injured tissues, platelets release molecules that help the clotting process.

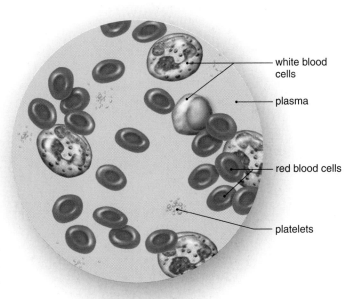

Figure 22.6   Blood, a liquid tissue.

Blood is classified as connective tissue because the cells are separated by a matrix—plasma, the liquid portion of blood. The cellular components of blood are red blood cells, white blood cells, and platelets (which are actually fragments of larger cells).

Check Your Progress

1. **Compare and contrast loose fibrous connective tissue with dense fibrous connective tissue.**
2. **Contrast the function of tendons with that of ligaments.**
3. **Contrast the structure and function of red blood cells with those of white blood cells.**

Answers: 1. Both loose and dense fibrous connective tissue are involved in binding organs together and providing support and protection. Loose fibrous connective tissue contains fewer collagen fibers, and these are less closely packed than those in dense fibrous connective tissue. 2. Tendons connect muscles to bones, while ligaments connect bones to other bones. 3. Red blood cells are smaller and lack a nucleus. The presence of hemoglobin for carrying oxygen makes them red. White blood cells contain a nucleus and without staining would appear translucent. They play a role in fighting infection.

Muscular Tissue Moves the Body

Muscular tissue and nervous tissue account for the ability of animals and their parts to move. **Muscle tissue** is also sometimes called *contractile tissue* because it contains contractile protein filaments, called actin and myosin filaments, that interact to produce movement. The three types of vertebrate muscles are skeletal, cardiac, and smooth.

Skeletal muscle, also called voluntary muscle, is attached by tendons to the bones of the skeleton, and when it contracts, bones move. Contraction of skeletal muscle is under voluntary control and occurs faster than in the other muscle types. The cells of skeletal muscle, called fibers, are cylindrical and quite long—sometimes they run the length of the muscle (**Fig. 22.7***a*). They arise during development when several cells fuse, resulting in one fiber with multiple nuclei. The nuclei are located at the edge of the cell, just inside the plasma membrane. The fibers have alternating light and dark bands running across the cells that give them a **striated** appearance. These bands are due to the placement of actin filaments and myosin filaments in the cell.

Cardiac muscle is found only in the walls of the heart, and its contraction pumps blood and accounts for the heartbeat. Like skeletal muscle, cardiac muscle has striations, but the contraction of the heart is autorhythmic and involuntary (Fig. 22.7*b*). Cardiac muscle cells also differ from skeletal muscle cells in that they have a single, centrally placed nucleus. The cells are branched and seemingly fused one with the other. The heart appears to be composed of one large interconnecting mass of muscle cells. Actually, cardiac muscle cells are separate and individual, but they are bound end to end at **intercalated disks,** areas where folded plasma membranes allow the contraction impulse to spread from one cell to the other.

Smooth muscle is so named because the cells lack striations. The spindle-shaped cells form layers in which the thick middle portion of one cell is opposite the thin ends of adjacent cells. Consequently, the nuclei form an irregular pattern in the tissue (Fig. 22.7*c*). Like cardiac muscle, smooth muscle is involuntary. Smooth muscle is also sometimes called *visceral* muscle because it is found in the walls of the viscera (intestine, stomach, and other internal organs) and blood vessels. Smooth muscle contracts more slowly than skeletal muscle but can remain contracted for a longer time. When the smooth muscle of the intestine contracts, food moves along its lumen (central cavity). When the smooth muscle of the blood vessels contracts, blood vessels constrict, helping to raise blood pressure.

Figure 22.7   Types of muscular tissue.

×250   intercalated disk   ×100   ×100

a. **Skeletal muscle**
- has striated, tubular, multinucleated fibers.
- is usually attached to skeleton.
- is voluntary.

b. **Cardiac muscle**
- has striated, branched, uninucleated fibers.
- occurs in walls of heart.
- is involuntary.

c. **Smooth muscle**
- has spindle-shaped, nonstriated, uninucleated fibers.
- occurs in walls of internal organs.
- is involuntary.

Nervous Tissue Communicates

Nervous tissue coordinates body parts and allows an animal to respond to the environment. The nervous system depends on (1) sensory input, (2) integration of data, and (3) motor output to carry out its functions. Nerves conduct impulses from sensory receptors such as pain receptors in the skin to the spinal cord and the brain, where integration occurs. The phenomenon called sensation occurs only in the brain, however. Nerves then conduct nerve impulses away from the spinal cord and brain to the muscles and glands, causing them to contract and secrete, respectively. In this way, a coordinated response to both internal and external stimuli is achieved.

A nerve cell is called a **neuron.** Every neuron has three parts: dendrites, a cell body, and an axon (**Fig. 22.8**). A *dendrite* is an extension that conducts signals toward the cell body. The *cell body* contains the major concentration of the cytoplasm and the nucleus of the neuron. An *axon* is an extension that conducts nerve impulses to other cells. The brain and spinal cord contain many neurons, whereas **nerves** contain only axons of neurons. The dendrites and cell bodies of these neurons are located in the spinal cord or brain, depending on whether it is a spinal nerve or a cranial nerve. A nerve is like a land-based telephone trunk cable, because just like telephone wires, each axon is a communication channel independent of the others.

In addition to neurons, nervous tissue contains **neuroglia,** cells that support and nourish neurons. They outnumber neurons nine to one and take up more than half the volume of the brain. Although their primary function is support, research is currently being conducted to determine how much neuroglia directly contribute to brain function. Schwann cells are the type of neuroglia that encircle long nerve fibers within nerves, forming a protective myelin sheath. The presence of myelin sheaths allows nerve impulses to travel much more quickly down the axon.

Check Your **Progress**

1. **Contrast the functions of skeletal, cardiac, and smooth muscle tissues.**
2. **Describe the components of a neuron, and give a function for each.**

Answers: 1. Skeletal muscle causes bones to move when it contracts. Cardiac muscle causes the heart to beat. Contraction of smooth muscle causes the walls of internal organs to constrict. 2. The dendrite is an extension that conducts signals toward the cell body. The cell body contains most of the cytoplasm and the nucleus. It carries on the usual functions of the cell. The axon is an extension that conducts nerve impulses.

dendrite

cell body

nucleus

nucleus of Schwann cell

axon

impulse

myelin sheath

×200

Nervous tissue
• Brain; spinal cord; nerves
• Conduction of nerve impulses

Figure 22.8   Nervous tissue.

Neurons are surrounded by neuroglia, such as Schwann cells, which envelope axons and form the myelin sheath. Only axons conduct nerve impulses.

22.2 | Organs and Organ Systems

Organs are composed of a number of tissues, and the structure and function of an organ are dependent on the tissues that comprise it. Since an organ has many types of tissues, it can perform a function that none of the tissues can do alone. To take an example, the function of the bladder is to store urine and to expel it when convenient. But this function is dependent on the individual tissues making up the bladder. The epithelium lining the bladder helps the organ store urine by stretching while still preventing urine from leaking into the body proper. The muscles of the bladder propel the urine forward so that it is voided from the body. Similarly, the functions of an organ system are dependent on its organs. The function of the urinary system is to produce urine, store it, and then transport it out of the body. The kidneys produce urine, the bladder stores it, and various tubes transport it from the kidneys to the bladder and also out of the body.

This text divides the systems of the body into those involved in (1) transport of fluids about the body, (2) maintenance of the body, (3) control of the body's systems, (4) sensory input and motor output, and (5) reproduction. All these systems have functions that contribute to homeostasis, the relative constancy of the internal environment (see Section 22.3).

Transport

The **cardiovascular system** (**Fig. 22.9***a*) consists of blood, the heart, and the blood vessels that carry blood throughout the body. The body's cells are surrounded by a liquid called **tissue fluid;** blood transports nutrients and oxygen to tissue fluid for the cells, and removes waste molecules excreted by cells from the tissue fluid. Recall that the internal environment of the body consists of the blood within the blood vessels and the tissue fluid that surrounds the cells.

The **lymphatic system** (Fig 22.9*b*) consists of lymphatic vessels, lymph, lymph nodes, and other lymphatic organs. Lymphatic vessels absorb fat from the digestive system and collect excess tissue fluid, which is returned to the blood in the cardiovascular system.

These two systems are also involved in defense against disease. Along with the thymus and spleen, certain blood cells in the lymph and blood are part of the **immune system,** which specifically protects the body from disease.

Maintenance of the Body

Three systems (respiratory, urinary, and digestive) add substances to and/or remove substances from the blood. If the composition of the blood remains constant, so does that of the tissue fluid.

The **respiratory system** (**Fig. 22.10***a*) consists of the lungs, the trachea, and other tubes that take air to and from the lungs. The respiratory system brings oxygen into the body and takes carbon dioxide out of the body through the lungs. It also exchanges gases with the blood.

The **urinary system** (Fig. 22.10*b*) contains the kidneys and the urinary bladder along with tubes that transport urine. This system rids blood of wastes and also helps regulate the fluid level and chemical content of the blood.

The **digestive system** (Fig. 22.10*c*) consists of the various organs along the digestive tract together with the associated organs, such as teeth, salivary glands, the liver, and the pancreas. This system receives food and digests it into nutrient molecules that enter the blood.

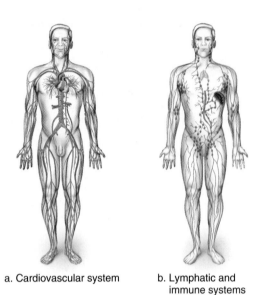

a. Cardiovascular system   b. Lymphatic and immune systems

Figure 22.9   **The body's transport systems.**

Aside from transport, the cardiovascular and lymphatic systems are involved in defense against disease.

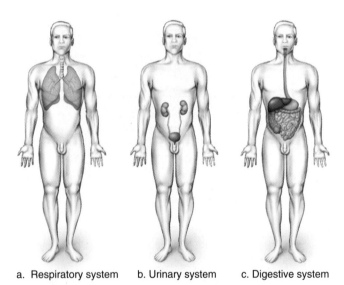

a. Respiratory system   b. Urinary system   c. Digestive system

Figure 22.10   **The body's maintenance systems.**

The respiratory, urinary, and digestive systems keep the internal environment constant.

Control

The **nervous system** (**Fig. 22.11**a) consists of the brain, spinal cord, and associated nerves. The nerves conduct nerve impulses from receptors to the brain and spinal cord. They also conduct nerve impulses from the brain and spinal cord to the muscles and glands, allowing us to respond to both external and internal stimuli. Sensory receptors and sense organs are sometimes considered a part of the nervous system.

The **endocrine system** (Fig. 22.11b) consists of the hormonal glands such as the thyroid and adrenal glands, which secrete chemicals that serve as messengers between body parts. These chemicals are known as hormones. Both the nervous and endocrine systems coordinate and regulate the functions of the body's other systems. The nervous system tends to cause quick responses in the body, while the body's responses to hormones released by the endocrine system tend to last much longer. The endocrine system also helps maintain the proper functioning of the male and female reproductive organs.

Sensory Input and Motor Output

The **integumentary system** (**Fig. 22.12**a) consists of the skin and its accessory structures. The sensory receptors in the skin, and in organs such as the eyes and ears, respond to specific external stimuli and communicate with the brain and spinal cord by way of nerve fibers. These messages may cause the brain to bring about a response to a stimulus.

The **skeletal system** (Fig. 22.12b) and the **muscular system** (Fig. 22.12c) enable the body and its parts to move as a result of motor output. The skeleton, as a whole, serves as a place of attachment for the skeletal muscles. Contraction of muscles in the muscular system actually accounts for movement of body parts.

These systems, along with the integumentary system, also protect and support the body. The skeletal system, consisting of the bones of the skeleton, protects body parts. For example, the skull forms a protective encasement for the brain, as does the rib cage for the heart and lungs. The skin serves as a barrier between the outside world and the body's tissues.

Reproduction

The **reproductive system** (**Fig. 22.13**) involves different organs in the male and female. The male reproductive system consists of the testes, other glands, and various ducts like the ductus deferens that conduct semen to and through the penis. The testes produce sex cells called sperm. The female reproductive system consists of the ovaries, oviducts, uterus, vagina, and external genitals. The ovaries produce sex cells called eggs. When a sperm fertilizes an egg, an offspring begins development.

Check Your Progress

1. Compare and contrast the function of the cardiovascular system with that of the lymphatic system.
2. Describe the common function of the nervous and endocrine systems.

Answers: 1. Both are transport systems. The cardiovascular system carries blood throughout the body, while the lymphatic system absorbs fat from the digestive system and collects excess tissue fluid and returns it to the blood.   2. Both the nervous and endocrine systems coordinate and regulate the functions of other systems.

a. Nervous system   b. Endocrine system

Figure 22.11 **The control systems.**

The nervous and endocrine systems coordinate the other systems of the body.

a. Integumentary system   b. Skeletal system   c. Muscular system

Figure 22.12 **Sensory input and motor output.**

Sensory receptors in the skin (integumentary system) and the sense organs send input to the skeletal and muscular systems, the control systems that cause a response to stimuli.

Reproductive system

Figure 22.13 **The reproductive system.**

The reproductive organs in the male and female differ. The reproductive system functions to ensure survival of the species.

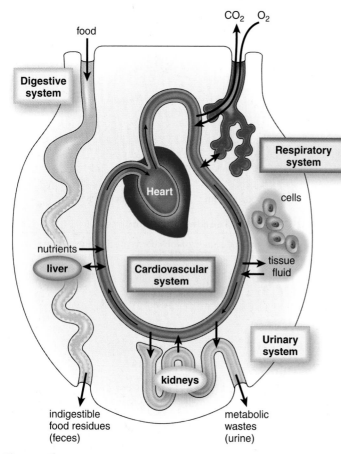

food

CO_2 O_2

Digestive system

Respiratory system

Heart

cells

nutrients

liver

Cardiovascular system

tissue fluid

Urinary system

kidneys

indigestible food residues (feces)

metabolic wastes (urine)

Figure 22.14 **Constancy of the internal environment (blood and tissue fluid).**

The respiratory system exchanges gases with the external environment and with the blood. The digestive system takes in food and adds nutrients to the blood, and the urinary system removes metabolic wastes from the blood and excretes them. When blood exchanges nutrients and oxygen for carbon dioxide and other wastes with tissue fluid, the composition of the tissue fluid stays within normal limits.

Check Your Progress

1. Why are blood and tissue fluid considered the internal environment?
2. How do the digestive, urinary, and respiratory systems contribute to homeostasis?

Answers: 1. Cells are bathed in a liquid environment called tissue fluid, and tissue fluid exchanges substances with the blood. Therefore, both are considered the internal environment of the body.   2. The digestive system adds nutrients to the blood, the urinary system removes metabolic wastes from the blood, and the respiratory system takes away carbon dioxide and adds oxygen from the blood.

22.3 | Homeostasis

It is likely you are most familiar with the use of the word *environment* to mean the external environment. There is much concern these days about our external environment because of the need to keep pollution to a minimum in order to maintain the health of ecosystems and organisms, including ourselves, that live in them. Your body, however, is composed of many cells, and therefore its internal environment is the environment of the cells. Cells live in a liquid environment called tissue fluid. Tissue fluid is constantly renewed by exchanges with the blood (**Fig. 22.14**). Therefore, blood and tissue fluid are the internal environment of the body. The volume and composition of tissue fluid remains relatively constant only as long as blood composition remains near normal levels. Relatively constant means that the composition of both tissue fluid and blood usually falls within a certain range of normality. Maintenance of the relatively constant condition of the internal environment is homeostasis.

Because of homeostasis, even though external conditions may change dramatically, internal conditions stay within a narrow range. For example, the temperature of the body is maintained near 37°C (97° to 99°F), even if the surrounding temperature is lower or higher. If you eat acidic foods, the pH of your blood still stays about 7.4, and even if you eat a candy bar, the amount of sugar in your blood remains at just about 0.1%.

All the systems of the body contribute to maintaining **homeostasis.** The cardiovascular system conducts blood to and away from capillaries, where exchange occurs. The heart pumps the blood, and thereby keeps it moving toward the capillaries. Red blood cells transport oxygen and participate in the transport of carbon dioxide. White blood cells fight infection, and platelets participate in the clotting process. The lymphatic system is accessory to the cardiovascular system. Lymphatic capillaries collect excess tissue fluid and return it via lymphatic vessels to the cardiovascular system. Lymph nodes help purify lymph and keep it free of pathogens.

The digestive system takes in and digests food, providing nutrient molecules that enter the blood to replace those that are constantly being used by the body's cells. The respiratory system removes carbon dioxide from and adds oxygen to the blood. The chief regulators of blood composition are the kidneys and the liver. Urine formation by the kidneys is extremely critical to the body, not only because it rids the body of metabolic wastes, but also because the kidneys carefully regulate blood volume, salt balance, and pH. The liver, among other functions, regulates the glucose concentration of the blood. Immediately after glucose enters the blood, the liver removes the excess for storage as glycogen. Later, the glycogen is broken down to replace the glucose that was used by body cells. In this way, the glucose composition of the blood remains relatively constant. The liver also removes toxic chemicals, such as ingested alcohol and other drugs. The liver makes urea, a nitrogenous end product of protein metabolism.

The nervous system and the endocrine system regulate the other systems of the body. They work together to control body systems so that homeostasis is maintained. In negative feedback mechanisms involving the nervous system, sensory receptors send nerve impulses to control centers in the brain, which then direct effectors to become active. Effectors can be muscles or glands. Muscles bring about an immediate change. Endocrine glands secrete hormones that bring about a slower, more lasting change that keeps the internal environment relatively stable.

Negative Feedback

Negative feedback is the primary homeostatic mechanism that allows the body to keep the internal environment constant. A negative feedback mechanism has at least two components: a sensor and a control center (**Fig. 22.15**). The sensor detects a change in the internal environment (a stimulus); the control center initiates an effect that brings conditions back to normal again. Now, the sensor is no longer activated. In other words, a negative feedback mechanism is present when the output of the system dampens the original stimulus.

Let's take a simple example. When the pancreas detects that the blood glucose level is too high, it secretes insulin, a hormone that causes cells to take up glucose. Now the blood sugar level returns to normal, and the pancreas is no longer stimulated to secrete insulin.

When conditions exceed their limits and feedback mechanisms cannot compensate, illness results. For example, if the pancreas is unable to produce insulin as in diabetes mellitus, the blood sugar level becomes dangerously high, and the individual can become seriously ill. The study of homeostatic mechanisms is therefore medically important.

Mechanical Example

A home heating system is often used to illustrate how a more complicated negative feedback mechanism works (**Fig. 22.16**). You set the thermostat at, say, 20°C (68°F). This is the *set point*. The thermostat contains a thermometer, a sensor that detects when the room temperature is above or below the set point. The thermostat also contains a control center; it turns the furnace off when the room is warm and turns it on when the room is cool. When the furnace is off, the room cools a bit, and when the furnace is on, the room warms a bit. In other words, a negative feedback system results in fluctuation above and below the set point.

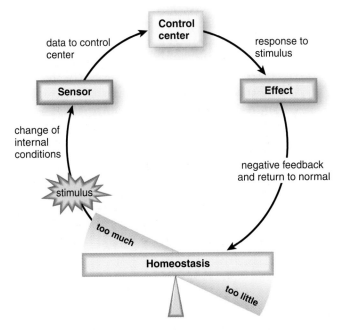

Figure 22.15   Negative feedback mechanism.

This diagram shows how the basic elements of a negative feedback mechanism work.

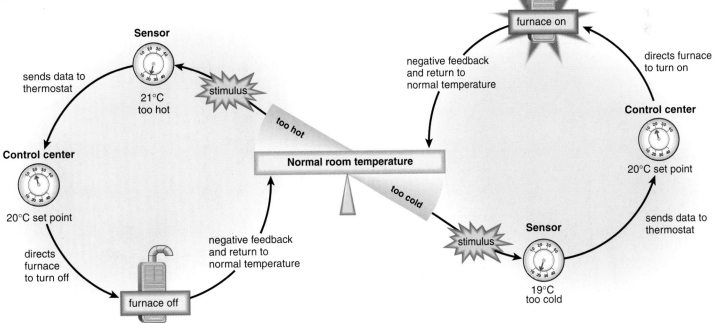

Figure 22.16   Regulation of room temperature.

This diagram shows how room temperature is returned to normal when the room becomes too hot. A contrary cycle in which the furnace turns on and gives off heat returns the room temperature to normal when the room becomes too cold.

Check
Your **Progress**

1. Describe the components of a negative feedback mechanism.
2. Describe how the control center in the brain responds to a drop in internal temperature.
3. Describe how the control center in the brain responds to a rise in internal temperature.

Answers: 1. A sensor detects a change in the internal environment; the control center initiates an effect that brings conditions back to normal and the sensor is no longer activated.   2. First, the control center brings about constriction of the blood vessels of the skin. If the temperature continues to drop, shivering begins in order to generate heat.   3. Blood vessels of the skin dilate, and sweat glands are activated.

Human Example: Regulation of Body Temperature

The thermostat for body temperature is located in a part of the brain called the hypothalamus. When the core body temperature falls below normal, the control center directs (via nerve impulses) the blood vessels of the skin to constrict (**Fig. 22.17**). This action conserves heat. If the core body temperature falls even lower, the control center sends nerve impulses to the skeletal muscles, and shivering occurs. Shivering generates heat, and gradually body temperature rises to 37°C (98.6°F). When the temperature rises to normal, the control center is inactivated.

When the body temperature is higher than normal, the control center directs the blood vessels of the skin to dilate. More blood is then able to flow near the surface of the body, where heat can be lost to the environment. In addition, the nervous system activates the sweat glands, and the evaporation of sweat helps lower body temperature. Gradually, body temperature decreases to 37°C (98.6°F).

Notice that a negative feedback mechanism prevents continued change in the same direction; body temperature does not get warmer and warmer, because warmth stimulates changes that decrease body temperature. Also, body temperature does not get colder and colder, because a body temperature below normal causes changes that bring body temperature up.

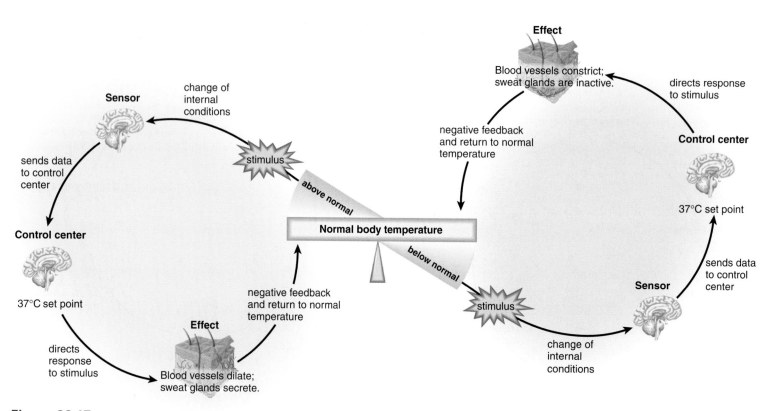

Figure 22.17   Regulation of body temperature.

This diagram shows how normal body temperature in maintained by a negative feedback system.

The Chapter in Review

Summary

22.1 | The Body's Organization

Levels of animal organization:

Four groups of vertebrate tissues:

Epithelial Tissue Protects

Epithelial tissue covers the body and lines its cavities.
- Types of epithelial tissue are squamous, cuboidal, and columnar.
- Epithelial cells sometimes form glands that secrete either into ducts or into the blood.

Connective Tissue Connects and Supports

In connective tissue, cells are separated by a matrix that contains fibers (e.g., collagen fibers). The four types are:
- Loose fibrous connective tissue, including adipose tissue
- Dense fibrous connective tissue (tendons and ligaments)
- Cartilage and bone (matrix for cartilage is more flexible than that for bone)
- Blood (matrix is a liquid called plasma)

Muscular Tissue Moves the Body

Muscular tissue is of three types: skeletal, cardiac, and smooth.
- Both skeletal muscle (attached to bones) and cardiac muscle (forms the heart wall) are striated.
- Both cardiac muscle (wall of heart) and smooth muscle (wall of internal organs) are involuntary.

Nervous Tissue Communicates

- Nervous tissue is composed of neurons and several types of neuroglia.
- Each neuron has dendrites, a cell body, and an axon. Axons conduct nerve impulses.

22.2 | Organs and Organ Systems

Organs make up organ systems, which are summarized in the table at the bottom of the page.

22.3 | Homeostasis

Homeostasis is the relative constancy of the internal environment. The internal environment consists of blood and tissue fluid. Due to exchange of nutrients and wastes with the blood, tissue fluid stays constant:

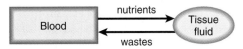

All organ systems contribute to the relative constancy of tissue fluid and blood.
- The cardiovascular system transports nutrients to cells and wastes from cells.
- The lymphatic system absorbs excess tissue fluid and functions in immunity.
- The digestive system takes in food and adds nutrients to the blood.
- The respiratory system carries out gas exchange with the external environment and the blood.
- The urinary system (i.e., the kidneys) removes metabolic wastes and regulates the pH and salt content of the blood.
- The nervous system and endocrine system regulate the other systems.

Negative Feedback

Negative feedback mechanisms keep the internal environment relatively stable. When a sensor detects a change above or below a set point, a control center brings about an effect that reverses the change, brings conditions back to normal again, and shuts off the response. Examples include:
- Regulation of blood glucose level by insulin.
- Regulation of room temperature by a thermostat and furnace.
- Regulation of body temperature by the brain and sweat glands.

ORGAN SYSTEMS

Transport	Sensory
Cardiovascular (heart and blood vessels) Lymphatic (lymphatic vessels)	Integumentary (skin)
Maintenance	**Motor**
Digestive (e.g., stomach, intestines) Respiratory (tubes and lungs) Urinary (tubes, urinary bladder, and kidneys)	Skeletal (bones and cartilage) Muscular (muscles)
Control	**Reproduction**
Nervous (brain, spinal cord, and nerves) Endocrine (glands, e.g., thyroid and adrenal glands)	Reproductive (tubes and testes in males; tubes, uterus, and ovaries in females)

Key Terms

blood 387
bone 387
cardiac muscle 388
cardiovascular system 390
cartilage 387
columnar epithelium 384
compact bone 387
connective tissue 386
cuboidal epithelium 384
dense fibrous connective tissue 386
digestive system 390
endocrine system 391
epithelial tissue 384
fibroblast 386
homeostasis 392
hyaline cartilage 387
immune system 390
integumentary system 391
intercalated disks 388
lacuna 387
ligament 387
loose fibrous connective tissue 386
lumen 384
lymphatic system 390

matrix 386
muscle tissue 388
muscular system 391
negative feedback 393
nerve 389
nervous system 391
nervous tissue 389
neuroglia 389
neuron 389
organ 382
platelet 387
red blood cell 387
reproductive system 391
respiratory system 390
skeletal muscle 388
skeletal system 391
smooth (visceral) muscle 388
squamous epithelium 384
striated 388
tendon 387
tissue 382
tissue fluid 390
urinary system 390
white blood cell 387

Testing Yourself

Choose the best answer for each question.

1. Label the levels of biological organization in the following illustration.

a.   b.   c.   d.   e.

2. Which of the following is not a feature of epidermis?
   a. contains keratin for strength
   b. stratified
   c. contains a matrix
   d. composed of epithelial cells

For questions 3–9, identify the tissue in the key that matches the description. Each answer may be used more than once. Each question may have more than one answer.

Key:

   a. epithelial tissue
   b. connective tissue
   c. muscular tissue
   d. nervous tissue

3. contains cells that are separated by a matrix
4. has a protective role
5. covers the body surface and forms linings of organs
6. composed of a group of cells with a similar function
7. best capable of replacing its cells
8. allows animals to respond to their environment
9. contains actin and myosin filaments
10. Label the blood smear in the following illustration.

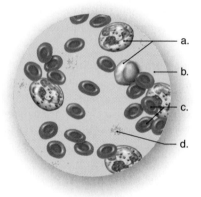

11. Which of the following systems does not add or remove substances from the blood?
    a. digestive   c. urinary
    b. cardiovascular   d. respiratory

12. Identify the type of muscle associated with each of the following illustrations.

a.   b.   c.

13. A major function of tissue fluid is to
    a. provide nutrients and oxygen to cells and remove wastes.
    b. keep cells warm.
    c. prevent cells from touching each other.
    d. provide flexibility by allowing cells to slide over each other.

14. Label the parts of a neuron in the following illustration.

a.

b.

c.

d.

e.

15. Which of the following is a function of skin?
    a. temperature regulation
    b. protection against water loss
    c. collection of sensory input
    d. protection from invading pathogens
    e. All of these are correct.

16. When a human being is cold, the blood vessels
    a. dilate, and the sweat glands are inactive.
    b. dilate, and the sweat glands are active.
    c. constrict, and the sweat glands are inactive.
    d. constrict, and the sweat glands are active.
    e. contract so that shivering occurs.

17. Blood is a(n) _____ tissue because it has a _____.
    a. connective, gap junction   c. epithelial, gap junction
    b. muscle, matrix   d. connective, matrix

18. The skeletal system functions in
    a. blood cell production.   c. movement.
    b. mineral storage.   d. All of these are correct.

19. The correct order for a negative feedback mechanism is
    a. sensory detection, control center, effect brings about change in environment.
    b. control center, sensory detection, effect brings about change in environment.
    c. sensory detection, control center, effect causes no change in environment.
    d. None of these are correct.

20. Which of the following is an example of negative feedback?
    a. Air conditioning goes off when room temperature lowers.
    b. Insulin decreases blood sugar levels after eating a meal.
    c. Heart rate increases when blood pressure drops.
    d. All of these are examples of negative feedback.

21. Which of these is not a type of epithelial tissue?
    a. cuboidal and columnar   d. pseudostratified
    b. bone and cartilage   e. All of these are epithelial
    c. squamous   tissue.

22. Which tissue is more apt to line a lumen?
    a. epithelial tissue   d. muscular tissue
    b. connective tissue   e. only smooth muscle
    c. nervous tissue

23. Tendons and ligaments
    a. are connective tissue.
    b. are associated with the bones.
    c. are found in vertebrates.
    d. contain collagen.
    e. All of these are correct.

24. Which of these components of blood fight infection?
    a. red blood cells   d. hydrogen ions
    b. white blood cells   e. All of these are correct.
    c. platelets

25. Which of these body systems contribute to homeostasis?
    a. digestive and urinary systems
    b. respiratory and nervous systems
    c. nervous and endocrine systems
    d. immune and cardiovascular systems
    e. All of these are correct.

26. Which type of epithelial tissue is composed of flat cells?
    a. squamous   c. columnar
    b. cuboidal   d. striated

27. When body temperature rises, sweat glands become _____ and blood vessels _____.
    a. active; constrict   c. active; dilate
    b. inactive; dilate   d. inactive; constrict

28. Which of these is a loose connective tissue?
    a. tendons   c. adipose tissue
    b. ligaments   d. All of these are correct

29. Cells working together form
    a. organs.   c. systems.
    b. tissues.   d. organisms.

30. Which of these is located in the abdominal cavity?
    a. heart
    b. digestive system
    c. much of the reproductive system
    d. Both b and c are correct.

31. Which of these is a connective tissue?
    a. bone   c. blood
    b. cartilage   d. All of these are correct.

32. Which of these is an example of homeostasis?
    a. Muscle tissue is specialized to contract.
    b. Normal body temperature is always about 37°C.
    c. There are more red blood cells than white blood cells.
    d. All of these are correct.

33. Which of these is true of humans?
    a. They have a vertebral column.
    b. The ventral cavity is divided into two cavities.
    c. The endocrine and nervous systems help coordinate the internal organs.
    d. Muscle and bone make up most of the body weight.
    e. All of these are true.

34. What type of epithelial cells are found in the epidermis?
    a. squamous
    c. cuboidal
    b. columnar
    d. Both *b* and *c* are correct.

35. Which of these types of muscles helps maintain posture?
    a. skeletal
    c. cardiac
    b. smooth
    d. involuntary

36. In a negative feedback control system,
    a. homeostasis is impossible.
    b. there is a constancy of the internal environment.
    c. there is a fluctuation about a mean.
    d. Both *a* and *b* are correct.

37. Which of these is a nerve cell?
    a. neuroglia
    c. cell body
    b. dendrite
    d. neuron

38. A muscle tissue that is both striated and involuntary is
    a. smooth.
    c. cardiac.
    b. skeletal.
    d. All of these are correct.

Thinking Scientifically

1. Patients with muscular dystrophy often develop heart disease. However, heart disease in these patients may not be detected until it has progressed past the point at which treatment options are effective. New research indicates that the link between muscular dystrophy and heart disease is so strong that muscular dystrophy patients should be screened for heart disease early, so that treatment may be begun. What do you suppose is the link between muscular dystrophy and heart disease?

2. According to a recent study, African Americans are 34% more likely to die from a liver transplant than are white Americans. What might be potential causes for this differential survival rate? How would you test your hypotheses?

 3. Not all animals have organ systems or even organs. What are the advantages to having specific tissues, organs, and organ systems that perform specific functions in the body? In other words, what might have been some of the selective pressures that led to the evolution of more complex organisms?

Bioethical Issue

Organ Insurance

Despite widespread efforts to convince people of the need for organ donors, supply continues to lag far behind demand. One proposed strategy to bring supply and demand into better balance is to develop an "insurance" program for organs. In this program, participants would pay the "premium" by promising to donate their organs at death. They, in turn, would receive priority for transplants as their "benefit." To avoid the problem of too many high-risk people applying for this type of insurance, a medical exam would be required so that only people with a normal risk of requiring a transplant would be accepted. Do you suppose this system would be an improvement over the current system in which organ donation is voluntary, with no tangible benefit? Can you think of any other strategy that would be more effective for increasing organ donations?

Essentials of Biology Website

The companion website for *Essentials of Biology* provides a wealth of information organized and integrated by chapter. You will find practice tests, animations, videos, and much more that will complement your learning and understanding of general biology.

http://www.mhhe.com/maderessentials2

Giving blood, which carries oxygen, can save a life.

An artificial pacemaker emits electrical signals that keep the heart contracting.

When we exercise, a red blood cell can move through the body in 20 seconds.

Our Heart's Pacemaker

THE BEAT OF the heart is under involuntary control, so we rarely consider the complex series of events that occurs when the heart moves blood to the lungs and the body tissues. Ordinarily, contraction of the heart is controlled by its natural pacemaker (referred to as the SA [sinoatrial] node), which automatically triggers the heart's rhythmic beating. At rest, the heart contracts about 70 times per minute, and each contraction moves the entire blood volume of approximately 4–6 liters through the system once every minute. The nervous system gets

Continued next page

23
The Transport Systems

Outline

involved when the body requires an increased amount of oxygen, as when you exercise vigorously. A particular nerve releases an adrenaline-like chemical in order to increase the activity of the SA node and therefore the heart rate. This action of the nervous system is mimicked whenever adrenaline is injected directly into the heart muscle, as is sometimes shown in the movies.

When damage occurs to the SA node, physicians often implant an artificial pacemaker. The heart responds to electrical signals generated by the SA node, and the artificial pacemaker emits electrical signals at the proper rate in order to duplicate the activity of a normal SA node. It is amazing to think that a simple electrical device can keep the entire cardiovascular system running.

The cardiovascular system is one of the body's transport systems. It must move blood to the tissues at a rate appropriate to meet the oxygen and nutrient demands of the cells. In this chapter, you will learn how the cardiovascular system works and how its functioning can be adjusted to meet the body's needs. Some of the consequences of a malfunctioning system are also discussed.

23.1 | Open and Closed Circulatory Systems

Most animals have a circulatory system to deliver oxygen and nutrients to cells and to take carbon dioxide and wastes away from cells. In some animals, the body plan makes a circulatory system unnecessary (**Fig. 23.1**). In a hydra, cells are either part of a single layer of external cells, or they line the gastrovascular cavity. In either case, each cell is exposed to water and can independently exchange gases and rid itself of wastes. The cells that line the gastrovascular cavity are specialized to carry out digestion. They pass nutrient molecules to other cells by diffusion. In a planarian, a trilobed gastrovascular cavity branches throughout the small, flattened body. No cell is very far from one of the digestive branches, so nutrient molecules can diffuse from cell to cell. Similarly, diffusion meets the respiratory and excretory needs of the cells.

Some other animals, such as nematodes and echinoderms, rely on movement of fluid within a coelom to circulate gases, nutrients, and wastes.

Open Circulatory Systems

More complex animals have a circulatory system in which a pumping **heart** moves fluid into blood vessels. The grasshopper is an example of an invertebrate animal that has an **open circulatory system** (**Fig. 23.2a**). A tubular heart pumps a fluid called hemolymph through a network of channels and cavities in the body. Collectively known as the **hemocoel,** these cavities or **sinuses** contain

Figure 23.1 **No circulatory systems.**

In hydras and planarians, each cell makes exchanges directly with the fluid in the gastrovascular cavity or the external environment. Therefore, there is no need for a circulatory system.

the animal's organs. Eventually, hemolymph (a combination of blood and tissue fluid) drains back to the heart. When the heart contracts, openings called ostia (sing., **ostium**) are closed; when the heart relaxes, the hemolymph is sucked back into the heart by way of the ostia. The hemolymph of a grasshopper is colorless because it does not contain hemoglobin or any other respiratory pigment that combines with and carries oxygen. Oxygen is taken to cells, and carbon dioxide is removed from them by way of air tubes, called tracheae, which are found throughout the body. The tracheae provide efficient transport and delivery of respiratory gases, while at the same time restricting water loss.

Closed Circulatory Systems

All vertebrates and some invertebrates have a **closed circulatory system** (Fig. 23.2*b*), which is more properly called a **cardiovascular system** because it consists of a strong muscular heart and blood vessels. In humans, the heart has two receiving chambers, called atria (sing., atrium), and two pumping chambers, called ventricles. There are three kinds of vessels: arteries, which carry blood away from the heart; capillaries, which exchange materials with tissue fluid; and veins, which return blood to the heart. Blood is always contained within these vessels and never runs freely into the body unless an injury occurs.

As blood passes through capillaries, the pressure of blood forces some water out of the blood and into the tissue fluid. Some of this fluid returns directly to a capillary, and some is picked up by lymphatic capillaries in the vicinity. The fluid, now called lymph, is returned to the cardiovascular system by lymphatic vessels. The function of the lymphatic system is discussed in Section 23.2.

Figure 23.2 **Open versus closed circulatory systems.**

a. The grasshopper has an open circulatory system. Hemolymph freely bathes the internal organs. The heart, a pump, keeps the hemolymph moving, but an open system probably could not rapidly supply oxygen to wing muscles. These muscles receive oxygen directly from tracheae (air tubes). **b.** Vertebrates and some invertebrates have a closed circulatory system. The heart pumps blood into the arteries, which take blood away from the heart to the capillaries where exchange takes place. Veins then return blood to the heart.

a. Open circulatory system

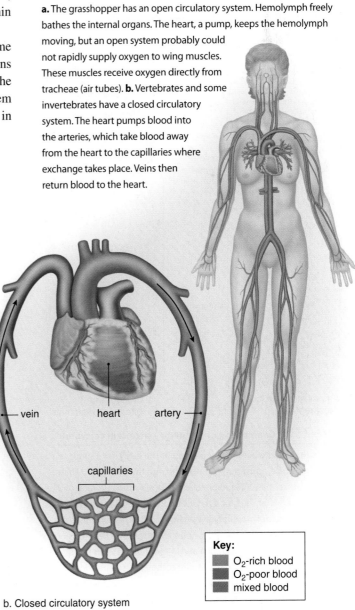

Key:
- O$_2$-rich blood
- O$_2$-poor blood
- mixed blood

b. Closed circulatory system

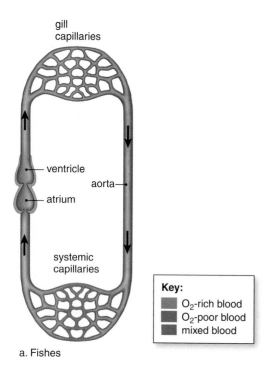

a. Fishes

Key:
- ⬛ O₂-rich blood
- ⬛ O₂-poor blood
- ⬛ mixed blood

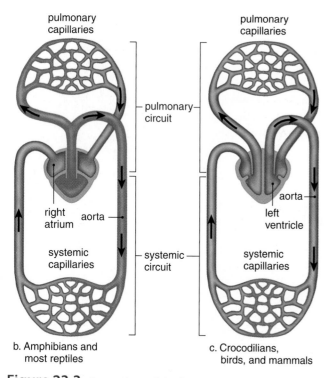

b. Amphibians and most reptiles

c. Crocodilians, birds, and mammals

Figure 23.3 Comparison of circulatory circuits in vertebrates.
a. In a fish, the blood moves in a single circuit. The heart has a single atrium and ventricle, and it pumps the blood into the gill region, where gas exchange takes place. Blood pressure created by the pumping of the heart is dissipated after the blood passes through the gill capillaries.
b. Amphibians and most reptiles have a two-circuit system in which the heart pumps blood to both the lungs and the body itself. There is a single ventricle, and some mixing of O₂-rich and O₂-poor blood takes place.
c. The pulmonary and systemic circuits are completely separate in crocodilians, birds, and mammals. The right side pumps blood to the lungs, and the left side pumps blood to the body proper.

Comparison of Vertebrate Circulatory Pathways

Two different types of circulatory pathways are seen among vertebrate animals. In fishes, blood follows a one-circuit (single-loop) pathway through the body. The heart has a single atrium and a single ventricle (**Fig. 23.3***a*). The pumping action of the ventricle sends blood under pressure to the gills, where gas exchange occurs. After passing through the gills, blood is under reduced pressure and flow. However, this single circulatory loop has advantages in that the gill capillaries receive O₂-poor blood and the systemic capillaries receive O₂-rich blood.

As a result of evolutionary changes, the other vertebrates have a two-circuit (double-loop) circulatory pathway. The heart pumps blood to the tissues through the **systemic circuit,** and also pumps blood to the lungs through the **pulmonary circuit.** This double pumping action is seen in terrestrial animals that utilize lungs to breathe air.

In amphibians, the heart has two atria, but only a single ventricle (Fig. 23.3*b*), and some mixing of O₂-rich and O₂-poor blood does occur. The same holds true for most reptiles, except that the ventricle has a partial septum, and so this mixing is reduced. The hearts of some reptiles (e.g., crocodilians and birds) and mammals are divided into right and left halves (Fig. 23.3*c*). The right ventricle pumps blood to the lungs, and the left ventricle, which is larger than the right ventricle, pumps blood to the rest of the body. This arrangement provides adequate blood pressure for both the pulmonary and systemic circuits.

Check Your **Progress**

1. Compare and contrast an open circulatory system with a closed circulatory system.
2. Contrast a one-circuit circulatory pathway with a two-circuit pathway.
3. Compare and contrast the amphibian heart with the mammalian heart.

Answers: 1. Both use a heart to pump fluid. An open system pumps hemolymph through channels and cavities. The hemolymph eventually drains back to the heart. A closed system pumps blood through vessels that carry blood both away from and back to the heart.   2. In fishes the one-circuit pathway utilizes a heart with one atrium and one ventricle to send blood to the gill capillaries and then to the systemic capillaries in a single loop. The two-circuit pathway pumps blood to both the pulmonary and systemic capillaries simultaneously.   3. The amphibian heart only has three chambers, one atrium and two ventricles. This causes the O₂-poor and O₂-rich blood to mix. The mammalian heart is four-chambered. The right side of the heart receives the O₂-poor blood and pumps it all to the lungs, where it picks up O₂. The left side receives O₂-rich blood and sends it out to the body.

23.2 | Transport in Humans

In the human cardiovascular system, like that of other vertebrates, the heart pumps blood into blood vessels that take it to capillaries where exchanges take place. In the lungs, carbon dioxide is exchanged for oxygen, and in the tissues, nutrients and oxygen are exchanged for carbon dioxide and other wastes. These exchanges in the lungs and tissues are so important that if the heart stops pumping, death results.

The Human Heart

In humans, the heart is a double pump: The right side of the heart pumps O_2-poor blood to the lungs, and the left side of the heart pumps O_2-rich blood to the tissues (**Fig. 23.4**). (The "right side of the heart" refers to how the heart is positioned in your body and not to the right side of the diagram.) The heart can be a double pump because a **septum** separates the right from the left side. Further, the septum is complete and prevents O_2-poor blood from mixing with O_2-rich blood.

Each side of the heart has two chambers. The upper, thin-walled chambers are called atria (sing., **atrium**), and they receive blood. The lower chambers are the thick-walled **ventricles,** which pump the blood away from the heart.

Valves occur between the atria and the ventricles, and between the ventricles and attached vessels. Because these valves close after the blood moves through, they keep the blood moving in the correct direction. The valves between the atria and ventricles are called the **atrioventricular valves,** and the valves between the ventricles and their attached vessels are called **semilunar valves** because their cusps look like half-moons.

This sequence traces the path of blood through the heart. The right atrium receives blood from attached veins (the **venae cavae**) that are returning O_2-poor blood to the heart from the tissues. After the blood passes through an atrioventricular valve (also called the *tricuspid valve* because of its three flaps), the right ventricle pumps it through the *pulmonary semilunar valve* into the **pulmonary trunk and arteries** that take it to the lungs. The **pulmonary veins** bring O_2-rich blood to the left atrium. After this blood passes through an atrioventricular valve (also called the *bicuspid valve*), the left ventricle pumps it through the *aortic semilunar valve* into the **aorta,** which takes it to the tissues.

O_2-poor blood is often associated with all veins and O_2-rich blood with all arteries, but this idea is incorrect: Pulmonary arteries and pulmonary veins are just the reverse. That is why the pulmonary arteries are colored blue and the pulmonary veins are colored red in Figure 23.4. The correct definitions are that an **artery** is a vessel that takes blood away from the heart, and a **vein** is a vessel that takes blood to the heart.

Like mechanical valves, the heart valves are sometimes leaky; they may not close properly, and there is a backflow of blood. A **heart murmur** is often due to leaky atrioventricular valves, which allow blood to pass back into the atria after the valves have closed. Rheumatic fever is a bacterial infection that begins in the throat and spreads throughout the body. The bacteria attack various organs, including the heart valves. When damage is severe, the valve can be replaced with a synthetic valve or one taken from a pig's heart.

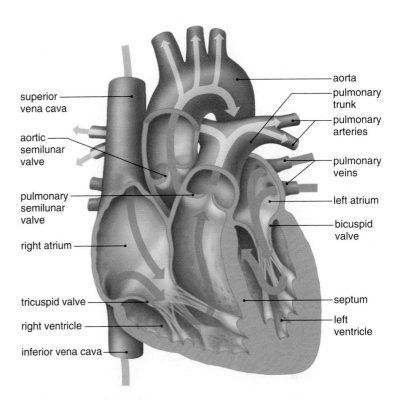

Figure 23.4   Heart anatomy and path of blood through the heart.

The right side of the heart receives and pumps O_2-poor blood to the lungs, and therefore is colored blue. The left side of the heart receives and pumps O_2-rich blood to tissues.

Time	Atria	Ventricles
0.15 sec	Systole	Diastole
0.30 sec	Diastole	Systole
0.40 sec	Diastole	Diastole

Phase 1: atrial systole

Phase 2: ventricular systole

Phase 3: atrial and ventricular diastole

Figure 23.5   The heartbeat.

A heartbeat is a cycle of events. **Phase 1:** The atria contract and pass blood to the ventricles. **Phase 2:** The ventricles contract and blood moves into the attached arteries. **Phase 3:** Both the atria and ventricles relax while the heart fills with blood.

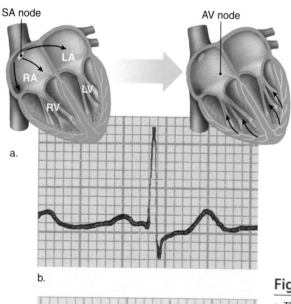

Figure 23.6   Control of the heartbeat.

a. The heart beats regularly because the SA node (called the pacemaker) automatically sends out an impulse that causes the atria (RA, LA) to contract and is picked up by the AV node. Thereafter, the ventricles (RV, LV) contract. **b.** An electrocardiogram records the electrical changes that occur as the heart beats. The large spike is associated with ventricular activation. **c.** Ventricular fibrillation produces an irregular ECG.

The Heartbeat

The heart's pumping action, known as the heartbeat or **cardiac cycle,** consists of a series of events: First the atria contract, then the ventricles contract, and then they both rest. **Figure 23.5** lists and depicts the events of a heartbeat, using the term **systole** to mean contraction and **diastole** to mean relaxation. When the heart beats, the familiar *lub-dub* sound is caused by the closing of the heart valves. The longer and lower-pitched *lub* occurs when the atrioventricular valves close, and the shorter and sharper *dub* is heard when the semilunar valves close. The **pulse** is a wave effect that passes down the walls of the arterial blood vessels when the aorta expands and then recoils following the ventricular systole. Because there is one pulse per ventricular systole, the pulse rate can be used to determine the heart rate. The heart beats about 70 times per minute, although a normal adult heart rate can vary from 60 to 100 beats per minute.

The beat of the heart is regular because it has an intrinsic pacemaker, called the **SA (sinoatrial) node.** The nodal tissue of the heart, located in two regions of the atrial wall, is a unique type of cardiac muscle tissue. Every 0.85 seconds, the SA node automatically sends out an excitation impulse that causes the atria to contract (**Fig. 23.6***a*). When this impulse is picked up by the **AV (atrioventricular) node,** it passes to fibers that cause the ventricles to contract. If the SA node fails to work properly, the ventricles still beat, due to impulses generated by the AV node, but the beat is slower (40–60 beats per minute). To correct this condition, it is possible to implant an artificial pacemaker, which automatically gives an electrical stimulus to the heart every 0.85 seconds. Pacemakers are available today that can adjust to the person's level of physical activity. Activity sensors in these self-adjusting pacemakers generate variable electrical signals, depending on the level of activity of the person. The pacemakers then change their output, depending on these signals.

Although the beat of the heart is intrinsic, it is regulated by the nervous system and various hormones. Activities such as yoga and meditation lead to activation of the vagus nerve, which slows the heart rate. Exercise or anxiety leads to release of the hormones norepinephrine and epinephrine by the adrenal glands, which cause it to speed up.

An **electrocardiogram (ECG)** is a recording of the electrical changes that occur in the wall of the heart during a cardiac cycle. Body fluids contain ions that conduct electrical currents, and therefore the electrical changes in heart muscle can be detected on the skin's surface. When an electrocardiogram is being taken, electrodes placed on the skin are connected by wires to an instrument that detects these electrical changes (Fig. 23.6*b*). Various types of abnormalities can be detected by an electrocardiogram. One of these, called ventricular fibrillation, is due to uncoordinated contraction of the ventricles (Fig. 23.6*c*). Ventricular fibrillation is of special interest because it can be caused by an injury or a drug overdose. It is the most common cause of sudden cardiac death in a seemingly healthy person. To stop fibrillation, a defibrillator can be used to apply a strong electrical current for a short period. Then, the SA node may be able to reestablish a coordinated beat. Easy to use defibrillators are becoming increasingly available in public places such as airports and college campuses.

Figure 23.7 Blood vessels.

a. Arteries have well-developed walls with a thick middle layer of elastic fibers and smooth muscle. **b.** Capillary walls are composed of an epithelium only one cell thick. **c.** Veins have flabby walls, particularly because the middle layer is not as thick as in arteries.

a. Artery

b. Capillary

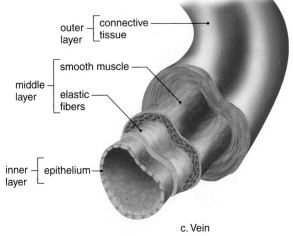

c. Vein

Blood Vessels

Arteries transport blood away from the heart. When the heart contracts, blood is sent under pressure into the arteries; thus, **blood pressure** accounts for the flow of blood in the arteries. Arteries have a much thicker wall than veins because of a well-developed middle layer composed of smooth muscle and elastic fibers. The elastic fibers allow arteries to expand and accommodate the sudden increase in blood volume that results after each heartbeat. The smooth muscle strengthens the wall and prevents overexpansion (**Fig. 23.7***a*).

Arteries branch into arterioles. **Arterioles** are small arteries just visible to the naked eye, and their diameter can be regulated by the nervous system, depending on the needs of the body. When arterioles are dilated, more blood flows through them, and when they are constricted, less blood flows. The constriction of arterioles can also raise blood pressure. Arterioles branch into **capillaries,** which are extremely narrow, microscopic tubes with a wall composed of only epithelium, often called endothelium (Fig. 23.7*b*). Capillaries, which are usually so narrow that red blood cells pass through in single file, allow exchange of nutrient and waste molecules across their thin walls. Capillary beds (many interconnected capillaries) are so widespread that in humans, all cells are less than a millimeter from a capillary. Because the number of capillaries is so extensive, blood pressure drops and blood flows slowly along. The slow movement of blood through the capillaries also facilitates efficient exchange of substances between the blood and tissue fluid. The entrance to a capillary bed is controlled by bands of muscle called precapillary sphincters. During muscular exercise, sphincters relax, and the capillary beds of the muscles are open. Also, after an animal has eaten, the capillary beds in the digestive tract are open. Otherwise, blood moves through a shunt that takes the blood from arteriole to venule (Fig. 23.9*b*).

Venules collect blood from capillary beds and join as they deliver blood to veins. Veins carry blood back to the heart. Blood pressure is much reduced by the time blood reaches the veins. The walls of veins are much thinner and their diameter wider than those of arteries (Fig. 23.7*c*). This allows skeletal muscle contraction to push on the veins, forcing the blood past a valve (**Fig. 23.8**). Valves within the veins point, or open, toward the heart, preventing a backflow of blood when they close. When inhalation occurs and the chest expands, the thoracic pressure falls and abdominal pressure rises. This action also aids the flow of venous blood back to the heart because blood flows in the direction of reduced pressure.

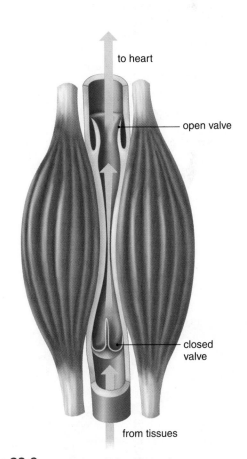

to heart

open valve

closed valve

from tissues

Figure 23.8 Movement of blood in a vein.

Veins have valves, which point toward the heart. Pressure on the walls of a vein, exerted by skeletal muscles, increases blood pressure within the vein and forces a valve open. When the muscles relax, gravity pulls the blood downward, closing the valves and preventing the blood from flowing in the opposite direction.

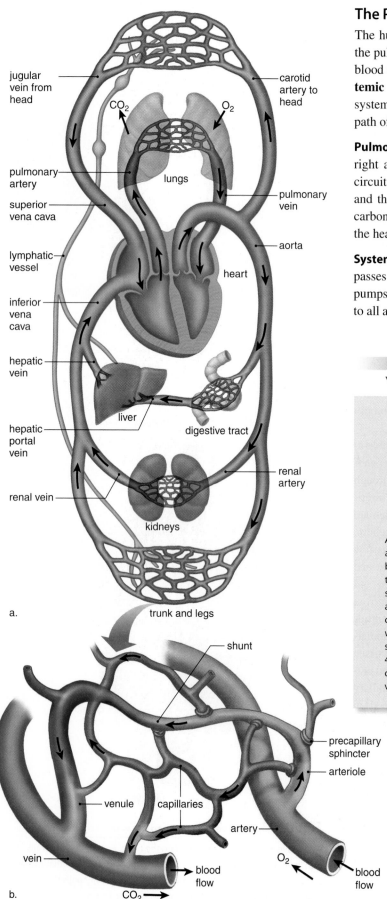

jugular
vein from
head

carotid
artery to
head

CO_2 O_2

pulmonary
artery

lungs

superior
vena cava

pulmonary
vein

lymphatic
vessel

aorta

heart

inferior
vena
cava

hepatic
vein

liver

hepatic
portal
vein

digestive tract

renal
artery

renal vein

kidneys

a.

trunk and legs

shunt

precapillary
sphincter

arteriole

venule   capillaries

artery

vein

blood
flow

O_2

blood
flow

CO_2

b.

The Pulmonary and Systemic Circuits

The human cardiovascular system includes two major circulatory pathways, the pulmonary circuit and the systemic circuit. The **pulmonary circuit** moves blood to and from the lungs, where O_2-poor becomes O_2-rich blood. The **systemic circuit** moves blood to and from the body proper. The function of the systemic circuit is to serve the needs of the body's cells. **Figure 23.9** traces the path of blood in both circuits.

Pulmonary Circuit O_2-poor blood from all regions of the body collects in the right atrium and then passes into the right ventricle (RV). The pulmonary circuit begins when the RV pumps blood to the lungs via the pulmonary trunk and the pulmonary arteries. As blood passes through pulmonary capillaries, carbon dioxide is given off and oxygen is picked up. O_2-rich blood returns to the heart via the pulmonary veins. Pulmonary veins enter the left atrium.

Systemic Circuit O_2-rich blood enters the left atrium from the lungs and passes into the left ventricle (LV). The systemic circuit begins when the LV pumps the blood into the **aorta.** Arteries branching from the aorta carry blood to all areas and organs of the body, where it passes through capillaries and col-

Check Your **Progress**

1. How do atria and ventricles differ in structure and function?
2. Trace the path of blood through the heart from the venae cavae to the aorta.
3. What are the events, including the valves, of a heartbeat, and what makes the familiar *lub-dub* sounds?
4. Outline the flow of blood through blood vessels away from and back to the heart.

Answers: 1. Atria, which are thin-walled, receive blood from the lungs (left atrium) and from the rest of the body (right atrium). Ventricles, which are thick-walled, pump blood to the lungs (right ventricle) and the rest of the body (left ventricle).   2. From the body: venae cavae, right atrium, tricuspid valve, right ventricle, pulmonary semilunar valve, pulmonary trunk and arteries. From the lungs: pulmonary veins, left atrium, bicuspid valve, left ventricle, aortic semilunar valve, aorta.   3. First the atria contract, then the ventricles contract, and then they both rest. The *lub* sound occurs when the atrioventricular valves close when the ventricles contract, and the *dub* sound occurs when the semilunar valves close as the heart begins to relax.
4. Arteries carry blood away from the heart to arterioles. Arterioles branch to form capillaries. Venules collect blood from capillaries and join as they deliver blood to veins. Veins carry blood back to the heart.

Figure 23.9   Path of blood.

a. Overview of cardiovascular system. When tracing blood from the right to the left side of the heart in the pulmonary circuit, you must include the pulmonary vessels. When tracing blood from the digestive tract to the right atrium in the systemic circuit, you must include the hepatic portal vein, the hepatic vein, and the inferior vena cava. **b.** To move from an artery to a vein, blood must move through a capillary bed where exchange occurs between blood and tissue fluid. When precapillary sphincters shut down a capillary bed, blood moves through a shunt from arteriole to venule.

lects in veins. Veins converge on the **venae cavae,** which return the O_2-poor blood to the right atrium. In the systemic circuit, arteries contain O_2-rich blood and have a bright red color, and veins contain O_2-poor blood and appear dull red or, when viewed through the skin, blue.

A **portal system** begins and ends in capillaries. For example, the hepatic portal takes blood from the intestines to the liver. The liver, an organ of homeostasis, modifies substances absorbed by the intestines and monitors the normal composition of the blood. The hepatic vein leaves the liver and enters the inferior vena cava (see Fig. 24.27).

Lymphatic System

The **lymphatic system** consists of lymphatic vessels and various lymphatic organs. Just now we are interested in the lymphatic vessels that take up excess tissue fluid and return it to cardiovascular veins in the shoulders, namely the subclavian veins (**Fig. 23.10**). The lymphatic vessels also take up fat in the form of lipoproteins at the digestive tract and transport it to these veins. Refer to Figure 23.9a to see how the lymphatic vessels fit into the organization of the cardiovascular system. As you will see in Chapter 26, the lymphatic system works with the immune system to help defend the body against disease.

Lymphatic vessels are quite extensive; most regions of the body are richly supplied with lymphatic capillaries. The construction of the larger lymphatic vessels is similar to that of cardiovascular veins, including the presence of valves. Also, the movement of lymph within these vessels is dependent upon skeletal muscle contraction. When the muscles contract, the lymph is squeezed past a valve that closes, preventing the lymph from flowing backward.

The lymphatic system is a one-way system that begins at the lymphatic capillaries. These capillaries take up fluid that has diffused from and not been reabsorbed by the blood capillaries. Once tissue fluid enters the lymphatic vessels, it is called **lymph.** The lymphatic capillaries join to form larger lymphatic vessels that merge before entering one of two ducts: the thoracic duct or the right lymphatic duct. The thoracic duct is much larger than the right lymphatic duct. It serves the lower limbs, the abdomen, the left arm, and the left side of both the head and the neck. The right lymphatic duct serves the right arm, the right side of both the head and the neck, and the right thoracic area. The lymphatic ducts enter the subclavian veins.

Cardiovascular Disorders

Cardiovascular disease is the leading cause of untimely death in Western countries. In the United States, it is estimated that about 20% of the population suffers from **hypertension,** which is high blood pressure. Normal blood pressure is 120/80 mm Hg. The top number is called *systolic* blood pressure because it is due to the contraction of the ventricles, and the bottom number is called *diastolic* blood pressure because it is measured during the relaxation of the ventricles. Hypertension occurs when blood pressure readings are higher than these numbers—say, 160/100. Hypertension is sometimes called a silent killer because it may not be detected until a stroke or heart attack occurs.

Inheritance and lifestyle contribute to hypertension. For example, hypertension is often seen in individuals who have atherosclerosis (formerly called arteriosclerosis), an accumulation of soft masses of fatty materials, particularly

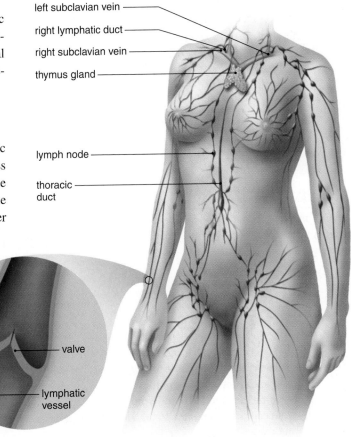

left subclavian vein
right lymphatic duct
right subclavian vein
thymus gland
lymph node
thoracic duct
valve
lymphatic vessel

Figure 23.10   Lymphatic vessels.

Lymphatic vessels drain excess fluid from the tissues and return it to the cardiovascular system. The enlargement shows that lymphatic vessels, like cardiovascular veins, have valves to prevent backward flow. Lymph nodes filter lymph and remove impurities.

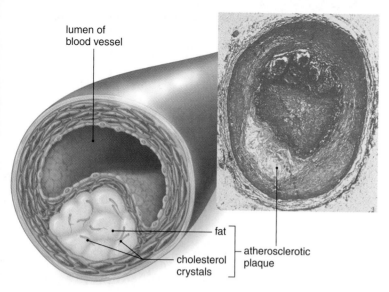

lumen of
blood vessel

fat

cholesterol
crystals

atherosclerotic
plaque

Figure 23.11 Plaque.

Plaque is an irregular accumulation of cholesterol and fat in the wall of an artery that closes off blood vessels. Of special concern is the closure of coronary arteries.

Check Your Progress

1. The blockage of which vessels in the systemic circuit lead to a heart attack? Describe the two surgical procedures used to correct blockage of these arteries.
2. Describe the function of the lymphatic system.
3. What conditions might occur as a result of hypertension and plaque?

Answers: 1. The coronary arteries that serve the heart. In balloon angioplasty, a balloon is inserted into the artery and inflated to break the blockage. In a coronary bypass operation, a piece of a blood vessel from another part of the body is attached to the artery so that blood flows through the new section rather than through the blocked area. 2. The lymphatic vessels return excess tissue fluid from the body and also fat from the digestive tract to cardiovascular veins. 3. Thromboembolism, stroke, heart attack.

cholesterol, beneath the inner linings of arteries (**Fig. 23.11**). Such deposits, called **plaque,** tend to protrude into the lumen of the vessel, interfering with the flow of blood. Atherosclerosis begins in early adulthood and develops progressively through middle age, but symptoms may not appear until an individual is 50 or older. To prevent its onset and development, the American Heart Association and other organizations recommend a diet low in saturated fat and cholesterol and rich in fruits and vegetables. Smoking, drug or alcohol abuse, obesity, and lack of exercise contribute to the risk of atherosclerosis.

Plaque can cause a clot to form on the irregular arterial wall. As long as the clot remains stationary, it is called a *thrombus,* but when and if it dislodges and moves along with the blood, it is called an *embolus.* If *thromboembolism* is not treated, serious health problems can result. A cardiovascular accident, also called a **stroke,** often occurs when a small cranial arteriole bursts or is blocked by an embolus. Lack of oxygen causes a portion of the brain to die, and paralysis or death can result. A person is sometimes forewarned of a stroke by a feeling of numbness in the hands or the face, difficulty in speaking, or temporary blindness in one eye. If a coronary artery becomes completely blocked due to thromboembolism, a heart attack can result.

The coronary arteries bring O_2-rich blood from the aorta to capillaries in the wall of the heart, and the cardiac veins return O_2-poor blood from the capillaries to the right ventricle. If the coronary arteries narrow due to cardiovascular disease, the individual may first suffer from angina pectoris, chest pain that is often accompanied by a radiating pain in the left arm. When a coronary artery is completely blocked, a portion of the heart muscle dies due to a lack of oxygen. This is known as a **heart attack.** Two surgical procedures are frequently performed to correct a blockage or facilitate blood flow. In a coronary bypass operation, a portion of a blood vessel from another part of the body is sutured from the aorta to the coronary artery, past the point of obstruction (**Fig. 23.12**a). Now blood flows normally again from the aorta to the wall of the heart. In balloon angioplasty, a plastic tube is threaded through an artery to the blockage, and a balloon attached to the end of the tube is inflated to break through the blockage. A stent is often used to keep the vessel open (Fig. 23.12b).

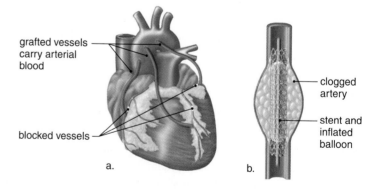

grafted vessels carry arterial blood

blocked vessels

clogged artery

stent and inflated balloon

a.

b.

Figure 23.12 Treatment for clogged coronary arteries.

(*left*) Many heart procedures, such as coronary bypass and stent insertion, can be performed using robotic surgery techniques. **a.** During a coronary bypass operation, blood vessels (usually veins from the leg) are stitched to the heart, taking blood past the region of obstruction. **b.** During stenting, a cylinder of expandable metal mesh is positioned inside the coronary artery by using a catheter. Then, a balloon is inflated so that the stent expands and opens the artery.

TV screen shows operation in progress.

Doctor manipulates instruments.

Moveable arms hold instruments.

23.3 | Blood: A Transport Medium

The **blood** of mammals helps maintain homeostasis. Blood's numerous functions include the following:

1. Transports substances to and from the capillaries, where exchange with tissue fluid take place.
2. Helps defend the body against invasion by pathogens (e.g., disease-causing viruses and bacteria).
3. Helps regulate body temperature.
4. Forms clots, preventing a potentially life-threatening loss of blood.

In humans, blood has two main portions: the liquid portion, called plasma, and the formed elements, consisting of various cells and platelets (**Fig. 23.13**).

Plasma

Plasma is composed mostly of water (90–92%) and proteins (7–8%), but it also contains smaller quantities of many types of molecules, including nutrients, wastes, and salts. The salts and proteins are involved in buffering the blood, effectively keeping the pH near 7.4, slightly basic. They also maintain the blood's osmotic pressure so that water has an automatic tendency to enter blood capillaries. Several plasma proteins are involved in blood clotting, and others transport large organic molecules in the blood. Albumin, the most plentiful of the plasma proteins, transports bilirubin, a breakdown product of hemoglobin. Lipoproteins transport cholesterol.

Formed Elements

The **formed elements** are red blood cells, white blood cells, and platelets. Among the formed elements, **red blood cells,** also called erythrocytes, transport oxygen. Red blood cells are small, biconcave disks that at maturity lack a nucleus and contain the respiratory pigment hemoglobin (**Fig. 23.14**). There are 6 million red blood cells in a small drop of whole blood, and each one of these cells contains about 250 million hemoglobin molecules. **Hemoglobin** contains iron, which combines loosely with oxygen; in this way, red blood cells transport oxygen. If the number of red blood cells is insufficient, or if the cells do not have enough hemoglobin, the individual suffers from anemia and has a tired, run-down feeling. At high altitudes, where oxygen levels are lower, red blood cell formation is stimulated. If an individual lives and works in a high altitude long enough they will have a greater number of red blood cells.

Red blood cells are manufactured continuously within certain bones, namely the skull, the ribs, the vertebrae, and the ends of the long bones. The hormone *erythropoietin* stimulates the production of red blood cells. The kidneys produce erythropoietin when they act on a precursor made by the liver. Now available as a drug, erythropoietin is helpful to persons with anemia and has also been abused by athletes to enhance their performance.

Before they are released from the bone marrow into the blood, red blood cells synthesize hemoglobin and lose their nuclei. After living about 120 days, they are destroyed chiefly in the liver and the spleen, where they are engulfed by large phagocytic cells. When red blood cells are destroyed, hemoglobin is released. The iron is recovered and returned to the red bone marrow for reuse.

Blood is spinning in a centrifuge.

plasma
(55% of whole blood)

leukocytes and platelets
(<1% of whole blood)

erythrocytes
(45% of whole blood)

formed elements

Figure 23.13 **Components of blood.**

After blood is centrifuged and settling occurs, it is apparent that blood is composed of plasma and formed elements.

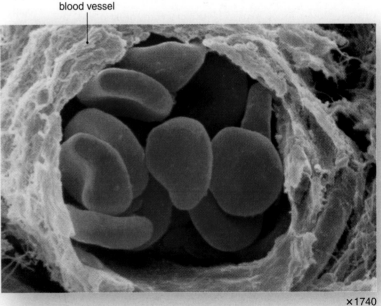

blood vessel

×1740

Figure 23.14 **Micrograph of red blood cells.**

Red blood cells contain hemoglobin, a red pigment that accounts for the red color of blood.

Formed Elements	Function
Red Blood Cells (erythrocytes) 4 million–6 million per mm³ blood	Transport O_2 and help transport CO_2.
White Blood Cells (leukocytes) 5,000–11,000 per mm³ blood	Fight infection.
Granular leukocytes • Neutrophils 40–70%	Phagocytize pathogens and cellular debris.
• Eosinophils 1–4%	Use granule contents to digest large pathogens, such as worms, and reduce inflammation.
• Basophils 0–1%	Promote blood flow to injured tissues and the inflammatory response.
Agranular leukocytes • Lymphocytes 20–45%	Responsible for specific immunity; B cells produce antibodies; T cells destroy cancer and virus-infected cells.
• Monocytes 4–8%	Become macrophages that phagocytize pathogens and cellular debris.
Platelets (thrombocytes) 150,000–300,000 per mm³ blood	Aid clotting.

• Appearance with Wright's stain.

Figure 23.15   Formed elements.

White blood cells are quite varied and have different functions that are all associated with defense of the body against infections.

Another portion of the molecules (i.e., heme) undergoes chemical degradation and is excreted by the liver as bile pigments in the bile. The bile pigments are primarily responsible for the color of feces.

White blood cells, also called leukocytes, help fight infections. White blood cells differ from red blood cells in that they are usually larger and have a nucleus, they lack hemoglobin, and without staining, they appear translucent. **Figure 23.15** shows the appearance of the various types of white blood cells. With staining, white blood cells appear light blue unless they have granules that bind with certain stains. The white blood cells that have granules also have a lobed nucleus. The agranular leukocytes have no granules and a spherical or indented nucleus. There are approximately 5,000–11,000 white blood cells in a small drop of whole blood. Growth factors are available to increase the production of all white blood cells, and these are helpful to people with low immunity, such as AIDS patients.

Red blood cells are confined to the blood, but white blood cells are able to squeeze between the cells of a capillary wall. Therefore, they are found in tissue fluid and lymph and in lymphatic organs. When an infection is present, white blood cells greatly increase in number. Many white blood cells live only a few days—they probably die while engaging pathogens. Others live months or even years.

When microorganisms enter the body due to an injury, the body's response is called an *inflammatory response* because swelling, reddening, heat, and pain occur at the injured site. Damaged tissue releases kinins, which dilate capillaries, and histamines, which increase capillary permeability. White blood cells called **neutrophils,** which are amoeboid, squeeze through the capillary wall and enter the tissue fluid, where they phagocytize foreign material. White blood cells called **monocytes** appear and are transformed into **macrophages,** large phagocytizing cells that release white blood cell growth factors. Soon the number of white blood cells increases explosively. A thick, yellowish fluid called pus contains a large proportion of dead white blood cells that have fought the infection.

Lymphocytes, another type of white blood cell, also play an important role in fighting infection. Certain lymphocytes called T cells attack infected cells that contain viruses. Other lymphocytes, called B cells, produce antibodies. Each B cell produces just one type of antibody, which is specific for one type of antigen. An **antigen,** which is most often a protein but sometimes a polysaccharide, causes the body to produce an **antibody** because the antigen does not belong to the body. Antigens are present in the outer covering of parasites or in their toxins. When antibodies combine with antigens, the complex is often phagocytized by a macrophage. An individual is actively immune when a large number of B cells are all producing the antibody needed for a particular infection.

Platelets and Blood Clotting

Platelets (also called thrombocytes) result from fragmentation of certain large cells, called megakaryocytes, in the red bone marrow. Platelets are produced at a rate of 200 billion a day, and a small drop of whole blood contains 150,000–300,000. These formed elements are involved in blood clotting, or coagulation.

Blood contains at least 12 clotting factors that participate in the formation of a blood clot. *Hemophilia* is an inherited clotting disorder in which the

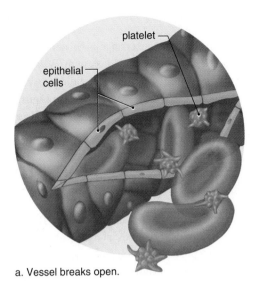

epithelial
cells

platelet

a. Vessel breaks open.

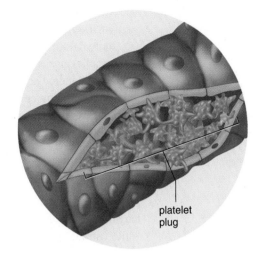

platelet
plug

b. Platelet plug forms.

Figure 23.16 Blood clotting.

a. When a capillary is injured, blood begins leaking out. **b.** Platelets congregate to form a platelet plug, and this temporarily seals the leak. **c.** Platelets and damaged tissue cells release an activator that sets in motion a series of reactions that end with (**d**) a blood clot.

Platelets and damaged
tissue cells release
prothrombin activator

Prothrombin $\xrightarrow{Ca^{2+}}$ Thrombin

Fibrinogen $\xrightarrow{Ca^{2+}}$ Fibrin threads
(Red blood cells
are trapped among
fibrin threads.)

c. Clotting occurs.

liver is unable to produce one of the clotting factors. The slightest bump can cause the affected person to bleed into the joints, and this leads to degeneration of the joints. Bleeding into muscles can lead to nerve damage and muscular atrophy. The most frequent cause of death due to hemophilia is bleeding into the brain.

Prothrombin and fibrinogen, two proteins involved in blood clotting, are manufactured and deposited in blood by the liver. Vitamin K, found in green vegetables and also formed by intestinal bacteria, is necessary for the production of prothrombin, and if by chance this vitamin is missing from the diet, hemorrhagic disorders can develop.

A series of reactions leads to the formation of a blood clot (**Fig. 23.16***a–c*). When a blood vessel in the body is damaged, platelets clump at the site of the puncture and form a plug that temporarily seals the leak. Platelets and the injured tissues release a clotting factor called prothrombin activator that converts prothrombin to thrombin. This reaction requires calcium ions (Ca^{2+}). **Thrombin,** in turn, acts as an enzyme that severs two short amino acid chains from each fibrinogen molecule. These activated fragments then join end to end, forming long threads of fibrin. Fibrin threads wind around the platelet plug in the damaged area of the blood vessel and provide the framework for the clot. Red blood cells also are trapped within the fibrin threads; these cells make a clot appear red (Fig. 23.16*d*). Clot retraction follows, during which the clot gets smaller as platelets contract. A fluid called serum is squeezed from the clot. A fibrin clot is present only temporarily. As soon as blood vessel repair is initiated, an enzyme called plasmin destroys the fibrin network and restores the fluidity of plasma.

fibrin
threads

d. Blood clot is present.   ×1430

Tissue fluid

Arterial end

tissue cell

salt

plasma protein

water ←

← oxygen

amino acids ←

← glucose

wastes →

← carbon dioxide

← water

Venous end

Figure 23.17 **Capillary exchange.**

A capillary, illustrating the exchanges that take place and the forces that aid the process. At the arterial end of a capillary (*top*), the blood pressure is higher than the osmotic pressure; therefore, water tends to leave the bloodstream. In the midsection, molecules, including oxygen and carbon dioxide, follow their concentration gradients. At the venous end of a capillary (*bottom*), the osmotic pressure is higher than the blood pressure; therefore, water tends to enter the bloodstream. Notice that the red blood cells and the plasma proteins are too large to exit a capillary.

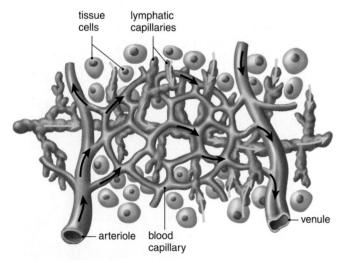

tissue cells

lymphatic capillaries

venule

arteriole

blood capillary

Figure 23.18 **Lymphatic capillary bed.**

A lymphatic capillary bed lies near a blood capillary bed. The heavy black arrows show the flow of blood. The yellow arrows show that lymph is formed when lymphatic capillaries take up excess tissue fluid.

Capillary Exchange in the Tissues

In Figure 23.1, we noted that in some animals, exchange is carried out by each cell individually because there is no cardiovascular system. When an animal does have a cardiovascular system, tissue fluid (the fluid between the cells) makes exchanges with blood within a capillary. Notice in **Figure 23.17** that amino acids, oxygen, and glucose exit a capillary and enter tissue fluid to be used by cells. On the other hand, carbon dioxide and wastes exit tissue fluid and enter a capillary to be taken away and excreted from the body. A chief purpose of the cardiovascular system is to take blood to the capillaries where exchange occurs. Without this exchange, homeostasis is not maintained, and the cells of the body perish.

Figure 23.17 illustrates certain mechanics of capillary exchange. Blood pressure and osmotic pressure are two opposing forces at work along the length of a capillary. Blood pressure is caused by the beating of the heart, while osmotic pressure is due to the salt and protein content of the blood. Blood pressure holds sway at the arterial end of a capillary and water exits. Blood pressure is reduced by the time blood reaches the venous end of a capillary, and osmotic pressure now causes water to enter. Midway between the arterial and venous ends of a capillary, blood pressure pretty much equals osmotic pressure, and passive diffusion alone causes nutrients to exit and wastes to enter. This works because tissue fluid always has the lesser amount of nutrients and the greater amount of wastes. After all, cells use nutrients and thereby create wastes.

The exchange of water at a capillary is not exact, and the result is always excess tissue fluid. Excess tissue fluid is collected by lymphatic capillaries, and in this way it becomes lymph (**Fig. 23.18**). Lymph contains all the components of plasma except much lesser amounts of protein. Lymph is returned to the cardiovascular system when the major lymphatic vessels enter the subclavian veins in the shoulder region (see Fig. 23.10).

In addition to nutrients and wastes, blood distributes heat to body parts. When you are warm, many capillaries that serve the skin are open, and your face is flushed. This helps rid the body of excess heat. When you are cold, skin capillaries close, conserving heat.

■ Check
Your **Progress**

1. List the functions of blood.
2. Contrast red blood cells with white blood cells.
3. Describe how a blood clot forms.

Answers: 1. Blood transports substances to and from the capillaries, defends against pathogen invasion, helps regulate body temperature, and forms clots to prevent excessive blood loss.   2. Red blood cells are smaller, lack a nucleus, contain hemoglobin, and are red in color. White blood cells are larger, have a nucleus, do not contain hemoglobin, and are translucent in appearance.   3. Platelets accumulate at the site of injury and release a clotting factor that results in the synthesis of thrombin. Thrombin synthesizes fibrin threads that provide a framework for the clot.

The Chapter in Review

Summary

23.1 | Open and Closed Circulatory Systems

- Some invertebrates do not have a transport system because their body plan allows each cell to exchange molecules with the external environment.
- Other invertebrates do have a transport system. Some have an open circulatory system, and some have a closed circulatory system.
- All vertebrates have a closed circulatory system in which arteries carry blood away from the heart to capillaries, where exchange takes place, and veins carry blood to the heart.

Comparison of Vertebrate Circulatory Pathways

- Fishes have a one-circuit pathway because the heart, with a single atrium and ventricle, pumps blood only to the gills.
- Other vertebrates have both pulmonary and systemic circuits. Amphibians have two atria but a single ventricle. Crocodilians, birds, and mammals, including humans, have a heart with two atria and two ventricles, in which O_2-rich blood is always separate from O_2-poor blood.

23.2 | Transport in Humans

The cardiovascular system consists of the heart and the blood vessels.

The Human Heart

The heart has a right and a left side. Each side has an atrium and a ventricle. Valves keep the blood moving in the correct direction.

Right side of the heart: The atrium receives O_2-poor blood from the tissues, and the ventricle pumps it to the lungs.

Left side of the heart: The atrium receives O_2-rich blood from the lungs, and the ventricle pumps it to the tissues.

Heartbeat: During a heartbeat, first the atria contract and then the ventricles contract. The heart sounds, *lub-dub*, are caused by the closing of valves.

SA node: The SA node (pacemaker) causes the two atria to contract. The SA node also stimulates the AV node.

AV node: The AV node causes the two ventricles to contract.

Blood Vessels

Arteries, with thick walls, take blood away from the heart to arterioles, which take blood to capillaries that have walls composed only of epithelial cells. Venules take blood from capillaries and merge to form veins, which have thinner walls than arteries have.

- Blood pressure, created by the beat of the heart, accounts for the flow of blood in the arteries.
- Skeletal muscle contraction is largely responsible for the flow of blood in the veins, which have valves preventing backward flow.

Blood Has Two Circuits

- In the pulmonary circuit, blood travels to and from the lungs.
- In the systemic circuit, the aorta divides into blood vessels that serve the body's cells. Venae cavae return O_2-poor blood to the heart.

Lymphatic System

The lymphatic system is a one-way system that consists of lymphatic vessels and lymphatic organs. The lymphatic vessels receive fat at the digestive tract and excess tissue fluid at blood capillaries, and carry these to the subclavian veins (cardiovascular veins in the shoulders).

Cardiovascular Disorders

Hypertension and atherosclerosis are two conditions that lead to heart attack and stroke. Following a heart-healthy diet, getting regular exercise, maintaining a proper weight, and not smoking cigarettes are protective against these conditions.

23.3 | Blood: A Transport Medium

Blood has two main parts: plasma and formed elements.

Plasma

Plasma contains mostly water (90–92%) and proteins (7–8%), but also nutrients, wastes, and salts.

- The proteins and salts buffer the blood and maintain its osmotic pressure.
- The proteins also have specific functions, such as participating in blood clotting and transporting molecules.

Formed Elements

The red blood cells contain hemoglobin and function in oxygen transport. Defense against disease depends on the various types of white blood cells:

Red blood cells

- Neutrophils and monocytes are phagocytic and are especially responsible for the inflammatory response.
- Lymphocytes are involved in the development of specific immunity to disease.

White blood cells

The platelets and two plasma proteins, prothrombin and fibrinogen, function in blood clotting, an enzymatic process that results in fibrin threads.

Platelets

Capillary Exchange in the Tissues

Capillary exchange in the tissues helps keep the internal environment constant. When blood reaches a capillary:

- Water moves out at the arterial end due to blood pressure.
- Water moves in at the venous end due to osmotic pressure.
- Nutrients diffuse out of and wastes diffuse into the capillary between the arterial end and the venous end.
- Lymphatic capillaries in the vicinity pick up excess tissue fluid and return it to cardiovascular veins.

Key Terms

antibody 410
antigen 410
aorta 403, 406
arteriole 405
artery 403, 405
atrioventricular valve 403
atrium (pl., atria) 403
AV (atrioventricular) node 404
blood 409
blood pressure 405
capillary 405
cardiac cycle 404
cardiovascular system 401
closed circulatory system 401
diastole 404
electrocardiogram (ECG) 404
formed elements 409
heart 400
heart attack 408
heart murmur 403
hemocoel 400
hemoglobin 409
hypertension 407
lymph 407
lymphatic system 407
lymphatic vessel 407
lymphocyte 410
macrophage 410

monocyte 410
neutrophil 410
open circulatory system 400
ostium (pl., ostia) 401
plaque 408
plasma 409
platelet 410
portal system 407
pulmonary artery 403
pulmonary circuit 402, 406
pulmonary trunk 403
pulmonary vein 403
pulse 404
red blood cell 409
SA (sinoatrial) node 404
semilunar valve 403
septum 403
sinus 400
stroke 408
systemic circuit 402, 406
systole 404
thrombin 411
vein 403
vena cava (pl., venae cavae) 403, 407
ventricle 403
venule 405
white blood cell 410

Testing Yourself

Choose the best answer for each question.

1. In insects with an open circulatory system, oxygen is taken to cells by
   a. blood.
   b. hemolymph.
   c. tracheae.
   d. capillaries.

2. In small invertebrates such as hydra, circulation of oxygen is accomplished by
   a. diffusion to each individual cell.
   b. the blood.
   c. the gills.
   d. hemolymph.

3. Label the components of the cardiovascular system in the following diagram.

4. Which of the following statements is true?
   a. Arteries carry blood away from the heart, and veins carry blood to the heart.
   b. Arteries carry blood to the heart, and veins carry blood away from the heart.
   c. All arteries carry O$_2$-rich blood, and veins carry O$_2$-poor blood.
   d. Arteries usually carry O$_2$-poor blood, and veins carry O$_2$-rich blood.

5. Label the following diagram of the heart.

6. In amphibians
   a. the heart pumps blood through one circuit.
   b. the heart has one atrium and two ventricles.
   c. the structure of the heart allow blood O$_2$-poor and O$_2$-rich blood to mix.
   d. the atria are responsible for pumping the blood to the body.

7. In humans, blood returning to the heart from the lungs returns to
   a. the right ventricle.
   b. the right atrium.
   c. both the right and left sides of the heart.
   d. the left ventricle.
   e. the left atrium.

8. The _____ circuit takes blood to and from the tissues.
   a. systemic
   b. pulmonary
   c. coronary
   d. atrial

9. Atria are
   a. the main pumps of the heart.
   b. the lower chambers of the heart.
   c. the holes in the heart of an invertebrate.
   d. the upper chambers of the heart.

10. An artificial pacemaker replaces the effect of the
   a. vagus nerve.
   b. SA (sinoatrial) node.
   c. AV (atrioventricular) node.
   d. ventricular systole.

11. Which of the following lists the events of the cardiac cycle in the correct order?
   a. contraction of atria, rest, contraction of ventricles
   b. contraction of ventricles, rest, contraction of atria
   c. contraction of atria, contraction of ventricles, rest
   d. contraction of ventricles, contraction of atria, rest

12. The *lub*, the first heart sound, is produced by the closing of
   a. the aortic semilunar valve.
   b. the pulmonary semilunar valve.
   c. the atrioventricular (tricuspid) valve.
   d. the atrioventricular (bicuspid) valve.
   e. both atrioventricular valves.

13. A _____ is a condition generally caused by leaky heart valves.
    a. heart murmur
    b. hypertension
    c. fibrillation
    d. plaque

14. An electrocardiogram measures
    a. chemical signals in the brain and heart.
    b. electrical activity in the brain and heart.
    c. chemical signals in the heart.
    d. electrical changes in the wall of the heart.

15. The _____ is the heart's intrinsic natural pacemaker.
    a. atrioventricular node
    b. sinoatrial node
    c. vagus nerve
    d. superior vena cava

16. Place the following blood vessels in order, from largest to smallest in diameter.
    a. arterioles, capillaries, arteries
    b. arteries, arterioles, capillaries
    c. capillaries, arteries, arterioles
    d. arterioles, arteries, capillaries
    e. arteries, capillaries, arterioles

17. _____ are the smallest of the blood vessels.
    a. Arteries
    b. Venules
    c. Veins
    d. Capillaries
    e. Arterioles

18. The lymphatic system
    a. returns excess tissue fluid to cardiovascular veins.
    b. is found in limited regions of the body.
    c. relies most extensively on diffusion to move lymph.
    d. is a two-way system.
    e. More than one of these are correct.

For questions 19–22, identify the cardiovascular disorder in the key that matches the description. Each answer may be used more than once. Each question may have more than one answer.

Key:

    a. hypertension
    b. atherosclerosis
    c. stroke
    d. heart attack

19. often not detected until after a stroke or heart attack

20. sometimes called a silent killer

21. results from the accumulation of plaque

22. may be caused by an embolism

23. Exchange of the gases oxygen and carbon dioxide occurs across the _____ of the _____.
    a. veins, lungs
    b. capillaries, tissues
    c. arteries, tissues
    d. All of these are correct.

24. The _____ and the _____ are the main artery and veins in the systemic circuit.
    a. aorta, venae cavae
    b. pulmonary trunk, pulmonary veins
    c. vena cava, aortas
    d. atrium, ventricles

25. _____ is tissue fluid that has entered lymphatic vessels.
    a. Plasma
    b. Blood
    c. Saliva
    d. Lymph

26. The average heart rate is about _____ beats per minute.
    a. 100
    b. 45
    c. 60
    d. 70

27. If the SA node fails and the AV node takes over, the result will be a heart rate that is
    a. slower.
    b. faster.
    c. the same.
    d. None of these are correct.

28. Which association is incorrect?
    a. white blood cells—infection fighting
    b. red blood cells—blood clotting
    c. plasma—water, nutrients, and wastes
    d. red blood cells—hemoglobin
    e. platelets—blood clotting

29. Plasma is
    a. composed mostly of proteins.
    b. has a pH of 12.
    c. the liquid portion of blood.
    d. approximately 10% water.

30. _____ are phagocytic white blood cells derived from monocytes.
    a. Erythrocytes
    b. Lymphocytes
    c. Platelets
    d. Macrophages

31. In the tissues, nutrients and _____ are exchanged for _____ and other wastes.
    a. blood, oxygen
    b. oxygen, carbon dioxide
    c. hemoglobin, tissue fluid
    d. None of these are correct.

32. Lymph is formed from
    a. urine.
    b. fats.
    c. excess tissue fluid.
    d. All of these are correct.

33. A decrease in lymphocytes would result in problems associated with
    a. clotting.
    b. immunity.
    c. oxygen transportation.
    d. All of these are correct.

34. The best explanation for the slow movement of blood in capillaries is
    a. skeletal muscles press on veins, not capillaries.
    b. capillaries have much thinner walls than arteries.
    c. there are many more capillaries than arterioles.
    d. venules are not prepared to receive so much blood from the capillaries.
    e. All of these are correct.

35. Which of the following assist in the return of venous blood to the heart?
    a. valves
    b. skeletal muscle contractions
    c. respiratory movements
    d. blood flow in the direction of reduced pressure
    e. All of these are correct.

36. Water enters the venous end of capillaries because
    a. osmotic pressure is higher than blood pressure.
    b. of an osmotic pressure gradient.
    c. of higher blood pressure on the venous side.
    d. of higher blood pressure on the arterial side.
    e. of higher red blood cell concentration on the venous side.

37. One-way conduction in lymphatic vessels is aided by
    a. valves.
    b. skeletal muscle contractions.
    c. Both a and b are correct.
    d. None of these are correct.

38. Which of the following is not a function of the lymphatic system?
    a. produces blood cells
    b. returns excess tissue fluid to the blood
    c. transports fat absorbed from the digestive system
    d. defends the body against disease

39. Red blood cells
    a. reproduce themselves by mitosis.
    b. live for several years.
    c. continually synthesize hemoglobin.
    d. are destroyed in the liver and spleen.
    e. More than one of these are correct.

40. Which of the following is not a formed element of blood?
    a. white blood cells   c. fibrinogen
    b. red blood cells   d. platelets

41. The last step in blood clotting
    a. is the only step that requires calcium ions.
    b. occurs outside the bloodstream.
    c. is the same as the first step.
    d. converts prothrombin to thrombin.
    e. converts fibrinogen to fibrin.

42. Hemorrhagic disorders can result when which vitamin is missing from the diet?
    a. A   d. D
    b. C   e. thiamine
    c. K

43. _____ are the individual cavities into which blood is pumped in invertebrates with open circulatory systems.
    a. Ostia   c. Atria
    b. Sinuses   d. Trachea

44. Water
    a. exits the bloodstream at the arterial and venous ends of a capillary.
    b. enters the bloodstream at the arterial and venous ends of a capillary.
    c. enters the bloodstream at the arterial end and exits cells at the venous end of a capillary.
    d. exits the bloodstream at the arterial end and enters cells at the venous end of a capillary.

45. In the following diagram, label arrows a–d as either blood pressure or osmotic pressure.

Thinking Scientifically

1. Explain why the evolution of the four-chambered heart was critical for the development of an endothermic lifestyle—the generation of internal heat—in birds and mammals.

2. Provide a physiological explanation for the benefit gained by athletes who train at high altitudes.

Bioethical Issue

Blood Doping

In the highly competitive world of professional athletics, individuals are constantly striving to gain an edge over the competition. With seven-figure salaries, international recognition, and multimillion-dollar endorsements as their incentives, some athletes resort to life-threatening measures to enhance their performance.

In recent years, various means of increasing the red blood cell count in athletes have been employed. By increasing the number of red blood cells in the body, more oxygen can be delivered to the muscles. This enhances athletic ability, especially in events requiring a high level of endurance.

In the past, blood doping was commonly used. Red blood cells (RBCs) were removed from the athlete's body and stored for several weeks. During this time, the body would release increased levels of the hormone erythropoietin to stimulate RBC production. When the RBC count returned to normal, the blood cells previously removed were injected back into the body. This was done a few days before the competition to get maximum effect. In more recent times, injections of artificially produced erythropoietin are being used to stimulate RBC production far above normal levels.

The artificial form of erythropoietin is known as epoetin alfa and is structurally identical to the naturally occurring form. Therefore, it is very difficult to determine when it is being used.

However, many athletes have paid a high price for the competitive advantage offered by epoetin alfa. The use of this hormone causes the blood to become thicker than normal. Hypertension and an increased risk of heart attack and stroke result. In recent decades, the suspected use of epoetin alfa has been linked to the deaths of numerous competitive cyclists.

What measures if any do you think should be taken to prevent the use of epoetin alfa by athletes? If you were a professional athlete and knew you would benefit from using epoetin alfa, how would you weigh the benefits against the risks?

Essentials of Biology Website

The companion website for *Essentials of Biology* provides a wealth of information organized and integrated by chapter. You will find practice tests, animations, videos, and much more that will complement your learning and understanding of general biology.

http://www.mhhe.com/maderessentials2

24

The Maintenance Systems

A urinalysis can involve any of 35 different tests.

small intestine

The small intestine of humans is about 6 meters (19.7 feet) long.

Outline

The respiratory system is one of the last organ systems to develop in the human fetus.

Respiratory Distress Syndrome

DURING HUMAN DEVELOPMENT, organ systems begin to function on an "as needed" basis. Within the first month of embryonic development, many organ systems (such as the cardiovascular and nervous systems) are present and functioning. However, some organ systems (such as the respiratory and digestive systems) that appear may not be refined until much later because they are not critical to early development. When babies are born prematurely (before the 37th week

—Continued next page

417

Continued—

of gestation), numerous health problems can occur, particularly with regard to these systems. One condition that is common in very premature babies (those born at 32 weeks or earlier) is respiratory distress syndrome (RDS). Because the lungs of a fetus do not normally function until full term (40 weeks), lung development is not complete when a baby is born prematurely. Usually, a natural surfactant that helps keep the air sacs of the lungs inflated so that gas exchange can occur is not being produced. Without treatment, RDS is often fatal because the blood is not being oxygenated. Now, an artificial surfactant is available, and premature babies are given the surfactant they lack. This treatment greatly increases the likelihood that premature babies will survive. Other medical breakthroughs are also now available to encourage the functioning of a premature baby's organ systems.

The maintenance systems have vital roles in the overall functioning of the body, and a malfunction in one of these systems can lead to problems in other systems. In this chapter, you will learn about the structure and function of the respiratory, urinary, and digestive systems and how they support the body's other organ systems.

24.1 | Respiratory System

The cells of your body are bathed in tissue fluid. They acquire oxygen and nutrients and get rid of carbon dioxide and other wastes through exchanges with tissue fluid. In turn, tissue fluid makes these same exchanges with the blood. Blood is refreshed because the respiratory, urinary, and digestive systems make exchanges with the external environment. Only in this way is blood supplied with the oxygen and nutrients cells require and cleansed of waste molecules. Notice in **Figure 24.1** that when blood enters the lungs of the respiratory system, it gives up carbon dioxide and picks up oxygen. Carbon dioxide exits the body and the oxygen is delivered to the body's cells. In this way, respiration contributes to homeostasis. **Respiration** in terrestrial vertebrates (including humans) requires these steps:

1. Breathing: inspiration (entrance of air into the lungs) and expiration (exit of air from the lungs).
2. External exchange of gases between the air and the blood within the lungs.
3. Internal exchange of gases between blood and tissue fluid; the body's cells exchange gases with tissue fluid.

Regardless of the particular gas-exchange surfaces of animals and the manner in which gases are delivered to the cells, in the end oxygen enters mitochondria, where aerobic cellular respiration takes place. Without the delivery of oxygen to the body's cells, ATP is not produced, and life ceases. Carbon dioxide (CO_2), a waste molecule given off by cells, is a by-product of cellular respiration.

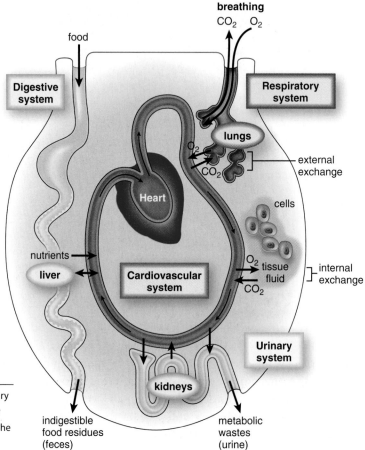

Figure 24.1 Keeping the internal environment steady.

The respiratory system functions in gas exchange. CO_2-laden blood enters the pulmonary capillaries and then CO_2 diffuses into the lungs and exits the body by way of respiratory passages. O_2-laden air enters the respiratory passages and lungs. Then O_2 diffuses into the blood at the pulmonary capillaries.

The Human Respiratory Tract

The human respiratory system includes all of the structures that conduct air in a continuous pathway to and from the **lungs** (**Fig. 24.2**). As air moves through the respiratory tract, it is filtered so that it is free of debris, warmed, and humidified. By the time the air reaches the lungs, it is at body temperature and saturated with water. In the nose, hairs and cilia act as screening devices. In the respiratory passages, cilia beat upward, carrying mucus, dust, and occasional bits of food that "went down the wrong way" into the throat, where the accumulation may be swallowed or spit out (**Fig. 24.3**). Smoking cigarettes and cigars inactivates and eventually destroys these cilia, so that the lungs become laden with soot and debris. This is the first step toward various lung disorders.

Conversely, as air moves out of the tract, it cools and loses its moisture. As air cools, it deposits its moisture on the lining of the tract, and the nose may even drip as a result of this condensation. However, the air still retains so much moisture that on a cold day, it forms a small cloud when we breathe out.

The Upper Respiratory Tract

The upper respiratory tract consists of the nasal cavities, pharynx, and larynx (Fig. 24.2). The nose, a prominent feature of the face, is the only external portion of the respiratory system. The nose contains the **nasal cavities,** narrow canals separated from one another by a septum composed of bone and cartilage. Tears from the eyes drain into the nasal cavities by way of tear ducts. For this reason, crying produces a runny nose. The nasal cavities communicate with the **sinuses,** air-filled spaces that reduce the weight of the skull and act as resonating chambers for the voice. If the ducts leading from the sinuses become inflamed, fluid may accumulate, causing a sinus headache. The nasal cavities are separated from the mouth by a partition called the palate. The palate has two portions. Anteriorly, the hard palate is supported by bone, and posteriorly, the soft palate is made solely of soft tissue and muscle.

The **pharynx** is a funnel-shaped passageway that connects the nasal cavity and mouth to the **larynx,** or voice box. The tonsils form a protective ring of lymphatic tissue at the junction of the mouth and the pharynx. *Tonsillitis* occurs when the tonsils become inflamed and enlarged. If tonsillitis occurs frequently and enlargement makes breathing difficult, the tonsils can be removed surgically, a procedure called a tonsillectomy. In the pharynx, the air passage and food passage cross because the larynx, which receives air, is anterior to the esophagus, which receives food. This arrangement may seem inefficient, since there is danger of choking if food accidentally enters the trachea, but it does have the advantage of letting you breathe through your mouth in case your nose is plugged. In addition, it permits greater intake of air during heavy exercise, when greater gas exchange is required. When swallowing occurs, the epiglottis, an elastic flap of cartilage, covers the glottis, the opening into the larynx, and this helps prevent choking.

Air passes from the pharynx through the **glottis.** The larynx is always open because it is formed by a complex of cartilages, among them the one that forms the "Adam's apple." At the edges of the glottis, embedded in mucous membrane, are the **vocal cords.** These flexible bands of connective tissue vibrate and produce sound when air is expelled past them through the glottis from the larynx. *Laryngitis* is an infection of the larynx with accompanying hoarseness, leading to the inability to speak audibly.

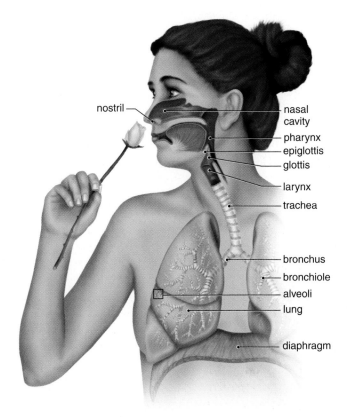

Figure 24.2 The human respiratory tract.

To reach the lungs, air moves from the nasal cavities, through the pharynx, larynx, trachea, bronchi, and finally the bronchioles, which end in the lungs.

×800

Figure 24.3 Ciliated cells of the respiratory passages.

These cilia sweep impurities up away from the lungs toward the throat where they can be swallowed. Smoking first inactivates and then destroys these cilia.

Lower Respiratory Tract

The lower respiratory tract contains the respiratory tree, consisting of the trachea, bronchi, and bronchioles (Fig. 24.2). The **trachea,** commonly called the windpipe, is a tube connecting the larynx to the bronchi. The trachea is held open by a series of C-shaped, cartilaginous rings that do not completely meet in the rear. This arrangement allows food to pass down through the esophagus, which lies right behind the trachea in the neck. The trachea divides into two primary **bronchi,** which enter the right and left lungs. *Bronchitis* is an infection of the bronchi. As bronchitis develops, a nonproductive cough becomes a deep cough that produces mucus and perhaps pus. The deep cough of smokers indicates that they have bronchitis and that the respiratory tract is irritated. When a person stops smoking, this progression reverses, and the airways become healthy again. Chronic bronchitis is the second step toward emphysema and lung cancer caused by smoking cigarettes. Lung cancer often begins in the bronchi, and from there it spreads to the lungs.

The bronchi continue to branch until there are a great number of smaller passages called **bronchioles.** The two bronchi resemble the trachea in structure, but as the passages divide and subdivide, their walls become thinner, the rings of cartilage disappear, and the amount of smooth muscle increases. During an attack of *asthma,* the smooth muscle of the bronchioles contracts, causing constriction of the bronchioles and characteristic wheezing. Each bronchiole terminates in an elongated space enclosed by a multitude of little air pockets, or sacs, called **alveoli** (sing., alveolus), which make up the lungs (see Fig. 24.7).

Respiration in Insects   While humans have one trachea, insects have many **tracheae,** little air tubes supported by rings of chitin that branch into every part of the body (**Fig. 24.4**). The tracheal system begins at spiracles, openings that perforate the insect's body wall, and ends in very fine, fluid-filled tubules, which may actually indent the plasma membranes of cells to come close to mitochondria. Ventilation is assisted by the presence of air sacs that help to draw in the air. Since no individual cell is very far from the site of gas exhange, the hemolymph in an insect does not transport oxygen, and no oxygen-carrying pigment is required.

Breathing

In humans the **diaphragm** is a muscular membranous partition that divides the **thoracic** (upper) **cavity** from the **abdominal** (lower) **cavity** of the body. When humans breathe, the volume of the thoracic cavity and lungs is increased by muscle contractions that both lower the diaphragm and raise the ribs (**Fig. 24.5**). Reptiles also ventilate this way except they lack a diaphragm. These movements create a negative pressure in the thoracic cavity and lungs, and air then flows into the lungs, a process called *inspiration.* It is important to realize that air comes in because the lungs have already opened up; air does not force the lungs open. When the rib and diaphragm muscles relax, the lungs recoil, and air moves out as a result of increased pressure in the lungs, a process called *expiration.*

In mammals and most reptiles, air moves in and out by the same route, and therefore, some residual air, low in oxygen, is always left in the lungs of humans. Birds, on the other hand, use a one-way ventilation mechanism (**Fig. 24.6**). Incoming air is carried past the lungs by a trachea that takes it to a set of abdominal air sacs. Then air passes forward through the lungs into a set of

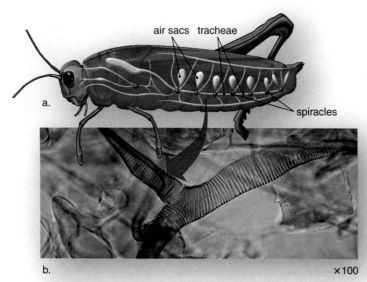

air sacs   tracheae

a.

spiracles

b.   ×100

Figure 24.4 Tracheae of insects.

A system of air tubes extending throughout the body of an insect carries oxygen to the cells. Air enters the tracheae at openings called spiracles. From here, the air moves to smaller tubes, which take it to the cells where gas exchange takes place. **a.** Diagram of tracheae. **b.** The photomicrograph shows how the walls of the trachea are stiffened with bands of chitin.

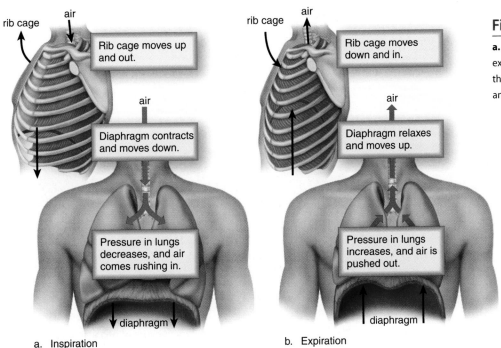

Figure 24.5 **Inspiration versus expiration.**

a. During inspiration, the thoracic cavity and lungs expand so that air is drawn in. **b.** During expiration, the thoracic cavity and lungs resume their original positions and pressures, forcing air out.

thoracic air sacs. Fresh air never mixes with used air in the lungs of birds, and thereby gas-exchange efficiency is greatly improved.

Increased concentrations of hydrogen ion (H^+) and carbon dioxide (CO_2) in the blood are the primary stimuli that increase the breathing rate in humans. The chemical content of the blood is monitored by chemoreceptors called the **aortic bodies** and **carotid bodies,** which are specialized structures in the walls of arteries, the aorta and common carotid arteries. These receptors are very sensitive to changes in H^+ and CO_2 concentrations, but they are only minimally sensitive to lower oxygen (O_2) concentrations. Information from the chemoreceptors goes to the *breathing center* in the brain, which increases the breathing rate when concentrations of hydrogen ions and carbon dioxide rise (see Fig. 24.10). The breathing center is also directly sensitive to the chemical content of the blood, including its oxygen content.

Check Your **Progress**

1. List the three steps of respiration in a terrestrial vertebrate.
2. Contrast the respiratory system in insects with that in humans.
3. Explain how air moves into the lungs of humans during breathing.

Answers: 1. The three steps of respiration are breathing, external exchange, and internal exchange. 2. Insects have many tracheae that branch into many smaller tubes, which deliver oxygen to the cells. Humans have a single trachea that leads to the bronchi and then to the bronchioles, which enter the lungs where gas exchange occurs. 3. Muscle contractions that lower the diaphragm and raise the ribs create a negative pressure that pulls air into the lungs.

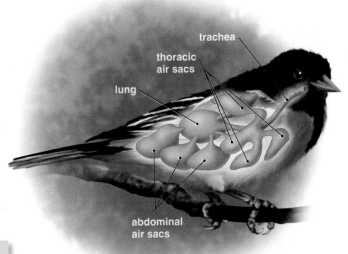

Figure 24.6 **Breathing in birds.**

Two types of air sacs are attached to the lungs of birds. When a bird inhales, most of the air enters the abdominal air sacs, and when a bird exhales, air moves through the lungs to the thoracic air sacs before exiting the trachea. This one-way flow of air through the lungs allows more fresh air to be present in the lungs with each breath, and this leads to greater uptake of oxygen from one breath of air.

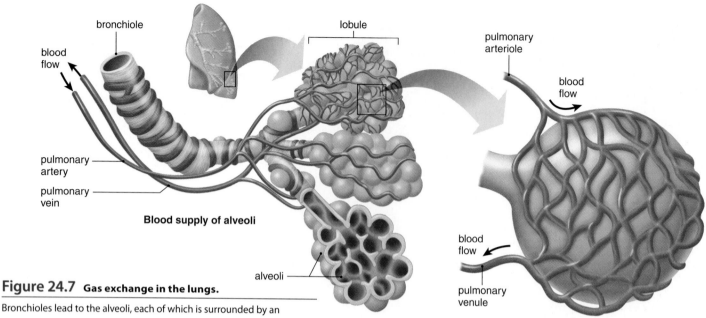

Blood supply of alveoli

Capillary network of one alveolus

Figure 24.7 Gas exchange in the lungs.

Bronchioles lead to the alveoli, each of which is surrounded by an extensive capillary network. The pulmonary artery and arteriole carry O_2-poor blood (colored blue) and the pulmonary venule and vein carry O_2-rich blood (colored red).

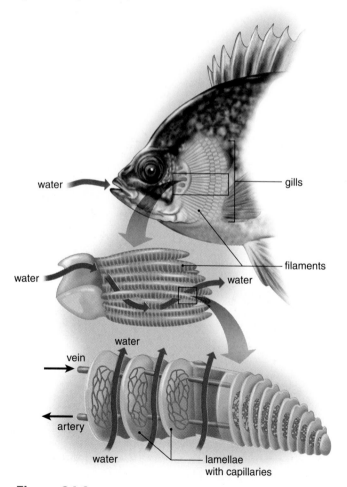

Figure 24.8 Gills of bony fishes.

Gills are finely divided into filaments, and each filament has many thin, platelike lamellae. Gases are exchanged between the capillaries inside the lamellae and the water that flows between the lamellae.

Lungs and External Exchange of Gases

The lungs of humans and other mammals are more elaborately subdivided than those of amphibians and reptiles. Frogs and salamanders have a moist skin that allows them to use the surface of their body for gas exchange in addition to the lungs. It has been estimated that human lungs have a total surface area at least 50 times the skin's surface area because of the presence of alveoli (**Fig. 24.7**).

An alveolus, like the capillary that surrounds it, is bounded by squamous epithelium. Diffusion alone accounts for gas exchange between the alveolus and the capillary. Carbon dioxide, being more plentiful in the pulmonary venule, diffuses from a pulmonary capillary to enter an alveolus, while oxygen, being more plentiful in the lungs, diffuses from an alveolus into a pulmonary capillary. The process of diffusion requires a gas-exchange region to be not only large, but also moist and thin. The alveoli are lined with surfactant, a film of lipoprotein that lowers the surface tension of water, thereby preventing the alveoli from collapsing. Some newborns, especially if premature, lack this film. Surfactant replacement therapy is now thankfully available to treat this condition.

The blood within pulmonary capillaries is indeed spread thin, and the red blood cells are pressed up against their narrow walls. The alveolar epithelium and the capillary epithelium are so close that together they are called the *respiratory membrane*. Hemoglobin in the red blood cells quickly picks up oxygen molecules as they diffuse into the blood.

Emphysema is a serious lung condition in which the walls of many alveoli have been destroyed. The lungs have less recoil, and there is less surface area for gas exchange to occur.

Gills of Fish   In contrast to the lungs of terrestrial vertebrates, fish and other aquatic animals use gills as their respiratory organ (**Fig. 24.8**). In fish, water is drawn into the mouth and out from the pharynx across the gills. The flow of blood in gill capillaries is opposite the flow of water across the gills, and therefore the blood is always exposed to water having a higher oxygen content. In the end, about 80–90% of the dissolved oxygen in water is absorbed.

Transport and Internal Exchange of Gases

Recall from Chapter 23 that hemoglobin molecules within red blood cells carry oxygen to the body's tissues. If hemoglobin did not transport oxygen, it is estimated it would take three years for an oxygen molecule to move from your lungs to your toes. Each hemoglobin molecule contains four polypeptide chains: two α and two β chains. Each polypeptide is folded around an iron-containing group called **heme.** It is actually the iron that bonds with oxygen and carries it to the tissues (**Fig. 24.9**). Since there are about 250 million hemoglobin molecules in each red blood cell, each cell is capable of carrying at least one billion molecules of oxygen. Hemoglobin gives up its oxygen in the tissues during internal exchange primarily because tissue fluid always has a lower oxygen concentration than blood does. This difference occurs because cells take up and utilize oxygen when they carry on cellular respiration. Another reason hemoglobin gives up oxygen is due to the warmer temperature and lower pH in the tissues, environmental conditions that are also caused by cellular respiration. When cells respire, they give off heat and carbon dioxide as by-products.

Carbon dioxide enters the blood during internal exchange because the tissue fluid always has a higher concentration of carbon dioxide. Most of the carbon dioxide is transported in the form of the **bicarbonate ion** (HCO_3^-). First, carbon dioxide combines with water, forming carbonic acid, and then this acid dissociates to a hydrogen ion (H^+) and HCO_3^-:

$$CO_2 + H_2O \longrightarrow H_2CO_3 \longrightarrow H^+ + HCO_3^-$$

carbon dioxide   water   carbonic acid   hydrogen ion   bicarbonate ion

The H^+ does cause the pH to lower, but only slightly because much of the H^+ is absorbed by the globin portions of hemoglobin. The HCO_3^- is carried in the plasma.

What happens to the above equation in the lungs? As blood enters the pulmonary capillaries, carbon dioxide diffuses out of the blood into the alveoli. Hemoglobin gives up the H^+ it has been carrying as this reaction occurs:

$$H^+ + HCO_3^- \longrightarrow H_2CO_3 \longrightarrow H_2O + CO_2$$

Now, much of the carbon dioxide diffuses out of the blood into the alveoli of the lungs. Should the blood level of H^+ rise, the breathing center in the brain increases the breathing rate, and as more CO_2 leaves the blood, the pH of blood is corrected (**Fig. 24.10**).

Check Your Progress

1. Explain what causes oxygen to diffuse into the blood from the lungs during external exchange but causes it to diffuse out of the blood into tissue fluid during internal exchange.
2. Describe what happens when carbon dioxide enters the blood.

Answers: 1. Oxygen follows its concentration gradient; there is more oxygen in the lungs than in the blood and there is more oxygen in the blood than in tissue fluid. Also, environmental conditions in the tissues cause hemoglobin to give up oxygen.   2. Carbon dioxide combines with water to form carbonic acid. Carbonic acid breaks down to the bicarbonate ion (HCO_3^-) and H^+, which combines with hemoglobin.

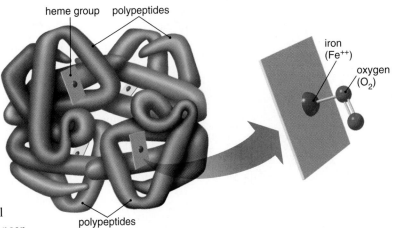

Figure 24.9   Hemoglobin.

Hemoglobin has two each of two different polypeptides. Each one is associated with a heme group. Oxygen bonds with the central iron atom of a heme group.

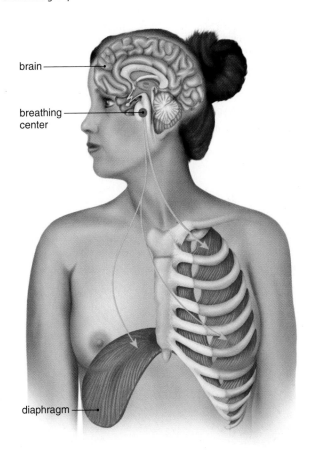

Figure 24.10   Neural control of breathing rate.

The brain regulates the breathing rate by controlling contraction of the rib cage muscles and the diaphragm. When the breathing rate increases, [H^+] lowers as CO_2 is removed from the blood, and the pH of the blood returns to normal.

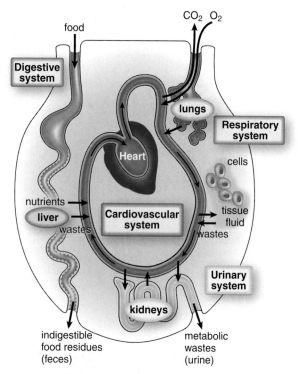

Figure 24.11 Keeping the internal environment steady.

Excretion carried out by the kidneys serves three functions: excretion of urea, produced by the liver, and other nitrogenous wastes, maintenance of water-salt balance, and maintenance of pH.

Figure 24.12 The human urinary system.

a. The human urinary system produces, stores, and expels urine from the body. **b.** The human kidney has three sections, which can be correlated with (**c**) the presence of about one million nephrons. Nephrons do the work of the kidney.

24.2 | Urinary System and Excretion

In complex animals, **kidneys** excrete nitrogenous wastes and are involved in regulating the water-salt balance of the body (**Fig. 24.11**). In addition, the mammalian kidney helps regulate the pH of blood. To summarize, these three functions can be associated with the kidney:

1. Excretion of nitrogenous wastes such as urea and uric acid.
2. Maintenance of the water-salt balance of the blood.
3. Maintenance of the acid-base balance of the blood.

In humans, the kidneys are bean-shaped, reddish-brown organs, each about the size of a fist. They are located on either side of the vertebral column just below the diaphragm, where they are partially protected by the lower rib cage. **Urine** made by the kidneys is conducted from the body by the other organs in the urinary system. Each kidney is connected to a **ureter,** a tube that takes urine from the kidney to the **urinary bladder,** where it is stored until it is voided from the body through the single **urethra** (**Fig. 24.12***a*). In males, the urethra passes through the penis, and in females, it opens in front of the opening of the vagina. In females, there is no connection between the genital (reproductive) and urinary systems, but there is a connection in males—that is, the urethra also carries sperm during ejaculation.

In amphibians, birds, reptiles, and some fishes, the bladder empties into the cloaca, a common chamber and outlet for the digestive, urinary, and genital tracts.

Kidneys

If a kidney is sectioned longitudinally, three major parts can be distinguished (Fig. 24.12*b*). The **renal cortex,** the outer region of a kidney, has a somewhat granular appearance. The **renal medulla** consists of the cone-shaped renal pyramids, which lie inside the renal cortex. The innermost part of the kidney is a hollow chamber called the **renal pelvis.** Urine collects in the renal pelvis and then is carried to the bladder by a ureter. Microscopically, each kidney is composed of about one million tiny tubules called **nephrons** (Fig. 24.12*c*). The nephrons of a kidney produce urine.

a. Urinary system   b. Kidney   c. Two nephrons

Nephrons

Each nephron is made of several parts (**Fig. 24.13**). The blind end of a nephron is pushed in on itself to form the **nephron capsule.** Filtration of the blood occurs at this portion of the nephron. The inner layer of the capsule is composed of specialized cells that allow easy passage of molecules. Leading from the capsule is a portion of the nephron called the **proximal tubule,** which is lined by cells with many mitochondria and tightly packed microvilli. Then comes the **nephron loop** with a descending limb and an ascending limb. This is followed by the **distal tubule.** Several distal tubules enter one **collecting duct.** A collecting duct delivers urine to the renal pelvis. The nephron loop and the collecting duct give the pyramids of the renal medulla their striped appearance (see Fig. 24.12*b*).

Each nephron has its own blood supply, and various exchanges occur between parts of the nephron and a blood capillary as urine forms.

Urine Formation

Urine formation requires three steps: filtration, reabsorption, and secretion (Fig. 24.13).

Filtration Filtration occurs whenever small substances pass through a filter and large substances are left behind. During urine formation, **filtration** is the movement of small molecules from a blood capillary to the inside of the capsule as a result of adequate blood pressure. Small molecules, such as water, nutrients, salts, and urea, move to the inside of the capsule. Plasma proteins and blood cells are too large to be part of this filtrate, so they remain in the blood.

If the composition of the filtrate were not altered in other parts of the nephron, death from loss of nutrients (starvation) and loss of water (dehydration) would quickly follow. The next step, reabsorption, helps prevent this from happening.

Reabsorption of Solutes **Reabsorption** takes place when substances from the proximal tubule move into the blood. Nutrients such as glucose and amino acids also return to the blood. This process is selective because some molecules such as glucose are both passively and actively reabsorbed. The cells of the proximal tubule have numerous microvilli, which increase the surface area, and numerous mitochondria, which supply the energy needed for active transport. However, if there is more glucose in the filtrate than there are carriers to handle it, glucose will appear in the urine. Glucose in the urine is a sign of diabetes mellitus, sometimes caused by lack of the hormone insulin.

Sodium ions (Na^+) are also actively pumped into the peritubular capillary, and then chloride ions (Cl^-) follow passively. Now water moves by osmosis from the tubule into the blood. About 60–70% of salt and water are reabsorbed at the proximal tubule.

Some of the urea, the primary nitrogenous waste product of human metabolism, and other types of nitrogenous wastes excreted by humans are passively reabsorbed, but most remain in the filtrate.

Secretion **Secretion** refers to the transport of substances into the nephron by means other than filtration. For our purposes, secretion may be particularly associated with the distal tubule. Substances such as uric acid, hydrogen ions, ammonia, and penicillin are eliminated by secretion. The process of secretion helps rid the body of potentially harmful compounds that were not filtered into the capsule.

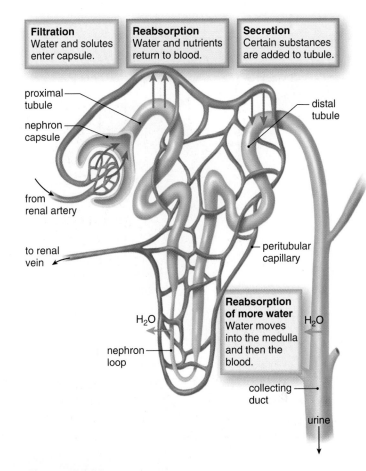

Filtration
Water and solutes enter capsule.

Reabsorption
Water and nutrients return to blood.

Secretion
Certain substances are added to tubule.

proximal tubule

nephron capsule

from renal artery

to renal vein

distal tubule

peritubular capillary

Reabsorption of more water
Water moves into the medulla and then the blood.

nephron loop

H_2O

H_2O

collecting duct

urine

Figure 24.13 Urine formation.

Urine formation requires three steps: filtration, reabsorption, and secretion. Production of a hypertonic urine occurs when water is reabsorbed into the blood along the length of the nephron and especially at the nephron loop and the collecting duct.

Check Your Progress

1. **List the three functions of the kidney.**
2. **List the parts of a nephron.**
3. **List the three steps involved in urine formation.**

Answers: 1. The kidney functions are excretion of nitrogenous wastes, maintenance of the water-salt balance of blood, and maintenance of the acid-base balance of blood.   2. A nephron consists of the renal capsule, proximal tubule, nephron loop, and distal tubule. Many nephrons empty into the same collecting duct.   3. Urine formation: *filtration* of small molecules from a blood capillary to the inside of the nephron capsule; *reabsorption* of water and nutrients from the nephron into the blood; and *secretion* from the blood into the nephron of potentially harmful substances.

Figure 24.14 Malpighian tubules.

The Malpighian tubules of insects are attached to the gut and surrounded by the hemolymph of the open circulatory system. Wastes diffuse into tubules. K^+ is secreted into these tubules, drawing in water by osmosis. Much of the water and K^+ is reabsorbed across the wall of the rectum.

Regulation of Water-Salt Balance and pH Typically, animals have some means of regulating the osmolarity of the internal environment so that the water-salt balance stays within normal limits. Insects have a unique excretory system consisting of long, thin tubules called **Malpighian tubules** attached to the gut (**Fig. 24.14**). Uric acid, their primary nitrogenous waste product, simply diffuses from the surrounding hemolymph into these tubules, and water follows a salt gradient established by active transport of potassium (K^+). Water and other useful substances are reabsorbed at the rectum, but the uric acid leaves the body at the anus. Insects that live in water or eat large quantities of moist food reabsorb little water. But insects in dry environments reabsorb most of the water and excrete a dry, semisolid mass of uric acid. Most animals can regulate the blood level of both water and salt. For example, freshwater fishes take up salt in the digestive tract and gills and produce large amounts of dilute urine. Saltwater fishes take in salt water by drinking, but then they pump ions out at the gills, and produce only small amounts of concentrated urine.

In mammals, the long nephron loop allows secretion of a hypertonic urine (see Fig. 24.13). The ascending limb of the nephron loop pumps salt and also urea into the renal medulla, and water follows by osmosis both at the descending limb of the nephron loop and at the collecting duct. As you will see in Chapter 27, at least three hormones are involved in regulating water-salt reabsorption by the kidneys. Drinking coffee interferes with one of these hormones, and that is why coffee is a *diuretic*, a substance that causes the production of more urine.

Most mammals can also regulate the pH of the blood. Both the bicarbonate (HCO_3^-) buffer system of the blood and the regulation of breathing rate (see Fig. 24.10) helps rid the body of CO_2 and contribute to maintaining blood pH.

As helpful as these mechanisms might be, only the kidneys can excrete a wide range of acidic and basic substances. The kidneys are slower acting than the buffer/breathing mechanism, but they have a more powerful effect on pH. For the sake of simplicity, we can think of the kidneys as reabsorbing bicarbonate ions and excreting hydrogen ions as needed to maintain the normal pH of the blood (**Fig. 24.15**). If the blood is acidic, hydrogen ions are excreted and bicarbonate ions are reabsorbed. If the blood is basic, hydrogen ions are not excreted and bicarbonate ions are not reabsorbed. The fact that urine is most often acidic shows that usually an excess of hydrogen ions is produced by the body and is excreted. Ammonia (NH_3) provides a means for buffering these hydrogen ions in urine: ($NH_3 + H^+ \longrightarrow NH_4^+$). Ammonia (the presence of which is quite obvious in the diaper pail or kitty litter box) is produced in tubule cells by the breakdown of amino acids. Phosphate provides another means of buffering hydrogen ions in urine.

Figure 24.15 Acid-base balance.

In the kidneys, bicarbonate ions (HCO_3^-) are reabsorbed and hydrogen ions (H^+) are excreted as needed to maintain the pH of the blood. Excess hydrogen ions are buffered, for example, by ammonia (NH_3), which becomes ammonium (NH_4^+). Ammonia is produced in tubule cells by the breakdown of amino acids.

Problems with Kidney Function

Many types of illnesses, especially diabetes, hypertension, and inherited conditions, cause progressive renal disease and renal failure. Urinary tract infections, an enlarged prostate gland, pH imbalances, or simply an intake of too much calcium can lead to kidney stones. Kidney stones form in the renal pelvis and usually pass unnoticed in the urine flow. If they grow to several centimeters and block the renal pelvis or ureter, back pressure builds up and destroys nephrons. One of the first signs of nephron damage is the presence of albumin, white blood cells, or even red blood cells in the urine, as detected by a urinalysis. If damage is so extensive that more than two-thirds of the nephrons are inoperative, urea and other waste substances accumulate in the blood. Although nitrogenous waste in the blood is a threat to homeostasis, the retention of water and salts is of even greater concern. **Edema,** fluid accumulation in the body tissues, may occur. Imbalance in the ionic composition of body fluids can lead to loss of consciousness and even to heart failure.

Hemodialysis and Kidney Transplants

Patients with renal failure can undergo **hemodialysis,** utilizing either an artificial kidney machine or **continuous ambulatory peritoneal dialysis (CAPD).** *Dialysis* is defined as the diffusion of dissolved molecules through a semipermeable membrane with pore sizes that allow only small molecules to pass through. In an artificial kidney machine, the patient's blood is passed through a membranous tube, which is in contact with a dialysis solution, or **dialysate (Fig. 24.16).** Substances more concentrated in the blood diffuse into the dialysate, and substances more concentrated in the dialysate diffuse into the blood. The dialysate is continuously replaced to maintain favorable concentration gradients. In this way, the artificial kidney can be utilized either to extract substances from the blood, including waste products or toxic chemicals and drugs, or to add substances to the blood—for example, bicarbonate ions (HCO_3^-) if the blood is acidic. In the course of a three- to six-hour hemodialysis procedure, from 50 to 250 grams of urea can be removed from a patient, which greatly exceeds the amount excreted by our kidneys within the same time frame. Therefore, a patient needs to undergo treatment only about twice a week.

CAPD is so named because the peritoneum, the epithelium that lines the abdominal cavity, is the dialysis membrane. A fresh amount of dialysate is introduced directly into the abdominal cavity from a bag that is temporarily attached to a permanently implanted plastic tube. The dialysate flows into the peritoneal cavity by gravity. Waste and salt molecules pass from the blood vessels in the abdominal wall into the dialysate before the fluid is collected four or eight hours later. The solution is drained into a bag from the abdominal cavity by gravity, and then it is discarded. One advantage of CAPD over an artificial kidney machine is that the individual can go about his or her normal activities during CAPD.

Patients with renal failure may undergo a transplant operation in which a functioning kidney from a donor is received. A person only needs one functioning kidney. The possibility of organ rejection exists as with all organ transplants. Receiving a kidney from a close relative has the highest chance of success. The current one-year survival rate is 97% if the kidney comes from a relative, and 90% if it is from a nonrelative. In the future, it may be possible to use kidneys from pigs or kidneys created in the laboratory for transplant operations.

Figure 24.16 **An artificial kidney machine.**

As the patient's blood is pumped through dialysis tubing, the tubing is exposed to a dialysate (dialysis solution). Wastes exit from blood into the solution because of a preestablished concentration gradient. In this way, blood is not only cleansed, but its water-salt and acid-base balances can also be adjusted.

Check Your **Progress**

1. Explain how the kidneys regulate blood pH.
2. Explain how an artificial kidney machine works.

Answers: 1. The kidneys reabsorb bicarbonate ions and excrete hydrogen ions as needed to maintain blood pH.   2. The patient's blood passes through a membranous tube filled with dialysate. The dialysate is continuously replaced to maintain favorable concentration gradients so that appropriate materials diffuse into and out of the blood.

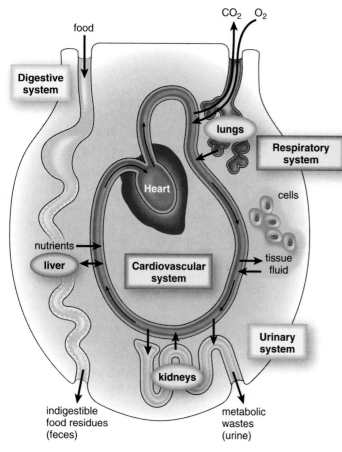

Figure 24.17   Keeping the internal environment steady.

The digestive system takes in food and digests it to nutrient molecules that enter the blood. The blood transports nutrients to the tissues where exchange occurs with tissue fluid.

Figure 24.18   Complete digestive system.

The earthworm (**a**) and humans (**b**) have complete digestive systems. A complete digestive system leads to specialization of organs along the digestive tract.

a.

24.3 | Digestive System

Digestion contributes to homeostasis by providing the body's cells with the nutrients they need to continue living. A digestive tract performs the following functions (**Fig. 24.17**):

1. Ingests food.
2. Breaks food down into small molecules that can cross plasma membranes.
3. Absorbs nutrient molecules.
4. Eliminates indigestible remains.

Tube-Within-a-Tube Body Plan

Hydras and planarians (see Fig. 23.17) have a sac body plan—that is, the mouth, like the top of a sack, serves as both an entrance and an exit. Most other animals, such as the earthworm, have a tube-within-a-tube plan, so called because the inner tube has both an entrance and an exit—the mouth and the anus. Therefore, it is called a complete digestive tract (**Fig. 24.18a**). Notice that the inner tube (the digestive tract, sometimes called the alimentary canal) is separated from the outer tube (the body wall) by the **coelom.** The digestive tract of humans and other vertebrates consists of so many specialized organs

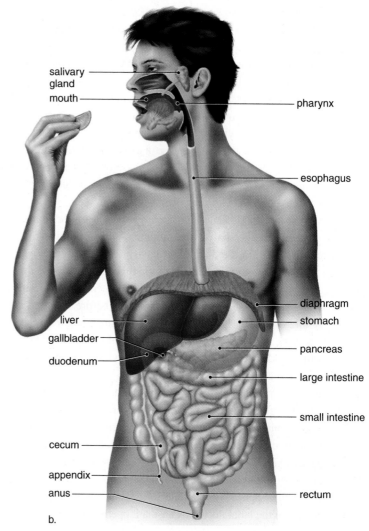

b.

that it might be hard to realize that the basic plan of vertebrates is the same as that of the earthworm (Fig. 24.18*b*). The tube-within-a-tube plan and a complete tract result in this specialization of parts.

Digestion of food in earthworms and humans is an extracellular process. Digestive enzymes, produced by glands in the wall of the tract or by accessory glands that lie nearby, are released into the tract. Food is never found within these accessory glands, only within the tract itself.

Mouth

In humans, the digestive system begins with the **mouth,** which chews food into pieces, beginning mechanical digestion. Many vertebrates have teeth, an exception being birds, which lack teeth and depend on the churning of small pebbles within a gizzard to break up their food. The teeth (dentition) of mammals reflect their diet (**Fig. 24.19**). **Carnivores** eat meat, which is easily digestible because the cells do not have a cellulose wall. **Herbivores** eat plant material, which needs a lot of chewing and other processing to break up the cellulose walls. Humans are **omnivores;** they eat both meat and plant material. The four front teeth (top and bottom) of humans are sharp, chisel-shaped incisors used for biting. On each side of the incisors are the pointed canines used for tearing food. The premolars and molars grind and crush food. It is as though humans are carnivores in the front of their mouths and herbivores in the back.

Food contains cells composed of molecules of carbohydrates, proteins, nucleic acids, and lipids. During digestion, hydrolytic enzymes break down these large molecules to smaller molecules. In the mouth, three pairs of **salivary glands** send saliva by way of ducts to the mouth, beginning chemical digestion. Saliva contains the enzyme **salivary amylase,** which breaks down starch, a carbohydrate, to maltose, a disaccharide. While in the mouth, food is manipulated by a muscular tongue, mixing the chewed food with saliva and then forming this mixture into a mass called a **bolus** that is swallowed. Taste buds that allow us to "taste" our food are also on the tongue as well as on the surface of the mouth.

Key:
- incisors
- canines
- premolars
- molars

a. Carnivores, such as cats, have enlarged canines for killing their prey, short incisors for scraping bones, and jagged molars for tearing apart flesh.

b. Herbivores, such as a horse, have sharp incisors for clipping grasses and large, flat premolars and molars for grinding.

c. The front teeth of humans resemble those of carnivores, and the back teeth are like those of herbivores.

Figure 24.19   Dentition among mammals.

nasal cavity
air
hard palate
tongue
soft palate
pharynx
epiglottis
glottis
larynx
esophagus

a. Anatomy of the pharynx

food mass
tongue
epiglottis

soft palate
food mass
epiglottis

soft palate
food mass
esophagus

b. Swallowing

Figure 24.20 The human mouth, pharynx, esophagus, and larynx.

a. The palates (both hard and soft) separate the mouth from the nasal cavities. The pharynx leads to the esophagus and the larynx. **b.** When food is swallowed, the soft palate closes off the nasal cavities, and the epiglottis closes off the larynx.

Swallowing

The human digestive and respiratory passages come together in the **pharynx** and then separate (**Fig. 24.20**a). When food is swallowed, the soft palate, the rear portion of the mouth's roof, moves up to close off the nasal cavities (Fig. 24.20b). A flap of tissue called the **epiglottis** covers the glottis, an opening into the larynx. Ordinarily the bolus must move through the pharynx and into the esophagus because the air passages are blocked. Unfortunately, we have all had the unpleasant experience of having food "go the wrong way". The wrong way may be either into the nasal cavities or into the trachea. If it is the latter, coughing will most likely force the food up out of the trachea and into the pharynx again.

The **esophagus** is a muscular tube that takes food to the stomach, which lies below the diaphragm. When food enters the esophagus, peristalsis begins. **Peristalsis** is a rhythmic contraction that moves the contents along in tubular organs—in this case, those of the digestive tract.

Stomach

The human **stomach** is a thick-walled, J-shaped organ (**Fig. 24.21**) that lies on the left side of the abdominal cavity below the liver. The stomach is continuous with the esophagus above and the duodenum of the small intestine below. The stomach is about 25 cm (10 in) long, regardless of the amount of food it holds, but the diameter varies, depending on how full it is. The stomach receives food from the esophagus, stores food, starts the digestion of proteins, and moves food into the small intestine.

The wall of the stomach has deep folds, which disappear as the stomach fills to an approximate capacity of 1 liter. Therefore, humans can periodically eat relatively large meals and spend the rest of their time at other activities. But the stomach is much more than a mere storage organ. The muscular walls of the stomach contract vigorously and mix food with juices that are secreted whenever food enters the stomach. The epithelial lining of the stomach, called a mucosa, has millions of gastric glands. These gastric glands produce gastric juice containing so much hydrochloric acid that the stomach routinely has a pH of about 2. Such strong acidity is usually sufficient to kill any microbes that might be in food. This low pH also promotes the activity of **pepsin,** a hydrolytic enzyme that acts on protein to produce peptides. In addition, strong acidity causes *heartburn* and *gastric reflux disease* when gastric juice backs up into the esophagus.

Check Your Progress

1. Compare and contrast the human digestive system with that of an earthworm.
2. List the functions of a digestive tract.
3. Explain why carnivores do not need teeth for grinding food.
4. Describe the function of peristalsis in digestion.

Answers: 1. Both are tube-within-a-tube plans, but the human system has many more specialized organs than that of an earthworm.   2. Ingest and break down food, absorb nutrients, and eliminate indigestible material.   3. Meat is easy to break down because it does not contain cellulose, so extensive chewing is not necessary.   4. Peristalsis moves materials through the digestive tract.

As with the rest of the digestive tract, a thick layer of mucus protects the wall of the stomach from enzymatic action. Still, an ulcer, which is an open sore in the wall caused by the gradual destruction of tissues, may occur in some individuals. Ulcers are due to an infection by an acid-resistant bacterium, *Helicobacter pylori*, which is able to attach to the epithelial lining. Wherever the bacterium attaches, the lining stops producing mucus, and the area becomes exposed to digestive action. Then an ulcer develops.

The stomach acts chemically but also acts physically on food. The wall of the stomach contains three muscle layers: one is longitudinal, another is circular, and the third layer is obliquely arranged. The muscular wall not only moves food along but it also churns, mixing the food with gastric juice and breaking it down into small pieces.

Alcohol and other liquids are absorbed in the stomach but food particles are not. Peristalsis pushes food along in the stomach as it does in other digestive organs (Fig. 24.21*b*). At the base of the stomach is a narrow opening controlled by a *sphincter*, a muscle that surrounds a tube and closes or opens it by contracting and relaxing. When food leaves the stomach, it is a thick, soupy liquid called chyme. Whenever the sphincter relaxes, a small quantity of chyme squirts through the opening into the small intestine.

Ruminants   Ruminants, a type of mammal that includes cattle, sheep, goats, deer, and buffalo, are named for a part of their stomach, the rumen (**Fig. 24.22**). The rumen contains symbiotic bacteria and protozoans that, unlike the mammal, produce enzymes that can digest cellulose. After these herbivores feed on grass, it goes to the rumen, where it is broken down by the enzymes released by the microbes and then it becomes small balls of cud. The cud returns to the mouth where the animal "chews the cud." The cud may return to the rumen for a second go-round before passing through the other chambers of the stomach. The rumen is an adaptation to a diet rich in fiber that may have been promoted by competition among the many types of animals that feed on grass. The last chamber in ruminants is analogous to the human stomach, being the place where protein is digested to peptides.

Figure 24.21   Anatomy of the human stomach.

a. The stomach has a thick wall that expands as it fills with food. The wall contains three layers of muscle, and their presence allows the stomach to churn and mix food with gastric juices. The mucosa of the stomach wall secretes mucus and contains gastric glands, which secrete gastric juices active in the digestion of protein. **b.** Peristalsis, a rhythmic contraction, occurs along the length of the digestive tract.

Check Your **Progress**

1. Describe the functions of the stomach.
2. Describe the specializations of the stomach of ruminants.

Answers: 1. The stomach stores food and continues digestion, kills microbes, and moves partially digested material into the intestines.   2. The stomachs of ruminants are divided into several chambers, including the rumen. The rumen houses bacteria and protozoans which secrete enzymes that digest cellulose.

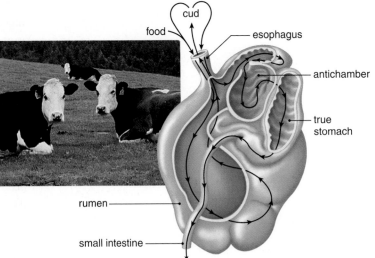

Figure 24.22   A ruminant's stomach.

Ruminants eat grass, which is made of cells with strong cellulose walls. The first chamber of a ruminant's stomach, called the rumen, contains symbiotic bacteria and protozoans that release enzymes that digest cellulose. After a first pass through the rumen, the "cud" returns to the mouth where it is leisurely chewed. Then, it may return to the rumen for a second go-round of digestion before passing through to the true stomach.

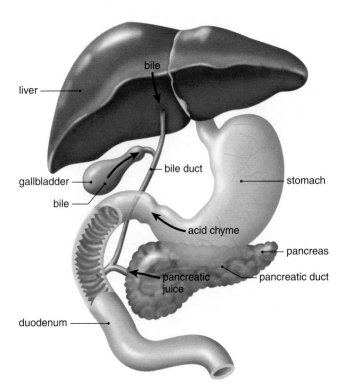

Figure 24.23 **The pancreatic and bile ducts empty into the duodenum.**

Bile, made by the liver and stored in the gallbladder, and pancreatic juice, which contains enzymes, enter the duodenum by way of ducts.

Small Intestine

Processing food in humans is more complicated than one might think. So far, the food has been chewed in the mouth and worked on by the enzyme salivary amylase, which digests starch to maltose. In addition, the digestion of protein has begun in the stomach as pepsin digests proteins to peptides. By now, the contents of the digestive tract are called **chyme.** Chyme passes from the stomach to the **small intestine,** a long, coiled tube that has two functions: (1) digestion of all the molecules in chyme, including polymers of carbohydrate, protein, nucleic acid, and fat; and (2) absorption of the nutrient molecules into the body.

The first part of the small intestine is called the duodenum. Two important accessory glands, the **liver,** the largest organ in the body, and the **pancreas,** located behind the stomach, send secretions to the duodenum by way of ducts (**Fig. 24.23**). The liver produces **bile,** which is stored in the **gallbladder.** Bile looks green because it contains pigments that are the products of hemoglobin breakdown. This green color is familiar to anyone who has observed how bruised tissue changes color. Hemoglobin within the bruise is breaking down into the same types of pigments found in bile. Bile also contains bile salts, which break up fat into fat droplets by a process called **emulsification.** Fat droplets mix with water and have more surface area for digestion by enzymes.

The pancreas produces pancreatic juice, which contains sodium bicarbonate ($NaHCO_3$) and digestive enzymes. $NaHCO_3$ neutralizes chyme and makes the pH of the small intestine slightly basic. The higher pH helps prevent autodigestion of the intestinal lining by pepsin and optimizes the pH for pancreatic enzymes. **Pancreatic amylase** digests starch to maltose; **trypsin** digests proteins to peptides; **lipase** digests fat droplets to glycerol and fatty acids; and **nuclease** digests nucleic acids to nucleotides.

Still more digestive enzymes are present in the small intestine. The wall of the small intestine contains fingerlike projections called **villi** (**Fig. 24.24***b*). The epithelial cells of the villi produce **intestinal enzymes,** which remain attached to them. These enzymes complete the digestion of peptides and sugars. Peptides, which result from the first step in protein digestion, are digested by peptidase to amino acids. Maltose, which results from the first step in starch digestion, is digested by maltase to glucose. Other disaccharides, each of which is acted upon by a specific enzyme, are digested to simple sugars also.

Finally, these small nutrient molecules can be absorbed into the body. Our cells use these molecules as a source of energy and as building blocks to make their own macromolecules.

Absorption by Villi The wall of the small intestine is adapted to absorbing nutrient molecules because it has an extensive surface area—approximately that of a tennis court! First, the mucous membrane layer of the small intestine has circular folds that give it an almost corrugated appearance (Fig. 24.24*a*). Second, on the surface of these circular folds are the villi. Finally, the cells on the surface of the villi have minute projections called **microvilli** (Fig. 24.24*c*). If the human small intestine were simply a smooth tube, it would have to be 500 to 600 meters long to have a comparable surface area for absorption. Carnivores have a much shorter digestive tract than herbivores because meat is easier to process than plant material (**Fig. 24.25**).

The villi of the small intestine absorb small nutrient molecules into the body. Each villus contains an extensive network of blood capillaries and a lymphatic capillary called a **lacteal.** As discussed in Chapter 23, the lymphatic system is an adjunct to the cardiovascular system—its vessels carry fluid, called lymph to the cardiovascular veins. Sugars and amino acids enter the

villi

lacteal
(lymphatic
capillary)

blood
capillaries

nucleus

columnar
epithelial cells

microvilli

a. Intestinal wall has
circular folds.

b. Circular folds
have villi.

×33,000

c. Cells of villi have microvilli.

Figure 24.24 **Small intestine and absorption of nutrients.**

The surface area of the small intestine is increased by three modifications: (**a**) circular folds,
(**b**) villi, and (**c**) microvilli. The blood vessels of the villi absorb amino acids and sugars.
Lacteals, which are lymphatic capillaries, absorb glycerol and fatty acids.

blood capillaries of a villus and are carried to the liver by way of the hepatic
portal system. In contrast, glycerol and fatty acids (digested from fats) enter
the epithelial cells of the villi, and within them are joined and packaged as
lipoprotein droplets, which enter a lacteal. Absorption continues until almost
all nutrient molecules have been absorbed. Absorption occurs by diffusion, as
well as by active transport, which requires an expenditure of cellular energy.
Lymphatic vessels transport lymph to cardiovascular
veins. Eventually, the bloodstream carries the nutrients absorbed by the digestive system to all the cells
of the body.

Large Intestine

The word *bowel* technically means the part of the digestive tract between the stomach and the anus, but it is sometimes used to mean only the large intestine. The **large
intestine** absorbs water, salts, and some vitamins. It also
stores indigestible material until it is eliminated at the anus.
The large intestine takes its name from its diameter rather
than its length, which is shorter than that of the small intestine. The large intestine has a blind pouch, the cecum, below
the entry of the small intestine with a small projection containing lymphatic tissue called the **appendix.** In humans, the
appendix may play a role in fighting infections by acting as a

a.

b.

Figure 24.25 **Digestive tract of a carnivore compared to a ruminant herbivore.**

The digestive tract of a carnivore (**a**) is much shorter than that of a ruminant herbivore (**b**) because
proteins can be more easily digested than plant matter.

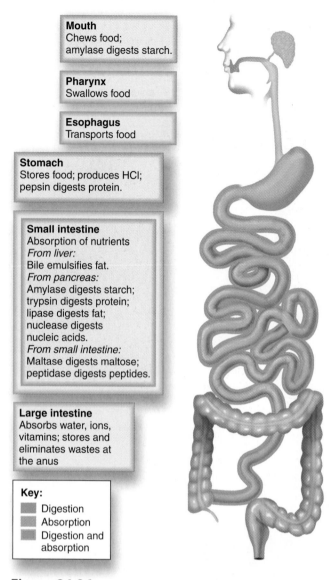

Mouth
Chews food;
amylase digests starch.

Pharynx
Swallows food

Esophagus
Transports food

Stomach
Stores food; produces HCl;
pepsin digests protein.

Small intestine
Absorption of nutrients
From liver:
Bile emulsifies fat.
From pancreas:
Amylase digests starch;
trypsin digests protein;
lipase digests fat;
nuclease digests
nucleic acids.
From small intestine:
Maltase digests maltose;
peptidase digests peptides.

Large intestine
Absorbs water, ions,
vitamins; stores and
eliminates wastes at
the anus

Key:
■ Digestion
■ Absorption
■ Digestion and
  absorption

Figure 24.26 **The digestive organs and their functions.**

A review of the processing of food to nutrient molecules and their absorption into the body.

reservoir of beneficial bacteria. In the condition called *appendicitis,* the appendix becomes infected and so filled with fluid that it may burst. If an infected appendix bursts before it can be removed, a serious, generalized infection of the abdominal lining, called peritonitis, may result.

The large intestine has a large population of bacteria, notably *Escherichia coli.* The bacteria break down indigestible material, and they also produce some vitamins, including vitamin K. Vitamin K is necessary for blood clotting. Digestive wastes (feces) eventually leave the body through the **anus,** the opening of the anal canal. Feces are about 75% water and 25% solid matter. Almost one-third of this solid matter is made up of intestinal bacteria. The remainder is indigestible plant material (also called fiber), fats, waste products (such as bile pigments), inorganic material, mucus, and dead cells from the intestinal lining. A diet that includes fiber adds bulk to the feces, improves regularity of elimination, and prevents *constipation.*

About 1.5 liters of water enter the digestive tract daily as a result of eating and drinking. An additional 8.5 liters also enter the digestive tract each day, carrying the various substances secreted by the digestive glands. About 95% of this water is absorbed by the small intestine, and much of the remaining portion is absorbed by the large intestine. If this water is not reabsorbed, **diarrhea** can occur, leading to serious dehydration and ion loss, especially in children.

The large intestine (also called the colon) is subject to the development of **polyps,** which are small growths arising from the mucosa. Polyps, whether they are benign or cancerous, can be removed surgically.

Figure 24.26 reviews the digestive process.

Accessory Organs

The pancreas and the liver are accessory organs of digestion, along with the teeth, salivary glands, and gallbladder.

Pancreas

The pancreas (see Fig. 24.23) functions as both an endocrine gland and an exocrine gland. **Endocrine glands** are ductless and secrete their products into the blood. The pancreas is an endocrine gland when it produces and secretes the hormones **insulin** and **glucagon** into the bloodstream. **Exocrine glands** secrete into ducts. The pancreas is an exocrine gland when it produces and secretes pancreatic juice into the **duodenum** of the small intestine through the common bile duct.

Liver

The liver has numerous functions, including the following:

1. Detoxifies the blood by removing and metabolizing poisonous substances.
2. Produces the plasma proteins, such as albumin and fibrinogen.
3. Destroys old red blood cells and converts hemoglobin to the breakdown products in bile (bilirubin and biliverdin).
4. Produces bile, which is stored in the gallbladder before entering the small intestine, where it emulsifies fats.
5. Stores glucose as glycogen and breaks down glycogen to glucose between meals to maintain a constant glucose concentration in the blood.
6. Produces urea from amino acids and ammonia.

Blood vessels from the large and small intestines as well as others merge to form the hepatic portal vein, which leads to the liver (**Fig. 24.27**). The liver helps maintain the glucose concentration in blood at about 0.1% by removing excess glucose from the hepatic portal vein and storing it as glycogen. Between meals, glycogen is broken down to glucose, and glucose enters the hepatic veins. Like plant-made starch, glycogen is made up of glucose molecules, and thus it is sometimes called animal starch. If the supply of glycogen and glucose runs short, the liver converts amino acids to glucose molecules.

Amino acids contain nitrogen in the form of amino groups, whereas glucose contains only carbon, oxygen, and hydrogen. Therefore, before amino acids can be converted to glucose molecules, deamination, the removal of amino groups from amino acids, must occur. By a complex metabolic pathway, the liver converts the amino groups to urea, the most common nitrogenous (nitrogen-containing) waste product of humans. After urea is formed in the liver, it is transported by the bloodstream to the kidneys, where it is excreted.

Liver Disorders   When a person is jaundiced, the skin has a yellowish tint due to an abnormally large quantity of bile pigments in the blood. In hemolytic jaundice, red blood cells are broken down in abnormally large amounts; in obstructive jaundice, the bile duct is obstructed, or the liver cells are damaged. Obstructive jaundice often occurs when crystals of cholesterol precipitate out of bile and form gallstones.

Jaundice can also result from viral infection of the liver, called *hepatitis*. Hepatitis A is most often caused by eating contaminated food. Hepatitis B and C are commonly spread by blood transfusions, kidney dialysis, and injection with unsterilized needles. These three types of hepatitis can also be spread by sexual contact.

Cirrhosis is a chronic liver disease in which the organ first becomes fatty, and later liver tissue is replaced by inactive fibrous scar tissue. Alcoholics often get cirrhosis, most likely due at least in part to the excessive amounts of alcohol the liver is forced to break down.

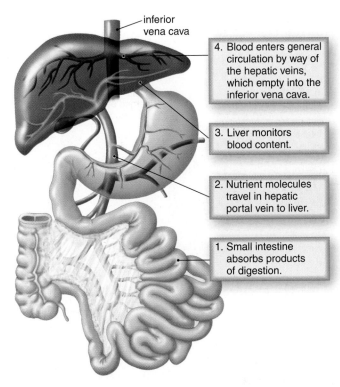

Figure 24.27   **Hepatic portal system.**

The hepatic portal vein takes the products of digestion from the digestive system to the liver, where they are processed before entering hepatic veins.

Digestive Juices

Does your mouth water when you smell food cooking? Even the thought of food can sometimes cause the nervous system to order the secretion of digestive juices. The secretion of these juices is also under the influence of several peptide hormones, so called because each is a small sequence of amino acids. When you eat a meal rich in protein, the stomach wall produces a peptide hormone that enters the bloodstream and doubles back to cause the stomach to produce more gastric juices. When protein and fat are present in the small intestine, another peptide hormone made in the intestinal wall stimulates the secretion of bile and pancreatic juices. In this way, the organs of digestion regulate their own needs.

The various digestive enzymes present in the digestive juices help break down carbohydrates, proteins, nucleic acids, and fats, the major components of food.

Check Your **Progress**

1. Contrast the anatomy and functions of the small intestine and the large intestine.
2. Describe the relationship between the duodenum and the liver and pancreas.
3. List the functions of the liver.

Answers: 1. The small intestine digests all types of food, and it produces enzymes that complete the breakdown of food to small molecules that can be absorbed across its villi. The villi are folds in the wall of the small intestine, and they are lined with cells that have microvilli formed by their plasma membranes. The small intestine is relatively long and skinny. No digestion occurs in the large intestine, but it does store digestive remains until they are eliminated. The large intestine also absorbs water, salts, and some vitamins. The large intestine lacks microvilli and villi. It is shorter in length and larger in diameter than the small intestine.   2. The liver and pancreas secrete bile and pancreatic juice, respectively, into the duodenum. These substances contribute to the digestion and absorption of nutrients.   3. The liver detoxifies the blood, produces plasma proteins, produces bile, stores glucose as glycogen, and produces urea.

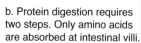

a. Carbohydrate (starch) digestion requires two steps. Only glucose molecules are absorbed at intestinal villi.

b. Protein digestion requires two steps. Only amino acids are absorbed at intestinal villi.

c. Fat digestion requires emulsification before digestion occurs. Then monoglycerides (glycerol bonded to one fatty acid) and free fatty acids are absorbed at intestinal villi.

Figure 24.28 Digestion and absorption of nutrients in the small intestine.

Starch is a carbohydrate, and its digestion begins in the mouth. Saliva from the salivary glands has a neutral pH and contains salivary amylase, the first enzyme to act on starch.

$$\text{starch} + H_2O \xrightarrow{\text{salivary amylase}} \text{maltose}$$

Maltose, a disaccharide, cannot be absorbed by the intestine; additional digestive action in the small intestine converts maltose to glucose, which can be absorbed.

Protein digestion begins in the stomach. Gastric juice secreted by gastric glands has a very low pH—about 2—because it contains hydrochloric acid (HCl). Pepsin, which is also present in gastric juice, acts on protein to produce peptides.

$$\text{protein} + H_2O \xrightarrow{\text{pepsin}} \text{peptides}$$

Peptides are usually too large to be absorbed by the intestinal lining, but later they are broken down to amino acids in the small intestine.

Starch, proteins, fats, and nucleic acids are all enzymatically broken down in the small intestine (Fig. 24.28). Pancreatic juice, which enters the duodenum, has a basic pH because it contains sodium bicarbonate ($NaHCO_3$). One pancreatic enzyme, pancreatic amylase, digests starch (Fig. 24.28a).

$$\text{starch} + H_2O \xrightarrow{\text{pancreatic amylase}} \text{maltose}$$

Another pancreatic enzyme, trypsin, digests protein (Fig. 24.28b).

$$\text{protein} + H_2O \xrightarrow{\text{trypsin}} \text{peptides}$$

Maltase and peptidases, enzymes produced by the small intestine, complete the digestion of starch to glucose and protein to amino acids, respectively. Glucose and amino acids are small molecules that cross into the cells of the villi and enter the blood (Fig. 24.28a, b).

Maltose, a disaccharide that results from the first step in starch digestion, is digested to glucose by **maltase.**

$$\text{maltose} + H_2O \xrightarrow{\text{maltase}} \text{glucose} + \text{glucose}$$

Other disaccharides have their own enzyme and are digested in the small intestine. The absence of any one of these enzymes can cause illness.

Peptides, which result from the first step in protein digestion, are digested to amino acids by **peptidases.**

$$\text{peptides} + H_2O \xrightarrow{\text{peptidases}} \text{amino acids}$$

Lipase, a third pancreatic enzyme, digests fat molecules in fat droplets after they have been emulsified by bile salts.

$$\text{fat droplets} + H_2O \xrightarrow{\text{lipase}} \text{glycerol} + 3 \text{ fatty acids}$$

Specifically, the end products of lipase digestion are monoglycerides (glycerol + one fatty acid) and fatty acids. These enter the cells of the villi, and within these cells, they are rejoined and packaged as lipoprotein droplets, called chylomicrons. Chylomicrons enter the lacteals (Fig. 24.28c).

The Chapter in Review

Summary

24.1 | Respiratory System

A respiratory system has these functions: (1) breathing; (2) external exchange of gases; and (3) internal exchange of gases.

The respiratory tract consists of several parts:

- **Nasal cavities:** In the nasal cavities of the nose, air is moistened, and hairs trap debris.
- **Pharynx:** Air crosses from front to back.
- **Larynx:** Contains the vocal cords.
- **Respiratory tree:** Consists of the trachea, bronchi, and bronchioles, which terminate in the alveoli of the lungs.

Insects have many tracheae. A tracheal system has no need of a respiratory pigment because air is delivered directly to cells.

- **Lungs:** Contain many alveoli, air sacs surrounded by a capillary network.

Breathing

- **Inspiration:** The diaphragm lowers, and the rib cage moves up and out; the lungs expand and air rushes in.
- **Expiration:** The diaphragm relaxes and moves up; the rib cage moves down and in; pressure in the lungs increases; air is pushed out of the lungs.

Lungs and External Exchange of Gases

- Alveoli are surrounded by pulmonary capillaries.
- CO_2 diffuses from the blood into the alveoli, and O_2 diffuses into the blood from the alveoli, because of their respective concentration gradients.
- Hemoglobin takes up oxygen.

Transport and Internal Exchange of Gases

Oxygen for cellular respiration follows this path:

- Transported by the iron portion of hemoglobin.
- Exits blood at tissues by diffusion.
- Given up by hemoglobin in capillaries because cellular respiration makes tissues warmer and more acidic.

Carbon dioxide, from cellular respiration, follows this path:

- Enters blood by diffusion.
- Taken up by red blood cells or joins with water to form carbonic acid.
- Carbonic acid breaks down to H^+ and bicarbonate ion (HCO_3^-):

$$CO_2 + H_2O \longrightarrow H_2CO_3 \longrightarrow H^+ + HCO_3^-$$

| carbon dioxide | water | carbonic acid | hydrogen ion | bicarbonate ion |

Bicarbonate ions are carried in plasma. H^+ combines with globin of hemoglobin.

- In the lungs, H^+ joins with bicarbonate ion to form carbonic acid, which breaks down to water and carbon dioxide:

$$H^+ + HCO_3^- \longrightarrow H_2CO_3 \longrightarrow H_2O + CO_2$$

24.2 | Urinary System and Excretion

The kidneys perform the following functions:

- Excrete nitrogenous wastes, such as urea and uric acid.
- Maintain the normal water-salt balance of blood.
- Maintain the acid-base balance of blood.

The urinary system consists of these parts:

- **Kidneys:** Produce urine.
- **Ureters:** Take urine to the bladder.
- **Urinary bladder:** Stores urine.
- **Urethra:** Releases urine to the outside.

Kidneys

Macroscopically, the kidneys have three parts: renal cortex, renal medulla, and renal pelvis. Microscopically, they contain the nephrons. A nephron has a nephron capsule, proximal tubule, nephron loop, and distal tubule.

Urine formation requires three steps:

- **Filtration:** Water, nutrients, and wastes move from the blood to the inside of the nephron capsule.
- **Reabsorption:** Primarily salts, water, and nutrients are reabsorbed at the proximal tubule.
- **Secretion:** Certain substances (e.g., hydrogen ions) are transported into the distal tubule from blood.

Regulation of Water-Salt Balance and pH Animals regulate their osmolarity; examples are the Malpighian tubules of insects and the specialized organs of saltwater versus freshwater fishes. The reabsorption of water and the production of a hypertonic urine involve establishment of a solute gradient that pulls water from the descending limb of the nephron loop and from the collecting duct. The kidneys keep blood pH at about 7.4 by reabsorbing HCO_3^- and excreting H^+ as needed. Ammonia buffers H^+ in the urine.

Problems with Kidney Function

Various medical conditions, including diabetes, kidney stones, and kidney infections, can lead to renal failure. Renal failure can be treated by hemodialysis using a kidney machine or CAPD, or by a kidney transplant.

24.3 | Digestive System

The digestive system, along with the respiratory and urinary systems, makes exchanges with the external environment and blood, thereby supplying the blood with nutrients and oxygen and cleansing it of waste molecules. Then the blood makes exchanges with the tissue fluid, and in this way cells acquire nutrients and oxygen and rid themselves of wastes.

Tube-Within-a-Tube Body Plan

Like the earthworm, humans have a complete digestive system that (1) ingests food, (2) breaks food down to small molecules that can cross plasma membranes, (3) absorbs these nutrient molecules, and (4) eliminates indigestible remains.

The digestive tract consists of several specialized parts:

- **Mouth:** Teeth chew the food, saliva contains salivary amylase for digesting starch, and the tongue forms a bolus for swallowing.
- **Pharynx:** The air and food passages cross in the pharynx. During swallowing, the air passage is blocked off by the soft palate and epiglottis; peristalsis begins.
- **Stomach:** The stomach expands and stores food and also churns, mixing food with the acidic gastric juices. This juice contains pepsin, an enzyme that digests protein. The stomach of ruminants has a special chamber, the rumen, where symbiotic bacteria and protozoans release enzymes that digest grass.
- **Small intestine:** The duodenum of the small intestine receives bile from the liver and pancreatic juice from the pancreas. Pancreatic juice contains trypsin (digests protein), lipase (digests fat), and pancreatic amylase (digests starch). The small intestine produces enzymes that finish digestion, breaking food down to small molecules that cross the villi. Amino acids and glucose enter blood capillaries. Glycerol and fatty acids are joined and packaged as lipoproteins before entering lymphatic vessels called lacteals.
- **Large intestine:** The large intestine stores the remains of digestion until they can be eliminated. It also absorbs water, salts, and some vitamins. Reduced water absorption results in diarrhea. The intake of water and fiber helps prevent constipation.

Accessory Organs

The pancreas is both an exocrine gland that produces pancreatic juice and an endocrine gland that produces the hormones insulin and glucagon. The liver produces bile, which is stored in the gallbladder. Pancreatic juice and bile enter the small intestine. The nervous system and the peptide hormones regulate the secretion of digestive juices and bile.

The hepatic portal vein carries absorbed molecules from the small intestine to the liver, an organ that performs many important functions.

▬ Key Terms

▬ Testing Yourself

Choose the best answer for each question.

1. Label the components of the human respiratory system in the following illustration.

2. Infection of one of the two branches just below the trachea is called
   a. tonsillitis.
   b. meningitis.
   c. a sinus infection.
   d. bronchitis.
   e. laryngitis.

3. In insects, _____ are small air tubes that carry air into every part of the body.
   a. bronchi
   b. spiracles
   c. tracheae
   d. alveoli

4. Birds have a higher gas-exchange capacity than other vertebrates because they
   a. have more alveoli in their lungs.
   b. have a stronger diaphragm.
   c. have a shorter trachea.
   d. never mix fresh air with used air.
   e. breathe at a more rapid rate.

5. The _____ or windpipe connects the larynx to the bronchi.
   a. glottis
   b. trachea
   c. pharynx
   d. epiglottis

6. Gas exchange occurs in the _____, small pockets in the bronchioles.
   a. alveoli
   b. tracheae
   c. tonsils
   d. nephrons

7. The larynx is
   a. the site of gas exchange in the human respiratory system.
   b. located between the trachea and the bronchi.
   c. a flap of cartilage which keeps food out of the respiratory tract during swallowing.
   d. the voice box.

8. Heme
   a. is the protein portion of a hemoglobin molecule.
   b. is the lead-containing group in a hemoglobin molecule.
   c. is the portion of a hemoglobin molecule which directly combines with oxygen molecules.
   d. floats free in blood plasma.

9. Label the components of the human urinary system in the following illustration.

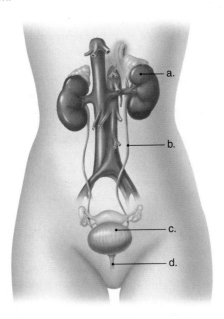

For questions 10–13, identify the kidney component in the key that matches the description.

Key:
   a. nephron capsule
   b. proximal tubule
   c. nephron loop
   d. distal tubule

10. pumps salt into the renal medulla, and water follows by osmosis
11. uric acid is eliminated by secretion here
12. microvilli reabsorb molecules such as glucose here
13. filtration occurs here
14. The presence of albumin in the urine is indicative of
   a. kidney stones.
   b. exposure to diuretics.
   c. the inability to adjust blood pH.
   d. hypertension.

15. _____ are the excretory organs of insects.
   a. Malpighian tubules
   b. Tracheae
   c. Kidneys
   d. Nephrons

16. Cleansing inhaled air involves
   a. hairs.
   b. cilia.
   c. mucus.
   d. All of these are correct.

17. Food and air both travel through the
   a. lungs.
   b. pharynx.
   c. larynx.
   d. trachea.

In questions 18–21, match the functions to the structures in the key.

Key:
   a. glottis
   b. larynx
   c. bronchi
   d. lungs

18. passage of air into larynx
19. passage of air into lungs
20. sound production
21. gas exchange
22. The use of an artificial kidney is known as
   a. filtration.
   b. excretion.
   c. hemodialysis.
   d. None of these are correct.

23. Which of the following materials would not be filtered from the blood at the nephron capsule?
   a. water
   b. urea
   c. protein
   d. nutrients
   e. salts

24. The renal medulla has a striped appearance due to the presence of which structures?
   a. nephron loop
   b. collecting duct
   c. capillaries
   d. Both a and b are correct.

25. Label the components of the human digestive system in the following illustration.

26. In the pharynx
    a. the digestive and respiratory passages come together.
    b. the glottis covers the epiglottis during swallowing.
    c. the digestive and urinary passages come together.
    d. chyme passes into the esophagus.

27. If plants did not have cellulose in their cell walls,
    a. an important nutrient would be missing in the diet of omnivores.
    b. the teeth of herbivores would more closely resemble those of carnivores.
    c. more birds would be carnivores.
    d. herbivores would have a longer digestive tract.

28. The stomach
    a. is lined with a thick layer of mucus.
    b. contains sphincter glands.
    c. has a pH of about 6.
    d. digests pepsin.
    e. More than one of these are correct.

29. A rumen in cows
    a. provides them with water.
    b. provides a home for symbiotic bacteria.
    c. helps them digest plant material.
    d. Both b and c are correct.

30. Which of the following is not a digestive enzyme?
    a. amylase   d. nuclease
    b. trypsin   e. isomerase
    c. lipase

31. Pepsin
    a. breaks down protein in the small intestine.
    b. breaks down protein in the stomach.
    c. is found in saliva.
    d. breaks down fats in the stomach.

32. The contents of the digestive tract are in the form of _____ when they pass from the stomach to the small intestine.
    a. chyme   c. feces
    b. a bolus   d. a small round mass

33. Which of the following is not a function of the liver?
    a. removal of poisonous substances from the blood
    b. secretion of digestive juices
    c. production of albumin
    d. storage of glucose
    e. production of bile

34. _____ are animals which eat meat.
    a. Omnivores   c. Frugivores
    b. Herbivores   d. Carnivores

Thinking Scientifically

1. Currently the only way people with diabetes mellitus can take their insulin is as an injection; no insulin pill is available. Taking into account what happens in the stomach and small intestine in terms of digestion and absorption, and that insulin is a protein hormone, explain why insulin can not be taken in pill form.

2. Individuals who have had their gallbladders removed are told to lower the fat intake in their diet after surgery. What effect will removal of the gallbladder have on their ability to digest and absorb fats?

3. Grasshoppers, like most other insects, are highly adapted to life on land. What adaptations to life on land are present in their respiratory and urinary systems?

Bioethical Issue

When to Terminate Life

As a result of serious injury or illness, a person may enter a persistent vegetative state. In this condition, the person does not recognize anyone and feels no pain. Advances in medical care now allow us to use ventilators and feeding tubes to keep vegetative patients alive for years. A living will allows patients to declare that they do not wish to be maintained in this way, giving doctors the legal authority to discontinue life-sustaining treatments. But for a patient without a living will, family members are left with a difficult decision and sometimes a court battle over what they all think is best. One recent case involved a woman who was kept in a vegetative state since suffering a heart attack in 1990. If her feeding tube were removed, she would die within two weeks. Her husband asked to have the tube removed, saying that would have been his wife's wish. Her parents, on the other hand, wanted her to be kept alive. They believed she was trying to communicate with noises and facial expressions, although her doctors said those activities were simply reflexes. What would you do if you were a judge and had to determine whether the husband's or the parents' wishes should be carried out?

Essentials of Biology Website

The companion website for *Essentials of Biology* provides a wealth of information organized and integrated by chapter. You will find practice tests, animations, videos, and much more that will complement your learning and understanding of general biology.

http://www.mhhe.com/maderessentials2

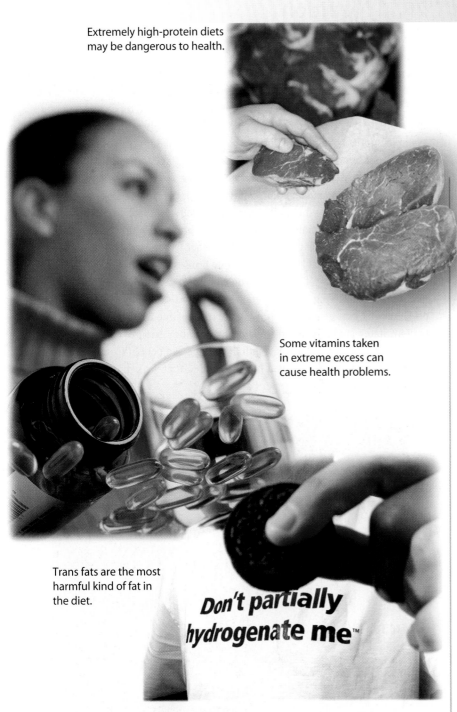

Extremely high-protein diets may be dangerous to health.

Some vitamins taken in extreme excess can cause health problems.

Trans fats are the most harmful kind of fat in the diet.

Don't partially hydrogenate me™

25

Human Nutrition

Outline

Trans Fat and Your Health

DID YOU KNOW that an extremely high-protein diet may eventually cause liver and kidney disease? That some vitamins can harm the body if taken in extreme excess? Or that trans fats are the worst possible kind of fat you can eat?

For some time now, we have known that cardiovascular health problems can result from eating high-fat and high-cholesterol foods. Further, saturated fats had to be listed on the Nutrition Facts panel of all

—*Continued next page*

441

Continued—

food packages, and many consumers became smart enough to check nutrition labels to avoid eating saturated fats. Enter trans fats. *Trans fats* are formed from liquid unsaturated fats when hydrogen has been added to them. This process, called hydrogenation, makes a liquid fat resemble a solid saturated fat. Why did manufacturers go through the trouble of creating trans fats when they could have just used a natural saturated fat? Well, we know that, until recently, the trans fat content of food did not have to be listed on nutrition labels. Therefore, a food product could appear healthier than it actually was if trans fats, instead of saturated fats, were used in processed foods.

But it turns out that trans fats are worse offenders than saturated fats when it comes to cardiovascular health. In 2006, the Food and Drug Administration (FDA) decided that trans-fat content *must* be disclosed on Nutrition Facts panels. Now, many manufacturers of products high in trans fats have lowered the trans-fat content of their foods but consumers should still check the nutrition labels for trans-fat content.

In this chapter, you will learn about the components of a healthy, balanced diet, as well as some of the consequences of an unhealthy one.

25.1 | Nutrition

The vigilance of your immune system, the strength of your muscles and bones, the ease with which your blood circulates—all aspects of your body's functioning—depend on proper **nutrition** (**Fig. 25.1**). A **nutrient** is a component of food that performs a physiological function in the body. Nutrients provide us with energy, promote growth and development by providing the material for cellular structures, and regulate cellular metabolism. They are also involved in homeostasis. For example, nutrients help maintain the fluid balance and proper pH of blood. Your body can make up for a nutrient deficiency to a degree, but eventually signs and symptoms of a **deficiency disorder** will appear. As an example, vitamin C is needed to synthesize and maintain *collagen*, the protein that holds tissues together. When the body lacks vitamin C, collagen weakens, and capillaries break easily. Gums may bleed, especially when the teeth are brushed, or tiny bruises may form under the skin when it is gently pressed. In other words, early signs of vitamin C deficiency are gums that bleed and skin that bruises easily (**Fig. 25.2**).

By learning about nutrition, you can improve your diet and increase the likelihood of enjoying a longer, more active and productive life. Conversely, poor diet and lack of physical activity are responsible for 400,000 deaths in the United States annually. Such lifestyle factors may soon overtake smoking as the major cause of preventable death. Everyone can benefit from learning what constitutes a poor diet versus a healthy diet so that we can choose foods that supply all the nutrients in proper balance.

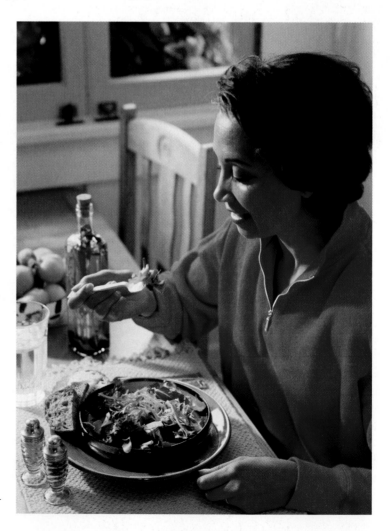

Figure 25.1 A healthy diet.

Making the right food choices leads to better health.

Introducing the Nutrients

Several factors, including cultural and ethnic backgrounds, financial situations, environmental conditions, and psychological states, influence what we eat. A **diet** is composed of a person's typical food choices. A *balanced diet* supplies all the nutrients in the proper proportions necessary for a healthy, functioning body.

There are six classes of nutrients: carbohydrates, lipids, proteins, minerals, vitamins, and water. An **essential nutrient** must be supplied by the diet because the body is not able to produce it or at least not in sufficient quantity to meet the body's needs. For example, amino acids are needed for protein synthesis and the body is unable to produce certain of them. Therefore, there are essential amino acids needed in the diet.

Carbohydrates, lipids, and proteins are called **macronutrients** because the body requires relatively large quantities of them. **Micronutrients,** such as vitamins and minerals, are needed in small quantities only. Macronutrients, not micronutrients, supply our energy needs. Although advertisements often imply that people can boost their energy levels by taking vitamin or mineral supplements, the body does not metabolize these nutrients for energy. Water does not provide energy either. Therefore, foods with high water content, such as fruits and vegetables, are usually lower in energy content than foods with less water and more macronutrient content.

Nearly every food is a mixture of nutrients. A slice of bread, for example, is about 50% carbohydrates, 35% water, 10% proteins, and 4% lipids. Vitamins and minerals make up less than 1% of the bread's nutrient content (**Fig. 25.3**). No single, naturally occurring food contains enough essential nutrients to meet all of our nutrient needs.

Contrary to popular belief, so-called "bad" foods or "junk foods" do have nutritional value. If a food contains water, sugar, or fat, it has nutritional value. However, such foods as sugar-sweetened soft drinks, cookies, and pastries have high amounts of fat and/or sugar in relation to their vitamin and mineral content. Therefore, these foods are more appropriately called *empty-calorie foods*, rather than junk foods. Diets that contain too many empty-calorie foods will assuredly lack enough vitamins and minerals.

a.

b.

Figure 25.2   Vitamin C deficiency.

a. Pinpoint hemorrhages (tiny bruises that appear as red spots in the skin) are an early indication of vitamin C deficiency. **b.** Bleeding gums are another early sign of vitamin C deficiency.

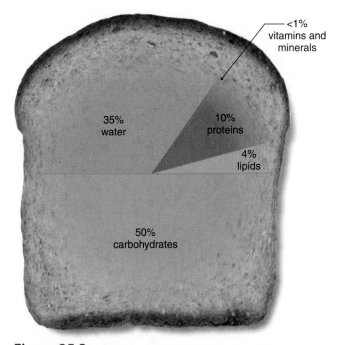

—<1% vitamins and minerals

35% water

10% proteins

4% lipids

50% carbohydrates

Figure 25.3   Nutrient composition of a slice of bread.

Most foods are mixtures of nutrients.

Check Your Progress

1. Which classes of nutrients do not provide energy?
2. Explain why the expression "junk food" may be inappropriate.
3. Give an example of an empty-calorie food, and explain why it fits this description.

Answers: 1. Vitamins, minerals, and water do not provide energy.   2. So-called "junk foods" do contain nutrients.   3. Empty-calorie foods that are rich sources of simple sugars and/or fat include sugar-sweetened soft drinks, pastries, and cookies. These are called empty-calorie foods because they provide little else besides calories.

Figure 25.4 **Major food sources.**

Carbohydrates, lipids, and proteins are represented.

25.2 | The Classes of Nutrients

Each type of nutrient usually has more than one function in the body and can be supplied by several food sources (**Fig. 25.4** and **Table 25.1**).

Carbohydrates

Carbohydrates are present in food as sugars, starch, and fiber. Fruits, vegetables, milk, and honey are natural sources of sugars. Glucose and fructose are monosaccharide sugars, and lactose (milk sugar) and sucrose (table sugar) are disaccharides. After absorption into the body, all sugars are converted to glucose for transport in the blood and use by cells. Glucose is the preferred direct energy source in cells.

Plants store glucose as starch, and animals store glucose as glycogen. High-starch foods are beans, peas, cereal grains, and potatoes. Starch is digested to glucose in the digestive tract, and any excess glucose is stored as glycogen. Although other animals likewise store glucose as glycogen in liver or muscle tissue (meat), it is gone by the time an animal is eaten for food. Except for honey and milk, which contain sugars, animal foods do not contain carbohydrates.

Fiber

Fiber includes various nondigestible carbohydrates derived from plants. Food sources rich in fiber include beans, peas, nuts, fruits, and vegetables. Whole-grain products are also a good source of fiber, and are therefore more nutritious than food products made from refined grains. During *refinement*, fiber and also vitamins and minerals are removed from grains, so that primarily starch remains. A slice of bread made from whole-wheat flour, for example, contains 3 grams of fiber; a slice of bread made from refined wheat flour contains less than a gram of fiber.

Table **25.1** | Summarizing the Classes of Nutrients

Class of Nutrient	Major Food Sources	Primary Physiological Roles	Macronutrient or Micronutrient
Carbohydrates	Sugars and starches in fruits, vegetables, cereals and other grains	Glucose is metabolized for energy; fiber adds to fecal bulk, preventing constipation.	Macronutrient
Lipids	Oil, margarine, salad dressings, meat, fish, poultry, nuts, fried foods, dairy products made with whole milk	Triglycerides are metabolized for energy and stored for insulation and protection of organs. Cholesterol is used to make certain hormones, bile, and vitamin D.	Macronutrient
Proteins	Meat, fish, poultry, eggs, nuts, dried beans and peas, soybeans, cereal products, milk, cheese, yogurt	Proteins are needed for building, repairing, and maintaining tissues; synthesizing enzymes and certain hormones; and producing antibodies.	Macronutrient
Minerals	Widespread in vegetables and other foods	Regulation of energy metabolism, maintenance of fluid balance, and production of certain structures, enzymes, and hormones	Micronutrient
Vitamins	Widespread in foods; plant foods are good sources of antioxidants.	Regulation of metabolism and physiological development	Micronutrient
Water	Widespread in foods, beverages	Participant in many chemical reactions; needed for maintenance of body fluids, temperature regulation, joint lubrication, and transportation of material in cells and the body	(Not applicable)

Technically, fiber is not a nutrient for humans because it cannot be digested to small molecules that enter the bloodstream. Insoluble fiber, however, adds bulk to fecal material, which stimulates movements of the large intestine, preventing constipation. Soluble fiber combines with bile acids and cholesterol in the small intestine and prevents them from being absorbed. In this way, high-fiber diets may protect against heart disease. The typical American consumes only about 15 grams of fiber each day; the recommended daily intake of fiber is 25 grams for women and 38 grams for men. To increase your fiber intake, eat whole-grain foods, snack on fresh fruits and raw vegetables, and include nuts and beans in your diet (**Fig. 25.5**).

Can Carbohydrates Be Harmful?

If you or someone you know has lost weight by following the Atkins or South Beach diet, you may think "carbs" are unhealthy and should be avoided. According to nutritionists, however, carbohydrates should supply a large portion of your energy needs. Evidence suggests that Americans are not eating the right kind of carbohydrates. In some countries, the traditional diet is 60–70% high-fiber carbohydrates, and these people have a low incidence of the diseases that plague Americans.

Obesity is associated with *type 2 diabetes* and *cardiovascular disease*, as discussed in Section 25.3. Some nutritionists hypothesize that the high intake of foods, such as those rich in refined carbohydrates and fructose sweeteners processed from cornstarch, may be responsible for the prevalence of obesity in the United States. These foods are empty-calorie foods that provide sugars but no vitamins or minerals. **Table 25.2** tells you how to reduce dietary sugars. Other nutritionists point out that consuming too much energy from any source contributes to body fat, which increases a person's risk of obesity and associated illnesses. Because many foods, such as doughnuts, cakes, pies, and cookies, are high in both refined carbohydrates and fat, it is difficult to determine which dietary component is responsible for the current epidemic of obesity among Americans.

Many people mistakenly believe children become hyperactive after eating sugar. There is no scientific basis for this belief because sucrose is broken down into glucose and fructose in the small intestine, and these sugars are absorbed into the bloodstream. The spike caused by their absorption does not last long. Excess glucose and fructose enters the liver, and fructose is converted to glucose. As you know, the liver stores glucose as glycogen, and glycogen is broken down to maintain the proper glucose level.

Lipids

Like carbohydrates, **triglycerides** (fats and oils) supply energy for cells, but **fat** is stored for the long term in the body. Fat deposits under the skin, called *subcutaneous fat*, insulate the body from cold temperatures; deeper fat deposits in the trunk protect organs against injury.

Nutritionists generally recommend that unsaturated rather than saturated fats be included in the diet. Two unsaturated fatty acids (alpha-linolenic and linoleic acids) are *essential dietary fatty acids*. Delayed growth and skin problems can develop when the diet lacks these essential fatty acids, which can be supplied by eating fatty fish and by including plant oils, such as canola and soybean oils, in the diet.

Figure 25.5   Fiber-rich foods.

Plants provide a good source of carbohydrates. They also provide fiber when they are not processed (refined).

Table 25.2 | Reducing Dietary Sugars

TO REDUCE DIETARY SUGAR:

1. Eat fewer sweets, such as candy, soft drinks, ice cream, and pastry.

2. Eat fresh fruits or fruits canned without heavy syrup.

3. Use less sugar—white, brown, or raw—and less honey and syrups.

4. Avoid sweetened breakfast cereals.

5. Eat less jelly, jam, and preserves.

6. Drink pure fruit juices, not imitations.

7. When cooking, use spices such as cinnamon, instead of sugar, to flavor foods.

8. Do not put sugar in tea or coffee.

9. Avoid potatoes and processed foods made from refined carbohydrates, such as white bread, rice, and pasta. These foods are immediately broken down to sugar during digestion.

Table 25.3 | Reducing Certain Lipids

TO REDUCE DIETARY SATURATED FAT:

1. Choose poultry, fish, or dry beans and peas as a protein source.

2. Remove skin from poultry and trim fat from red meats before cooking, and place on a rack so that fat drains off.

3. Broil, boil, or bake rather than frying.

4. Limit your intake of butter, cream, trans fats, shortening, and tropical oils (coconut and palm oils).*

5. Use herbs and spices to season vegetables instead of butter, margarine, or sauces. Use lemon juice instead of salad dressing.

6. Drink skin milk instead of whole milk, and use skim milk in cooking and baking.

TO REDUCE DIETARY CHOLESTEROL:

1. Eat white fish and poultry in preference to cheese, egg yolks, liver, and certain shellfish (shrimp and lobster).

2. Substitute egg whites for egg yolks in both cooking and eating.

3. Include soluble fiber in the diet. Oat bran, oatmeal, beans, corn, and fruits, such as apples, citrus fruits, and cranberries are high in soluble fiber.

*Although coconut and palm oils are from plant sources, they are mostly saturated fats.

Figure 25.6 Beans as food.

Beans are a good source of protein, but they do not supply all nine of the essential amino acids. To ensure a complete source of protein in the diet, they should be eaten in combination with a grain such as rice.

Animal foods such as butter, meat, whole milk, and cheeses contain saturated fats. Plant oils contain unsaturated fats. Each type of oil has a particular percentage of monounsaturated and polyunsaturated fatty acids (see Fig. 3.13).

Cholesterol, a lipid, can be synthesized by the body. Cells use cholesterol to make various compounds, including bile, steroid hormones, and vitamin D. Plant foods do not contain cholesterol; only animal foods such as cheese, egg yolks, liver, and certain shellfish (shrimp and lobster) are rich in cholesterol.

Can Lipids Be Harmful?

Elevated blood cholesterol levels are associated with an increased risk of cardiovascular disease, the number one killer of Americans. A diet rich in cholesterol and saturated fats increases the risk of cardiovascular disease (see also Section 25.3). Statistical studies suggest that trans fats are even more harmful than saturated fats. Trans fats arise when unsaturated oils are hydrogenated to produce a solid fat found largely in processed foods. Trans fatty acids may reduce the function of plasma membrane receptors that clear cholesterol from the bloodstream. Trans fatty acids are found in commercially packaged foods such as cookies and crackers, in commercially fried foods such as french fries, in packaged snacks such as microwave popcorn, and in vegetable shortening and some margarines. **Table 25.3** tells you how to reduce harmful lipids in the diet.

Proteins

Dietary **proteins** are digested to amino acids, which cells use to synthesize hundreds of cellular proteins. Of the 20 different amino acids, nine are *essential amino acids* that must be present in the diet. Children will not grow if their diets lack the essential amino acids. Eggs, milk products, meat, poultry, and most other foods derived from animals contain all nine essential amino acids and are "complete" or "high-quality" protein sources.

Foods derived from plants generally do not have as much protein per serving as those derived from animals, and each type of plant food generally lacks one or more of the essential amino acids. Therefore, most plant foods are "incomplete" or "low-quality" protein sources. Vegetarians, however, do not have to rely on animal sources of protein. To meet their protein needs, total vegetarians (vegans) can eat grains, beans, and nuts in various combinations (**Fig. 25.6**). Also, tofu, soymilk, and other foods made from processed soybeans are complete protein sources. A balanced vegetarian diet is quite possible with a little planning.

Can Proteins Be Harmful?

According to nutritionists, protein should not supply the bulk of dietary calories. The average American eats about twice as much protein as he or she needs, and some people may be on a diet that encourages the intake of proteins instead of carbohydrates as an energy source. Also, body builders, weight lifters, and other athletes may include amino acid or protein supplements in the diet because they think these supplements will increase muscle mass. However, excess amino acids are not always converted into muscle tissue. When they are used as an energy source, the liver removes the nitrogen portion (*deamination*) and uses it to form urea, which is excreted in urine. The water needed for excretion of urea can cause dehydration when a person is exercising and losing water by sweating. High-protein diets can also increase calcium loss in the urine and encourage the formation of kidney stones. Furthermore, high-protein foods often contain a high amount of fat.

Minerals

The body needs about 20 elements called **minerals** for numerous physiological roles, including regulation of biochemical reactions, maintenance of fluid balance, and incorporation into certain structures and compounds. **Major minerals** are needed in the body at a higher level than **trace minerals**. More than 100 milligrams (mg) per day of each major mineral is required in the diet and less than 100 mg per day of each trace mineral is required in the diet. **Table 25.4** lists the major minerals and a sample of the trace minerals. Their functions and food sources are given and as well as the health effects of too little or too much intake.

Occasionally, individuals (especially women) do not receive enough iron, calcium, magnesium, or zinc in their diets. Adult females need more iron in the diet than males (18 mg compared to 10 mg) if they are menstruating each month. *Anemia*, characterized by a run-down feeling due to insufficient red blood cells, results when the diet lacks sufficient iron. Many people take calcium supplements as directed by a physician to counteract *osteoporosis*, a

Table 25.4 | Minerals

Mineral	Functions	Food Sources	Too Little	Too Much
Major Minerals				
Calcium (Ca^{2+})	Strong bones and teeth, nerve conduction, muscle contraction	Dairy products, leafy green vegetables	Stunted growth in children, low bone density in adults	Kidney stones; interferes with iron and zinc absorption
Phosphorus (PO_4^{3-})	Bone and soft tissue growth; part of phospholipids, ATP, and nucleic acids	Meat, dairy products, sunflower seeds, food additives	Weakness, confusion, pain in bones and joints	Low blood and bone calcium levels
Potassium (K^+)	Nerve conduction, muscle contraction	Many fruits and vegetables, bran	Paralysis, irregular heartbeat, eventual death	Vomiting, heart attack, death
Sodium (Na^+)	Nerve conduction, pH and water balance	Table salt	Lethargy, muscle cramps, loss of appetite	High blood pressure, calcium loss
Chloride (Cl^-)	Water balance	Table salt	Not likely	Vomiting, dehydration
Magnesium (Mg^{2+})	Part of various enzymes for nerve and muscle contraction, protein synthesis	Whole grains, leafy green vegetables	Muscle spasm, irregular heartbeat, convulsions, confusion, personality changes	Diarrhea
Sulfur (S^{2-})	Part of some amino acids and some vitamins	Meats, legumes, and whole grains	Not likely	Not likely
Trace Minerals				
Zinc (Zn^{2+})	Protein synthesis, wound healing, fetal development and growth, immune function	Meats, legumes, whole grains	Delayed wound healing, night blindness, diarrhea, mental lethargy	Anemia, diarrhea, vomiting, renal failure, abnormal cholesterol levels
Iron (Fe^{2+})	Hemoglobin synthesis	Whole grains, meats, prune juice	Anemia, physical and mental sluggishness	Iron toxicity disease, organ failure, eventual death
Copper (Cu^{2+})	Iron metabolism; part of various antioxidant enzymes	Meat, nuts, legumes	Anemia, stunted growth in children	Damage to internal organs if not excreted
Iodine (I^-)	Thyroid hormone synthesis	Iodized table salt, seafood	Thyroid deficiency	Depressed thyroid function, anxiety
Selenium (SeO_4^{2-})	Part of antioxidant enzyme	Seafood, meats, eggs	Vascular collapse, possible cancer development	Hair and fingernail loss, discolored skin

Table 25.5 | Reducing Dietary Sodium

TO REDUCE DIETARY SODIUM:

1. Use spices instead of salt to flavor foods.

2. Add little or no salt to foods at the table, and add only small amounts of salt when you cook.

3. Eat unsalted crackers, pretzels, potato chips, nuts, and popcorn.

4. Avoid hot dogs, ham, bacon, luncheon meats, smoked salmon, sardines, and anchovies.

5. Avoid processed cheese and canned or dehydrated soups.

6. Avoid brine-soaked foods, such as pickles or olives.

7. Read labels to avoid high-salt products.

Figure 25.7   Rickets.

Bowing of bones (rickets) due to vitamin D deficiency.

Check Your Progress

1. **What two minerals are needed for strong bones and teeth?**
2. **List the vitamins that are antioxidants.**
3. **Name the nutrient in your body that is present in the largest quantity.**

Answers: 1. Calcium and phosphorus are needed for strong bones and teeth.   2. Vitamins C, E, and A (beta-carotene) are antioxidants.   3. Water is present in the largest quantity.

degenerative bone disease that affects an estimated one-quarter of older men and one-half of older women in the United States.

One mineral that people consume too much of is sodium. The recommended amount of sodium intake per day is 500 mg, although the average American takes in 4,000–4,700 mg each day. About one-third of the sodium we consume occurs naturally in foods; another one-third is added during commercial processing; and we add the last one-third either during home cooking or at the table in the form of table salt. **Table 25.5** gives recommendations for reducing the amount of sodium in the diet.

Vitamins

Vitamins are organic compounds (other than carbohydrates, fats, and proteins including amino acids) that regulate various metabolic activities. **Tables 25.6** and **25.7** list the major vitamins and some of their functions and food sources.

Although many people think vitamins can enhance health dramatically, prevent aging, and cure diseases such as arthritis and cancer, there is no scientific evidence that vitamins are "wonder drugs." However, vitamins C, E, and A have been shown to defend the body against free radicals, and therefore they are termed **antioxidants.** These vitamins are especially abundant in fruits and vegetables, and so it is suggested that we eat about 4½ cups of fruits and vegetables per day. To achieve this goal, we should consume salad greens, raw or cooked vegetables, dried fruit, and fruit juice in addition to traditional apples and oranges and other fresh foods.

Vitamin deficiencies can lead to disorders, and even death, in humans. Although many foods in the United States are now enriched, or fortified, with vitamins, some individuals, especially the elderly, young children, alcoholics, and low-income people, are still at risk for vitamin deficiencies, generally as a result of poor food choices. For example, skin cells normally contain a precursor cholesterol molecule that is converted to vitamin D after UV exposure. But a vitamin D deficiency leads to a condition called rickets, in which defective mineralization of the skeleton causes bowing of the legs (**Fig. 25.7**). Most milk today is fortified with vitamin D, which helps prevent the occurrence of rickets. Another example is vitamin C deficiency, the effects of which are illustrated in Figure 25.2. If a diet involves high alcohol consumption, then vitamin deficiencies may occur even if the intake of vitamins is adequate. Deficiencies occur because alcohol interferes with the absorption of certain vitamins such as vitamin B_{12}, folacin, and vitamin A and it increases the excretion of other vitamins such as vitamin C.

Water

Water comprises about 60% of an adult's body. Water participates in many chemical reactions; in addition, watery fluids lubricate joints, transport other nutrients, and help maintain body temperature. Beverages, soups, fruits, and vegetables are sources of water, and most solid foods also contain some water. The amount of *total* water (water from beverages and foods) that you need to consume depends on your physical activity level, diet, and environmental conditions. On average, men should consume about 125 ounces and women should consume about 90 ounces of total water each day. Thirst is a healthy person's best guide for meeting water needs and avoiding dehydration.

Table 25.6 | Water-Soluble Vitamins

Vitamin	Functions	Food Sources	Conditions Caused By:	
			Too Little	Too Much
Vitamin C	Antioxidant; needed for forming collagen; helps maintain capillaries, bones, and teeth	Citrus fruits, leafy green vegetables, tomatoes, potatoes, cabbage	Scurvy, delayed wound healing, infections	Gout, kidney stones, diarrhea, decreased copper
Thiamine (vitamin B$_1$)	Part of coenzyme needed for cellular respiration; also promotes activity of the nervous system	Whole-grain cereals, dried beans and peas, sunflower seeds, nuts	Beriberi, muscular weakness, enlarged heart	Can interfere with absorption of other vitamins
Riboflavin (vitamin B$_2$)	Part of coenzymes such as FAD; aids cellular respiration, including oxidation of protein and fat	Nuts, dairy products, wholegrain cereals, poultry, leafy green vegetables	Dermatitis, blurred vision, growth failure	Unknown
Niacin (nicotinic acid)	Part of coenzymes NAD and NADP; needed for cellular respiration, including oxidation of protein and fat	Peanuts, poultry, whole-grain cereals, leafy green vegetables, beans	Pellagra, diarrhea, mental disorders	High blood sugar and uric acid, vasodilation, etc.
Folacin (folic acid)	Coenzyme needed for production of hemoglobin and formation of DNA	Dark leafy green vegetables, nuts, beans, whole-grain cereals	Megaloblastic anemia, spina bifida	May mask B$_{12}$ deficiency
Vitamin B$_6$	Coenzyme needed for synthesis of hormones and hemoglobin; CNS control	Whole-grain cereals, bananas, beans, poultry, nuts, leafy green vegetables	Rarely, convulsions, vomiting, seborrhea, muscular weakness	Insomnia, neuropathy
Pantothenic acid	Part of coenzyme A needed for oxidation of carbohydrates and fats; aids in the formation of hormones and certain neurotransmitters	Nuts, beans, dark green vegetables, poultry, fruits, milk	Rarely, loss of appetite, mental depression, numbness	Unknown
Vitamin B$_{12}$	Complex, cobalt-containing compound; part of the coenzyme needed for synthesis of nucleic acids and myelin	Dairy products, fish, poultry, eggs, fortified cereals	Pernicious anemia	Unknown
Biotin	Coenzyme needed for metabolism of amino acids and fatty acids	Generally in foods, especially eggs	Skin rash, nausea, fatigue	Unknown

Table 25.7 | Fat-Soluble Vitamins

Vitamin	Functions	Food Sources	Conditions Caused By:	
			Too Little	Too Much
Vitamin A	Antioxidant needed for normal vision, immune system function and cellular growth	Yellow-orange and dark green fruits and vegetables, milk and cereals	Night blindness, poor immune system functioning	Nausea, headaches, bone fractures, hair loss
Vitamin D	A group of steroids needed for development and maintenance of bones and teeth	Milk fortified with vitamin D, fish liver oil; also made in the skin when exposed to sunlight	Rickets, bone decalcification and weakening	Calcification of soft tissues, diarrhea, possible renal damage
Vitamin E	Antioxidant that prevents oxidation of vitamin A and polyunsaturated fatty acids	Leafy green vegetables, fruits, vegetable oils, nuts, whole-grain breads and cereals	Unknown	Diarrhea, nausea, headaches, fatigue, muscle weakness
Vitamin K	Needed for synthesis of substances active in clotting of blood	Leafy green vegetables, cabbage, cauliflower	Easy bruising and bleeding	Can interfere with anticoagulant medication

25.3 | Nutrition and Health

Many serious disorders in Americans are linked to a diet that results in excess body fat. In the United States, the number of people who are overweight or obese has reached epidemic proportions. Nearly two-thirds of adult Americans have too much body fat.

Excess body fat increases the risk of type 2 diabetes, cardiovascular disease, and possibly certain cancers. These conditions are among the leading causes of disability and death in the United States. Therefore, it is important for us all to stay within the recommended weight for our height.

Are You Overweight?

Medical researchers use the **body mass index (BMI)** to determine if a person is overweight or obese. On the whole, the height to which we grow is determined genetically, while our weight is influenced by other factors as well, particularly diet and lifestyle. BMI is the relationship between a person's weight and height. Here's how to calculate your BMI:

$$\frac{\text{weight (pounds)} \times 703.1}{\text{height}^2 \text{ (inches)}}$$

Example: A woman whose height is 63 inches and weighs 133 pounds has a BMI of 23.56.

$$\frac{133 \times 703.1}{63^2} = \frac{93512.3}{3969} = 23.56 \text{ (BMI)}$$

Healthy BMIs	=	18.5 to 24.9
Overweight BMIs	=	25.0 to 29.9
Obese BMIs	=	30.0 to 39.9
Morbidly obese BMIs	=	40.0 or more

Energy Intake Versus Energy Output

While genetics is a factor in being overweight, a person cannot become fat without taking in more food energy (calories) than are expended.

Energy Intake   Scientists use a bomb calorimeter to measure a food's caloric value (**Fig. 25.8**). A portion of food is placed inside the chamber and then ignited. As the food burns, it raises the temperature of the water that surrounds the chamber. Scientists measure the change in water temperature and calculate the number of calories in the food. The energy value of food is often reported in kilocalories (kcal)[1]. A kilocalorie is the amount of heat that raises the temperature of a liter of water by 1°C.

For practical purposes, you can estimate a food's caloric value if you know how many grams of carbohydrate, fat, protein, and alcohol it contains. Each gram of carbohydrate or protein supplies 4 kcal, and each gram of fat supplies 9 kcal. Although alcohol is not a nutrient, it is considered a food, and each gram supplies 7 kcal. Therefore, if a serving of food contains 30 grams of carbohydrate, 9 grams of fat, and 5 grams of protein, it supplies 221 kcal.

carbohydrate	30 g × 4 kcal	=	120 kcal
fat	9 g × 9 kcal	=	81 kcal
protein	5 g × 4 kcal	=	20 kcal
Total			221 kcal

Figure 25.8   **Measuring the energy content of food.**

Scientists use a bomb calorimeter to determine the amount of calories supplied by a particular food.

wire to ignite food

thermometer

insulation around chamber

air space

water

chamber for food

---

[1] Nutritionists use the term Calorie, while scientists prefer kcal.

Energy Output   The body expends energy primarily for (1) metabolic functions; (2) physical activity; and (3) digesting, absorbing, and processing nutrients from food. Scientists can access a person's energy expenditure for a particular physical activity by measuring oxygen intake and carbon dioxide output during performance of the activity (**Fig. 25.9**).

For practical purposes, here's how to estimate your daily energy needs:

1. **Kcal needed daily for metabolic functions** Multiply your weight in kilograms (weight in pounds divided by 2.2) times 1.0 if you are a man, and times 0.9 if you are a woman. Then multiply that number by 24.
   Example: Meghan, a woman who weighs 130 pounds (about 59 kg), would use the following steps to calculate her daily caloric use:

   0.9 kcal × 59 kg × 24 hours = approximately 1,274 kcal/day

2. **Kcal needed daily for physical activity** Choose a multiplication factor from one of the follow categories:
   Sedentary (little or no physical activity) = .20 to .40
   Light (walk daily) = .55 to .65
   Moderate (daily vigorous exercise) = .70 to .75
   Heavy (perform physical labor/endurance training) = .80 to 1.20

   Multiply this factor times the kcal value you obtained in step 1.
   Example: Meghan performs light physical activity daily. She multiplies .55 × 1,274 kcal to determine her physical activity use, which is 701 kcal.

3. **Kcal needed for digestion, absorption, and processing of nutrients**
   Example: Meghan adds 1,274 kcal and 701 kcal to calculate her total energy needs so far. The answer is 1,975 kcal.
   Multiply this value by 0.1, and add it to the total kcal from steps 1 and 2 to get your total daily energy needs.
   Example: Meghan multiplies 1,975 kcal by 0.1 and adds that value (197.5) to 1,975 to obtain her total daily energy needs of 2,172 kcal. Therefore, Meghan will maintain her weight of 130 pounds if she continues to consume about 2,170 kcal a day and to perform light physical activities.

a.

b.

Figure 25.9 **Measuring energy needed for a physical activity.**

a. A sedentary job requires much less energy than (**b**) a physical job. **c.** Scientists measure oxygen intake and carbon dioxide output to determine the kcal required for a particular physical activity.

c.

Check
Your Progress

1. Ted's BMI is 32.5. According to this information, in what range is his weight?
2. Name the device used to determine the number of calories in a food.
3. Daily caloric needs are based on what three types of requirements?

Answers: 1. Ted is obese.   2. A bomb calorimeter is used to determine the calories in food.
3. Daily caloric needs are based on metabolic functions; physical activity; and digestion, absorption, and processing of food.

Intake	Output	Weight Change

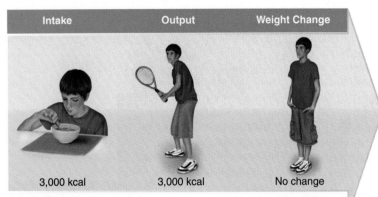

3,000 kcal   3,000 kcal   No change

Energy balance (equilibrium)

4,000 kcal   2,000 kcal   Increase

Positive energy balance

2,000 kcal   3,000 kcal   Decrease

Negative energy balance

Figure 25.10   How body weight changes.

These illustrations show the relationships between food intake, energy use, and weight change.

Maintaining a Healthy Weight

To maintain weight at an appropriate level, the kcal intake (eating) should not exceed the daily kcal output (metabolism + physical activity + processing food). For many Americans, this ratio is out of sync; they take in more calories than they need. The extra energy is converted to fat stored in *adipose tissue*, and they become overweight. To lose weight, an overweight person needs to lower the kcal intake and increase the kcal output in the form of physical activity. Only then does the body metabolize its stored fat for energy needs, allowing the person to lose weight. **Figure 25.10** illustrates how body weight changes in relation to kcal intake and kcal output.

Dieting   Fad weight-reduction diets—high-protein, low-carb, high-fiber, and even cabbage soup diets—come and go. During the first few weeks of following a fad diet, overweight people often lose weight rapidly, because they consume fewer calories than usual, and excess body fat is metabolized for energy needs. In most cases, however, dieters become bored with eating the same foods and avoiding their favorite foods, which may be high in fat and sugars. When most dieters "go off" their diets, they regain the weight that was lost, and they often feel frustrated and angry at themselves for failing to maintain the weight loss.

There are no quick and easy solutions for losing weight. The typical fad diet is nutritionally unbalanced and difficult to follow over the long term. Weight loss and weight maintenance require making permanent lifestyle changes, such as increasing the level of physical activity and reducing portion sizes. Behavior modification allows an overweight person to lose weight safely, generally at a reasonable rate of about ½ to 2 pounds per week. Once body weight is under control, it needs to be maintained by continuing to eat sensibly.

Disorders Associated with Obesity

Type 2 diabetes and cardiovascular disease are often seen in people who are obese.

Type 2 Diabetes

As discussed in Chapter 27, diabetes comes in two forms, type 1 diabetes and type 2 diabetes. When a person has type 1 diabetes, the pancreas does not produce insulin, and the patient has to have daily insulin injections.

In contrast to type 1 diabetes, children and more often adults with type 2 diabetes are usually obese and display impaired insulin production and insulin resistance. Normally, the presence of insulin causes the cells of the body to take up and metabolize glucose. In a person with insulin resistance, the body's cells fail to take up glucose even when insulin is present. Therefore, the blood glucose value exceeds the normal level, and glucose appears in the urine.

Type 2 diabetes is increasing rapidly in most industrialized countries of the world. Because type 2 diabetes is most often seen in people who are obese, dietary factors are generally believed to contribute to the development of type

2 diabetes. Further, a healthy diet, increased physical activity, and weight loss have been seen to improve the ability of insulin to function properly in type 2 diabetics (**Fig. 25.11**). How might diet contribute to the occurrence of type 2 diabetes? Simple sugars in foods, such as candy and ice cream, immediately enter the bloodstream, as do sugars from the digestion of starch within white bread and potatoes. When the blood glucose level rises rapidly, the pancreas produces an overload of insulin to bring the level under control. Chronically high insulin levels apparently lead to insulin resistance, increased fat deposition, and a high blood fatty acid level. Over the years, the body's cells become insulin resistant, and thus type 2 diabetes can occur. In addition, high fatty acid levels can lead to increased risk of cardiovascular disease.

It is well worth the effort to control type 2 diabetes because all diabetics, whether type 1 or type 2, are at risk for blindness, kidney disease, and cardiovascular disease.

Cardiovascular Disease

In the United States, cardiovascular disease, which includes hypertension, heart attack, and stroke, is among the leading causes of death. Cardiovascular disease is often due to arteries blocked by plaque, which contains saturated fats and cholesterol. Cholesterol is carried in the blood by two types of lipoproteins: low-density lipoprotein (LDL) and high-density lipoprotein (HDL). LDL is thought of as "bad" because it carries cholesterol from the liver to the cells, while HDL is thought of as "good" because it carries cholesterol from the cells to the liver, which takes it up and converts it to bile salts.

Saturated fats, including trans fats, tend to raise LDL cholesterol levels, while unsaturated fats lower LDL cholesterol levels. Beef, dairy foods, and coconut oil are rich sources of saturated fat. Foods containing partially hydrogenated oils (e.g., vegetable shortening and stick margarine) are sources of trans fats (**Fig. 25.12**). Unsaturated fatty acids in olive and canola oils, most nuts, and coldwater fish tend to lower LDL-cholesterol levels. Furthermore, coldwater fish (e.g., herring, sardines, tuna, and salmon) contain polyunsaturated fatty acids and especially *omega-3 unsaturated fatty acids* that can reduce the risk of cardiovascular disease. Taking fish oil supplements to obtain omega-3s is not recommended without a physician's approval, because too much of these fatty acids can interfere with normal blood clotting.

The American Heart Association recommends limiting total cholesterol intake to 300 mg per day. Eggs are very nutritious, but yolks contain about 210 mg of cholesterol. Most healthy people can eat a couple of whole eggs each week without experiencing an increase in their blood cholesterol levels. Overall, dietary saturated fats and trans fats raise LDL-cholesterol levels more than dietary cholesterol.

A physician can determine if blood lipid levels are normal. If a person's cholesterol and triglyceride levels are elevated, modifying the fat content of your diet, losing excess body fat, and exercising regularly can reduce them. If lifestyle changes do not lower blood lipid levels enough to reduce the risk of cardiovascular disease, a physician may prescribe special medications.

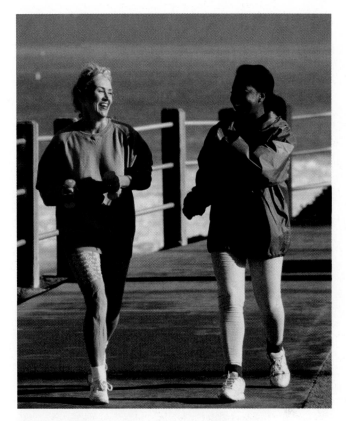

Figure 25.11   Exercising for good health.

Regular exercise helps prevent and control type 2 diabetes.

Figure 25.12   French fries.

Most people enjoy eating french fries, but fast-food fries are apt to contain saturated fats and trans fats.

a.

b.

Figure 25.13 Eating disorders.

a. People with anorexia nervosa have a mistaken body image and think they are fat even though they are thin. **b.** Those with bulimia nervosa overeat and then purge their bodies of the food they have eaten.

Eating Disorders

People with eating disorders are dissatisfied with their body image. Social, cultural, emotional, and biological factors all contribute to the development of an eating disorder. These serious conditions can lead to malnutrition, disability, and death. Regardless of the eating disorder, early recognition and treatment are crucial. Treatment usually includes psychological counseling and anti-depressant medications.

Anorexia nervosa is a severe psychological disorder characterized by an irrational fear of getting fat that results in the refusal to eat enough food to maintain a healthy body weight (**Fig. 25.13***a*). A self-imposed starvation diet is often accompanied by occasional binge eating that is followed by purging and extreme physical activity to avoid weight gain. Binges usually include large amounts of high-calorie foods, and purging episodes involve self-induced vomiting and laxative abuse. About 90% of people suffering from anorexia nervosa are young women; an estimated 1 in 200 teenage girls is affected.

A person with **bulimia nervosa** binge-eats, and then purges to avoid gaining weight (Fig. 25.13*b*). The binge-purge cyclic behavior can occur several times a day. People with bulimia nervosa can be difficult to identify because their body weights are often normal and they tend to conceal their binging and purging practices. Women are more likely than men to develop bulimia; an estimated 4% of young women suffer from this condition.

Other abnormal eating practices include binge-eating disorder and muscle dysmorphia. Many obese people suffer from **binge-eating disorder,** a condition characterized by episodes of overeating that are not followed by purging. Stress, anxiety, anger, and depression can trigger food binges. A person suffering from **muscle dysmorphia** thinks his or her body is underdeveloped. Body-building activities and a preoccupation with diet and body form accompany this condition. Each day, the person may spend hours in the gym working out on muscle-strengthening equipment. Unlike anorexia nervosa and bulimia, muscle dysmorphia affects more men than women.

Check Your Progress

1. What causes a person to gain weight?
2. Why are fad diets of limited value?
3. Salmon, tuna, and other coldwater fish are good sources of certain polyunsaturated fatty acids. What are these fatty acids called?
4. Sally diets because she thinks she is fat, even though her friends tell her she is already too thin. Sally is in danger of what eating disorder?

Answers: 1. Taking in more calories than required for daily energy needs causes weight gain.   2. Typical fad diets are prone to be nutritionally unbalanced and hard to follow for the long term.   3. Omega-3 unsaturated fatty acids.   4. Sally is in danger of anorexia nervosa.

25.4 | How to Plan Nutritious Meals

Planning nutritious meals and snacks involves making daily food choices based on a wide variety of information about recommended amounts of nutrients. A day's food intake should provide the proper balance of nutrients—neither too much nor too little of each nutrient. Food guides can be helpful in planning your diet. Additionally, reading the Nutrition Facts panel on packaged foods can help you choose a healthier source of nutrients.

The Food Pyramid

The U.S. Department of Agriculture (USDA) has a new updated food guide pyramid (**Fig. 25.14**). The food guide pyramid used prior to 2005 had been criticized for not emphasizing the need for whole-grain breads and cereals, low-fat dairy products, low-sugar fruits and vegetables, and unsaturated fats such as olive and canola oil in the diet. It also did not place an emphasis on exercise as a means to control weight gain.

The new pyramid, like the old one, groups foods with similar nutritional content and ranks food groups according to dietary emphasis so the diet will be balanced. The new pyramid emphasizes foods that should be eaten often and omits foods that should not be eaten on a regular basis. Additionally, the USDA provides recommendations concerning the minimum quantity of foods in each group that should be eaten daily. The sample menu in **Table 25.8** shows how the food guide pyramid can be used to plan a day's meals and snacks that supply about 2,200 kcal.

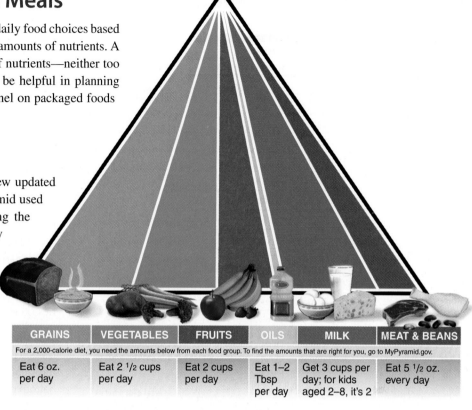

GRAINS	VEGETABLES	FRUITS	OILS	MILK	MEAT & BEANS
Eat 6 oz. per day	Eat 2 ½ cups per day	Eat 2 cups per day	Eat 1–2 Tbsp per day	Get 3 cups per day; for kids aged 2–8, it's 2	Eat 5 ½ oz. every day

For a 2,000-calorie diet, you need the amounts below from each food group. To find the amounts that are right for you, go to MyPyramid.gov.

Figure 25.14 **The 2008 USDA food guide pyramid.**

In 2005, the U.S. Department of Agriculture (USDA) first presented the current food guide pyramid. This food guide pyramid can be helpful when planning nutritious meals and snacks.

Source: Data from the U.S Department of Agriculture.

Table 25.8 | Preparing Meals and Snacks using the Food Guide Pyramid

Breakfast	Snack	Lunch	Snack	Dinner	Dessert
½ cup oat flakes cereal	Whole-wheat muffin	Toasted American cheese sandwich (whole-grain bread)	Apple	3 oz. hamburger patty	½ cup frozen vanilla yogurt topped with sliced strawberries
8 oz. nonfat milk	2 Tbsp peanut butter mixed with 1 tsp sunflower seeds	1 cup vegetable salad	4 graham crackers	Whole-wheat bun	
½ cup blueberries	8 oz. nonfat milk	1 Tbsp olive oil and vinegar salad dressing	½ cup baked beans	1 cup steamed broccoli spears	
1 slice whole-grain toast		4 oz. tomato juice		1 tsp soft margarine	
1 tsp soft margarine					
4 oz. orange juice					

serving size: how much people usually eat at one sitting

total number of kcal per serving size

total number of kcal from fat per serving size

% daily value for protein not usually given

kcal for each type of macronutrient

% of amount needed daily based on a 2,000 kcal diet

% daily value for a few vitamins and minerals

recommended intake for a 2,000 kcal and a 2,500 kcal diet

Figure 25.15 **Nutrition Facts panel.**

The Nutrition Facts panel provides information about the amounts of certain nutrients and other substances in a serving of food.

Making Sense of Nutrition Labels

A Nutrition Facts panel, such as the one shown in **Figure 25.15** provides specific dietary information about the product and general information about the nutrients the product contains.

Serving Size The serving size is based on the typical serving size for the product and not necessarily what the food guide pyramid would consider a serving size. If you are comparing Calorie (kcal)[1] and other data about products of the same type, you definitely want to be sure the serving size is the same for each product.

Calories The total number of kcal is based on the serving size. Obviously, if you eat twice the serving size, you have taken in twice the number of calories.

The bottom of the panel lists the Calories (kcal) per gram (g) of fat, carbohydrate, and protein. You can use this information to calculate the calories per serving for each type of nutrient. The total for all nutrients should agree with the figure given for Calories (kcal) at the top of the panel, namely 260.

% Daily Value The % daily value (the % of the total amount needed in a 2,000 kcal diet) is calculated by comparing the specific information about this product with the information given at the bottom of the panel. For example, the product in Figure 25.15 has a fat content of 13 g, and the total daily recommended amount is less than 65 g, so 13/65 = 20%. *These % daily values are not applicable for people who require more or less than 2,000 Calories (kcal) per day.*

A % daily value for protein is generally not given because determining % daily value would require expensive testing of the protein quality of the product by the manufacturer. Also, notice there is a % daily value for carbohydrates but not sugars because there is no daily value for sugar.

How to Use the Panel If the serving sizes are the same, you can use Nutrition Facts panels to compare two products of the same type. For example, if you wanted to reduce your caloric intake and increase your fiber and vitamin C intakes, comparing the panels from two different food products would allow you to see which one is lowest in calories and highest in fiber and vitamin C.

Dietary Supplements

Dietary supplements are nutrients and plant products (such as herbal teas) that are used to enhance health. The U.S. government does not require dietary supplements to undergo the same safety and effectiveness testing that new prescription drugs must complete before they are approved. Therefore, many herbal products have not been tested scientifically to determine their benefits. Although people often think herbal products are safe because they are "natural," many plants, including lobelia, comfrey, and kava kava, can be poisonous.

Dietary supplements that contain nutrients can also cause harm. Most fat-soluble vitamins are stored in the body and can accumulate to toxic levels,

---

[1] Nutritionists use the term Calorie, while scientists prefer kcal.

particularly vitamins A and D. Although excesses of water-soluble vitamins can be excreted, cases involving toxic amounts of vitamins B_6, thiamin, and vitamin C have been reported. Minerals can be harmful, even deadly, when ingested in amounts that exceed the body's needs.

Healthy people can take a daily supplement that contains recommended amounts of vitamins and minerals. Some people have metabolic diseases or physical conditions that interfere with their ability to absorb or metabolize certain nutrients. These individuals may need to add certain nutrient supplements to their diet. However, people should not take high doses of dietary supplements without checking with their physicians.

The Bottom Line

While food guide pyramids differ in various ways, most nutritionists can agree that a healthy diet does the following:

- Has a moderate total fat intake and is low in saturated fat, trans fats, and cholesterol. (See Table 25.3 for help in achieving this goal.)
- Is rich in whole-grain products, vegetables, and legumes (e.g., beans and peas) as sources of complex carbohydrates and fiber.
- Is low in refined carbohydrates, such as starches and sugars. (See Table 25.2 for help in achieving this goal.)
- Is low in salt and sodium intake. (See Table 25.5 for help in achieving this goal.)
- Contains only adequate amounts of protein, largely from poultry, fish, and plant origins.
- Includes only moderate amounts of alcohol.
- Contains adequate amounts of minerals and vitamins, but avoids questionable food additives and supplements.

Check Your Progress

1. What should Jim consult to compare the caloric value and vitamin A content of canned peaches and canned pears?
2. For the product in Figure 25.15, how many Calories are in a serving? How many grams of total fat are in a serving?
3. Is it always safe to take high doses of nutritional supplements?

Answers: 1. Jim should read the Nutrition Facts panels on the labels.   2. Total Calories = 260; total fat = 13 grams   3. No, vitamins, especially fat-soluble vitamins and minerals can be toxic in large doses.

The Chapter in Review

Summary

25.1 | Nutrition

Nutrients perform varied physiological functions in the body. A healthy diet can lead to a longer and more active life.

Introducing the Nutrients

Nearly every food is a mixture of nutrients. Balanced diets supply nutrients in proportions necessary for health. Essential nutrients must be supplied by the diet, or else deficiency disorders result. Macronutrients (carbohydrates, lipids, and proteins) supply energy. Micronutrients (vitamins and minerals) and water do not supply energy.

<1% vitamins and minerals

35% water

50% carbohydrates

10% proteins

4% lipids

25.2 | The Classes of Nutrients

The six classes of nutrients are carbohydrates, lipids, proteins, minerals, vitamins, and water. Their functions and food sources follow.

Carbohydrates (Macronutrient)

- Carbohydrates in the form of sugars and starch provide energy for cells. Glucose is a simple carbohydrate the body uses directly for energy.
- Sources of carbohydrates are fruits, vegetables, cereals, and other grains.
- Fiber includes nondigestible carbohydrates from plants. Fiber in the diet can prevent constipation and may protect against cardiovascular disease, diabetes, and colon cancer. Refined carbohydrates from which fiber, vitamins, and minerals have been removed are a poor dietary choice.
- Nutritionists debate whether dietary fat or refined carbohydrates are responsible for the prevalence of obesity today.

Lipids (Macronutrient)

- Triglycerides (fats and oils) supply energy and are stored for insulation and protection of organs. Alpha-linolenic and linoleic acids are essential fatty acids.
- Cholesterol is used to make bile, steroid hormones, and vitamin D. The only food sources of cholesterol are foods derived from animals. Elevated blood cholesterol levels can lead to cardiovascular disease.
- Sources of lipids include oils, fats, whole-milk dairy products, meat, fish, poultry, and nuts.
- High intake of saturated fats, trans fats, and cholesterol is harmful to health.

Proteins (Macronutrient)

- The body uses the 20 different amino acids to synthesize hundreds of proteins. Nine amino acids are essential to the diet. Most animal foods are complete sources of protein because they contain all nine essential amino acids.
- Sources of proteins include meat, fish, poultry, eggs, nuts, soybeans, and cheese.
- Healthy vegetarian diets rely on various sources of plant proteins.

- Consumption of excess proteins can be harmful because the excretion of excess urea taxes the kidneys and can lead to kidney stones.

Minerals (Micronutrient)

- Minerals regulate metabolism and are incorporated into structures and compounds.
- About 20 minerals, obtained from most foods, are needed by the body.
- Lack of the mineral calcium can lead to osteoporosis.
- Most people consume too much sodium.

Vitamins (Micronutrient)

- Vitamins, obtained from most foods, regulate metabolism and physiological development. Vitamins C, E, and A also serve as antioxidants.
- Lack of any of the vitamins can lead to certain disorders.

Water

- Water participates in chemical reactions, lubricates joints, transports other nutrients, and helps maintain body temperature.

25.3 | Nutrition and Health

Excess body fat increases the risk of type 2 diabetes, cardiovascular disease, and certain cancers.

Body Mass Index

Medical researchers use the body mass index (BMI) to determine if a person has a healthy weight, is overweight, or is obese.

Energy Intake Versus Energy Output

- Energy intake: The number of daily calories (kcal) consumed is based on grams of carbohydrate, fat, and protein in the foods eaten.
- Energy output: The number of daily calories (kcal) used is based on the amounts needed for metabolic functions, physical activity, and digesting, absorbing, and processing nutrients.

Maintaining a Healthy Weight

- To maintain a healthy weight, kcal intake should not exceed kcal output. To lose weight, decrease caloric intake and increase physical activity.
- Fad diets, in general, are nutritionally unbalanced and difficult to follow over the long term.

Disorders Associated with Obesity

- Type 2 diabetes: A healthy diet, increased physical activity, and weight loss improve insulin function in type 2 diabetics.
- Cardiovascular disease: Saturated fats and trans fats are associated with high blood LDL-cholesterol levels. To reduce risk of cardiovascular disease, cholesterol intake should be limited, and the diet should include sources of unsaturated, polyunsaturated, and omega-3 fatty acids to reduce cholesterol levels.

Eating Disorders

People suffering from eating disorders are dissatisfied with their body image. Anorexia nervosa and bulimia nervosa are serious psychological disturbances that can lead to malnutrition and death.

25.4 | How to Plan Nutritious Meals

Planning nutritious meals and snacks involves making healthy and informed food choices.

The Food Pyramid

Food guide pyramids group foods according to similar nutrient content. Such guides can be helpful when planning nutritious meals and snacks. The new USDA food guide pyramid also emphasizes the need for exercise to prevent weight gain.

Making Sense of Nutrition Labels

The information in the Nutrition Facts panel on packaged foods can be useful when comparing foods for nutrient content, especially serving size, total calories, and % daily value.

Dietary Supplements

Dietary supplements are nutrients and plant products that are taken to enhance health. A multiple vitamin and mineral supplement that provides recommended amounts of nutrients can be taken daily, but herbal and nutritional supplements can be harmful if misused.

▬▬ Key Terms

anorexia nervosa   454
antioxidant   448
binge-eating disorder   454
body mass index (BMI)   450
bulimia nervosa   454
carbohydrate   444
cholesterol   446
deficiency disorder   442
diet   443
dietary supplement   456
essential nutrient   443
fat   445
fiber   444

macronutrient   443
major mineral   447
micronutrient   443
mineral   447
muscle dysmorphia   454
nutrient   442
nutrition   442
obesity   445
protein   446
trace mineral   447
triglyceride   445
vitamin   448

▬▬ Testing Yourself

Choose the best answer for each question.

1. Vitamins are considered
   a. micronutrients because they are small in size.
   b. micronutrients because they are needed in small quantities.
   c. macronutrients because they are large in size.
   d. macronutrients because they are needed in large quantities.

For questions 2–9, choose the class of nutrient from the key that matches the description. Each answer may be used more than once.

Key:
   a. carbohydrates
   b. lipids
   c. proteins
   d. minerals
   e. vitamins
   f. water

2. preferred source of direct energy for cells
3. comprises the majority of human body mass

4. not present in most animal foods
5. include the essential fatty acids alpha-linolenic and linoleic
6. include antioxidants
7. generally found in higher levels in animal than in plant sources
8. an example is cholesterol
9. includes calcium, phosphorus, and potassium
10. The refining of whole-grain foods removes
    a. fiber.
    b. vitamins.
    c. minerals.
    d. More than one of these are correct.
    e. All of these are correct.
11. Cholesterol is
    a. a lipid that has only negative effects on the body.
    b. is a carbohydrate.
    c. an essential vitamin.
    d. a lipid necessary for the synthesis of bile, steroid hormones, and vitamin D.
12. What is the BMI (body mass index) of a person who is 5 feet 10 inches tall and weighs 160 pounds?
    a. 31   d. 18
    b. 27   e. 12
    c. 23
13. A serving of graham crackers contains 24 grams of carbohydrate, 3 grams of fat, and 2 grams of protein. How many calories are in a serving?
    a. 131   d. 261
    b. 236   e. 141
    c. 116
14. _____ is a psychological disorder in which a person refuses to eat because of an irrational fear of becoming fat.
    a. Bulimia   c. Anorexia nervosa
    b. Muscle dysmorphia   d. Binge-eating disorder
15. With reference to the food pyramid, fill in each component and the recommended daily dietary amount based on a 2,000 Calorie (kcal) diet.

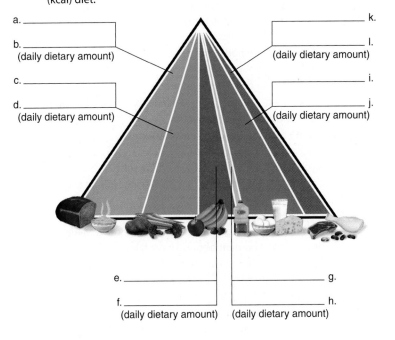

a. _____ / b. _____ (daily dietary amount) / c. _____ / d. _____ (daily dietary amount) / e. _____ / f. _____ (daily dietary amount) / g. _____ / h. _____ (daily dietary amount) / i. _____ / j. _____ (daily dietary amount) / k. _____ / l. _____ (daily dietary amount)

16. Vitamins
    a. are inorganic compounds necessary in the diet.
    b. are found in high levels in the body and used as building blocks for body tissues.
    c. are organic compounds, found in low levels in the body, that regulate metabolic activities.
    d. can all be synthesized by the human body.

17. _____ are compounds that defend the body against free radicals.
    a. Macronutrients.
    b. Proteins
    c. Minerals
    d. Antioxidants.

18. The body mass index (BMI) is
    a. a measure of height relative to weight to determine whether or not a person is overweight.
    b. a measure of height relative to age to determine whether or not a person is of normal height.
    c. relatively low if a person is overweight.
    d. relatively high if a person is underweight.

19. You are comparing two breakfast cereals, and one has fewer calories than the other per serving. What should you check before deciding that one will lead to weight loss rather than the other?
    a. the percentage of vitamins provided by each
    b. the grams of protein provided by each
    c. the grams of sugar provided by each
    d. the suggested serving size
    e. the size of the box

20. A percent daily value for sugar is not included in a nutrition label because
    a. sugar is not a nutrient.
    b. it is too difficult to determine the caloric value of sugar.
    c. there is a daily value given for carbohydrates but not for sugars.
    d. sugar quality varies from product to product.

21. A _____ is a component of food that provides a physiological function.
    a. macromolecule
    b. nutrient
    c. calorie
    d. chemical

22. The amino acids that must be consumed in the diet are called essential. Nonessential amino acids
    a. can be produced by the body.
    b. are only needed occasionally.
    c. are stored in the body until needed.
    d. can be taken in by supplements.

23. _____ is a nondigestible carbohydrate derived from plants.
    a. Fiber   b. Starch   c. Glycogend   d. Vitamin A

24. Bulimia nervosa is characterized by
    a. a restrictive diet, but often includes excessive exercise.
    b. binge eating followed by purging.
    c. an obsession about body shape and weight.
    d. a distorted body image so that the person feels fat even when emaciated.

25. Which of these is a true statement?
    a. Technically speaking, junk foods have nutritional value.
    b. Water is not a nutrient because it contains no kcal.
    c. Even though fiber can be digested, it is not a nutrient.
    d. Cholesterol should be avoided because it plays no role in the body.
    e. A vegetarian diet is sure not to provide all the essential amino acids.

26. Fats and oils are examples of
    a. sugars.   b. proteins.   c. triglycerides.   d. low-calorie foods.

Thinking Scientifically

1. One reason obesity is such a pervasive problem in the United States is that serving sizes have increased dramatically in the past two decades. For example, 20 years ago, a typical soda was 6.5 ounces, while today a 20-ounce soda is considered average. A serving of french fries was 2.6 ounces in the past, but is at least 6 ounces today. Bagels have doubled in size. Complete the following table to determine the caloric values of today's servings, and then compare them.

	Soda Then 6.5 oz.	Soda Now 20 oz.	Calorie Increase
Fat (g) 0	0	_____	
Carbohydrate (g)	21	_____	
Protein (g)	0	_____	
Total calories	_____	_____	_____
	Fries Then 2.6 oz.	**Fries Now 6 oz.**	**Calorie Increase**
Fat (g) 0	11	_____	
Carbohydrate (g)	28	_____	
Protein (g)	3	_____	
Total calories	_____	_____	_____
	Bagels Then 3 oz.	**Bagels Now 6 oz.**	**Calorie Increase**
Fat (g) 0	0.5	_____	
Carbohydrate (g)	40	_____	
Protein (g)	6	_____	
Total calories	_____	_____	_____

2. Did humans evolve as herbivores, carnivores, or omnivores? Our dendition has both carnivore and herbivore characteristics. What about the data regarding the nine essential amino acids in our diet (see page 446)? Does that help conclude what type diet we were originally adapted to? Why or why not?

Bioethical Issue

Misleading Food Advertisements

Some people believe that food and drink manufacturers are at least partly to blame for the obesity epidemic in the United States because of misleading advertising. For example, their advertisements show portion sizes that encourage excess consumption and make people believe that certain unhealthy substitutes are the equivalent of a well-balanced meal. Advertisements also lead children to desire sugar-coated cereals and fructose-loaded drinks. Do you think the government should regulate such advertising? If so, how extensive should the regulations be? If not, how might parents counter the effects of such advertising?

Essentials of Biology Website

The companion website for *Essentials of Biology* provides a wealth of information organized and integrated by chapter. You will find practice tests, animations, videos, and much more that will complement your learning and understanding of general biology.

http://www.mhhe.com/maderessentials2

26

Defenses Against Disease

AIDS eventually leads to death because the HIV virus (red) destroys immune system cells (green).

People who have had organ transplants must take immuno-suppressive drugs for the rest of their lives.

A fever increases metabolism, enabling the immune system to fight infection more efficiently.

Outline

Can Fever Be Beneficial?

IF YOU WERE asked the purpose of a fever, what would your answer be? Many people think the heat generated by a fever actually "burns out" the pathogens in the body, thus killing them. In reality, the heat from increased body temperature does not directly kill the pathogens, but instead increases your metabolism so that your immune system defense mechanisms can work better to fight the pathogens. In fact, an increase

—Continued next page

461

Continued—

of 1°C in body temperature causes about a 10% increase in metabolism. While a low-grade fever can be beneficial, a high-grade fever can be detrimental, as body temperature moves further away from normal. If a fever becomes too high and is not reduced, death will occur. For this reason, we often take medications to reduce fevers in case they start to elevate to potentially dangerous levels.

An interesting manipulation of body temperature is the basis for weight-loss supplements that claim to be "thermogenic." These supplements raise the body temperature slightly (as would a natural fever), increasing metabolism as a way to bring about weight loss. However, some people have suffered serious consequences, including death, after taking such supplements. For this reason, the FDA has banned one of the major thermogenic supplemental ingredients—ephedra.

The immune system has an incredible array of defenses to help keep you free of infection, as well as to protect you from cancer. Most of these defenses work remarkably well and only rarely fail. In this chapter, you will learn about how the immune system protects you on a daily basis.

26.1 | Organs, Tissues, and Cells of the Immune System

The **immune system** plays an important role in keeping us healthy because it fights infections and cancer. First and foremost, the immune system contains the **lymphatic organs:** red bone marrow, the thymus gland, the lymph nodes, and the spleen (**Fig. 26.1**). The **tonsils**, which are located in the pharynx, and the **appendix**, which is attached to a portion of the large intestine, are areas of lymphatic tissue that also belong to the immune system. The immune system not only contains a network of lymphatic organs and lymphatic tissues, but it also includes cells. In particular, we will be discussing the cell types mentioned in **Table 26.1**. You will want to refer to this table as you read the chapter.

Immunity is the body's capability to repel foreign substances, pathogens, and cancer cells. Nonspecific immunity indiscriminately repels pathogens, while specific immunity requires that a certain antigen be present. As previously discussed on page 410, an **antigen** is any molecule, usually a protein or carbohydrate, that stimulates the immune system.

Lymphatic Organs

Each of the lymphatic organs has a particular function in immunity, and each is rich in lymphocytes, one of the types of white blood cells.

Red Bone Marrow

In a child, most bones have red bone marrow, and in an adult it is still present in the bones of the skull, the sternum (breastbone), the ribs, the clavicle, the pelvic

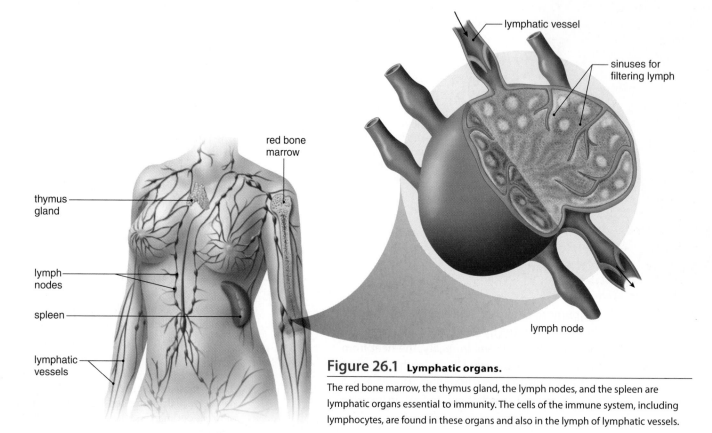

Figure 26.1 Lymphatic organs.

The red bone marrow, the thymus gland, the lymph nodes, and the spleen are lymphatic organs essential to immunity. The cells of the immune system, including lymphocytes, are found in these organs and also in the lymph of lymphatic vessels.

bones, the vertebral column and the ends of the humerus and femur nearest their attachment to the body. **Red bone marrow** produces all types of blood cells, but in this chapter we are interested in the cells listed in Table 26.1.

Lymphocytes differentiate into either **B lymphocytes (B cells)** or **T lymphocytes (T cells),** which are discussed at length later in the chapter. Bone marrow is not only the source of B lymphocytes, but also the place where B lymphocytes mature. T lymphocytes mature in the thymus gland. We shall learn that B cells produce antibodies and that T cells kill antigen-bearing cells outright.

Thymus Gland

The soft, bilobed **thymus gland** varies in size, but it is larger in children and shrinks as we get older. The thymus gland plays a role in the maturity of T lymphocytes. Immature T lymphocytes migrate from the bone marrow through the bloodstream to the thymus, and then they mature. Only about 5% of T lymphocytes ever leave the thymus. These T lymphocytes have survived a critical test: If any show the ability to react with "self" cells, they die. If they have potential to attack a foreign cell, they leave the thymus and enter lymphatic vessels and organs.

The thymus gland also produces thymic hormones thought to aid in the maturation of T lymphocytes.

Lymph Nodes

Lymph nodes are small, ovoid structures occurring along lymphatic vessels (Fig. 26.1). Lymph nodes filter lymph and keep it free of pathogens and antigens. Lymph is filtered as it flows through a lymph node because its many sinuses (open spaces) are lined by **macrophages,** large phagocytic cells that engulf and then devour as many as a hundred pathogens and still survive (see Fig. 26.4a). Lymph nodes are also instrumental in fighting infections and cancer because they contain many lymphocytes.

Some lymph nodes are located near the surface of the body, and such nodes are named for their location. For example, inguinal nodes are in the groin, and axillary nodes are in the armpits. Physicians often feel for the presence of swollen, tender lymph nodes in the neck as evidence that the body is fighting an infection. This method is a noninvasive, preliminary way to help make a diagnosis.

Spleen

The **spleen,** which is about the size of your fist, is in the upper left abdominal cavity. The spleen's unique function is to filter the blood. This soft, spongy organ contains so-called red pulp and white pulp. Blood passing through the many sinuses in the red pulp is filtered of pathogens and debris, including worn-out red blood cells, because the sinuses are lined by macrophages. The white pulp contains lymphocytes that are actively engaged in fighting infections and also cancer.

The spleen's outer capsule is relatively thin, and an infection or a severe blow can cause the spleen to burst. The spleen's functions can be fulfilled by other organs, but a person without a spleen is often slightly more susceptible to infections. As a result they will receive certain vaccinations and may have to receive antibiotic therapy indefinitely.

Table 26.1 | Immunocell Type and Function

Cell	Function
Macrophages	Phagocytize pathogens; inflammatory response and specific immunity
Mast cells	Release histamine, which promotes blood flow to injured tissues; inflammatory response
Neutrophils	Phagocytize pathogens; inflammatory response
Natural killer cells	Kill virus-infected and tumor cells by cell-to-cell contact
Lymphocytes	Responsible for specific immunity
B cells	Produce plasma cells and memory cells
Plasma cells	Produce specific antibodies
Memory cells	Ready to produce antibodies in the future
T cells	Regulate immune response; produce cytotoxic T cells and helper T cells
Cytotoxic T cells	Kill virus-infected and cancer cells
Helper T cells	Regulate immunity

Check Your Progress

1. List the various components of the immune system.
2. List the lymphatic organs, and identify their functions.
3. Define immunity.

Answers: 1. The immune system is composed of lymphatic organs, tissues, and cells as well as the products of those cells, including antibodies.   2. Red bone marrow: produces blood cells, including white blood cells; thymus: aids in maturation of T lymphocytes and testing of their ability to function; spleen: filters blood of pathogens and debris; lymph nodes: filter lymph of pathogens and debris.   3. Immunity is the body's ability to repel foreign substances, pathogens, and cancer cells.

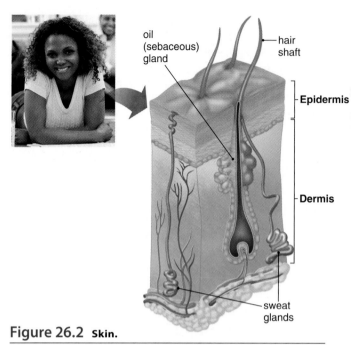

oil
(sebaceous)
gland

hair
shaft

Epidermis

Dermis

sweat
glands

Figure 26.2 Skin.

The cells of the epidermis harden and die as they progress to the outer layer of skin. These outer dead cells form a protective barrier against invasion by pathogens. The sweat and oil glands are acidic enough to inhibit invasion by bacteria.

histamine

mast
cell

free nerve
ending (pain)

capillary

neutrophil

pathogens

macrophage

Figure 26.3 The inflammatory response.

If you cut your skin, the inflammatory response occurs immediately. As blood flow increases, the area gets red and warm. As mast cells, a type of white blood cell, release chemicals such as histamine, the capillary becomes more permeable, localized swelling occurs, and pain receptors are stimulated. Neutrophils and macrophages begin phagocytizing the bacteria.

26.2 | Nonspecific Defenses

The body has an **innate immunity** composed of the various types of nonspecific defenses that are our first line of defense against most types of infections. The nonspecific defenses are barriers to entry, the inflammatory response, the complement system, and natural killer cells.

Barriers to Entry

Skin and the mucous membranes lining the respiratory, digestive, reproductive, and urinary tracts serve as mechanical barriers to entry by pathogens. Oil gland secretions contain chemicals that weaken or kill certain bacteria on the skin (**Fig. 26.2**). The upper respiratory tract is lined by ciliated cells that sweep mucus and trapped particles up into the throat, where they can be swallowed or expectorated (spit out). The stomach has an acidic pH, which inhibits growth or kills many types of bacteria. The various bacteria that normally reside in the intestine and other areas, such as the vagina, prevent pathogens from taking up residence.

The Inflammatory Response

Whenever tissue is damaged, a series of events occurs that is known as the **inflammatory response** (**Fig. 26.3**). The inflamed area has four outward signs: redness, warmth, swelling, and pain. The inflammatory response involves the first three cell types listed in Table 26.1.

When an injury occurs, damaged tissue cells and **mast cells** release chemical mediators, such as **histamine**, which cause the capillaries to dilate and become more permeable. Excess blood flow due to enlarged capillaries causes the skin to redden and become warm. Increased permeability allows proteins and fluids to escape into the tissues, resulting in swelling. The swollen area stimulates free nerve endings, causing the sensation of pain.

Neutrophils are phagocytic white blood cells that migrate to the site of injury. They are amoeboid and can change shape to squeeze through capillary walls and enter tissue fluid. When neutrophils phagocytize pathogens, they are enclosed within a vesicle. The pathogens are later destroyed by hydrolytic enzymes when this vesicle combines with a lysosome, one of the cellular organelles.

Also present are *macrophages* (**Fig. 26.4a**). As stated, lymph nodes contain macrophages, but so do some tissues, particularly connective tissues, which have resident macrophages that routinely act as scavengers, devouring old blood cells, bits of dead tissue, and other debris. During the inflammatory response, macrophages can also bring about an explosive increase in the number of leukocytes by liberating growth factors, which pass by way of blood to the red bone marrow. There, they stimulate the production and release of white blood cells, primarily neutrophils. As the infection is being overcome, some neutrophils may die. These—along with dead cells, dead bacteria, and living white blood cells—form pus, a whitish material. The presence of pus indicates that the body is trying to overcome an infection. When a blood vessel ruptures or is torn, the blood forms a clot to seal the break and prevent any further pathogen invasion.

The macrophages that have been fighting the infection move through the tissue fluid and lymph to the lymph nodes. Now lymphocytes are activated to react to the threat of an infection. Sometimes inflammation persists, and the result is chronic inflammation that is often treated by administering anti-inflammatory agents such as aspirin, ibuprofen, or cortisone. These medications act against the chemical mediators such as histamine released by the white blood cells in the area.

The Complement System

The **complement system**, often simply called complement, is composed of a number of blood plasma proteins designated by the letter C and a subscript. A limited amount of activated complement protein is needed because a domino effect occurs: As soon as a protein is activated, it in turn stimulates many more.

The complement proteins are activated when pathogens enter the body. The proteins "complement" certain immune responses, which accounts for their name. For example, they are involved in and amplify the inflammatory response because complement proteins attract phagocytes to the scene. Some complement proteins bind to the surfaces of pathogens already coated with antibodies, which ensures that the pathogens will be phagocytized by a neutrophil or macrophage.

Certain other complement proteins join to form a membrane attack complex that produces holes in the surfaces of microbes (Fig 26.4*b*). Fluids and salts then enter the pathogen to the point that it bursts.

Natural Killer Cells

Natural killer (NK) cells are large, granular lymphocytes that kill virus-infected cells and tumor (cancer) cells by cell-to-cell contact (see Table 26.1).

What makes an NK cell attack and kill a cell? The cells of your body ordinarily have "self" proteins on their surface that bind to receptors on NK cells. Sometimes virus-infected cells and cancer cells undergo alterations and lose their ability to produce self proteins. When NK cells can find no self proteins to bind to, they kill the cell by the same method as T lymphocytes (see Fig. 26.8).

NK cells are not specific—their numbers do not increase when exposed to a particular antigen, and they have no means of "remembering" the antigen from previous contact.

cytoplasmic extension from macrophage

bacteria

×6,000

a.

membrane attack complex

complement proteins

plasma membrane of pathogen

inside

b.

Figure 26.4 **Ways to get rid of pathogens.**

a. Macrophages, the body's scavengers, engulf pathogens and chop them up inside lysosomes. **b.** Complement proteins come together and form a membrane attack complex in the surface of a pathogen. Water and salts enter, and the pathogen bursts.

Check Your **Progress**

1. Describe the steps in the inflammatory response.
2. Describe the function of the complement system.
3. Explain how natural killer cells "know" what to attack.

Answers: 1. Damaged cells and mast cells release chemicals that cause capillaries to dilate and become more permeable. Neutrophils and macrophages migrate to the site of injury and phagocytize pathogens.   2. The complement system produces proteins that complement other forms of immunity. Certain complement proteins also poke holes in invading pathogens, causing them to burst.   3. Natural killer cells attack cells that have lost their ability to produce self proteins, often because of a viral infection.

Table 26.2 | Characteristics of B Cells

- Provide an antibody response to a pathogen
- Produced and become mature in bone marrow
- Reside in lymph nodes and spleen; circulate in blood and lymph
- Directly recognize antigen and then undergo cell division
- Cell division produces antibody-secreting plasma cells as well as memory B cells

26.3 | Specific Defenses and Aquired Immunity

When nonspecific defenses have failed to prevent an infection, specific defenses or **acquired immunity** come into play. Specific defenses respond to antigens. Pathogens have antigens, and antigens can be part of a foreign cell or a cancer cell. When our body learns to destroy a particular antigen, we have become immune to it. Because our immune system does not ordinarily respond to the proteins on the surface of our own cells (as if they were antigens), the immune system is said to be able to distinguish "self" from "nonself." Only in this way can the immune system aid, rather than disrupt, homeostasis.

Lymphocytes are capable of recognizing an antigen because they have plasma membrane receptor proteins shaped to allow them to combine with a specific antigen. It is often said that the receptor and the antigen fit together like a lock and a key. Because we encounter a million different antigens during our lifetime, we need a great diversity of lymphocytes to protect us against them. Remarkably, diversification occurs to such an extent during the maturation process that a lymphocyte type exists for any possible antigen.

Immunity usually lasts for some time. For example, once we recover from the measles, we usually do not get the illness a second time. Immunity is primarily the result of the action of the B lymphocytes and the T lymphocytes. B lymphocytes mature in the bone marrow, and T lymphocytes mature in the thymus gland. B lymphocytes, also called B cells, give rise to plasma cells, which produce antibodies (Table 26.2). These antibodies are secreted into the blood, lymph, and other body fluids. In contrast, T lymphocytes, also called T cells, do not produce antibodies. Some T cells regulate the immune response, and other T cells directly attack cells that bear antigens (see Table 26.1).

B Cells and the Antibody Response

Each B cell can bind only to a specific antigen—the antigen that fits the binding site of its receptor. The receptor is called a **B-cell receptor (BCR)**. Some B cells never have anything to do because an antigen that fits the binding site of their type of receptor never shows up. But if an antigen does bind to a BCR, that particular B cell is activated, and it divides, producing many plasma cells and also memory B cells (**Fig. 26.5**). A B cell is stimulated to divide and produce plasma cells by **cytokines,** which are signaling proteins that stimulate various immune cells to perform their functions. Plasma cells are larger than regular B cells because they have extensive rough endoplasmic reticulum for the mass production and secretion of antibodies specific to the antigen. The antibodies produced by plasma cells and secreted into the blood and lymph are identical to the BCR of the activated B cell.

Memory B cells are the means by which long-term immunity is possible. If the same antigen enters the system again, memory B cells quickly divide and give rise to more plasma cells capable of producing the correct antibodies.

Once the threat of an infection has passed, the development of new plasma cells ceases, and those present undergo apoptosis. Apoptosis is a process of programmed cell death (PCD) involving a cascade of specific cellular events leading to the death and destruction of the cell (see Fig. 8.11, p. 122).

Figure 26.5 **B cell and the antibody response.**

When an antigen combines with a BCR, the B cell divides to produce plasma cells and memory B cells. Plasma cells produce antibodies but eventually disappear. Memory B cells remain in the body ready to produce the same antibody in the future.

The Function of Antibodies

Antibodies are immunoglobulin proteins that are capable of combining with a specific antigen (**Fig. 26.6**). The antigen-antibody reaction can take several forms, but quite often the reaction produces complexes of antigens combined with antibodies. Such antigen-antibody complexes, sometimes called immune complexes, mark the antigens for destruction (see Fig. 26.5). An antigen-antibody complex may be engulfed by neutrophils or macrophages, or it may activate complement. Complement makes pathogens more susceptible to phagocytosis, as discussed previously.

ABO Blood Type   Most likely you know your ABO blood type—whether you have type A, B, AB, or O blood. These letters stand for antigens on your red blood cells. If you have type O blood, you do not have either antigen A or B on your red cells. Some blood types have antibodies in the plasma (**Table 26.3**). As an example, type O blood has both anti-A and anti-B antibodies in the plasma. You cannot give a person with type O blood a transfusion from an individual with type A blood. If you do, the antibodies in the recipient's plasma will react to type A red blood cells, and agglutination will occur. **Agglutination,** the clumping of red blood cells, causes blood to stop circulating and red blood cells to burst. On the other hand, you can give type O blood to a person with any blood type because type O red blood cells bear neither A nor B antigens.

It is possible to determine who can give blood to whom based on the ABO system. However, there are other red blood cell antigens in addition to A and B that are used in typing blood. Therefore, it is best to physically put the donor's blood on a slide with the recipient's blood to observe whether the two types match (no agglutination occurs) before blood can be safely transfused from one person to another.

a.

b.

Figure 26.6   **Why an antibody is specific.**

a. During a lifetime, a person will encounter a million different antigens that differ in shape. An antibody has two variable regions that end in an antigen-binding site. The variable regions vary so much that the binding site of each antibody has a shape that will fit only one specific antigen. **b.** Computer model of an antibody molecule. The antigen combines with the two side branches.

Table 26.3 | Blood Types

Blood Type	Antigens on Red Blood Cells	Antibodies in Plasma	Compatible Donors
A	A	Anti-B	A, O
B	B	Anti-A	B, O
AB	A, B	Neither anti-A nor anti-B	A, B, AB, O
O	—	Both anti-A and anti-B	O

Check Your **Progress**

1. **Explain how antibodies are produced in response to an antigen.**
2. **Explain what the letters in an ABO blood type refer to.**

Answers: 1. When a B-cell receptor binds to an antigen, a B cell is activated. Now the B cell divides to produce many plasma cells. The plasma cells secrete antibodies against the antigen.   2. The letters A, B, and O refer to the antigens carried on red blood cells. Type A produces antigen A, type B produces antigen B, type AB produces both antigens A and B, and type O does not produce antigen A or B.

Table 26.4 | Characteristics of T Cells

- Provide a cellular response to virus-infected cells and cancer cells
- Produced in bone marrow, mature in thymus gland
- Antigen must be presented in groove of an MHC[1] molecule
- Cytotoxic T cells destroy nonself protein-bearing cell
- Helper T cells secrete cytokines that control the immune response

[1] MHC = major histocompatibility complex. The MHC genes encode MHC proteins that are displayed on the cell surface and define an individual's tissue type. When tissues/organs have like MHC proteins, they are histocompatible, and transplants between individuals are possible.

T Cells and the Cellular Response

When T cells leave the thymus gland, they have unique **T-cell receptors (TCRs)** just as B cells do. Unlike B cells, however, T cells are unable to recognize an antigen without help. The antigen must be presented to them by an **antigen-presenting cell (APC),** such as a macrophage. After ingesting and destroying a pathogen, APCs travel to a lymph node or the spleen, where T cells also congregate. When a macrophage phagocytizes a virus, it is digested in a lysosome. An antigen from the virus is combined with an MHC "self" protein, and the complex appears on the cell surface. Then the (MHC + antigen) is presented to a T cell. The importance of self proteins in plasma membranes was first recognized when it was discovered that they contribute to the specificity of tissues and make it difficult to transplant tissue from one human to another.

The two main types of T cells are **helper T cells (T_H cells)** and **cytotoxic T cells (T_C cells)** (Table 26.4). Each of these types has a TCR that can recognize an antigen fragment in combination with an MHC molecule. However, the major difference in antigen recognition by these two types of cells is that the T_H cells only recognize antigens presented by APCs with MHC class II molecules on their surface, while T_C cells only recognize antigens presented by APCs with MHC class I molecules on their surface. In **Figure 26.7**, the antigen is represented by a red triangle, and the helper T cell that binds to the antigen has the specific TCR that can combine with this particular MHC II + antigen. Now the helper T cell is activated and divides to produce more helper T cells.

As the illness disappears, the immune reaction wanes, and the activated T cells become susceptible to apoptosis. Apoptosis contributes to homeostasis by regulating the number of cells present in an organ, or in this case, the immune system. When apoptosis does not occur as it should, T-cell cancers (e.g., lymphomas and leukemias) can result. Also, in the thymus gland, any T cell that has the potential to destroy the body's own cells undergoes apoptosis.

Functions of Cytotoxic T Cells and Helper T Cells

Cytotoxic T cells specialize in cell to cell combat. They have storage vacuoles containing perforins or enzymes called granzymes. After a cytotoxic T cell binds to a virus-infected or cancer cell presenting the same antigen it has learned to recognize, it releases perforin molecules that perforate the plasma membrane,

Macrophage (APC)
virus
MHC II + antigen
virus inside lysosome
antigen
Helper T cell
T cell receptor (TCR) that fits the antigen + MHC II
cytokines
Activation
Helper T cells

Figure 26.7   Activation of a T cell.

For a T cell to be activated, the antigen must be presented along with an MHC protein to the T cell by an APC, often a macrophage. Each type of T cell bears a specific receptor, and if its TCR fits the MHC II + antigen complex, it is activated to divide and produce more T cells—in this case, helper T cells, which also have this type TCR. The helper T cells then produce and release cytokines.

forming a pore. Cytotoxic T cells then deliver a supply of granzymes into the pore, and these cause the cell to undergo apoptosis and die. Once cytotoxic T cells have released the perforins and granzymes, they move on to the next target cell. Cytotoxic T cells are responsible for a cellular response to virus-infected and cancer cells (**Fig. 26.8**).

Helper T cells specialize in regulating immunity by secreting cytokines that in particular stimulate B cells and cytotoxic T cells. Similar to B cells, cloned T cells include memory T cells that live for many years and can jump start an immune response to an antigen dealt with before. Because HIV, the virus that causes AIDS, infects helper T cells and other cells of the immune system, it inactivates the immune response and makes HIV-infected individuals susceptible to opportunistic infections. Infected macrophages serve as reservoirs for the HIV virus. AIDS is discussed in Section 26.5.

Tissue Rejection

Certain organs, such as the skin, the heart, and the kidneys, could be transplanted easily from one person to another if the body did not attempt to reject them. Rejection occurs because cytotoxic T cells and also antibodies bring about destruction of foreign tissues in the body. When rejection occurs, the immune system is correctly distinguishing between self and nonself.

Organ rejection can be controlled by carefully selecting the organ to be transplanted and administering immunosuppressive drugs. It is best if the transplanted organ has the same type of MHC proteins as those of the recipient because otherwise the transplanted organ will be antigenic to T cells. Two well-known immunosuppressive drugs, cyclosporine and tacrolimus, act by inhibiting the response of T cells to cytokines. Without cytokines, all types of immune responses are weak.

Figure 26.8

Cytotoxic T cells and the cellular response.

a. Scanning electron micrograph of cytotoxic T cell attacking a target cell, which is either a virus-infected or cancer cell. **b.** A cytotoxic T cell attacks any cell that presents it with an MHC I + antigen it has learned to recognize. First, vesicles release perforin, which forms a pore in the target cell. Then vesicles release granzymes, which cause the cell to undergo apoptosis.

Check Your **Progress**

1. Compare and contrast B cells with T cells.
2. What happens to plasma cells and cytotoxic T cells when an infection has passed?

Answers: 1. Both are involved with specific defenses against disease. B cells are produced and mature in the bone marrow and give rise to plasma cells that produce antibodies. T cells are produced in the bone marrow and mature in the thymus as cytotoxic or helper T cells.   2. They undergo programmed cell death, called apoptosis.

a.

b.

Figure 26.9 **Active immunity due to immunizations.**

a. Immunization often requires more than one injection. **b.** A minimal primary response occurs after the first vaccine injection, but after a second injection, the secondary response usually shows a dramatic rise in the amount of antibody present in plasma.

26.4 | Immunizations

After you have had an infection, you are sometimes immune to getting it again. Good examples of diseases against which you can acquire immunity are the childhood diseases measles and mumps. Unfortunately, few sexually transmitted diseases stimulate lasting immunity; for example, a person can get gonorrhea over and over again. If lasting immunity is possible, a vaccine for the disease likely exists or can be developed. **Vaccines** are substances that usually do not cause illness, but even so, the immune system responds to them. Traditionally, vaccines are the pathogens themselves, or their products, that have been treated so that they are no longer virulent (able to cause disease). Today, it is possible to genetically engineer bacteria to mass-produce a protein from pathogens, and this protein can be used as a vaccine. This method has now produced a vaccine against hepatitis B, a virus-induced disease, and it is being used to prepare a vaccine against malaria, a protozoan-induced disease.

Immunization promotes **active immunity.** After a vaccine is given, it is possible to follow an active immune response by determining the amount of antibody present in a sample of plasma; this is called the antibody titer. After the first exposure to a vaccine, a primary response occurs. For a period of several days, no antibodies are present; then their concentration rises slowly, levels off, and gradually declines as the antibodies bind to the antigen or simply break down (**Fig. 26.9**). After a second exposure to the vaccine, a secondary response is expected. The concentration now rises rapidly to a level much greater than before; then it slowly declines. The second exposure is called a "booster" because it boosts the antibody concentration to a high level. The high antibody concentration now is expected to help prevent disease symptoms if the individual is exposed to the disease-causing antigen.

Active immunity is dependent upon the presence of memory B cells and possibly memory T cells that are capable of responding to lower doses of antigen. Active immunity is usually long-lasting, but a booster may be required after a certain number of years.

Although the body usually makes its own antibodies, it is sometimes possible to give an individual prepared antibodies (immunoglobulins) to combat a disease. Because these antibodies are not produced by the individual's plasma cells, this so-called **passive immunity** is temporary. For example, newborn infants are passively immune to some diseases because antibodies have crossed the placenta from the mother's blood. These antibodies soon disappear, however, so that within a few months, infants become more susceptible to infections. Breast-feeding prolongs the natural passive immunity an infant receives from the mother because antibodies are present in the mother's milk.

Even though passive immunity does not last, it is sometimes used to prevent illness in a patient who has been unexpectedly exposed to an infectious disease. Usually, the patient receives an injection of gamma globulin, a portion of blood that contains antibodies, preferably taken from an individual who has recovered from the illness. In the past, horses were immunized, and gamma globulin was taken from them to provide the needed antibodies against such diseases as diphtheria, botulism, and tetanus. Unfortunately, patients who received these antibodies became ill about 50% of the time, because the serum contained proteins that the individual's immune system recognized as foreign. This condition was called serum sickness.

Check Your **Progress**

1. Explain why you cannot be vaccinated for most sexually transmitted diseases.
2. Describe the function of a booster shot.
3. Explain the function of gamma globulin injections to prevent illness in a patient who is unexpectedly exposed to an infectious disease.

Answers: 1. Vaccinations only work if acquired immunity is possible. People cannot develop immunity to most sexually transmitted diseases, and so a vaccine cannot be developed.   2. A booster shot is an injection of the same vaccine to increase the antibody concentration to a high level.   3. Gamma globulin from someone who had the same disease will contain antibodies against the pathogen in question.

26.5 | Immune System Problems

The immune system usually protects us from disease because it can distinguish self from nonself. Sometimes, however, it responds in a manner that harms the body, as when individuals develop allergies or have an autoimmune disease.

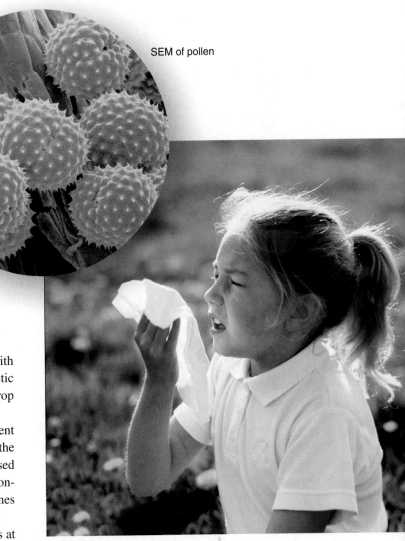

SEM of pollen

Allergies

Allergies are hypersensitivities to substances in the environment, such as pollen, food, or animal hair, that ordinarily would not cause an immune reaction. The response to these antigens, called **allergens**, usually includes some unpleasant symptoms (**Fig. 26.10**). An allergic response is regulated by cytokines secreted by both T cells and macrophages.

    Immediate allergic responses are caused by receptors attached to the plasma membrane of mast cells in the tissues. When an allergen attaches to receptors on mast cells, they release histamine and other substances that bring about the symptoms.

    An immediate allergic response can occur within seconds of contact with the antigen. The symptoms can vary, but a dramatic example, anaphylactic shock, is a severe reaction characterized by a sudden and life-threatening drop in blood pressure.

    Allergy shots, injections of the allergen in question, sometimes prevent the onset of an allergic response. It has been suggested that injections of the allergen may cause the body to build up high quantities of antibodies released by plasma cells, and these combine with allergens received from the environment before they have a chance to reach the receptors located in the membranes of mast cells.

    Delayed allergic responses are probably initiated by memory T cells at the site of allergen contact in the body. A classic example of a delayed allergic response is the skin test for tuberculosis (TB). When the test result is positive, the tissue where the antigen was injected becomes red and hardened. This shows that the person has been previously exposed to tubercle bacillus, the cause of TB. Contact dermatitis, which occurs when a person is allergic to poison ivy, jewelry, cosmetics, and so forth, is also an example of a delayed allergic response.

Figure 26.10   Allergies.

When people are allergic to pollen, they develop symptoms that include watery eyes, sinus headaches, increased mucus production, labored breathing, and sneezing.

Autoimmune Diseases

When cytotoxic T cells or antibodies mistakenly attack the body's own cells as if they bear antigens, the resulting condition is known as an **autoimmune disease.** Exactly what causes autoimmune diseases is not known. However, sometimes they occur after an individual has recovered from an infection.

    In the autoimmune disease *myasthenia gravis*, neuromuscular junctions do not work properly, and muscular weakness results. In *multiple sclerosis* (*MS*), the myelin sheath of nerve fibers breaks down, causing various neuromuscular disorders. A person with *systemic lupus erythematosus* has various symptoms prior to death due to kidney damage. In *rheumatoid arthritis*, the joints are affected (**Fig. 26.11**). Researchers suggest that rheumatic fever and also type 1 diabetes are autoimmune illnesses. As yet, there are no cures for autoimmune diseases, but they can be controlled with drugs.

Figure 26.11   Rheumatoid arthritis.

Rheumatoid arthritis is due to recurring inflammation in skeletal joints. Complement proteins, T cells, and B cells all participate in deterioration of the joints, which eventually become immobile.

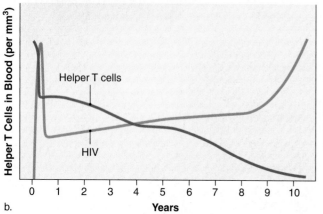

a.

b.

Figure 26.12 HIV infection.

a. An HIV infection leads to AIDS, characterized by a number of associated illnesses, including a cancer that results in skin lesions. **b.** At first the body produces enough helper T cells to keep the HIV infection under control, but then as the number of helper T cells declines, the HIV infection takes over.

AIDS

Understanding why AIDS patients are so sick gives us a whole new level of appreciation for the workings of a healthy immune system. An immune system ravaged by **AIDS (acquired immunodeficiency syndrome)** can no longer fight off the onslaught of viruses, fungi, and bacteria that the body encounters every day. **HIV (human immunodeficiency virus),** which causes AIDS, lives in and destroys helper T cells, which promote the activity of all the other cells in the immune system. For years, the body is able to maintain an adequate number of T cells, but finally the virus gains the upper hand. Without drug therapy, the number of T cells eventually drops from thousands to hundreds as the immune system becomes helpless (**Fig. 26.12**). The symptoms of AIDS begin with weight loss, chronic fever, cough, diarrhea, swollen glands, and shortness of breath, and progress to those of rare diseases. *Pneumocystis pneumonia*, a respiratory disease found in cats, and *Kaposi's sarcoma*, a very rare type of cancer, are often observed in patients with advanced AIDS. Death approaches rapidly and certainly.

HIV is transmitted by sexual contact with an infected person, including vaginal or rectal intercourse and oral/genital contact. Also, needle-sharing among intravenous drug users is a high-risk behavior. Babies born to HIV-infected women may become infected before or during birth, or through breast-feeding after birth. In 2007, approximately 33.2 million people worldwide were estimated to be living with HIV, with 6,800 people becoming newly infected and 5,700 people dying every day. Even though male-to-male sexual contact still accounts for the greatest number of new HIV cases in the United States, heterosexual contact or intravenous drug use accounts for the greatest percentage of increase in new cases.

Advances in treatment have reduced the serious complications of an HIV infection and have prolonged life. The sooner drug therapy begins after infection, the better the chances that the immune system will not be destroyed by HIV. Also, medication must be continued indefinitely. Unfortunately, new strains of the virus have emerged that are resistant to the new drugs used for treatment. The likelihood of transmission from mother to child at birth can be lessened if the mother receives medication prior to birth and the child is delivered by cesarean section.

Many investigators are working on a vaccine for AIDS. Some are trying to develop a vaccine in the traditional way, using the entire virus. Others are working on vaccines that utilize just a single HIV protein as the vaccine. So far, no method has resulted in sufficient antibodies to keep an infection at bay. After many clinical trials, none too successful, most investigators agree that a combination of various vaccines may be the best strategy to bring about a response by both B cells and T cells.

HIV infection is preventable. Suggestions for preventing an infection are to (1) abstain from sexual intercourse or develop a long-term monogamous (always same partner) sexual relationship with a person who is free of HIV; (2) be aware that having relations with an intravenous drug user is risky behavior; (3) avoid anal-rectal intercourse because the lining of the rectum is thin, and infected T cells easily enter the body there; (4) always use a latex condom during sexual intercourse if you do not know that your partner has been free of HIV for the past five years; (5) avoid oral sex because this can be a means of transmission; and (6) be cautious about the use of alcohol or any drug that may prevent you from being able to control your behavior.

Check Your Progress

1. Describe the relationship between allergies and allergens.
2. Contrast an immediate allergic response with a delayed allergic response.
3. Explain why AIDS patients cannot fight pathogens.

Answers: 1. Allergies are hypersensitive responses to antigens called allergens.   2. An immediate allergic response occurs within seconds of exposure to an allergen and is caused by chemicals released by mast cells. Delayed allergic responses take longer to develop and are probably initiated by memory T cells.   3. The HIV virus lives in and destroys helper T cells, which are necessary for the activity of all other immune system cells.

The Chapter in Review

Summary

26.1 | Organs, Tissues, and Cells of the Immune System

The immune system consists of lymphatic organs, tissues, and cells, as well as the products of these cells. The lymphatic organs are:

- Red bone marrow, where all blood cells are made and the B lymphocytes mature.
- Thymus gland, where T lymphocytes mature (see Table 26.1).
- Lymph nodes, where lymph is cleansed of pathogens and debris.
- Spleen, where blood is cleansed of pathogens and debris.
- Other organs, such as the tonsils and appendix, are patches of lymphatic tissue.

26.2 | Nonspecific Defenses

Immunity involves nonspecific and specific defenses. Nonspecific defenses include:

- Barriers to entry (e.g., skin).
- Inflammatory response involving mast cells, which release histamine to increase capillary permeability, resulting in redness, warmth, swelling, and pain. Neutrophils and macrophages enter tissue fluid and engulf pathogens.
- The complement system, which has many functions. One is to attack bacteria outright by forming a pore in the surface of a bacterium. Water and salts then enter, and the bacterium bursts.
- Natural killer cells that can tell self proteins from nonself and cause virus-infected cells to undergo apoptosis.

26.3 | Specific Defenses

Specific defenses require B lymphocytes and T lymphocytes, also called B cells and T cells.

B Cells and the Antibody Response

The BCR (B-cell receptor) of each type B cell is specific to a particular antigen. When the antigen binds to a BCR, that B cell divides to produce plasma cells and memory B cells.

- Plasma cells secrete antibodies and eventually undergo apoptosis. Plasma cells are responsible for antibody response to an antigen.

- Memory B cells remain in the body and produce antibodies if the same antigen enters the body at a later date.

T Cells and the Cellular Response

T cells are responsible for cellular response to an infection. Each TCR (T-cell receptor) is specific to a particular antigen. For a T cell to recognize an antigen, the antigen must be presented to it, usually by a macrophage, along with an MHC protein.

> Cytotoxic T cell (activated by MHC I + antigen) ⟶ Kill virus-infected cell
> Helper T cells (activated by MHC II + antigen) ⟶ Regulate immune response

- Cytotoxic T cells kill virus-infected or cancer cells on contact because these cells bear the MHC I antigen it has learned to recognize. First, perforin is secreted, and these molecules form a pore; then granzymes are secreted and cause the cell to undergo apoptosis.
- Helper T cells produce cytokines and stimulate other immune cells.

26.4 | Immunizations

Immunity occurs after an infection or a vaccination.

- A vaccine brings about immunity to a particular infection. Two injections may be required because the number of antibodies is higher after the second injection, called a booster shot. A booster shot in the future also increases the number of antibodies.

- Passive immunity (receiving preformed antibodies) is short-lived because the antibodies are administered to, not made by, the individual.

26.5 | Immune System Problems

Allergies

Allergic responses to various substances can be immediate or delayed. Anaphylactic shock is a dangerous immediate allergic response.

Autoimmune Diseases

In an autoimmune disease (e.g., rheumatoid arthritis, multiple sclerosis, and perhaps type 1 diabetes), the immune system mistakenly attacks the body's tissues.

AIDS

The HIV virus lives in and destroys helper T cells. Without drug treatment, the number of T cells falls off, and the individual dies from infections that are rare in healthy individuals. Sexual contact and needle-sharing transmit AIDS from person to person. Combined drug therapy has prolonged the lives of some people infected with HIV. So far, vaccine research has met with limited success. However, AIDS is a preventable disease if certain behaviors are avoided.

Key Terms

acquired immunity   466

active immunity   470

agglutination   467

AIDS (acquired immunodeficiency syndrome)   472

allergen   471

allergy   471

antibody   467

antigen   462

antigen-presenting cell (APC)   468

appendix   462

autoimmune disease   471

B-cell receptor (BCR)   466

B lymphocyte (B cell)   463

complement system   465

cytokine   466

cytotoxic T cell   468

delayed allergic response   471

helper T cell   469

histamine   464

HIV (human immunodeficiency virus)   472

immediate allergic response   471

immune system   462

immunity   462

inflammatory response   464

innate immunity   464

lymphatic organs   462

lymph node   463

macrophage   463

mast cell   464

natural killer (NK) cell   465

neutrophil   464

passive immunity   470

red bone marrow   463

spleen   463

T-cell receptor (TCR)   468

thymus gland   463

T lymphocyte (T cell)   463

tonsils   462

vaccine   470

Testing Yourself

Choose the best answer for each question.

1. This lymphatic tissue is associated with the respiratory system.
   a. red bone marrow
   b. spleen
   c. tonsil
   d. appendix
   e. lymph node

2. Label the lymphatic organs in the following illustration.

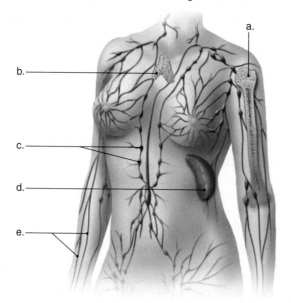

3. Which of the following is not a barrier to pathogen entry?
   a. oil gland secretions of the skin
   b. acidic pH in the stomach
   c. cilia in the upper respiratory tract
   d. nonpathogenic bacteria in the digestive tract
   e. saliva in the mouth

For questions 4–10, identify the lymphatic organ in the key that matches the description. Some answers may be used more than once.

Key:

   a. red bone marrow
   b. thymus gland
   c. lymph nodes
   d. spleen
   e. appendix

4. produces stem cells

5. located in every body cavity except the dorsal cavity

6. produces lymphocytes

7. site of maturation of T cell

8. contains red pulp and white pulp

9. region of lymphatic tissue attached to the large intestine

10. shrinks with age

11. During the inflammatory response,
    a. T cells move to the site of injury.
    b. capillaries become constricted.
    c. histamine is produced.
    d. capillaries become less permeable.
    e. More than one of these are correct.

12. _____ is the chemical released by certain blood cells, causing capillaries to dilate and become more permeable.
    a. Granzymes   c. Allergen
    b. Histamine   d. Cytokine

13. Which is a lymphatic organ?
    a. spleen   c. thymus gland
    b. tonsil   d. All of these are correct.

For questions 14–18, identify the type of blood cell in the key that matches the description. Some answers may be used more than once.

Key:

   a. macrophage
   b. B lymphocyte
   c. cytotoxic T cell
   d. helper T cell
   e. mast cell

14. large white blood cell that phagocytizes pathogens and presents antigens to T cells

15. white blood cell produced in the bone marrow that gives rise to plasma cells

16. secretes granzymes into infected cells to cause apoptosis

17. responsible for production of histamine

18. release cytokines and activate several components of the inmmue system

19. The _____ cleanses _____, and in its absence, problems associated with _____ may occur.
    a. spleen, lymph, oxygen transport
    b. liver, lymph, infection
    c. spleen, blood, infection
    d. spleen, blood, oxygen transport

20. Which cells will phagocytize pathogens?
    a. neutrophils   c. mast cells
    b. macrophages   d. Both a and b are correct.

21. During agglutination
    a. red blood cells clump.
    b. white blood cells attach pathogens.
    c. it is unlikely that a blood clot will form.
    d. platelets form a plug within a blood vessel.

22. Complement
    a. is a general defense mechanism.
    b. is involved in the inflammatory response.
    c. is a series of proteins present in the plasma.
    d. plays a role in destroying bacteria.
    e. All of these are correct.

23. Antibodies
    a. cause specific illnesses.
    b. result in immunity for a specific pathogen.
    c. are produced by helper T cells.
    d. can bond with many different types of pathogens.

24. _____ release histamines.
    a. Mast cells   c. Monocytes
    b. Neutrophils   d. All of these are correct.

25. Which applies to T lymphocytes?
    a. mature in bone marrow
    b. mature in thymus gland
    c. Both a and b are correct.
    d. None of these are correct.

26. Allergens
    a. cause an immune response in most people.
    b. are antigens to those people who are allergic to them.
    c. reduce the production of histamine by mast cells.
    d. are molecules that are usually rare in the environment.

27. AIDS is caused by which of the following viruses?
    a. SIV   d. AIDS
    b. HSV   e. HIV
    c. HPV

28. Plasma cells are
    a. the same as memory cells.
    b. formed from blood plasma.
    c. B cells that are actively secreting antibody.
    d. inactive T cells carried in the plasma.
    e. a type of red blood cell.

29. _____ is a condition that results when cytotoxic T cells attack the body's own cells.
    a. An allergic response   c. Passive immunity
    b. Autoimmune disease   d. Active immunity

30. Vaccines are associated with
    a. active immunity.   c. passive immunity.
    b. long-lasting immunity. d. Both a and b are correct.

31. MHC proteins play a role in
    a. aquired immunity.   c. tissue transplantation.
    b. histamine release.   d. All of these are correct.

32. Programmed cell death is known as
    a. activation.   c. clonal selection.
    b. apoptosis.   d. None of these are correct.

33. Type O blood has _____ antibodies in the plasma.
    a. A   c. Both a and b are correct.
    b. B   d. Neither a nor b is correct.

34. Label the structures in the following diagram of skin which contribute to the skin's protection of the body.

35. Which of the following is a nonspecific defense of the body?
    a. immunoglobulin   d. vaccine
    b. B cell   e. inflammatory response
    c. T cell

36. Antibodies combine with antigens
    a. at specific regions of the antibodies.
    b. at multiple regions of the antibodies.
    c. only if macrophages are present.
    d. Both a and c are correct.

37. Which one of these pairs is mismatched?
    a. helper T cells—help complement react
    b. cytotoxic T cells—active in tissue rejection
    c. macrophages—activate T cells
    d. memory T cells—long-living line of T cells
    e. T cells—mature in thymus

38. An antigen-presenting cell (APC)
    a. presents antigens to T cells.
    b. secretes antibodies.
    c. marks each human cell as belonging to that particular person.
    d. secretes cytokines only.

39. Which of the following is nonphagocytic and participates in a specific defense against an infection?
    a. mast cells
    b. macrophages
    c. neutrophils
    d. histamine
    e. plasma cells

40. Vaccines are
    a. the same as monoclonal antibodies.
    b. treated bacteria or viruses, or one of their proteins.
    c. short-lived.
    d. MHC proteins.
    e. All of these are correct.

41. Which of the following is not an example of an autoimmune disease?
    a. multiple sclerosis
    b. myasthenia gravis
    c. contact dermatitis
    d. systemic lupus erythematosus
    e. rheumatoid arthritis

42. Natural killer cells attack cells that
    a. have been determined to contain a virus.
    b. have lost their self proteins.
    c. are malformed.
    d. are dividing rapidly.

43. A person with type A blood can receive a transfusion from someone with type
    a. A blood only.
    b. A or B blood.
    c. A or AB blood.
    d. A or O blood.
    e. A, B, AB, or O blood.

44. Unlike B cells, T cells
    a. require help to recognize an antigen.
    b. are components of a specific defense response.
    c. are types of lymphocytes.
    d. contribute to homeostasis.

45. Passive immunity
    a. is permanent.
    b. may result from immunoglobulin injections.
    c. requires memory B cells.
    d. may be induced by vaccines.

46. Anaphylactic shock is an example of
    a. an immediate allergic response.
    b. an autoimmune disease.
    c. a type of serum shock.
    d. a delayed allergic response.

47. The HIV virus lives in and destroys
    a. B cells.
    b. cytotoxic T cells.
    c. helper T cells.
    d. All of these are correct.

48. AIDS patients generally die due to which of the following causes?
    a. rare diseases and infections
    b. HIV virus
    c. strokes
    d. heart attacks
    e. cancer

Thinking Scientifically

1. At one time, tonsillectomies (removal of the tonsils) were more commonplace than they are today. Why are the tonsils seemingly so susceptible to infection? Why should they not be removed at the first sign of infection?

2. The transplantation of organs from one person to another was impossible until the discovery of immunosuppressive drugs. Now, with the use of drugs such as cyclosporine, organs can be transplanted without rejection. Transplant patients must take immunosuppressive drugs for the remainder of their lives. What are the risks associated with long-term use of immunosuppressive drugs?

3. What role does the human immune system play in the evolution of viruses and bacteria? How does the evolution of pathogens affect our ability to achieve and maintain immunity against them?

Bioethical Issue

Access to AIDS Medications

Over 33 million people worldwide are living with AIDS. This disease is deadly without proper medical care, but can be chronic if treated. Drug companies typically charge a high price for AIDS medications because Americans and their insurance companies can afford them. However, these drugs are out of reach in many countries, such as those in Africa, where AIDS is a widespread problem. Some people argue that drug companies should use the profits from other drugs (such as those for heart disease, depression, and impotence) to make AIDS drugs affordable to those who need them. That has not happened yet. In some countries, governments have allowed companies to infringe on foreign patents held by major drug companies so that they can produce affordable AIDS drugs. Do drug companies have a moral obligation to provide low-cost AIDS drugs, even if they have to do so at a loss of revenue? Is it right for governments to ignore patent laws in order to provide their citizens with affordable drugs?

Essentials of Biology Website

The companion website for *Essentials of Biology* provides a wealth of information organized and integrated by chapter. You will find practice tests, animations, videos, and much more that will complement your learning and understanding of general biology.

http://www.mhhe.com/maderessentials2

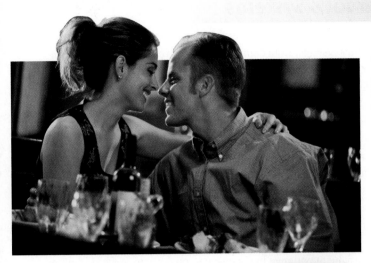
The nervous and endocrine systems control our behavior.

The Control Systems

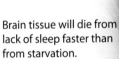
Brain tissue will die from lack of sleep faster than from starvation.

If brain tissue is removed, it is possible to live with half a brain.

Outline

Life with Half a Brain?

OBVIOUSLY, THE BRAIN'S role in the human body is critical to survival. The brain controls nearly all the activities in the body, including those of the other organ systems. When particular areas of the brain malfunction, a multitude of problems, or even death, can occur. Certain conditions, such as epilepsy, inherited disease, or infection, can cause unusual electrical activity in areas of the brain, leading to severe and

—Continued next page

Continued—

debilitating seizures. When the condition is serious enough and cannot be controlled by medication, physicians may consider performing a surgery termed a hemispherectomy. In this procedure, a portion of the brain (anywhere from a lobe up to a complete half) is completely removed. Unbelievably, many of the functions linked to the side of the brain that was removed can be "re-mapped" to the remaining side. In fact, most children who have undergone this procedure have experienced minimal lasting side effects, other than some minor paralysis on one side of the body. Learning ability does not seem to be significantly damaged by the procedure. Such drastic surgeries have taught us that the brain is a highly adaptable organ that can recover from significant trauma.

Understanding the structure and function of the nervous system is of critical importance to understanding the functioning of other organs and organ systems in the body. In this chapter, you will learn how the central and peripheral nervous systems operate, as well as how the central nervous system communicates with the endocrine system, which in turn regulates the body's internal environment.

27.1 | Nervous Systems

The nervous system and the endocrine system work together to regulate the activities of the other systems. Both control systems use chemical signals when they respond to changes that might threaten homeostasis, but they have different means of delivering these signals (**Fig. 27.1**). The nervous system quickly sends a message along a nerve fiber directly to a target organ, such as skeletal or smooth muscle. Once a chemical signal is released, the muscle brings about an appropriate response.

The endocrine system uses the blood vessels of the cardiovascular system to send a hormone as a messenger to a target organ, such as the liver. The endocrine system is slower-acting because it takes time for the hormone to move through the bloodstream to a target organ. Also, hormones change the metabolism of cells, and this takes time; however, the response is longer-lasting. Cellular metabolism tends to remain the same for at least a limited period of time.

Let's first turn our attention to the human nervous system. As we examine it, we will also compare it to the nervous systems of other animals.

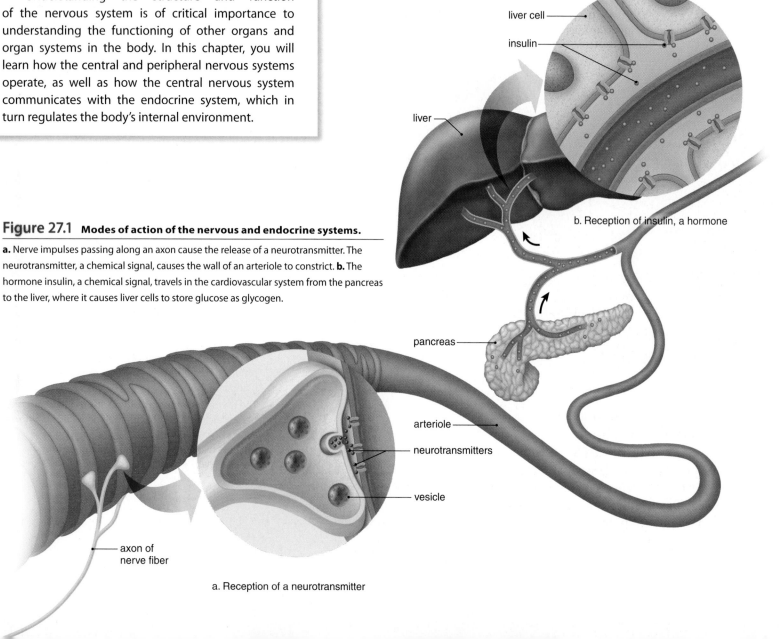

Figure 27.1 **Modes of action of the nervous and endocrine systems.**

a. Nerve impulses passing along an axon cause the release of a neurotransmitter. The neurotransmitter, a chemical signal, causes the wall of an arteriole to constrict. **b.** The hormone insulin, a chemical signal, travels in the cardiovascular system from the pancreas to the liver, where it causes liver cells to store glucose as glycogen.

liver cell

insulin

liver

b. Reception of insulin, a hormone

pancreas

arteriole

neurotransmitters

vesicle

axon of nerve fiber

a. Reception of a neurotransmitter

The Human Nervous System

The nervous system is always involved in an animal's ability to move around. In fact, about the only function of the nervous system in some animals, such as planarians, is movement, particularly to feed. Some planarians are predators, and the worm wraps itself around its prey, entangling it in slime and pinning it down. Then a muscular pharynx extends out the mouth, and by a sucking motion, the prey is torn up and swallowed. The bilateral symmetry of planarians is reflected in the organization of their nervous system. They have two lateral nerve cords (bundles of nerves) joined together by transverse nerves. The arrangement is called a *ladderlike* nervous system. A "brain" receives sensory information from the eyespots and sensory cells in the auricles. The two lateral nerve cords allow a rapid transfer of information from the cerebral ganglia to the posterior end, and the transverse nerves between the nerve cords keep the movement of the two sides coordinated. The nervous organization in planarians is a foreshadowing of the central and peripheral nervous systems that are found in more complex invertebrates, such as an earthworm, and in vertebrates, including humans (**Fig. 27.2**).

In humans, the nervous system controls the muscular system and works along with the endocrine system to maintain homeostasis. The **central nervous system (CNS)** includes the brain and spinal cord, which have a central location—they lie in the midline of the body. The **peripheral nervous system (PNS)** consists of nerves that lie outside the central nervous system. The brain gives off paired cranial nerves (one on each side of the body), and the spinal cord gives off paired spinal nerves. The division between the central nervous system and the peripheral nervous system is arbitrary; the two systems work together and are connected to one another.

What makes the human nervous system more complex than the planarian system? Five trends during the evolution of the vertebrate nervous system can be identified:

- A CNS developed that is able to summarize incoming messages before ordering outgoing messages.
- Nerve cells (neurons) became specialized to send messages to the central nervous system (CNS), between neurons in the CNS, or away from the CNS.
- A brain evolved that has special centers for receiving input from various regions of the body and for directing their activity.
- The CNS became connected to all parts of the body by peripheral nerves. Therefore, the central nervous system can respond to both external and internal stimuli.
- Complex sense organs, such as the human eye and ear, arose that can detect changes in the external environment.

This chapter discusses the first four aspects of the human nervous system. Sensory input and muscle response are the topics of Chapter 28.

Figure 27.2 Comparison of nervous systems.

Invertebrates, such as a planarian and an earthworm, as well as vertebrates, such as humans, have a central nervous system (e.g., brain) and a peripheral nervous system (nerves).

Check Your **Progress**

1. Describe the structure and function of the nervous system in planarians, and explain how it is anatomically different from that of humans.
2. List the five trends toward complexity of the nervous system in animals.

Answers: 1. The major function of the planarian nervous system is to bring about movement. The planarian nervous system is composed of two lateral nerve cords and transverse nerves that join them to form a ladderlike system. The human nervous system controls the muscular system and all the other systems. Humans have a well-developed central nervous system, which works in conjunction with the peripheral nervous system.   2. The central nervous system processes incoming messages before sending out messages; neurons are specialized for carrying messages; the brain has specialized regions for receiving information from different parts of the body; the CNS can respond to both internal and external stimuli; and sense organs detect changes in the external environment.

Figure 27.3 Types of neurons.

A sensory neuron, an interneuron, and a motor neuron are drawn here to show their arrangement in the body. Only axons conduct nerve impulses. In a sensory neuron, a process that extends from the cell body divides into an axon that takes nerve impulses all the way from the dendrites to the CNS. In a motor neuron and interneuron the axon extends directly from the cell body. The axon of sensory and motor neurons is covered by a myelin sheath. All long axons have a myelin sheath.

Neurons

The shape of a nerve cell, or **neuron,** is suitable to its function. The **cell body** contains the nucleus and other organelles that allow a cell to function. The neuron's many short **dendrite** fibers fan out to receive signals from sensory receptors or other neurons. These signals can result in nerve impulses carried by an axon. The **axon,** a nerve fiber that is typically longer than a dendrite, is the portion of a neuron that conducts nerve impulses. An axon can reach all the way from the end of your spinal cord to the tip of your big toe or from your big toe to the spinal cord, depending on which way messages are being conducted. Long axons are covered by a white **myelin sheath** formed from the membranes of tightly spiraled cells that leave gaps called **nodes of Ranvier.** Myelin sheaths account for our impression that nerves are white and glistening.

The nervous system has three types of neurons specific to its three functions (**Fig. 27.3**):

1. The nervous system receives sensory input. **Sensory neurons** perform this function. They take nerve impulses from sensory receptors to the CNS. The sensory receptor, which is the distal end of the axon of a sensory neuron, may be as simple as a naked nerve ending (a pain receptor), or may be built into a highly complex organ, such as the eye or ear. In any case, the axon of a sensory neuron can be quite long if the sensory receptor is far from the CNS.

2. The nervous system performs integration—in other words, the CNS sums up the input it receives from all over the body. **Interneurons** occur entirely within the CNS and take nerve impulses between various parts of the CNS. Some interneurons lie between sensory neurons and motor neurons, and some take messages from one side of the spinal cord to the other or from the brain to the spinal cord, and vice versa. They also form complex pathways in the brain where processes that account for thinking, memory, and language occur.

3. The nervous system generates motor output. **Motor neurons** take nerve impulses from the CNS to muscles or glands. Motor neurons cause muscle fibers to contract or glands to secrete, and therefore they are said to *innervate* these structures.

Check Your Progress

1. List the three major types of neurons and their functions.
2. What part of a neuron might be covered by a myelin sheath?
3. What part or parts of a neuron increase the available surface for receiving signals from other neurons or sensory receptors?

Answers: 1. Sensory neurons take nerve impulses from sensory receptors to the central nervous system. Interneurons carry nerve impulses between parts of the central nervous system. Motor neurons carry nerve impulses from the central nervous system to muscles or glands.   2. The axon.   3. Dendrites.

The Nerve Impulse

Like some other cellular processes, the nerve impulse is also dependent on concentration gradients. In neurons, these concentration gradients are maintained by the sodium-potassium pump. This pump actively transports sodium ions (Na^+) to the outside of the axon and actively transports potassium ions (K^+) inside. Aside from ion concentration differences across the axon's membrane, a charge difference also exists. *The inside of an axon is negative* compared to the outside. This charge difference is due in part to an unequal distribution of ions across the membrane, but is also due to the presence of large, negatively charged proteins in the axon cytoplasm.

The charge difference across the axon's membrane offers a potential for change, or an **action potential,** as the nerve impulse is also called. The nerve impulse is a rapid, short-lived, self-propagating reversal in the charge difference across the axon's membrane. **Figure 27.4** shows how it works. A nerve impulse involves two types of gated channel proteins in the axon's membrane. In contrast to ungated channel proteins, which constantly allow ions across the membrane, gated channel proteins open and close in response to a stimulus such as a signal from another neuron. One type of gated channel protein allows sodium (Na^+) to pass through the membrane, and the other allows potassium (K^+) to pass through the membrane. As an axon is conducting a nerve impulse, the Na^+ gates open at a particular location, and the inside of the axon becomes positive as Na^+ moves from outside the axon to the inside. The Na^+ gates close, and then the K^+ gates open. Now K^+ moves from inside the axon to outside the axon, and the charge reverses back again.

In Figure 27.4, the axon is unmyelinated, and the action potential at one locale stimulates an adjacent part of the axon's membrane to produce an action potential. In myelinated axons, an action potential at one node of Ranvier causes an action potential at the next node (**Fig. 27.5**). This type of conduction, called **saltatory conduction,** is much faster than otherwise. Imagine running down a long hallway as quickly as you can; then picture yourself able to get to the end of the very same hall in just a few leaping bounds. Leaping would enable you to travel the same distance in a much shorter time; likewise, saltatory conduction greatly speeds the conduction of nerve impulses. In thin, unmyelinated axons, the nerve impulse travels about 1.0 meter/second, but in thick, myelinated axons, the rate is more than 100 meters/second due to saltatory conduction. In any case, action potentials are self-propagating; each action potential generates another along the length of an axon.

The conduction of a nerve impulse (action potential) is an all-or-none event—that is, either an axon conducts a nerve impulse or it does not. The intensity of a message is determined by how many nerve impulses are generated within a given time span. An axon can conduct a volley of nerve impulses because only a small number of ions are exchanged with each impulse. As soon as an impulse has passed by each successive portion of an axon, it undergoes a short refractory period during which it is unable to conduct an impulse. During a refractory period, the sodium gates cannot yet open. This period ensures that nerve impulses travel in only one direction and do not reverse.

Figure 27.5   Conduction of a nerve impulse in a myelinated axon.

Action potentials can occur only at gaps in the myelin sheath called nodes of Ranvier. This makes the speed of conduction much faster than in unmyelinated axons. In humans, all long axons are myelinated.

a.

b.

Figure 27.4   **Conduction of action potentials in an unmyelinated axon.**

a. Na^+ and K^+ each have their own gated channel protein by which they cross the axon's membrane. **b.** During an action potential, Na^+ enters the axon, and the charge difference between inside and outside reverses (blue); then K^+ exits, and the charge difference is restored (red). The action potential moves from section to section in an unmyelinated axon. Only a few ions are exchanged at a time.

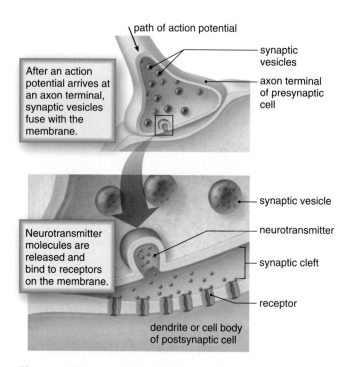

After an action potential arrives at an axon terminal, synaptic vesicles fuse with the membrane.

path of action potential

synaptic vesicles

axon terminal of presynaptic cell

synaptic vesicle

Neurotransmitter molecules are released and bind to receptors on the membrane.

neurotransmitter

synaptic cleft

receptor

dendrite or cell body of postsynaptic cell

Figure 27.6 Synapse structure and function.

Transmission across a synapse from one neuron (the presynaptic cell) to another occurs when a neurotransmitter is released, diffuses across a synaptic cleft, and binds to a receptor in the plasma membrane of the next neuron (the postsynaptic cell). Each axon releases only one type of neurotransmitter, symbolized here by a red ball.

Check Your Progress

1. Sequentially list the steps in an action potential.
2. Explain why saltatory conduction is rapid.
3. Explain how a signal is carried across the synaptic cleft.

Answers: 1. First: Na^+ gates open, and Na^+ moves to the inside; the inside of the axon becomes positive. Second: K^+ gates open, and K^+ moves to the outside; the inside of the axon becomes negative again.   2. In an unmyelinated axon, the action potential must travel the entire span of the membrane. Saltatory conduction occurs when axons are myelinated, allowing an action potential at one node to cause an action potential at the next node, skipping the membrane between nodes.   3. Nerve impulses at the axon terminal cause synaptic vesicles to release a neurotransmitter into the synaptic cleft. The neurotransmitters then diffuse across the cleft and bind to a receptor proteins.

The Synapse

Each axon has many axon terminals (**Fig. 27.6**). In the CNS, a terminal of one neuron, known as the *presynaptic cell*, lies very close to the dendrite (or the cell body) of another neuron, the *postsynaptic cell*. This region of close proximity is called a **synapse.** In the PNS, when the postsynaptic cell is a muscle cell, the region is called a neuromuscular junction. A small gap exists at a synapse, and this gap is called the **synaptic cleft.** How is it possible to excite the next neuron or a muscle cell across this gap? Nerve impulses cannot themselves pass across the synaptic cleft. Instead, transmission across a synaptic cleft is carried out by chemical signals called **neurotransmitters,** which are stored in synaptic vesicles. When nerve impulses traveling along an axon reach an axon terminal, synaptic vesicles release a neurotransmitter into the synaptic cleft. Neurotransmitter molecules diffuse across the cleft and bind to a specific receptor protein on the postsynaptic cell.

Depending on the type of neurotransmitter and/or the type of receptor, the response of the postsynaptic cell can be toward excitation or toward inhibition. At least 25 different neurotransmitters have been identified, but two very well-known neurotransmitters are **acetylcholine (ACh)** and **norepinephrine (NE).**

Once a neurotransmitter has been released into a synaptic cleft and has initiated a response, the neurotransmitter is removed from the cleft. In some synapses, the postsynaptic cell produces enzymes that rapidly inactivate the neurotransmitter. For example, the enzyme **acetylcholinesterase (AChE)** breaks down acetylcholine. In other synapses, the presynaptic cell rapidly reabsorbs the neurotransmitter. For example, norephinephrine is reabsorbed by the axon terminal. The short existence of neurotransmitters at a synapse prevents continuous stimulation (or inhibition) of the postsynaptic cell.

A single neuron has many dendrites plus the cell body, and both can have synapses with many other neurons. One thousand to ten thousand synapses per single neuron are not uncommon. Therefore, a neuron is on the receiving end of many signals. An excitatory neurotransmitter produces a potential change that drives the neuron closer to an action potential, and an inhibitory neurotransmitter produces a potential change that drives the neuron farther from an action potential. Neurons integrate these incoming signals. **Integration** is the summing up of both excitatory and inhibitory signals. If a neuron receives many excitatory signals (either at different synapses or at a rapid rate from one synapse), chances are the axon will transmit a nerve impulse. On the other hand, if a neuron receives both inhibitory and excitatory signals, the integration of these signals may prohibit the axon from firing.

Drug Abuse

Many drugs that affect the nervous system act by interfering with or promoting the action of neurotransmitters. A drug can either enhance or block the release of a neurotransmitter, mimic the action of a neurotransmitter or block the receptor for it, or interfere with the removal of a neurotransmitter from a synaptic cleft. *Stimulants* are drugs that increase activity of the CNS, and *depressants* decrease the activity of the CNS. Increasingly, researchers believe that dopamine, a neurotransmitter in the brain, is responsible for mood. Many of the new medications developed to counter drug dependence and mental illness affect the release, reception, or breakdown of dopamine.

Drug abuse is apparent when a person takes a drug at a dose level and under circumstances that increase the potential for a harmful effect. A drug abuser often takes more of the drug than was intended. Drug abusers are apt to display a psychological and/or physical dependence on the drug. With physical dependence, formerly called "addiction," more of the drug is needed to get the same effect, and withdrawal symptoms occur when the user stops taking the drug.

Cocaine

Cocaine is an alkaloid derived from the shrub *Erythroxylon coca*. It is sold in powder form and as crack, a more potent extract. Because cocaine prevents the synaptic uptake of dopamine, the neurotransmitter remains in the synapse for a prolonged period of time and continues to stimulate the postsynaptic cell. As a result, the user experiences a "rush" sensation. The epinephrine-like effects of dopamine account for the state of arousal that lasts for several minutes after the rush experience.

A cocaine binge can go on for days, after which the individual suffers a crash. During the binge period, the user is hyperactive and has little desire for food or sleep but has an increased sex drive. During the crash period, the user is fatigued, depressed, and irritable, has memory and concentration problems, and displays no interest in sex.

Cocaine causes extreme physical dependence. With continued cocaine use, the postsynaptic cells become increasingly desensitized to dopamine. The user, therefore, experiences physical dependence withdrawal symptoms, and an intense craving for cocaine. These are indications that the person is highly dependent upon the drug.

Overdosing on cocaine can cause seizures and cardiac and respiratory arrest. It is possible that long-term cocaine abuse causes brain damage (**Fig. 27.7**). Babies born to addicts suffer withdrawal symptoms and may have neurological and developmental problems.

Heroin

Heroin is derived from morphine, an alkaloid of opium. Once heroin is injected into a vein (**Fig. 27.8**), a feeling of euphoria, along with relief of any pain, occurs within 3 to 6 minutes. Side effects can include nausea, vomiting, restlessness, anxiety, mood disorders, and respiratory and circulatory depression. Heroin binds to receptors meant for the endorphins, special neurotransmitters that ease pain and produce a feeling of tranquility. With time, the body's production of endorphins decreases. Physical dependence develops, and the euphoria originally experienced upon injection is no longer felt. Heroin withdrawal symptoms are quite severe, and infants born to women who are physically dependent also experience these withdrawal symptoms.

Marijuana

The dried flowering tops, leaves, and stems of the Indian hemp plant *Cannabis sativa* contain and are covered by a resin that is rich in THC (tetrahydrocannabinol). The names *cannabis* and *marijuana* apply to either the plant or THC. Usually, marijuana is smoked in a cigarette form called a "joint."

Recently, researchers have found that marijuana binds to a receptor for anandamide, a neurotransmitter that seems to create a feeling of peaceful contentment. The occasional marijuana user experiences a mild euphoria along with alterations in vision and judgment, which result in distortions of space and time. Motor incoordination, including the inability to speak coherently, takes place. Heavy use can result in hallucinations, anxiety, depression, rapid flow of ideas, body image distortions, paranoid reactions, and similar psychotic symptoms. Craving and difficulty in stopping usage can occur as a result of regular use.

brain inactivity →

a. Before cocaine use, brain is more active.

b. After cocaine use, brain is less active.

Figure 27.7 **Effect of cocaine on brain.**

In a cocaine user, PET scans show that (**a**) the usual activity of the brain is (**b**) reduced. The color red indicates brain tissue is active.

Figure 27.8 **Drug abuse.**

Blood-borne diseases such as AIDS and hepatitis B pass from one drug abuser to another when they share needles.

The Central Nervous System

The organization of the brains in certain vertebrates—namely, reptiles (alligator and goose), and mammals (horse and human)—is compared in **Figure 27.9***a*. If we divide the brain into a hindbrain, midbrain, and forebrain, we can see that the forebrain is most prominent in humans. The forebrain of mammals also has an altered function in that it becomes the last depository for sensory information. This change accounts for why the forebrain carries on much of the integration for the entire nervous system before it sends out motor instructions to glands and muscles. In humans, the **spinal cord** provides a means of communication between the brain and the spinal nerves, which are a part of the PNS. (Spinal nerves leave the spinal cord and take messages to and from the skin, glands, and muscles in all areas of the body, except for the head and face.) Myelinated long fibers of interneurons in the spinal cord run together in bundles called **tracts.** These tracts connect the spinal cord to the brain. Because certain of the tracts cross over at one point, the left side of the brain controls the right side of the body, and vice versa. Also, as discussed on page 487, the spinal cord is involved in reflex actions, which are programmed, built-in circuits that allow for protection and survival. They are present at birth and require no conscious thought to take place.

The Brain

Our discussion will center on these parts of the brain: the cerebrum, the diencephalon, the cerebellum, and the brain stem (Fig. 27.9*b*).

Cerebrum   The cerebrum communicates with and coordinates the activities of the other parts of the brain. The cerebrum has two halves, and each half has a number of lobes, which are color coded in **Figure 27.10.** Most of the cerebrum is white matter where the long axons of interneurons are taking nerve impulses to and from the cerebrum. The highly convoluted outer layer of gray matter that covers the cerebrum is called the **cerebral cortex.** The cerebral cortex contains over one billion cell bodies, and it is the region of the cerebrum that interprets and initiates sensation, voluntary movement, and higher thought processes.

Investigators have found that each part of the cerebrum has specific functions. To take an example, the **primary sensory area** located in the parietal lobe receives information from the skin, skeletal muscles, and joints. Each part of the body has its own particular receiving area. The **primary motor area,** on the other hand, is in the frontal lobe just before the small cleft that divides the frontal lobe from the parietal lobe. Voluntary commands to skeletal muscles involve the primary motor area, and the muscles in each part of the body are controlled by a certain section of the primary motor area.

Figure 27.9   Vertebrate brains.

a. A comparison of reptile and mammalian brains shows that the forebrain increased in size and complexity among these animals. **b.** The human brain.

The lobes of the cerebral cortex have a number of specialized centers to receive information from the sensory receptors for sight, hearing, and smell. The lobes also have *association areas* where integration occurs. The **prefrontal area,** an association area in the frontal lobe, receives information from the other association areas and uses this information to reason and plan our actions. Integration in this area accounts for our most cherished human abilities to think critically and to formulate appropriate behaviors.

Diencephalon Beneath the cerebrum is the diencephalon, which contains the hypothalamus and the thalamus (see Fig. 27.9*b*). The **hypothalamus** is an integrating center that helps maintain homeostasis by regulating hunger, sleep, thirst, body temperature, and water balance. The hypothalamus controls the pituitary gland, and thereby serves as a link between the nervous and endocrine systems. The **thalamus** is on the receiving end for all sensory input except smell. Information from the eyes, ears, and skin arrives at the thalamus via the cranial nerves and tracts from the spinal cord. The thalamus integrates this information and sends it on to the appropriate portions of the cerebrum. The thalamus is involved in arousal of the cerebrum; it participates in motor functions and higher mental processes such as memory and emotions.

The pineal gland, which secretes the hormone melatonin, is located in the diencephalon. Presently, there is much popular interest in melatonin because it is released at night when we are sleeping. Some researchers hypothesize that it can be used to help prevent jet lag or insomnia.

Cerebellum The **cerebellum** has two portions that are joined by a narrow median strip. Each portion is primarily composed of white matter, which in longitudinal section has a treelike pattern (see Fig. 27.9*b*). Overlying the white matter is a thin layer of gray matter that forms a series of complex folds.

The cerebellum receives sensory input from the eyes, ears, joints, and skeletal muscles about the present position of body parts, and it also receives motor output from the cerebral cortex that specifies where these parts should be located. After integrating this information, the cerebellum sends motor impulses by way of the brain stem to the skeletal muscles. In this way, the cerebellum maintains posture and balance. It also ensures that all of the muscles work together to produce smooth, coordinated voluntary movements. The cerebellum assists the learning of new motor skills such as playing the piano or hitting a baseball.

Brain Stem The **brain stem,** which contains the midbrain, the pons, and the medulla oblongata, connects the rest of the brain to the spinal cord (see Fig. 27.9*b*). It contains tracts that ascend or descend between the spinal cord and higher brain centers. The **midbrain** contains important visual and auditory reflex centers, and coordinates responses such as the startle reflex. This occurs when you automatically turn your head in response to a sudden loud noise, trying to see its source. The **medulla oblongata** contains a number of reflex centers for regulating heartbeat, breathing, and vasoconstriction (blood pressure). It also contains the reflex centers for vomiting, coughing, sneezing, hiccuping, and swallowing. In addition, the medulla oblongata helps control various internal organs. The **pons** links the medulla with the midbrain, and is vital for the control of breathing.

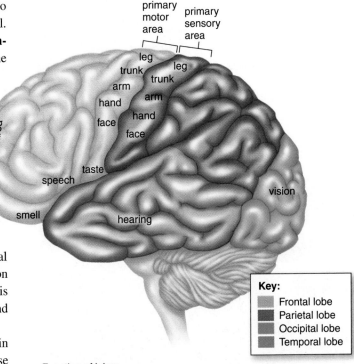

Key:
- Frontal lobe
- Parietal lobe
- Occipital lobe
- Temporal lobe

Function of lobes
- Frontal lobe—reasoning, planning, speech, movement, emotions, and problem solving
- Parietal lobe—integration of sensory input from skin and skeletal muscles, understanding speech
- Occipital lobe—seeing, perception of visual stimuli
- Temporal lobe—hearing, perception of auditory stimuli

Figure 27.10 Functional regions of the cerebral cortex.

Specific areas of the cerebral cortex receive sensory input from particular sensory receptors, integrate various types of information, or send out motor commands to particular areas of the body.

Check Your Progress

1. Compare and contrast the primary sensory area with the primary motor area of the cerebrum.
2. List the three parts of the brain stem.

Answers: 1. The primary sensory area receives information from the skin, skeletal muscles, and joints. The primary motor area sends voluntary commands to skeletal muscles. 2. Midbrain, pons, and medulla oblongata.

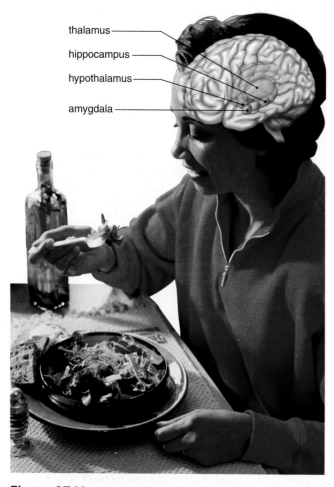

Figure 27.11   The limbic system.

The limbic system includes the diencephalon and parts of the cerebrum. It joins higher mental functions, such as reasoning, with more instinctive feelings, such as fear and pleasure. Eating is a pleasurable activity for most people.

Figure 27.12   A nerve.

A nerve contains the axons of many neurons.

The Limbic System

The **limbic system** is a complex network that includes the diencephalon and areas of the cerebrum (**Fig. 27.11**). The limbic system blends higher mental functions and primitive emotions into a united whole. It accounts for why activities such as sexual behavior and eating seem pleasurable and also why, say, mental stress can cause high blood pressure.

Two significant structures within the limbic system are the hippocampus and the amygdala, which are essential for learning and memory. The hippocampus, a seahorse-shaped structure that lies deep in the temporal lobe, is well situated in the brain to make the prefrontal area aware of past experiences stored in sensory association areas. The amygdala, in particular, can cause these experiences to have emotional overtones. A connection between the frontal lobe and the limbic system means that reason can keep us from acting out strong feelings.

Learning and Memory   Memory is the ability to hold a thought in mind or recall events from the past, ranging from a word we learned only yesterday to an early emotional experience that has shaped our lives. Learning takes place when we retain and utilize past memories.

The prefrontal area in the frontal lobe is active during short-term memory, as when we temporarily recall a telephone number. Some telephone numbers go into long-term memory. Think of a telephone number you know by heart, and see if you can bring it to mind without also thinking about the place or person associated with that number. Most likely you cannot, because typically long-term memory is a mixture of what is called *semantic memory* (numbers, words, and so on) and *episodic memory* (persons, events, and other associations). *Skill memory* is a type of memory that can exist independent of episodic memory. Skill memory is what allows us to perform motor activities like riding a bike or playing ice hockey. A person who has **Alzheimer disease (AD)** experiences a progressive loss of memory, particularly for recent events. Gradually the person loses the ability to perform any type of daily activity and becomes bedridden. In AD patients, clusters of abnormal tissue develop among degenerating neurons, especially in the hippocampus and amygdala. Major research efforts are devoted to seeking a cure for AD.

What parts of the brain are functioning when we remember something from long ago? Our long-term memories are stored in bits and pieces throughout the sensory association areas of the cerebral cortex. The hippocampus gathers this information together for use by the prefrontal area of the frontal lobe when we remember Uncle George or our summer holiday. Why are some memories so emotionally charged? The amygdala is responsible for fear conditioning and for associating danger with sensory information received from the thalamus and the cortical sensory areas.

The Peripheral Nervous System

The peripheral nervous system (PNS) lies outside the central nervous system and contains **nerves,** which are bundles of axons (**Fig. 27.12**). The cell bodies of neurons are found in the CNS—that is, the brain and spinal cord—or in ganglia. Ganglia (sing., **ganglion**) are collections of cell bodies within the PNS.

Humans have 12 pairs of **cranial nerves** attached to the brain. Cranial nerves are largely concerned with the head, neck, and facial regions of the body. However, the vagus nerve is a cranial nerve that has branches not only to the pharynx and larynx, but also to most of the internal organs.

Humans have 31 pairs of **spinal nerves,** and each contains many sensory and motor axons. The dorsal root of a spinal nerve contains the axons of sensory neurons, which conduct impulses to the spinal cord from sensory receptors. The cell body of a sensory neuron is in the **dorsal root ganglion.** The ventral root contains the axons of motor neurons, which conduct impulses away from the cord, largely to skeletal muscles (**Fig. 27.13**). Each spinal nerve serves the particular region of the body in which it is located.

The Somatic System

The **somatic system** of the PNS includes the nerves that take sensory information from external sensory receptors to the CNS and motor commands away from the CNS to skeletal muscles. Voluntary control of skeletal muscles always originates in the brain. Involuntary responses to stimuli, called **reflexes,** can involve either the brain or just the spinal cord. Flying objects cause our eyes to blink, and sharp pins cause our hands to jerk away even without our having to think about it.

Figure 27.13 illustrates the path of a reflex that involves only the spinal cord. If your hand touches a sharp pin, sensory receptors in the skin generate nerve impulses that move along sensory axons toward the spinal cord. Sensory neurons that enter the spinal cord pass signals on to many interneurons. Some of these interneurons synapse with motor neurons. The short dendrites and the cell bodies of motor neurons are in the spinal cord, but their axons leave the cord. Nerve impulses travel along motor axons to an effector, which brings about a response to the stimulus. In this case, a muscle contracts so that you withdraw your hand from the pin. Various other reactions are possible—you will most likely look at the pin, wince, and cry out in pain. This whole series of responses is explained by the fact that some of the interneurons involved carry nerve impulses to the brain. Also, sense organs send messages to the brain that make us aware of our actions. The brain makes you aware of the stimulus and directs these other reactions to it.

Figure 27.13   **A reflex arc showing the path of a spinal reflex.**

A stimulus (e.g., a pinprick) causes sensory receptors in the skin to generate nerve impulses that travel in sensory axons to the spinal cord. Interneurons integrate data from sensory neurons and then relay signals to motor neurons. Motor axons convey nerve impulses from the spinal cord to a skeletal muscle, which contracts. Movement of the hand away from the pin is the response to the stimulus.

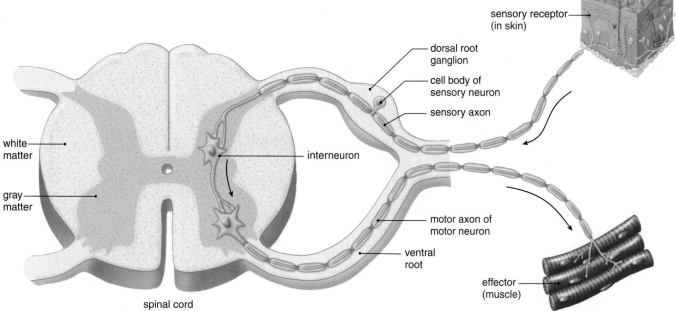

The Autonomic System

The **autonomic system** of the PNS automatically and involuntarily regulates the activity of glands and cardiac and smooth muscle. The system is divided into the parasympathetic and sympathetic divisions (**Fig. 27.14**). Reflex actions, such as those that regulate blood pressure and breathing rate, are especially important to the maintenance of homeostasis. These reflexes begin when the sensory neurons in contact with internal organs send information to the CNS. They are completed by motor neurons within the autonomic system.

The **parasympathetic division** includes a few cranial nerves (e.g., the vagus nerve) and also axons that arise from the last portion of the spinal cord. The parasympathetic division, sometimes called the "housekeeper division," promotes all the internal responses we associate with a relaxed state (see Fig. 27.14, photograph). For example, it causes the pupil of the eye to constrict, promotes digestion of food, and retards the heartbeat. The neurotransmitter utilized by the parasympathetic division is acetylcholine (ACh).

Axons of the **sympathetic division** arise from portions of the spinal cord. The sympathetic division is especially important during emergency situations and is associated with "fight or flight." If you need to fend off a foe or flee from danger, active muscles require a ready supply of glucose and oxygen. On the one hand, the sympathetic division accelerates the heartbeat and dilates the airways. On the other hand, the sympathetic division inhibits the digestive tract, since digestion is not an immediate necessity if you are under attack. The sympathetic nervous system utilizes the neurotransmitter norepinephrine, which has a structure like that of epinephrine (adrenaline) released by the adrenal medulla.

Parasympathetic

- constrict pupils
- secrete saliva
- constrict airways
- slow down heartbeat
- adrenal gland
- stomach
- increase secretion
- large intestine
- increase motility
- empty colon
- empty bladder
- bladder

Sympathetic

- dilate pupils
- stop saliva secretion
- spinal cord
- dilate airways
- speed up heartbeat
- secrete adrenaline
- decrease secretion
- decrease motility
- small intestine
- retain colon contents
- delay emptying

Figure 27.14
Autonomic system.

The parasympathetic and sympathetic motor axons go to the same organs, but they have opposite effects. The parasympathetic division is active when we, as mammals, feel warm and cozy in the arms of someone who loves us (see photograph). The sympathetic division is active when we are stressed and feel threatened.

Check Your **Progress**

1. **Explain why the parasympathetic division of the autonomic system is called the "housekeeper division."**

Answer: 1. The parasympathetic division keeps the "house" in order by maintaining the internal responses associated with a relaxed state, such as pupil contraction, food digestion, and slow heartbeat.

27.2 | Endocrine Systems

The endocrine system consists of glands and tissues that secrete chemical signals we call hormones (**Fig. 27.15**). **Endocrine glands** do not have ducts; they secrete their hormones directly into the bloodstream for distribution throughout the body. They can be contrasted with exocrine glands, which have ducts and secrete their products into these ducts for transport to body cavities. For example, the salivary glands send saliva into the mouth by way of the salivary ducts.

The endocrine system and the nervous system are intimately involved in homeostasis, the relative stability of the internal environment. Hormones directly affect blood composition and pressure, body growth, and many more life processes. Certain hormones are involved in the maturation and function of the reproductive organs, and these are discussed in Chapter 29.

The Action of Hormones

The cells that can respond to a hormone have receptor proteins that bind to the hormone. Hormones cause these cells to undergo a metabolic change. The type of change is dependent on the chemical structure of the hormone. **Steroid hormones** (see Fig. 3.15, p. 39) are lipids, and they can pass through the plasma membrane. The hormone-receptor complex then binds to DNA, and gene expression follows—for example, a protein such as an enzyme is made by the cell. The enzyme goes on to produce a change in the target cell's function (**Fig. 27.16***a*).

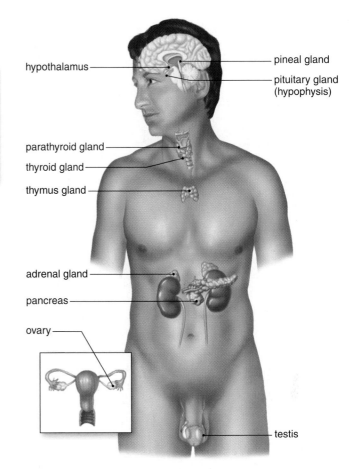

Figure 27.15   The endocrine system.

Anatomical location of major endocrine glands in the body.

a. Action of steroid hormone

b. Action of peptide hormone

Figure 27.16   How hormones work.

a. A steroid hormone (S) is a chemical signal that is able to enter the cell. Reception of this messenger causes the cell to synthesize a product by way of the cellular machinery for protein synthesis. **b.** A peptide hormone (P) is a "first messenger" that is received by a cell at the plasma membrane. Reception of the first messenger and a signal transduction pathway lead to a "second messenger" that changes the metabolism of the cell.

Figure 27.17   Negative feedback inhibition.

The hormones secreted by the thyroid (and also the adrenal cortex and gonads) feed back to inhibit the anterior pituitary and hypothalamic-releasing hormones so that their blood levels stay relatively constant.

Peptide hormones comprise a category that includes peptides, proteins, glycoproteins, and modified amino acids. Peptide hormones can't pass through the plasma membrane so they bind to a receptor protein in the plasma membrane (Fig. 27.16*b*). The peptide hormone is called the "first messenger" because a signal transduction pathway leads to a second molecule—that is, the "second messenger," that changes the metabolism of the cell. The second messenger sets in motion an enzyme pathway that is sometimes called an enzyme cascade because each enzyme in turn activates another. Because enzymes work over and over, every step in an enzyme cascade leads to more reactions—the binding of a single peptide hormone molecule can result in as much as a thousandfold response.

Hypothalamus and Pituitary Gland

The **hypothalamus,** a part of the brain (see Fig. 27.9), helps regulate the internal environment. For example, it is on the receiving end of information about the heartbeat and body temperature. And to correct any abnormalities, the hypothalamus communicates with the medulla oblongata, where the brain centers that control the autonomic system are located. The hypothalamus is also a part of the endocrine system, containing specialized hormone-secreting neurons. It controls the glandular secretions of the **pituitary gland,** a small gland connected to the brain by a stalklike structure. The pituitary has two portions, the anterior pituitary and the posterior pituitary, which are distinct from each other.

Anterior Pituitary

The hypothalamus controls the anterior pituitary by producing **hypothalamic-releasing hormones,** most of which are stimulatory. The **anterior pituitary** in turn stimulates other glands:

1. **Thyroid-stimulating hormone (TSH)** stimulates the thyroid to produce triiodothyronine (T_3) and thyroxine (T_4).
2. **Adrenocorticotropic hormone (ACTH)** stimulates the adrenal cortex to produce the glucocorticoids.
3. **Gonadotropic hormones (FSH** and **LH)** stimulate the gonads—the testes in males and the ovaries in females—to produce gametes and sex hormones.

In these instances, a three-tiered control system develops that is described in **Figure 27.17.** For example, the secretion of thyroid-releasing hormone (TRH) by the hypothalamus stimulates the thyroid to produce the thyroid-stimulating hormone (TSH), and the thyroid produces its hormones (T_3 and T_4), which feed back to inhibit the release of the first two hormones mentioned.

Two other hormones produced by the anterior pituitary do not affect other endocrine glands (**Fig. 27.18**). **Prolactin (PRL)** is produced in quantity during pregnancy and after childbirth. It causes the mammary glands in the breasts to develop and produce milk. It also plays a role in carbohydrate and fat metabolism. **Growth hormone (GH)** promotes skeletal and muscular growth. It stimulates the rate at which amino acids enter cells and protein synthesis occurs. Underproduction of growth hormone leads to pituitary dwarfism, and overproduction can lead to pituitary gigantism.

Check Your Progress

1. Explain how steroid and peptide hormones induce metabolic changes in cells.
2. Name the only two hormones produced by the anterior pituitary that do not target other endocrine glands.

Answers: 1. Steroid hormones enter the cell and stimulate genes to produce enzymes that alter cell activity. Peptide hormones induce an enzyme cascade that alters cell activity.   2. Prolactin (PRL) and growth hormone (GH).

Figure 27.18 The hypothalamus and the pituitary.

(*left*) The hypothalamus controls the secretions of the anterior pituitary, and the anterior pituitary controls the secretions of the thyroid gland, adrenal cortex, and gonads, which are also endocrine glands. Growth hormone and prolactin are also produced by the anterior pituitary. (*Right*) The hypothalamus produces two hormones, ADH and oxytocin, which are stored and secreted by the posterior pituitary.

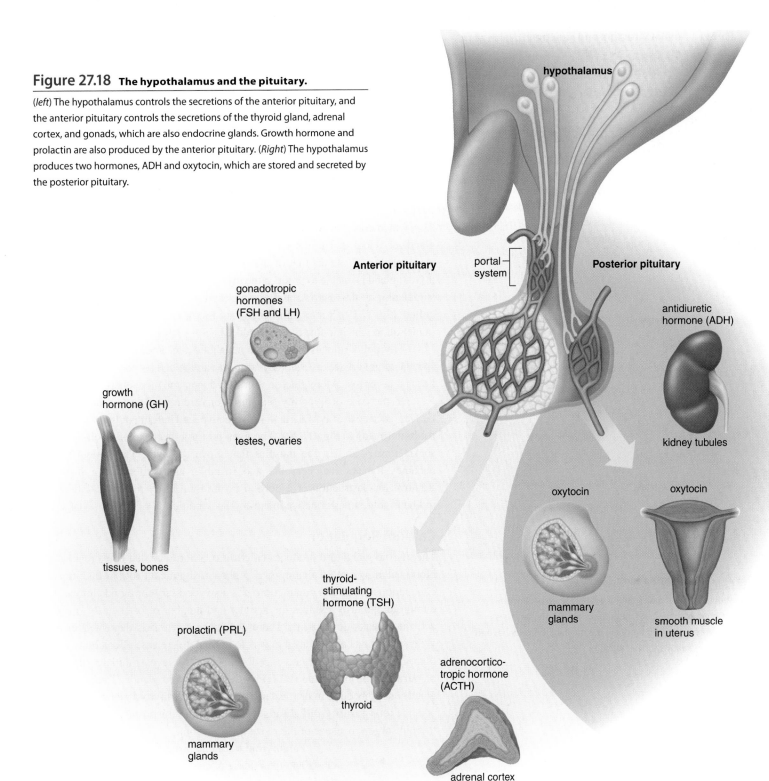

hypothalamus

Anterior pituitary

portal system

Posterior pituitary

gonadotropic hormones (FSH and LH)

antidiuretic hormone (ADH)

growth hormone (GH)

testes, ovaries

kidney tubules

oxytocin

oxytocin

tissues, bones

thyroid-stimulating hormone (TSH)

mammary glands

smooth muscle in uterus

prolactin (PRL)

thyroid

adrenocortico-tropic hormone (ACTH)

mammary glands

adrenal cortex

Posterior Pituitary

The hypothalamus produces two hormones, **antidiuretic hormone (ADH)** and **oxytocin** (Fig. 27.18). The axons of hypothalamic secretory neurons extend into the **posterior pituitary,** where hormones are stored in the axon terminals. When the hypothalamus determines that the blood is too concentrated, ADH is released from the posterior pituitary. Upon reaching the kidneys, ADH causes water to be reabsorbed. As the blood becomes dilute, ADH is no longer released. This is also an example of control by **negative feedback** because the effect of the hormone (to dilute blood) acts to shut

Figure 27.19
Simple goiter.

An enlarged thyroid gland can result from too little iodine in the diet. The thyroid is under constant stimulation to produce more of its hormones and so it enlarges, resulting in a simple goiter.

affected eye

Figure 27.20   **Exophthalmos.**

Enlargement of the thyroid gland can cause the eyes to protrude, a condition called exophthalmos. In this individual only the left eye was affected. The overactive thyroid produces too much of its hormones and the individual is hyperactive and nervous.

Check Your Progress

1. List the two hormones that are produced by the hypothalamus, then stored in and released from the posterior pituitary.
2. Compare and contrast calcitonin with parathyroid hormone (PTH).
3. Explain why a simple goiter results from a diet low in iodine.

Answers: 1. Antidiuretic hormone (ADH) and oxytocin.   2. Both hormones regulate blood calcium level. Calcitonin is produced by the thyroid gland, and reduces blood calcium. PTH is produced by the parathyroid glands, and increases blood calcium.   3. Iodine is a component of hormones produced by the thyroid gland. These hormones cannot be made unless iodine is present. Consequently, the thyroid receives constant stimulation from the anterior pituitary, which causes it to enlarge. The result is a simple goiter.

down the release of the hormone. Negative feedback, as discussed in Chapter 22, maintains homeostasis.

Oxytocin, the other hormone made in the hypothalamus, causes uterine contraction during childbirth and milk letdown (ejection) when a baby is nursing.

Thyroid and Parathyroid Glands

The **thyroid gland** is a large gland located in the neck that produces two hormones: triiodothyronine (T_3), which contains three iodine atoms, and **thyroxine (T_4),** which contains four iodine atoms. To produce these hormones, the thyroid gland needs iodine. The concentration of iodine in the thyroid gland can increase to as much as 25 times that of the blood. If iodine is lacking in the diet, the thyroid gland is unable to produce the thyroid hormones. In response to constant stimulation by the anterior pituitary, the thyroid enlarges, resulting in a **simple goiter (Fig. 27.19).** Some years ago, it was discovered that the use of iodized salt (table salt to which iodine has been added) helps prevent simple goiter.

Thyroid hormones increase the metabolic rate. They do not have a single, specific target organ; instead, they stimulate all the cells of the body to metabolize at a faster rate. More glucose is broken down, and more energy is utilized.

In the case of hyperthyroidism (oversecretion of thyroid hormone), or Graves disease, the thyroid gland is overactive, and bulging of the eyes (**exophthalmos**) results (**Fig. 27.20**). The eyes protrude because of swelling in eye socket tissues and in the muscles that move the eyes. The patient usually becomes hyperactive, nervous, and irritable, and suffers from insomnia. Removal or destruction of a portion of the thyroid by means of radioactive iodine is sometimes effective in curing the condition.

Calcium Regulation

The thyroid gland also produces **calcitonin,** a hormone that helps regulate the blood calcium level. Calcium (Ca^{2+}) plays a significant role in both nervous conduction and muscle contraction. It is also necessary for blood clotting. Calcitonin temporarily reduces the activity and number of osteoclasts, a type of cell that breaks down bone. Therefore, more calcium is deposited in bone. When the blood calcium level lowers to normal, the release of calcitonin by the thyroid is inhibited by negative feedback, but a low level stimulates the release of **parathyroid hormone (PTH)** by the parathyroid glands. The parathyroid glands are embedded in the posterior surface of the thyroid gland. Many years ago, the four **parathyroid glands** were sometimes mistakenly removed during thyroid surgery because of their size and location.

Parathyroid hormone promotes the activity of osteoclasts and the release of calcium from the bones. PTH also promotes the reabsorption of calcium by the kidneys, where it activates vitamin D. Vitamin D, in turn, stimulates the absorption of calcium from the intestine. These effects bring the blood calcium level back to the normal range so that the parathyroid glands no longer secrete PTH.

When insufficient parathyroid hormone production leads to a dramatic drop in the blood calcium level, tetany results. In **tetany,** the body shakes from continuous muscle contraction. This effect is brought about by increased excitability of the nerves, which initiate nerve impulses spontaneously and without rest.

Adrenal Glands

Two **adrenal glands** sit atop the kidneys. Each adrenal gland consists of an inner portion called the **adrenal medulla** and an outer portion called the **adrenal cortex.** These portions, like the anterior pituitary and the posterior pituitary, have no functional connection with one another.

The hypothalamus exerts control over the activity of both portions of the adrenal glands. It initiates nerve impulses that travel by way of the brain stem, spinal cord, and sympathetic nerve fibers to the adrenal medulla, which then secretes its hormones. The hypothalamus, by means of ACTH-releasing hormone, controls the anterior pituitary's secretion of ACTH, which in turn stimulates the adrenal cortex. Stress of all types, including emotional and physical trauma, and even vigorous exercise, prompts the hypothalamus to stimulate both the adrenal medulla and the adrenal cortex.

Adrenal Medulla

Epinephrine (adrenaline) and **norepinephrine** (noradrenaline) produced by the adrenal medulla rapidly bring about all the body changes that occur when an individual reacts to an emergency situation. In so doing, these two hormones complement the actions of the sympathetic autonomic system. The effects of these hormones are short-term. In contrast, the hormones produced by the adrenal cortex provide a long-term response to stress.

Adrenal Cortex

The two major types of hormones produced by the adrenal cortex are the **mineralocorticoids,** such as aldosterone, and the **glucocorticoids,** such as cortisol. Aldosterone acts on the kidneys and thereby regulates salt and water balance, leading to increases in blood volume and blood pressure. Cortisol regulates carbohydrate, protein, and fat metabolism, leading to an increase in the blood glucose level. It is also an anti-inflammatory agent. The adrenal cortex also secretes a small amount of both male and female sex hormones in both sexes.

When the level of adrenal cortex hormones is low due to hyposecretion, a person develops **Addison disease.** ACTH may build up as more is secreted to attempt to stimulate the adrenal cortex. The excess can cause bronzing of the skin because ACTH in excess stimulates melanin production. Without cortisol, glucose cannot be replenished when a stressful situation arises. Even a mild infection can lead to death. The lack of aldosterone results in the loss of sodium and water by the kidneys and the development of low blood pressure, and possibly severe dehydration. Left untreated, Addison disease can be fatal.

When the level of adrenal cortex hormones is high due to hypersecretion, a person develops **Cushing syndrome (Fig. 27.21).** The excess cortisol results in a tendency toward diabetes mellitus as muscle protein is metabolized and subcutaneous fat is deposited in the midsection. The trunk is obese, while the arms and legs remain a normal size. An excess of aldosterone and reabsorption of sodium and water by the kidneys lead to a basic blood pH and hypertension. The face swells and takes on a moon shape. Masculinization may occur in women because of excess adrenal male sex hormones.

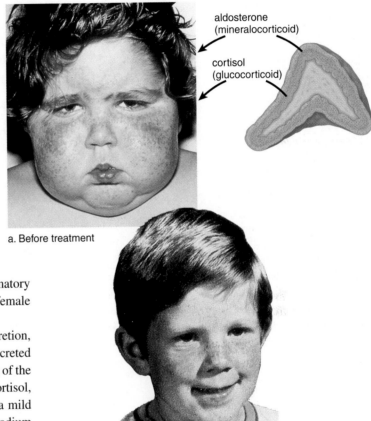

aldosterone (mineralocorticoid)

cortisol (glucocorticoid)

a. Before treatment

b. After treatment

Figure 27.21   Cushing syndrome.

Cushing syndrome results from hypersecretion of hormones by the adrenal cortex possibly due to a tumor. **a.** Patient first diagnosed with Cushing syndrome. **b.** Four months later, after treatment.

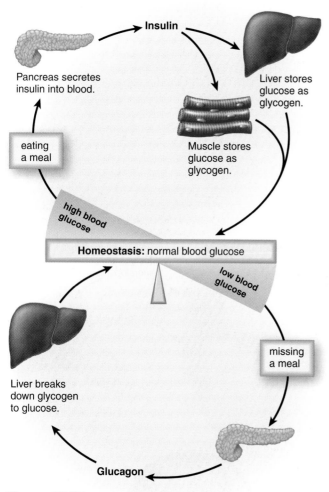

Figure 27.22   Regulation of blood glucose level.

Typical of hormones, insulin is regulated by negative feedback—once the blood glucose level is low, insulin is no longer secreted. The effect of insulin is countered by glucagon, which raises the blood glucose level. The two hormones work together to keep the blood glucose level relatively constant.

Pancreas

The **pancreas** is composed of two types of tissue. Exocrine tissue produces and secretes digestive juices that pass through ducts to the small intestine. Endocrine tissue, called the **pancreatic islets** (islets of Langerhans), produces and secretes the hormones **insulin** and **glucagon** directly into the blood (**Fig. 27.22**).

Insulin is secreted when there is a high blood glucose level, which usually occurs just after eating. Insulin stimulates the uptake of glucose by cells, especially liver cells, muscle cells, and adipose tissue cells. In liver and muscle cells, glucose is then stored as glycogen. In muscle cells, glucose supplies energy for ATP production, leading to protein metabolism and muscle contraction. In fat cells, the breakdown of glucose supplies glycerol and acetyl groups for the formation of fat. In these ways, insulin lowers the blood glucose level.

Glucagon is usually secreted between meals, when the blood glucose level is low. The major target tissues of glucagon are the liver and adipose tissue. Glucagon stimulates the liver to break down glycogen to glucose and to use fat and protein in preference to glucose as energy sources. The latter spares glucose and makes more available to enter the blood. In these ways, glucagon raises the blood glucose level.

Diabetes Mellitus

Diabetes mellitus is a fairly common hormonal disease in which liver cells, and indeed all body cells, do not take up and/or metabolize glucose. Therefore, the cells are in need of glucose even though there is plenty in the blood. As the blood glucose level rises, glucose, along with water, is excreted in the urine. The loss of water in this way causes the diabetic person to be extremely thirsty.

Two types of diabetes mellitus have been distinguished. In *type 1 diabetes* (insulin-dependent diabetes), the pancreas is not producing insulin. The condition is believed to be brought on by exposure to an environmental agent, most likely a virus, whose presence causes cytotoxic T cells to destroy the pancreatic islets. The cells turn to the breakdown of protein and fat for energy. The metabolism of fat leads to acidosis (acid blood) that can eventually cause coma and death. As a result, the individual must have daily insulin injections. These injections control the diabetic symptoms but can still cause inconveniences, since either taking too much insulin or failing to eat regularly can bring on the symptoms of hypoglycemia (low blood sugar). These symptoms include perspiration, pale skin, shallow breathing, and anxiety. Because the brain requires a constant supply of glucose, unconsciousness can result. The treatment is quite simple: Immediate ingestion of a sugar cube or fruit juice can very quickly counteract hypoglycemia.

Of the 16 million people who now have diabetes in the United States, most have *type 2 diabetes* (non-insulin-dependent diabetes). This type of diabetes mellitus usually occurs in people of any age who are obese and inactive, as discussed in Chapter 25. The pancreas produces insulin, but the liver and muscle cells do not respond to it in the usual manner. They are said to be insulin resistant. If type 2 diabetes is untreated, the results can be as serious as those of type 1. People with diabetes of either type are prone to blindness, kidney disease, and circulatory disorders. It is usually possible to prevent or at least control type 2 diabetes by adhering to a low-fat and low-sugar diet and exercising regularly.

Check Your Progress

1. Compare and contrast the adrenal medulla with the adrenal cortex.
2. Contrast the function of insulin with that of glucagon.
3. Contrast type 1 diabetes with type 2 diabetes.

Answers: 1. Both are components of the adrenal glands and are controlled by the hypothalamus. The adrenal medulla provides rapid, short-term responses to stress, while the adrenal cortex provides long-term responses to stress.   2. Insulin lowers blood sugar levels by stimulating the uptake of glucose by cells. Glucagon raises blood sugar levels by stimulating the breakdown of glycogen to glucose.   3. People with type 1 diabetes do not produce insulin, so they require daily insulin injections. People with type 2 diabetes produce insulin, but the liver and muscle cells do not respond to it as they should.

The Chapter in Review

Summary

27.1 | Nervous System

Both the nervous and endocrine systems utilize chemical signals. The nervous system is organized into a central nervous system (CNS) and a peripheral nervous system (PNS).

Neurons

- Are composed of a cell body, an axon, and dendrites. Only axons conduct nerve impulses.
- Exist as three types: sensory (takes nerve impulses to the CNS), interneuron (takes nerve impulses between neurons of the CNS), and motor (takes nerve impulses away from the CNS).

The Nerve Impulse

- The sodium-potassium pump transports Na^+ ions out of the axon and K^+ ions into the axon. The inside of the axon has a negative charge; the outside has a positive charge.
- The nerve impulse is an action potential—there is a reversal of charge as Na^+ ions flow in, and then a return to the previous charge difference as K^+ flows out of an axon.
- The nerve impulse is much faster in myelinated axons because the impulse jumps from node to node.

The Synapse

- The synapse is a region of close proximity between an axon terminal and the next neuron (CNS), or between an axon terminal and a muscle cell (PNS).
- A nerve impulse causes the release of a neurotransmitter (can be excitatory or inhibitory) into the synaptic cleft.
- Neurotransmitters are ordinarily removed quickly from the synaptic cleft. AChE breaks down acetylcholine, for example.
- Drugs affect the action of neurotransmitters.

The Central Nervous System

The Brain

- The cerebrum functions in sensation, reasoning, learning and memory, language, and speech. The cerebral cortex has a primary sensory area in the parietal lobe that receives sensory information from each part of the body and a primary motor area in the frontal lobe that sends out motor commands to skeletal muscles. Association areas carry on integration.
- In the diencephalon, the hypothalamus helps control homeostasis; the thalamus specializes in sending sensory input on to the cerebrum.
- The cerebellum primarily coordinates skeletal muscle contractions.
- In the brain stem, the medulla oblongata has centers for vital functions, such as breathing and the heartbeat, and helps control the internal organs.

The Limbic System

The limbic system blends higher functions into a united whole. The hippocampus and amygdala have roles in learning and memory and appear to be affected in Alzheimer disease.

The Peripheral Nervous System

- Somatic system: Reflexes (automatic responses) involve a sensory receptor, sensory neuron, interneurons in the spinal cord, and a motor neuron.
- Autonomic system: The sympathetic division is active during times of stress, and the parasympathetic division is active during times of relaxation. Both divisions control the same internal organs.

27.2 | Endocrine System

Endocrine glands secrete hormones into the bloodstream for distribution to target organs or tissues. Hormones are either steroids or peptides.

Hypothalamus and Pituitary Gland

Hypothalamus secretes

releasing and inhibiting hormones	control anterior pituitary
antidiuretic hormone (ADH)	released by posterior pituitary; causes water uptake by kidneys
oxytocin	released by posterior pituitary; causes uterine contractions

Anterior Pituitary secretes

gonadotropic hormones (FSH and LH)	stimulate gonads
thyroid-stimulating hormone (TSH)	stimulates thyroid
adrenocorticotropic hormone (ACTH)	stimulates adrenal cortex
prolactin	causes milk production
growth hormone (GH)	causes cell division, protein synthesis, bone growth; too little = *pituitary dwarfism*; too much = *pituitary gigantism*

Thyroid and Parathyroid Glands

Thyroid Gland secretes

thyroxine (T_4) and triiodothyronine (T_3)	increase metabolic rate; *simple goiter* when iodine is lacking, *exophthalmos* when overactive
calcitonin	lowers blood calcium level

Parathyroid Glands secrete

parathyroid hormone (PTH)	raises blood calcium level

Adrenal Glands

Adrenal Medulla secretes

epinephrine and norepinephrine	response to emergency situations

Adrenal Cortex secretes

mineralocorticoids (aldosterone)	causes kidneys to reabsorb Na^+; too little as in *Addison disease* = low blood pressure; too much as in *Cushing syndrome* = high blood pressure

glucocorticoids (cortisol)	raises blood glucose level; too little as in *Addison disease* = body can't respond to stress; too much as in *Cushing syndrome* = diabetes
sex hormones	male/female differences
Pancreas secretes	
insulin	causes cells to take up and liver to store glucose as glycogen; too little = *diabetes mellitus*
glucagon	causes liver to break down glycogen

Key Terms

acetylcholine (ACh) 482
acetylcholinesterase (AChE) 482
action potential 481
Addison disease 493
adrenal cortex 493
adrenal gland 493
adrenal medulla 493
adrenocorticotropic hormone (ACTH) 490
Alzheimer disease (AD) 486
anterior pituitary 490
antidiuretic hormone (ADH) 491
autonomic system 488
axon 480
brain stem 485
calcitonin 492
cell body 480
central nervous system (CNS) 479
cerebellum 485
cerebral cortex 484
cranial nerve 486
Cushing syndrome 493
dendrite 480
diabetes mellitus 494
dorsal root ganglion 487
endocrine gland 489
epinephrine 493
exophthalmos 492
ganglion 486
glucagon 494
glucocorticoid 493
gonadotropic hormones (FSH and LH) 490
growth hormone (GH) 490
hypothalamic-releasing hormone 490
hypothalamus 485, 490
insulin 494
integration 482

interneuron 480
limbic system 486
medulla oblongata 485
memory 486
midbrain 485
mineralocorticoid 493
motor neuron 480
myelin sheath 480
negative feedback 491
nerve 486
neuron 480
neurotransmitter 482
nodes of Ranvier 480
norepinephrine (NE) 482, 493
oxytocin 491
pancreas 494
pancreatic islets 494
parasympathetic division 488
parathyroid gland 492
parathyroid hormone (PTH) 492
peptide hormone 490
peripheral nervous system (PNS) 479
pituitary gland 490
pons 485
posterior pituitary 491
prefrontal area 485
primary motor area 484
primary sensory area 484
prolactin (PRL) 490
reflex 487
saltatory conduction 481
sensory neuron 480
simple goiter 492
somatic system 487
spinal cord 484
spinal nerve 487
steroid hormone 489
sympathetic division 488

synapse 482
synaptic cleft 482
tetany 492
thalamus 485
thyroid gland 492

thyroid-stimulating hormone (TSH) 490
thyroxine (T_4) 492
tract 484

Testing Yourself

Choose the best answer for each question.

1. Unlike the nervous system, the endocrine system
   a. uses chemical signals as a means of communication.
   b. helps maintain equilibrium.
   c. sends messages to target organs.
   d. changes the metabolism of cells.

2. Pain receptors are at the distal ends of
   a. interneurons.   c. sensory neurons.
   b. intraneurons.   d. motor neurons.

3. Glucagon causes
   a. use of fat for energy.
   b. glycogen to be converted to glucose.
   c. use of amino acids to form fats.
   d. Both a and b are correct.
   e. None of these are correct.

4. When comparing the interior of an axon at rest to the exterior, there is
   a. a charge difference.
   b. an ion concentration difference.
   c. both a charge and ion concentration difference.
   d. neither a charge nor an ion concentration difference.

5. Long-term use of cocaine causes an intense craving for the drug because the body has begun to make
   a. less dopamine.   c. less anandamide.
   b. more dopamine.   d. more anandamide.

6. Myelin is formed from
   a. dendrites.   c. axons.
   b. Schwann cells.   d. motor neurons.

7. Label the parts of the brain in the following illustration.

For questions 8–12, identify the part of the brain in the key that matches the description. Some answers may be used more than once. Some questions may have more than one answer.

Key:

a. cerebrum
b. diencephalon
c. cerebellum
d. brain stem

8. regulates hunger, thirst, and sleep
9. receives sensory information from eyes and ears
10. composed of two parts; mostly white matter
11. regulates heartbeat, breathing, and blood pressure
12. responsible for sensation, voluntary movement, and higher thought processes
13. In contrast to exocrine glands, endocrine glands
    a. secrete products.
    b. induce a response by the body.
    c. utilize ducts.
    d. produce hormones.
14. Both the adrenal medulla and the adrenal cortex are
    a. endocrine glands.
    b. in the same organ.
    c. involved in our response to stress.
    d. All of these are correct.
15. Type 1 diabetes is thought to result from a virus that
    a. interferes with gene expression in pancreatic cells.
    b. causes T cells to destroy the pancreatic islets.
    c. breaks down insulin.
    d. prevents the secretion of insulin by the pancreatic islets.
16. Label the parts of the endocrine system in the following illustration.

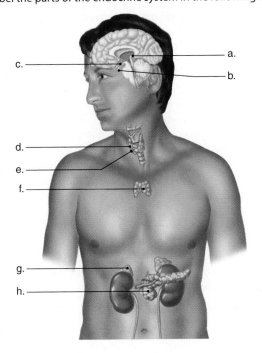

17. The cerebellum
    a. coordinates skeletal muscle movements.
    b. receives sensory input from the joints and muscles.
    c. receives motor input from the cerebral cortex.
    d. All of these are correct.
18. An interneuron can relay information
    a. from a sensory neuron to a motor neuron.
    b. only from a motor neuron to another motor neuron.
    c. only from a sensory neuron to another sensory neuron.
    d. None of these are correct.
19. The sympathetic division of the autonomic nervous system will
    a. increase heart rate and digestive activity.
    b. decrease heart rate and digestive activity.
    c. cause pupils to constrict.
    d. None of these are correct.
20. Effects of drugs can include
    a. prevention of neurotransmitter release.
    b. prevention of reuptake by the presynaptic membrane.
    c. blockages to a receptor.
    d. All of these are correct.
    e. None of these are correct.
21. When the action potential begins, sodium gates open, allowing Na$^+$ to cross the membrane. Now the polarity changes to
    a. negative outside and positive inside.
    b. positive outside and negative inside.
    c. neutral outside and positive inside.
    d. There is no difference in charge between outside and inside.
22. Transmission of the nerve impulse across a synapse is accomplished by the
    a. movement of Na$^+$ and K$^+$.
    b. release of a neurotransmitter by a dendrite.
    c. release of a neurotransmitter by an axon.
    d. release of a neurotransmitter by a cell body.
    e. Any one of these is correct.
23. Long-term complications of diabetes include
    a. blindness.
    b. kidney disease.
    c. circulatory disorders.
    d. All of these are correct.
    e. None of these are correct.
24. The autonomic system has two divisions, called the
    a. CNS and PNS.
    b. somatic and skeletal systems.
    c. efferent and afferent systems.
    d. sympathetic and parasympathetic divisions.
25. The limbic system
    a. involves portions of the cerebral lobes and the diencephalon.
    b. is responsible for our deepest emotions, including pleasure, rage, and fear.
    c. is a system necessary to memory storage.
    d. is not directly involved in language and speech.
    e. All of these are correct.

26. Which of these correctly describes the distribution of ions on either side of an axon when it is not conducting a nerve impulse?
    a. more sodium ions (Na^+) outside and more potassium ions (K^+) inside
    b. more K^+ outside and less Na^+ inside
    c. charged protein outside and Na^+ and K^+ inside
    d. Na^+ and K^+ outside and water only inside
    e. chloride ions (Cl^-) outside and K^+ and Na^+ inside

27. Growth hormone is produced by the
    a. posterior adrenal gland.
    b. posterior pituitary.
    c. anterior pituitary.
    d. kidneys.
    e. None of these are correct.

28. Which of the following is not true of the nerve impulse? The nerve impulse
    a. is subject to a short refractory period before it can occur again.
    b. is slower in thick myelinated fibers.
    c. is also called an action potential.
    d. moves from node to node in myelinated fibers.

29. PTH causes the blood level of calcium to _____, and calcitonin causes it to _____.
    a. increase, not change
    b. increase, decrease
    c. decrease, also decrease
    d. decrease, increase
    e. not change, increase

30. The portion of a neuron that conducts nerve impulses is the
    a. axon.   c. dendrite.
    b. cell body.   d. nucleus.

31. A bundle of axons in the PNS is called a
    a. ganglion.   d. node of Ranvier.
    b. tract.   e. synapse.
    c. nerve.

32. The inner portion of an adrenal gland is the
    a. adrenal cortex.   d. medulla oblongata.
    b. synaptic cleft.   e. adrenal medulla.
    c. cerebral cortex.

Thinking Scientifically

1. Recent research indicates that Parkinson disease damages the sympathetic division of the peripheral nervous system. One test for sympathetic division function, called the Valsalva maneuver, requires the patient to blow against resistance. A functional nervous system will compensate for the decrease in blood output from the heart by constricting blood vessels. How do you suppose Parkinson patients respond to the Valsalva maneuver? How does this relate to a common condition in Parkinson patients, called orthostatic hypotension, in which blood pressure falls suddenly when the person stands up, leading to dizziness and fainting?

2. Researchers have been trying to determine the reason for a dramatic rise in type 2 diabetes in recent decades. A sedentary lifestyle and poor eating habits certainly contribute to the risk for developing the disease. In addition, however, some researchers have observed a connection between childhood vaccination and type 1 diabetes. Epidemiological data from countries that have recently initiated mass immunization programs indicate that the incidence of type 1 diabetes has increased there as well. What might be the connection between vaccination and diabetes?

3. Scientists studying orangutans on the Indonesian islands of Borneo and Sumatra recently found that the subspecies inhabiting the food-scarce region of Borneo have smaller brains than those on Sumatra where food is abundant. What can we infer from this observation about the relationship between brain size and diet in orangutan evolution?

Bioethical Issue

Human Growth Hormone

Recently, the Food and Drug Administration (FDA) approved the use of human growth hormone (hGH) to treat "short stature." This decision implies that short stature is a medical condition with a legitimate need to be treated. In this case, medical care is not treating a disease, but enhancing a feature of an otherwise healthy person. Proponents of the FDA decision say that administering hGH to short people will help them avoid discrimination and live a more normal life in a world designed for taller people. Opponents say that this decision may lead to slippery-slope scenarios in which medical treatments that make us smarter or faster are assumed to make us better. Do you think hGH should be used to "cure shortness"? Typically, parents will need to make decisions about hGH treatments for their children, since the treatments are generally administered to preschoolers. Should parents be allowed to make decisions about treatments that will determine their children's heights?

At the other end of the life span, hGH has been widely advertised as a cure for aging. Levels of hGH naturally decline with age, and there is evidence that treating older adults with hGH boosts muscle mass and reduces body fat. However, the only effective way to administer hGH is by injection; the hGH pills hyped on numerous websites do not work. Injections of hGH are by prescription only and very expensive, around $1,000 a month. Furthermore, hGH therapy can result in undesirable side effects such as elevated blood sugar, fluid retention, and joint pain. Nevertheless, there is keen interest in hGH therapy not only from those who wish to delay or avoid the effects of aging, but also by athletes who wish to improve their performance. Should the physical decline of aging be accepted, or should we try to maintain a youthful body using hGH?

Essentials of Biology Website

The companion website for *Essentials of Biology* provides a wealth of information organized and integrated by chapter. You will find practice tests, animations, videos, and much more that will complement your learning and understanding of general biology.

http://www.mhhe.com/maderessentials2

Sensory Input and Motor Output

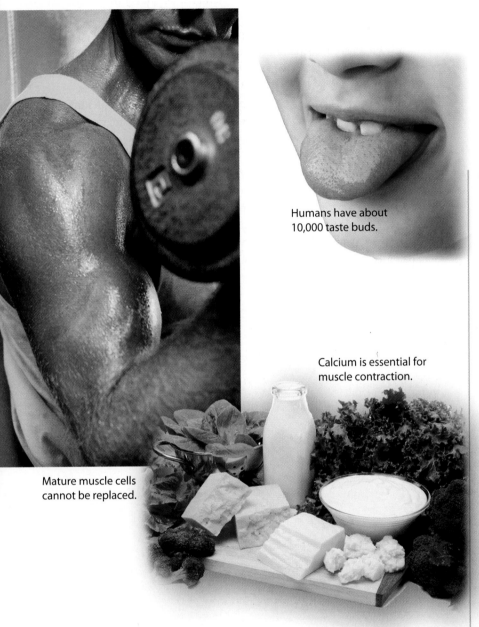

Humans have about 10,000 taste buds.

Calcium is essential for muscle contraction.

Mature muscle cells cannot be replaced.

Outline

Your Body Needs Calcium

WHEN YOU THINK of the importance of calcium in nutrition, you most likely think about bone. Bone serves as a depository for calcium so that there is always a reserve in the body, and as an added benefit, storing calcium makes bone tissue strong. However, calcium plays other important roles in the body as well.

One of the most essential functions of calcium involves muscle contraction. No muscle in the body (including the heart and the diaphragm)

—Continued next page

Continued—

can contract without the presence of calcium. If you were dependent solely on obtaining calcium from your daily diet and you ran out of calcium, your heart and diaphragm wouldn't contract. Then, the inability to circulate your blood and to breathe would cause you to die. Luckily, our bones do stockpile calcium, so when we don't get enough in our diet, we get it from the reserve supply in our bones. The downside is that in order to tap into the reserve of calcium in our bones, bone tissue must be degraded. This is why obtaining enough calcium in the diet is critical. When there is an imbalance between the amount of calcium taken from bones and the amount of calcium deposited in bones, conditions such as osteoporosis often result. In osteoporosis, the bones are fragile, and a broken hip, spine, or wrist often occurs. This is why we hear so much about including plenty of calcium in the diet.

The functions of bones and muscle are closely related to each other. In this chapter, in addition to learning about these motor output systems, you will learn about sensory input systems.

28.1 | The Senses

All living things respond to stimuli. Stimuli are environmental signals that tell us about the external environment or the internal environment. In Chapter 21 we learned that plants often respond to external stimuli, such as light, by changing their growth pattern. An animal's response often results in motion. Complex animals rely on sensory receptors to provide information to the central nervous system (brain and spinal cord), which integrates sensory input before directing a motor response (**Fig. 28.1**).

Sense organs, as a rule, are specialized to receive one kind of stimulus. The eyes ordinarily respond to light, ears to sound waves, pressure receptors to pressure, and chemoreceptors to chemical molecules. Sensory receptors transform the stimulus into nerve impulses that reach a particular section of the cerebral cortex. Those from the eye reach the visual areas, and those from the ears reach the auditory areas. The brain, not the sensory receptor, is responsible for sensation and perception, and each part of the brain interprets impulses in only one way. For example, if by accident the photoreceptors of the eye are stimulated by pressure and not light, the brain causes us to see "stars" or other visual patterns.

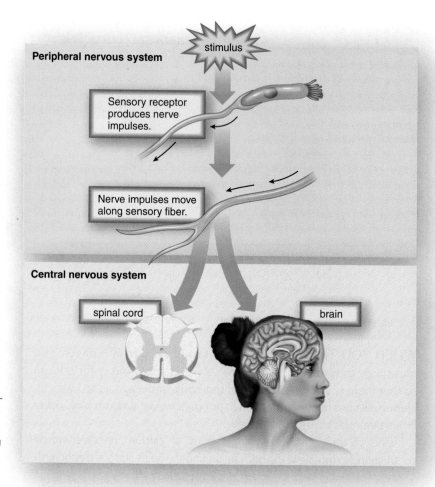

Figure 28.1 **Sensory input.**

After detecting a stimulus, sensory receptors initiate nerve impulses within the PNS. These impulses give the central nervous system (CNS) information about the external and internal environment. The CNS integrates all incoming information and then initiates a motor response to the stimulus.

Chemical Senses

The fundamental functions of sensory receptors include helping animals to stay safe, find food, and find mates. *Chemoreceptors* give us the ability to detect chemicals in the environment, which is believed to be our most primitive sense. Chemoreceptors occur almost universally in animals. For example, they are present all over the body of planarians (flatworms), but concentrated on the auricles at the sides of the head. Male moths have receptors for a sex attractant on their antennae. The receptors on the antennae of the male silkworm moth are so sensitive that only 40 out of 40,000 receptor proteins need to be activated in order for the male to respond to a chemical released by the female (**Fig. 28.2***a*). Other insects, such as the housefly, have chemoreceptors largely on their feet—a fly tastes with its feet instead of its mouth. In mammals, the receptors for taste are located in the mouth, and the receptors for smell are located in the nose.

Taste and Smell

In humans, **taste buds,** located primarily on the tongue, contain taste receptor cells, and the nose contains olfactory receptor cells (Fig. 28.2*b,c*). Receptor proteins for chemicals are located on the microvilli of taste receptor cells and on the cilia of olfactory receptor cells. When molecules bind to these receptor proteins, nerve impulses are generated in sensory nerve fibers that go to the brain. When they reach the appropriate cortical areas, they are interpreted as taste and smell, respectively.

There are at least five primary types of tastes: bitter, sour, salty, sweet, and umami (Japanese for "savory" or "delicious"). Foods rich in certain amino acids. such as the common seasoning monosodium glutamate (MSG), as well as certain flavors of cheese, beef broth, and some seafood, produce the taste of umami. Taste buds for each primary taste are located throughout the tongue, but may be concentrated in particular regions. A particular food can stimulate more than one of these types of taste buds. In this way, the response of taste buds can result in a range of sweet, sour, salty, umami, and bitter tastes. The brain appears to survey the overall pattern of incoming sensory impulses and to take a "weighted average" of their taste messages as the perceived taste. Similarly, an odor contains many odor molecules, which activate a characteristic combination of receptor proteins. When this complex information is communicated to the cerebral cortex, we know we have smelled a rose—or an onion!

An important role of taste and also smell is to trigger reflexes that start the digestive juices flowing. A revolting or repulsive substance in the mouth can initiate the gag reflex or even vomiting. Smell is even more important to our survival. Smells associated with danger, such as smoke, can trigger the fight-or-flight reflex. Unpleasant smells can cause us to sneeze or choke.

Have you ever noticed that a certain aroma vividly brings to mind a certain person or place? A person's perfume may remind you of someone else, or the smell of boxwood may remind you of your grandfather's farm. The olfactory bulbs have direct connections with the limbic system and its centers for emotions and memory. One investigator showed that when subjects smelled an orange while viewing a painting, they not only remembered the painting when asked about it later, but they also had many deep feelings about it.

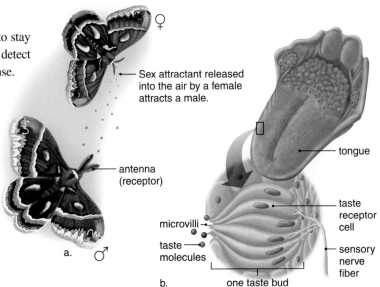

Sex attractant released into the air by a female attracts a male.

antenna (receptor)

a. ♂   ♀

tongue

microvilli

taste molecules

taste receptor cell

sensory nerve fiber

b.   one taste bud

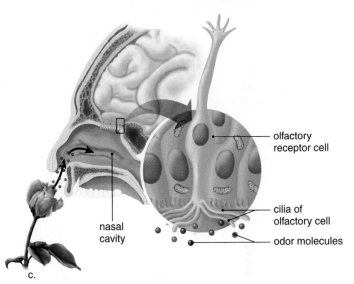

olfactory receptor cell

cilia of olfactory cell

nasal cavity

odor molecules

c.

Figure 28.2   Chemoreceptors.

Chemoreceptors are common in the animal kingdom. **a.** Male moths have receptor proteins on their antennae for a sex attractant released by the female. **b.** Humans have receptor proteins for taste on the microvilli of taste receptor cells and (**c**) receptor proteins for smell on the cilia of olfactory receptor cells.

Check Your Progress

1. **Summarize the function of taste receptor cells.**
2. **List the five primary types of tastes.**
3. **Explain why a certain odor can often evoke a strong memory of a person or place.**

Answers: 1. After chemicals bind to the receptor proteins of taste receptor cells, nerve impulses go to the brain, which interprets the nerve impulses as tastes.   2. Bitter, sour, salty, sweet, and umami.   3. The olfactory system is directly tied to the centers for emotion and memory in the limbic system, so an odor can help us recall an emotion or memory.

Hearing and Balance

The human ear has two sensory functions: hearing and balance (equilibrium). The sensory receptors for both of these consist of *hair cells* with long microvilli called stereocilia. These microvilli, unlike those of taste cells, are sensitive to mechanical stimulation. Therefore, they are termed *mechanoreceptors*.

The similarity of the sensory receptors for balance and hearing and their presence in the same organ suggest an evolutionary relationship between them. In fact, the sense organs of the mammalian ear may have evolved from a type of sense organ in fishes.

Hearing

Most invertebrates cannot hear. Some arthropods, including insects, do have sound receptors, but they are quite simple. In insects, the ear consists of a pair of air pockets, each enclosed by a membrane, called the tympanic membrane, that passes sound vibrations to sensory receptors. The human ear has a tympanic membrane also, but it is between the outer and middle ear (**Fig. 28.3***a*). The outer ear collects sound waves that cause the tympanic membrane to move back and forth (vibrate) ever so slightly. Three tiny bones in the middle ear (the ossicles) amplify the sound about 20 times as it moves from one to the other. The last of the ossicles (the stapes) strikes the membrane of the oval window, causing it to vibrate, and in this way, the pressure is passed to a fluid within the hearing portion of the inner ear called the cochlea (Fig. 28.3*b*). The term cochlea means snail shell. Specifically, the sensory receptors of hearing are located in the cochlear canal of the cochlea. The sensory receptors for hearing are hair cells whose stereocilia are embedded in a gelatinous membrane. Collectively, they are called the **spiral organ** (organ of Corti) (Fig. 28.3*c,d*).

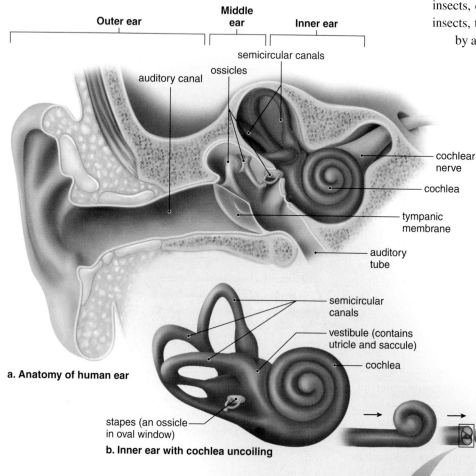

a. Anatomy of human ear

b. Inner ear with cochlea uncoiling

c. Spiral organ

d. Sensory receptors for hearing

Figure 28.3   The human ear.

a. The outer ear collects sound waves, the middle ear amplifies sound waves, and the cochlea contains the sensory receptors for hearing. **b.** The inner ear contains the cochlea, and also the semicircular canals and the vestibule. **c.** The sensory receptors for hearing are hair cells within the spiral organ. **d.** We hear when pressure waves within the canals of the cochlea cause the hair cells to vibrate and their stereocilia to bend.

The outer ear and middle ear, which collect and amplify sound waves, are filled with air. The auditory tube functions to relieve pressure in the middle ear. But the inner ear is filled with fluid; therefore, fluid pressure waves actually stimulate the spiral organ. When the stapes strikes the membrane of the oval window, pressure waves cause the hair cells to move up and down, and the stereocilia of the hair cells embedded in the gelatinous membrane bend. The hair cells of the spiral organ synapse with the cochlear nerve, and when their stereocilia bend, nerve impulses begin in the cochlear nerve and travel to the brain stem. When these impulses reach the auditory areas of the cerebral cortex, they are interpreted as sound.

Each part of the spiral organ is sensitive to different wave frequencies, or *pitch*. Near the tip, the spiral organ responds to low pitches, such as a tuba, and near the base, it responds to higher pitches, such as a bell or a whistle. The nerve fibers from each region along the length of the spiral organ lead to slightly different areas in the brain. The pitch sensation we experience depends upon which region of the spiral organ vibrates and which area of the brain is stimulated.

Volume is a function of the amplitude of sound waves. Loud noises cause the spiral organ to vibrate to a greater extent. The resulting increased stimulation is interpreted by the brain as volume. It is believed that the brain interprets the *tone* of a sound based on the distribution of the hair cells stimulated.

Hearing Loss Especially when we are young, the middle ear is subject to infections that can lead to hearing impairment. Today, it is quite common for youngsters to have "tubes" put into the tympanic membrane to allow the middle ear to drain in an effort to prevent this type of hearing loss. The mobility of ossicles decreases with age, and if bone grows over the stapes, the only remedy is implantation of an artificial stapes that can move.

Deafness due to middle ear damage is called **conduction deafness.** Deafness due to spiral organ damage is called **nerve deafness.** In today's society, noise pollution is common, and even city traffic can be loud enough to damage the stereocilia of hair cells (**Fig. 28.4**). It stands to reason, then, that frequently attending rock concerts, constantly playing a stereo loudly, or using earphones at high volume can also damage hearing. The first hint of danger can be temporary hearing loss, a "full" feeling in the ears, muffled hearing, or tinnitus (ringing in the ears). If exposure to noise is unavoidable, specially designed noise-reduction earmuffs are available, and it is also possible to purchase earplugs made from compressible, spongelike material at the drugstore or sporting-goods store. These earplugs are not the same as those worn for swimming, and they should not be worn interchangeably. Finally, you should be aware that some medicines may damage the ability to hear. Anyone taking anticancer drugs, such as cisplatin, and certain antibiotics, such as streptomycin, should be especially careful to protect the ears from loud noises.

Check Your **Progress**

1. Explain how the ear can distinguish between sounds with different pitches.
2. How do we know if a noise is loud?
3. Distinguish between nerve deafness and conduction deafness.

Answers: 1. Each part of the spiral organ is sensitive to a different pitch. The nerve fibers in each part connect to and stimulate a different part of the brain. 2. The greater the amplitude of sound waves, the greater the degree of stimulation of the spiral organ. The brain interprets the high level of stimulation as a high volume, i.e. a loud noise. 3. Nerve deafness is caused by damage to the spiral organ. Conduction deafness is due to middle ear damage.

a. Normal hair cells

b. Damaged hair cells

Figure 28.4 Effect of noise on hearing.

a. Normal hair cells in the spiral organ of a guinea pig. **b.** Spiral organ damaged by 24-hour exposure to a noise level equivalent to that of a heavy-metal rock concert. Damaged cells cannot be replaced, so hearing is permanently impaired.

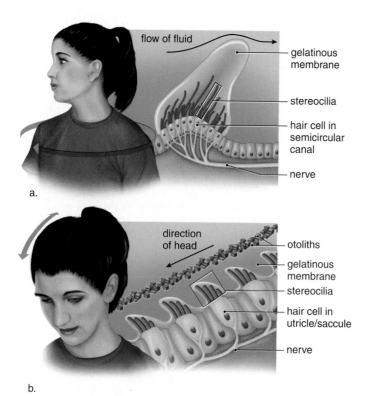

a.

b.

Figure 28.5 **Sense of balance in humans.**

a. Sensory receptors for rotational equilibrium. When the head rotates, the gelatinous membrane is displaced, bending the stereocilia. **b.** Sensory receptors for gravitational equilibrium. When the head bends, otoliths are displaced, causing the gelatinous membrane to sag and the stereocilia to bend.

Statocyst

a.

Lateral line

b.

Balance

Humans have two senses of balance (equilibrium): rotational and gravitational. We are able to detect the rotational and/or angular movement of the head as well as the straight-line movement of the head with respect to gravity.

Rotational equilibrium involves the semicircular canals (see Fig. 28.3b). In the base of each canal, hair cells have stereocilia embedded within a gelatinous membrane (**Fig. 28.5a**). Because there are three semicircular canals, each responds to head movement in a different plane of space. As fluid within a semicircular canal flows over and displaces the gelatinous membrane, the stereocilia of the hair cells bend, and the pattern of impulses carried to the central nervous system (CNS) changes. These data, usually supplemented by vision, tell the brain how the head is moving. *Vertigo* is dizziness and a sensation of rotation. It is possible to bring on a feeling of vertigo by spinning rapidly and stopping suddenly. Now the person feels like the room is spinning because of sudden stimulation of stereocilia in the semicircular canals.

Gravitational equilibrium refers to the position of the head with relation to gravity. It depends on the **utricle** and **saccule,** two membranous sacs located in the inner ear (see Fig. 28.3b). Both of these sacs also contain hair cells with stereocilia in a gelatinous membrane (Fig. 28.5b). Calcium carbonate ($CaCO_3$) granules (the otoliths) rest on this membrane. When the head moves forward or back, up or down, the otoliths are displaced, and the membrane moves, bending the stereocilia of the hair cells. This movement alters the frequency of nerve impulses to the CNS. These data, usually supplemented by vision, tell the brain the direction of the movement of the head.

Similar Receptors in Other Animals

Gravitational equilibrium organs, called **statocysts,** are found in several types of invertebrates, from cnidarians to molluscs to crustaceans. These organs give information only about the position of the head; they are not involved in the sensation of movement (**Fig. 28.6a**). When the head stops moving, a small particle called a statolith stimulates the cilia of the closest hair cells, and these cilia generate impulses, indicating the position of the head.

The **lateral line** system of fishes utilizes sense organs similar to those we have been studying in the human inner ear (Fig. 28.6b). In bony fishes, the system consists of sense organs located within a canal that has openings to the outside. As you might expect, the sense organ is a collection of hair cells with cilia embedded in a gelatinous membrane. Water currents and pressure waves from nearby objects cause the membrane and the cilia of the hair cells to bend. Thereafter the hair cells initiate nerve impulses that go to the brain. Fishes use these data, not for hearing or balance, but to locate other fish, including predators, prey, and mates.

Figure 28.6 **Sensory receptors in other animals.**

a. Invertebrates utilize statocysts to determine their position. When the statolith stops moving, cilia of the nearest hair cells are stimulated, telling the position of the head. **b.** The lateral line of fishes is not for hearing or balance. It is for knowing the location of other fishes.

Vision

Sensory receptors that are sensitive to light are called **photoreceptors.** In planarians, eyespots allow these animals to determine the direction of light only. Other photoreceptors form actual images. Image-forming eyes provide information about an object and how far away it is. Such detailed information is invaluable to an animal.

Among invertebrates, arthropods have **compound eyes** composed of many independent visual units, each of which possesses a lens to focus light rays on photoreceptors (**Fig. 28.7***a*). The image that results from all the stimulated visual units is crude because the small size of compound eyes limits the number of visual units, which still might number as many as 28,000.

Vertebrates (including humans) and certain molluscs, such as the squid and the octopus, have a **camera-type eye** (Fig. 28.7*b*). A single lens focuses light on photoreceptors, which number in the millions and are closely packed together within a retina. Since molluscs and vertebrates are not closely related, the camera-type eye evolved independently in each group.

Insects have color vision, but they make use of a slightly shorter range of the electromagnetic spectrum compared to humans. However, they can see the longest of the ultraviolet rays, and this enables them to be especially sensitive to the reproductive parts of flowers, which have particular ultraviolet patterns. Some fishes and most reptiles are believed to have color vision, but among mammals, only humans and other primates have expansive color vision. It would seem, then, that this trait was adaptive for a diurnal habit (active during the day), which accounts for its retention in only a few mammals.

The Human Eye

When looking straight ahead, each of our eyes views the same object from a slightly different angle. This slight displacement of the images permits *binocular vision*, the ability to perceive three-dimensional images and to sense depth. Like the human ear, the human eye has numerous parts, many of which are involved in preparing the stimulus for the sensory receptors. In the case of the ear, sound wave energy is magnified before it reaches the sensory receptors. In the case of the eye, light rays are brought to a focus on the photoreceptors located within the **retina.** In **Figure 28.8,** the **cornea** and especially the **lens** are involved in focusing light rays on the photoreceptors. The **iris,** the colored part of the eye, regulates the amount of light that enters the eye by way of the **pupil.** The retina generates nerve impulses that are sent to the visual part of the cerebral cortex, and from this information, the brain forms an image of the object.

The shape of the lens is controlled by the ciliary muscles. When we view a distant object, the lens remains relatively flat, but when we view a near object, the lens rounds up. With normal aging, the lens loses its ability to accommodate for near objects; therefore, many people need reading glasses when they reach middle age. Aging, or possibly exposure to the sun, also makes the lens subject to cataracts; the lens becomes opaque, and therefore incapable of transmitting light rays. Currently, surgery is the only viable treatment for cataracts.

a. Compound eye

b. Camera-type eye

Figure 28.7 **Eyes.**

a. A compound eye has several visual units. Each visual unit has a lens that focuses light onto photoreceptor cells. **b.** A camera-type eye has one lens that focuses light onto a retina containing many photoreceptors.

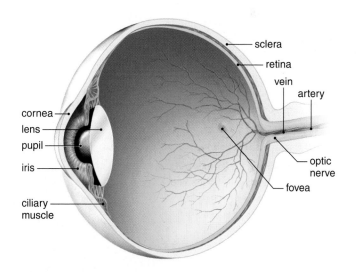

Figure 28.8 **The human eye.**

Light passes through the cornea and through the pupil (a hole in the iris), and is focused by the lens on the retina, which houses the photoreceptors.

Check Your Progress

1. Which structures of the inner ear are responsible for gravitational equilibrium? Rotational equilibrium?
2. Contrast a compound eye with a camera-type eye.
3. List, in order, the structures a ray of light encounters on its way to the retina of the human eye.

Answers: 1. The utricle and saccule are responsible for gravitational equilibrium, and the semicircular canals for rotational equilibrium. 2. A compound eye contains many lenses that focus light rays on photoreceptors, while a camera-type eye uses a single lens to focus light rays on the photoreceptors. 3. Cornea, pupil, lens, retina.

Figure 28.9 Photoreceptors of the eye.

In rods, the membrane of each disk contains rhodopsin, a complex molecule containing the protein opsin and the pigment retinal. When rhodopsin absorbs light energy, it splits, releasing opsin, which sets in motion a cascade of reactions that ends in nerve impulses.

Figure 28.10 The retina.

The retina contains a layer of rods and cones, a layer of intermediate cells, and a layer of ganglion cells. Integration of signals occurs at the synapses between the layers, and much processing occurs before nerve impulses are sent to the brain.

Photoreceptors of the Eye

Figure 28.9 illustrates the structure of the photoreceptors in the human eye, which are called **rods** and **cones.** Both types of photoreceptors contain a visual pigment similar to that found in all types of eyes throughout the animal kingdom. The visual pigment in rods is a deep-purple pigment called rhodopsin. **Rhodopsin** is a complex molecule made up of the protein opsin and a light-absorbing molecule, called *retinal*, which is a derivative of vitamin A. When a rod absorbs light, rhodopsin splits into opsin and retinal, leading to a cascade of reactions that ends in the generation of nerve impulses. Rods are very sensitive to light, and therefore are suited to night vision. (Because carrots are rich in vitamin A, it is true that eating carrots can improve your night vision.) Rod cells are plentiful throughout the entire retina; therefore, they also provide us with peripheral vision and perception of motion.

The cones, on the other hand, are located primarily in a part of the retina called the **fovea.** Cones are activated by bright light; they allow us to detect the fine detail and the color of an object. Color vision depends on three different kinds of cones, which contain pigments called the B (blue), G (green), and R (red) pigments. Each pigment is made up of retinal and opsin, but a slight difference in the opsin structure of each accounts for their specific absorption patterns. Various combinations of cones are believed to be stimulated by in-between shades of color.

Retina The retina has three layers of cells and light has to penetrate through the first two layers to reach the photoreceptors (**Fig. 28.10**). The intermediate cells of the middle layer process and relay visual information from the photoreceptors to the ganglion cells that have axons forming the optic nerve. The sensitivity of cones versus rods is mirrored by how directly they connect to ganglion cells. Information from several hundred rods may converge on a single ganglion cell, while cones show very little convergence. As signals pass through the layers of the retina, integration occurs. Integration improves the overall contrast and quality of information sent to the brain, which uses the information to form an image of the object.

No rods and cones occur where the optic nerve exits the retina. Therefore, no vision is possible in this area. You can prove this to yourself by putting a very small dot to the right of center on a piece of paper. Close your left eye, then use your right hand to move the paper slowly toward your right eye while you look straight ahead. The dot will disappear at one point—this point is your **blind spot.**

Seeing in the Dark

Some nocturnal animals rely on vision, but others use sonar (sound waves) to find their way in the dark. Bats flying in a dark room easily avoid obstacles in their path because they echolocate, like submarines. When searching for food, they emit ultrasonic sound (above the range humans can hear) in chirps that bounce back off their prey. They are able to determine the distance to dinner by timing the echo's return—a long delay means the prey is far away.

A bat-inspired sonar walking stick is being perfected to help visually impaired people sense their surroundings. It too emits ultrasonic chirps and picks up the echoes from nearby objects. Buttons on the cane's handle vibrate gently to warn a user to dodge a low ceiling or to sidestep objects blocking the path. A fast, strong signal means an obstacle is close by.

Cutaneous Receptors and Proprioceptors

Cutaneous receptors are located in the skin, and **proprioceptors** are located in the muscles and joints. These sensory receptors are connected to the primary sensory area of the cerebral cortex where each part of the body is represented. Here, sensory input is received from the skin, muscles, and joints in each part of the body (**Fig. 28.11**). Thus, these receptors are important to our well-being.

Cutaneous Receptors

Skin, the outermost covering of our body, contains numerous sensory receptors that help us respond to changes in our environment, be aware of dangers, and communicate with others. The sensory receptors in skin are for touch, pressure, pain, and temperature.

Skin has two regions, called the epidermis and the dermis (**Fig. 28.12**). The epidermis is packed with cells that become keratinized as they rise to the surface. Among these cells are free nerve endings responsive to cold or to warmth. Cold receptors are far more numerous than warmth receptors, but there are no known structural differences between the two. Also in the epidermis are pain receptors sensitive to extremes in temperature or pressure and to chemicals released by damaged tissues. Sometimes, stimulation of internal receptors is felt as pain in the skin. This is called *referred pain*. For example, pain from the heart may be felt in the left shoulder and arm. This effect most likely happens when nerve impulses from the pain receptors of internal organs travel to the spinal cord and synapse with neurons also receiving impulses from the skin.

The dermis of the skin contains sensory receptors for pressure and touch (Fig. 28.12). The Pacinian corpuscle is an onion-shaped pressure receptor that lies deep inside the dermis. Several other cutaneous receptors detect touch. A free nerve ending, called a root hair plexus, winds around the base of a hair follicle and produces nerve impulses if the hair is touched. Touch receptors are concentrated in parts of the body essential for sexual stimulation: the fingertips, the palms, the lips, the tongue, the nipples, the penis, and the clitoris.

Proprioceptors

Proprioceptors help the body maintain equilibrium and posture, despite the force of gravity always acting upon the skeleton and muscles. A **muscle spindle** consists of sensory nerve endings wrapped around a few muscle cells within a connective tissue sheath. **Golgi tendon** organs and other sensory receptors are located in the joints. The rapidity of nerve impulses by proprioceptors is proportional to the stretching of the organs they occupy. A motor response results in contraction of muscle fibers adjoining the proprioceptor.

Check
Your Progress

1. Compare and contrast rod and cone photoreceptors.
2. Where are cutaneous receptors and proprioceptors located?

Answers: 1. Both contain a visual pigment consisting of retinal and opsin. All rods have the same visual pigment rhodopsin. The opsin portion of the pigment varies slightly from one type of cone to the next. Rods are distributed throughout the retina, while cones are concentrated in the fovea. Rods are sensitive to low levels of light, and cones are activated by bright light. Rods enable us to see in dim light, have peripheral vision, and detect motion; and cones provide us with detailed color vision.   2. Cutaneous receptors are found in the skin, proprioceptors in skeletal muscles and joints.

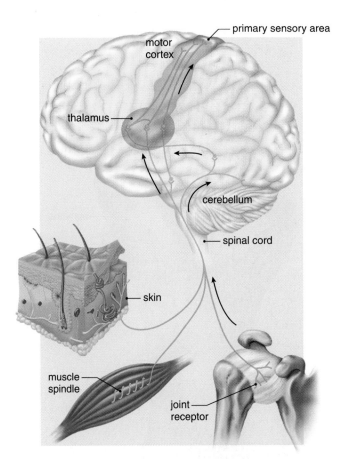

Figure 28.11   Sensory input to primary sensory area of brain.

Sensory receptors in the skin, muscles, and joints send nerve impulses to the CNS. Sensation and perception occur when these reach the primary sensory area of the cerebral cortex. A motor response can be initiated by the spinal cord and cerebellum without the involvement of the cerebrum.

Figure 28.12   Cutaneous receptors.

Numerous receptors are in the skin. Free nerve endings (yellow) in the epidermis detect pain, heat, and cold. Various touch and pressure receptors (red) are in the dermis.

Figure 28.13   Musculoskeletal system.

Major skeletal muscles and bones of the body are labeled. The human skeleton contains bones that belong to the axial skeleton (red labels) and those that belong to the appendicular skeleton (black labels).

28.2 | The Motor Systems

In this section, we consider the motor systems of humans—the muscles and the bones—as forming a single system, namely the musculoskeletal system (**Fig. 28.13**). Combining the two seems appropriate because in many cases the functions of muscles and bones overlap:

Both skeletal muscles and bones support the body and make movement of body parts possible.

Both skeletal muscles and bones protect internal organs. Skeletal muscles pad the bones that protect the heart and lungs, the brain, and the spinal cord.

Both muscles and bones aid the functioning of other systems. Without the movement of the rib cage, breathing would not occur. As an aid to digestion, the jaws have sockets for teeth; skeletal muscles move the jaws so food can be chewed, and smooth muscle moves food along the digestive tract (see Fig. 24.21*b*). Red bone marrow supplies the red blood cells that carry oxygen, and the pumping of cardiac muscle in the wall of the heart moves the blood to the tissues, where exchanges with tissue fluid occur.

In addition to these shared functions, the muscles and bones have individual functions:

Skeletal muscle contraction assists movement of blood in the veins and lymphatic vessels (see Fig. 23.8). Without the return of lymph to the cardiovascular system and blood to the heart, circulation could not continue.

Skeletal muscles help maintain a constant body temperature (see Fig. 28.17). Skeletal muscle contraction causes ATP to break down, releasing heat that is distributed about the body.

Bones store fat and calcium. Fat is stored in yellow bone marrow (see Fig. 28.15), and the extracellular matrix of bone contains calcium. Calcium ions play a major role in muscle contraction and nerve conduction.

The Human Skeleton

The **endoskeleton** of vertebrates can be contrasted with the **exoskeleton** of arthropods. Both skeletons are jointed, which has helped members of both groups of animals successfully live on land. The endoskeleton of humans is composed of bone, which is living material and capable of growth. Like an exoskeleton, an endoskeleton protects vital internal organs, but unlike an exoskeleton, it need not limit the space available for internal organs because it grows as the animal grows. The soft tissues that surround an endoskeleton protect it, and injuries to soft tissues are apt to be easier to repair than is the skeleton itself.

The exoskeleton of arthropods is composed of chitin, a strong, flexible, nitrogenous polysaccharide. Besides providing protection against wear and tear and against enemies, an exoskeleton also prevents drying out. Although an arthropod exoskeleton provides support for muscle contractions, it does not grow with the animal, and arthropods molt to rid themselves of an exoskeleton that has become too small (**Fig. 28.14***a*). This process makes them vulnerable to predators.

One other type of skeleton is seen in the animal kingdom. In animals, such as worms, that lack a hard skeleton, a fluid-filled internal cavity can act as a **hydrostatic skeleton** (Fig. 28.14*b*). A hydrostatic skeleton offers support and resistance to the contraction of muscles so that the animal can move. As an analogy, consider that a garden hose stiffens when filled with water, and that a water-filled balloon changes shape when squeezed at one end. Similarly, an animal with a hydrostatic skeleton can change shape and perform a variety of movements.

Axial and Appendicular Skeletons

The 206 bones of the human skeleton are arranged into an axial skeleton and an appendicular skeleton.

Axial Skeleton   The labels for the bones of the **axial skeleton** are in color in Figure 28.13. The **skull** consists of the **cranium,** which protects the brain, and the facial bones. The most prominent of the **facial bones** are the lower and upper jaws, the cheekbones, and the nasal bones. The **vertebral column** (spine) extends from the skull to the **sacrum** (tailbone). It consists of a series of vertebrae separated by pads of fibrocartilage called the **intervertebral disks.** On occasion, disks can slip or even rupture. A damaged disk pressing against the spinal cord or spinal nerves causes pain. Removal of the disk may be required.

The **rib cage,** composed of the ribs and **sternum** (breastbone), demonstrates how the skeleton can be protective and flexible at the same time. In the United States, automobile accidents are the most common cause of sternum fractures; people with such injuries are typically examined for signs of heart damage. The rib cage protects the heart and lungs but moves when we breathe.

Appendicular Skeleton   The **appendicular skeleton** contains the bones of two girdles and their attached limbs. The **pectoral (shoulder) girdle** and upper limbs are specialized for flexibility; the **pelvic (hip) girdle** and lower limbs are specialized for strength. The pelvic girdle also protects internal organs.

A **clavicle** (collarbone) and a **scapula** (shoulder blade) make up the shoulder girdle. The shoulder pads worn by football players are designed to cushion direct blows that could cause clavicle fractures. The **humerus** of the arm articulates only with the scapula; the joint is stabilized by tendons and ligaments that form a **rotator cuff.** Vigorous circular movements of the arm can lead to rotator cuff injuries. Two bones (the **radius** and **ulna**) contribute to the easy twisting motion of the forearm. The flexible hand contains bones of the wrist, palm, and five fingers.

The pelvic girdle contains two massive **coxal bones** that form a bowl, called the pelvis, and articulate with the longest and strongest bones of the body, the **femurs** (thighbones). The strength of the femurs allows these massive bones to support the weight of the upper half of the body. The kneecap protects the knee. The **tibia** of the leg is the shinbone. The **fibula** is the more slender bone in the leg. Each foot contains bones of the ankle, instep, and five toes.

a.

fluid-filled internal cavity

b.

muscles

muscles

epidermis

cuticle

Figure 28.14   Types of skeletons.

a. Arthropods have an exoskeleton that must be shed as they grow.
b. Worms have a hydrostatic skeleton, in which muscle contraction pushes against a fluid-filled internal cavity.

Check Your Progress

1. Describe the functions of bones and muscles when they work together.
2. List the components of the axial skeleton.
3. List the components of the appendicular skeleton.

Answers: 1. Bones and muscles support the body and allow body parts to move, protect internal organs, and help other systems to function.   2. The axial skeleton is composed of the skull, vertebral column, ribs, and sternum.   3. The appendicular skeleton is composed of the shoulder girdle, the pelvic girdle, and their attached limbs.

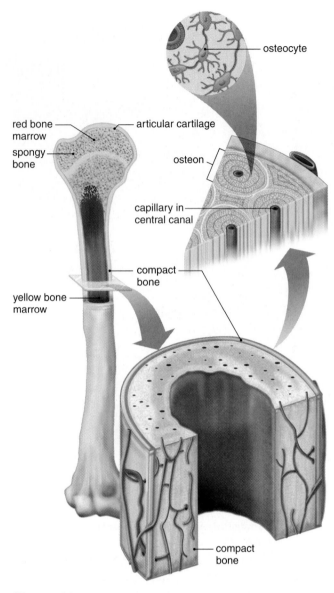

Figure 28.15   Structure of bone.

A bone has a central cavity that is usually filled with yellow bone marrow. Both spongy bone and compact bone are living tissues composed of bone cells within a matrix that contains calcium. The spaces of spongy bone contain red bone marrow.

Structure of a Bone

When a long bone is split open, as in **Figure 28.15**, a cavity is revealed that is bounded on the sides by compact bone. **Compact bone** contains many **osteons** (Haversian systems) where bone cells called **osteocytes** lie in tiny chambers arranged in concentric circles around central canals. The cells are separated by an extracellular matrix that contains mineral deposits, primarily calcium and phosphorous salts. Two other cell types are constantly at work in bones. **Osteoblasts** deposit bone, and **osteoclasts** secrete enzymes that digest the matrix of bone and release calcium into the bloodstream. (Recall from page 492 that the endocrine system regulates the activity of osteoclasts in order to maintain a homeostatic level of calcium in the bloodstream.) When a person has **osteoporosis** (weak bones subject to fracture), osteoclasts are working harder than osteoblasts. Intake of high levels of dietary calcium, especially when a person is young and more active, encourages denser bones and lessens the chance of getting osteoporosis later in life.

A long bone has **spongy bone** at each end. Spongy bone has numerous bony bars and plates separated by irregular spaces. The spaces are often filled with **red bone marrow,** a specialized tissue that produces red blood cells. The cavity of a long bone is filled with yellow bone marrow and stores fat. Beyond the spongy bone are a thin shell of compact bone and a layer of hyaline cartilage, which is important to healthy joints.

Figure 28.16   Skeletal muscle fiber structure and function.

A muscle fiber contains many myofibrils divided into sarcomeres, which are contractile. When innervated by a motor neuron, the myofibrils contract, and the sarcomeres shorten because actin filaments slide past the myosin filaments.

Skeletal Muscle Structure and Physiology

The three types of muscle—smooth, cardiac, and skeletal—have different structures (see Fig. 22.7). Smooth muscles contain sheets of long, spindle-shaped cells, each with a single nucleus. Cardiac cells are striated (having a striped appearance), each with a single nucleus. Cardiac muscle tissue contains branched chains of cells that interconnect, forming a lattice network. Skeletal muscle cells, called muscle fibers, are also striated. They are quite elongated, and run the length of a skeletal muscle. Skeletal muscle fibers arise during development when several cells fuse, resulting in one long, multinucleated cell. Skeletal muscle contraction has been extensively studied.

Skeletal Muscle Contraction

When skeletal muscle fibers contract, they shorten. Let's look at details of the process, beginning with the motor axon (**Fig. 28.16**). When nerve impulses travel down a motor axon and arrive at an axon terminal, synaptic vesicles release ACh (acetylcholine) into a synaptic cleft. ACh quickly diffuses across the cleft and binds to receptors in the plasma membrane of a muscle fiber, called the sarcolemma. The sarcolemma generates impulses that travel along its T tubules to the endoplasmic reticulum, which is called the *sarcoplasmic reticulum* in muscle fibers. The release of calcium from calcium storage sites causes muscle fibers to contract.

The contractile portions of a muscle fiber are many parallel, threadlike **myofibrils** (Fig. 28.16). An electron microscope shows that myofibrils (and therefore skeletal muscle fibers) are striated because of the placement of protein filaments within contractile units called **sarcomeres.** A sarcomere extends between two dark lines called the Z lines. Sarcomeres contain thick filaments made up of the protein **myosin** and thin filaments made up of the protein **actin.** As a muscle fiber contracts, the sarcomeres within the myofibrils shorten because actin (thin) filaments slide past the myosin (thick) filaments and approach one another. The movement of actin filaments in relation to myosin filaments is called the **sliding filament model** of muscle contraction. During the sliding process, the sarcomere shortens, even though the filaments themselves remain the same length.

Why the Filaments Slide   The thick filament is a bundle of myosin molecules, each having a globular head with the capability of attaching to the actin filament when calcium (Ca^{2+}) is present. First, myosin binds to and hydrolyzes ATP. Thus, energized myosin heads attach to an actin filament. The release of ADP + P causes myosin to shift its position and pull the actin filaments to the center of the sarcomere. The action is similar to the movement of your hand when you flex your forearm (**Fig. 28.17**).

In the presence of another ATP, a myosin head detaches from actin. Now the heads attach at a location further along the actin filament. The cycle occurs again and again, and the actin filaments move nearer and nearer the center of the sarcomere each time the cycle is repeated. Contraction continues until nerve impulses cease and calcium ions are returned to their storage sites. The membranes of the sarcoplasmic reticulum contain active transport proteins that pump calcium ions back into the calcium storage sites, and the muscle relaxes. When a person or animal dies, ATP production ceases. Without ATP, the myosin heads cannot detach from actin, nor can calcium be pumped back into the sarcoplasmic reticulum. As a result, the muscles remain contracted, a phenomenon called *rigor mortis*.

1 Myosin head has combined with ATP.

2 ATP is hydrolyzed to ADP + P as myosin head attaches to actin.

3 ADP + P release causes head to change position and actin filament to move.

Figure 28.17   Why a muscle shortens when it contracts.

The presence of calcium (Ca^{2+}) sets in motion a chain of events (1–3) that causes myosin heads to attach to and pull an actin filament toward the center of a sarcomere. After binding to other ATP molecules, myosin heads will return to their resting position. Then, the chain of events (1–3) occurs again, except that the myosin heads reattach further along the actin filament.

Check Your Progress

1. Contrast the functions of osteoblasts and osteoclasts.
2. List the three types of muscle.
3. What are the individual contractile units of a myofibril?

Answers: 1. Osteoblasts deposit bone, and osteoclasts break down bone.   2. Smooth muscle, cardiac muscle, and skeletal muscle.
3. Sarcomeres.

Figure 28.18   Muscles and bones.

Graceful movements are possible because tendons attach muscles to bones across a joint.

Figure 28.19   Action of muscles.

When muscles contract, they shorten. Therefore, muscles only pull—they cannot push. This causes them to work in antagonistic pairs; each member of the pair pulls on a bone in the opposite direction. For example, (**a**) when the biceps contracts, the forearm flexes (raises), and (**b**) when the triceps contracts, the forearm extends (lowers).

Skeletal Muscles Move Bones at Joints

Joints are classified as immovable, like those of the cranium; slightly movable, like those between the vertebrae; and freely movable (synovial joints), like those in the knee and hip. In **synovial joints, ligaments** bind the two bones together, providing strength and support, and forming a capsule containing lubricating synovial fluid. All of our movements, from those of graceful and agile ballet dancers to those of aggressive and skillful football players, occur because muscles are attached to bones by tendons that span movable joints (**Fig. 28.18**). Because muscles shorten when they contract, they have to work in antagonistic pairs. If one muscle of an antagonistic pair flexes the joint and raises the limb, the other extends the joint and straightens the limb. **Figure 28.19** illustrates this principle with regard to the movement of the forearm at the elbow joint.

Figure 28.20*b* illustrates the anatomy of a freely movable synovial joint. Note that, in addition to a cavity filled with synovial fluid, a synovial joint may include additional structures, namely menisci and bursae. Menisci (sing., **meniscus**), are C-shaped pieces of hyaline cartilage between the bones. These give added stability and act as shock absorbers. Fluid-filled sacs called bursae (sing., **bursa**) ease friction between bare areas of bone and overlapping muscles, or between skin and tendons.

The **ball-and-socket joints** at the hips and shoulders are synovial joints that allow movement in all planes, even rotational movement (Fig. 28.20*c*). The elbow and knee joints are synovial joints called **hinge joints** because, like a hinged door, they largely permit movement in one direction only (Fig. 28.20*d*).

Joint Disorders   *Sprains* occur when ligaments and tendons are overstretched at a joint. For example, a sprained ankle can result if you turn your ankle too far. Overuse of a joint may cause inflammation of a bursa, called *bursitis*. Tennis elbow is a form of bursitis. A common knee injury is a torn meniscus. Because fragments of menisci can interfere with joint movements, most physicians believe they should be removed. Today, arthroscopic surgery is possible to remove cartilage fragments or to repair ligaments or cartilage. A small instrument bearing a tiny lens and light source is inserted into a joint, as are the surgical instruments. Fluid is then added to distend the joint and allow visualization of its structure. Usually, the surgery is displayed on a monitor so that the whole operating team can see the operation. Arthroscopy is much less traumatic than surgically opening the knee with long incisions. The benefits of arthroscopy include small incisions, faster healing, a more rapid recovery, and less scarring. Because arthroscopic surgical procedures are often performed on an outpatient basis, the patient is able to return home on the same day.

Rheumatoid arthritis, discussed on page 471, is not as common as osteoarthritis (OA), which is the deterioration of an overworked joint. Constant compression and abrasion continually damage articular cartilage, and eventually it softens, cracks, and wears away entirely in some areas. As the disease progresses, the exposed bone thickens and forms spurs that cause the bone ends to enlarge and restrict joint movement. Weight loss can ease arthritis. Taking off 3 pounds can reduce the load on a hip or knee joint by 9 to 15 pounds. A sensible exercise program helps build up muscles, which stabilize joints. Low-impact activities, such as biking and swimming, are best.

Today, replacement of damaged joints with a prosthesis (artificial substitute) is often possible. Some people have found glucosamine-chondroitin supplements beneficial as an alternative to joint replacement. Glucosamine, an amino sugar, is thought to promote the formation and repair of cartilage. Chondroitin, a carbohydrate, is a cartilage component that is thought to promote

Figure 28.20 Synovial joints.

a. The synovial joints of the human skeleton allow the body to be flexible and move with precision even when bearing a weight.
b. Generalized synovial joint. Problems arise when menisci or ligaments are torn, bursae become inflamed, and articular cartilage wears away. **c.** The shoulder is a ball-and-socket joint that permits movement in three planes. **d.** The elbow is a hinge joint that permits movement in a single plane.

b. Generalized synovial joint

c. Ball-and-socket joint

d. Hinge joint

a. A gymnast depends on flexible joints.

water retention and elasticity and to inhibit enzymes that break down cartilage. Both compounds are naturally produced by the body.

Exercise A sensible exercise program has many benefits. Exercise improves muscular strength, muscular endurance, and flexibility. It improves cardiorespiratory endurance and may lower the blood cholesterol level. People who exercise are less likely to develop various types of cancer. Exercise promotes the activity of osteoblasts, and therefore helps prevent osteoporosis (see page 510). It helps prevent weight gain, not only because of increased activity, but also because as muscle mass increases, the body is less likely to accumulate fat. Exercise even relieves depression and enhances the mood.

Check Your Progress

1. **Explain why a muscle shortens when it contracts.**
2. **Contrast a hinge joint with a ball-and-socket joint.**

Answers: 1. The globular heads of myosin molecules attach to actin filaments and pull them toward the center of the sarcomere. Then ATP allows the heads to be released, and they reattach at a new location along the actin filament, resulting in contraction. 2. A hinge joint, such as a knee or elbow, allows movement mainly in one direction. A ball-and-socket joint, such as a hip or shoulder, allows movement in all planes.

The Chapter in Review

Summary

28.1 | The Senses

All living things respond to stimuli. In animals, stimuli generate nerve impulses that often go to a CNS, which integrates the information before initiating motor responses:

Stimulus ⟶ Nerve impulse ⟶ CNS ⟶ Motor response

Chemical Senses

Chemoreception is found universally in animals and is therefore believed to be the most primitive sense.
Human taste buds and olfactory receptor cells are chemoreceptors.
- Taste buds have microvilli with receptor proteins that bind to chemicals in food.
- Olfactory receptor cells have cilia with receptor proteins that bind to odor molecules.

Hearing and Balance

Mammals have an ear that may have evolved from the lateral line of fishes. The sensory receptors for hearing are hair cells with stereocilia that respond to pressure waves.
- Hair cells respond to stimuli that have been received by the outer ear and amplified by the ossicles in the middle ear.
- Hair cells are found in the spiral organ and are located in the cochlear canal of the cochlea. The spiral organ generates nerve impulses that travel to the brain.

The sensory receptors for balance (equilibrium) are also hair cells with stereocilia.
- Hair cells in the base of the semicircular canals provide rotational equilibrium.
- Hair cells in the utricle and saccule provide gravitational equilibrium.

Vision

Arthropods have a compound eye; squids and humans have a camera-type eye. In humans, the photoreceptors:
- Respond to light that has been focused by the cornea and lens.
- Consist of two types, rods and cones. In rods, rhodopsin splits into opsin and retinal.
- Communicate with the next layer of cells in the retina. Integration occurs in the three layers of the retina before nerve impulses go to the brain.

Cutaneous Receptors and Proprioceptors

These receptors communicate with the primary sensory area of the brain. They consist of receptors for hot, cold, pain, touch, and pressure (cutaneous) and stretching (proprioceptors).

28.2 | The Motor Systems

Together, the muscles and bones:
- Support the body and allow parts to move.
- Help protect internal organs.
- Assist the functioning of other systems.

In addition:
- Skeletal muscle contraction assists movement of blood in cardiovascular veins and lymphatic vessels.
- Skeletal muscles provide heat that warms the body.
- Bones are storage areas for calcium and phosphorous salts, as well as sites for blood cell formation.

The Human Skeleton

Arthropods have an exoskeleton, and molting is needed to replace it as the animal grows. Humans have an endoskeleton that grows with them. Worms have a hydrostatic skeleton.
The human skeleton is divided into two parts:
- The axial skeleton is made up of the skull, vertebral column, sternum, and ribs.
- The appendicular skeleton is composed of the girdles and their appendages.

Bone contains:
- Compact bone (site of calcium storage) and spongy bone (site of red bone marrow).
- A cavity (site of fat storage).

Skeletal Muscle Structure and Physiology

Skeletal muscle is one of the three types of muscles:
- Smooth muscle is composed of spindle-shaped cells that form a sheet.
- Cardiac muscle has striated cells that form a lattice network.
- Skeletal muscle has striated, tubular cells (fibers) that run the length of the muscle.

In a skeletal muscle cell:
- Myofibrils, myosin filaments, and actin filaments are arranged in a sarcomere.
- Myosin filaments pull actin filaments, and sarcomeres shorten.

Skeletal muscles:
- Move bones at joints.
- Work in antagonistic pairs.
- Move bones at synovial joints that are subject to a number of disorders, such as sprains, bursitis, and torn ligaments and cartilage.

Key Terms

actin   511

appendicular skeleton   509

axial skeleton   509

ball-and-socket joints   512

blind spot   506

bursa   512

camera-type eye   505

clavicle   509

compact bone   510

compound eye   505

conduction deafness   503

cone   506

cornea   505

coxal bones   509

cranium   509

cutaneous receptor   507

endoskeleton   508

exoskeleton   508

facial bones   509

femurs   509

fibula   509

fovea   506

Golgi tendon   507

gravitational equilibrium   504

hinge joint   512

humerus   509

hydrostatic skeleton   509

intervertebral disks   509

iris   505

lateral line   504

lens   505

ligament   512

meniscus   512

muscle spindle   507

myofibril   511

myosin   511

nerve deafness   503

osteoblast   510

osteoclast   510

osteocyte   510

osteon   510

osteoporosis   510

pectoral (shoulder) girdle   509

pelvic (hip) girdle   509

photoreceptor   505

proprioceptor   507

pupil   505

radius   509

red bone marrow   510

retina   505

rhodopsin   506

rib cage   509

rod   506

rotational equilibrium   504

rotator cuff   509

saccule   504

sacrum   509

sarcomere   511

scapula   509

skull   509

sliding filament model   511

spiral organ   502

spongy bone   510

statocyst   504

sternum   509

synovial joint   512

taste buds   501

tibia   509

ulna   509

utricle   504

vertebral column   509

Testing Yourself

Choose the best answer for each question.

1. The most primitive sense is probably the ability to detect
   a. light.
   c. chemicals.
   b. sound.
   d. pressure.

2. Label the parts of the ear in the following illustration.

3. Loud noises generally lead to hearing loss due to damage to the
   a. outer ear.
   b. middle ear.
   c. inner ear.

4. The function of the lateral line system in fishes is to
   a. detect sound.
   b. locate other fish.
   c. maintain gravitational equilibrium.
   d. maintain rotational equilibrium.

5. The human eye focuses by
   a. changing the thickness of the lens.
   b. changing the shape of the lens.
   c. opening and closing the pupil.
   d. rotating the lens.

6. A blind spot occurs where the
   a. iris meets the pupil.
   b. retina meets the lens.
   c. optic nerve meets the retina.
   d. cornea meets the retina.

7. Label the parts of the eye in the following illustration.

8. Cold receptors differ from warmth receptors in
   a. abundance.   c. structure.
   b. shape.   d. All of these are correct.

9. Unlike an exoskeleton, an endoskeleton
   a. grows with the animal.
   b. is composed of chitin.
   c. is jointed.
   d. protects internal organs.

10. A component of the appendicular skeleton is the
    a. rib cage.   c. femur.
    b. skull.   d. vertebral column.

For questions 11–14, identify the type of muscle in the key that matches the description. Some answers may be used more than once. Some questions may have more than one answer.

Key:

   a. smooth
   b. cardiac
   c. skeletal

11. composed of cells with one nucleus each

12. composed of fibers that result when many cells fuse

13. composed of a lattice network of branched cells

14. striated

15. The deterioration of a joint over time can cause
    a. osteoarthritis.   c. bursitis.
    b. a sprain.   d. tendonitis.

16. The _____ contains the sensory receptors for hearing.
    a. utricle   d. osteon
    b. fovea   e. spiral organ
    c. saccule

17. The thick filaments in a sarcomere are composed mainly of
    a. calcium.   d. rhodopsin.
    b. actin.   e. Both b and c are correct.
    c. myosin.

18. A freely movable, fluid-filled joint is a _____ joint.
    a. synovial   d. otolithic
    b. proprioceptive   e. hydrostatic
    c. Golgi

19. The _____ is the region of the eye which controls the amount of light that enters.
    a. cornea   d. fovea
    b. pupil   e. retina
    c. lens

20. _____ are gravitational equilibrium organs found in several types of invertebrates.
    a. Utricles   d. Statocysts
    b. Rods   e. Spiral organs
    c. Saccules

21. C-shaped pieces of hyaline cartilage lying between the bones in a synovial joint are termed
    a. bursae.   c. intervertebral disks.
    b. menisci.   d. cutaneous receptors.

22. The _____ is the portion of the retina containing most of the cones.
    a. blind spot   d. lens
    b. cornea   e. fovea
    c. iris

23. The contractile portion of a skeletal muscle fiber is the
    a. myofibril.   c. thick filament.
    b. thin filament.   d. sarcoplasmic reticulum.

24. Which cell type deposits bone tissue?
    a. osteocyte   c. osteoblast
    b. osteoclast   d. None of these are correct.

25. The portion of the axial skeleton that extends from the skull to the sacrum is the
    a. rib cage.   d. vertebral column.
    b. sternum.   e. coxal bone.
    c. scapula.

Thinking Scientifically

1. The two leading causes of blindness are age-related macular degeneration and diabetic retinopathy. Both are characterized by the development of an abnormally high number of blood vessels (angiogenesis) in the retina. Why do you suppose angiogenesis impairs vision? Progress in cancer research has led to new strategies for the treatment of these eye diseases. Do you see a connection between the causes of blindness and cancer?

 2. Human genome researchers have found a family of approximately 80 genes that encode receptors for bitter-tasting compounds. The genes encode proteins made in taste receptor cells of the tongue. Typically, many types of taste receptors are expressed per cell on the tongue. However, in the olfactory system, each cell expresses only one of the 1,000 olfactory receptor genes. Different cells express different genes. How do you suppose this difference affects our ability to taste versus smell different chemicals? Why might the two systems have evolved such different patterns of gene expression?

Bioethical Issue

Deafness: Disability, or Cultural Identity?

A lesbian couple, both of whom are deaf, sought out a deaf sperm donor in order to have a deaf child. The women considered deafness to be a cultural identity rather than a disability. They felt that they would be better parents to a deaf child because they understood the "culture" of deafness. Should this couple have been allowed to choose a father based on his ability to contribute genes for deafness? In a situation like this, whose rights should prevail, the parents' rights to choose their mate or the child's rights to have a chance for normal hearing? Is deafness a cultural identity or a disability?

Essentials of Biology Website

The companion website for *Essentials of Biology* provides a wealth of information organized and integrated by chapter. You will find practice tests, animations, videos, and much more that will complement your learning and understanding of general biology.

http://www.mhhe.com/maderessentials2

29

Reproduction and Development

For a male to be fertile, the ejaculate usually needs to contain 40 million sperm per milliliter.

A human fetus can be viable after only 22 weeks of development (compared to the normal 40 weeks) but a premature infant has a greater risk of illness.

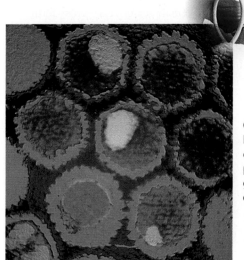

One out of five young women may have a sexually transmitted disease caused by a herpes virus. Babies born prematurely to these women are at greater risk of contracting the disease during birth.

Outline

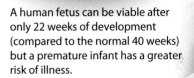

STDs and Infertility

MOST PEOPLE KNOW how sexually transmitted diseases (STDs) are contracted, but many do not understand how they are treated or their long-term consequences. Bacterial STDs, such as chlamydia and gonorrhea, are very easy to treat, but when untreated can have lasting effects, including permanent infertility. Many people are surprised to find that they have become infertile as the result of an STD. If these diseases are easy to treat, why would anyone not be treated and take the risk of

—Continued next page

Continued—

becoming infertile? There are many reasons, but one of the most important considerations is that many people with an STD do not have any symptoms of the infection. As long as everything appears normal, they do not seek treatment; meanwhile, they may pass the disease to others, and run the risk of long-term complications. This is the primary reason that testing for STDs (even when everything "seems OK") is so important. In addition to the bacterial STDs, several viral STDs are of concern—not only the well-known human immunodeficiency virus (HIV) but also human papillomavirus (HPV). HPV is the primary cause of cervical cancer (the cervix is at the entrance to the uterus). There are usually no noticeable signs of early cervical cancer, but it can be detected by having yearly checkups. Treatment for cervical cancer can result in the inability to have children if the cervix and the uterus must be surgically removed.

Reproduction is a fascinating topic with relevance not only to STDs, but also to birth control, reproductive technology, and development. In this chapter, you will learn about reproduction, including some unique issues faced by humans. You will also learn about the portion of human development that occurs before birth and about the birth process itself.

29.1 | How Animals Reproduce

Animals usually reproduce sexually, but some, on occasion, can reproduce asexually. In asexual reproduction, there is only one parent, and in sexual reproduction there are two parents.

Asexual Versus Sexual Reproduction

Hydras can reproduce by budding. A new individual arises as an outgrowth (bud) of the parent (**Fig. 29.1**). Many flatworms can constrict into two halves; each half regenerates to become a new individual. Fragmentation, followed by **regeneration,** is also seen among sponges, echinoderms, and corals. Chopping up a sea star does not kill it; instead, each fragment grows into another animal.

Parthenogenesis is a modification of sexual reproduction in which an unfertilized egg develops into a complete individual. In honeybees, the queen bee can fertilize eggs or allow eggs to pass unfertilized as she lays them. The fertilized eggs become diploid females called workers, and the unfertilized eggs become haploid males called drones. Parthenogenesis is also observed in some fish, amphibian, and reptile species.

In sexual reproduction, animals usually produce gametes in specialized organs called **gonads.** The gonads are **testes,** which produce sperm, and **ovaries,** which produce eggs. Eggs or sperm are derived from germ cells that become specialized for this purpose during early development. During sexual reproduction, the egg of one parent is usually fertilized by the sperm of another, and a **zygote** (fertilized egg) results. Even among earthworms, which are **hermaphroditic**—each worm has both male and female sex organs—cross-fertilization occurs. In coral reef fishes called wrasses, a male has a harem of several females. If the male dies, the largest female becomes a male. Many aquatic animals practice **external fertilization**—that is, eggs and sperm join outside the body in the water (**Fig. 29.2**).

Figure 29.1 **Asexual reproduction.**

A new hydra can bud from an adult hydra. This is a form of asexual reproduction. Two hydras (*right*) are now present whereas before there was only one hydra (*left*).

Copulation is sexual union to facilitate the reception of sperm by a female. In terrestrial animals, males typically have a penis for depositing sperm into the vagina of females (**Fig. 29.3**). Aquatic animals also have copulatory organs. Lobsters and crayfish have modified swimmerets. Among terrestrial animals, birds lack a penis and vagina. They have a cloaca, a chamber that receives products from the digestive, urinary, and reproductive tracts. A male transfers sperm to a female after placing his cloacal opening against hers.

Reproduction in Water Versus on Land

Many aquatic animals have a larval stage, an immature form capable of feeding. Since the larva has a different lifestyle, it is able to use a different food source than the adult. In sea stars, the bilaterally symmetrical larva undergoes metamorphosis to become a radially symmetrical juvenile. Crayfish, on the other hand, do not have a larval stage; the egg hatches into a tiny juvenile with the same form as the adult. Egg-laying animals are *oviparous*.

Reptiles, particularly birds, provide their eggs with plentiful yolk, a rich nutrient material, and they have no larval stage. Complete development takes place within a shelled egg containing **extraembryonic membranes** to serve the needs of the embryo and prevent drying out. The shelled egg frees these animals from the need to reproduce in the water and is a significant adaptation to the terrestrial environment.

Birds, in particular, tend their eggs, and newly hatched birds usually have to be fed before they are able to fly away and seek food for themselves. Complex hormones and neural regulation are involved in the reproductive behavior of parental birds. Other animals take a different course. They do not deposit and tend their eggs; instead, they are *ovoviviparous*, meaning that their eggs are retained in the body until they hatch, releasing fully developed offspring that have a way of life like the parent. Oysters, which are molluscs, retain their eggs in the mantle cavity, and male sea horses, which are vertebrates, have a special brood pouch in which the eggs develop. Garter snakes, water snakes, and pit vipers retain their eggs in their bodies until they hatch, and thus give birth to living young.

Finally, most mammals are *viviparous*, meaning that they produce living young. After offspring are born, the mother supplies the nutrients needed for further growth. Viviparity represents the ultimate in caring for the unborn. How did viviparity among certain mammals come about? Some mammals, such as the duckbill platypus and the spiny anteater, lay eggs. In contrast, marsupial offspring are born in a very immature state; they finish their development within a pouch, where they are nourished on milk. Only among primates, including humans, does a single embryo usually develop. The **placenta** is a complex structure derived, in part, from the chorion, which first appeared in a shelled egg. The evolution of the placenta allowed the developing offspring to exchange materials with the mother internally.

Figure 29.2   Reproducing in the water.
Animals that reproduce in the water have no need to protect their eggs and embryos from drying out. Here, male and female frogs are mating. They deposit their gametes in the water, where fertilization takes place. The egg contains yolk that nourishes the embryo until it is a free-swimming larva that can feed for itself.

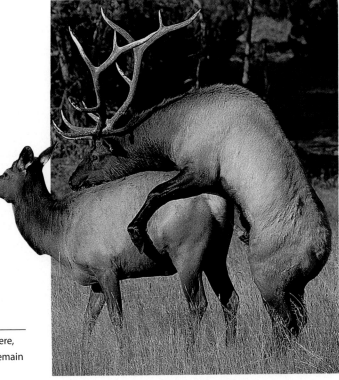

Figure 29.3   Reproducing on land.
Animals that reproduce on land need to protect their gametes and embryos from drying out. Here, the male passes sperm to the female by way of a penis, and the developing embryo/fetus will remain in the female's body until it is capable of living independently.

29.2 | Human Reproduction

In human males and females, the reproductive system consists of two components: (1) the gonads, either testes or ovaries, which produce gametes and sex hormones; and (2) accessory organs that conduct gametes, and in the case of the female, house the embryo/fetus.

Male Reproductive System

The human male reproductive system includes the testes (sing., testis), the epididymis (pl., epididymides), the vas deferens (pl., vasa deferentia), and the urethra (**Fig. 29.4**). (The urethra in males is a part of both the urinary system and the reproductive system.) The paired **testes,** which produce sperm, are suspended within the scrotum. The testes begin their development inside the abdominal cavity, but they descend into the scrotum as development proceeds. If the testes do not descend—and the male does not receive hormone therapy or undergo surgery to place the testes in the scrotum—sterility (the inability to produce offspring) results. This type of sterility occurs because normal sperm production is inhibited at body temperature; a slightly cooler temperature is required.

Sperm produced by the testes mature within the **epididymis,** a coiled tubule lying just outside each testis. Maturation seems to be required for the sperm to swim to the egg. Once the sperm have matured, they are propelled into the **vas deferens** (pl., vasa deferentia) by muscular contractions. The vasa deferentia are severed or blocked in a surgical form of birth control called a vasectomy (see Table 29.1). Sperm are stored in both the epididymis and the vas deferens. When a male becomes sexually aroused, sperm enter first the ejaculatory duct and then the urethra, part of which is located within the penis.

The **penis** is a cylindrical organ that hangs in front of the scrotum. Three cylindrical columns of spongy, erectile tissue containing distensible blood spaces extend through the shaft of the penis (Fig. 29.4b). During sexual arousal, nervous reflexes cause an increase in arterial blood flow to the penis. This

a. Frontal View

b. Side View

Figure 29.4   Male reproductive system.

a. Frontal view. **b.** Side view. The testes produce sperm. The seminal vesicles, the prostate gland, and the bulbourethral glands provide a fluid medium for the sperm. Circumcision is the removal of the foreskin.

increased blood flow fills the blood spaces in the erectile tissue, and the penis, which is normally limp (flaccid), stiffens and increases in size. These changes are called an **erection.** If the penis fails to become erect, the condition is called erectile dysfunction. Drugs such as Viagra, Levitra, and Cialis work by increasing blood flow to the penis so that when a man is sexually excited, he can achieve and keep an erection.

Semen (seminal fluid) is a thick, whitish fluid that contains sperm and secretions from three glands: the seminal vesicles, prostate gland, and bulbourethral glands. The **seminal vesicles** lie at the base of the bladder. Each joins a vas deferens to form an ejaculatory duct that enters the urethra. As sperm pass from the vas deferens into the ejaculatory duct, these vesicles secrete a thick, viscous fluid containing nutrients for use by the sperm. Just below the bladder is the **prostate gland,** which secretes a milky alkaline fluid believed to activate or increase the motility of the sperm. In older men, the prostate gland may become enlarged, thereby constricting the urethra and making urination difficult. Also, prostate cancer is the most common form of cancer in men. Slightly below the prostate gland, on either side of the urethra, is a pair of small glands called **bulbourethral glands,** which have mucous secretions with a lubricating effect. Notice from Figure 29.4 that the urethra also carries urine from the bladder during urination.

If sexual arousal reaches its peak, ejaculation follows an erection. The first phase of ejaculation is called emission. During emission, the spinal cord sends nerve impulses via appropriate nerve fibers to the epididymides and vasa deferentia. Their muscular walls contract, causing sperm to enter the ejaculatory ducts, whereupon the seminal vesicles, prostate gland, and bulbourethral glands release their secretions. Secretions from the bulbourethral glands occur first and may or may not contain sperm.

During the second phase of ejaculation, called expulsion, rhythmic contractions of muscles at the base of the penis and within the urethral wall expel semen in spurts from the opening of the urethra. These contractions are an example of release from muscle tension. An erection lasts for only a limited amount of time. The penis then returns to its normal flaccid state. Following ejaculation, a male may typically experience a time, called the refractory period, during which stimulation does not bring about an erection. The contractions that expel semen from the penis are a part of male **orgasm,** the physiological and psychological sensations that occur at the climax of sexual stimulation.

The Testes

A longitudinal section of a testis shows that it is composed of compartments called lobules, each of which contains one to three tightly coiled **seminiferous tubules** (Fig. 29.5a,b). A microscopic cross section of a seminiferous tubule reveals that it is packed with cells undergoing spermatogenesis, a process that involves reducing the chromosome number from diploid (2n) to haploid (n). Also present are *Sertoli cells*, which support, nourish, and regulate the production of sperm (Fig. 29.5c). A sperm has three distinct parts: a head, a middle piece, and a tail. The head contains a nucleus and is capped by a membrane-bounded acrosome. The tail is a flagellum that allows sperm to swim toward the egg, and the middle piece contains energy-producing mitochondria.

The ejaculated semen of a normal human male contains 40 million sperm per milliliter, ensuring an adequate number for fertilization to take place. Fewer than 100 sperm ever reach the vicinity of the egg, however, and only one sperm normally enters an egg.

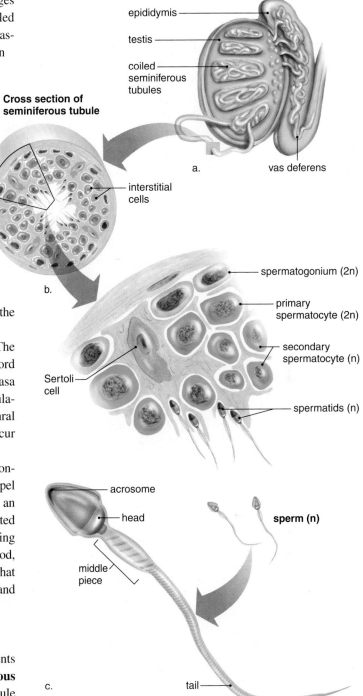

Cross section of seminiferous tubule

epididymis

testis

coiled seminiferous tubules

a.

vas deferens

b.

interstitial cells

Sertoli cell

spermatogonium (2n)

primary spermatocyte (2n)

secondary spermatocyte (n)

spermatids (n)

acrosome

head

sperm (n)

middle piece

c.

tail

Figure 29.5   Seminiferous tubules.

a. The testes contain seminiferous tubules, where sperm are produced. **b.** Cross section of a tubule. As spermatogenesis occurs, the chromosome number is reduced to the haploid number. **c.** A sperm has a head, a middle piece, and a tail. The nucleus is in the head, which is capped by an acrosome.

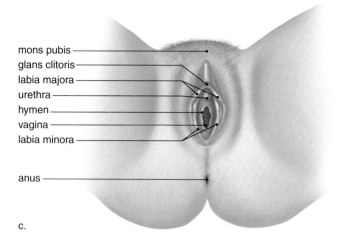

Figure 29.6 **Female reproductive system.**

a. Frontal view of the female reproductive system. The ovaries produce one oocyte (egg) per month. Fertilization occurs in the oviduct, and development occurs in the uterus. The vagina is the birth canal and organ of sexual intercourse. **b.** Side view of female reproductive system plus nearby organs. **c.** Female external genitals, the vulva. At birth, the opening of the vagina is partially occluded by a membrane called the hymen. Physical activities and sexual intercourse disrupt the hymen.

Hormonal Regulation in Males

The hypothalamus has ultimate control of the testes' sexual function because it secretes a hormone called gonadotropin-releasing hormone, or GnRH, that stimulates the anterior pituitary to produce the gonadotropic hormones. Both males and females have two gonadotropic hormones—follicle-stimulating hormone (FSH) and luteinizing hormone (LH). In males, FSH promotes spermatogenesis in the seminiferous tubules. LH in males was formerly called interstitial cell-stimulating hormone (ICSH) because it controls the production of testosterone by the interstitial cells, which are scattered in the spaces between the seminiferous tubules.

Testosterone, the main sex hormone in males, is essential for the normal development and functioning of the sexual organs. Testosterone is also necessary for the maturation of sperm. In addition, testosterone brings about and maintains the male **secondary sex characteristics** that develop at the time of **puberty,** the time of life when sexual maturity is attained. Males are generally taller than females and have broader shoulders and longer legs relative to trunk length. The deeper voice of males compared to females is due to males having a larger larynx with longer vocal cords. Because the so-called "Adam's apple" is a part of the larynx, it is usually more prominent in males than in females.

Testosterone causes males to develop noticeable hair on the face, chest, and occasionally other regions of the body, such as the back. Testosterone also leads to the receding hairline and pattern baldness that occur in males. Testosterone is responsible for the greater muscular development in males, which is why steroids are sometimes abused.

Check Your Progress

1. Describe the changes that occur in the penis during an erection.
2. Name the glands that contribute fluids to the semen.
3. What is the main sex hormone in males?

Answers: 1. Blood flow increases to the blood spaces in three columns of erectile tissue in the penis, causing it to stiffen and enlarge.   2. Seminal vesicles, prostate gland, and bulbourethral glands.   3. Testosterone.

Female Reproductive System

The human female reproductive system includes the ovaries, the oviducts, the uterus, and the vagina (**Fig. 29.6**a,b). The oviducts, also called uterine or fallopian tubes, extend from the ovaries to the uterus; however, the oviducts are not attached to the ovaries. Instead, the oviducts have fingerlike projections called **fimbriae (sing., fimbria)** that sweep over the ovaries. When an oocyte bursts from an ovary during ovulation, it usually is swept into an oviduct by the combined action of the fimbriae and the beating of cilia that line the oviducts. Fertilization, if it occurs, normally takes place in the first one-third of an oviduct, and the developing embryo is propelled slowly by ciliary movement and tubular muscle contraction to the uterus. The **uterus** is a thick-walled muscular organ about the size and shape of an inverted pear. The narrow end of the uterus is called the **cervix.** An embryo completes its development after embedding itself in the uterine lining, called the **endometrium.** A small opening at the cervix leads to the vaginal canal. The vagina is a tube at a 45° angle with the body's vertical axis. The mucosal lining of the vagina lies in folds, and therefore the vagina can expand. This ability to expand is especially important when the

oviduct

Ovary

developing follicle

corpus luteum

ruptured follicle

fimbriae

a. Ovarian cycle

oogonium (2n)

primary oocyte (2n) in primary follicle

first polar body

secondary oocyte (n) released by ruptured follicle

second polar body

If fertilization occurs, the secondary oocyte becomes an egg (n).

b. Oogenesis

Figure 29.7 **Ovarian cycle and oogenesis.**

As a follicle matures in the ovary, the oocyte enlarges and is surrounded by layers of follicular cells and fluid. Eventually, ovulation occurs, the mature follicle ruptures, and the secondary oocyte is released. If fertilization occurs, it is usually in the first third of the oviduct. A single follicle actually goes through all the stages in one place within the ovary.

vagina serves as the birth canal, and it also can facilitate sexual intercourse, when the penis is inserted into the vagina.

The external genital organs of a female are known collectively as the vulva (Fig. 29.6c). The mons pubis and two folds of skin called labia minora and labia majora are on either side of the urethral and vaginal openings. At the juncture of the labia minora is the clitoris, which is homologous to the penis of males. The clitoris has a shaft of erectile tissue and is capped by a pea-shaped glans. The many sensory receptors of the clitoris allow it to function as a sexually sensitive organ. Most females are born with a thin membrane called the hymen, which partially obstructs the vaginal opening and has no apparent biological function. The hymen is typically ruptured by physical activities including sexual intercourse, tampon insertion, or even athletics.

The Ovaries

An oogonium in the ovary gives rise to an oocyte surrounded by epithelium (**Fig. 29.7**). This is called a primary **follicle.** An ovary contains many primary follicles, each containing an oocyte. At birth, a female has as many as 2 million primary follicles, but the number is reduced to 300,000–400,000 by the time of puberty. Only a small number of primary follicles (about 400) ever mature and produce a secondary oocyte. When mature, the follicle balloons out on the surface of the ovary and bursts, releasing the secondary oocyte surrounded by follicle cells. The release of a secondary oocyte from a mature follicle is termed **ovulation.** Oogenesis is completed when and if the secondary oocyte is fertilized by a sperm. A follicle that has lost its oocyte develops into a **corpus luteum.** If fertilization and pregnancy do not occur, the corpus luteum begins to degenerate after about ten days.

The ovarian cycle is controlled by the gonadotropic hormones FSH and LH from the anterior pituitary gland (**Fig. 29.8**). During the first half, or **follicular phase** (pink), of the cycle, FSH promotes the development of follicles that primarily secrete estrogens, called *estrogen* for the sake of simplicity. As the blood level of estrogen rises, it exerts feedback control over FSH secretion, and ovulation occurs. Ovulation marks the end of the follicular phase. During the second half, or **luteal phase** (yellow), of the ovarian cycle, LH promotes the development of a corpus luteum, which primarily secretes progesterone. As the blood level of progesterone rises, it exerts feedback control over LH secretion so

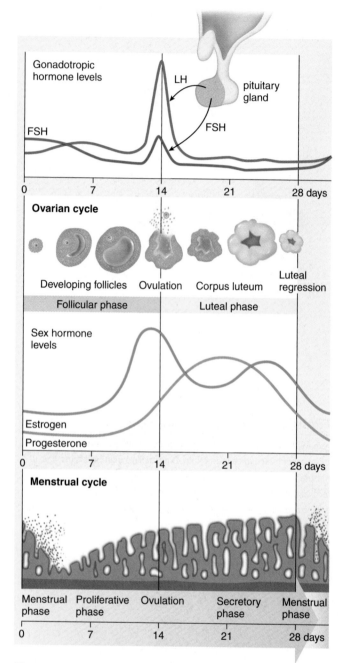

Figure 29.8 Ovarian and menstrual cycles.

During the follicular phase of the ovarian cycle (pink), FSH released by the anterior pituitary promotes the maturation of follicles in the ovary. Ovarian follicles produce increasing levels of estrogen, which causes the endometrium to thicken during the proliferative phase of the menstrual cycle (bottom). After ovulation and during the luteal phase of the ovarian cycle (yellow), LH promotes the development of a corpus luteum. This structure produces increasing levels of progesterone, which causes the endometrium to become secretory. Menstruation begins when progesterone production declines to a low level.

that the corpus luteum begins to degenerate if fertilization does not occur. As the luteal phase comes to an end, menstruation occurs.

Notice that the female sex hormones estrogen and progesterone affect the endometrium of the uterus, causing the series of events known as the **menstrual cycle** (Fig. 29.8, *bottom*). The twenty-eight-day menstrual cycle in the nonpregnant female is divided as follows:

During *days 1–5*, female sex hormones are at a low level in the body, causing the endometrium to disintegrate and its blood vessels to rupture. A flow of blood, mucus, and degenerating endometrium known as the **menses,** passes out of the vagina during **menstruation,** also known as the menstrual period.

During *days 6–13*, increased production of estrogen by ovarian follicles causes the endometrium to thicken and to become vascular and glandular. This is called the proliferative phase of the menstrual cycle. Ovulation usually occurs on *day 14* of the 28-day cycle.

During *days 15–28*, increased production of progesterone by the corpus luteum causes the endometrium to double in thickness and the uterine glands to mature, producing a thick mucoid secretion. This is called the secretory phase of the menstrual cycle. The endometrium now is prepared to receive the developing embryo. But if fertilization does not occur and no embryo embeds itself, the corpus luteum degenerates, and the low level of sex hormones in the female body causes the endometrium to break down. Menses begins, marking day one of the next cycle. Even while menstruation is occurring, the anterior pituitary begins to increase its production of FSH, and new follicles begin to mature.

Estrogen and Progesterone

Estrogen and **progesterone** are the female sex hormones. Estrogen in particular, is essential for the normal development and functioning of the female reproductive organs. Estrogen is also largely responsible for the secondary sex characteristics in females, including body hair and fat distribution. In general, females have a more rounded appearance than males because of a greater accumulation of fat beneath their skin. Also, the pelvic girdle realigns so that females have wider hips than males, and the thighs converge at a greater angle toward the knees. Both estrogen and progesterone are required for breast development as well.

Menopause, which usually occurs between ages 45 and 55, is the time in a woman's life when the ovarian and menstrual cycles cease. Menopause is not complete until menstruation has been absent for a year.

Check Your Progress

1. List the components of the male and female reproductive systems and state their functions.
2. Relate the phases of the ovarian cycle to the phases of the menstrual cycle.

Answers: 1. Testes in males and ovaries in females produce gametes; vasa deferentia in males and oviducts in females transport gametes; the penis in males and the vagina in females are organs of intercourse.   2. (a) During the follicular phase of the ovarian cycle, a follicle is producing an egg and releasing estrogen. Estrogen causes the uterine lining to thicken (proliferative phase of menstrual cycle). (b) Ovulation occurs. (c) During the luteal phase of the ovarian cycle, the corpus luteum is producing progesterone. Progesterone causes the uterine lining to become secretory (secretory phase of menstrual cycle). (d) If pregnancy does not occur, the corpus luteum degenerates, and menstruation occurs.

Aspects of Reproduction

Important topics related to human reproduction include control of reproduction, infertility, and sexually transmitted diseases.

Control of Reproduction

Table 29.1 lists various means of birth control and gives their rate of effectiveness, assuming consistent and correct usage. **Figure 29.9** illustrates some birth control devices. **Abstinence**—that is, not engaging in sexual intercourse—is very reliable and has the added advantage of avoiding sexually transmitted diseases. Oral contraception (the **birth control pill**) often involves taking a combination of estrogen and progesterone on a daily basis. These hormones effectively shut down the pituitary production of both FSH and LH so that no follicle in the ovary begins to develop in the ovary; since ovulation does not occur, pregnancy cannot take place. Because of possible side effects, including headaches, blurred vision, chest or abdominal pain, and swollen legs, women taking birth control pills should see a physician regularly.

Contraceptive implants utilize a synthetic progesterone to prevent ovulation by disrupting the ovarian cycle. The older version of the implant consists of six match-sized, time-release capsules that are surgically implanted under the skin of a woman's upper arm. The newest version consists of a single capsule that remains effective for about three years.

Contraceptive injections are available as progesterone only or a combination of estrogen and progesterone. The length of time between injections can vary from three months to a few weeks.

oral contraception (birth control pill)

contraceptive implant

female condom

spermicidal jelly and diaphragm

intrauterine device

Depo-Provera injection

Figure 29.9 **Contraceptive devices.**

Table 29.1 | Common Birth Control Methods

Name	Procedure	Effectiveness*	Name	Procedure	Effectiveness*
Abstinence	Refrain from sexual intercourse	100%	Diaphragm	Latex cap inserted into vagina to cover cervix before intercourse	With jelly, about 90%

Sterilization
  Vasectomy
  Tubal ligation
	Vas deferens cut and tied
	Oviducts cut and tied
		Almost 100%
		Almost 100%
			Cervical cap	Latex cap held by suction over cervix	Almost 85%

Combined estrogen/progesterone available as a pill, injectable, or vaginal ring and patch	Pill is taken daily; injectable and ring last a month; patch is replaced weekly	About 100%			
			Male condom	Latex sheath fitted over erect penis	About 85%
			Female condom	Polyurethane liner fitted inside vagina	About 85%
Progesterone only available as a tube implant and injectable	Implant lasts three years; injectable lasts three weeks	About 95%	Coitus interruptus	Penis withdrawn before ejaculation	About 75%
			Jellies, creams, foams	Spermicidal products inserted before intercourse	About 75%
Intrauterine device (IUD)	Newest hormone-free device contains progesterone and lasts up to ten years	More than 90%	Natural family planning	Day of ovulation determined by record keeping; various methods of testing	About 70%
Vaginal sponge	Sponge permeated with spermicide is inserted into vagina	About 90%	Douche	Vagina cleansed after intercourse	Less than 70%

*The percentage of sexually active women per year who will not get pregnant using this method.

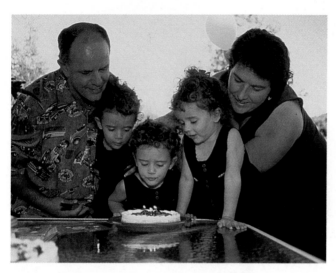

Figure 29.10 Control of reproduction.

The medical profession can offer help to control reproduction. The number of births can be decreased by the proper use of birth control, and the number of births can be increased through assisted reproductive technologies.

Interest in barrier methods of birth control has recently increased because they offer some protection against sexually transmitted diseases. A **female condom,** now available, consists of a large polyurethane tube with a flexible ring that fits onto the cervix. The open end of the tube has a ring that covers the external genitals. A **male condom** is most often a latex sheath that fits over the erect penis. The ejaculate is trapped inside the sheath, and thus does not enter the vagina. When used in conjunction with a spermicide, the protection is better than with the condom alone. The **diaphragm** is a soft latex cup with a flexible rim that lodges behind the pubic bone and fits over the cervix. Each woman must be properly fitted by a physician, and the diaphragm can be inserted into the vagina no more than two hours before sexual relations. Also, it must be used with spermicidal jelly or cream and should be left in place at least six hours after sexual relations. The **cervical cap** is a minidiaphragm.

An **intrauterine device (IUD)** is a small piece of molded plastic that is inserted into the uterus by a physician or other qualified healthcare practitioner. IUDs are believed to alter the environment of the uterus and oviducts so that fertilization probably will not occur—but if it should occur, implantation cannot take place. One type of IUD has copper wire wrapped around the plastic.

Contraceptive vaccines are now being developed. For example, a vaccine intended to immunize women against human chorionic gonadotropin (hCG), a hormone necessary to maintaining the implantation of the embryo, was successful in a limited clinical trial. Since hCG is not normally present in the body, no autoimmune reaction is expected, but the immunization does wear off with time. Other researchers believe that it would also be possible to develop a safe antisperm vaccine that could be used in women.

A morning-after pill, or emergency contraception, refers to a medication that will prevent pregnancy after unprotected intercourse. One type, a kit called Preven, contains four synthetic progesterone pills; two are taken up to 72 hours after unprotected intercourse, and two more are taken 12 hours later. The medication upsets the normal menstrual cycle, making it difficult for an embryo to implant itself in the endometrium. In a recent study, it was estimated that the medication was 75% effective in preventing unintended pregnancies.

Mifepristone, better known as RU-486, is a pill that is presently used to cause the loss of an implanted embryo by blocking the progesterone receptor proteins of endometrial cells. When taken in conjunction with a prostaglandin to induce uterine contractions, RU-486 is 95% effective. It is possible that someday this medication will also be a "morning-after pill," taken when menstruation is late without evidence that pregnancy has occurred.

Much research today is devoted to developing safe and effective hormonal birth control for males. Implants, pills, patches, and injections are being explored as ways to deliver testosterone and/or progesterone at adequate levels to suppress sperm production. Even the most successful formulations are still in the experimental stage and are unlikely to be available outside of clinical trials for at least a few more years.

Infertility

Control of reproduction does not only mean preventing pregnancy. It also involves methods of treating infertility (**Fig. 29.10**). **Infertility** is the failure of a couple to achieve pregnancy after one year of regular, unprotected intercourse. The American Medical Association estimates that 15% of all couples are infertile. The cause of infertility can be attributed to the male (40%), the female (40%), or both (20%).

Check Your Progress

1. Name two barrier methods of birth control and their benefits.
2. How are the pill, contraceptive implants, contraceptive injections, and the morning-after pill similar?

Answers: 1. Condoms and the diaphragm can be used at the last minute, and condoms in particular offer some protection against STDs.   2. They all work by interfering with the release of and/or the normal balance of sex hormones in the body of a female.

The most frequent cause of infertility in males is low sperm count and/or a large proportion of abnormal sperm, which can be due to environmental influences. Physicians advise that a sedentary lifestyle coupled with smoking and alcohol consumption can lead to male infertility. When males spend most of the day sitting in front of a computer or television or driving, the testes' temperature remains too high for adequate sperm production.

In females, body weight is an important fertility factor. Only if a woman is of normal weight, do fat cells produce a hormone called leptin that stimulates the hypothalamus to release GnRH so that follicle formation begins in the ovaries. If a woman is overweight, the ovaries may contain many small ineffective follicles and ovulation does not occur. About 10% of woman of childbearing age have an endocrine condition that may be due to an inability to utilize insulin properly. The condition is called polycystic ovary syndrome because the ovaries contain many cysts (small fluid-filled sacs) but no functioning follicles and ovulation does not occur. An impaired ability to respond to insulin is implicated because many women with PCOS eventually develop type 2 diabetes. Other causes of infertility in females are blocked oviducts due to pelvic inflammatory disease (see page 529), and endometriosis, the presence of uterine tissue outside the uterus, particularly in the oviducts and on the abdominal organs.

Sometimes the causes of infertility can be corrected by medical intervention so that couples can have children. It is also possible to give females fertility drugs, gonadotropic hormones that stimulate the ovaries and bring about ovulation. Such hormone treatments have been known to cause multiple ovulations and multiple births.

When reproduction does not occur in the usual manner, many couples adopt a child. Others sometimes try the assisted reproductive technologies discussed in the following paragraphs.

Assisted Reproductive Technologies

Assisted reproductive technologies (ART) consist of techniques used to increase the chances of pregnancy. Often, sperm and/or eggs are retrieved from the testes and ovaries, and fertilization takes place in a clinical or laboratory setting.

Artificial Insemination by Donor (AID)   During artificial insemination, harvested sperm are placed in the vagina by a physician. Sometimes a woman is artificially inseminated by her partner's sperm. This technique is especially helpful if the partner has a low sperm count, because the sperm can be collected over a period of time and concentrated so that the sperm count is sufficient to result in fertilization. Often, however, a woman is inseminated by sperm acquired from a donor who is a complete stranger to her. At times, a combination of partner and donor sperm is used.

In Vitro Fertilization (IVF) and Intracytoplasmic Sperm Injection (ICSI)   During IVF and ICSI (**Fig. 29.11**), conception occurs outside the body in a laboratory. Ultrasound machines can now spot follicles in the ovaries that hold immature eggs; therefore, the latest method is to forgo the administration of fertility drugs and retrieve immature eggs by using a needle. In IVF, the immature eggs are then brought to maturity in glassware before concentrated sperm are added. In ICSI, a single sperm is injected into an egg, usually because a male has severe infertility problems. After about two to four days, embryos are ready to be transferred to the uterus of the woman, who is now in the secretory phase of her menstrual cycle. If desired, embryos can be tested for a genetic disease, and only those found to be free of disease will be used. If implantation is successful and development is normal, pregnancy continues to term.

Figure 29.11   Intracytoplasmic Sperm Injection (ICSI).

A microscope connected to a television screen (at left) is used to carry out ICSI. On the screen, note that a pipette (at left of egg) holds the egg steady while a needle (not visible) introduces the sperm into the egg, ensuring fertilization.

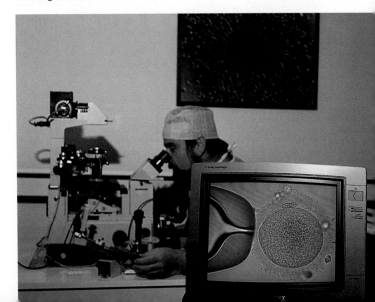

If several embryos are produced and transferred at a time, multiple births are common. Or there may be excess embryos left over and these may be frozen for transfer later or they may be donated to other couples or used in research.

Gamete Intrafallopian Transfer (GIFT)   The term **gamete** refers to a sex cell, either a sperm or an egg. Gamete intrafallopian transfer was devised to overcome the low success rate (15–20%) of in vitro fertilization. The method is exactly the same as in vitro fertilization, except the eggs and the sperm are placed in the oviducts immediately after they have been brought together. GIFT has the advantage of being a one-step procedure for the woman—the eggs are removed and reintroduced all in the same time period. A variation on this procedure is to fertilize the eggs in the laboratory and then place the zygotes in the oviducts.

Sexually Transmitted Diseases

Abstinence is the best protection against the spread of sexually transmitted diseases. For those who are sexually active, a latex condom offers some protection. Among STDs caused by viruses, treatment is available for AIDS and genital herpes, but these conditions are not curable. Only STDs caused by bacteria (e.g., chlamydia, gonorrhea, and syphilis) are curable with antibiotics.

STDs Caused by Viruses   Acquired **immunodeficiency syndrome (AIDS)** is caused by a virus called **human immunodeficiency virus (HIV).** HIV attacks the type of lymphocyte known as helper T cells. Helper T cells, you will recall, stimulate the activities of B lymphocytes, which produce antibodies. After an HIV infection sets in, helper T cells begin to decline in number, and the person becomes debilitated and more susceptible to other types of infections (**Fig. 29.12**). AIDS has three stages of infection, called categories A, B, and C.

During the category A stage, which may last about a year, the individual is an asymptomatic carrier. He or she may exhibit no symptoms, but can pass on the infection. Immediately after infection and before the blood test becomes positive, a large number of infectious viruses are present in the blood, and these could be passed on to another person. Even after the blood test becomes positive, the person remains well as long as the body produces sufficient helper T cells to keep the count higher than 500 per mm³. With a combination therapy of several drugs, AIDS patients may remain in this stage indefinitely.

During the category B stage, which may last six to eight years, the lymph nodes swell, and the person may experience weight loss, night sweats, fatigue, fever, and diarrhea. Infections such as thrush (white sores on the tongue and in the mouth) and herpes recur.

Finally, the person may progress to category C, which is full-blown AIDS characterized by nervous disorders and the development of an opportunistic disease, such as an unusual type of pneumonia or skin cancer. Opportunistic diseases are those that occur only in individuals who have little or no capability of fighting an infection. Without intensive medical treatment, the AIDS patient dies about seven to nine years after infection.

Genital warts are caused by the human papillomaviruses (HPVs). According to the CDC (Centers for Disease Control and Prevention), genital HPV is the most common STD in the United States today; about half of all sexually active individuals will acquire an HPV infection at some point in their lives. Many times, carriers either do not have any sign of warts or merely have flat lesions. When

a. AIDS patient, Tom Moran, July 1987

b. AIDS patient, Tom Moran, early January 1988

c. AIDS patient, Tom Moran, late January 1988

Figure 29.12   The course of AIDS.

This individual went through all the stages of AIDS before he died. Everyone should protect themselves from sexually transmitted diseases (STDs). If you have an STD, you are more susceptible to getting another one.

present, the warts commonly are seen on the penis and foreskin of men and near the vaginal opening in women. Genital warts are associated with cancer of the cervix, as well as tumors of the vulva, the vagina, the anus, and the penis.

Genital herpes is characterized by painful blisters on the genitals. Once the blisters rupture, they leave painful ulcers that may take as long as three weeks or as little as five days to heal. The blisters may be accompanied by fever, pain on urination, swollen lymph nodes in the groin, and in women, a copious discharge. After the ulcers heal, the disease is only latent, and blisters can recur, although usually at less frequent intervals and with milder symptoms. Fever, stress, sunlight, and menstruation are associated with recurrence of symptoms.

Hepatitis is an infection of the liver and can lead to liver failure, liver cancer, and death. Several types of hepatitis exist, some of which may be transmitted sexually. The type of hepatitis and the virus that causes it are designated by the same letter. Hepatitis A is usually acquired from sewage-contaminated drinking water, but this infection can also be sexually transmitted through oral/anal contact. Hepatitis B, which is spread in the same manner as AIDS, is even more infectious. Fortunately, a vaccine is now available for hepatitis B. Hepatitis C is spread when a person comes in contact with the blood of an infected person.

STDs Caused by Bacteria   **Chlamydia** is a bacterial infection of the lower reproductive tract that is usually mild or asymptomatic, especially in women. About 8 to 21 days after infection, men may experience a mild burning sensation on urination and a mucoid discharge. Women may have a vaginal discharge along with the symptoms of a urinary tract infection. Chlamydia also causes cervical ulcerations, which increase the risk of acquiring AIDS. If the infection is misdiagnosed or if a woman does not seek medical help, there is a particular risk of the infection spreading from the cervix to the oviducts so that **pelvic inflammatory disease (PID)** results. This very painful condition can result in blockage of the oviducts with the possibility of sterility and infertility.

Gonorrhea is easier to diagnose in males than in females because the male is more likely to experience painful urination and a thick, greenish-yellow urethral discharge. In males and females, a latent infection leads to PID, which affects the vasa deferentia or oviducts. As the inflamed tubes heal, they may become partially or completely blocked by scar tissue, resulting in sterility or infertility.

Syphilis has three stages, which are typically separated by latent periods. During the final stage, syphilis may affect the cardiovascular and/or nervous systems. An infected person may become mentally retarded, become blind, walk with a shuffle, or show signs of insanity. Gummas, which are large, destructive ulcers, may develop on the skin or within the internal organs. Syphilitic bacteria can cross the placenta, causing birth defects or stillbirth. Syphilis is easily diagnosed with a blood test.

STDs Caused by Other Organisms   Females very often have vaginitis, or infection of the vagina, caused by either the flagellated protozoan *Trichomonas vaginalis* or the yeast *Candida albicans*. **Trichomoniasis** is most often acquired through sexual intercourse, and the asymptomatic male is usually the reservoir of infection. *Candida albicans*, however, is a normal organism found in the vagina; its growth simply increases beyond normal under certain circumstances. For example, women taking birth control pills are sometimes prone to yeast infections. Also, the legitimate and indiscriminate use of antibiotics for infections elsewhere in the body can alter the normal balance of organisms in the vagina so that a yeast infection flares up.

Figure 29.13 lists the precautions everyone should take to avoid sexually transmitted diseases.

Figure 29.13   **Condom use can help reduce the spread of STDs.**

Some guidelines for preventing the spread of STDs are:

1. Abstain from sexual intercourse or develop a long-term monogamous (always the same person) relationship with a person who is free of STDs.

2. Refrain from multiple sex partners or having a relationship with a person who does have multiple sex partners.

3. Be aware that having relations with an intravenous drug user is risky because the behavior of this group puts them at risk for AIDS and hepatitis B.

4. Avoid anal intercourse because HIV has easy access through the lining of the rectum.

5. Always use a latex condom if your partner has not been free of STDs for the past five years.

6. Avoid oral sex because this may be a means of transmitting AIDS and other STDs.

7. Stop, if possible, the habit of injecting drugs, and if you cannot stop, at least always use a sterile needle.

Check Your Progress

1. Contrast artificial insemination with in vitro fertilization.
2. List the sexually transmitted diseases that are caused by viruses.
3. Why are STDs caused by viruses more problematic than those caused by bacteria?

Answers: 1. During artificial insemination, sperm are placed by a physician into a recipient's vagina, and fertilization occurs internally. During in vitro fertilization, donor eggs are fertilized by donor sperm in laboratory glassware. An embryo is then implanted in the uterus of a woman.   2. AIDS, genital warts, genital herpes, and hepatitis.   3. Bacterial infections are treatable with antibiotics, but viral infections are not.

chorion
becomes part of the placenta where the embryo/fetus receives oxygen and nutrient molecules and rids itself of waste molecules

yolk sac
first site of blood cell formation

allantois
its blood vessels become the blood vessels of the umbilical cord

amnion
contains the amniotic fluid, which cushions and protects the embryo

Figure 29.14   The extraembryonic membranes.

Humans, like other animals that reproduce on land, are dependent on these membranes to protect and nourish the embryo (and later the fetus).

29.3 | Human Development

Development encompasses all the events that occur from the time of fertilization until the animal is fully formed—in this case, a human being. Humans and other mammals have stages similar to those of all animals but with some marked differences, chiefly because of the presence of extraembryonic membranes (outside the embryo). Developing mammalian embryos (and fetuses in placental mammals such as humans), like those of reptiles and birds, depend on these membranes to protect and nourish them. **Figure 29.14** shows what these membranes are and what they do in a mammal.

Fertilization

Fertilization, which results in a zygote, requires that the sperm and the secondary oocyte interact (**Fig. 29.15**). The plasma membrane of the secondary oocyte is surrounded by an extracellular material termed the zona pellucida. In turn, the zona pellucida is surrounded by a few layers of adhering follicle cells.

During fertilization, a sperm moves past the leftover follicle cells, and acrosomal enzymes released by exocytosis digest a route through the zona pellucida. A sperm binds and fuses to the secondary oocyte's plasma membrane, and its nucleus enters. Only then does the secondary oocyte complete meiosis II and become an egg. Finally, the haploid egg and sperm nuclei fuse. Only one sperm should fertilize an egg, or else the zygote will have too many chromosomes and development will not be normal. Changes in the zona pellucida prevent the binding and penetration of additional sperm (called polyspermy).

Figure 29.15   Fertilization.

1 The sperm makes its way through the adhering follicle cells. **2** Acrosomal enzymes digest a portion of the zona pellucida. **3** Sperm binds to and fuses with egg plasma membrane. **4** Sperm nucleus enters cytoplasm of egg. **5** Sperm and egg nuclei fuse to produce a zygote.

Early Embryonic Development

The first two months of development are considered the embryonic period. Approximately six days of development occur in the oviduct before the embryo implants itself in the uterine lining (endometrium) (**Fig. 29.16**). During the first stage of development, the embryo becomes multicellular. Following fertilization, the zygote undergoes **cleavage,** which is cell division without growth. DNA replication and mitotic cell division occur repeatedly, and the cells get smaller with each division. Notice that cleavage only increases the number of cells; it does not change the original volume of the egg cytoplasm. The resulting tightly packed ball of cells is called a **morula.**

The cells of the morula continue to divide, but they also secrete a fluid into the center of a ball of cells. A hollow ball of cells, called the **blastocyst,** is formed, surrounding a fluid-filled cavity called the blastocoel. Within the ball is an inner cell mass that will go on to become the embryo. The outer layer of cells is the first sign of the chorion, the extraembryonic membrane that will contribute to the development of the placenta. As the embryo implants itself in the uterine lining (endometrium), the placenta begins to form and to secrete the hormone hCG (human chorionic gonadotropin). This hormone is the basis for the pregnancy test, and it serves to maintain the corpus luteum past the time it normally disintegrates. Because of hCG, the endometrium is maintained until this function is taken over by estrogen and progesterone, produced by the placenta. Ovulation and menstruation do not occur during pregnancy.

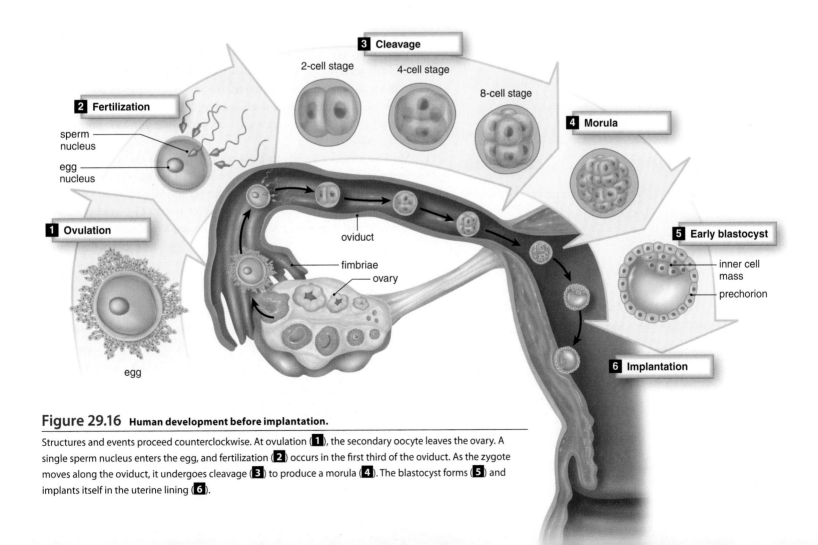

Figure 29.16 **Human development before implantation.**

Structures and events proceed counterclockwise. At ovulation (**1**), the secondary oocyte leaves the ovary. A single sperm nucleus enters the egg, and fertilization (**2**) occurs in the first third of the oviduct. As the zygote moves along the oviduct, it undergoes cleavage (**3**) to produce a morula (**4**). The blastocyst forms (**5**) and implants itself in the uterine lining (**6**).

Figure 29.17 **Gastrulation and neurulation.**

a. Gastrulation is the movement of cells to establish the three germ layers: ectoderm, mesoderm, and endoderm. **b.** During neurulation, the neural tube, which becomes the central nervous system, develops right above the notochord. Later, the notochord is replaced by the vertebral column. The neural crests contain ectodermal cells that did not participate in forming the neural tube. They develop into a number of different structures, including the autonomic nervous system.

a. Gastrulation

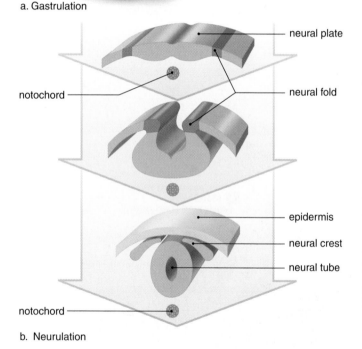

b. Neurulation

Later Embryonic Development

Gastrulation

During the third week of development, a slit called the primitive streak appears. The **gastrula** stage is evident when certain cells begin to push, or invaginate, at the primitive streak, creating three layers of cells (**Fig. 29.17***a*). Also evident is the **amnion,** the extraembryonic membrane that surrounds the developing embryo and later the fetus. The amnion encloses the amniotic fluid, which offers protection from sudden blows and movements, allows freedom of fetal movement, helps maintain a relatively constant temperature, and prevents drying out, which assists lung development.

Gastrulation is complete when the three layers of cells that will develop into adult organs have been produced. The outer layer is the **ectoderm;** the inner layer is the **endoderm;** and the third, or middle, layer of cells is called the **mesoderm.** Ectoderm, mesoderm, and endoderm are called the embryonic **germ layers.** As shown in **Figure 29.18,** the organs of an animal's body develop from these three germ layers.

Neurulation

The first organs to form are those of the central nervous system (Fig. 29.17*b*). The newly formed mesoderm cells that lie along the main longitudinal axis of the animal coalesce to form a dorsal supporting rod called the **notochord.** The central nervous system develops from midline ectoderm located just above the notochord. During **neurulation,** thickening of cells, called the **neural plate,** is seen along the dorsal surface of the embryo. Then, neural folds develop on either side of a neural groove, which becomes the **neural tube** when these folds fuse. The anterior portion of the neural tube becomes the brain, and the posterior portion becomes the spinal cord. Development of the neural tube is an example of **induction,** the process by which one tissue or organ influences the development of another. Induction occurs because the tissue initiating the induction releases a chemical that turns on genes in the tissue being induced.

Midline mesoderm cells that did not contribute to the formation of the notochord now become two longitudinal masses of tissue. These two masses become blocked off into somites, which are serially arranged on both sides along the length of the notochord. Somites give rise to the vertebrae and to muscles associated with the axial skeleton. The sequential order of the verte-

Figure 29.18 The germ layers and associated organs.

Embryonic Germ Layer	Vertebrate Adult Structures
Ectoderm (outer layer)	Nervous system; epidermis of skin; epithelial lining of oral cavity and rectum
Mesoderm (middle layer)	Musculoskeletal system; dermis of skin; cardiovascular system; urinary system; reproductive system; outer layers of respiratory and digestive systems
Endoderm (inner layer)	Epithelial lining of digestive tract and respiratory tract; associated glands of these systems; epithelial lining of urinary bladder

brae and the muscles of the trunk testify that chordates are segmented animals. Lateral to the somites, the mesoderm splits and forms the mesodermal lining of the coelom. In addition, the neural crest consists of a band of cells that develops where the neural tube pinches off from the ectoderm. These cells migrate to various locations where they contribute to the formation of skin and muscles, in addition to the adrenal medulla and the ganglia of the peripheral nervous system.

By the end of the third week, over a dozen somites are evident, and the blood vessels and gut have begun to develop. At this point, the embryo is about 2 millimeters (mm) long.

Organ Formation Continues

A human embryo at five weeks has little flippers called limb buds (**Fig. 29.19**). Later, the arms and the legs develop from the limb buds, and even the hands and the feet become apparent. During the fifth week, the head enlarges, and the sense organs become more prominent. It is possible to make out the developing eyes, ears, and even the nose.

The umbilical cord has developed from a bridge of mesoderm called the body stalk, which connects the caudal (tail) end of the embryo with the chorion. A fourth extraembryonic membrane, the allantois, is contained within this stalk, and its blood vessels become the umbilical blood vessels. The head and the tail then lift up as the body stalk moves anteriorly by constriction. Once this process is complete, the **umbilical cord,** which connects the developing embryo to the placenta, is fully formed (**Fig. 29.20**).

A remarkable change in external appearance occurs during the sixth through eighth weeks of development—the embryo becomes easily recognized as human. Concurrent with brain development, the neck region develops, making the head distinct from the body. The nervous system is developed well enough to permit reflex actions, such as a startle response to touch. At the end of this period, the embryo is about 38 mm (1.5 inches) long and weighs no more than an aspirin tablet, even though all its organ systems are established.

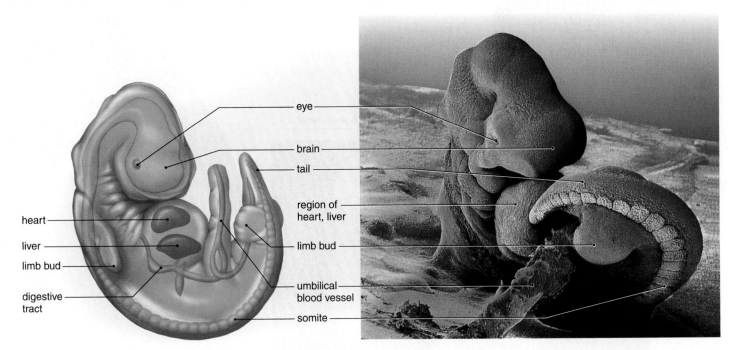

Figure 29.19   **Human embryo at the beginning of the fifth week.**

Endometrium

developing placenta

chorionic villi

maternal blood vessels

amniotic cavity

umbilical cord

amnion

chorion

extraembryonic cavity

Figure 29.20 **Placenta.**

Blood vessels within the umbilical cord lead to the placenta, where exchange takes place between fetal blood and maternal blood.

Figure 29.21 **9-week fetus.**

Placenta

The placenta has a fetal side contributed by the chorion, the outermost extraembryonic membrane, and a maternal side consisting of uterine tissues. Notice in Figure 29.20 that the chorion has treelike projections called the **chorionic villi.** The chorionic villi are surrounded by maternal blood; yet maternal and fetal blood never mix because exchange always takes place across the walls of the villi. Carbon dioxide and other wastes move from the fetal side to the maternal side of the placenta, and nutrients and oxygen move from the maternal side to the fetal side.

The umbilical cord stretches between the placenta and the fetus. The umbilical blood vessels are an extension of the fetal circulatory system and simply take fetal blood to and from the placenta. Harmful chemicals can also cross the placenta. This is of particular concern during the embryonic period, when various structures are first forming. Each organ or part seems to have a sensitive period during which a substance can alter its normal development. For example, if a woman takes the drug thalidomide, a tranquilizer, between days 27 and 40 of her pregnancy, the infant is likely to be born with deformed limbs. After day 40, however, the limbs will develop normally.

Fetal Development and Birth

Fetal development encompasses the third to the ninth months (**Figs. 29.21–29.24**). Fetal development is marked by an extreme increase in size. Weight multiplies 600 times, going from less than 28 grams to 3 kilograms. During this time, too, the fetus grows to about 50 centimeters in length. The genitalia appear in the third month, so it is possible to tell if the fetus is male or female.

Soon after the third month, hair, eyebrows, and eyelashes add finishing touches to the face and head. In the same way, fingernails and toenails complete the hands and feet. Later, during the fifth through seventh months, a fine, downy hair (lanugo) covers the limbs and trunk, only to later disappear. The fetus looks very old because the skin is growing so fast that it wrinkles. A waxy, almost cheeselike, substance (called vernix caseosa) protects the wrinkly skin from the watery amniotic fluid.

Figure 29.22 **12- to 16-week fetus.**

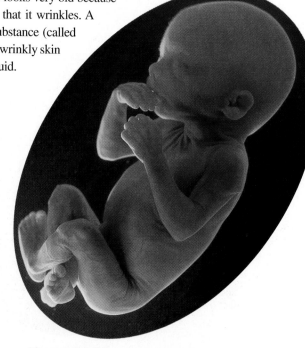

Figure 29.23 **20- to 28-week fetus.**

The fetus at first only flexes its limbs and nods its head, but later it can move its limbs vigorously to avoid discomfort. The mother feels these movements from about the fourth month on. The other systems of the body also begin to function. As early as 10 weeks, the fetal heartbeat can be heard through a stethoscope. A fetus born at 22 weeks has a chance of surviving, although the lungs are still immature and often cannot capture oxygen adequately. Weight gain during the last couple of months increases the likelihood of survival.

The Stages of Birth

The latest findings suggest that when the fetal brain is sufficiently mature, the hypothalamus causes the pituitary to stimulate the adrenal cortex so that androgens are released into the bloodstream. The placenta utilizes androgens as a precursor for estrogen, a hormone that stimulates the production of oxytocin and prostaglandin (a molecule produced by many cells that acts as a local hormone). All three of these molecules—estrogen, oxytocin, prostaglandin—cause the uterus to contract and expel the fetus.

The process of birth (parturition) includes three stages (**Fig. 29.25**). During the first stage, the cervix dilates to allow passage of the baby's head and body. The amnion usually bursts about this time. During the second stage, the baby is born, and the umbilical cord is cut (**Fig. 29.26**). During the third stage, the placenta is delivered.

ruptured amniotic sac

placenta

a. First stage of birth: cervix dilates

b. Second stage of birth: baby emerges

placenta
uterus
umbilical cord

c. Third stage of birth: expelling afterbirth

Figure 29.25 **Three stages of birth (parturition).**

a. Dilation of cervix. **b.** Birth of baby. **c.** Expulsion of afterbirth (placenta).

Check Your **Progress**

1. **What is the significance of germ layer formation during gastrulation?**
2. **Explain how induction occurs.**
3. **Contrast embryonic development with fetal development.**
4. **Of what use is the afterbirth during development?**

Answers: 1. Each of the germ layers gives rise to particular structures and organs. 2. One tissue releases a chemical that turns on genes in another tissue, influencing its development. 3. Embryonic development occurs during months 0–2 and is characterized by the establishment of organ systems. Fetal development occurs during months 3–9 and is characterized by an increase in size. 4. The afterbirth contains the placenta, the region of exchange with the mother's bloodstream.

Figure 29.24 **24-week fetus.**

Figure 29.26 **Newborn (40 weeks).**

The Chapter in Review

Summary

29.1 | How Animals Reproduce

Animals usually reproduce sexually, but some can also reproduce asexually.

- In asexual reproduction, a single parent produces offspring.
- In sexual reproduction, gametes produced by gonads join to form a zygote that develops into an offspring.
- Animals adapted to reproducing in the water shed their gametes into water; fertilization and development occur in water. Animals that reproduce on land protect their gametes and embryos from drying out.
- Reptiles and mammals have extraembryonic membranes.

29.2 | Human Reproduction

Male Reproductive System

- Sperm are produced in the testis, mature in the epididymis, and are stored in the epididymis and the vas deferens.
- Sperm enter the urethra (in the penis) prior to ejaculation along with seminal fluid (produced by seminal vesicles, the prostate gland, and bulbourethral glands).
- Spermatogenesis occurs in the seminiferous tubules of the testes, which also produce testosterone in interstitial cells.

Hormone Regulation in Males Testosterone brings about the maturation of the sex organs during puberty and promotes the secondary sex characteristics of males.

Female Reproductive System

- Oocytes are produced in the ovary and move through the oviduct to the uterus. The uterus opens into the vagina. The external genital area of women includes the clitoris, the labia minora, the labia majora, and the vaginal opening.
- A follicle produces a secondary oocyte and becomes a corpus luteum. The follicle and later the corpus luteum produce estrogen and progesterone.

The ovarian and menstrual cycles last 28 days. The events of these cycles are as follows:

- Days 1–13: Menstruation occurs for five days; the anterior pituitary produces FSH, and follicles produce primarily estrogen. Estrogen causes the endometrium to increase in thickness.
- Ovulation occurs on day 14.
- Days 15–28: LH from the anterior pituitary causes the corpus luteum to produce progesterone. Progesterone causes the endometrium to become secretory.

Estrogen and Progesterone Estrogen and progesterone bring about the maturation of the sex organs during puberty and promote the secondary sex characteristics of females.

Aspects of Reproduction

Numerous birth control methods and devices are available for those who wish to prevent pregnancy. Infertile couples are increasingly making use of assisted methods of reproduction.

Sexually transmitted diseases include the following:

- AIDS, an epidemic disease that destroys the immune system
- Genital warts, which can lead to cancer of the cervix
- Genital herpes, which repeatedly flares up
- Hepatitis, especially types A and B
- Chlamydia and gonorrhea, which cause pelvic inflammatory disease (PID)
- Syphilis, which leads to cardiovascular and neurological complications if untreated

29.3 | Human Development

Development encompasses all the events that occur from fertilization to a fully formed animal, in this case a human being.

Fertilization

The acrosome of a sperm releases enzymes that digest a pathway for the sperm through the zona pellucida. The sperm nucleus enters the egg and fuses with the egg nucleus.

Early Embryonic Development (months 1 and 2)

- Cleavage, which occurs in the oviduct, is cell division and formation of a morula and blastocyst. The blastocyst implants itself in the endometrium.
- Gastrulation is invagination of cells into the blastocoel, which results in formation of the germ layers. The germ layers are ectoderm, mesoderm, and endoderm.
- Organ formation can be related to the germ layers. Organ development begins with neural tube formation. Induction helps account for the steady progression of organ formation during embryonic development.
- The placenta has a fetal side and a maternal side. Gases and nutrients are exchanged at the placenta. The umbilical blood vessels are an extension of the embryo/fetal cardiovascular system and carry blood to and from the placenta.

Fetal Development and Birth (months 3–9)

- During fetal development, refinement of organ systems occurs, and the fetus adds weight.
- Birth has three stages: The cervix dilates, the baby is born, and the placenta is delivered.

Key Terms

abstinence   525

acquired immunodeficiency syndrome (AIDS)   528

amnion   532

birth control pill   525

blastocyst   531

bulbourethral glands   521

cervical cap   526

cervix   522

chlamydia   529

Testing Yourself

Choose the best answer for each question.

1. The evolutionary significance of the placenta is that it allowed
   a. animals to retain their eggs until they hatched.
   b. offspring to be born in an immature state and then develop outside the mother.
   c. offspring to exchange materials with the mother while developing inside the mother.
   d. mammals to develop the ability to produce milk.

2. The union of an egg and sperm is
   a. fertilization.
   b. parthenogenesis.
   c. insemination.
   d. contraception.
   e. implantation.

3. Sperm are never found in the
   a. prostate gland.
   b. epididymides.
   c. urethra.
   d. vasa deferentia.
   e. ejaculatory duct.

4. The function of the prostate gland is to
   a. cause sperm to reproduce.
   b. improve sperm motility.
   c. provide nutrients for sperm.
   d. extend the life of sperm.

5. Spermatogenesis produces cells that are
   a. diploid and genetically identical to each other.
   b. diploid and genetically different from each other.
   c. haploid and genetically identical to each other.
   d. haploid and genetically different from each other.

6. The time in a woman's life when the ovarian and menstrual cycles cease is
   a. puberty.
   b. copulation.
   c. menopause.
   d. parturition.
   e. adolescence.

7. Label the parts of the male reproductive system in the following illustration.

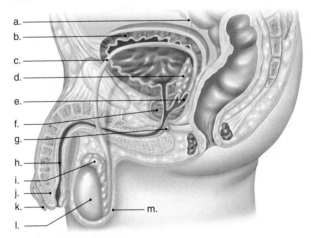

8. Label the parts of the female reproductive system in the following illustration.

9. Secondary sex characteristics in women are due mainly to the effects of
   a. estrogen.
   b. progesterone.
   c. follicle-stimulating hormone.
   d. luteinizing hormone.
   e. testosterone.

10. The process of birth begins when
    a. the fetal heart is sufficiently developed.
    b. the umbilical cord can no longer support the growth of the fetus.
    c. the fetal brain is mature enough to release androgens.
    d. all organ systems have completed development.

For questions 11–14, identify the stage of embryonic development in the key that matches the description.

Key:

   a. cleavage   c. neurulation
   b. gastrulation   d. organ formation

11. The notochord is formed.
12. The zygote divides without increasing in size.
13. Ectoderm and endoderm are formed.
14. Reflex actions, such as response to touch, develop.
15. Hepatitis is caused by a
    a. bacterium.   c. protozoan.
    b. fungus.   d. virus.
16. In human males, sterility results when the testes are located inside the abdominal cavity instead of the scrotum because
    a. the sperm cannot pass through the vas deferens.
    b. sperm production is inhibited at body temperature.
    c. the sperm cannot travel the extra distance.
    d. digestive juices destroy the sperm.
17. In the human menstrual cycle, ovulation usually occurs
    a. during days 1–5.   c. on day 14.
    b. during days 6–13.   d. during days 15–28.
18. In the testes, the _____ produce(s) the sperm.
    a. seminiferous tubules   c. seminal vesicles
    b. bulbourethral glands   d. prostate gland
19. Exchange of gases and waste between maternal and fetal blood occurs across the walls of the
    a. corpus luteum.   c. chorionic villi.
    b. seminiferous tubules.   d. germ layers.
20. The release of a secondary oocyte from a follicle is
    a. fertilization.   d. copulation.
    b. menses.   e. ovulation.
    c. parturition.
21. The assisted reproductive technology (ART) in which harvested sperm are placed in the vagina by a physician is
    a. in vitro fertilization (IVF).
    b. artificial insemination by donor (AID).
    c. gamete intrafallopian transfer (GIFT).
    d. intracytoplasmic sperm injection (ICSI).
22. Which pair is matched incorrectly?
    a. HIV—AIDS
    b. HPVs—genital warts
    c. chlamydia—pelvic inflammatory disease (PID)
    d. gonorrhea—liver failure
23. The process by which one tissue or organ influences the development of another is
    a. induction.   d. cleavage.
    b. insemination.   e. gastrulation.
    c. neurulation.

24. Human fetal development encompasses
    a. the third to the ninth months.
    b. the first to the third months.
    c. the first to the ninth months.
    d. the sixth to the ninth months.
25. The _____ is a dorsal supporting rod that forms from mesoderm during embryonic development.
    a. neural plate   d. notochord
    b. neural tube   e. morula
    c. umbilical cord

Thinking Scientifically

1. Female athletes who train intensively often stop menstruating. The important factor appears to be the reduction of body fat below a certain level. Give a possible evolutionary explanation for a relationship between body fat in females and reproductive cycles.

2. Human females undergo menopause between the ages of 45 and 55. In most other animal species, both males and females maintain their reproductive capacity throughout their lives. Why do you suppose human females lose that ability in midlife? Why don't human males go through menopause? (Hint: Think of the traditional roles of women and men in child-rearing.)

Bioethical Issue

The HPV Vaccine

In June 2006, the FDA (Food and Drug Administration) licensed Gardasil, a vaccine that protects against four different types of human papillomaviruses (HPVs), including the two known to cause 90% of genital warts and 70% of cervical cancers. The HPV vaccine is intended for use before the onset of sexual activity, and thus is recommended for 11- and 12-year-old girls, although it can be given to any female between the ages of 9 and 26. Gardasil typically has no side effects other than temporary soreness in the injection site. The vaccine is nearly 100% effective against four specific types of HPVs; however, because there are over 40 different types of sexually transmitted HPVs, Gardasil does not provide absolute protection against either genital warts or cervical cancer.

When the HPV vaccine was first made available, some parents were eager to have their daughters vaccinated. But enthusiasm for Gardasil has not been universal. In 2007, the governor of Texas signed an executive order (later overturned by the state legislature) requiring all schoolgirls to receive the HPV vaccine. Controversy ensued, as some parents were upset by the idea of vaccinating their young daughters against an STD. Some opponents argued that requiring Gardasil would encourage promiscuity, perhaps lulling young women into a false sense of security, so they might believe themselves perfectly immune to genital warts and cervical cancer.

Should parents be required to have their daughters vaccinated with Gardasil? Or should such vaccination be up to the discretion of the parents? Does the HPV vaccine encourage reckless sexual behavior, or is it a practical means of preventing a potentially life-threatening disease?

Essentials of Biology Website

The companion website for *Essentials of Biology* provides a wealth of information organized and integrated by chapter. You will find practice tests, animations, videos, and much more that will complement your learning and understanding of general biology.

http://www.mhhe.com/maderessentials2

30

Ecology of Populations

The world population increases by 85 million each year.

Over half the world's population lives in Asia.

To control population growth, China has imposed the one-child-per-couple policy on its citizens.

Controlling Population Growth

THE WORLD'S HUMAN population is growing at an alarming rate. Some estimates predict it will exceed 9 billion people within 50 years. If this rate of growth continues, the planet will eventually be unable to support the human population. Different countries have found ways to encourage their citizens to limit population growth. One of the most scrutinized strategies is China's one-child

—Continued next page

Outline

Continued—

policy. This policy allows each Chinese couple to have only one child. Punishments for those who do not follow the policy include imposing fines of as much as eight times the annual household income, destroying a family's home, or even forcing a pregnant woman to terminate her pregnancy. While some officials believe that such a drastic policy is necessary to decrease the growth of the Chinese population, others feel the policy is too harsh and no more effective than the much less restrictive policies, such as family planning and birth control education, that are being practiced in other overpopulated countries. The situation in China poses a major ethical dilemma. Is it more important to control population growth or to respect individuals' reproductive rights?

In this chapter, you will learn about ecology and population dynamics in all sorts of populations, including humans. One important lesson to be learned is that there are no easy solutions when it comes to ecological problems.

30.1 | The Human Population

The human population practically covers the face of the Earth. However, on both a large scale and a small scale, the human population has a clumped distribution. Fifty-six percent of the world's people live in Asia, and most Asian populations live in China and India. Mongolia has a population density of only 0.25 persons per square kilometer, while Bangladesh has a density of over 1,000 persons per square kilometer. On every continent, human population densities are highest along the coasts.

Present Population Growth

For most of the twentieth century, the growth of the global human population was relatively slow, but as more individuals achieved reproductive age, the growth rate began to increase. The Industrial Revolution (1800s) brought an increase in production of food and jobs. At this time, the growth curve for the human population began to increase steeply, so that currently the population is undergoing rapid growth (**Fig. 30.1**).

Just as with populations in nature, the growth rate of the human population is determined by the difference between the number of people born and the number of people who die each year. For example, the current global birthrate is an estimated 21 per 1,000 per year, while the global death rate is approximately 8 per 1,000 per year. From these numbers, we can calculate that the current annual growth rate of the human population is:

$$(21 - 8)/1,000 = (13)/1,000 = 1.3\%$$

Figure 30.1 Human population growth.

It is predicted that the world's population size may level off at 9 billion or increase to 12 billion by 2050, depending on the future growth rate. Close examination of the curve shows changes in growth that occurred in the 1300s during the black plague, in the 1800s with the Industrial Revolution, and in the 1900s with advances in science and medicine.

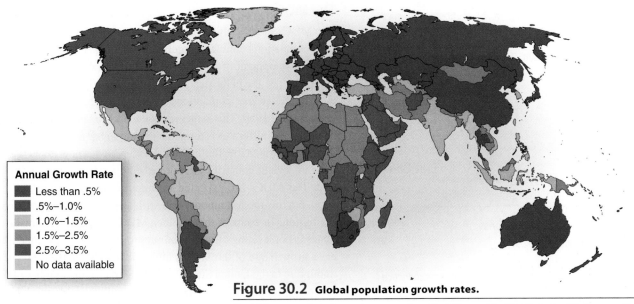

Annual Growth Rate
- Less than .5%
- .5%–1.0%
- 1.0%–1.5%
- 1.5%–2.5%
- 2.5%–3.5%
- No data available

Figure 30.2   **Global population growth rates.**

The highest population growth rates are found in Africa and the Middle East, while the growth rates throughout Europe are much lower.

Source: United Nations Population Division, 1993. Note: Data refer to 1990–95 (World Resource Institute).

Future Population Growth

Future growth of the human population is of great concern. It's possible that population increases will exceed the rate at which resources can be supplied. The potential for dire consequences can be appreciated by considering the *doubling time*—the length of time it takes for a population to double in size. The doubling time of the human population is now estimated to be 35 years. Will we be able to meet the extreme demands for resources by such a large population increase within such a short period of time? Already there are areas across the globe where people have inadequate access to fresh water, food, and shelter. Yet, in just 35 years, the world would need to double the amount of food, jobs, water, energy, and other resources in order to maintain the present standard of living.

Rapid growth usually begins to decline when resources, such as food and space, become scarce. Then population growth declines to zero, and the population levels off at a size appropriate to the carrying capacity of its environment. The carrying capacity reflects the number of individuals the environment can sustain for an indefinite period of time. The Earth's carrying capacity for humans has not been determined. Some authorities think the Earth is potentially capable of supporting far more people than currently inhabit the planet, perhaps as many as 50–100 billion people. Others believe that we may have already exceeded the number of humans the Earth can support, and that the human population may undergo a catastrophic crash to a much smaller size.

More-Developed Versus Less-Developed Countries

Complicating the issue of future human population growth is the fact that not all countries have the same growth rate (**Fig. 30.2**). The countries of the world today can be divided into two groups (**Fig. 30.3**). In the **more-developed countries (MDCs)**, such as those in North America and Europe, population growth is modest, and the people enjoy a fairly good standard of living. In the **less-developed countries (LDCs)**, such as those in Latin America, Asia, and Africa, population growth is dramatic, and the majority of the people live in poverty.

a.

b.

Figure 30.3   **More-developed versus less-developed countries.**

People in the (**a**) more-developed countries have a high standard of living and will contribute the least to world population growth, while people in the (**b**) less-developed countries have a low standard of living and will contribute the most to the world population growth.

a. More-developed countries (MDCs)

b. Less-developed countries (LDCs)

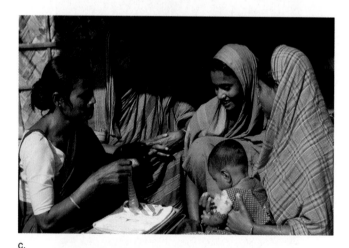

c.

Figure 30.4 **Age structure diagrams (1998).**

The diagrams predict that (**a**) the MDCs are approaching stabilization, whereas (**b**) the LDCs will expand rapidly due to their age distributions. **c.** Improved women's rights and increasing contraceptive use could change this scenario. Here a community health worker is instructing women in Bangladesh about the use of contraceptives.

(a, b) Data from the United Nations Population Division, 1998, p. 536.

The MDCs

The MDCs did not always have a low growth rate. Between 1850 and 1950, their populations doubled, largely because of the decline in the death rate due to the development of modern medicine and improved socioeconomic conditions. The decline in the death rate was followed shortly thereafter by a decline in birthrate, so that populations in the MDCs have experienced only modest growth since 1950. This sequence of events (i.e., decreased death rate followed by decreased birthrate) is termed a demographic transition.

The growth rate for the MDCs as a whole is now about 0.2%, but populations in several countries are not growing at all, or are actually decreasing in size. The MDCs are expected to increase by 52 million between 2002 and 2050, but this modest amount will still keep their total population at just about 1.2 billion. In contrast to the other MDCs, population growth in the United States is not leveling off and continues to increase. The United States currently has a growth rate of 0.6%. This higher rate is due to the fact that many people immigrate to the United States each year, and a large number of the women are still of reproductive age. Therefore, it is unlikely that the United States will experience a decline in its growth rate in the near future.

The LDCs

Death rates began to decline sharply in the LDCs following World War II, with the introduction of modern medicine, but the birthrate remained high. The *collective* growth rate of the LDCs peaked at 2.5% between 1960 and 1965, and since that time the rate has declined to 1.5%. Unfortunately, the growth rates in 46 of the less-developed countries have not declined. Thirty-five of these countries are in sub-Saharan Africa, where women on average are presently bearing more than five children each.

Between 2002 and 2050, the population of the LDCs is expected to jump from 5 billion to at least 8 billion. Africa will not share appreciably in this increase because of the many deaths there due to AIDS. The majority of the increase is expected to occur in Asia. Asia, with 31% of the world's arable (farmable) land, is already home to 59% of the human population. Twelve of the world's most polluted cities are located in Asia. If the human population increases as expected, Asia will experience even more urban pollution, acute water scarcity, and a significant loss of wildlife over the next 50 years.

▬ Check ▬
Your **Progress**

1. **Usually, why does rapid population growth begin to decline?**
2. **Contrast MDCs with LDCs.**

Answers: 1. Historically, when resources such as food and space become scarce, population size levels off at the carrying capacity of the environment.   2. MDCs, such as those in Europe and North America, have modest population growth and a good standard of living. LDCs, such as those in Latin America, Asia, and Africa, have a fast rate of population growth, and most of the people live in poverty.

Comparing Age Structures

An age structure diagram is divided into three groups: dependency, reproductive, and postreproductive (**Fig. 30.4**). Many MDCs have a stable age structure, meaning that the number of individuals in each group is just about the same. Therefore, MDC populations are expected to remain about the same size if couples have two children, and to decline if each couple has fewer than two children. In contrast, the age structure diagram of most LDCs has a pyramid shape, with the dependency group the largest. Therefore, LDC populations will continue to expand, even after replacement reproduction is attained. However, the more quickly replacement reproduction is achieved, the sooner zero population growth will result.

Replacement reproduction occurs when each couple in a population produces only two children. It might seem at first that replacement reproduction would cause a population to undergo zero population growth and therefore no increase in population size. However, if there are more young women entering the reproductive years than there are older women leaving them, replacement reproduction would still result in population growth.

Population Growth and Environmental Impact

Population growth is putting extreme pressure on each country's social organization, the Earth's resources, and the biosphere. Since the population of the LDCs is still growing at a significant rate, it might seem that their population increase will be the main cause of future environmental degradation. But this is not necessarily the case because the MDCs consume a much larger proportion of the Earth's resources than do the LDCs. This consumption leads to environmental degradation, which is also of the utmost concern.

Environmental Impact

The environmental impact (E.I.) of a population is measured not only in terms of population size, but also in terms of resource consumption per capita and the pollution resulting from this consumption. In other words:

$$E.I. = \text{population size} \times \text{resource consumption per capita}$$
$$= \text{pollution per unit of resource used}$$

Therefore, there are two possible causes of overpopulation: population size and resource consumption. Overpopulation is more obvious in LDCs; resource consumption is more obvious in MDCs, because the per capita consumption is so much higher in the MDCs. For example, an average American family, in terms of per capita resource consumption and water requirements, is the equivalent of 30 people in India. We need to realize that only a limited number of people can be sustained anywhere near the standard of living enjoyed in the MDCs.

The comparative environmental impacts of MDCs and LDCs are shown in **Figure 30.5**. The MDCs account for only about one-fourth of the world population. However, compared to the LDCs, MDCs account for 90% of the hazardous waste production due their high rate of consumption of such resources as fossil fuel, metal, and paper.

LDCs 78%   MDCs 22%

a. Population

LDCs 10%   MDCs 90%

b. Hazardous wastes

LDCs 40%   MDCs 60%

c. Consumption: fossil fuels

LDCs 20%   MDCs 80%

d. Consumption: metals

LDCs 25%   MDCs 75%

e. Consumption: paper

Figure 30.5   Environment impacts.

a. The combined population of the MDCs is much smaller than that of the LDCs. **b.** MDCs produce most of the world's hazardous wastes. **c–e.** MDCs consume much more fossil fuel, metals, and paper than do LDCs.

Figure 30.6 Patterns of distribution within a population.

a. A population of mature desert shrubs. **b.** Young, small desert shrubs are clumped. **c.** Medium shrubs are randomly distributed. **d.** Large shrubs are uniformly distributed.

30.2 | Characteristics of Populations

Various characteristics of all populations change over time. Populations are often periodically subject to environmental instability. During such times of environmental stability, individuals are under pressures of natural selection. Individuals better adapted to the environment will leave behind more offspring than those that are not as well adapted. Once the population becomes adapted to the new environment, an increase in size will occur again. So when an ecologist lists the characteristics of a population, this is a snapshot of the characteristics at a particular time and place.

Distribution and Density

Resources are the components of an environment that support its organisms. Light, water, space, mates, and food are some important resources for wildlife populations. The availability of resources influences the spatial distribution of individuals in a population. Three terms—**clumped, random,** and **uniform** distributions—are used to describe the observed patterns of distribution. In a study of desert shrubs, it was found that the distribution changes from clumped to random to uniform as the plants mature (**Fig. 30.6**). After sufficient study, competition for belowground resources was found to cause the distribution pattern to become uniform.

Suppose we were to step back and consider the distribution of individuals, not within a single desert or forest or pond, but within a species' range. A **range** is that portion of the globe where the species can be found. On this scale, all the members of a species within the range make up the population. Members of a population are clumped within the range because they are located in areas suitable for their growth.

Population density is the number of individuals per unit area or volume. Population density tends to be higher in areas with plentiful resources than in areas with scarce resources. Other factors can be involved, however. In general, population density declines with increasing body size. Consider that tree seedlings live at a much higher population density than do mature trees. Therefore, more than one factor must be taken into account when explaining population densities.

Population Growth

As mentioned on page 540, the growth rate is dependent on the number of individuals born and the number of individuals that die each year. Populations grow when the number of births exceeds the number of deaths. To take another example, if the number of births is 30 per year, and the number of deaths is 10 per year per 1,000 individuals, the growth rate is 2%.

In cases where the number of deaths exceeds the number of births, the value of the growth rate is negative—the population is shrinking.

Check Your Progress

1. The spatial distribution of individuals in a population can be described in what three ways?
2. List two major factors that influence population density.

Answers: 1. Clumped, random, and uniform.   2. Resource availability and body size.

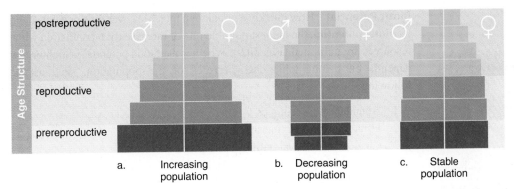

Figure 30.7 **Age structure diagrams.**

Typical age structure diagrams for hypothetical populations that are (**a**) increasing, (**b**) decreasing, or (**c**) stable. Different numbers of individuals in each age class create these distinctive shapes. In each diagram, the left half represents males while the right half represents females.

Demographics and Population Growth

Availability of resources and certain characteristics of a population—called **demographics**—influence its growth rate.

One demographic characteristic of interest for all populations is the **age structure** of a population. It is customary to use bar diagrams to depict the age structure of a population. The population is divided into members that are *prereproductive, reproductive,* or *postreproductive* based on their age. If the prereproductive and reproductive individuals outnumber those that are postreproductive, the birthrate will exceed the death rate (**Fig. 30.7***a*). In contrast, if the postreproductive individuals exceed those that are prereproductive and/or reproductive, the number of individuals in the population will decrease over time because the death rate will exceed the birthrate (Fig. 30.7*b*). If the numbers of prereproductive, reproductive, and postreproductive individuals are approximately equal, the birthrate and death rate will be equal, and the number of individuals in the population will remain relatively stable over time (Fig. 30.7*c*).

Ecologists also study patterns of **survivorship**—how the age at death influences population size. For example, if members of a population die young, the number of reproductive individuals is lower than otherwise. First, an ecologist constructs a life table, which lists the number of individuals that die or survive at each age. The best way to arrive at a life table is to identify a large number of individuals that are born at about the same time and keep records on them from birth until death.

A life table for Dall sheep is given in **Figure 30.8***a*. This table was constructed by gathering a large number of skulls and counting the growth rings on the horns to estimate each sheep's age at death, given that one growth ring is produced annually. Plotting the number of survivors per 1,000 births against age produces a *survivorship curve* (Fig. 30.8*b*). While each species has its own particular pattern, three types of survivorship curves are common. In a type I curve, typical of Dall sheep and humans, survival is high until old age, when deaths increase due to illness. In a type II curve, the possibility of death is equally likely over all age groups. In a type III curve, death is likely among the young, and few individuals reach old age.

Age (years)	Number of survivors at beginning of year	Number of deaths during year
0–1	1,000	199
1–2	801	12
2–3	789	13
3–4	776	12
4–5	764	30
5–6	734	46
6–7	688	48
7–8	640	69
8–9	571	132
9–10	439	187
10–11	252	156
11–12	96	90
12–13	6	3
13–14	3	3
14–15	0	

1,000–199

801–12

etc.

a.

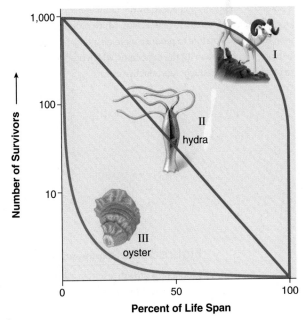

b.

Figure 30.8 Life table and survivorship curves.

a. A life table for the Dall sheep. **b.** Three typical survivorship curves. Among Dall sheep, with a type I curve, most individuals survive until old age, when they gradually die off. Among hydras, with a type II curve, there is an equal chance of death at all ages. Among oysters, with a type III curve, most die when they are young, and few become adults that are able to survive until old age.

Biotic Potential   How quickly a population increases over time depends upon its **biotic potential,** the highest possible rate of increase for a population when resources are unlimited (**Fig. 30.9**). Whether the biotic potential is high or low depends on demographic characteristics of the population, such as the following:

1. Abundance of resources
2. Usual number of offspring per reproduction
3. Chances of survival until age of reproduction
4. How often each individual reproduces
5. Age at which reproduction begins

Patterns of Population Growth

The patterns of population growth are dependent on (1) the biotic potential of the species combined with other demographics, and (2) the availability of resources. The two fundamental patterns of population growth are exponential growth and logistic growth.

Exponential Growth

Suppose ecologists are studying the growth of a population of insects that are capable of infesting and taking over an area. Under these circumstances, **exponential growth** is expected. An exponential pattern of population growth results in a J-shaped curve (**Fig. 30.10**). This growth pattern can be likened to compound interest at the bank: The more your money increases, the more interest you will get. If the insect population has 2,000 individuals and the per capita rate of increase is 20% per month, there will be 2,400 insects after one month, 2,880 after two months, 3,456 after three months, and so forth.

Notice that a J-shaped curve has these phases:

Lag phase   Growth is slow because the number of individuals in the population is small.
Exponential growth phase   Growth is accelerating.

Usually, exponential growth can continue only as long as resources in the environment are unlimited. When the number of individuals in the population approaches the number that can be supported by available resources, competition among individuals for resources increases.

a.

b.

Figure 30.9   Biotic potential.

A population's maximum growth rate under ideal conditions—that is, its biotic potential—is greatly influenced by the number of offspring produced in each reproductive event. **a.** Pigs, which produce many offspring that quickly mature to produce more offspring, have a much higher biotic potential than (**b**) the rhinoceros, which produces only one or two offspring per infrequent reproductive events.

Figure 30.10   Exponential growth.

Exponential growth results in a J-shaped growth curve because the growth rate is positive.

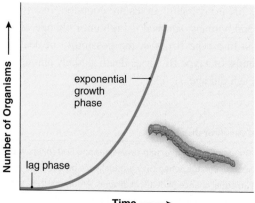

Logistic Growth

As resources decrease, population growth levels off, and a pattern of population growth called **logistic growth** is expected. Logistic growth results in an S-shaped growth curve (**Fig. 30.11**).

Notice that an S-shaped curve has these phases:

Lag phase Growth is slow because the number of individuals in the population is small.

Exponential growth phase Growth is accelerating due to biotic potential (see Fig. 30.9).

Deceleration phase The rate of population growth slows because of increased competition among individuals for available resources.

Stable equilibrium phase Little if any growth takes place because births and deaths are about equal.

The stable equilibrium phase is said to occur at the carrying capacity of the environment. Recall from Section 30.1 (p. 541) that the **carrying capacity** is the total number of individuals that the resources can support. This number is not a constant and varies with the circumstances. For example, a large island can support a larger population of penguins than a small island, because the smaller island has a limited amount of space for nesting sites.

Applications Our knowledge of logistic growth has practical applications. The model predicts that exponential growth will occur only when population size is much lower than the carrying capacity. So, if humans are using a fish population as a continuous food source, it would be best to maintain that population size in the exponential phase of growth when biotic potential is having its full effect and the birthrate is the highest. If people overfish, the fish population will sink into the lag phase, and it may be years before exponential growth recurs (**Fig. 30.12**). On the other hand, if people are trying to limit the growth of a pest, it is best to reduce the carrying capacity rather than to reduce the population size. Reducing the population size only encourages exponential growth to begin once again. Farmers can reduce the carrying capacity for a pest by alternating rows of different crops instead of growing one type of crop throughout the entire field.

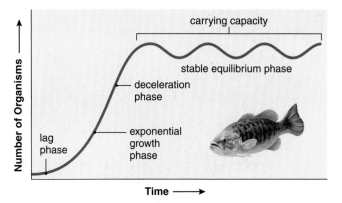

Figure 30.11 Logistic growth.

Logistic growth produces an S-shaped growth curve because competition for resources increases as the population approaches the carrying capacity.

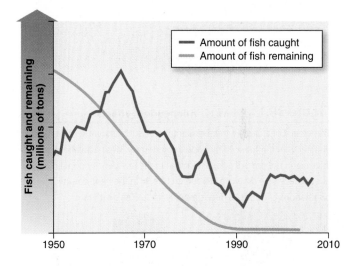

Figure 30.12 Sustainable population density.

As the amount of fish caught increased over the 1950s and 1960s, the remaining amount decreased. Because fishing still continues, fish population growth is maintained in its lag phase. Fishing must be reduced to allow the fish population to enter an exponential growth phase. Now a balance between fish population growth and fish caught could be maintained regularly at a higher amount.

Check Your Progress

1. What is the relationship between survivorship and population size?
2. List the four demographic characteristics of a population that determine its biotic potential.
3. Contrast exponential growth with logistic growth.

Answers: 1. If members of a population typically die young, the size of the population will be lower than expected. 2. Usual number of offspring per reproduction; chances of survival until age of reproduction; how often each individual reproduces; age at which reproduction begins. 3. Exponential growth produces a J-shaped curve and occurs when resources are unlimited. Logistic growth produces an S-shaped curve and occurs when resources are limited.

a.

survivors
40%

deaths
60%

b.

Figure 30.13 **Density-independent effects.**

The impact of a density-independent factor, such as weather or a natural disaster, is not influenced by population density. Two populations of field mice are in the path of a flash flood. (**a**) In the low-density population, 3 out of the 5 mice drown, a 60% death rate. **b.** In the high-density population, 12 out of 20 mice drown—also a 60% death rate.

a.

b.

Figure 30.14 **Density-dependent effects—competition.**

The impact of competition is directly proportional to the density of a population. **a.** When density is low, every member of the population has access to the resource. **b.** When density is high, members of the population must compete to gain access to available resources, and some fail to gain access.

Factors That Regulate Population Growth

Ecologists have long recognized that the environment contains both biotic (living) and abiotic (nonliving) components, and that these two components play an important role in regulating population size in nature.

Density-Independent Factors

Abiotic factors, such as weather or natural disasters, are typically **density-independent factors.** Abiotic factors can cause sudden and catastrophic reductions in population size. However, density-independent factors cannot in and of themselves regulate population size because the effect is not influenced by the number of individuals in the population. In other words, the intensity of the effect does not increase with increased population size. For example, the proportion of a population killed in a flash flood is independent of density—floods don't necessarily kill a larger percentage of a dense population than of a less dense population. If a mouse population has only five members and three drown due to a flash flood, the death rate is 60%. Also, if a mouse population has twenty mice and twelve drown, the death rate is still 60% (**Fig. 30.13**).

Density-Dependent Factors

Biotic factors are considered **density-dependent factors** because the percentage of the population affected does increase as the density of the population increases. Competition, predation, and parasitism are all biotic factors that increase in intensity as the density increases. We will discuss these interactions between populations again in Chapter 31 because they influence community composition and diversity.

Competition **Competition** can occur when members of the same species attempt to utilize resources (such as light, food, or space) that are in limited supply but are necessary for survival. As a result, not all members of the population can have access to the resource to the degree necessary to ensure survival or reproduction or some other aspect of their life history.

As an example, let's consider a woodpecker population in which members have to compete for nesting sites. Each pair of birds requires a tree hole in order to raise offspring. If there are more holes than breeding pairs, each pair can have a hole in which to lay eggs and rear young birds (**Fig. 30.14***a*). But if there are fewer holes than breeding pairs, each pair must compete to acquire a nesting site (Fig. 30.14*b*). Pairs that fail to gain access to holes will be unable to contribute new members to the population.

A well-known example of competition for food concerns the reindeer on St. Paul Island. Overpopulation led to an insufficiency of food to sustain all members, causing the population to crash—that is, it became drastically reduced.

Resource partitioning among different age groups is a way to reduce competition for food. During the life cycle of butterflies, the caterpillar stage requires different food than the adult stage. Caterpillars graze on leaves, while adult butterflies feed on nectar produced by flowers. Therefore, the parents are less apt to compete with their offspring for food.

Predation   **Predation** occurs when one organism, the predator, eats another, the prey. In the broadest sense, predation can include not only animals such as lions that kill zebras, but also filter-feeding blue whales that strain krill from the ocean waters, and even herbivorous deer that browse on trees and bushes.

The effect of predation generally increases as the prey population increases in density, because prey are easier to find when hiding places are limited. Consider a field inhabited by a population of mice (**Fig. 30.15**). Each mouse must have a hole in which to hide to avoid being eaten by a hawk. If there are 102 mice, but only 100 holes, two mice will be left out in the open. It might be hard for the hawk to find only two mice in the field. If neither mouse is caught, the predation rate is 0/2 = 0%. However, if there are 200 mice and only 100 holes, there is a greater chance that the hawk will be able to find some of these 100 mice without holes. If half of the exposed mice are caught, the predation rate is 50/100 = 50%. Therefore, increasing the density of the available prey has increased the proportion of the population preyed upon.

Predator-Prey Population Cycles   Instead of remaining steady, some predator and prey populations experience an increase, and then a decrease, in population size. A famous example of such a cycle occurs between the snowshoe hare and the Canada lynx, a type of small cat (**Fig. 30.16**). The snowshoe hare is a common herbivore in the coniferous forests of North America, where it feeds on terminal twigs of various shrubs and small trees. Investigators first assumed that the predatory lynx brings about a decrease in the hare population, and this decrease in prey brings about a subsequent decrease in the lynx population. Once the hare population recovers, so does the lynx population, and the result is a boom–bust cycle. But some biologists noted that the decline in snowshoe hare abundance is accompanied by low growth and reproductive rates, which could be signs of food shortage. A field experiment showed that if the lynx predator was denied access to the hare population, the cycling of the hare population still occurred based on food availability. The results suggested that a hare–food cycle and a predator–hare cycle have combined to produce the pattern observed in Figure 30.16.

b.

Figure 30.15   **Density-dependent effects—predation.**

The impact of predation on a population is directly proportional to the density of the population. **a.** In a low-density population, the chances of a predator finding the prey are low, resulting in little predatioin (mortality rate of 0/2, or 0%). **b.** In the higher-density population, there is a greater likelihood of the predator locating potenntial prey, resulting in a greater predation rate (mortality rate of 50/100, or 50%)

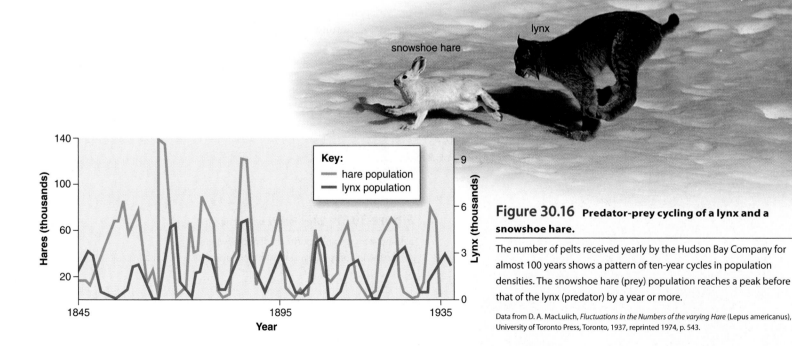

Figure 30.16   **Predator-prey cycling of a lynx and a snowshoe hare.**

The number of pelts received yearly by the Hudson Bay Company for almost 100 years shows a pattern of ten-year cycles in population densities. The snowshoe hare (prey) population reaches a peak before that of the lynx (predator) by a year or more.

Data from D. A. MacLuiich, *Fluctuations in the Numbers of the varying Hare* (Lepus americanus), University of Toronto Press, Toronto, 1937, reprinted 1974, p. 543.

Opportunistic Pattern

Small individuals
Short life span
Fast to mature
Many offspring
Little or no care of offspring

a.

30.3 | Life History Patterns and Extinction

A **life history** consists of a particular mix of the characteristics we have been discussing. After studying many different types of populations, from mayflies to humans, ecologists have discovered two fundamental and contrasting patterns: the opportunistic life history pattern and the equilibrium life history pattern.

Opportunistic species tend to exhibit exponential growth. The members of the population are small in size, mature early, have a short life span, and provide limited parental care for a great number of offspring (**Fig. 30.17a**). Density-independent effects tend to regulate the population size, which is large enough to survive an event that threatens to annihilate it. A population has a high dispersal capacity. Various types of insects and weeds provide the best examples of opportunistic species.

Equilibrium species exhibit logistic growth, and the population size remains close to or at the carrying capacity (Fig. 30.17b). Resources are relatively scarce, and individuals best able to compete—those with phenotypes best suited to the environment—have the largest number of offspring. They allocate energy to their own growth and survival and to the growth and survival of their few offspring. Therefore, they are fairly large, slow to mature, and have a fairly long life span. The growth of equilibrium species tends to be regulated by density-dependent effects. Various birds and mammals provide the best examples of equilibrium species.

Equilibrium Pattern

Large individuals
Long life span
Slow to mature
Few offspring
Much care of offspring

Figure 30.17 Life history patterns.

a. Dandelions are an opportunistic species, whereas (**b**) mountain gorillas are an equilibrium species.

b.

Extinction

Extinction is the total disappearance of a species or higher group. Which species shown in Figure 30.17, the dandelion or the mountain gorilla, is apt to become extinct? Because the dandelion matures quickly, produces many offspring at one time, and has seeds dispersed widely by wind, it can withstand a local decimation more easily than the mountain gorilla.

A study of equilibrium species shows that three other factors—namely, size of geographic range, degree of habitat tolerance, and size of local populations—can help determine whether an equilibrium species is in danger of extinction. **Figure 30.18** compares several equilibrium species on the basis of these three factors. The mountain gorilla has a restricted geographic range, narrow habitat tolerance (few preferred places to live), and small local population. This combination of characteristics makes the mountain gorilla very vulnerable to extinction. The possibility of extinction increases depending on whether a species is similar to the gorilla in one, two, or three ways. Such population studies can assist conservationists and others trying to preserve the biosphere's biodiversity.

▬ Check ▬
Your **Progress**

1. Contrast opportunistic populations with equilibrium populations.
2. List five factors that can determine whether an equilibrium population is in danger of extinction.

Answers: 1. Opportunistic populations exhibit exponential growth and are regulated by density-independent effects. Equilibrium populations exhibit logistic growth and remain near carrying capacity.   2. Time to maturity; number of offspring produced; size of geographic range; degree of habitat tolerance; and size of local populations.

Figure 30.18   Vulnerability to extinction.

Vulnerability is particularly tied to range, habitat, and size of population. The amount of white-highlighted text in the boxes indicates the chances and causes of extinction.

Organism

Population

Community

Ecosystem

Figure 30.19 Levels of organization.
The study of ecology encompasses the organism, population, community, and ecosystem levels of biological organization.

30.4 | The Scope of Ecology

Ecology is the scientific study of the interactions of organisms with one another and with their physical environment. Ecology is one of the two biological sciences of most interest to the public today, the other being genetics. Ecological studies offer information key to the survival of all species, present and future. Understanding ecology will help us make informed decisions, ranging from what kind of car to drive and whether to use pesticides on our lawns, to how to support the preservation of a forested area in our towns. Current ecological decisions will affect not only our lives, but also the lives of generations to come.

Ecology involves the study of several levels of biological organization (**Fig. 30.19**). Some ecologists study **organisms,** focusing on their adaptations to a particular environment. For example, organismal ecologists might investigate what attributes of a clownfish in a coral reef allow it to live in warm tropical waters.

In this chapter, we learned that organisms are usually part of a **population,** a group of individuals of the same species in a given location at the same time. At the population level of study, ecologists describe changes in population size over time. For example, a population ecologist might compare the number of clownfishes living in a given location today to data she obtained from that same location 20 years ago.

A **community** consists of all the various populations at a particular locale. A coral reef contains numerous populations of fishes, crustaceans, corals, algae, and so forth. At the community level, ecologists study how the interactions between populations affect the populations' well-being. For example, a community ecologist might study how a decrease in algal populations affects the population sizes of crustaceans and fishes living on the coral reef.

An **ecosystem** consists of a community of living organisms as well as their physical environment. As an example of how the physical environment affects a community, consider that the presence of suspended particles in the water decreases the amount of sunlight reaching algae living on the coral reef. Without solar energy, algae cannot produce the organic nutrients they and the other populations require.

Finally, the **biosphere** is the portion of the Earth's surface—air, water, land—where living things exist. Having information about the many levels of organization within a coral reef allows ecologists to understand how a coral reef contributes to the biodiversity and dynamics of the biosphere.

Ecology: A Biological Science

Ecology began as a part of **natural history,** the discipline of observing and describing organisms in their environment, but today ecology is also an experimental science. A central goal of ecology is to develop models that explain and predict the distribution and abundance of populations based on their interactions within an ecosystem. Achieving such a goal involves testing hypotheses. For example, ecologists might formulate and test hypotheses about the role fire plays in maintaining a lodgepole pine forest (**Fig. 30.20**). To test these hypotheses, ecologists could compare the characteristics of a community before and after a prescribed burn of the area by professionals under controlled conditions. Ultimately, ecologists wish to develop models about the distribution and abundance of ecosystems within the biosphere.

Ecology and Environmental Science

The field of environmental science applies ecological principles to practical human concerns. Comprehension of ecological principles helps us understand why a functioning biosphere is critical to our survival and why our population of 6 billion people and our overconsumption of resources pose threats to the biosphere.

  Conservation biology is a new discipline that studies all aspects of biodiversity with the goal of conserving natural resources, including wildlife, for the benefit of this generation and future generations. Conservation biology recognizes that wildlife species are an integral part of a well-functioning biosphere on which human life depends.

Check
Your Progress

1. List the levels of biological organization above organ systems.
2. How is environmental science helpful to conservation biology?

Answers: 1. Organisms, populations, communities, ecosystems, and the biosphere.
2. Environmental science involves understanding biodiversity; conservation biology uses this information to preserve biodiversity.

a. After a burn.

b. Before a burn.

Figure 30.20   Studying the effects of fire.

Fire is a natural occurrence in lodgepole pine forests, such as these in Yellowstone National Park. Over time, the recently burned forest (**a**) will mature and look like the unburned forest (**b**).

The Chapter in Review

 Summary

30.1 | The Human Population

The human population exhibits clumped distribution (both on a large scale and a small scale) and is undergoing rapid growth.

Future Population Growth

At the present growth rate, the doubling time for the human population is estimated to be 35 years. This rate of growth will put extreme demands on resources, and growth will decline due to resource scarcity. Eventually the population will most likely level off at its carrying capacity.

More-Developed Countries (MDCs) The growth rate for MDCs is about 0.1%. The total MDC population between 2002 and 2050 will remain at around 1.2 billion.

Less-Developed Countries (LDCs) The growth rate for LDCs is about 1.6%. The LDC population between 2002 and 2050 is expected to increase from 5 billion to 8 billion.

Comparing Age Structures

The age structure of populations is divided into three age groups: dependency, reproductive, and postreproductive. The MDCs are approaching stabilization with just about equal numbers in each group. The LDCs will continue to expand because their prereproductive group is the largest.

Population Growth and Environmental Impact

Two types of environmental impact can occur: The LDCs put stress on the biosphere due to population growth, and the MDCs put stress on the biosphere due to resource consumption and waste production.

30.2 | Characteristics of Populations

Distribution patterns and population density are both dependent on resource availability.

Distribution Patterns Clumped, random, uniform.

Population Density The number of individuals per unit area is higher in areas with abundant resources and lower when resources are limited.

Demographics and Population Growth Population growth can be calculated based on the annual birthrate and death rate. Population growth is determined by:
- Resource availability
- Demographics (age structures, survivorship, and biotic potential—the highest rate of increase possible)

Patterns of Population Growth

The two patterns of population growth are exponential growth and logistic growth.

Exponential Growth Exponential growth results in a J-shaped curve. The two phases of exponential growth are lag phase (slow growth) and exponential growth phase (accelerating growth).

Logistic Growth Logistic growth results in an S-shaped curve. The four phases of logistic growth are lag phase (slow growth), exponential growth phase (accelerating growth), deceleration phase (slowing growth), and stable equilibrium phase (relatively no growth).

Factors That Regulate Population Growth

The two factors that regulate population growth are density-independent and density-dependent factors.

Density-independent factors include abiotic factors, such as weather and natural disasters. The effect of the factor is not dependent on density.

Density-dependent factors include biotic factors, such as competition and predation. The effect of the factor is dependent on density.

30.3 | Life History Patterns and Extinction

The two fundamental life history patterns are exhibited by opportunistic populations and equilibrium populations.

Opportunistic populations are characterized by individuals with short life spans, who mature quickly, produce many offspring, have strong dispersal ability, provide little or no care of offspring, and exhibit exponential growth.

Equilibrium populations are characterized by individuals with long life spans, who mature slowly, produce few offspring, provide much care of offspring, and exhibit logistic growth.

Extinction

Extinction is the total disappearance of a species or higher group. Opportunistic populations are less likely than equilibrium populations to become extinct. Three factors in particular influence vulnerability of equilibrium populations to extinction: size of geographic range, degree of habitat tolerance, and size of local populations.

30.4 | The Scope of Ecology

Ecology is an experimental science that studies the interactions of organisms with each other and with the physical environment. The levels of biological organization studied by ecologists are:
- Organisms
- Population
- Community
- Ecosystem
- Biosphere

Key Terms

age structure   545
biosphere   552
biotic potential   546
carrying capacity   547
clumped   544

community   552
competition   548
demographics   545
density-dependent factor   548
density-independent factor   548

Testing Yourself

Choose the best answer for each question.

1. Decreased death rate followed by decreased birthrate has occurred in
   a. MDCs.
   b. LDCs.
   c. MDCs and LDCs.
   d. neither MDCs nor LDCs.

2. Human societies at present are characterized by
   a. ever-increasing population growth.
   b. unsustainable practices.
   c. concentrations along coastlines.
   d. All of these are correct.

3. A population's maximum growth rate is also called its
   a. carrying capacity.
   b. biotic potential.
   c. growth curve.
   d. replacement rate.

4. Which of these levels of ecological study involves both abiotic and biotic components?
   a. organisms
   b. populations
   c. communities
   d. ecosystem
   e. All of these are correct.

5. The biological level of organization subject to evolution by natural selection is called
   a. an organism.
   b. a population.
   c. a community.
   d. a ecosystem.
   e. the biosphere.

6. Globally, if a population lives only along a lake's shoreline, members of this population exhibit which type of spatial distribution?
   a. variable
   b. clumped
   c. random
   d. uniform

7. Which are likely to have the highest population density?
   a. zebras on the African savanna during the dry season
   b. mice in the frozen tundra
   c. moose in a Canadian forest
   d. earthworms in an organically rich soil

8. Calculate the growth rate of a population of 500 individuals in which the birthrate is 10 per year and the death rate is 5 per year.
   a. 5%
   b. −5%
   c. 1%
   d. −1%

9. If the human birthrate was reduced to 15 per 1,000 per year and the death rate remained the same (9 per 1,000), what would be the growth rate?
   a. 9%
   b. 6%
   c. 10%
   d. 0.6%
   e. 15%

10. Replacement reproduction in a population with a pyramid-shaped age structure diagram results in
    a. no population growth.
    b. population growth.
    c. a decline in the population.

11. Label the following age structure diagrams to indicate whether the population is stable, increasing, or decreasing.

a. _____   b. _____   c. _____

12. If the number of prereproductive and reproductive members of a population exceeds the number of postreproductive members, the population will
    a. grow.
    b. remain stable.
    c. decline.

13. An S-shaped growth curve indicates
    a. logistic growth.
    b. logarithmic growth.
    c. exponential growth.
    d. additive growth.

14. Exponential growth occurs
    a. when population size is increasing to an ever-higher amount.
    b. at the carrying capacity of the environment.
    c. when people are poor and don't have enough to eat.

For statements 15–18, indicate the interaction in the key exemplified by the scenario.

Key:
   a. density-independent factor
   b. competition
   c. predation
   d. predator-prey cycle

15. A severe drought destroys the entire food supply of a herd of gazelle.

16. A population of feral cats increases in size as the mouse population increases and then crashes regularly after the mouse population enters periods of decline.

17. Swift coyotes are able to catch rabbits and live to reproduce. Other coyotes lack a good food source and are not strong enough to reproduce.

18. Deer prefer to feed on a dense thicket of oak saplings in preference to more widely spaced young oak trees.

19. Which of the following is not an adaptive feature of an opportunistic life history pattern?
    a. many offspring
    b. little or no care of offspring
    c. long life span
    d. small individuals
    e. fast to mature

20. Natural selection favors a population of organisms living in a frequently disturbed environment to
    a. have a long life.
    b. invest much time and energy into parental care for young.
    c. have many offspring.
    d. mature slowly.

21. When the carrying capacity of the environment is exceeded, the population typically
    a. increases, but at a slower rate.
    b. decreases and stabilizes at the highest level possible.
    c. reverts to an opportunistic life history strategy.
    d. experience extinction immediately.

22. A pyramid-shaped age distribution means that
    a. the prereproductive group is the largest group.
    b. the population will grow for some time in the future.
    c. the country is more likely an LDC than an MDC.
    d. fewer women are leaving the reproductive years than entering them.
    e. All of these are correct.

23. Which of these is a density-independent factor?
    a. competition   c. weather
    b. predation   d. resource availability

24. Which of the following would not limit the biotic potential of an organism?
    a. a decrease in predators of this population
    b. an increase in parasites of this population
    c. maintaining a high degree of competition between this population and other species
    d. an decrease in natural resources available to this population
    e. All of these would limit a population's biotic potential.

Thinking Scientifically

1. In Sri Lanka, the death rate is 6 per 1,000, while the birthrate is 19 per 1,000. Calculate the current population growth rate. The formula for the doubling time of a population is:

   $t = 0.69/r$, where t is the doubling time and r is the growth rate.

Determine the doubling rate (the number of years it will take to double the population size) in Sri Lanka. Be sure to convert your growth rate from a percentage (e.g., 2.1%) to a decimal value (0.021). If the birthrate drops to 10 per 1,000, what is the doubling rate?

2. Assume that a population of dandelions grows exponentially so that it doubles in size every week. The population (beginning with one plant) expands to fill a field in 20 weeks. How many weeks will it take to fill one-quarter of the field? How long will it take to fill half the field?

 3. What type data would you collect to show that a population is undergoing stabilizing selection, as per Figure 16.9?

Bioethical Issue

Saving Species from Extinction

Species that are more prone to certain risk factors are more likely than others to become extinct. For example, species with a unique lineage, such as the giant panda, are likely to be at severe risk for extinction. Should our limited resources for species protection be focused on species that are at the highest risk for extinction? Some argue that high-risk species are less successful products of evolution and should not receive extraordinary protection. Consequently, all species at risk for extinction should be equally protected. Which camp are you in? Support your position.

Essentials of Biology Website

The companion website for *Essentials of Biology* provides a wealth of information organized and integrated by chapter. You will find practice tests, animations, videos, and much more that will complement your learning and understanding of general biology.

http://www.mhhe.com/maderessentials2

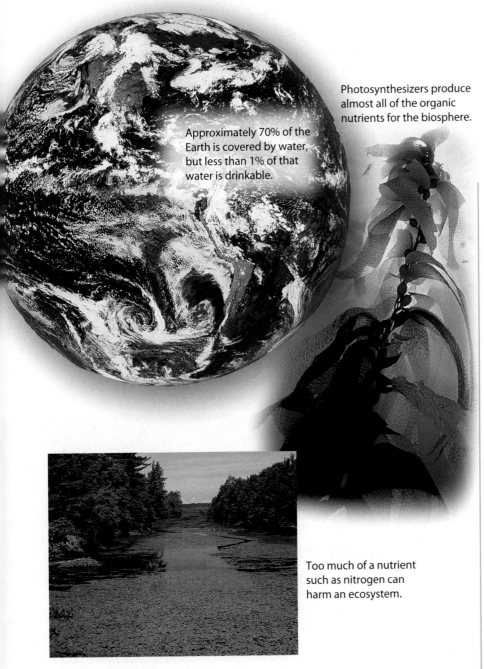

Photosynthesizers produce almost all of the organic nutrients for the biosphere.

Approximately 70% of the Earth is covered by water, but less than 1% of that water is drinkable.

Too much of a nutrient such as nitrogen can harm an ecosystem.

31
Communities and Ecosystems

Outline

Chemical Cycling

ESSENTIAL ELEMENTS are elements necessary for all living things. Nitrogen is one of these elements, and even though it is the most abundant gas in Earth's atmosphere, it is not biologically available to plants until it has been "fixed." This "fixing" process is part of the nitrogen cycle, and much of it is performed by soil bacteria. Once this process is complete, plants can absorb the nitrogen from

—Continued next page

Continued—

the soil in the form of ammonium and nitrate ions, and other organisms receive this nitrogen when they feed on plants. Carbon dioxide, another atmospheric gas, is the source of all carbon used in life's organic molecules. Plants and other autotrophs convert carbon dioxide into usable organic molecules through the process of photosynthesis. Animals that then feed on plants, acquire the necessary organic molecules, which get broken down through the process of cellular respiration. Carbon dioxide, a waste product of cellular respiration, is then exhaled back into the atmosphere, thus completing the carbon cycle.

In order for nutrients to cycle and energy to flow through both the environment and its organisms, many levels of interactions among species and their habitats must exist. In this chapter, you will learn about communities; types of interactions among species that influence natural selection; characteristics of aquatic and terrestrial ecosystems; and chemical cycling, including the nitrogen and carbon cycles.

31.1 | Ecology of Communities

A **community** is an assemblage of populations of multiple species, interacting with one another within a single environment. For example, the various species living on and within a fallen log, such as plants, fungi, worms, and insects, interact with one another and form a community. The fungi break down the log and provide food for the earthworms and insects living in and on the log. Those insects may feed on one another, too. If birds flying throughout the forest feed on the insects and worms living in and on the log, they are also part of the larger forest community.

Communities come in different sizes, and it is sometimes difficult to decide where one community ends and another one begins. The relationships and interactions between species in a community form over time. Some of the relationships between species are products of **coevolution,** by which an evolutionary change in one species results in an evolutionary change in another. **Figure 31.1** gives examples of how flowering plants and their pollinators are co-adapted. The Australian orchid, *Chiloglottis trapeziformis*, resembles the body of a wasp, and its odor mimics the pheromones of a female wasp. Therefore, male wasps are attracted to the flower, and when they attempt to mate with it, they become covered with pollen, which they transfer to the next orchid flower (Fig. 31.1*b*). The orchid is dependent upon wasps for pollination because neither wind nor other insects pollinate these flowers.

a.

b.

c.

Figure 31.1 Coevolution.

Flowers and pollinators have evolved to be suited to one another. **a.** Hummingbird-pollinated flowers are usually red, a color that these birds can see, and the petals are recurved to allow the stamens to dust the birds' heads. **b.** The reward offered by the flower is not always food. This orchid looks and smells like the female of this wasp's species. The male tries to copulate with flower after flower and in the process transfers pollen. **c.** Bats are nocturnal, and the flowers they pollinate are white or light-colored making them visible in moonlight. The flowers smell like bats and are large and sturdy, enabling them to withstand insertion of the bat's head as it uses its long, bristly tongue to lap up nectar and pollen.

All species in a community possess adaptations suitable to the conditions of the particular physical environment. An **ecosystem** consists of these species interacting with each other and with the physical environment. If the physical environment of an ecosystem changes, corresponding changes will most likely occur in the species comprising the community and in the relationships among these species. Extinction of species can occur when environmental change is too rapid for suitable adaptations to evolve.

Rapid environmental changes can be detrimental to humans too, even though technology increases our ability to adapt. Sometimes the economy of an area is dependent upon the climate and, in some cases, the species composition, of an aquatic or terrestrial ecosystem. Therefore, human activities that negatively impact the community of the area can also negatively affect the economy of that area. Knowledge of community and ecosystem ecology will help you better understand how human can, in the end, be detrimental to ourselves.

Community Composition and Diversity

Two characteristics of communities—species composition and diversity—allow us to compare communities. The species composition of a community, also known as **species richness,** is simply a listing of the various species found in that community. The **diversity** includes both species richness and species distribution.

Species Composition

It is apparent by comparing the photographs in **Figure 31.2** that a coniferous forest has a different species composition than a tropical rain forest. Narrow-leaved evergreen tree species are prominent in the coniferous forest, whereas broad-leaved evergreen tree species are numerous in the tropical rain forest. As the list of mammals demonstrates, a coniferous community and a tropical rain forest community contain different species of mammals. Ecologists comparing these two communities would also find differences in the number of other plants and animal species. In the end, ecologists would conclude that not only are the species compositions of these two communities different, but the tropical rain forest supports more species and therefore higher species richness.

Diversity

The diversity of a community goes beyond species richness to include distribution and relative abundance, the number of individuals of each species per unit area. For example, suppose a deciduous forest in West Virginia has, among individuals of other species, 76 poplar trees but only one American elm. If you were simply walking through this forest, you might miss the lone American elm. If, instead, the forest had 36 poplar trees and 41 American elm trees, the forest would seem more diverse to you, and indeed would have a higher diversity value, although both would have the same species richness value. The greater the species richness and the more even the distribution of the species in the community, the greater the diversity.

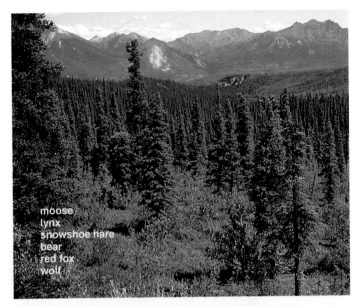

moose
lynx
snowshoe hare
bear
red fox
wolf

a.

monkey
sloth
anteater
kinkajou
jaguar
tapir
bat

b.

Figure 31.2 **Community species composition.**

Communities differ in their species composition, as exemplified by the predominant plants and animals in (**a**) a coniferous forest and (**b**) a tropical rain forest. Some mammals found in coniferous forests and in tropical rain forests are listed.

a.

b.

Figure 31.3 **Climax communities.**

Does succession in a particular area always lead to the same climax community? For example, temperate forests (**a**) only occur where there is adequate rainfall, and deserts (**b**) occur where rainfall is minimal. Even so, the exact same mix of plants and animals may not always arise because the assemblage of organisms depends on which organisms, by chance, migrate to the area.

Ecological Succession

Community species composition and diversity do change over time, although they may seem to remain static because it can take decades—often longer than the human life span—for noticeable changes to occur. Natural forces, such as glaciers, volcanic eruptions, lightning-ignited forest fires, hurricanes, tornadoes, and floods are considered disturbances and bring about community changes. Communities also change because of human activities, such as logging, road building, sedimentation, and farming. A more or less orderly process of community change is known as **ecological succession.**

Ecologists have developed models to explain why succession occurs and to predict patterns of succession following a disturbance. The climax-pattern model says that the climate of an area always leads to the same stable **climax community,** a specific assemblage of bacterial, fungal, plant, and animal species (**Fig. 31.3**). For example, a coniferous forest community is expected in northern latitudes, a deciduous forest in temperate zones, and a tropical forest in areas with a tropical climate. Now that we know disturbances influence community composition and diversity, and that despite the climate, the composition of a community is not always the same, the climax model of succession is being modified.

Two Types of Succession

Ecologists define two types of ecological succession: primary and secondary (**Fig. 31.4**). **Primary succession** starts where soil has not yet formed. For example, hardened lava flows or the scraped bedrock that remains following a glacial retreat are subject to primary succession. **Secondary succession** begins, for example, in a cultivated field that is no longer farmed, where soil is already present. With both primary and secondary succession, a progression of species occurs over time as first spores of fungi and vascular seedless plants and then seeds of nonvascular plants, followed by seeds of gymnosperms and/or angiosperms, are carried into the area by wind, water, or animals from the surrounding regions (**Fig. 31.5**; see also Fig. 31.4).

rock	lichens/mosses	grass	low shrub	high shrub	shrub-tree	low tree	high tree
primary succession								
secondary succession								

Figure 31.4 **Primary and secondary succession.**

Primary succession begins on areas of bare rock. Secondary succession begins in areas where soil remains following natural or human-caused disturbance.

The first species to appear in an area undergoing either primary or secondary succession are called opportunistic pioneer species. These species are small in stature, short-lived, quick to mature, and they produce numerous offspring per reproductive event. The first pioneer species to arrive are photosynthetic organisms such as lichens and mosses. Lichens are two organisms from different kingdoms, living as one. The fungal partners of lichens play a critical role by breaking down rock or lava into usable mineral nutrition, not only for their algal partners but also for pioneer plants. The mycorrhizal fungal partners of plants (see Fig. 18.26) pass minerals directly to plants so that they can grow successfully in poor soil. Pioneer plant species that become established in an area are accompanied by pioneer herbivore species (e.g., insects) and then carnivore species (e.g., small mammals). As the community continues to change, equilibrium species become established in the area. Equilibrium species, such as deer, wolves, and bears, are larger in size, long-lived, and slow to mature, and they produce few offspring per reproductive event.

Check Your Progress

1. **Describe what is meant by coevolution.**
2. **Contrast a community with an ecosystem.**
3. **Contrast species richness with diversity.**
4. **Contrast primary succession with secondary succession.**

Answers: 1. An evolutionary change in one species results in an evolutionary change in another species. 2. A community is a group of populations, while an ecosystem is all the species in a community interacting with each other and with the environment. 3. Species richness is the list of species in a community (species composition), while diversity encompasses both composition and the relative abundance of each species. 4. Primary succession occurs where soil has not yet been formed, while secondary succession occurs where soil is present.

a. During the first year, only the remains of corn plants are seen.

b. During the second year, wild grasses have invaded the area.

c. By the fifth year, the grasses look more mature, and sedges have joined them.

d. After twenty years, the juniper trees are mature, and there are also birch and maple trees in addition to the blackberry shrubs.

Figure 31.5 Secondary succession.

Secondary successional changes in a western Pennsylvania field from (**a**) first year, (**b**) second year, (**c**) fifth year, and (**d**) after twenty years.

Table 31.1 | Species Interactions

Interaction	Expected Outcome
Competition (− −)	Abundance of both species decreases.
Predation (+ −)	Abundance of predator increases, and abundance of prey decreases.
Parasitism (+ −)	Abundance of parasite increases, and abundance of host decreases.
Commensalism (+ 0)	Abundance of one species increases, and the other is not affected.
Mutualism (+ +)	Abundance of both species increases.

Figure 31.6 **Niche of a backswimmer.**

Backswimmers require warm, clear pond water containing insects that they can eat and vegetation where they can hide from predators.

Interactions in Communities

Species interactions, especially competition for resources, fashion a community into a dynamic system of interspecies relationships. In **Table 31.1,** the plus and minus signs tell how the relationship affects the abundance of the two interacting species. **Competition** between two species for limited resources has a negative effect on the abundance of both species. In **predation,** one animal, the predator, feeds on another, the prey (see p. 549); in **parasitism,** one species obtains nutrients from another species, called the host, but usually does not kill the host. **Commensalism** is a relationship in which one species benefits while the second species is not harmed. Commensalism often occurs when one species provides a home or transportation for another. In **mutualism,** two species interact in such a way that both benefit.

Ecological Niche

Each species occupies a particular position in the community, both in a spatial and a functional sense. Spatially, species live in a particular area of the community, or **habitat,** such as underground, in the trees, or in shallow water. Functionally, each species plays a role, such as whether it is a photosynthesizer, predator, prey, parasite, or decomposer. The **ecological niche** of a species incorporates the role the species plays in its community, its habitat, and its interactions with other species. The niche includes the living and nonliving resources that individuals in the population need to meet their energy, nutrient, and survival demands. The habitat of an insect, called a backswimmer, is a pond or lake, where it eats other insects (**Fig. 31.6**). The pond or lake must contain vegetation where the backswimmer can hide from its predators, including fish and birds. The pond water must be clear enough for the backswimmer to see its prey and warm enough for it to maintain a good metabolic rate. Since it is difficult to describe and measure the total niche of a species in a community, ecologists often focus on a certain aspect of a species' niche, as with the birds featured in **Figure 31.7.**

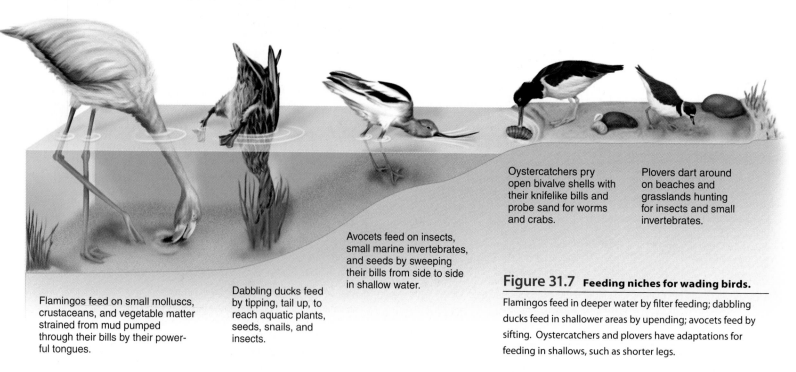

Flamingos feed on small molluscs, crustaceans, and vegetable matter strained from mud pumped through their bills by their powerful tongues.

Dabbling ducks feed by tipping, tail up, to reach aquatic plants, seeds, snails, and insects.

Avocets feed on insects, small marine invertebrates, and seeds by sweeping their bills from side to side in shallow water.

Oystercatchers pry open bivalve shells with their knifelike bills and probe sand for worms and crabs.

Plovers dart around on beaches and grasslands hunting for insects and small invertebrates.

Figure 31.7 **Feeding niches for wading birds.**

Flamingos feed in deeper water by filter feeding; dabbling ducks feed in shallower areas by upending; avocets feed by sifting. Oystercatchers and plovers have adaptations for feeding in shallows, such as shorter legs.

Competition

Competition for resources such as light, space, or nutrients contributes to the niche of each species and helps structure the community. Laboratory experiments helped ecologists formulate the **competitive exclusion principle,** which states that no two species can occupy the same niche at the same time. In the 1930s, G. F. Gause grew two species of *Paramecium* in one test tube containing a fixed amount of bacterial food. Although populations of each species survived when grown in separate test tubes, only one species, *Paramecium aurelia*, survived when the two species were grown together (**Fig. 31.8**). *P. aurelia* acquired more of the food resource and had a higher population growth rate than did *P. caudatum*. Eventually, as the *P. aurelia* population grew and obtained an increasingly greater proportion of the food resource, the number of *P. caudatum* individuals decreased, and the population died out.

Niche Specialization Competition for resources does not always lead to localized extinction of a species. Multiple species coexist in communities by partitioning, or sharing, resources. In another laboratory experiment using other species of *Paramecium*, Gause found that the two species could survive in the same test tube if one species consumed bacteria at the bottom of the tube and the other ate bacteria suspended in solution. This **resource partitioning** decreased competition between the two species, leading to increased niche specialization. What could have been one niche became two more specialized niches due to species differences in feeding behavior.

When three species of ground finches of the Galápagos Islands live on the same islands, their beak sizes differ, and each feeds on a different-sized seed (**Fig. 31.9**). When the finches live on separate islands, their beaks tend to be the same intermediate size, enabling each to feed on a wider range of seeds. Such **character displacement** often is viewed as evidence that competition and resource partitioning have taken place.

The niche specialization that permits coexistence of multiple species can be very subtle. Species of warblers that occur in North American forests are all about the same size, and all feed on budworms, a type of caterpillar found on spruce trees. Robert MacArthur recorded the length of time each warbler species spent in different regions of spruce canopies to determine where each species did most of its feeding. He discovered that each species primarily used different parts of the tree canopy and, in that way, had a more specialized niche. As another example, consider that three types of birds—swallows, swifts, and martins—all eat flying insects and parachuting spiders. These birds even frequently fly in mixed flocks. But each type of bird has a different nesting site and migrates at a slightly different time of year.

Mutualism

Mutualism, a symbiotic relationship in which both members benefit, is now recognized to be at least as important as competition in shaping community structure. The relationship between plants and their pollinators mentioned previously is a good example of mutualism. Perhaps the relationship began when herbivores feasted on pollen. The provision of nectar by the plant may have spared the pollen, and at the same time allowed the animal to become an agent of pollination. By now, pollinator mouthparts are adapted to gathering the nectar of a particular plant species, and this species is dependent on the pollinator for dispersing pollen. As also mentioned previously, lichens can grow on rocks because the fungal member leaches minerals that are provided to the algal partner. The algal partner, in turn, photosynthesizes and provides organic food for both members of the relationship.

Figure 31.8 Competitive exclusion principle demonstrated by *Paramecium*.

The competitive exclusion principle states that no two species occupy the exact same niche. When grown separately, *Paramecium caudatum* and *Paramecium aurelia* exhibit logistic growth. When grown together, *P. aurelia* excludes *P. caudatum*.

Data from G.F. Gause, *The Struggle for Existence*, 1934, Williams & Wilkins Company, Baltimore, MD, p. 557.

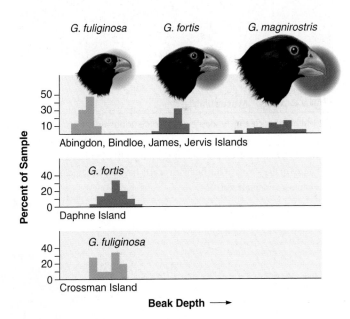

Figure 31.9 Character displacement in finches on the Galápagos Islands.

When *G. fuliginosa, G. fortis,* and *G. magnirostris* coexist on the same island, their beak sizes are appropriate for eating small-, medium-, and large-sized seeds, respectively. When *G. fortis* and *G. fuliginosa* are on separate islands, their beaks have the same intermediate size, which allows them to eat seeds of various sizes. Character displacement is evidence that resource partitioning has occurred.

In tropical America, ants form mutualistic relationships with certain plants. The bullhorn acacia tree is adapted to provide a home for ants of the species *Pseudomyrmex ferruginea* (**Fig. 31.10**). Unlike other acacias, this species has swollen thorns with a hollow interior where ant larvae can grow and develop. In addition to housing the ants, acacias provide them with food. The ants feed from nectaries at the base of the leaves and eat fat- and protein-containing nodules called Beltian bodies, which are found at the tips of the leaves. For 24 hours a day, ants constantly protect the tree from herbivores that would like to feed on it. The ants are so critical to the trees' survival that when the ants on experimental trees were poisoned, the trees died.

The outcome of mutualism is an intricate web of species interpendencies critical to the community. For example, in areas of the western United States, the branches and cones of whitebark pine are turned upward, meaning that the seeds do not fall to the ground when the cones open. Birds called Clark's nutcrackers eat the seeds of whitebark pine trees and store them in the ground (**Fig. 31.11**). Therefore, Clark's nutcrackers are critical seed dispersers for the trees. Also, grizzly bears find the stored seeds and consume them. Whitebark pine seeds do not germinate unless their seed coats are exposed to fire. When natural forest fires in the area are suppressed, whitebark pine trees decline in number, and so do Clark's nutcrackers and grizzly bears. When lightning-ignited fires are allowed to burn, or prescribed burning is used in the area, the whitebark pine populations increase, as do the populations of Clark's nutcrackers and grizzly bears.

Community Stability

As witnessed by the discussion of succession, community stability is fragile. However, some communities have one species that stabilizes the community, helps maintain its characteristics, and essentially helps hold its web of interactions together. Such a species is known as a **keystone species,** or species on which the existence of a large number of other species in the ecosystem depends. The term "keystone" comes from the name for the center stone of an arch that holds the other stones in place so that the arch can keep its shape.

Figure 31.10 **Mutualism.**

The bullhorn acacia tree is adapted to provide nourishment for a mutualistic ant species. **a.** The thorns are hollow, and the ants live inside. **b.** The bases of the leaves have nectaries (openings) where ants can feed. **c.** The tips of the leaves of the bullhorn acacia have Beltian bodies, which ants harvest for larval food.

Figure 31.11 **Interdependence of species.**

(**a**) The Clark's nutcracker feeds on the seeds of (**b**) the white bark pine. The nutcracker stores the seeds in the ground, and these seeds may be found and eaten by (**c**) the grizzly bear. Clark's nutcrackers and grizzly bears disperse the seed of the pine.

Keystone species are not always the most numerous in the community. However, the loss of a keystone species can lead to the extinction of other species and a loss of diversity. For example, bats are designated as keystone species in tropical forests. Bats are pollinators and also disperse the seeds of certain tropical trees. When bats are killed off or their roosts destroyed, these trees fail to reproduce. The grizzly bear is a keystone species in the northwestern United States and Canada. Grizzly bears disperse as many as 7,000 berry seeds in one dung pile. Grizzly bears also kill the young of many herbivorous mammals, such as deer, and thereby keep their populations under control.

The sea otter is a keystone species of a kelp forest ecosystem. Kelp forests, created by large brown seaweeds, provide a home for a vast assortment of organisms. The kelp forests occur just off the coast and protect coastline ecosystems from damaging wave action. Among other species, sea otters eat sea urchins, keeping their population size in check. Otherwise, sea urchins feed on the kelp, causing the kelp forest and its associated species to severely decline. Fishermen don't like sea otters because they also prey on abalone, a mollusc prized for its commercial value. Many do not realize that without the otters, abalone and many other species would not be around because their natural habitat, a kelp forest, would no longer exist.

Native Versus Exotic Species

Native species are species indigenous to an area. They colonize an area without intentional or accidental human assistance. For example, you naturally find maple trees in Vermont and many other states of the eastern United States. The introduction of **exotic species,** or nonnative species, into a community greatly disrupts normal interactions, and therefore a community's web of species. Populations of exotic species may grow exponentially because they are often better competitors, or because their population size is not controlled by their natural predators or diseases. The unique assemblage of native species on an island often cannot compete well against an exotic species. For example, myrtle trees, introduced into the Hawaiian Islands from the Canary Islands, are mutualistic with a type of bacterium capable of fixing atmospheric nitrogen. The bacterium provides the plant with a form of nitrogen it can utilize. This feature allows myrtle trees to become established on nitrogen-poor volcanic soil, a distinct advantage in Hawaii. Once established, myrtle trees halt the normal succession of native plants on volcanic soil.

Exotic species disrupt communities in continental areas as well. The red fox was deliberately imported into Australia to prey on the previously introduced European rabbit, but instead, the red fox has now reduced the populations of native small mammals (**Fig. 31.12**a). The brown tree snake was accidentally introduced onto a number of islands in the Pacific Ocean (Fig. 31.12b). The brown tree snake eats eggs, nestlings, and adult birds. On Guam, the snakes have reduced ten native bird species to the point of extinction. On the Galápagos Islands, black rats accidentally carried to the islands by ships have reduced populations of the giant tortoise. Goats and feral pigs have changed the vegetation on the islands from highland forest to pampas-like grasslands and destroyed stands of cacti. In the United States, gypsy moths, zebra mussels, the Chestnut blight fungus, fire ants, and African bees are well-known exotic species that have killed native species. At least two species, fire ants and African bees, have attacked humans with serious consequences.

a.

b.

Figure 31.12 Exotic species.

Human introduction of exotic species, such as (**a**) the red fox and (**b**) the brown tree snake, has disrupted communities in Australia and Guam, respectively.

Check Your Progress

1. List the five major types of species interactions in a community.
2. Distinguish between habitat and ecological niche.
3. Why is character displacement a form of resource partitioning?
4. Describe the role of a keystone species.

Answers: 1. Competition, predation, parasitism, commensalism, and mutualism. 2. A habitat is the special location of a species. It is one component of an ecological niche, which also includes the role the species plays in the community and its interactions with other species.   3. Due to character displacement, species are able to feed on different types of food.   4. A keystone species stabilizes the community, helps maintain its characteristics, and helps hold the web of interactions together.

Figure 31.13 Producers.

Green plants and algae are photoautotrophs.

a. Herbivores

b. Carnivores

Figure 31.14 Consumers.

a. Caterpillars and giraffes are herbivores. **b.** A praying mantis and an osprey are carnivores.

31.2 | Ecology of Ecosystems

An ecosystem is more inclusive than a community because community ecology only considers how species interact with one another. When studying ecosystem ecology, interactions with the physical environment are also considered. For example, one important aspect of ecological niche is how an organism acquires food. It is obvious that autotrophs interact with the physical environment but so do heterotrophs.

Autotrophs

Autotrophs take in only inorganic nutrients (e.g., CO_2 and minerals) and an outside energy source to produce organic nutrients for their own use and for all the other members of a community. They are called **producers** because they produce food. *Photoautotrophs*, often called photosynthetic organisms, produce most of the organic nutrients for the biosphere (**Fig. 31.13**). Algae of all types possess chlorophyll and carry on photosynthesis in freshwater and marine habitats. Algae make up the phytoplankton, which are photosynthesizing organisms suspended in water. Green plants are the dominant photosynthesizers on land. All photosynthesizing organisms release O_2 to the atmosphere.

Some bacteria are *chemoautotrophs*. They obtain energy by oxidizing inorganic compounds such as ammonia, nitrites, and sulfides, and they use this energy to synthesize organic compounds. Chemoautotrophs have been found to support communities in some caves and also at hydrothermal vents along deep-sea oceanic ridges.

Heterotrophs

Heterotrophs need a source of preformed organic nutrients, and they release CO_2 to the atmosphere. They are called **consumers** because they consume food. **Herbivores** are animals that graze directly on algae or plants (**Fig. 31.14***a*). In aquatic habitats, zooplankton, such as protozoans, are herbivores; in terrestrial habitats, insects, including caterpillars, play that role. Among larger herbivores, giraffes browse on trees. **Carnivores** eat other animals; for example, a praying mantis catches and eats caterpillars, and an osprey preys on and eats fish (Fig. 31.14*b*). These examples illustrate that there are primary consumers (e.g., giraffes), secondary consumers (e. g., praying mantis), and tertiary consumers (e.g., osprey). Sometimes tertiary consumers are called top predators. **Omnivores** are animals that eat both plants and animals. As you likely know, most humans are omnivores.

The **decomposers** are heterotrophic bacteria and fungi, such as molds and mushrooms, that break down dead organic matter, including animal wastes (**Fig. 31.15**). They perform a very valuable service because they release inorganic nutrients (CO_2 and minerals) that are then taken up by plants once more. Otherwise, plants would rely on minerals to be slowly released

Figure 31.15 Decomposers.

Mushrooms and bacteria are decomposers.

from rocks. **Detritus** is composed of the remains of dead organisms plus the bacteria and fungi that aid in decay. Fanworms feed on detritus floating in marine waters, while clams take detritus from the sea bottom. Earthworms and some beetles, termites, and maggots are soil detritus feeders.

Energy Flow and Chemical Cycling

The living components of ecosystems process energy and chemicals. Energy flow through an ecosystem begins when producers absorb solar energy, and chemical cycling begins when producers take in inorganic nutrients from the physical environment (**Fig. 31.16**). Thereafter, via photosynthesis, producers convert the solar energy and inorganic nutrients into chemical energy in the form of organic nutrients, such as carbohydrates. Producers synthesize organic nutrients directly for themselves and indirectly for the heterotrophic components of the ecosystem. Energy flows through an ecosystem because as organic nutrients pass from one component of the ecosystem to another, as when a herbivore eats a plant or a carnivore eats a herbivore, a portion is used as an energy source. Eventually, the energy dissipates into the environment as heat. Therefore, the vast majority of ecosystems cannot exist without a continual supply of solar energy.

Only a portion of the organic nutrients made by producers is passed on to consumers because plants use organic molecules to fuel their own cellular respiration. Similarly, only a small percentage of nutrients consumed by lower-level consumers, such as herbivores, is available to higher-level consumers, or carnivores. As **Figure 31.17** demonstrates, a certain amount of the food eaten by a herbivore is never digested and is eliminated as feces. Metabolic wastes are excreted as urine. Of the assimilated energy, a large portion is utilized during cellular respiration for the production of ATP and thereafter becomes heat. Only the remaining energy, which is converted into increased body weight or additional offspring, becomes available to carnivores.

The elimination of feces and urine by a heterotroph, and indeed the death of all organisms, does not mean that organic nutrients are lost to an ecosystem. Instead, they represent the organic nutrients made available to decomposers. Decomposers convert the organic nutrients back into inorganic chemicals and release them to the soil or atmosphere. Chemicals complete their cycle within an

Key:
energy
nutrients

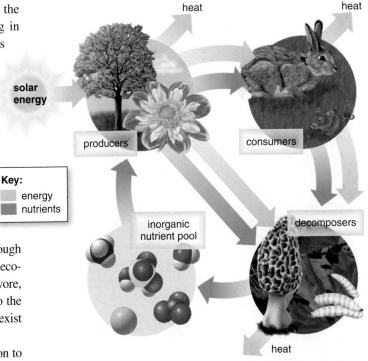

Figure 31.16   Chemical cycling and energy flow.

Chemicals cycle within, but energy flows through, an ecosystem. As energy is repeatedly passed from one component to another, all the chemical energy derived from solar energy dissipates as heat.

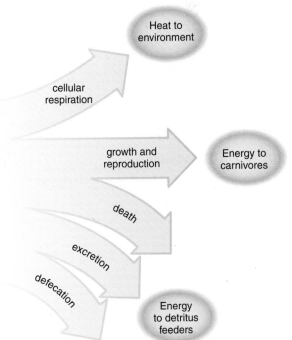

Figure 31.17   Energy balances.

Only about 10% of the nutrients and energy taken in by a herbivore is passed on to carnivores. A large portion goes to detritus feeders. Another large portion is used for cellular respiration.

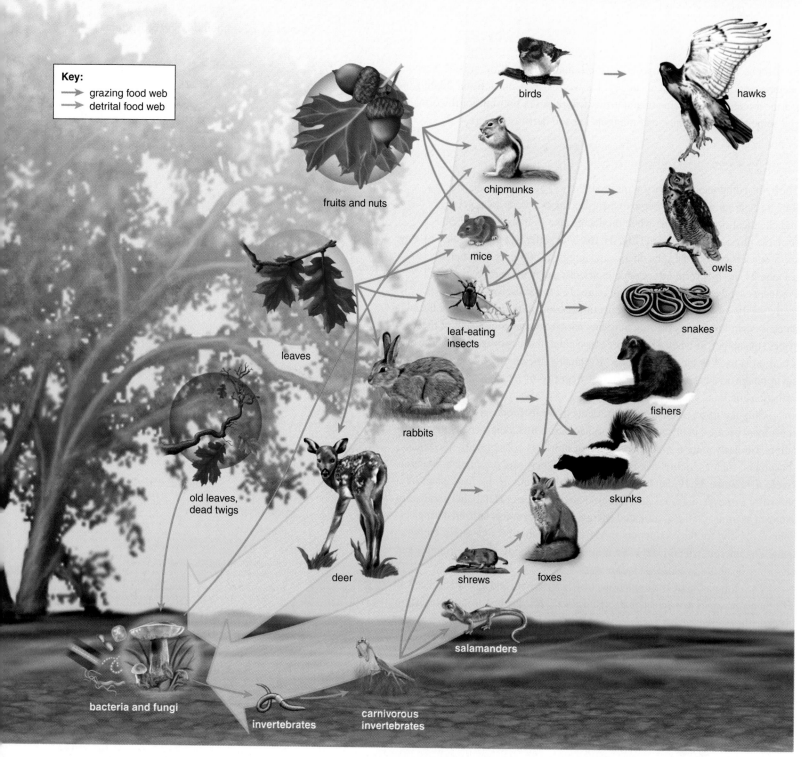

Key:
→ grazing food web
→ detrital food web

fruits and nuts

leaves

old leaves,
dead twigs

birds

chipmunks

mice

leaf-eating
insects

rabbits

deer

hawks

owls

snakes

fishers

skunks

shrews   foxes

salamanders

bacteria and fungi

invertebrates

carnivorous
invertebrates

Figure 31.18 Food webs.

The grazing and detrital food webs of ecosystems are linked.

ecosystem when inorganic chemicals are absorbed by the producers from the atmosphere or from soil.

Energy Flow

Applying the principles discussed so far to a temperate deciduous forest, ecologists can draw a **food web** to represent the interconnecting paths of energy flow between the component species of the ecosystem. In **Figure 31.18**, the green arrows are part of a **grazing food web** because the web begins with plants, such as the oak trees depicted. A **detrital food web** (brown arrows) begins with bacteria and fungi. In the grazing food web, caterpillars and other herbivorous insects feed on the leaves of the trees, while other herbivores, including mice, rabbits, and

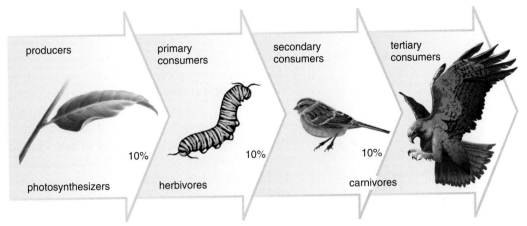

Figure 31.19   Food chain.

A food chain diagrams a single path of energy flow in an ecosystem. Ten percent of the energy goes toward growth and reproduction of each step in the path.

producers   primary consumers   secondary consumers   tertiary consumers

photosynthesizers   herbivores   carnivores

10%   10%   10%

deer, feed on leaves at or near the ground. Birds, chipmunks, and mice feed on fruits and nuts of the trees, but they are in fact omnivores because they also feed on caterpillars and other insects. These herbivores and omnivores all provide food for a number of different carnivores. In the detrital food web, detritus, which includes smaller decomposers such as bacteria and fungi, is food for larger organisms. Because some of these organisms, such as shrews and salamanders, become food for aboveground animals, the detrital and the grazing food webs are joined.

We tend to think that the aboveground parts of trees are the largest storage form of organic matter and energy, but this is not necessarily the case. In temperate deciduous forests, the organic matter lying on the forest floor and mixed into the soil, along with the belowground roots of the trees, contains over twice the energy of the leaves, branches, and trunks of living trees combined. Therefore, more energy and matter in a forest may be stored in or funneled through the detrital food web than the grazing food web.

Trophic Levels and Ecological Pyramids   The arrangement of component species in **Figure 31.19** suggests that organisms are linked to one another in a straight line according to feeding relationships, or who eats whom. Diagrams that show a single path of energy flow in an ecosystem are called **food chains** (Fig. 31.19). A **trophic level** is a level of nourishment within a food web or chain. In the grazing food web (see Fig. 31.18), going from left to right: The trees are producers (the first trophic level), the first series of animals are herbivores (the second trophic level), and many of the animals in the next series are carnivores (the third and possibly fourth trophic levels). Food chains are short because energy is lost between trophic levels. In general, only about 10% of the energy of one trophic level is available to the next trophic level. Therefore, if an herbivore population consumes 1,000 kg of plant material, only about 100 kg is converted to herbivore tissue, 10 kg to first-level carnivores, and 1 kg to second-level carnivores. The so-called 10% rule of thumb explains why few carnivores can be supported in a food web. The flow of energy with large losses between successive trophic levels is sometimes depicted as an **ecological pyramid** (Fig. 31.20).

A pyramid based on the number of organisms can run into problems because, for example, one tree can support many herbivores. Pyramids of **biomass,** which is the number of organisms multiplied by their weight, eliminate size as a factor. Even then, apparent inconsistencies can arise. In aquatic ecosystems, such as lakes and open seas where algae are the only producers, the herbivores at some point in time may have a greater biomass than the producers. Why? Because even though the algae reproduce rapidly, they are also consumed at a high rate.

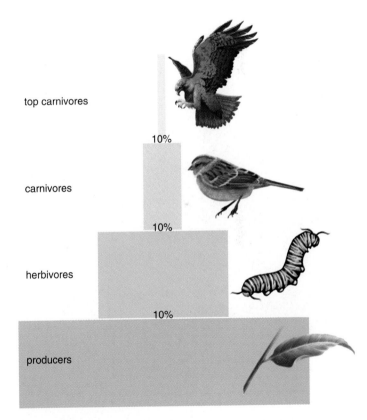

top carnivores

carnivores

herbivores

producers

10%

10%

10%

Figure 31.20   Ecological pyramid.

An ecological pyramid depicts the loss of nutrients and energy from one trophic level to the next.

Chemical Cycling

The pathways by which chemicals cycle within ecosystems involve both living (producers, consumers, decomposers) and nonliving (rock, inorganic nutrients, atmosphere) components, and therefore are known as **biogeochemical cycles.** Biogeochemical cycles can be sedimentary or gaseous. In a *sedimentary cycle,* such as the phosphorus cycle, the chemical is absorbed from the sediment by plant roots, passed to heterotrophs, and eventually returned to the soil by decomposers, usually in the same general area. In a *gaseous cycle,* such as the nitrogen and carbon cycles, the element returns to and is withdrawn from the atmosphere as a gas.

Chemical cycling of an element may involve **reservoirs** and exchange pools as well as the biotic community (**Fig. 31.21**). A *reservoir* is a source normally unavailable to organisms. For example, much carbon is found in calcium carbonate shells in ocean bottom sediments. An *exchange pool* is a source from which organisms generally take elements. For example, photosynthesizers can utilize carbon dioxide in the atmosphere for their carbon needs. The *biotic community* consists of the autotrophic and heterotrophic species of an ecosystem that feed on one another. Human activities, such as mining or burning fossil fuels, increase the amounts of chemical elements removed from reservoirs and cycling within ecosystems. As a result, the physical environment of the ecosystem contains excess chemicals that, in turn, may alter the species composition and diversity of the biotic community.

Check
Your **Progress**

1. **Contrast the two types of autotrophs.**
2. **Contrast herbivores with carnivores.**
3. **Contrast the first organisms in the grazing food web with those in the detrital food web.**
4. **Contrast a food chain with an ecological pyramid.**

Answers: 1. Photoautotrophs use photosynthesis to produce organic nutrients, while chemoautotrophs use energy from inorganic compounds to produce organic nutrients.   2. Herbivores are animals that feed on autotrophs, while carnivores feed on other animals.   3. The first organisms in the grazing food web are herbivorous insects, while those in the detrital food web are bacteria and fungi.   4. A food chain depicts a single path of energy flow in an ecosystem, while an ecological pyramid depicts the entire flow of energy between trophic levels.

Figure 31.21   Model for chemical cycling.

Chemical nutrients cycle between these components of ecosystems. Reservoirs, such as fossil fuels, minerals in rocks, and sediments in oceans, are normally relatively unavailable sources, but exchange pools, such as those in the atmosphere, soil, and water, are available sources of chemicals for the biotic community. When human activities (purple arrows) remove chemicals from reservoirs and make them available to the biotic community, pollution can result.

Phosphorus Cycle

On land, the very slow weathering of rocks fostered by fungi adds phosphates (PO_4^{2-} and HPO_4^{2-}) to the soil, some of which become available to terrestrial plants for uptake (**Fig. 31.22**). Phosphates made available by weathering also run off into aquatic ecosystems, where algae absorb phosphates from the water before they become trapped in sediments. Phosphates in sediments become available again only when a geological upheaval exposes sedimentary rocks to weathering once more.

Producers use phosphates in a variety of molecules, including phospholipids, ATP, and the nucleotides that become a part of DNA and RNA. Animals consume producers and incorporate some of the phosphates into teeth, bones, and shells that take many years to decompose. Decomposition of dead plant and animal material and animal wastes does, however, make phosphates available to producers at a faster rate than weathering. Because much of the available phosphates are utilized within food chains, phosphate is usually a limiting inorganic nutrient for ecosystems. In other words, the finite supply of phosphates limits plant growth, and therefore primary productivity.

The importance of phosphates and calcium to population growth is demonstrated by considering the fate of lemmings every four years (**Fig. 31.23**). You may have heard that lemmings dash mindlessly over cliffs into the sea; ecologists tell us that actually these lemmings are migrating to find food. What happened? Every four years or so, grasses and sedges of the tundra (see Fig. 31.28) become rich in minerals, and the lemming population starts to explode. Once the lemmings number in the millions, the grasses and sedges of the tundra suffer a decline caused by a lack of minerals. Now the lemming population suffers a crash, but it takes about four years before the animals decompose in this cold region and minerals return to the producers. Then the cycle begins again.

Human Activities   A **transfer rate** is defined as the amount of a nutrient that moves from one component of the environment to another within a specified period of time. Human activities affect the dynamics of a community by altering transfer rates. For example, humans mine phosphate ores and use them to make fertilizers, animal feed supplements, and detergents. Phosphate ores are slightly radioactive, and therefore mining phosphate poses a health threat to all organisms, including the miners. Animal wastes from livestock feedlots, fertilizers from lawns and cropland, and untreated and treated sewage discharged from cities all add excess phosphates to nearby waters. The result is **eutrophication,** or overenrichment, of a body of water, which causes an algal overpopulation called an algal bloom. When the algae die and decay, oxygen is consumed, causing fish kills. In the mid-1970s, Lake Erie was dying because of eutrophication. Control of nutrient phosphates, particularly in sewage effluent and household detergents, reversed the situation.

Figure 31.22   The phosphorus cycle.

The phosphorus cycle is a sedimentary biogeochemical cycle. Globally, phosphates flow into large bodies of water and become a part of sedimentary rocks. Thousands or millions of years later, the seafloor can rise; the phosphates are then exposed to weathering and become available. Locally, phosphates cycle within a community when plants on land and algae in the water take them up. Animals gain phosphates when they feed on plants or algae. Decomposers return phosphates to plants or algae, and the cycle within the community begins again.

Figure 31.23   Lemmings.

Lemmings, small rodents, feed on plant species in an Arctic community called the tundra.

Nitrogen Cycle

Nitrogen, in the form of nitrogen gas (N_2), comprises about 78% of the atmosphere by volume. But plants cannot make use of nitrogen gas. Instead, plants rely on various types of bacteria to make nitrogen available to them. Therefore, nitrogen, like phosphorus, is also a limiting inorganic nutrient of producers in ecosystems.

Plants can take up both ammonium (NH_4^+) and nitrate (NO_3^-) from the soil and incorporate the nitrogen into amino acids and nucleic acids. Two processes, nitrogen fixation and nitrification, convert nitrogen gas, N_2, into NH_4^+ and NO_3^-, respectively (**Fig. 31.24**). *Nitrogen fixation* occurs when nitrogen gas is converted to ammonium. Some cyanobacteria in aquatic ecosystems and some free-living, nitrogen-fixing bacteria in soil are able to fix nitrogen in this way. Other nitrogen-fixing bacteria live in nodules on the roots of legumes, plants such as peas, beans, and alfalfa. They make organic compounds containing nitrogen available to the host plant.

Nitrification is the production of nitrates. Ammonium in the soil is converted to nitrate by certain nitrifying soil bacteria in a two-step process. First, nitrite-producing bacteria convert ammonium to nitrite (NO_2^-), and then nitrate-producing bacteria convert nitrite to nitrate. In Figure 31.24, notice that the biotic community subcycle in the nitrogen cycle does not depend on the presence of nitrogen gas.

Denitrification is the conversion of nitrate to nitrogen gas, which enters the atmosphere. Denitrifying bacteria are chemoautotrophs living in the anaerobic mud of lakes, bogs, and estuaries that carry out this process as a part of their own metabolism. In the nitrogen cycle, denitrification counterbalanced nitrogen fixation until humans started making fertilizer.

Human Activities   Human activities significantly alter the transfer rates in the nitrogen cycle by producing fertilizers from N_2—in fact, humans nearly double the fixation rate. The nitrate in fertilizers, just like phosphate, can leach out of agricultural soils into surface waters, leading to eutrophication. Deforestation by humans also causes a loss of nitrogen to groundwater and makes regrowth of the forest difficult. The underground water supplies in farming areas today are more apt to contain excess nitrate. A high concentration of nitrates interferes with blood oxygen levels, and infants below the age of six months who drink water containing excessive amounts of nitrate can become seriously ill and, if untreated, may die.

To cut back on fertilizer use, it might be possible to genetically engineer soil bacteria with increased nitrogen fixation rates. Also, farmers could grow legumes that increase the nitrogen content of soil (**Fig. 31.25**). In one study, the rotation of legumes and winter wheat produced a better yield than fertilizers after several years.

Figure 31.24   The nitrogen cycle.

The nitrogen cycle is a gaseous biogeochemical cycle normally maintained by the work of several populations of soil bacteria.

Key:
- human activities
- natural events

N_2 in atmosphere

proteins in animals and plants

human activities

plants

nitrogen-fixing bacteria in nodules and soil

denitrifying bacteria

dead organisms and animal waste

Biotic community

nitrates (NO_3^-)

decomposers

nitrifying bacteria

ammonium (NH_4^+)

Figure 31.25   Root nodules.

Bacteria that live in nodules on the roots of plants in the legume family, such as pea plants, convert nitrogen in the air to a form that plants can use to make proteins.

Carbon Cycle

In the carbon cycle, organisms in both terrestrial and aquatic eco-systems exchange carbon dioxide with the atmosphere (**Fig. 31.26**). On land, plants take up carbon dioxide from the air, and through photosynthesis, they incorporate carbon into organic nutrients that are used by auto-trophs and heterotrophs alike. When aerobic organ-isms respire, a portion of this carbon is returned to the atmosphere as carbon dioxide, a waste product of cellular respiration.

In aquatic ecosystems, the exchange of carbon dioxide with the atmosphere is indirect. Carbon dioxide from the air combines with water to produce bicarbonate ion (HCO_3^-), a source of carbon for algae that produce food for themselves and for heterotrophs. Similarly, when aquatic organisms respire, the carbon dioxide they give off becomes bicarbonate ion. The amount of bicarbon-ate in the water is in equilibrium with the amount of carbon dioxide in the air.

Living and dead organisms contain organic carbon and serve as one of the reservoirs for the carbon cycle. The world's biotic components, particularly trees, contain billions of tons of organic carbon, and additional tons are estimated to be held in the remains of plants and animals in the soil. If dead plant and animal remains fail to decompose, they are sub-jected to extremely slow physical processes that transform them into coal, oil, and natural gas, the **fossil fuels.** Most of the fossil fuels were formed during the Carboniferous period, 286–360 MYA, when an exceptionally large amount of organic matter was buried before decomposing. Another reservoir is the calcium carbonate ($CaCO_3$) that accumulates in limestone and in shells. Many marine organisms have calcium carbonate shells that remain in bottom sedi-ments long after the organisms have died. Geological forces change these sedi-ments into limestone.

Human Activities   The transfer rates of carbon dioxide due to photosynthesis and cellular respiration are just about even. However, more carbon dioxide is being deposited in the atmosphere than is being removed. This increase is largely due to the burning of fossil fuels and the destruc-tion of forests to make way for farmland and pasture. Becasue of fossil fuel consumption and deforestation, CO_2 in the atmosphere is predicted to rise from 0.36% today to 0.75% by 2100. CO_2 concen-trations in the atmosphere would rise higher but the oceans take up CO_2, and so far they have been taking up more CO_2 than they vent.

The increased amount of carbon dioxide (and other gases) in the atmosphere are predicted to cause a rise in temperature called **global warming.** These gases allow the sun's rays to pass through, but they absorb and reradiate heat back to the Earth, a phenomenon called the **greenhouse effect.** Scientists predict that if the ice at the poles melts and sea levels rise as a result, many of the world's most populous cities will be flooded. Furthermore, weather pattern changes might cause the American Midwest to become a dust bowl.

Figure 31.26   The carbon cycle.

The carbon cycle is a gaseous biogeochemical cycle. Producers take in carbon dioxide from the atmosphere and convert it to organic molecules that feed all organisms. Fossil fuels arise when organisms die but do not decompose. The burning of fossil fuels releases carbon dioxide and causes environmental pollution.

Check Your **Progress**

1. Describe how phosphate enters the phosphorus cycle.
2. Describe the significance of nitrogen fixation and nitrification to the nitrogen cycle.
3. Explain how removing forests may affect the amount of atmospheric carbon dioxide.
4. Contrast a sedimentary biogeochemical cycle with a gaseous one.

Answers: 1. Most phosphate comes from decomposition of organisms, but some comes from weathering.   2. Both processes allow nitrogen to enter the nitrogen cycle.   3. Plants take in atmospheric carbon dioxide so the removal of trees reduces the amount of CO_2 removed from the atmosphere.   4. A sedimentary cycle involves the exchange of chemicals in the soil, while a gaseous cycle involves chemicals in the atmosphere.

31.3 | Ecology of Major Ecosystems

The **biosphere,** which encompasses all the ecosystems on planet Earth, is the final level of biological organization. **Aquatic ecosystems** are divided into those composed of fresh water and those composed of salt water (marine ecosystems) (**Fig. 31.27**). The ocean is a marine ecosystem that covers 70% of the Earth's surface. Two types of freshwater ecosystems are those with standing water, such as lakes and ponds, and those with running water, such as rivers and streams. The richest marine ecosystems lie near the coasts. Coral reefs are located offshore, while marshes occur where rivers meet the sea.

Scientists recognize several distinctive major types of **terrestrial ecosystems,** also called *biomes* (**Fig. 31.28**). Temperature and rainfall define the biomes, which contain communities adapted to the regional climate. The tropical rain forests, which occur at the equator, have a high average temperature and the greatest amount of rainfall of all the biomes. They are dominated by large evergreen, broad-leaved trees. The savanna is a tropical grassland with a high temperature and alternating wet and dry seasons. Temperate grasslands receive less rainfall than temperate forests (in which trees lose their leaves during the winter) and more water than deserts, which lack trees. The taiga is a very cold northern coniferous forest, and the tundra, which borders the North Pole, is also very cold, with long winters and a short growing season. A permafrost persists even during the summer in the tundra and prevents large plants from becoming established.

Figure 31.27   The major aquatic ecosystems.

Aquatic ecosystems are divided into those that have salt water, such as (**a**) the ocean and (**b**) those that have fresh water, such as a river. Saltwater, or marine, ecosystems also include (**c**) coral reefs and (**d**) marshes.

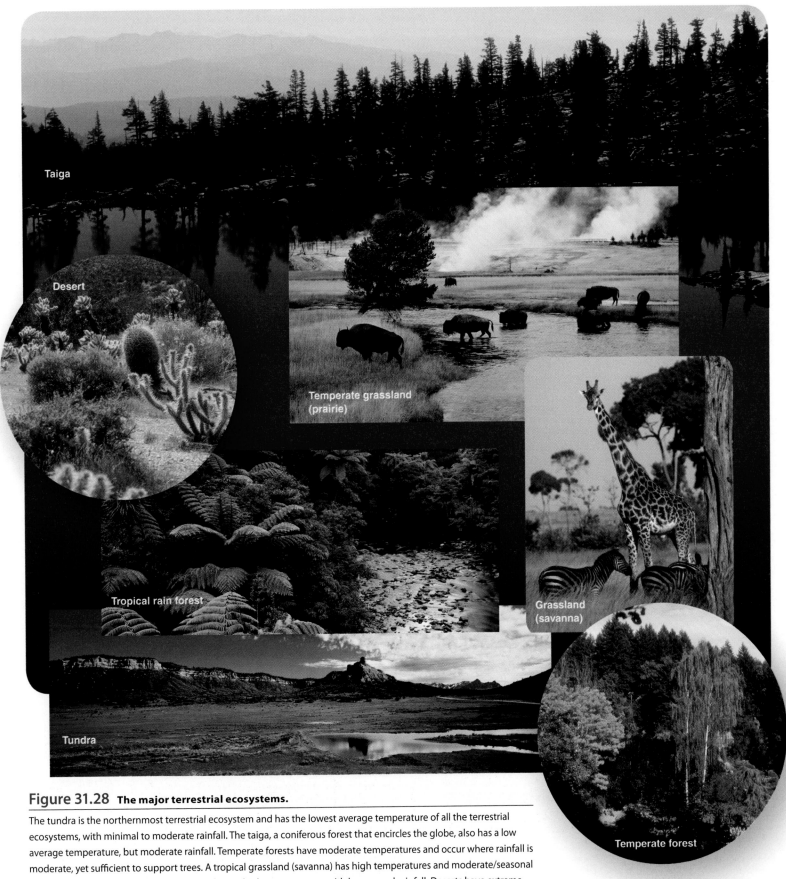

Figure 31.28 The major terrestrial ecosystems.

The tundra is the northernmost terrestrial ecosystem and has the lowest average temperature of all the terrestrial ecosystems, with minimal to moderate rainfall. The taiga, a coniferous forest that encircles the globe, also has a low average temperature, but moderate rainfall. Temperate forests have moderate temperatures and occur where rainfall is moderate, yet sufficient to support trees. A tropical grassland (savanna) has high temperatures and moderate/seasonal rainfall. A temperate grassland (prairie) has low to high temperatures, with low annual rainfall. Deserts have extreme and changeable temperatures with minimal rainfall. Tropical rain forests, which generally occur near the equator, have a high average temperature and the greatest amount of rainfall of all the terrestrial ecosystems.

Primary Productivity

One way to compare ecosystems is based on **primary productivity,** the rate at which producers capture and store energy as organic nutrients over a certain length of time. Temperature and moisture, and secondarily the nature of the soil, influence the primary productivity and, as already discussed, the assemblage of species in an ecosystem. In terrestrial ecosystems, primary productivity is generally lowest in high-latitude tundras and deserts, and highest at the equator where tropical forests occur (**Fig. 31.29**). The high productivity of tropical rain forests provides varied niches and much food for consumers. The number and diversity of species in tropical rain forests are the highest of all the terrestrial ecosystems. Therefore, conservation biologists are interested in preserving as much of this biome as possible.

The primary productivity of aquatic communities is largely dependent on the availability of inorganic nutrients. Estuaries, swamps, and marshes are rich in organic nutrients and in decomposers that convert those organic nutrients into their inorganic chemical components. Estuaries, swamps, and marshes also contain a large number of varied species, particularly in the early stages of their development before they venture forth into the sea. Therefore, all of these coastal regions are in great need of preservation. The open ocean has a productivity somewhere between that of a desert and the tundra because it lacks a concentrated supply of inorganic nutrients. Coral reefs exist near the coasts in warm tropical waters where currents and waves bring nutrients and where sunlight penetrates to the ocean floor. Coral reefs are areas of remarkable biological abundance, equivalent to that of tropical rain forests.

Check Your Progress

1. Describe the two major types of ecosystems of the biosphere.
2. List the terrestrial ecosystems of the world.
3. Explain why swamps have higher levels of primary productivity than open oceans.

Answers: 1. The biosphere is divided into aquatic ecosystems and terrestrial ecosystems. 2. Taiga, savanna, prairie, temperate forest, desert, tropical rain forest, and tundra.   3. Swamps have a more concentrated supply of nutrients and decomposers.

Figure 31.29   Primary productivity.

Ecologists can compare ecosystems based on primary productivity, the rate at which producers convert and store solar energy as chemical energy.

The Chapter in Review

Summary

31.1 | Ecology of Communities

Knowledge of community and ecosystem ecology is important for understanding the impacts of human alterations to the environment.

Community A community is an assemblage of the populations of different species interacting with each other in a given area.

Ecosystem An ecosystem consists of species interacting with one another and with the physical environment.

Ecological Succession

bare rock ⟶ lichens/mosses ⟶ grasses ⟶ shrubs ⟶ trees

The two types of ecological succession are primary succession (begins on bare rock) and secondary succession (following a disturbance; begins where soil is present). Ecological succession leads to a stable climax community.

Interactions in Communities

Species in communities interact with one another in the following ways:

- **Competition** Species vie with one another for resources, such as light, space, and nutrients. Aspects of competition are the competitive exclusion principle, resource partitioning, and character displacement.
- **Predation** One species (predator) eats another species (prey).
- **Parasitism** One species (parasite) obtains nutrients from another species (host) but does not kill the host species.
- **Commensalism** One species benefits from the relationship, while the other species is not harmed.
- **Mutualism** Two species interact in a way that benefits both.

Ecological Niche The ecological niche of a species is defined by the role it plays in its community, the habitat, and its interactions with other species.

Keystone Species The interactions of a keystone species in the community hold the community and its species together. Removal of a keystone species can lead to species extinctions and loss of diversity. An example of a keystone species is the grizzly bear.

Native Versus Exotic Species Native species are indigenous to a given area and thrive without assistance. Exotic species are introduced into an area, and may disrupt the balance and interactions among native species in that area's community.

31.2 | Ecology of Ecosystems

In the food chain of an ecosystem, some populations are autotrophs and some are heterotrophs.

Autotrophs

The autotrophs are the producers. They require only inorganic nutrients (e. g., CO_2 and minerals) and an outside energy source to produce organic nutrients for their own use and for the use of other members of the community. Examples of autotrophs are algae, cyanobacteria, and plants.

Heterotrophs

The heterotrophs are the consumers. They require a preformed source of organic nutrients and give off CO_2. Examples of consumers are herbivores (feed on plants), carnivores (feed on animals), and omnivores (feed on both plants and animals). Other heterotrophs are the decomposers (the bacteria and fungi that aid in decay).

Energy Flow and Chemical Cycling

Energy flows through an ecosystem, while chemicals cycle within an ecosystem.

Food Webs and Food Chains Energy flows in an ecosystem through food chains and detrital and grazing food webs.

Trophic Level A trophic level is a level of nourishment in a food web or chain.

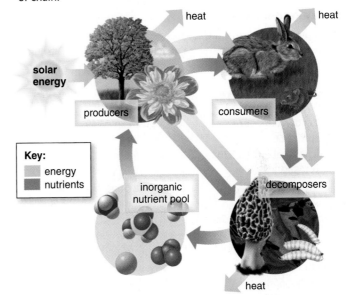

Ecological Pyramid An ecological pyramid illustrates the energy losses that occur between trophic levels.
- Only about 10% of the energy of one trophic level is available to the next trophic level.
- Top carnivores occupy the last and smallest trophic level.

Biogeochemical Cycle Chemicals cycle within an ecosystem through various biogeochemical cycles, such as the phosphorus cycle, the nitrogen cycle, and the carbon cycle. Human activities significantly alter the transfer rates in these cycles.

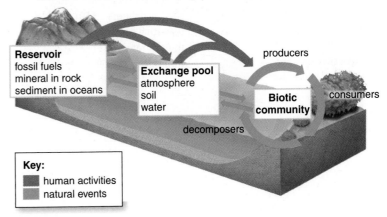

31.3 | Ecology of Major Ecosystems

The biosphere encompasses all the major ecosystems of the Earth.

Aquatic Ecosystems The aquatic ecosystems are classified as freshwater ecosystems (rivers, streams, lakes, ponds) and marine ecosystems (oceans, coral reefs, saltwater marshes).

Terrestrial Ecosystems The terrestrial ecosystems are called biomes. The major biomes are:
- Tundra
- Taiga
- Temperate forests
- Tropical grasslands (savanna)
- Temperate grasslands (prairie)
- Deserts
- Tropical rain forests

Primary Productivity

Primary productivity is the rate at which producers capture solar energy and convert it to chemical energy over a specified length of time. The number of species in an ecosystem is positively related to its primary productivity.

Key Terms

aquatic ecosystem   574
autotroph   566
biogeochemical cycle   570
biomass   569
biosphere   574
carnivore   566
character displacement   563
climax community   560
coevolution   558
commensalism   562
community   558
competition   562
competitive exclusion principle   563
consumer   566
decomposer   566
detrital food web   568
detritus   567
diversity   559
ecological niche   562
ecological pyramid   569
ecological succession   560
ecosystem   559
eutrophication   571
exotic species   565

food chain   569
food web   568
fossil fuel   573
global warming   573
grazing food web   568
greenhouse effect   573
habitat   562
herbivore   566
heterotroph   566
keystone species   564
mutualism   562
native species   565
omnivore   566
parasitism   562
predation   562
primary productivity   576
primary succession   560
producer   566
reservoir   570
resource partitioning   563
secondary succession   560
species richness   559
terrestrial ecosystem   574
transfer rate   571
trophic level   569

Testing Yourself

Choose the best answer for each question.

1. As diversity increases,
   a. species richness increases, and distribution of species becomes more even.
   b. species richness decreases, and distribution of species becomes more even.
   c. species richness increases, and distribution of species becomes less even.
   d. species richness decreases, and distribution of species becomes less even.

2. Which is not an adaptive trait of an opportunistic pioneer species?
   a. long life span   c. small size
   b. short time to maturity   d. high reproductive output

For statements 3–7, indicate the type of interaction described in each scenario.

Key:
   a. competition
   b. predation
   c. parasitism
   d. commensalism
   e. mutualism

3. An alfalfa plant gains fixed nitrogen from the bacterial species *Rhizobium* in its root system, while *Rhizobium* gains carbohydrates from the plant.

4. Both foxes and coyotes in an area feed primarily on a limited supply of rabbits.

5. Roundworms live and reproduce within a cat's digestive tract.

6. A fungus captures nematodes as a food source.

7. An orchid plant lives in the treetops, gaining access to sun and pollinators, but not harming the trees.

8. The abundance of both species is expected to increase as a result of which type of interaction?
   a. predation   d. competition
   b. commensalism   e. parasitism
   c. mutualism

9. According to the competitive exclusion principle,
   a. one species is always more competitive than another for a particular food source.
   b. competition excludes multiple species from using the same food source.
   c. no two species can occupy the same niche at the same time.
   d. competition limits the reproductive capacity of species.

10. Fungi are examples of
   a. autotrophs.   d. omnivores.
   b. herbivores.   e. decomposers.
   c. carnivores.

11. In the following diagram, fill in the components of chemical cycling and nutrient flow.

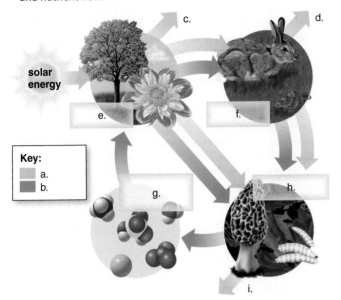

Key:
- a.
- b.

12. An ecological pyramid depicts the amount of _____ in various trophic levels.
    - a. food
    - b. organisms
    - c. energy
    - d. nutrients
    - e. both b and c are correct.

13. Which of the following would be a primary consumer in a vegetable garden?
    - a. aphid sucking sap from cucumber leaves
    - b. lady beetle eating aphids
    - c. songbird eating lady beetles
    - d. fox eating songbirds
    - e. All of these are correct.

14. Detritus always contains
    - a. bacteria and fungi.
    - b. leaf litter.
    - c. decaying logs.
    - d. animal carcasses.

15. Producers are
    - a. autotrophs.
    - b. herbivores.
    - c. omnivores.
    - d. carnivores.
    - e. Both a and b are correct.

16. The first trophic level in a food web is occupied by the
    - a. producers.
    - b. primary consumers.
    - c. secondary consumers.
    - d. tertiary consumers.

17. Which of the following represents a grazing food chain?
    - a. leaves → detritus feeders → deer → owls
    - b. birds → mice → snakes
    - c. nuts → leaf-eating insects → chipmunks → hawks
    - d. leaves → leaf-eating insects → mice → snakes

18. In a grazing food web, carnivores that eat herbivores are
    - a. producers.
    - b. primary consumers.
    - c. secondary consumers.
    - d. tertiary consumers.

19. Identify the components of the ecological pyramid in the following diagram.

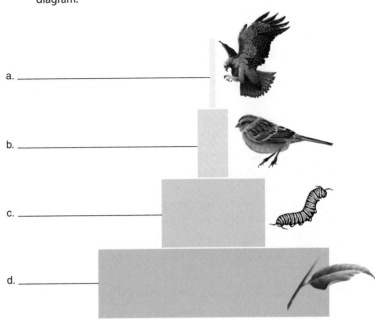

a. _____

b. _____

c. _____

d. _____

20. Minerals in rocks are considered members of this component of an ecosystem.
    - a. detritus
    - b. community
    - c. reservoir
    - d. decomposers

21. Which of the following is a sedimentary biogeochemical cycle?
    - a. carbon
    - b. nitrogen
    - c. phosphorus

22. Underground oil is an example of a carbon
    - a. cycle.
    - b. pathway.
    - c. reservoir.
    - d. exchange pool.

For questions 23–25, match the description to the process in the key.

Key:
- a. nitrogen fixation
- b. nitrification
- c. denitrification

23. nitrate to nitrogen gas

24. nitrogen gas to nitrate

25. nitrogen gas to ammonium

26. Which of the following is not a component of the nitrogen cycle?
    - a. proteins
    - b. ammonium
    - c. decomposers
    - d. photosynthesis
    - e. bacteria in root nodules

27. Which biome is characterized by a coniferous forest with low average temperature and moderate rainfall?
    - a. taiga
    - b. savanna
    - c. tundra
    - d. tropical rain forest
    - e. temperate forest

28. Which biome has the lowest primary productivity?
    - a. tundra
    - b. lake
    - c. sandy beach
    - d. prairie
    - e. temperate forest

For questions 29–34, match the description to the biome in the key.

Key:

    a. tundra

    b. taiga

    c. tropical rain forest

    d. temperate grassland

    e. tropical grassland

    f. desert

29. very cold northern coniferous forest

30. prairie

31. northernmost terrestrial ecosystem; persisting permafrost

32. changeable temperatures with minimal rainfall; low primary productivity

33. occurs near the equator; high temperatures, large amount of rainfall; high primary productivity

34. savanna, alternating wet and dry seasons

For questions 35–38, match the description to the type of aquatic ecosystem in the key.

Key:

    a. deep ocean waters

    b. marshes

    c. rivers, lakes

    d. coral reefs

35. lie near the coast; high primary productivity

36. lowest primary productivity

37. freshwater ecosystems

38. lie offshore; high primary productivity

39. Ecosystems include which of the following component that communities do not?

    a. energy transfer between members

    b. interaction with physical environment

    c. intricate food webs

    d. herbivores and carnivores

40. Character displacement evolves in response to

    a. predation.

    b. competiton among species for a single niche.

    c. loss of keystone species.

    d. primary succession.

Thinking Scientifically

1. One of the most striking examples of coevolution is between insects and flowers. The earliest angiosperms produced wind-pollinated flowers, which released large quantities of pollen. The ovules were partially exposed and exuded tiny droplets of sugary sap to catch passing pollen. Outline a course of events that could have resulted in the coevolution we observe today between a flower and its pollinator.

2. Over 200 wildlife species have been associated with prairie dog colonies in the Great Plains of the United States. The prairie dog, which burrows, forages, and feeds in the area, acts as a keystone species. If the prairie dogs are destroyed, many other species will suffer as well. How do you think the activities of the prairie dogs influence the survival of so many other species?

Bioethical Issue

Protection of Native Species

Many exotic species, such as zebra mussels and sea lampreys, are so obviously troublesome that most people do not object to programs aimed at controlling their populations. However, some exotic species' eradication programs meet with more resistance. For example, the mute swan, one of the world's largest flying birds, is beautiful, graceful, and makes an impressive presence. However, it is very aggressive and territorial. The mute swan was introduced to the United States from Asia and Europe in the nineteenth century as an ornamental bird but has since established a large wild population. The birds consume large amounts of aquatic vegetation and displace native birds from feeding and nesting areas. The U.S. Fish and Wildlife Service explains that it will be necessary to kill 3,000 mute swans in Maryland in the next two years in order to protect native bird populations. Attempts to limit the size of the mute swan populations in Maryland and other states have been met with opposition by citizens who find the birds beautiful.

Do you feel that native populations need not be protected as long as the exotic species serves a suitable human purpose? Or, do you feel that native species should be protected regardless?

Essentials of Biology Website

The companion website for *Essentials of Biology* provides a wealth of information organized and integrated by chapter. You will find practice tests, animations, videos, and much more that will complement your learning and understanding of general biology.

http://www.mhhe.com/maderessentials2

The Earth loses 4,000–6,000 species due to rain forest deforestation each year.

The Earth loses 12–17 million acres of farmland due to erosion each year.

The creation of a successful, self-sustaining ecosystem has been attempted.

32

Human Impact on the Biosphere

Outline

Ecosystems Under Glass

BETWEEN 1987 AND 1989, scientists attempted to develop a sustainable ecosystem from scratch on 1,650 acres of land in Arizona. This incredibly expensive Biosphere 2 project consisted of a series of airtight domes designed to house thousands of plant and animal species, including several humans, for two years. Some of the early supporters were interested in the project's potential for the future creation of

—Continued next page

Continued—

closed systems capable of supporting colonies of people on the moon or on Mars.

After careful planning and construction, Biosphere 2 was sealed, and the experiment began. For two years, Biosphere 2 recycled 100% of its wastes and water without any buildup of harmful chemicals in the atmosphere, and the rain forest more than doubled its size. Some unexpected results were documented as well, such as invasive plant and animal species destroying other species and low levels of photosynthesis due to cloudy weather producing high levels of carbon dioxide and decreasing levels of oxygen. Most of the Biosphere 2 experimenters concluded that in order to have a sustainable ecosystem, there is much more to learn about the abiotic and biotic factors that hang in delicate balance.

Today, it's now possible to visit Biosphere 2 and tour its rain forest, grassland, desert, marsh, and marine ecosystems. Visitors gain both an appreciation for the Earth's ecosystems and an understanding of the choices people must make in order to manage the planet well. In this chapter, you will learn about how humans negatively impact ecosystems, and some of the potential solutions that will help us move toward a sustainable society.

32.1 | Resources and Pollution

Human beings have certain basic needs, and they use resources to meet these needs. Land, water, food, energy, and minerals are the maximally used resources (**Fig. 32.1**).

Some resources are nonrenewable, and some are renewable. **Nonrenewable resources** are limited in supply. For example, the amount of land, fossil fuels, and minerals is finite and can be exhausted. Better extraction methods and efficient use can make the supply last longer, but eventually these resources will run out. **Renewable resources** are not limited in supply. We can use water and certain forms of energy (e.g., solar energy) or harvest plants and animals for food, and more supply will always be forthcoming. However, even though these resources are renewable, we must be careful not to squander them. Consider, for example, that most species have a population threshold below which they cannot recover or sustain their numbers, as when the huge herds of buffalo that once roamed western United States disappeared after being overexploited.

Unfortunately, a side effect of resource consumption can be pollution. **Pollution** is any alteration of the environment in an undesirable way. Pollution is often caused by human activities. The impact of humans on the environment is proportional to the size of the population. As the population grows, so does the need for resources and the amount of pollution generated by using these resources. Consider that six people adding waste to the ocean might not be alarming, but 6 billion people doing so would certainly affect its cleanliness. In modern times, the consumption of mineral and energy resources has grown faster than population size, most likely because people in the less-developed countries (LDCs) have increased their use of them.

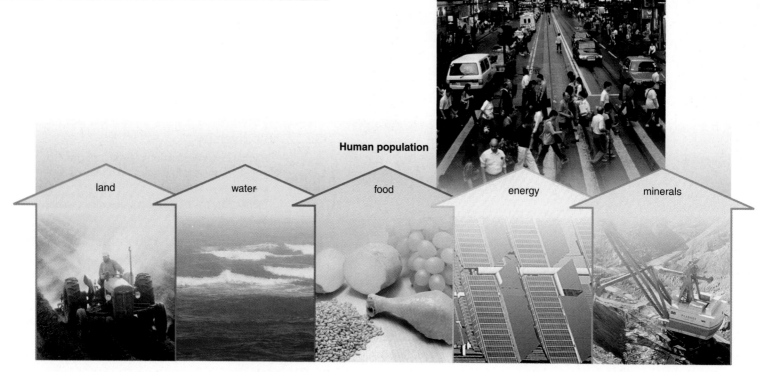

Figure 32.1 Resources.

Humans use land, water, food, energy, and minerals to meet their needs, including a place to live, food to eat, and products that make their lives easier.

Land

People require a physical place to live. Worldwide, there are currently more than 32 people for each km² (83 people per mi²) of all available land, including Antarctica, mountain ranges, jungles, and deserts. Naturally, land is also needed for a variety of uses aside from homes, such as agriculture, electric power plants, manufacturing plants, highways, hospitals, schools, and so on.

Beaches and Human Habitation

At least 40% of the world population lives within 100 km (60 mi) of a coastline, and this number is expected to increase. In the United States today, over one-half of the population lives within 80 km (50 mi) of the coasts (including the Great Lakes). Living right on the coast is an unfortunate choice because it accelerates natural beach erosion and loss of habitat for marine organisms.

Beach Erosion   An estimated 70% of the world's beaches are eroding; **Figure 32.2** shows how extensive this issue is in the United States. The seas have been rising for the past 12,000 years, ever since the climate turned warmer after the last Ice Age and the forces associated with waves, tides, and wind have consequently been amplified.

Humans often participate in activities that divert more water to the oceans, contributing to rising seas and beach erosion. For example, humans have filled in coastal wetlands, such as mangrove swamps in the southern United States and saltwater marshes in the northern United States formerly considered 'wastelands.' With growing recognition of the services provided by wetlands, wetland conservation has been growing in the United States during the past 40 years, but it is just starting in South America, where a project to straighten the Paraná River will drain the world's largest wetland. Reasons to protect coastal wetlands are that they serve as spawning areas for fish and other forms of marine life as well as buffer against hurricane storm surges and protect against shoreline erosion. They are also habitats for certain terrestrial species, including many types of birds.

Humans often try to stabilize beaches by building groynes (structures that extend from the beach into the water) and seawalls. Groynes trap sand and build beaches on one side as well as slowing the longshore currents, but erosion is increased on the other side. Seawalls, in the end, also increase erosion because ocean waves remove sand from in front of and to the side of the seawalls. Importing sand is a better solution, but it is very costly and can disturb plant and animal populations. It's estimated that today, the U.S. shoreline loses 40% more sediment than it receives, because the building of dams prevents sediment from reaching the coast.

Coastal Pollution   The coasts are particularly subject to pollution because toxic substances placed in freshwater lakes, rivers, and streams may eventually find their way to the coast. Oil spills at sea cause localized harmful effects also.

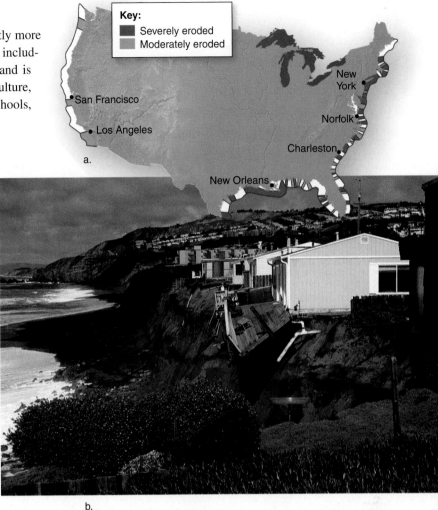

a.

b.

Figure 32.2   Beach erosion.

a. Most of the U.S. coastline is subject to beach erosion. **b.** Therefore, people who choose to live near the coast may eventually lose their homes.

Check Your Progress

1. **Distinguish between renewable and nonrenewable resources.**
2. **Describe two problems related to human habitation near oceans.**

Answers: 1. Renewable resources are not limited in supply, while nonrenewable resources are. 2. Beach erosion occurs due to land development on the shore, destruction of wetlands, and coastal pollution occurs due to human activities in the vicinity of the water.

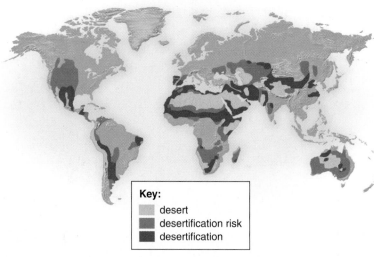

Figure 32.3 Desertification.

Desertification is a worldwide occurrence that reduces the amount of land suitable for human habitation.

Source: Data from A. Goudie and I. Wilkinson, *The Warm Desert Environment*. Copyright 1977 by Cambridge University Press, New York.

Figure 32.4 Deforestation.

a. Nearly half of the world's forest lands have been cleared for farming, logging, and urbanization. **b.** The soil of tropical rain forests is not suitable for long-term farming.

Sources: United Nations Environment Program, World Resources Institute.

Semiarid Lands and Human Habitation

Forty percent of the Earth's lands are already deserts, and any land adjacent to a desert is in danger of becoming desert if it is improperly managed by humans. **Desertification** is the conversion of semiarid land to desert-like conditions (**Fig. 32.3**).

Quite often, desertification begins when humans allow animals to overgraze the land. The soil can no longer hold rainwater, which runs off instead of keeping the remaining plants alive or recharging wells. Humans then remove whatever vegetation they can find to use as fuel or fodder for their animals. The end result is a desert unable to support agriculture, which is then abandoned as people move on to continue the process someplace else. Some estimate that nearly three-quarters of all rangelands worldwide are in danger of desertification. The famine in Ethiopia during 2002–2003 was due, at least in part, to degradation of the land to the point that it could no longer support human beings and their livestock.

Tropical Rain Forest and Human Habitation

Deforestation, the removal of trees, has long allowed humans to live in areas where forests once covered the land. The concern recently has been that people are settling in tropical rain forests, such as the Amazon, following the building of roads (**Fig. 32.4**). This land, too, is subject to desertification. Soil in the tropics is often thin and nutrient-poor because all the nutrients are tied up in the trees and other vegetation. When the trees are felled, removed, and the land is used for agriculture or grazing, it quickly loses its fertility and becomes subject to desertification.

Other consequences of deforestation include the loss of biodiversity and an increase in atmospheric carbon dioxide because trees that once took in carbon dioxide for photosynthesis have been removed. The destruction of tropical rain forests kills many unique plants and stresses the animals that live there because their supply of food is reduced and they no longer have homes to live in.

Water

Access to clean drinking water is considered a human right, but actually most fresh water is utilized by agriculture and industry (**Fig. 32.5**). Worldwide, 70% of all fresh water is used to irrigate crops! Much of a recent surge in demand for water stems from increased industrial activity and irrigation-intensive agriculture, the type of agriculture that now supplies about 40% of the world's food crops. Domestically, in the more-developed countries (MDCs), more water is usually used for bathing, flushing toilets, and watering lawns than for drinking and cooking. In the water-poor areas of the world, people may not have ready access to drinking water, and if they do, the water may be unclean.

Increasing Water Supplies

Although the needs of the human population overall do not exceed the renewable supply, this is not the case in certain regions of the United States and the world. As illustrated in Figure 32.3, about 40% of the world's land is desert, and deserts are bordered by semiarid land. When necessary to support human population growth, the supply of fresh water is increased by damming rivers and withdrawing water from aquifers.

Dams   The world's 45,000 large dams catch 14% of all precipitation runoff, provide water for up to 40% of irrigated land, and give some 65 countries more than half their electricity. Damming of certain rivers has been so extensive that they no longer flow as they once did. The Yellow River in China fails to reach the sea most years; the Colorado River barely makes it to the Gulf of California; and even the Rio Grande dries up before it can merge with the Gulf of Mexico. The first and third longest rivers on Earth, the Nile in Egypt and the Ganges in India, respectively, are also so overexploited that at some times of the year, they hardly make it to the ocean.

Dams have other drawbacks: (1) Reservoirs behind the dam lose water due to evaporation and seepage into underlying rock beds. The amount of water lost sometimes equals the amount made available! (2) The salt left behind by evaporation and agricultural runoff can make a river's water unusable farther downstream. (3) Sediment buildup causes dams to hold back less water; with time, dams may become useless for storing water. (4) The reduced water below the dam has a negative impact on the native wildlife.

Aquifers   To meet their freshwater needs, people are pumping vast amounts of water from **aquifers,** which are natural reservoirs found just below or as much as 1 km below the surface. Aquifers hold about 1,000 times the amount of water that falls on land as precipitation each year. This water accumulates from rain that fell in far-off regions as many as hundreds of thousands of years ago. In the past 50 years, groundwater depletion has become a problem in many areas of the world. In substantial portions of the High Plains Aquifer, which stretches from South Dakota to the Texas Panhandle, much water is pumped out for irrigation or for household water use. In the 1950s, India had 100,000 motorized pumps in operation; today, India has 20 million pumps, a huge increase in groundwater pumping.

Environmental Consequences   Removal of water is causing land **subsidence,** settling of the soil as it dries out. In California's San Joaquin Valley, an area of more than 13,000 square km has subsided, and in the worst spot, the surface has dropped more than 9 meters (m)! In some parts of Gujarat, India, the water table has dropped as much as 7 m. Subsidence damages canals, buildings, and underground pipes. Withdrawal of groundwater can cause sinkholes, as when an underground cavern collapses because water no longer holds up its roof.

Saltwater intrusion is another consequence of groundwater depletion. The flow of water from streams and aquifers usually keeps them fairly free of seawater. But as water is withdrawn, the water table can lower to the point that seawater backs up into streams and aquifers. Saltwater intrusion reduces the supply of fresh water along the coast.

Conservation of Water

By 2025, two-thirds of the world's population may be living in countries that are facing serious water shortages. Some solutions for expanding water supplies have been suggested. Using drip irrigation delivers more water to crops and saves about 50% over traditional methods while increasing crop yields as well (**Fig. 32.6**). Although the first drip systems were developed in 1960, they are used on less than 1% of irrigated land. Most governments subsidize irrigation so heavily that farmers have little incentive to invest in drip systems or other water-saving methods. Planting drought- and salt-tolerant crops would decrease the water required for agriculture. Recycling water and adopting conservation measures could help the world's industries cut their water demands by more than half.

a. Agriculture uses most of the fresh water consumed.

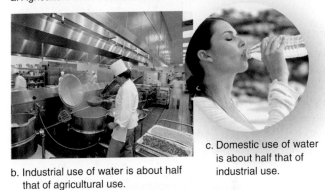

c. Domestic use of water is about half that of industrial use.

b. Industrial use of water is about half that of agricultural use.

Figure 32.5   Global water use.

a. Agriculture primarily uses water for irrigation. **b.** Industry uses water in various ways. **c.** People use water for drinking, showering, flushing toilets, and watering lawns.

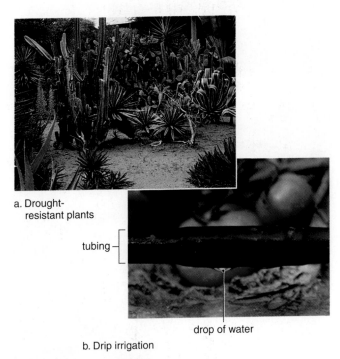

a. Drought-resistant plants

tubing —

drop of water

b. Drip irrigation

Figure 32.6   Conservation measures to save water.

a. Planting drought-resistant plants in parks and gardens and drought-resistant crops in the field cuts down on the need to irrigate. **b.** Drip irrigation delivers water directly to the roots.

1. Explain how desertification occurs.
2. List two problems associated with deforestation.
3. List problems associated with dams.
4. Describe two consequences of groundwater depletion.

Answers: 1. Humans allow animals to overgraze in semiarid regions. The loss of vegetation prevents soil from holding water, and so it runs off. Plants are also removed for fuel and additional animal food.   2. Increase in atmospheric CO_2 and loss of biodiversity.   3. Dams change the flow of rivers, lose water due to evaporation and seepage, accumulate sediment, and impact wildlife in estuaries.   4. Groundwater depletion causes subsidence and saltwater intrusion.

Figure 32.7   Conservation method.

a. Polyculture reduces the ability of one parasite to wipe out an entire crop and reduces the need to use a herbicide to kill weeds. This farmer has planted alfalfa in between strips of corn, which also replenishes the nitrogen content of the soil (instead of adding fertilizers). **b.** Contour farming with no-till conserves topsoil because water has less tendency to run off. **c.** Instead of pesticides, it is sometimes possible to use a natural predator. Here ladybugs are eating cottony-cushions scale insects on citrus trees.

a. Polyculture

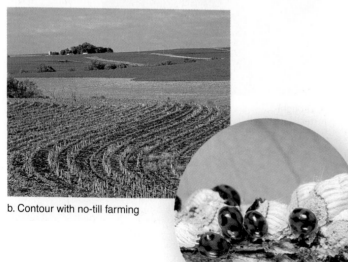

b. Contour with no-till farming

c. Biological pest control

Food

In 1950, the human population numbered 2.5 billion, and there was only enough food to provide less than 2,000 calories per person per day; now, with over 6.7 billion people on Earth, the world food supply provides more than 2,500 calories per person per day. Generally speaking, food comes from three activities: growing crops, raising animals, and fishing the seas. Unfortunately, modern farming methods, which have increased the food supply, include some harmful practices:

1. *Planting of a few genetic varieties.* The majority of farmers practice monoculture, meaning farmers plant a single type (strain) of a crop throughout their fields. Unfortunately, the resulting lack of genetic diversity means that a single type of parasite or disease would infect the entire monoculture. This scenario does not take into account any genetic variants that may exhibit disease resistance.
2. *Heavy use of fertilizers, pesticides, and herbicides.* Fertilizer production is energy-intensive, and fertilizer runoff contributes to water pollution. Pesticides reduce soil fertility because they kill beneficial soil organisms as well as pests, and some pesticides and herbicides are linked to the development of cancer.
3. *Generous irrigation.* As already discussed, water is sometimes taken from aquifers whose water content may in the future become so reduced that it could be too expensive to pump out any more.
4. *Excessive fuel consumption.* Irrigation pumps remove water from aquifers, and large farming machines are used to spread fertilizers, pesticides, and herbicides, as well as to sow and harvest the crops. In effect, modern farming methods transform fossil fuel energy into food energy.

Figure 32.7 shows ways to minimize the harmful effects of modern farming practices.

Soil Loss and Degradation

Land suitable for farming and for grazing animals is being degraded worldwide. Topsoil, the topmost portion of the soil, is the richest in organic matter and the most capable of supporting grass and crops. When bare soil is acted on by water and wind, soil erosion occurs and topsoil is lost. As a result, marginal rangeland becomes desert, and farmland loses its productivity.

The custom of planting the same crop in straight rows that facilitate the use of large farming machines has caused the United States and Canada to have one of the highest rates of soil erosion in the world. Conserving the nutrients now being lost could save farmers $20 billion annually in fertilizer costs. Much of the eroded sediment ends up in lakes and streams, where it reduces the ability of aquatic species to survive.

Since almost all water contains dissolved salts, this salt is left behind either when plants take up the water or a dry climate causes the water to evaporate. Between 25% and 35% of the irrigated western croplands are thought to have undergone **salinization,** an accumulation of mineral salts generated through irrigation. Salinization makes the land unsuitable for growing crops.

Green Revolutions

About 50 years ago, research scientists began to breed tropical wheat and rice varieties specifically for farmers in the LDCs. The dramatic increase in yield due to the introduction of these new varieties around the world was called "the green revolution." These plants helped the world food supply keep pace with the rapid increase in world population. Most green revolution plants are called "high responders" because they need high levels of fertilizer, water, and pesticides in order to produce a high yield. In other words, they require the same subsidies and create the same ecological problems as do modern farming methods.

Genetic Engineering   Genetic engineering can produce transgenic plants with new and different traits, among them, resistance to both insects and herbicides. When herbicide-resistant crops are planted, weeds are easily controlled, less tillage is needed, and soil erosion is minimized. Researchers also want to produce crops that tolerate salt, drought, and cold. In addition, some progress has been made in increasing the food quality of crops so that they will supply more of the proteins, vitamins, and minerals people need. Genetically engineered crops are resulting in another green revolution.

Nevertheless, some citizens are opposed to the use of genetically engineered crops, fearing that they will damage the environment and lead to health problems in humans.

Domestic Livestock

A low-protein, high-carbohydrate diet consisting only of grains such as wheat, rice, or corn can lead to malnutrition. In the LDCs, kwashiorkor, a condition caused by a severe protein deficiency, is seen in infants and children ages 1–3, usually after a new baby arrives in the family and the older children are no longer breast-fed. Such children are lethargic, irritable, and have bloated abdomens. Mental retardation is expected.

In the MDCs, many people tend to have more than enough protein in their diet. Almost two-thirds of U.S. cropland is devoted to producing livestock feed. This means that a large percentage of the fossil fuel, fertilizer, water, herbicides, and pesticides we use are actually for the purpose of raising livestock. Typically, cattle are range-fed for about four months, and then they are brought to crowded feedlots where they receive growth hormone and antibiotics while feeding on grain or corn. Most pigs and chickens spend their entire lives in crowded pens and cages (**Fig. 32.8**).

Figure 32.8 **Crowding of livestock.**

Hogs milling in a feedlot pen.

b. Drag net fishing

Figure 32.9 **Fishing.**

a. The world fish catch has declined in recent years (**b**) because modern fishing methods are overexploiting fisheries. **c.** Aquaculture, the farming of aquatic organisms, is decreasing the demand from the oceans.

Sources: Data from U.S. Marine Fisheries Service.

Figure 32.9 **Fishing.**

a. The world fish catch has declined in recent years (**b**) because modern fishing methods are overexploiting fisheries. **c.** Aquaculture, the farming of aquatic organisms, is decreasing the demand from the oceans.

Sources: Data from U.S. Marine Fisheries Service.

a. World fish catch

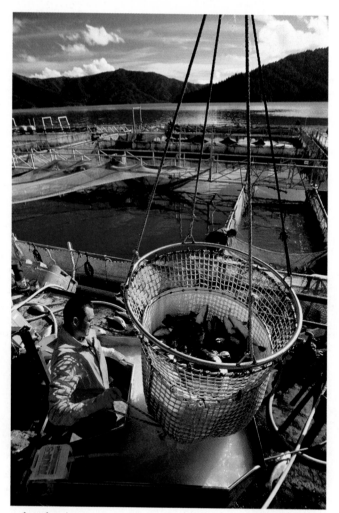

c. Aquafarming

Just as livestock eat a large proportion of the crops in the United States, raising livestock accounts for much of the pollution associated with farming. Consider also that presently, fossil fuel energy is needed not just to produce herbicides and pesticides and to grow food, but also to make the food available to the livestock. Raising livestock is extremely energy-intensive in the MDCs. In addition, water is used to wash livestock wastes into nearby bodies of water, where they add significantly to water pollution. Whereas human wastes are sent to sewage treatment plants, raw animal wastes are not.

For these reasons, it is prudent to recall the ecological energy pyramid (see Fig. 31.20), which shows that as you move up the food chain, energy is lost. As a rule of thumb, for every 10 calories of energy from a plant, only 1 calorie is available for the production of animal tissue in a herbivore. In other words, it is possible to feed ten times as many people on grain as on meat.

Fishing

Since 2000, the world fish catch has been on the decline (**Fig. 32.9***a*). Worldwide, between 1970 and 1990, the number of large boats devoted to fishing doubled to 1.2 million. The U.S. fishing fleet participated in this growth due to the availability of federal loans for building fishing boats. The new boats have sonar and depth recorders, and their computers remember the sites of previous catches so that the boats can go there again. Helicopters, planes, and even satellite data are used to help find fish. The number of North Atlantic swordfish caught in the United States declined 70% from 1980 to 1990, and the average weight fell from 115 to 60 pounds. The Atlantic bluefin tuna is so overfished that it may never recover and instead become extinct.

Modern fishing practices negatively impact biodiversity because a large number of marine animals are caught by chance in the huge nets some fishing boats use (Fig. 32.9*b*). These animals are discarded. The world's shrimp fishery has an annual catch of 1.8 million tons, but the other animals caught and discarded in the process amount to 9.5 million tons. Raising tuna, shrimp, and other aquatic organisms in controlled settings, called aquafarming, has reduced the fishing pressures on wild populations (Fig. 32.9*c*). In fact, over 90% of shrimp available in the United States is produced through aquaculture.

Energy

Presently, about 6% of the world's energy supply comes from nuclear power, and 75% comes from fossil fuels; both of these are finite, nonrenewable sources. Although it was once predicted that the nuclear power industry would fulfill a significant portion of the world's energy needs, this has not happened for two reasons: (1) People are very concerned about nuclear power dangers, such as the meltdown that occurred in 1986 at the Chernobyl nuclear power plant in Russia. (2) Radioactive wastes from nuclear power plants remain a threat to the environment for thousands of years, and we still have not determined the best way to safely store them.

As you learned in Chapter 31, oil, natural gas, and coal are **fossil fuels**— the compressed remains of organisms that died many thousands of years ago. The MDCs presently consume more than twice as much fossil fuel as the LDCs, yet there are many more people in the LDCs than in the MDCs. It has been estimated that each person in the MDCs uses approximately as much energy in one day as a person in an LDC does in one year.

Among the fossil fuels, oil burns more cleanly than coal, which may contain a considerable amount of sulfur. So despite the fact that the United States has a goodly supply of coal, imported oil is our preferred fossil fuel today. Even so, the burning of any fossil fuel contributes to environmental problems because as it burns, pollutants are emitted into the air.

Fossil Fuels and Global Climate Change

The percentage of CO_2 in the atmosphere is 0.03%. However, the level of carbon dioxide in the atmosphere in 1850 was about 280 parts per million (ppm), and today it is about 350 ppm. This increase is largely due to the burning of fossil fuels and the burning and clearing of forests to make way for farmland and pasture. Human activities are causing the emission of other gases as well. For example, the amount of methane given off by oil and gas wells, rice paddies, and all sorts of organisms, including domesticated cows, is increasing by about 1% a year. These gases are known as **greenhouse gases** because, just like the panes of a greenhouse, they allow solar radiation to pass through but hinder the escape of infrared heat back into space.

Today, data collected around the world show a steady rise in the concentration of the various greenhouse gases. These data are used to generate computer models that predict the Earth may warm to temperatures never before experienced by living things. The global climate has already warmed about 0.6°C since the Industrial Revolution. Computer models are unable to consider all possible variables, but the Earth's temperature may rise 1.5°–4.5°C by 2060 if greenhouse emissions continue at the current rates (**Fig. 32.10**).

Some models predict that as the oceans warm due to global warming, temperatures in the polar regions will rise to a greater degree than in other regions. If so, glaciers will melt, and sea levels will rise, not only due to this melting but also because water expands as it warms. Water evaporation will increase, and most likely precipitation will increase along the coasts, while conditions inland become drier. The occurrence of droughts will reduce agricultural yields, and will also cause trees to die off. Coastal agricultural lands, such as the deltas of China and Bangladesh, India, will be inundated, and billions will have to be spent to keep U.S. coastal cities, such as New York, Boston, Miami, and Galveston, from disappearing into the sea.

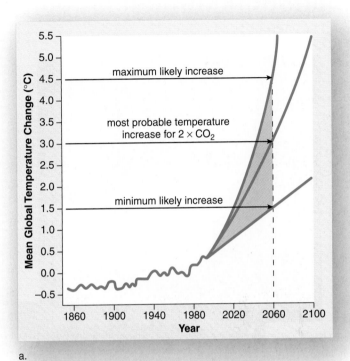

a.

b.

Figure 32.10   Global warming.

a. Mean global temperature is expected to rise due to the introduction of greenhouse gases into the atmosphere. Global warming has the potential to significantly affect the world's biodiversity. **b.** A temperature rise of only a few degrees causes coral reefs to "bleach" and become lifeless. As the oceans warm and land recedes, coral reefs could move northward.

Renewable Energy Sources

Renewable types of energy include hydropower, geothermal, wind, and solar (**Fig. 32.11**).

Wind Power   Wind power is expected to account for a significant percentage of our energy needs in the future. Despite the common belief that a huge amount of land is required for the "wind farms" that produce commercial electricity, the actual amount of space for a wind farm compares favorably to the amount of land required by a coal-fired power plant or a solar thermal energy system.

A community that generates its own electricity by using wind power can solve the problem of uneven energy production by selling electricity to a local public utility when an excess is available and buying electricity from the same facility when wind power is in short supply.

Hydropower   Hydroelectric plants convert the energy of falling water into electricity. Hydropower accounts for about 10% of the electric power generated in the United States and almost 98% of the total renewable energy used. Brazil, New Zealand, and Switzerland produce at least 75% of their electricity with water power, but Canada is the world's leading hydropower producer. Worldwide, hydropower presently generates 19% of all electricity utilized, but

Figure 32.11   Renewable energy sources.

a. Hydropower dams provide a clean form of energy but can be ecologically disastrous in other ways. **b.** Wind power requires land on which to place enough windmills to generate energy. **c.** Photovoltaic cells on rooftops and (**d**) sun-tracking mirrors on land can now collect diffuse solar energy more cheaply than in the past.

this percentage is expected to rise due to increased use in certain countries. For example, Iceland has an ambitious hydropower project under way because presently it uses only 10% of its potential capacity.

Much of the hydropower development in recent years has been due to the construction of enormous dams, which are known to have detrimental environmental effects. An alternative choice may be small-scale dams that generate less power per dam but do not have the same environmental impact.

Geothermal Energy   Elements such as uranium, thorium, radium, and plutonium undergo radioactive decay below the Earth's surface and then heat the surrounding rocks to hundreds of degrees Centigrade. When the rocks are in contact with underground streams or lakes, huge amounts of steam and hot water are produced. This steam can be piped up to the surface to supply hot water for home heating or to run steam-driven turbogenerators. The California's Geysers project is the world's largest geothermal electricity-generating complex.

Energy and the Solar-Hydrogen Revolution   Solar energy is diffuse energy that must be (1) collected, (2) converted to another form, and (3) stored if it is to compete with other available forms of energy. Passive solar heating of a house is successful when the windows of the house face the sun, the building is well insulated, and heat can be stored in water tanks, rocks, bricks, or some other suitable material.

In a **photovoltaic (solar) cell,** a wafer of an electron-emitting metal is in contact with another metal that collects the electrons and passes them along into wires in a steady stream. Ever since the oil shocks of the 1970s, the U.S. government has been supporting the development of photovoltaics. As a result, the price of buying a system has dropped from about $100 per watt to around $4. The photovoltaic cells placed on roofs, for example, generate electricity that can be used inside a building and/or sold back to a power company.

Several types of solar power plants are now operational in California. In one type, huge reflectors focus sunlight on a pipe containing oil. The heated pipes boil water, generating steam that drives a conventional turbogenerator. In another type, 1,800 sun-tracking mirrors focus sunlight onto a molten salt receiver mounted on a tower. The hot salt generates steam that drives a turbogenerator.

Scientists are working on the possibility of using solar energy to extract hydrogen from water via electrolysis. The hydrogen can then be used as a clean-burning fuel; when it burns, water is produced. Presently, cars have internal combustion engines that run on gasoline. In the future, vehicles are expected to be powered by fuel cells, which use hydrogen to produce electricity (**Fig. 32.12**), and many gasoline–fuel cell hybrid vehicles now exist. The electricity runs a motor that propels the vehicle. Fuel cells are now powering buses in Vancouver and Chicago, and more buses are planned.

Hydrogen fuel can be produced locally or in central locations, using energy from photovoltaic cells. If produced in central locations, hydrogen can be piped to filling stations using the natural gas pipes—already plentiful in the United States. The advantages of a solar-hydrogen revolution would be at least twofold: (1) the world would no longer be dependent on oil; and (2) environmental problems, such as acid rain and smog, would begin to lessen.

Figure 32.12   Solar-hydrogen revolution.

a. Hydrogen fuel cells. **b.** This bus is powered by hydrogen fuel. **c.** The use of fuel-cell hybrid vehicles, such as this prototype, will reduce air pollution and dependence on fossil fuels.

Figure 32.13   Modern mining capabilities.

Giant mining machines—some as tall as a 20-story building—can remove an enormous amount of the Earth's crust in one scoop in order to mine for coal or a metal ore.

Minerals

Minerals are nonrenewable raw materials in the Earth's crust that can be mined (extracted) and used by humans. Nonrenewable minerals include fossil fuels; nonmetallic raw materials, such as sand, gravel, and phosphate; and metals, such as aluminum, copper, iron, lead, and gold.

Nonrenewable resources are subject to depletion—that is, the supply will eventually run out. A depletion curve is dependent on how fast the resource is used, whether new reserves can be found, and whether recycling and reuse are possible. We can extend our supply of fossil fuels if we conserve our use and if we find new reserves. In addition to these possibilities, metals can be recycled.

Most of the metals mined each year are consumed by the United States, Japan, and Europe, despite the fact that they are primarily mined in South America, South Africa, and countries of the former Soviet Union. One of the greatest threats to the maintenance of ecosystems and biodiversity is surface mining, called strip mining. In the United States, huge machines can go as far as removing mountaintops in order to reach a mineral (**Fig. 32.13**). The land, devoid of vegetation, takes on a surreal appearance, and runoff from rain washes toxic waste deposits into nearby streams and rivers. Legislation now requires that strip miners restore the land to its original condition, a process that can take years to complete.

The most dangerous metals to human health are the heavy metals: lead, mercury, arsenic, cadmium, tin, chromium, zinc, and copper. They are used to produce batteries, electronics, pesticides, medicines, paints, inks, and dyes. In the ionic form, they enter the body and inhibit vital enzymes. That's why these items should be discarded carefully and taken to hazardous waste sites.

Other Sources of Pollution

Synthetic organic compounds and wastes are also pollutants of concern.

Synthetic Organic Compounds

In addition to metals, synthetic organic compounds are another area of considerable ecological concern due to their detrimental effects on the health of living things, including humans. Synthetic organic compounds play a role in the production of plastics, pesticides, herbicides, cosmetics, coatings, solvents, wood preservatives, and hundreds of other products.

Synthetic organic compounds include halogenated hydrocarbons, in which halogens (chlorine, bromine, fluorine) have replaced certain hydrogens. **Chlorofluorocarbons (CFCs)** are a type of halogenated hydrocarbon in which both chlorine and fluorine atoms replace some of the hydrogen atoms. CFCs have brought about a thinning of the Earth's ozone shield, which protects terrestrial life from the dangerous effects of ultraviolet radiation. In most MDCs, legislation has been passed to prevent the production of any more CFCs. Hydrofluorocarbons, which contain no chlorine, are expected to take their place in coolants and other products. The ozone shield is predicted to recover by 2050; in the meantime, many more cases of skin cancer are expected to occur.

Other synthetic organic chemicals pose a direct and serious threat to the health of living things, including humans. Rachel Carson's book *Silent Spring*, published in 1962, made the public aware of the deleterious effects of pesticides.

Wastes

Every year, the countries of the world discard billions of tons of solid wastes, some on land and some in fresh and marine waters. Solid wastes are visible wastes, some of which are hazardous to our health.

Industrial Wastes Industrial wastes are generated during the mining and production of a product. Clean-water and clean-air legislation in the early 1970s prevented venting industrial wastes into the atmosphere and flushing them into waterways. Industry turned to land disposal, which was unregulated at the time. Utilization of deep-well injection, pits with plastic liners, and landfills led to much water pollution and human illness, including cancer. An estimated 5 billion metric tons of highly toxic chemicals were improperly discarded in the United States between 1950 and 1975. The public's concern was so great that the Environmental Protection Agency (EPA) came into existence. Using an allocation of monies called the superfund, the EPA oversees the cleanup of hazardous waste disposal sites in the United States.

Among the most commonly found contaminants are heavy metals (lead, arsenic, cadmium, chromium) and organic compounds (trichloroethylene, toluene, benzene, polychlorinated biphenyls [PCBs], and chloroform). Some of the chemicals used in pesticides, herbicides, plastics, food additives, and personal hygiene products are classified as endocrine disrupters. These products can affect the endocrine system and interfere with reproduction. In the environment, they occur at a level 1,000 times greater than the hormone levels in human blood.

Decomposers are unable to break down these wastes. They enter and remain in the bodies of organisms because they are not excreted. Therefore, they become more concentrated as they pass along a food chain, a process termed **biological magnification (Fig. 32.14)**. This effect is most apt to occur in aquatic food chains, which have more links than terrestrial food chains. Humans are one of the final consumers in both types of food chains, and in some areas, human milk contains detectable amounts of the polychlorinated hydrocarbons DDT and PCBs.

Sometimes industrial wastes accumulate in the mud of deltas and estuaries of highly polluted rivers and cause environmental problems if disturbed. Industrial pollution is being addressed in many MDCs, but usually has low priority in LDCs.

Sewage Raw sewage can contribute to oxygen depletion in lakes and rivers. Sewage serves as a fertilizer for plants. This can lead to eutrophication; first there is an algal bloom, and then when the algae use up all the nutrients, the algae die. Decomposition of dead algae robs the water of oxygen, which can result in a massive fish kill. Also, human feces can contain pathogenic microorganisms that cause cholera, typhoid fever, and dysentery. In regions of the LDCs where sewage treatment is practically nonexistent, many children die each year from these diseases. Sewage treatment plants use bacteria to break down organic matter to inorganic nutrients, such as nitrates and phosphates, which then enter surface waters.

Check Your Progress

1. List two reasons why nuclear power is not a major fuel source.
2. Describe the effects of global warming.
3. List four types of renewable energy.
4. Explain why synthetic organic compounds are harmful.

Answers: 1. People are concerned about the dangers associated with nuclear power, and it is difficult to safely store nuclear wastes. 2. Global warming will cause sea levels to rise, evaporation to increase, and precipitation to increase near coastlines and decrease inland. 3. Hydropower, geothermal, wind, and solar. 4. Synthetic compounds can cause a depletion in the ozone layer, and can harm humans and other animals.

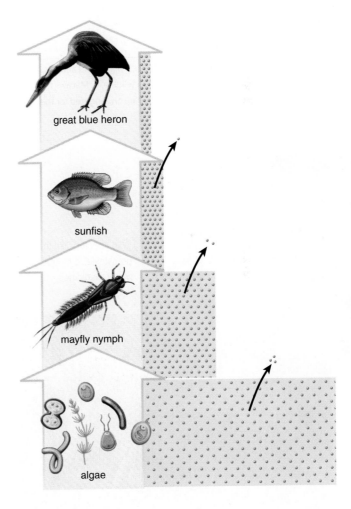

Figure 32.14 Biological magnification.

A poison (*dots*), such as DDT, that is excreted in relatively small amounts (*arrows*) becomes maximally concentrated as it passes along a food chain due to the reduced size of the trophic levels.

great blue heron

sunfish

mayfly nymph

algae

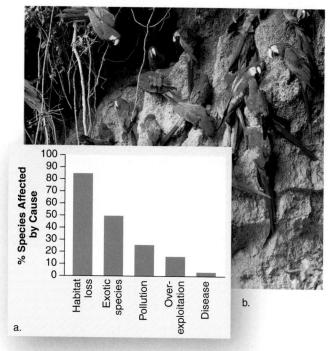

Figure 32.15 **Habitat loss.**

a. In a study examining records of imperiled U.S. plants and animals, habitat loss emerged as the greatest threat to wildlife. **b.** Macaws that reside in South American tropical rain forests are endangered for the reasons listed in the graph in (**a**).

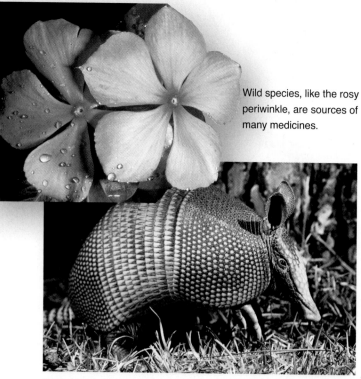

Wild species, like the rosy periwinkle, are sources of many medicines.

Wild species, like the nine-banded armadillo, play a role in medical research.

Figure 32.16 **Medicinal value of biodiversity.**

Many wildlife species are sources of medicines for today's ills.

32.2 | Biodiversity

Biodiversity can be defined as the variety of life on Earth, described in terms of the number of different species. We are presently in a biodiversity crisis—the number of extinctions (loss of species) expected to occur in the near future will, for the first time, be attributable to human activities. In the United States, researchers have found that habitat loss was involved in 85% of the extinction cases (**Fig. 32.15***a*). Exotic species had a hand in nearly 50%, pollution was a factor in 24%, overexploitation in 17%, and disease in 3%. The percentages add up to more than 100% because most of these species are imperiled for more than one reason. For example, not only have macaws had their habitat reduced by encroaching timber and mining companies, but they are also hunted for food and collected for the pet trade (Fig. 32.15*b*).

Conservation biology strives to reverse the trend toward the possible extinction of thousands of plants and animals. To bring this about, it is necessary to make all people aware that biodiversity is a resource of immense value—both direct and indirect.

Direct Values of Biodiversity

Figure 32.16 depicts some of the direct values of biodiversity, which include medicines, foods, and other products that benefit human beings.

Medicinal Value

Most of the prescription drugs used in the United States were originally derived from organisms. The rosy periwinkle from Madagascar is an excellent example of a tropical plant that has provided us with useful medicines (Fig. 32.16*a*). Potent chemicals from this plant are now used to treat two forms of cancer: leukemia and Hodgkin disease. Due to these drugs, the survival rate for childhood leukemia has gone from 10 to 90%, and Hodgkin disease is now usually curable. Although the value of saving a life cannot be calculated, it is still sometimes easier for us to appreciate the worth of a resource if it is explained in monetary terms. Thus, researchers tell us that, judging from the success rate in the past, an additional 328 types of drugs are yet to be found in tropical rain forests, and the value of this resource to society is probably $147 billion.

The popular antibiotic penicillin is derived from a fungus, and certain species of bacteria produce the antibiotics tetracycline and streptomycin. These drugs have proven to be indispensable in the treatment of diseases, including certain sexually transmitted diseases.

Leprosy is among the diseases for which there is, as yet, no cure. The bacterium that causes leprosy will not grow in the laboratory, but scientists discovered that it grows naturally in the nine-banded armadillo (Fig. 32.16*b*). Having a source for the bacterium may make it possible to find a potential cure for leprosy. The blood of horseshoe crabs contains a substance called limulus amoebocyte lysate, which is used to ensure that medical devices, such as pacemakers, surgical implants, and prostheses, are free of bacteria. Blood is taken from 250,000 horseshoe crabs a year, and then they are returned to the sea unharmed.

Agricultural Value

Crops such as wheat, corn, and rice are derived from wild plants that have been modified to be high producers. The same high-yield, genetically similar strains tend to be grown worldwide. When cultivated rice crops in Africa were being devastated by a virus, researchers grew wild rice plants from thousands of seed samples until they found one that contained a gene for resistance to the virus. These wild plants were then used in a breeding program to transfer the gene into high-yield rice plants. If this variety of wild rice had become extinct before its resistance could be discovered, rice cultivation in Africa might have collapsed.

Biological pest controls—specifically, natural predators and parasites— are often preferable to chemical pesticides (**Fig. 32.17***a*). When a rice pest called the brown planthopper became resistant to pesticides, farmers began to use natural enemies of the brown planthopper instead. The economic savings were calculated at well over $1 billion. Similarly, cotton growers in Cañete Valley, Peru, found that pesticides were no longer working against the cotton aphid due to resistance. Research identified natural predators that are now being used to an even greater degree by cotton farmers.

Most flowering plants are pollinated by animals, such as bees, wasps, butterflies, beetles, birds, and bats (Fig. 32.17*b*). The honeybee has been domesticated, and it pollinates almost $10 billion worth of food crops annually in the United States. The danger of this dependency on a single species is exemplified by mites that have now wiped out more than 20% of the commercial honeybee population in the United States. Where can we get resistant bees? From the wild, of course. The value of wild pollinators to the U.S. agricultural economy has been calculated at $4.1 to $6.7 billion a year.

Consumptive Use Value

We have had much success in cultivating crops, keeping domesticated animals, growing trees on plantations, and so on. However, the environment provides all sorts of other products that are sold in marketplaces worldwide, including wild fruits and vegetables, skins, fibers, beeswax, and seaweed. Also, some people obtain their meat directly from the environment. In one study, researchers calculated that the economic value of wild pig in the diet of native hunters in Sarawak, East Malaysia, was about $40 million per year.

Similarly, many trees are still felled in the natural environment for their wood. Researchers have calculated that a species-rich forest in the Peruvian Amazon is worth far more if the forest is used for fruit and rubber production than for timber production (Fig. 32.17*d*). Fruit and the latex needed to produce rubber can be brought to market for an unlimited number of years, whereas once the trees are gone, no more timber can be harvested.

Wild species, like ladybugs, play a role in biological control of agricultural pests.

a.

b.

Wild species, like the long-nosed bat, are pollinators of agricultural and other plants.

c.

Wild species, like many marine species, provide us with food.

Figure 32.17 Agriculture and consumptive value of wildlife.

Many wildlife species help account for our bountiful harvests and are sources of food for today's world.

d.

Wild species, like rubber trees, can provide a product indefinitely if the forest is not destroyed.

biogeochemical cycles
waste disposal
provision of fresh water
prevention of soil erosion
regulation of climate
ecotourism

Figure 32.18 **Indirect value of ecosystem.**

Forests and the oceans perform many of the functions listed as the
indirect value of ecosystems.

Indirect Values of Biodiversity

All wild species play important roles in the ecosystems to which they belong. If we want to preserve them, it is more economical to save the ecosystems than the individual species. Ecosystems perform many indirect services for modern humans, who increasingly live in cities. These services are said to be indirect because they are wide-ranging and not easily perceptible (**Fig. 32.18**).

Biogeochemical Cycles

Recall from Chapter 31 that ecosystems are characterized by energy flow and chemical cycling. The biodiversity within ecosystems contributes to the workings of the water, carbon, nitrogen, phosphorus, and other biogeochemical cycles. We are dependent on these cycles for fresh water, removal of carbon dioxide from the atmosphere, uptake of excess soil nitrogen, and provision of phosphate. When human activities upset one aspect of biogeochemical cycles, other parts within the cycles will also be affected. Technology is unable to artificially contribute to or create any of the biogeochemical cycles.

Waste Disposal

Decomposers break down dead organic matter and other types of wastes to inorganic nutrients that are used by the producers within ecosystems. This function aids humans immensely because we dump millions of tons of waste material into natural ecosystems each year. If not for decomposition, waste would soon cover the entire surface of our planet. We can build sewage treatment plants, but they are expensive, and few of them break down solid wastes completely to inorganic nutrients. It is less expensive and more efficient to water plants and trees with partially treated wastewater and let soil bacteria cleanse it completely.

Biological communities are also capable of breaking down and immobilizing pollutants, such as heavy metals and pesticides that humans release into the environment. A review of wetland functions in Canada assigned a value of $50,000 per hectare (2.471 acres or 10,000 square meters) per year to the ability of natural areas to purify water and take up pollutants.

Provision of Fresh Water

Few terrestrial organisms are adapted to living in a salty environment—they need fresh water. The water cycle continually supplies fresh water to terrestrial ecosystems. Humans use fresh water in innumerable ways, including drinking it and irrigating their crops. Freshwater ecosystems, such as rivers and lakes, also provide us with fish and other types of organisms for food.

Unlike other commodities, there is no substitute for fresh water. We can remove salt from seawater to obtain fresh water, but the cost of desalination is about four to eight times the average cost of fresh water acquired via the water cycle.

Forests and other natural ecosystems exert a "sponge effect." They soak up water and then release it at a regular rate. When rain falls in a natural area, plant foliage and dead leaves lessen its impact, and the soil slowly absorbs it, especially if the soil has been aerated by organisms. The water-holding capacity of forests reduces the possibility and degree of flooding. The value of a marshland outside Boston, Massachusetts, has been estimated at $72,000 per hectare per year solely on its ability to reduce floods. Forests release water slowly for days or weeks after the rains have ceased. Comparing rivers in West African coffee plantations, those flowing through forests release twice as much water halfway through the dry season and three to five times as much at the end of the dry season. These data show the water-retaining ability of forests.

Prevention of Soil Erosion

Intact terrestrial ecosystems naturally retain soil and prevent soil erosion. The importance of this ecosystem attribute is especially noticeable following deforestation. In Pakistan, the world's largest dam, the Tarbela Dam, is losing its storage capacity of 12 billion cubic meters many years sooner than expected because silt is building up behind the dam due to deforestation of areas upriver. At one time, the Philippines were exporting $100 million worth of oysters, mussels, clams, and cockles each year. Now, silt carried down rivers following deforestation is smothering the mangrove ecosystem that serves as a nursery for many marine species. Most coastal ecosystems are not as bountiful as they once were because of deforestation elsewhere and its consequences.

Regulation of Climate

At the local level, trees provide shade, block drying winds, and reduce the need for fans and air conditioners during the summer.

Globally, forests regulate the climate because they take up carbon dioxide, a greenhouse gas. The leaves of trees use carbon dioxide when they photosynthesize, release oxygen and the bodies of the trees store carbon. When trees are cut and burned, carbon is no longer released into the soil through natural decomposition, but instead, carbon dioxide is released into the atmosphere. Subsequently, the reduction of forests worldwide reduces the amount of carbon dioxide removed from the atmosphere, amplifying the issue. Deforestation often removes soil nutrients needed by future tree generations, which in turn limits reforestation.

Ecotourism

Almost everyone prefers to vacation in the natural beauty of an ecosystem (**Fig. 32.19**). In the United States, nearly 100 million people enjoy vacationing in a natural setting. To do so, they spend $5 billion each year on fees, travel, lodging, and food. Many tourists want to go sport fishing, whale watching, boat riding, hiking, birdwatching, and the like. Some merely want to immerse themselves in the beauty of a natural environment. Some lesser developed countries are realizing that there is more economic potential in ecotourism than in cutting forests for timber.

Check Your Progress

1. List the direct values of biodiversity.
2. List the indirect values of biodiversity.

Answers: 1. Medicines, agriculture, and human consumption.   2. Contributions to biogeochemical cycles, waste disposal, provision of fresh water, prevention of soil erosion, regulation of climate, and ecotourism.

Figure 32.19 Ecotourism.

Whale watchers experience the thrill of seeing an orca surfacing off the coast of Washington State.

a. Human society at present

b. Sustainable society

Figure 32.20   **Current human society versus a sustainable society.**

a. At present, our "throw-away" society is characterized by high input of energy and raw material, large output of waste materials and energy in the form of heat, and minimal recycling (*white arrows*). **b.** A sustainable society would be characterized by the use of only renewable energy sources, reuse of heat and waste materials, and maximal recycling of products (*blue arrows*).

32.3 | Working Toward a Sustainable Society

A **sustainable society** would be able to provide the same goods and services for future generations of human beings as it does now. At the same time, biodiversity would be preserved.

Today's Society

The following evidence indicates that at present, human society is most likely not sustainable (**Fig. 32.20***a*):

- A considerable proportion of land, and therefore natural ecosystems, is presently being used for human purposes (homes, agriculture, factories, etc.).
- Agriculture requires large inputs of nonrenewable fossil fuel energy, fertilizer, and pesticides, which create much pollution. More fresh water is used for agriculture than in homes.
- At least half of the agricultural yield in the United States goes toward feeding animals. According to the ten-to-one rule of thumb, it takes 10 pounds of grain to grow one pound of meat. Therefore, it is wasteful for citizens in MDCs to eat as much meat as they do. Also, animal sewage pollutes water.
- Even though fresh water is a renewable resource, we are running out of the available supply within a given time frame.
- Our society primarily utilizes nonrenewable fossil fuel energy, which leads to acid precipitation, smog, and various other pollutants entering the ecosystems.
- Minerals are nonrenewable, and the mining, processing, and use of minerals are responsible for much environmental pollution.

Characteristics of a Sustainable Society

A natural ecosystem can offer clues as to what a sustainable human society would be like. A natural ecosystem is characterized by use of renewable solar energy, and its materials cycle through the various populations back to the producers once again. It is clear that if we want to develop a sustainable society, we too should use renewable energy sources and recycle materials (Fig. 32.20*b*).

We can apply these principles to both rural areas and urban areas, as suggested in the next two sections.

Rural Sustainabililty

In rural areas, we should put the emphasis on preservation. We should preserve ecosystems, including terrestrial ecosystems—such as forests and prairies—and aquatic ecosystems—both freshwater and brackish ones along the coast. We should also preserve agricultural land, groves of fruit trees, and other areas that provide us with renewable resources.

It is imperative that we take all possible steps to preserve what remains of our topsoil and replant areas with native grasses, as necessary. Native grasses stabilize the soil and rebuild soil nutrients, while also serving as a source of renewable biofuel. Trees can be planted to break the wind and protect the soil from erosion, while providing a product as well. Creative solutions to today's ecological problems are very much needed.

Here are some other possible ways to help make rural areas sustainable:

- Plant *cover crops*, which often are a mixture of legumes and grasses, to stabilize the soil between rows of cash crops or between seasonal plantings of cash crops.

- Use *multiuse farming* by planting a variety of crops and use a variety of farming techniques to increase the amount of organic matter in the soil.
- Replenish soil nutrients through composting, organic gardening, or other self-renewable methods.
- Use low flow or trickle irrigation, retention ponds, and other water-conserving methods.
- Increase the planting of *cultivars* (plants propagated vegetatively), which are resistant to blight, rust, insect damage, salt, drought, and encroachment by noxious weeds.
- Use *precision farming (PF)* techniques that rely on accumulated knowledge to reduce habitat destruction, while improving crop yields.
- Use *integrated pest management (IPM),* which encourages the growth of competitive beneficial insects and uses biological controls to reduce the abundance of a pest.
- Plant a variety of species, including native plants, to reduce our dependence on traditional crops.
- Plant *multipurpose trees*—trees with the ability to provide numerous products and perform a variety of functions, in addition to serving as windbreakers (**Fig. 32.21**). Remember that mature trees can provide many different types of products: Mature rubber trees provide us with rubber, and tagua nuts are an excellent substitute for ivory, for example.
- Maintain and restore wetlands, especially in hurricane or tsunami-prone areas. Protect deltas from storm damage. By protecting wetlands, we protect the spawning ground for many valuable fish nurseries.
- Use renewable forms of energy, such as wind and biofuel.
- Support local farmers, fishermen, and feed stores by buying food products produced close to home.

Urban Sustainability

More and more people are moving to a city. Much thought needs to be given about how to serve the needs of new arrivals, without overexpansion of the city.
Here are some other possible ways to help make a city sustainable:
- Design an energy-efficient transportation system to rapidly move people about.
- Use solar or geothermal energy to heat buildings; cool them with an air-conditioning system that uses seawater; and, in general, use conservation methods to regulate the temperature of buildings.
- Utilize *green roofs*—a wild garden of grasses, herbs, and vegetables on the tops of buildings—to assist temperature control, supply food, reduce the amount of rainwater runoff, and be visually appealing (**Fig. 32.22**).
- Improve storm-water management by using sediment traps for storm drains, artificial wetlands, and holding ponds. Increase use of porous surfaces for walking paths, parking lots, and roads. These surfaces reflect less heat, while soaking up rainwater runoff.
- Instead of traditional grasses, plant native species that attract bees and butterflies and require less water and fertilizers.
- Create *greenbelts* that suit the particular urban setting. Include plentiful walking and bicycle paths.
- Revitalize old sections of a city, before developing new sections.
- Use lighting fixtures that hug the walls or ground and send light down; control noise levels by designing quiet motors.
- Promote sustainability by encouraging recycling of business equipment; use low-maintenance building materials, rather than wood.

Figure 32.21 Rural sustainability.

Trees planted by a farmer to break wind and prevent soil erosion can also have other purposes, such as supplying fruits and nuts.

Figure 32.22 Urban sustainability.

Green roofs are ecologically sound and can be visually beautiful.

Check Your Progress

1. List the characteristics of today's society that make it unsustainable.
2. List the characteristics of a sustainable society.

Answers: 1. Large proportions of natural ecosystems are being used for human purposes; agriculture requires large inputs; much agriculture is used for animal production; fresh water is being used up; most energy comes from fossil fuels; and minerals are causing pollution.   2. Use of renewable energy resources, reuse of heat and waste materials, and maximal recycling.

The Chapter in Review

Summary

32.1 | Resources and Pollution

Five resources are maximally used by humans:

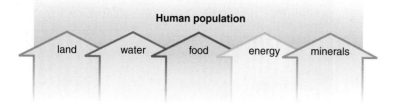

Resources are either nonrenewable or renewable.

Nonrenewable resources are limited in supply:
- Land
- Fossil fuel
- Minerals

Renewable resources are not limited in supply:
- Water
- Solar energy
- Food

Land

Human activities, such as habitation, farming, and mining, contribute to erosion, pollution, desertification, deforestation, and loss of biodiversity.

Water

Industry and agriculture use most of the freshwater supply. Water supplies are increased by damming rivers and drawing from aquifers. As aquifers are depleted, subsidence, sinkhole formation, and saltwater intrusion can occur. If used by industries, water conservation methods could cut world water consumption by half.

Food

Food comes from growing crops, raising animals, and fishing.
- Modern farming methods increase the food supply, but some methods harm the land, pollute water, and consume fossil fuels excessively.
- Transgenic plants can increase the food supply and reduce the need for chemicals.
- Raising livestock contributes to water pollution and uses fossil fuel energy.
- The increased number and high efficiency of fishing boats have caused the world fish catch to decline.

Energy

Most of the world's energy is supplied by the burning of fossil fuels, a nonrenewable resource. The burning of fossil fuels causes pollutants and gases to enter the air.
- Burning fossil fuels and burning to clear land for farming adds CO_2, a greenhouse gas, to the atmosphere. Other gases are emitted by various means. Greenhouse gases allow solar radiation to pass through the atmosphere to the Earth, but infrared heat cannot escape back into space.
- Renewable energy sources include hydropower, geothermal, wind, and solar power.

Minerals

Minerals are a nonrenewable resource that can be mined. These raw materials include sand, gravel, phosphate, and metals. The act of mining causes destruction of the land by erosion, loss of vegetation, and toxic runoff into bodies of water. Some metals are dangerous to human health. Land ruined by mining can take many years to recover.

Synthetic Organic Compounds Compounds such as chlorofluorocarbons (CFCs) are detrimental to the ozone layer and to the health of living things, including humans.

Wastes Raw sewage and industrial wastes can pollute land and bodies of water. Some industrial wastes cause biological magnification, by which toxins become more concentrated as they pass along a food chain.

32.2 | Biodiversity

Biodiversity is the variety of life on Earth. It has both direct value and indirect value.

Direct Values of Biodiversity

Direct values of biodiversity are:
- Medicinal value (medicines derived from living organisms)
- Agricultural value (crops derived from wild plants)
- Biological pest controls and animal pollinators
- Consumptive use values (food production)

Indirect Values of Biodiversity

Biodiversity in ecosystems contributes to:
- The function of biogeochemical cycles (water, carbon, nitrogen, phosphorus, and others)
- Waste disposal (through the action of decomposers and the ability of natural communities to purify water and take up pollutants)
- Fresh water provision through the water biogeochemical cycle
- Prevention of soil erosion, which occurs naturally in intact ecosystems
- Climate regulation (plants take up carbon dioxide)
- Ecotourism (human enjoyment of a beautiful ecosystem)

32.3 | Working Toward a Sustainable Society

A sustainable society would use only renewable energy sources, would reuse heat and waste materials, and would recycle almost everything. It would also provide the same goods and services presently provided and would preserve biodiversity.

Key Terms

aquifer 585	nonrenewable resources 582
biodiversity 594	photovoltaic (solar) cell 591
biological magnification 593	pollution 582
chlorofluorocarbons (CFCs) 592	renewable resources 582
deforestation 584	salinization 587
desertification 584	saltwater intrusion 585
fossil fuel 589	subsidence 585
greenhouse gases 589	sustainable society 598
mineral 592	

Testing Yourself

Choose the best answer for each question.

1. Beach erosion is caused by
   a. rising sea levels.
   b. mining for minerals.
   c. human development.
   d. preserving coastal wetlands.
   e. Both a and c are correct.

2. Desertification typically happens because
   a. deserts naturally expand in size.
   b. humans allow overgrazing.
   c. desert animals wander into adjacent areas for food.
   d. humans tap into limited water supplies for water needs in the nearby desert.

3. Soils in tropical rain forests are typically nutrient-poor because
   a. they receive more water than other ecosystems.
   b. they are sandy.
   c. nitrogen-fixing bacteria are absent.
   d. nutrients are tied up in vegetation.

4. Most fresh water in the world is used for
   a. drinking.
   b. supporting industry.
   c. irrigating crops.
   d. cooking.
   e. bathing.

5. Which of the following will help conserve fresh water?
   a. saltwater intrusion
   b. drought-tolerant crops
   c. increasing irrigation
   d. using water contained in aquifers

6. The first green revolution resulted from the development of
   a. high-responder wheat and rice varieties.
   b. effective irrigation systems.
   c. transgenic crop plants.
   d. crops designed for animal feed.

7. The preferred fossil fuel in the United States is
   a. coal because it produces less pollution than oil.
   b. oil because it produces less pollution than coal.
   c. coal because it is more abundant than oil.
   d. oil because it is more abundant than coal.

For questions 8–12, indicate the type of energy associated with each statement. Some questions may have more than one answer.

Key:
   a. fossil fuels
   b. hydropower
   c. geothermal energy
   d. wind power
   e. solar energy

8. may require the building of dams
9. environmental impact minimal
10. may lead to inland droughts
11. must be stored at times
12. may be used to extract hydrogen from water

13. A major negative effect of the dumping of raw sewage into lakes and rivers is
   a. oxygen depletion.
   b. the buildup of carbon.
   c. a reduction of light penetration into the water.
   d. an increase in populations of small fish.

14. Show how the following diagram must change in order to develop a sustainable society.

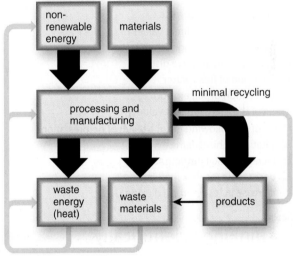

15. The preservation of ecosystems indirectly provides fresh water because
   a. trees produce water as a result of photosynthesis.
   b. animals excrete water-based products.
   c. forests soak up water and release it slowly.
   d. ecosystems promote the growth of bacteria that release water into the environment.

16. A renewable energy source is
   a. hydropower.
   b. natural gas.
   c. coal.
   d. solar energy.
   e. Both a and d are correct.

17. A nonrenewable energy source is
   a. wind.   c. solar energy.
   b. geothermal energy.   d. fossil fuel.

18. A transition to hydrogen fuel technology will
    a. be long in coming and not likely be of major significance.
    b. lessen many current environmental problems.
    c. not be likely since it will always be expensive and as polluting as fossil fuels.
    d. be of major consequence, but resource limitations for obtaining hydrogen will hinder its progress.

19. Most fresh water is held in
    a. groundwater.   d. ice and snow.
    b. lakes.   e. oceans.
    c. rivers.

20. Subsidence involves
    a. the overexploitation of land.
    b. the removal of water from aquifers.
    c. the removal of plant life.
    d. desertification or deforestation.

21. Evaporation of excess water from irrigation of farmland causes
    a. salinization.
    b. accumulation of mineral salts.
    c. subsidence.
    d. loss of topsoil.
    e. Both a and b are correct.

22. Which of these are indirect values of species?
    a. participation in biogeochemical cycles
    b. participation in waste disposal
    c. provision of fresh water
    d. prevention of soil erosion
    e. All of these are indirect values.

23. Which of the following is not a result of deforestation?
    a. siltation behind dams
    b. increase in soil nutrients
    c. decrease in local oxygen output
    d. increase in local carbon dioxide

Thinking Scientifically

1. In the early 1960s, Soviet officials built dams on two large Siberian rivers to support a developing agricultural economy based on cotton, a crop with a high demand for water. Since that time, the Aral Sea (which is fed by those two rivers) in central Asia has shrunk to 25% of its original size. Now plans are being discussed to build a canal to divert water from two other Siberian rivers into the dammed ones in order to keep up with agricultural water demands. What might be the positive and negative environmental impacts of connecting the two river systems for the purpose of increasing the supply of water for cotton production? Another option is to remove the dams from the first two rivers and let them flow freely again. What might be the positive and negative impacts of this strategy?

2. Every time a new species becomes extinct, we lose a potential source of human medicine. Since we do not know which species are likely to provide valuable medicines, it is difficult to know where to focus conservation efforts. If you were in charge of determining high priorities for species conservation in a country, with the goal of identifying plants with medicinal value, what factors would you consider in making your decisions?

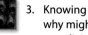

3. Knowing that organisms are closely adapted to their environment, why might global climate change cause the extinction of mammals even living in the tropics?

Bioethical Issue

Control of Beach Erosion

Beach erosion is a serious problem that is difficult to address. One control strategy has been to build hard structures such as seawalls and goynes to reduce the impact of waves. However, these structures are expensive to construct and actually accelerate beach erosion by concentrating the power of waves. Construction of new barriers has recently been banned in several states. The opposite approach to saving beaches is "managed retreat." This method allows coastal erosion to occur naturally by removing man-made beach structures. Several parks have adopted this strategy by removing all hard structures. Should private property owners on beaches be encouraged or forced to remove physical barriers as well? If so, should they be compensated for the loss of beach that will occur as a result of natural erosion? Should incentives be put in place to encourage the planting of native vegetation to reduce erosion?

Essentials of Biology Website

The companion website for *Essentials of Biology* provides a wealth of information organized and integrated by chapter. You will find practice tests, animations, videos, and much more that will complement your learning and understanding of general biology.

http://www.mhhe.com/maderessentials2

Appendix A   Answer Key

Chapter 1

Testing Yourself

1. d; **2. a.** small molecules; **b.** large molecules; **c.** cells; **d.** tissues; **e.** organs; **f.** organ systems; **g.** complex organisms; **3.** b; **4.** d; **5.** c; **6.** c; **7.** e; **8.** d; **9.** b; **10.** b; **11.** d; **12.** c; **13.** b; **14.** c; **15.** a; **16.** c; **17.** b; **18.** a; **19.** d; **20.** g; **21.** a; **22.** c

Thinking Scientifically

1. Domain Archaea gave rise to Eukarya. Kingdom Protista gave rise to plants, animals, and fungi. **2.** Nasal mucus contains an antibacterial chemical. Tears also contain an antibacterial chemical.

Chapter 2

Testing Yourself

1. a; **2.** c; **3.** c; **4.** d; **5.** b; **6.** a; **7.** a; **8.** a; **9.** b; **10.** c; **11.** b

Thinking Scientifically

1. When the bird enters the water, surface tension produced by hydrogen bonding keeps the surface smooth and continuous. When the bird flies out of the water, the cohesiveness of water molecules causes drops of water to be pulled out with the bird. **2.** Silicon, with one extra shell, is too big to form the same varied molecules as carbon.

Chapter 3

Testing Yourself

1. c; **2.** c; **3.** a; **4.** d; **5.** b; **6.** a; **7.** a; **8.** a; **9.** c; **10.** b; **11.** d; **12.** b; **13.** d; **14.** c; **15.** a; **16.** a; **17.** c; **18.** b; **19.** d; **20.** e; **21.** b; **22.** a; **23.** d; **24.** c; **25.** d; **26.** b; **27.** d; **28.** d; **29.** c; **30.** a; **31.** a; **32.** c; **33.** e; **34.** a; **35.** e; **36.** d; **37.** a

Thinking Scientifically

1. a. The plasma membrane. **b.** It increases, because the plant improves its ability to function at lower temperatures by increasing the proportion of linolenic acid in its plasma membrane. Chilling the plants results in injury due to reduced membrane function because the lipids in the plasma membrane solidify slightly. This does not allow the embedded proteins to float within the membrane as rapidly and impairs the membrane's function. **c.** Canola oil remains liquid, while butter (the saturated fat) is a solid. **d.** The fatty acids within the plasma membrane of cold-tolerant plants are mostly unsaturated, whereas the membrane fatty acids of cold-intolerant plants are more saturated and tend to solidify at lower temperatures. This explains why cold-intolerant plants, such as tropical plants, are often damaged by exposure to low temperatures. **2.** A protein's ability to function correctly is dependent upon its shape. In this case, the amino acid with the polar *R* group may interact with other polar *R* groups within the protein

and change the enzyme's shape. **3.** Mutations help drive evolution so that the DNA codes for a different sequence of amino acids in a protein. Different mutations occur as two species evolve separately.

Chapter 4

Testing Yourself

1. d; **2.** a; **3.** c; **4.** b; **5.** c; **6.** d; **7.** a; **8.** b; **9.** d; **10.** e; **11.** c; **12.** b; **13.** a; **14.** b; **15.** c; **16.** d; **17.** c; **18. a.** vesicle; **b.** centrioles; **c.** mitochondrion; **d.** rough endoplasmic reticulum; **e.** smooth endoplasmic reticulum; **f.** lysosome; **g.** Golgi apparatus; **h.** nucleus; see Fig. 4.7, p. 56, in text. **19.** a; **20.** d

Thinking Scientifically

1. a. A dietician may suggest a diet that would not include the offending long-chain fatty acids. **b.** Lysosomes are similar to peroxisomes in that they are single-membrane organelles containing enzymes that break down cellular compounds. **2.** The plastid must be necessary to the life of the parasite because the parasite dies if the plastid no longer functions. **3.** Just as a house cannot store its contents, so a cell cannot be different from its environment without a boundary.

Chapter 5

Testing Yourself

1. d; **2.** e; **3.** b; **4.** d; **5.** d; **6.** a, d; **7.** b, c; **8.** b, c; **9.** a, d; **10. a.** without enzyme; **b.** with enzyme; see Fig. 5.10, p. 78, in text; **11.** c; **12.** a, b, c; **13.** b, c, d; **14.** d; **15.** d; **16.** a; **17.** b; **18.** c; **19.** b; **20.** e; **21.** c; **22.** a; **23.** d; **24.** c; **25.** b; **26.** a; **27.** b; **28.** b; **29.** b; **30.** c; **31.** b

Thinking Scientifically

1. Using GTP instead of ATP as an energy source for a subset of enzymes allows the cell to maintain separate "pools" of energy available for these reactions. For example, using GTP to fuel protein synthesis allows a cell to continue making proteins even when energy levels (the ATP pool) are low. It also allows protein synthesis to continue when oxidative phosphorylation, the process by which most of a cell's ATP is produced, becomes blocked. Since GTP is produced directly from the citric acid cycle, energy for protein synthesis is still available. **2.** A normal channel protein allows chloride to leave the cells. When chloride leaves, water follows by osmosis, keeping the outer surfaces hydrated. Since the defective cystic fibrosis chloride channel does not allow chloride to leave the cell, water remains inside, and the cell surface will be covered with a thick mucus. **3.** The composition of the archaeal plasma membrane is different from that of bacteria and eukaryotes. Furthermore, archaeal enzymes have evolved to withstand this high temperature.

Chapter 6

Testing Yourself

1. c; **2.** a; **3.** d; **4.** a; **5.** a; **6.** a; **7.** b; **8.** d; **9. a.** stroma; **b.** electron transport chain; **c.** ATP; d. NADPH; **e.** ATP synthase complex; **f.** thylakoid membrane; **g.** thylakoid space; see Fig. 6.7, p. 91, in text; **10.** b; **11.** c; **12.** a, b, c; **13.** b; **14.** b; **15.** c; **16.** d; **17.** a; **18.** d; **19.** e; **20.** b; **21.** d; **22.** a; **23.** d

Thinking Scientifically

1. The alga was producing O_2 as a byproduct of photosynthesis wherever its chlorophyll was able to absorb the needed light energy. The bacteria were congregated wherever this O_2 was produced, and were using it for cellular respiration. **2.** Both photosynthesis and vision are adapted to utilizing this portion of the electronic spectrum, which is the most abundant high-energy radiation source available.

Chapter 7

Testing Yourself

1. a; **2.** c; **3.** a; **4.** a; **5.** d; **6.** b; **7.** a; **8.** c; **9.** b; **10. a.** Pyruvate is broken down to an acetyl group; **b.** Acetyl group is taken up and a C_6 molecule results; **c.** Oxidation results in NADH and CO_2; **d.** ATP is produced by substrate-level ATP synthesis; **e.** Oxidation produces more NADH and $FADH_2$; **11.** c; **12.** a, c, d; **13.** a; **14.** a, b, c; **15.** d; **16.** b, c; **17.** a; **18.** b, c, d; **19.** c; **20.** d; **21.** c; **22.** b; **23.** b; **24.** a; **25.** c; **26.** c; **27.** a; **28. a.** citric acid cycle; **b.** anaerobic; **c.** pyruvate; **d.** fermentation; **29.** c

Thinking Scientifically

1. High concentrations of malonate inhibit an enzyme called succinate dehydrogenase through a feedback mechanism. Malonate is the product of the reaction catalyzed by succinate dehydrogenase, and it competes with succinate for the active site of the enzyme. **2.** As we age, our mitochondria lose efficiency and do not break down fats as quickly. This causes fats to accumulate in cells. Since fats and oils enter the cellular respiration pathway as acetyl CoA, there are plenty of citric acid cycle substrates to create energy, and more glucose is not needed. Since the cells already have plenty of energy, they ignore the insulin signal and do not readily take up glucose. Thus, fatty acid accumulation can lead to insulin resistance. **3.** The inner membrane of mitochondria has invaginations and this membrane would have been the plasma membrane of a free-living prokaryote cell. The inner membrane of mitochondria bears a metabolic pathway.

Appendix A

A-1

Chapter 8

Testing Yourself

1. d; **2.** c; **3.** a; **4.** c; **5.** a; **6.** b; **7. a.** G_1; **b.** S; **c.** G_2; **d.** prophase; **e.** metaphase; **f.** anaphase; **g.** telophase; **h.** cytokinesis; see Fig. 8.3, p. 116, in text; **8.** a; **9.** d; **10.** c; **11.** c; **12.** b; **13.** a, b; **14.** b; **15.** b; **16.** c; **17.** b; Each chromosome has one chromatid; therefore, the answer is 42; **18.** d; Each chromosome has two chromatids; therefore, the answer is 84; **19.** a; **20.** a; **21.** c; **22.** d; **23.** e; **24.** e; **25.** d; **26.** b

Thinking Scientifically

1. Cancer results from a series of mutations ("hits") that accumulate in cells over a person's lifetime. People exposed to high levels of radiation are likely to have a higher number of somatic mutations than the general population. They would then require fewer additional random mutations to occur before they develop cancer. A reasonable explanation for the appearances of different cancers at different times is that some cancers require a larger number of mutations to occur before they develop. Cancers that require more mutations to occur would more likely occur later in life. **2.** Sister chromatids of chromosomes (inside the nucleus) must link to microtubules produced by centrosomes (outside the nucleus) in order to be pulled apart during mitosis. Therefore, microtubules cannot attach to and separate chromatids if the nuclear membrane is present. The current hypothesis is that cytoplasmic dynein, a molecular motor protein, tears the nuclear envelope and transports pieces of the membrane away from the area along microtubules. Once pieces are torn away, the nuclear envelope loses shape and cytoplasmic proteins flow into the nuclear region. The rush of cytoplasmic proteins is believed to stimulate chromosome condensation and spindle formation. So, dismantling the nuclear membrane is coordinated with chromosome behavior during mitosis. **3.** Checkpoints ensure that when the cell cycle is finished, the chromosomal content is normal. If the chromosomal content is not normal, then cancer can become prevalent and the species may become extinct.

Chapter 9

Testing Yourself

1. a; **2.** d; **3.** b; **4.** a; **5.** a; **6.** c; **7.** a; **8.** d; **9.** a; **10.** f; **11.** h; **12.** c; **13.** g; **14.** b; **15.** e; **16. a.** homologues condense and undergo synapsis, chromosomes condense and do not pair; **b.** tetrads at spindle equator, dyads at spindle equator; **c.** homologues separate, sister chromatids separate; **d.** (top) two daughter cells result from meiosis I; (bottom) four daughter cells following meiosis II, two daughter cells following mitosis; see Fig. 9.8, p. 137, in text; **17.** a; **18.** c; **19.** c; **20.** c; **21.** b

Thinking Scientifically

1. A mosaic such as (46,XY/47,XXY) may result from nondisjunction occurring during mitosis, especially at a very early stage of development. A fertile Klinefelter mosaic may arise from an individual in which karyotypically normal (46,XY) embryonic cells gave rise to the testes. **2.** Hypotheses to explain the relationship between the increase in Down syndrome with advancing maternal age include: (1) The older a woman is, the longer her oocytes have been arrested in meiosis, and the higher the risk that these oocytes have been exposed to mutagens which might cause nondisjunction; (2) A woman may have a pool of oocytes resulting from nondisjunction which take longer to mature, and thus are more commonly released as she ages; (3) Estrogen levels (which control the rate of meiosis in developing oocytes) drop with advancing maternal age and may slow down the rate of meiosis. This might allow nondisjunction to occur more frequently in older women. **3.** Large amounts of genetic variability in a population allow the population to quickly adapt to changing conditions. For example, if a population is suddenly exposed to a new pathogen, it is likely that at least a few individuals will be able to survive the pathogen and the species would continue.

Chapter 10

Testing Yourself

1. b; **2.** c; **3.** d; **4.** d; **5.** b; **6.** a; **7.** e; **8. a.** TG; **b.** tg; **c.** TtGg; **d.** TTGG; **e.** TtGg; **f.** TTgg; **g.** Ttgg; **h.** TtGG; **i.** ttGG; **j.** TtGg; **k.** ttGG; **l.** ttgg; see Fig. 10.6, p. 148, in text; **9.** b; **10.** c; **11.** c; **12.** d; **13.** a; **14.** e; **15.** c; **16.** d; **17.** e; **18.** b; **19.** e; **20.** d; **21.** a; **22.** a, b; **23.** a, b, c, d; **24.** c, d; **25.** a, b, d; **26.** a; **27.** b

Thinking Scientifically

1. a. CcPp x CcPp would produce 9/16 purple (C_P_) and 7/16 white offspring (C_pp, ccP_, or ccpp). **2.** The chromosomal segment contained the SRY gene. An XX man has an SRY gene on an X chromosome, and would show the male phenotype. XY women have a Y chromosome that is missing its SRY, so they would have the female phenotype. **3.** In a changing environment, a particular combination of alleles may occur to produce an advantageous polygenic trait. The individuals with this trait would be selected to reproduce more than other individuals.

Chapter 11

Testing Yourself

1. a. sugar-phosphate backbone; **b.** base; **c.** hydrogen bonds; **d.** sugar; see Fig. 11.5, p. 165, in text; **2.** a; **3.** a; **4.** d; **5.** b; **6.** b; **7.** a; **8.** c; **9.** b; **10.** b; **11.** c; **12.** b; **13.** c; **14.** c; **15.** c; **16.** a; **17.** c; **18.** d

Thinking Scientifically

1. DNA from the heat-killed virulent strain contains genes that make them virulent. Somehow, the living non-virulent bacteria obtained some of this DNA, converting them into a virulent strain. **2.** The priniples of evolution tell us that it would be too disadvantageous to maintain this DNA if it didn't have a necessary function in these animals.

Chapter 12

Testing Yourself

1. b; **2.** c; **3.** d; **4. a.** regulatory gene; **b.** promoter; **c.** operator; **d.** lactose metabolizing genes; **e.** mRNA; **f.** active repressor; **g.** RNA polymerase; see Fig. 12.5, p. 186, in text; **5.** b; **6.** c; **7.** d; **8.** c; **9.** a; **10.** b; **11.** c; **12.** a; **13.** d; **14.** b; **15.** e; **16.** c; **17.** c; **18.** e; **19.** e; **20.** d; **21.** c; **22.** a

Thinking Scientifically

1. If a mutation in cyclin D leads to its expression outside of the G_1 stage, or prevents it from being destroyed during the S stage, the tumor suppressor RB would be inhibited at other points during the cell cycle. This would tilt the balance of cell cycle regulation towards promotion of the cell cycle, and rapid growth might result. There are other possible explanations as well. **2.** A mutation outside a gene may alter the way in which it is regulated. For example, a mutation in an enhancer may hide the site from a transcriptional activator, so that the gene is not expressed when a growth factor binds to a receptor. In this way, altering the method in which a gene is regulated can sometimes lead to disease. There are other correct explanations as well. **3.** With alternative mRNA splicing, one gene can serve as the coding sequence for a number of different proteins.

Chapter 13

Testing Yourself

1. a; **2.** c; **3.** d; **4. a.** aa; **b.** A_; **c.** Aa; **d.** Aa; **e.** Aa; **f.** Aa; **g.** aa; **h.** aa; **i.** A_; see Fig. 13.7, p. 204, in text; **5. a.** Aa; **b.** Aa; **c.** aa; **d.** Aa; **e.** aa; **f.** Aa; **g.** Aa; **h.** aa; see Fig. 13.8, p. 204, in text; **6. a.** $X^b Y$; **b.** $X^B X^b$; **c.** $X^B Y$; **d.** $X^B Y$; **e.** $X^b Y$; see Fig. 13.9, p. 205, in text; **7.** b; **8.** a, e; **9.** b, d, e; **10.** a, c; **11.** b, d; **12.** e; **13.** b; **14.** e, f; **15.** a, b, c, d; **16.** f; **17.** h; **18.** b; **19.** g, h; **20.** a; **21.** d; **22.** b; **23.** Parents: Aa and Aa; child: aa; **24.** Yes, the son may inherit the father's recessive allele; **25.** Child: $X^{cb} X^{cb}$; Mother: $X^{CB} X^{cb}$ or $X^{cb} X^{cb}$; Father: $X^{cb} Y$; **26.** b; **27.** b; **28.** d; **29.** c

Thinking Scientifically

1. In many cases, gene dosage effects may cause disease. For example, you have learned that some genetic disorders are dominant because having a single mutant allele of the gene is enough to alter the phenotype. In these cases, the mutant allele may have an additional function, not possessed by the normal allele, which causes the disorder. Likewise, having a third or fourth copy of a gene may cause too much of the protein to be made, which may lead to a genetic disorder if the increased enzyme activity is harmful. **2.** The adenovirus may have inserted itself into or nearby a gene that is involved in controlling the cell cycle, such as a proto-oncogene or a tumor suppressor gene. In these cases, disrupting the gene may have inactivated it. In the case cited, the virus was found to have inserted itself near a proto-oncogene called LMO-2. **3.** When diseases are homozygous recessive, the recessive allele can remain hidden in a portion of the populace at all times. Only when two carriers mate is there a chance that the offspring will have the disease.

Chapter 14

Testing Yourself

1. b; **2.** b; **3.** d; **4.** e; **5.** c; **6.** b; **7.** b; **8. a.** Originally, giraffes had short necks. **b.** Giraffes stretched their necks in order to reach food. **c.** Today, most giraffes have long necks. **d.** Originally, giraffe neck length varied. **e.** Struggle to exist caused long-necked giraffes to have the most offspring. **f.** Today, most giraffes have long necks; **9.** e; **10.** d; **11.** d; **12.** c; **13.** b; **14.** a; **15.** c; **16.** b; **17.** a; **18.** c; **19.** e; **20.** e; **21.** e; **22.** e; **23.** d; **24.** a

Thinking Scientifically

1. Humans and other animals share this common structure because we share a common ancestor that had such a structure. The dietary history of humans shows one that had a higher degree of plants than the current diets. **2.** Different genes evolve at different rates; genes are under varying levels of evolutionary constraint. **3.** Islands have populations that are both isolated and with different environments from a mainland population.

Chapter 15

Testing Yourself

1. a; **2.** d; **3.** c; **4.** a; **5.** c; **6.** c; **7.** d; **8.** b; **9.** a; **10.** c; **11.** c; **12.** c; **13.** c; **14.** c; **15.** c; **16.** c; **17.** c; **18.** b; **19.** a; **20.** a; **21.** Curve is shifted to one extreme or the other; **22.** Curve narrows, but mean remains the same; **23.** The one curve splits into two curves, each curve having its mean at a value that is now seen as an extreme value

Thinking Scientifically

1. a. The most fit individuals of a population are the ones that reproduce more than other individuals. Therefore, it is adaptive for the male to be aggressive during nest construction because mating hasn't taken place yet. Once the egg is laid, the male has the assurance that the offspring is his and energy should be conserved. After the offspring has hatched, it would be disadvantageous to spend energy on aggression. **b.** If genetic drift of alleles was causing the male to be aggressive, he would be aggressive all the time without regard to the reproductive cycle.

Chapter 16

Testing Yourself

1. c; **2.** c; **3.** b; **4.** f; **5.** a; **6.** g; **7.** e; **8.** c; **9.** h; **10. a.** species 1; **b.** geographic barrier; **c.** genetic changes; **d.** species 2; **e.** genetic changes; **f.** species 3; **11.** b; **12.** a; **13.** c; **14.** a, b, c; **15.** d; **16.** b; **17.** e; **18.** e; **19. a.** 3; **b.** vertebrae, amniotic egg and internal fertilization; **c.** snake and lizard, they share the most derived traits; **20.** b; **21.** a; **22.** c

Thinking Scientifically

1. a. ferns; **b.** produce seed; **c.** naked seeds; **d.** conifers, needle-like leaves; **e.** Ginkgos, fan-like leaves; **f.** enclosed seeds; **g.** monocots, one embryonic leaf; **h.** eudicots, two embryonic leaves. **2.** Species get stranded on islands that are far away from a mainland and adapt to the island's environment, often different from the mainland

environment from which they came. The Florida Keys are too close together and too close to the Florida mainland. Species are able to move among the islands and thus do not speciate.

Chapter 17

Testing Yourself

1. c; **2. a.** capsid; **b.** protein subunit; **c.** DNA/nucleic acid; **d.** spike; **3.** d; **4.** b; **5.** a; **6.** b; **7.** d; **8.** c; **9.** c; **10.** a; **11.** c; **12.** a; **13.** b; **14.** a; **15.** d; **16.** c; **17.** c; **18.** d; **19.** a; **20. a.** contractile vacuole (partially full); **b.** pellicle; **c.** cilia; **d.** food vacuole; **e.** oral groove; **f.** macronucleus; **g.** micronucleus; **h.** gullet; **i.** anal pore; **j.** contractile vacuole (full)

Thinking Scientifically

1. As members of the domain Eukarya, protists are more closely related to archaea than to bacteria. Archaea and eukarya have certain ribosomal proteins in common that bacteria do not share; they initiate transcription in the same fashion, and have similar tRNAs. Viruses are not shown on the tree because they are acellular and are not considered to belong to any of the three domains of life. **2.** Bacteriophages could infect bacteria in a patient's body and initiate a lytic cycle that would kill the bacteria. The most important benefit would be the ability to treat infections of antibiotic-resistant bacteria against which drugs are ineffective. The main shortcoming is that bacteria would almost certainly evolve resistance to bacteriophage infection.

Chapter 18

Testing Yourself

1. a; **2.** e; **3.** e; **4.** b; **5. a.** sporophyte (2n); **b.** meiosis; **c.** spore (n); **d.** gametophyte (n); **e.** gametes (n); **f.** zygote (2n); **6.** c, d, e; **7.** a, b, c, d; **8.** d, e; **9.** a; **10.** d; **11.** a, b, c; **12.** e; **13.** a, b, c, d, e; **14. a.** anther; **b.** filament; **c.** stigma; **d.** style; **e.** ovary; **f.** ovule; **g.** sepals (calyx); **h.** petals (corolla); **15.** e; **16.** b; **17.** b; **18.** e; **19.** e; **20.** c; **21.** a; **22.** b; **23.** c; **24.** e; **25.** b; **26.** a; **27.** d; **28.** a; **29.** c

Thinking Scientifically

1. The original soil contains mycorrhizal fungi, which have mutualistic relationships with plant roots and enhance the growth of plants. **2.** Protists, plants, and fungi all belong to Domain Eukarya. The protist ancestors of plants resembled certain green algae, possessing chlorophylls *a* and *b* and accessory pigments for photosynthesis, storing carbohydrates as starch, having cellulose in their cell walls, and utilizing an alteration of generations life cycle. The protist ancestors of fungi probably resembled modern chytrids in having flagellated gametes and spores, a feature that has since been lost in the evolution of the other fungal groups. Cell walls may have contained chitin, a feature shared by all modern fungi, including chytrids. Fungal ancestors may have had an alternation of generations life cycle, as do some chytrids.

Chapter 19

Testing Yourself

1. b; **2. a.** ectoderm; **b.** mesoderm; **c.** endoderm; **d.** ectoderm; **e.** mesoderm; **f.** endoderm; **g.** pseudocoelom; **h.** ectoderm; **i.** mesentery; **j.** endoderm; **k.** coelom; **l.** mesoderm; see Fig. 19.6, p. 313, in text; **3.** d; **4.** g; **5.** a, b, c, d, e, g; **6.** a, b, c, f; **7.** a; **8.** e; **9.** c; **10.** f; **11.** b; **12.** e; **13.** c; **14.** b; **15.** a; **16.** a; **17.** c; **18.** a; **19.** d; **20.** d; **21.** a; **22.** b; **23.** e; **24.** e; **25.** a; **26.** e; **27.** a; **28.** b; **29.** d; **30.** d; **31.** b; **32.** b; **33.** a; **34.** e; **35.** a

Thinking Scientifically

1. The recent findings support the idea that a constant lineage exists in Africa from primitive to modern humans. However, it does not prove that this lineage exists only in Africa. This discovery of similar fossils in Asia and Europe would provide support for the multiregional continuity hypothesis. **2.** Animals that are sessile tend to be radially symmetrical because their food comes to them from all directions. There is no need to have anterior and posterior body regions. Animals that move through their environment are bilaterally symmetrical, with the anterior portion containing sensory organs. This allows the animal to sense and respond to the environment as it travels through it.

Chapter 20

Testing Yourself

1. c; **2.** b; **3.** c; **4.** c; **5. a.** terminal bud; **b.** blade (leaf); **c.** lateral bud; **d.** node; **e.** shoot system; **f.** vascular tissues; **g.** root system; see Fig. 20.1, p. 342, in text; **6.** b; **7.** a; **8.** a, b, c, d; **9.** b; **10.** a, b, c; **11.** b; **12.** c; **13.** a; **14.** d; **15. a.** cuticle; **b.** upper epidermis; **c.** palisade mesophyll; **d.** leaf vein; **e.** spongy mesophyll; **f.** lower epidermis; **g.** guard cell; see Fig. 20.9, p. 347, in text; **16. a.** epidermis; **b.** cortex; **c.** vascular bundle; **d.** xylem; **e.** phloem; **f.** pith; see Fig. 20.11*a*, p. 348, in text; **17. a.** endodermis; **b.** pericycle; **c.** xylem; **d.** phloem; **e.** cortex; **f.** epidermis; **g.** root hair; **h.** apical meristem; **i.** root cap; **j.** zone of cell division; **k.** zone of elongation; **l.** zone of maturation; see Fig. 20.13, p. 350, in text

Thinking Scientifically

1. There are unique cell recognition proteins on the surface of the nitrogen-fixing bacteria that the plant recognizes. **2.** When the stomata open at night, both CO_2 and water in the fog are available.

Chapter 21

Testing Yourself

1. d; **2.** a; **3.** e; **4.** a; **5.** a; **6.** e; **7.** b; **8.** c; **9.** a; **10.** b; **11.** e; **12.** d; **13. a.** sporophyte; **b.** meiosis; **c.** microspore; **d.** megaspore; **e.** male gametophyte; **f.** female gametophyte; **g.** egg; **h.** sperm; **i.** zygote; see Fig. 21.10, p. 366, in text; **14.** d; **15.** b; **16.** d; **17.** b; **18.** c; **19.** d; **20.** b; **21.** a; **22.** b; **23.** c; **24.** a; **25.** e; **26.** b; **27.** b; **28. a.** stigma; **b.** style; **c.** carpel; **d.** ovary; **e.** ovule; **f.** receptacle; **g.** peduncle; **h.** sepal; **i.** petal; **j.** stamen; **k.** filament; **l.** anther; see Fig. 21.11, p. 366, in text; **29.** a; **30.** b; **31.** c; **32.** d; **33.** e; **34.** b; **35.** a; **36.** d; **37.** d; **38.** b; **39.** a; **40. a.** seed coat; **b.** embryo; **c.** cotyledon; **d.** seed coat; **e.** first true leaves;

f. cotyledons (two); **g.** seed coat; see Fig. 21.19, p. 372, in text

Thinking Scientifically

1. When plants are shipped long distances, they produce ethylene as a result of the stress. Ethylene builds up within the shipping container, and leaf abscission is stimulated. The plastic sleeves promote ethylene buildup near the plant, amplifying the abscission response. **2.** You could determine whether flowers or stems are responsible by cutting the flowers off the stems and looking to see whether the stems still exhibit heliotropism. If so, then the stems rather than the flowers are responsible. You could determine which portion of the stem is responsible for heliotropism by shading different portions of the stem and determining which treatment blocks the response. **3.** Although there are few pollinators available, plants that bloom at this time of year generally do not have to compete with other species to attract pollinators that are available. The flowers of early-blooming species tend to be low-growing to protect them from cold snaps that may occur, often bear hairs that will keep frost away from the plant, and sometimes close up at night to preserve warmth.

Chapter 22

Testing Yourself

1. a.cell; **b.** tissue; **c.** organ; **d.** organ system; **e.** organism; **2.** c; **3.** b; **4.** a, b; **5.** a; **6.** a, b, c, d; **7.** a, b, c; **8.** c, d; **9.** c; **10. a.** white blood cell; **b.** plasma; **c.** red blood cells; **d.** platelets; **11.** b; **12. a.** skeletal; **b.** cardiac; **c.** smooth; **13.** a; **14. a.** dendrite; **b.** cell body; **c.** nucleus; **d.** nucleus of Schwann cell; **e.** axon; **15.** e; **16.** c; **17.** d; **18.** d; **19.** a; **20.** d; **21.** b; **22.** a; **23.** e; **24.** b; **25.** e

Thinking Scientifically

1. Even though the heart is made of cardiac muscle instead of skeletal muscle, the probable link between the two conditions is that types of muscle tissue are involved. The cardiac muscle may deteriorate in a similar way to that of skeletal muscle. **2.** There are several possibilities of the increased likelihood of death after a liver transplant in African Americans. Their general health may be worse before the transplant, which would affect survival after the transplant. They may not have the same support post-transplant as other populations. If their nutrition, taking of medications is not what it should be for best survival they will have more complications. There may also be genetic differences between African Americans and other populations that affect the success of liver transplants. For example, the gene for sickle cell anemia is more common in African Americans than in other populations. Traits like this may impact how well their bodies handle transplants. **3.** All cells need to carry out exchanges with the external environment either directly or indirectly. With an increase in size, specialized organs and organ systems carry out exchanges with the external environment so that cells are able to carry out exchanges with the internal environment, blood, and tissue fluid. Just as a large company uses specialized rooms to carry out its functions so does a large animal use specialized organs that work together within a system.

Chapter 23

Testing Yourself

1. c; **2.** a; **3. a.** pulmonary vein; **b.** aorta; **c.** renal artery; **d.** lymphatic vessel; **e.** pulmonary artery; **f.** superior vena cava; **g.** inferior vena cava; **h.** hepatic vein; **i.** hepatic portal vein; **j.** renal vein; **4.** a; **5. a.** superior vena cava; **b.** aortic semilunar valve; **c.** pulmonary semilunar valve; **d.** right atrium; **e.** tricuspid valve; **f.** right ventricle; **g.** inferior vena cava; **h.** aorta; **i.** pulmonary trunk; **j.** pulmonary arteries; **k.** pulmonary veins; **l.** left atrium; **m.** bicuspid valve; **n.** septum; **o.** left ventricle; **6.** c; **7.** e; **8.** a; **9.** d; **10.** b; **11.** c; **12.** e; **13.** a; **14.** d; **15.** b; **16.** b; **17.** d; **18.** a; **19.** a; **20.** a; **21.** b; **22.** c, d; **23.** b; **24.** a; **25.** d; **26.** d; **27.** a; **28.** b; **29.** c; **30.** d; **31.** b; **32.** c; **33.** b; **34.** c; **35.** e; **36.** a; **37.** c; **38.** a; **39.** d; **40.** c; **41.** e; **42.** c; **43.** b; **44.** d; **45. a.** blood pressure; **b.** osmotic pressure; **c.** blood pressure; **d.** osmotic pressure

Thinking Scientifically

1. The four-chambered heart keeps O_2-poor and O_2-rich blood separate. This allows the maximum amount of oxygen to be delivered to the cells. This in turn allows for the greatest amount of energy to be produced. In other words, a four-chambered heart supports the high metabolism of an endothermic animal. **2.** Lower oxygen levels found at higher altitudes stimulate an increased production of red blood cells and hemoglobin. Therefore, if an athlete trains at higher altitudes long enough for this to take affect, he/she will have the ability to carry a higher amount of oxygen in their blood stream. This effect will continue even at lower altitudes for a few months after the high altitude training has stopped. More oxygen means more energy through aerobic cellular respiration for muscle contraction.

Chapter 24

Testing Yourself

1. a. nostril; **b.** nasal cavity; **c.** pharynx; **d.** epiglottis; **e.** glottis; **f.** larynx; **g.** trachea; **h.** bronchus; **i.** bronchiole; **j.** alveoli; **k.** lung; **l.** diaphragm; **2.** d; **3.** c; **4.** d; **5.** b; **6.** a; **7.** d; **8.** c; **9. a.** kidney; **b.** ureter; **c.** bladder; **d.** urethra; **10.** c; **11.** d; **12.** b; **13.** a; **14.** d; **15.** a; **16.** d; **17.** b; **18.** a; **19.** c; **20.** b; **21.** d; **22.** c; **23.** c; **24.** d; **25. a.** salivary glands; **b.** mouth; **c.** liver; **d.** gallbladder; **e.** duodenum; **f.** cecum; **g.** appendix; **h.** anus; **i.** pharynx; **j.** esophagus; **k.** diaphragm; **l.** stomach; **m.** pancreas; **n.** large intestine; **o.** small intestine; **p.** rectum; **26.** a; **27.** b; **28.** a; **29.** d; **30.** e; **31.** b; **32.** a; **33.** b; **34.** d

Thinking Scientifically

1. Protein digestion begins in the stomach. The pepsin produced by glands in the stomach would breakdown the insulin, since it is a protein. The insulin would not make it to the small intestine where most substances are absorbed. **2.** Without normal amounts and concentrations of bile, the individual will have difficulty digesting and absorbing fats. Fats that remain in the intestines may cause abdominal discomfort. Individuals may also suffer from fat soluble vitamin deficiencies due to the difficulty they will have in absorbing them. **3.** Grasshoppers have a tracheal system consisting of air tubes to carry out gas exchange and their exoskeleton is both air and waterproof. They excrete uric acid which requires little water to excrete and their Malpighian tubules absorb most of the water back into their bodies.

Chapter 25

Testing Yourself

1. b; **2.** a ; **3.** f ; **4.** a; **5.** b; **6.** e; **7.** c; **8.** b; **9.** d; **10.** e; **11.** d; **12.** c; **13.** a; **14.** c; **15. a.** grains; **b.** 6 oz.; **c.** vegetables; **d.** 2.5 cups; **e.** fruits; **f.** 2 cups; **g.** oils; **h.** 1-2 tbsp; **i.** milk; **j.** 3 cups kids, 2 cups if older; **k.** meats and beans; **l.** 5.5 oz; **16.** c; **17.** d; **18.** a; **19.** d; **20.** c; **21.** b; **22.** a; **23.** a; **24.** b; **25.** a; **26.** c

Thinking Scientifically

1. See table below. **2.** All nine essential amino acids are easily obtainable from eating meat. However, our dendition and also the length of our digestive tracts suggests that we evolved as omnivores. Normally, we need both plant matter (for fiber) and animal matter (for amino acids) in our diet. A vegetarian diet takes careful planning.

Chapter 26

Testing Yourself

1. c; **2. a.** red bone marrow; **b.** thymus; **c.** lymph nodes; **d.** spleen; **e.** lymph vessels; **3.** d; **4.** a; **5.** c; **6.** a; **7.** b; **8.** d; **9.** e;

Ch. 25.1 Thinking Scientifically

	Soda then 6.5 oz	Soda Now 20 oz	Calorie Increase
Fat (g)	0	0	
Carbohydrate (g)	21	65	
Protein (g)	0	0	
Total Calories	84	260	176
	Fries then 2.6 oz	Fries Now 6 oz	Calorie Increase
Fat (g)	11	25	
Carbohydrate (g)	28	65	
Protein (g)	3	7	
Total Calories	22.3	513	290
	Bagels then 3 oz	Bagels Now 6 oz	Calorie Increase
Fat (g)	0.5	1.0	
Carbohydrate (g)	40	80	
Protein (g)	6	12	
Total Calories	188.5	377	188.5

10. b; **11.** c; **12.** b; **13.** d; **14.** a; **15.** b; **16.** c; **17.** e; **18.** d; **19.** c; **20.** d; **21.** a; **22.** e; **23.** b; **24.** a; **25.** b; **26.** b; **27.** e; **28.** c; **29.** b; **30.** d; **31.** d; **32.** b; **33.** d; **34. a.** hair shaft; **b.** oil gland; **c.** sweat glands; **d.** dermis; **e.** epidermis; **35.** e; **36.** a; **37.** a; **38.** a; **39.** e; **40.** b; **41.** c; **42.** a; **43.** d; **44.** a; **45.** b; **46.** a; **47.** c; **48.** a

Thinking Scientifically

1. The tonsils are composed of lymph tissue and help to capture pathogens as they enter the digestive and respiratory tracts. Because of their location and function, they often contain bacteria and that bacteria can overwhelm the tissue of the tonsils and cause an infection. Either through the body's usual defenses or through help from medication, individuals can usually overcome these infections. The tonsils, once healthy, will once again aid the body in the fight against infection. Only if they become chronically infected should they be removed. At that point, they no longer perform their function. **2.** Immunosuppressive drugs not only dampen the body's response to the transplanted tissue but to other foreign cells as well. Therefore, the body's ability to fight infections is weakened and the transplant patients become more susceptible to all types of infections. **3.** Our immune system recognizes pathogens by their antigens. If a pathogen changes its antigens then the immune system has a reduced ability to recognize that pathogen. Therefore, our immune system is a selective agent for pathogens that can change their antigens and thereby escape recognition. Also, the greater the ability of a pathogen to change its antigens, the greater difficulty it is for scientists to make an effective vaccine.

Chapter 27

Testing Yourself

1. d; **2.** c; **3.** d; **4.** c; **5.** a; **6.** b; **7. a.** cerebrum; **b.** skull; **c.** corpus callosum; **d.** diencephalon; **e.** pineal gland; **f.** pituitary gland; **g.** cerebellum; **h.** spinal cord; **i.** medulla oblongata; **j.** pons; **k.** midbrain; **l.** brain stem; see Fig. 27.9*b*, p. 484, in text; **8.** b; **9.** a, c; **10.** a; **11.** d; **12.** a; **13.** d; **14.** d; **15.** b; **16. a.** pineal gland; **b.** pituitary gland (hypophysis); **c.** hypothalamus; **d.** parathyroid gland; **e.** thyroid gland; **f.** thymus gland; **g.** adrenal gland; **h.** pancreas; see Fig. 27.15, p. 489, in text; **17.** d; **18.** a; **19.** d; **20.** d; **21.** a; **22.** c; **23.** d; **24.** d; **25.** e; **26.** a; **27.** c; **28.** b; **29.** b; **30.** a; **31.** c; **32.** e

Thinking Scientifically

1. Parkinson's patients do not exhibit a constriction of blood vessels, so their blood pressure slowly decreases. The loss of function of the sympathetic division prevents the body from regulating blood pressure, so rapid changes in blood pressure can occur, leading to orthostatic hypotension. **2.** Diabetes type 1 is thought to result when an environmental agent such as a virus causing T cells to destroy the pancreatic islets. Vaccines that use a live virus may introduce that agent. For example, the mumps and rubella viruses may remain in the body for years after vaccination and can infect pancreatic islet cells. Infection lowers the levels of insulin secreted by these cells. In addition, exposure to the viruses may lead to

an autoimmune disorder that causes the destruction of pancreatic islet cells. **3.** The brain is an energetically-expensive organ, accounting for a substantial proportion of our daily calorie intake. The selective pressure in an environment with scarce food, like that on Borneo, would be to sacrifice brain size to conserve energy. In an environment with plentiful food like Sumatra, the orangutans can afford to spend extra energy on nervous tissue, and thus reap the advantage of larger brain size.

Chapter 28

Testing Yourself

1. c; **2. a.** cochlear nerve; **b.** cochlea; **c.** tympanic membrane; **d.** auditory tube; **e.** auditory canal; **f.** ossicles; **g.** semicircular canals; **h.** outer ear; **i.** middle ear; **j.** inner ear; see Fig. 28.3*a*, p. 502; **3.** c; **4.** b; **5.** b; **6.** c; **7. a.** sclera; **b.** retina; **c.** optic nerve; **d.** vein; **e.** artery; **f.** fovea; **g.** ciliary muscle; **h.** iris; **i.** pupil; **j.** cornea; **k.** lens; see Fig. 28.8, p. 505; **8.** a; **9.** b; **10.** c; **11.** a; **12.** c; **13.** b; **14.** b, c; **15.** a; **16.** e; **17.** c; **18.** a; **19.** b; **20.** d; **21.** b; **22.** e; **23.** a; **24.** c; **25.** d

Thinking Scientifically

1. The blood vessels interfere with the transmission of light to the back of the eye, consequently impairing the signal sent to the brain. Both cancer and eye diseases involve the growth of new blood vessels. Anti-angiogenesis drugs that have been developed for cancer treatment are currently being studied for the treatment of eye diseases. **2.** Our olfactory system allows us to recognize and discriminate between many different odors. Therefore, we can detect the difference between a dangerous odor, such as smoke, and a harmless one, such as flower fragrance. On the other hand, every natural toxin tastes bitter, so it is more important to be able to sense bitterness than to discriminate between bitter tastes. Therefore, humans can identify bitter compounds by taste, but cannot distinguish between different compounds.

Chapter 29

Testing Yourself

1. c; **2.** a; **3.** a; **4.** b; **5.** d; **6.** c; **7. a.** ureter; **b.** bladder; **c.** vas deferens; **d.** seminal vesicle; **e.** ejaculatory duct; **f.** prostate gland; **g.** bulbourethral gland; **h.** urethra; **i.** epididymis; **j.** penis; **k.** foreskin; **l.** testis; **m.** scrotum; see Fig. 29.4*b*, p. 520, in text; **8. a.** oviduct; **b.** ovary; **c.** rectum; **d.** cervix; **e.** bladder; **f.** vagina; **g.** urethra; see Fig. 29.6*b*, p. 522, in text; **9.** a; **10.** d; **11.** c; **12.** a; **13.** b; **14.** d; **15.** d; **16.** b; **17.** c; **18.** a; **19.** c; **20.** e; **21.** b; **22.** d; **23.** a; **24.** a; **25.** d

Thinking Scientifically

1. Fetuses take a high caloric toll on their mothers. In modern times food supply is usually sufficient for our needs, but earlier in human history the food supply may not have been as dependable. Mothers with low percentages of body fat might starve during pregnancy or the prolonged period of breast-feeding that early human infants presumably required, and

thus would not be able to reproduce at all. By stopping the reproductive cycle when the body senses insufficient food reserves, the woman is perhaps better able to survive to a time when food is more plentiful. **2.** The most logical (although controversial) explanation is that menopause in women developed early in human evolution. It allowed women to channel their efforts into caring for their existing children, increasing the survival rate of these children. This would give the woman the best chance to pass her genes through generations, so it increased her fitness. Since men were not the main caregivers for their children, their fitness would not be increased if they evolved the ability to undergo menopause.

Chapter 30

Testing Yourself

1. a; **2.** d; **3.** b; **4.** d; **5.** b; **6.** b; **7.** d; **8.** c; **9.** d; **10.** b; **11. a.** increasing; **b.** decreasing; **c.** stable; **12.** a; **13.** a; **14.** a; **15.** a; **16.** d; **17.** b; **18.** c; **19.** c; **20.** c; **21.** c; **22.** e; **23.** c; **24.** a

Thinking Scientifically

1. 5.3 years; 17.25 years. **2.** ¼ filled takes 18 weeks; ½ filled takes 19 weeks. **3.** A burrow provides a safe, stable home for animals and, therefore, permits for all age classes to survive equally and maintain a stable population. Scorpions that do not burrow are exposed to the elements and predation. Most offspring do not survive to adulthood. A Type III life history strategy is necessary for survival in which many offspring are produced, but mortality is high. Varying environmental conditions cause fluctuating population size.

Chapter 31

Testing Yourself

1. a; **2.** a; **3.** e; **4.** a; **5.** c; **6.** b; **7.** d; **8.** c; **9.** c; **10.** e; **11. a.** energy; **b.** nutrients; **c.** heat; **d.** heat; **e.** producer; **f.** consumer; **g.** inorganic nutrients; **h.** decomposers; **i.** heat; **12.** e; **13.** a; **14.** a; **15.** a; **16.** a; **17.** d; **18.** c; **19. a.** tertiary consumer; **b.** secondary consumer; **c.** primary consumer; **d.** producer; **20.** c; **21.** c; **22.** c; **23.** c; **24.** b; **25.** a; **26.** d; **27.** a; **28.** a; **29.** b; **30.** d; **31.** a; **32.** f; **33.** c; **34.** d; **35.** b; **36.** a; **37.** c; **38.** d; **39.** a; **40.** b

Thinking Scientifically

1. Insects that are best adapted for obtaining the most food from flowers that are most accessible to them. Those best adapted flowers for attracting insect pollinators produce more offspring and thus the next generation, and likewise those insects that obtain the most food produce more offspring than others. This process continues throughout time, leading to very specific adaptations. **2.** The prairie dog is a keystone species because, for example, other species such as coyotes, hawks, and snakes all rely on prairie dogs for food. The black-footed ferret cannot survive unless the prairie dog remains. Other animals such as burrowing white-tailed rabbits live in abandoned prairie dog burrows, and bison, pronghorn antelope, and cattle prefer to graze near their towns because prairie dog foraging increases seed dispersal and the quality of grassland plants.

Chapter 32

Testing Yourself

1. e; **2.** b; **3.** d; **4.** c; **5.** b; **6.** a; **7.** d; **8.** b; **9.** c, d, e; **10.** a; **11.** a, d, e; **12.** e; **13.** a; **14.** Use more renewable resources than non-renewable resources, increase recycling, refine processing to reduce waste; **15.** c; **16.** e; **17.** d; **18.** b; **19.** a; **20.** b; **21.** e; **22.** e; **23.** b

Thinking Scientifically

1. Diverting the two rivers into one would create a large lake that could serve as habitat for animal and plant species. However, downstream from these two rivers, the aquatic habitat would be damaged, likely reducing the number of species. Removing the dams would replenish life downstream but the water level is still 25% of the original volume. **2.** How rare the species is, how difficult it will be to preserve it versus the cost of preserving it, the potential benefit to human medicine. **3.** The internal temperature of mammals must stay within the normal range. If the external temperature is too high for homeostatic mechanisms to control and their internal temperature rises above normal, they will die off.

Appendix B Metric System

Unit and Abbreviation	Metric Equivalent	Approximate English-to-Metric Equivalents	Units of Temperature
Length			
nanometer (nm)	$= 10^{-9}$ m $(10^{-3}\,\mu m)$		
micrometer (μm)	$= 10^{-6}$ m $(10^{-3}$ mm)		
millimeter (mm)	$= 0.001\ (10^{-3})$ m		
centimeter (cm)	$= 0.01\ (10^{-2})$ m	1 inch = 2.54 cm 1 foot = 30.5 cm	
meter (m)	$= 100\ (10^{2})$ cm $= 1{,}000$ mm	1 foot = 0.30 m 1 yard = 0.91 m	
kilometer (km)	$= 1{,}000\ (10^{3})$ m	1 mi = 1.6 km	
Weight (mass)			
nanogram (ng)	$= 10^{-9}$ g		
microgram (μg)	$= 10^{-6}$ g		
milligram (mg)	$= 10^{-3}$ g		
gram (g)	$= 1{,}000$ mg	1 ounce = 28.3 g 1 pound = 454 g	
kilogram (kg)	$= 1{,}000\ (10^{3})$ g	= 0.45 kg	
metric ton (t)	$= 1{,}000$ kg	1 ton = 0.91 t	
Volume			
microliter (μl)	$= 10^{-6}$ l $(10^{-3}$ ml)		
milliliter (ml)	$= 10^{-3}$ l $= 1$ cm^3 (cc) $= 1{,}000$ mm^3	1 tsp = 5 ml 1 fl oz = 30 ml	
liter (l)	$= 1{,}000$ ml	1 pint = 0.47 l 1 quart = 0.95 l 1 gallon = 3.79 l	
kiloliter (kl)	$= 1{,}000$ l		

°C	°F	
100	212	Water boils at standard temperature and pressure.
71	160	Flash pasteurization of milk
57	134	Highest recorded temperature in the United States, Death Valley, July 10, 1913
41	105.8	Average body temperature of a marathon runner in hot weather
37	98.6	Human body temperature
13.7	56.66	Human survival is still possible at this temperature.
0	32.0	Water freezes at standard temperature and pressure.

To convert temperature scales:

$$°C = \frac{(°F-32)}{1.8}$$

$$°F = 1.8(°C) + 32$$

Glossary

A

abdominal cavity Located inferior to the thoracic cavity and is separated from it by the diaphragm; contains most other internal organs except the heart and the lungs. 430

abscisic acid (ABA) Plant hormone that causes stomata to close and initiates and maintains dormancy. 362

abscission Dropping of leaves, fruits, or flowers from a plant. 362

abstinence Method of birth control; the practice of not engaging in sexual intercourse. 525

acetylcholine (ACh) Neurotransmitter active in both the peripheral and central nervous systems. 482

acetylcholinesterase (AChE) Enzyme that breaks down acetylcholine within a synapse. 482

acetyl-CoA Molecule made up of a 2-carbon acetyl group attached to coenzyme A. During cellular respiration, the acetyl group enters the citric acid cycle for further breakdown. 104

acid Molecules tending to raise the hydrogen ion concentration in a solution and to lower its pH numerically. 27

acoelomate Animals without a coelom, as in flatworms. 313

acquired immunodeficiency syndrome (see AIDS)

actin Muscle protein making up the thin filaments in a sarcomere; its movement shortens the sarcomere, yielding muscle contraction. Actin filaments play a role in the movement of the cell and its organelles. 511

actin filament Cytoskeletal filament of eukaryotic cells composed of the protein actin; also refers to the thin filaments of muscle cells. 64

action potential Electrochemical changes that take place across the axomembrane; the nerve impulse. 481

active immunity Resistance to disease due to the immune system's response to a microorganism or a vaccine. 470

active site Region on the surface of an enzyme where the substrate binds and where the reaction occurs. 77

active transport Use of a plasma membrane carrier protein to move a molecule or ion from a region of lower concentration to one of higher concentration; it opposes equilibrium and requires energy. 80

adaptation Organism's modification in structure, function, or behavior suitable to the environment. 4, 222

adaptive radiation Rapid evolution of several species from a common ancestor into new ecological or geographical zones. 253

Addison disease Condition resulting from a deficiency of adrenal cortex hormones; characterized by low blood glucose, weight loss, and weakness. 493

adenine (A) One of four nitrogen-containing bases in nucleotides composing the structure of DNA and RNA. 163

adenosine triphosphate (ATP) Nucleotide with three phosphate groups. The breakdown of ATP into ADP + Ⓟ makes energy available for energy-requiring processes in cells. 62

adhesion junction Junction between cells in which the adjacent plasma membranes do not touch but are held together by intercellular filaments attached to buttonlike thickenings. 67

adrenal cortex Located in the adrenal gland; produces the glucocorticoid and mineralocorticoid hormones. 493

adrenal gland Gland that lies atop a kidney; the *adrenal medulla* produces the hormones epinephrine and norepinephrine, and the *adrenal cortex* produces the glucocorticoid and mineralocorticoid hormones. 493

adrenal medulla Inner portion of the adrenal gland; secretes the hormones epinephrine and norepinephrine. 493

adrenocorticotropic hormone (ACTH) Hormone secreted by the anterior lobe of the pituitary gland that stimulates activity in the adrenal cortex. 490

adult stem cell Found in many organs of an adult's body, can become like embryonic stem cells by adding four genes. 185

age structure In demographics, a display of the age groups of a population; a growing population has a pyramid-shaped diagram. 545

agglutination Clumping of red blood cells due to a reaction between antigens on red blood cell plasma membranes and antibodies in the plasma. 467

AIDS (acquired immunodeficiency syndrome) Disease caused by a retrovirus and transmitted via body fluids; characterized by failure of the immune system. 472, 528

algae (sing., alga) Type of protist that carries on photosynthesis; unicellular forms are a part of phytoplankton, and multicellular forms are called seaweed. 279

alkaptonuria Rare autosomal genetic disorder in which the individual is unable to metabolize homogentisic acid. 206

allele Alternative form of a gene—alleles occur at the same locus on homologous chromosomes. 131

allergen Foreign substance capable of stimulating an allergic response. 471

allergy Immune response to substances that usually are not recognized as foreign. 471

allopatric speciation Origin of new species between populations that are separated geographically. 252

alternation of generations Life cycle, typical of plants, in which a diploid sporophyte alternates with a haploid gametophyte. 290, 366

alternative mRNA processing Lets cells produce multiple proteins from the same gene by changing the way exons are joined. 190

alveolus (pl., alveoli) In humans, terminal, microscopic, grapelike air sac found in lungs. 420

Alzheimer disease (AD) Brain disorder characterized by a general loss of mental abilities. 486

amino acid Organic molecule composed of an amino group and an acid group; covalently bonds to produce peptide molecules. 41

amniocentesis Procedure for removing amniotic fluid surrounding the developing fetus for testing of the fluid or cells within the fluid. 201

amnion Extraembryonic membrane of reptiles (including birds) and mammals that forms an enclosing, fluid-filled sac. 532

amoeboid Cell that moves and engulfs debris with pseudopods. 282

amphibian Member of a class of vertebrates that includes frogs, toads, and salamanders; they are still tied to a watery environment for reproduction. 327

analogous structure Structure that has a similar function in separate lineages but differs in anatomy and ancestry. 228, 259

analogy Similarity of characteristics due to convergent evolution. 259

anaphase Mitotic phase during which daughter chromosomes move toward the poles of the spindle. 117

angiogenesis Formation of new blood vessels; one mechanism by which cancer spreads. 123, 190

angiosperm Flowering plant; the seeds are borne within a fruit. 290, 296

animal Multicellular, heterotrophic organism belonging to the animal kingdom. 7

annelid The segmented worms, such as the earthworm and the clam worm. 319

annual ring Layer of wood (secondary xylem) usually produced during one growing season. 349

anorexia nervosa Eating disorder characterized by a morbid fear of gaining weight. 454

anterior pituitary Portion of the pituitary gland that is controlled by the hypothalamus and produces six types of hormones, some of which control other endocrine glands. 484, 490

anther In flowering plants, pollen-bearing portion of stamen. 296, 367

anthropoid Group of primates that includes monkeys, apes, and humans. 332

antibody Protein produced in response to the presence of an antigen; each antibody combines with a specific antigen. 410, 467

anticodon Three-base sequence in a transfer RNA molecule base that pairs with a complementary codon in mRNA. 170

antidiuretic hormone (ADH) Hormone secreted by the posterior pituitary that increases the permeability of the collecting ducts in a kidney. 491

antigen Foreign substance, usually a protein or a polysaccharide, that stimulates the immune system to react, such as to produce antibodies. 410, 462

antigen-presenting cell (APC) Cell that displays the antigen to certain cells of the immune system so they can defend the body against that particular antigen. 468

antioxidant Substances, such as vitamins C, E, and A, which defend the body against free radicals. 448

anus Outlet of the digestive tube. 434

aorta In humans, the major systemic artery that takes blood from the heart to the tissues. 403, 406

aortic body Structure located in the walls of the aorta; contains chemoreceptors sensitive to hydrogen ion and carbon dioxide concentrations in the blood. 421

apical dominance Influence of a terminal bud in suppressing the growth of lateral buds. 361

apical meristem In vascular plants, masses of cells in the root and shoot that reproduce and elongate as primary growth occurs. 345

apoptosis Programmed cell death involving a cascade of specific cellular events leading to death and destruction of the cell. 122

appendicular skeleton Part of the vertebrate skeleton forming the appendages, shoulder girdle, and hip girdle. 509

appendix In humans, small, tubular appendage containing lymphatic tissue, that extends outward from the cecum of the large intestine; a part of the immune system. 433, 462

aquaporin Membrane channel through which water moves by simple diffusion. 79

aquatic ecosystem Ecosystems that are freshwater ecosystems (rivers, streams, lakes, ponds) and saltwater (marine) ecosystems (oceans, coral reefs, saltwater marshes). 574

aquifer Rock layers that contain water and will release it in appreciable quantities to wells or springs. 585

arachnid Group of arthropods that contains spiders, scorpions, and ticks. 321

archaea Prokaryotic members of the domain Archaea; often live in unique habitats such as swamps, highly saline environments, and submarine thermal vents. 278

arteriole Vessel that takes blood from an artery to capillaries. 405

artery Blood vessel that transports blood away from the heart. 403, 405

arthropod Invertebrates with an exoskeleton and jointed appendages, such as crustaceans and insects. 320

artificial selection Breeders selecting individuals with particular traits, or combinations of traits, to reproduce over others. 224

assortative mating Individuals tend to mate with those that have the same phenotype with respect to certain characteristics. 238

aster Short, radiating fibers produced by the centrosomes in animal cells. 117

atom Smallest particle of an element that displays the properties of the element. 18

atomic mass Mass of an atom equal to the number of protons plus the number of neutrons within the nucleus. 18

atomic number Number of protons within the nucleus of an atom. 19

atomic symbol One or two letters that represent the name of an element—e.g., H stands for a hydrogen atom, and Na stands for a sodium atom. 18

ATP synthase Enzyme that is part of an *ATP synthase complex* and functions in the production of ATP in chloroplasts and mitochondria. 91

atrioventricular valve Heart valve located between an atrium and a ventricle. 403

atrium (pl., atria) Chamber; particularly an upper chamber of the heart lying above a ventricle. 403

australopithecine One of several species of *Australopithecus,* a genus that contains the first generally recognized hominins. 334

autoimmune disease Disease that results when the immune system mistakenly attacks the body's own tissues. 471

autonomic system Portion of the peripheral nervous system that regulates internal organs. 488

autosome Any chromosome other than the sex-determining pair. 130

autotroph Organism that can capture energy and synthesize organic molecules from inorganic nutrients. 566

auxin Plant hormone regulating growth, particularly cell elongation; also called indoleacetic acid (IAA). 360

AV (atrioventricular) node Small region of neuromuscular tissue that transmits impulses received from the SA node to the ventricular walls. 404

axial skeleton Part of the vertebrate skeleton forming the vertical support or axis, including the skull, the rib cage, and the vertebral column. 509

axon Elongated portion of a neuron that conducts nerve impulses, typically from the cell body to the synapse. 480

B

bacteria Prokaryotic members of the domain Bacteria; most diverse and prevalent organisms on Earth. 273

bacteriophage Virus that infects bacteria. 269

ball-and-socket joint Synovial joint (hip and shoulder joints) that allows movement in all directions. 512

bark External part of a tree, containing cork, cork cambium, and phloem. 349

Barr body Dark-staining body (discovered by M. Barr) in the nuclei of female mammals that contains a condensed, inactive X chromosome. 139

base Molecules tending to lower the hydrogen ion concentration in a solution and raise the pH numerically. 27

B-cell receptor (BCR) Complex on the surface of a B cell that binds an antigen and stimulates the B cell. 466

benign tumor Tumor that does not invade adjacent tissue and stays at the site or origin. 123

bicarbonate ion Ion that participates in buffering the blood, and the form in which carbon dioxide is transported in the bloodstream. 423

bilateral symmetry Body plan having two corresponding or complementary halves. 311

bile Secretion of the liver that is temporarily stored and concentrated in the gallbladder before being released into the small intestine, where it emulsifies fat. 432

binary fission Splitting of a parent cell into two daughter cells; serves as an asexual form of reproduction in bacteria. 274

binge-eating disorder Condition characterized by overeating episodes that are not followed by purging. 454

binomial name Scientific name of an organism, the first part of which designates the genus and the second part of which designates the specific epithet. 6

biodiversity Total number of species, the variability of their genes, and the communities in which they live. 14, 594

bioethics The branch of ethics concerned with the development and implementation of biological technology. 14

biogeochemical cycle Circulating pathway of elements such as carbon and nitrogen involving exchange pools, storage areas, and biotic communities. 570

biogeography Study of the geographical distribution of organisms. 220

bioinformatics Computer technologies used to study the genome and other molecular data. 179

biological magnification Process by which substances become more concentrated in organisms in the higher trophic levels of a food web. 593

biological molecule Organic molecules in cells: carbohydrates, lipids, proteins, and nucleic acids. 34

biology Scientific study of life. 10

biomass Weight of one or more organisms. 569

bioremediation Cleanup of the environment using bacteria to break down pollutants such as oil spills and Agent Orange. 276

biosphere Zone of air, land, and water at the surface of the Earth in which organisms are found. 9, 552, 574

biotechnology Term that encompasses genetic engineering and other techniques that make use of natural biological systems to create a product or achieve a particular result desired by humans. 176

biotic potential Maximum population growth rate under ideal conditions. 546

bipedalism Walking erect on two feet. 333

bird An endothermic reptile that has feathers and wings, is often adapted for flight, and lays hard-shelled eggs. 327

birth control pill Oral contraception containing estrogen and progesterone. 525

bivalve Type of mollusc with a shell composed of two valves; includes clams, oysters, and scallops. 318

blade Broad, expanded portion of a plant leaf that may be single or compound leaflets. 346

blastocyst Early stage of animal embryonic development that consists of a hollow, fluid-filled ball of cells. 531

blind spot Region of the retina, lacking rods or cones, where the optic nerve exits the retina. 506

blood Fluid circulated by the heart through a closed system of vessels. 387, 409

blood pressure Force of blood pushing against the inside wall of blood vessels. 405

B lymphocyte (B cell) Lymphocyte that matures in the bone marrow and, when stimulated by the presence of a specific antigen, gives rise to antibody-producing plasma cells. 463

body mass index (BMI) Calculation used to determine whether or not a person is overweight or obese. 450

bolus Mass of chewed food mixed with saliva. 429

bone Connective tissue having protein fibers and a hard matrix of inorganic salts, notably calcium salts. 387

bony fishes Fishes with a bony rather than cartilaginous skeleton. 326

bottleneck effect An evolutionary event in which a significant percentage of a population is prevented from reproducing, therefore reducing variation. Population bottlenecks increase the influence of *genetic drift*. 239

brain stem In mammals; portion of the brain consisting of the medulla oblongata, pons, and midbrain. 485

bronchiole In terrestrial vertebrates, small tube that conducts air from a bronchus to the alveoli. 420

bronchus (pl., bronchi) In terrestrial vertebrates, branch of the trachea that leads to the lungs. 420

brown algae Marine photosynthetic protists with a notable abundance of xanthophyll pigments; this group includes well-known seaweeds of northern rocky shores. 273

bryophyte The nonvascular plants—the mosses, liverworts, and hornworts; these plants have no vascular tissue and occur in moist locations. 290

buffer Substance or group of substances that tend to resist pH changes of a solution, thus stabilizing its relative acidity and basicity. 28

bulbourethral glands Two small structures located below the prostate gland in males; adds secretions to semen. 521

bulimia nervosa Eating disorder characterized by binge eating followed by purging via self-induced vomiting or use of a laxative. 454

bursa Saclike, fluid-filled structure, lined with synovial membrane, that occurs near a synovial joint. 512

C

C$_3$ plant Plant that fixes carbon dioxide via the Calvin cycle; the first stable product of C$_3$ photosynthesis is a 3-carbon compound. 94

C$_4$ plant Plant that fixes carbon dioxide to produce a C$_4$ molecule that releases carbon dioxide to the Calvin cycle. 94

calcitonin Hormone secreted by the thyroid gland that increases the blood calcium level. 492

calorie Amount of heat energy required to raise the temperature of 1 gram of water 1°C. 72

Calvin cycle reaction Portion of photosynthesis that takes place in the stroma of chloroplasts and can occur in the dark; it uses the products of the light reactions to reduce CO$_2$ to a carbohydrate. 88

calyx The sepals collectively; the outermost flower whorl. 296

CAM Crassulacean-acid metabolism; plant fixes carbon dioxide at night to produce a C$_4$ molecule that releases carbon dioxide to the Calvin cycle during the day. 95

camera-type eye Type of eye found in vertebrates and certain molluscs; a single lens focuses an image on closely packed photoreceptors. 505

cancer Malignant tumor whose nondifferentiated cells exhibit loss of contact inhibition, uncontrolled growth, and the ability to invade tissue and metastasize. 123

capillary Microscopic blood vessel; gases and other substances are exchanged across the walls of a capillary between blood and tissue fluid. 405

capsid Protective protein container of the genetic material of a virus. 268

capsule Gelatinous layer surrounding the cells of green algae and certain bacteria. 52

carbohydrate Class of organic compounds that includes monosaccharides, disaccharides, and polysaccharides; present in food as sugars, starch, and fiber. 35, 444

carbon dioxide (CO$_2$) fixation Photosynthetic reaction in which carbon dioxide is attached to an organic compound. 92

carcinogenesis Development of cancer. 123

cardiac cycle One complete cycle of systole and diastole for all heart chambers. 404

cardiac muscle Striated, involuntary muscle tissue found only in the heart. 388

cardiovascular system Organ system in which blood vessels distribute blood under the pumping action of the heart. 390, 401

carnivore Consumer in a food chain that eats other animals. 429, 566

carotenoid Yellow or orange pigment that serves as an accessory to chlorophyll in photosynthesis. 89

carotid body Structure located at the branching of the carotid arteries; contain chemoreceptors sensitive to hydrogen ion and carbon dioxide concentrations in blood. 421

carpel Ovule-bearing unit that is a part of a pistil. 296

carrier Heterozygous individual who has no apparent abnormality but can pass on an allele for a recessively inherited genetic disorder. 155

carrying capacity Largest number of organisms of a particular species that can be maintained indefinitely by a given environment. 547

cartilage Connective tissue in which the cells lie within lacunae embedded in a flexible, proteinaceous matrix. 387

cartilaginous fishes Fishes with a cartilaginous rather than bony skeleton; includes sharks, rays, and skates. 326

cell Smallest unit that displays the properties of life; composed of cytoplasm surrounded by a plasma membrane. 2, 50

cell body Portion of a neuron that contains a nucleus and from which dendrites and an axon extend. 480

cell cycle Repeating sequence of events in eukaryotes that involves cell growth and nuclear division; consists of the stages G$_1$, S, G$_2$, and M. 116

cell plate Structure across a dividing plant cell that signals the location of new plasma membranes and cell walls. 120

cell-signaling pathway Cellular pathway that includes a receptor for a signal molecule and a signal transduction pathway leading to a specific response in the cell. 191

cell theory One of the major theories of biology, which states that all organisms are made up of cells; cells are capable of self-reproduction and come only from preexisting cells. 52

cellular respiration Metabolic reactions that use the energy from carbohydrates, fatty acid, or amino acid breakdown to produce ATP molecules. 63

cellulose Polysaccharide that is the major complex carbohydrate in plant cell walls. 36

cell wall Structure that surrounds a plant, protistan, fungal, or bacterial cell and maintains the cell's shape and rigidity. 52, 66

central nervous system (CNS) Portion of the nervous system consisting of the brain and spinal cord. 479

centriole Cell organelle, existing in pairs, that occurs in the centrosome and may help organize a mitotic spindle for chromosome movement during animal cell division. 65

centromere Constricted region on a chromosome joining two sister chromatids. 115

centrosome Central microtubule organizing center of cells. In animal cells, it contains two centrioles. 64, 117

cephalization Having a well-recognized anterior head with a brain and sensory receptors. 313

cephalopod Type of mollusc in which a modified foot develops into the head region; includes squids, cuttlefish, octopuses, and nautiluses. 318

cerebellum In terrestrial vertebrates, portion of the brain that coordinates skeletal muscles to produce smooth, graceful motions. 485

cerebral cortex Outer layer of cerebral hemispheres; receives sensory information and controls motor activities. 484

cervical cap Birth control device made of latex or rubber in the shape of a cup which covers the cervix, considered a minidiaphragm. 526

cervix Narrow end of the uterus leading into the vagina. 522

character displacement Tendency for characteristics to be more divergent when similar species belong to the same community than when they are isolated from one another. 563

charophytes A group of freshwater green algae thought to be the closest relatives of land plants. 288

checkpoint In the cell cycle, one of several points where the cell cycle can stop or continue on, depending on the internal signal it receives; ensures that each step of the cell cycle is completed before the next one begins. 121

chemoautotroph Organism able to synthesize organic molecules by using carbon dioxide as the carbon source and the oxidation of an inorganic substance (such as hydrogen sulfide) as the energy source. 275

chemoheterotroph Organism that is unable to reproduce its own organic molecules, and therefore requires organic nutrients in its diet. 275

chitin Strong but flexible nitrogenous polysaccharide found in the exoskeleton of arthropods. 36, 320

chlamydia Bacterial STD of the lower reproductive tract that can result in pelvic inflammatory disease if not treated. 529

chlorofluorocarbons (CFCs) Organic compounds containing carbon, chlorine, and fluorine atoms. CFCs such as Freon can deplete the ozone shield by releasing chlorine atoms in the upper atmosphere. 592

chlorophyll Green pigment that absorbs solar energy and is important in algal and plant photosynthesis; occurs as chlorophyll *a* and chlorophyll *b*. 87

chloroplast Membrane-bounded organelle in algae and plants with chlorophyll-containing membranous thylakoids; where photosynthesis takes place. 62, 87

cholesterol One of the major lipids found in animal plasma membranes; makes the membrane impermeable to many molecules. 39, 446

chordate Animal that has a dorsal tubular nerve cord, a notochord, pharyngeal pouches, and postanal tail at some point in its life cycle. 324

chorionic villi Treelike extensions of the chorion (fetal portion of the placenta) that project into the maternal portion of the placenta. 534

chorionic villus sampling (CVS) Prenatal test in which a sample of chorionic villi cells is removed for diagnostic purposes. 201

chromatid A single DNA strand of a chromosome. Chromosomes may consist of a pair of sister chromatids. 116

chromatin Network of fibrils consisting of DNA and associated proteins observed within a nucleus that is not dividing. 58, 115

chromosome Structure consisting of DNA complexed with proteins that transmits genetic information from the previous generation of cells and organisms to the next generation. 58, 115

chromosome map Sequence that shows the relative distance between gene loci on a chromosome. 157

chyme Thick, semiliquid food material that passes from the stomach to the small intestine. 432

ciliate Complex, unicellular protist that moves by means of cilia and digests food in food vacuoles. 281

cilium (pl., cilia) Short, hairlike projection from the plasma membrane, occurring usually in larger numbers. 65

citric acid cycle Cycle of reactions in mitochondria that begins with citric acid. It breaks down an acetyl group and produces CO_2, ATP, NADH, and $FADH_2$; also called the Krebs cycle. 101

clade A group of organisms that includes a common ancestor and all of its descendants has its own particular derived traits. 261

cladistics School of systematics that uses derived characters to determine monophyletic groups and construct cladograms. 261

cladogram Branching diagrammatic tree used to depict evolutionary history of a group of organisms. 261

class One of the categories, or taxa, used by taxonomists to group species; class is the taxon above the order level. 6, 258

clavicle Collarbone. 509

cleavage Cell division without cytoplasmic addition or enlargement; occurs during the first stage of animal development. 531

climax community In ecology, community that results when succession has come to an end. 560

closed circulatory system Cardiovascular system in all vertebrates and some invertebrates that is composed of a muscular heart and blood vessels. 401

club mosses Lycophyges, a group of seedless vascular plants also known as ground pines. 292

clumped Spatial distribution of individuals in a population in which individuals are more dense in one area than in another. 544

cnidarian Invertebrates existing as either a polyp or medusa with two tissue layers and radial symmetry. 314

codominance Inheritance pattern in which both alleles of a gene are equally expressed in a heterozygote. 151

codon Three-base sequence in messenger RNA that causes the addition of a particular amino acid into a protein, or termination of translation. 169

coelom Body cavity lying between the digestive tract and body wall that is completely lined by mesoderm. 428

coenzyme Nonprotein organic molecule that aids the action of the enzyme to which it is loosely bound. 91, 101

coenzyme A (CoA) Molecule that helps oxidate pyruvate in the preparatory (prep) reaction during cellular respiration. 101

coevolution Joint evolution in which one species exerts selective pressure on the other species. 558

cohesion-tension model Explanation for upward transport of water in xylem based upon transpiration-created tension and the cohesive properties of water molecules. 353

collecting duct Duct within the kidney that receives fluid from several nephrons; the reabsorption of water occurs here. 425

collenchyma Plant tissue composed of cells with unevenly thickened walls; supports growth of stems and petioles. 345

color blindness Deficiency in one or more of the three kinds of cone cells responsible for color vision; a common X-linked recessive disorder. 208

columnar epithelium Type of epithelial tissue in which the cells are rectangular with the nuclei typically located at the bottom of each cell. 384

commensalism Symbiotic relationship in which one species is benefited, and the other is neither harmed nor benefited. 562

common ancestor Ancestor held in common by at least two lines of descent. 249

community Assemblage of populations interacting with one another within the same environment. 9, 552, 558

compact bone Type of bone that contains osteons consisting of concentric layers of matrix and osteocytes in lacunae. 387, 510

competition Interaction between two organisms in which both require the same limited resource, which results in harm to both. 548, 562

competitive exclusion principle Theory that no two species can occupy the same niche. 563

complementary base pairing Hydrogen bonding between particular purines and pyrimidines in DNA. 166

complement system Series of proteins in plasma that form a nonspecific defense mechanism against a pathogen invasion; it complements the antigen-antibody reaction. 465

compound Substance having two or more different elements united chemically in a fixed ratio. 21

compound eye Type of eye found in arthropods; it is composed of many independent visual units. 505

conclusion Statement made following an experiment as to whether or not the results support the hypothesis. 11

condom, female Birth control method that blocks the entrance of sperm to the uterus; also prevents STDs. 526

condom, male Sheath used to cover the penis during sexual intercourse; used as a contraceptive and, if latex, to minimize the risk of transmitting infection. 526

conduction deafness Deafness due to middle ear damage. 503

cone Photoreceptor cell in vertebrate eyes that responds to bright light and makes color vision possible. In conifers, structure that bears either pollen (male gametophyte) or seeds (female gametophyte); the sporangia-bearing structures of certain seedless vascular plants are also termed cones. 295, 506

conifer Member of a group of cone-bearing gymnosperm plants that includes pine, cedar, and spruce trees. 295

conjugation Transfer of genetic material from one cell to another. 275

conjugation pili Tubular, rigid bacterial structures that enable the transmission of DNA from cell to cell. 53

connective tissue Type of animal tissue that binds structures together, provides support and protection, fills spaces, stores fat, and forms blood cells; adipose tissue, cartilage, bone, and blood are types of connective tissue. 386

consumer Organism that feeds on another organism in a food chain; primary consumers eat plants, and secondary consumers eat animals. 566

contact inhibition In cell culture, the point where cells stop dividing when they become a one-cell-thick sheet. 122

continuous ambulatory peritoneal dialysis (CAPD) Dialysis that takes place inside the body using the peritoneum, the natural lining of the abdomen, as the dialysis membrane. 427

contraceptive implant Birth control method utilizing synthetic progesterone; prevents ovulation by disrupting the ovarian cycle. 525

contraceptive injection Birth control method utilizing progesterone or estrogen and progesterone; prevents ovulation by disrupting the ovarian cycle. 525

contraceptive vaccine Under development, this birth control method immunizes against the hormone HCG, crucial to maintaining implantation of the embryo. 526

control group Sample that goes through all the steps of an experiment but does not contain the experimental variable being tested; a standard against which the results of an experiment are checked. 11

convergent evolution Similarity in structure in distantly related groups due to adaptation to the environment. 259

copulation Sexual union between a male and a female. 519

cork cambium Lateral meristem that produces cork. 349

cork cell Dead cells found in the outer covering of the bark of trees; may be sloughed off. 349

cornea Transparent, anterior portion of the outer layer of the eyeball. 505

corolla Petals, collectively; usually the conspicuously colored flower whorl. 296

corpus luteum Follicle that has released an egg and increases its secretion of progesterone. 523

cortex In plants, ground tissue bounded by the epidermis and vascular tissue in stems and roots; in animals, outer layer of an organ, such as the cortex of the kidney or adrenal gland. 348

cotyledon Seed leaf for embryo of a flowering plant; provides nutrient molecules for the developing plant before photosynthesis begins. 297, 344, 370

coupled reaction Reaction that occurs simultaneously; one is an exergonic reaction that releases energy, and the other is an endergonic reaction that requires an input of energy in order to occur. 75

covalent bond Chemical bond in which atoms share one pair of electrons. 22

coxal bones Two massive bones that form the pelvis and articulate with the femurs (thighbones). 509

cranial nerve Nerve that arises from the brain. 486

cranium Portion of the skull that protects the brain. 509

cristae Short, fingerlike projections formed by the folding of the inner membrane of mitochondria. 63

Cro-Magnon Common name for the first fossils to be designated *Homo sapiens*. 336

crossing-over Exchange of segments between nonsister chromatids of a tetrad during meiosis. 133

crustacean Member of a group of marine arthropods that contains, among others, shrimps, crabs, crayfish, and lobsters. 320

cuboidal epithelium Type of epithelial tissue in wich the cells are cube-shaped. 384

culture Total pattern of human behavior; includes technology and the arts, and is dependent upon the capacity to speak and transmit knowledge. 334

Cushing syndrome Condition resulting from hypersecretion of glucocorticoids; characterized by thin arms and legs and a "moon face," and accompanied by high blood glucose and sodium levels. 493

cutaneous receptor Sensory receptors for pressure and touch found in the dermis of the skin. 507

cuticle Waxy layer covering the epidermis of plants that protects the plant against water loss and disease-causing organisms. 345

cyanobacteria Photosynthetic bacteria that contain chlorophyll and release O_2; formerly called blue-green algae. 275

cyclin Protein that cycles in quantity as the cell cycle progresses; combines with and activates the kinases that function to promote the events of the cycle. 121

cystic fibrosis Generalized, autosomal recessive disorder of infants and children in which there is widespread dysfunction of the exocrine glands. 206

cytokine Type of protein secreted by a T cell that attacks viruses, virally infected cells, and cancer cells. 469

cytokinesis Division of the cytoplasm following mitosis and meiosis. 116

cytokinin Plant hormone that promotes cell division; often works in combination with auxin during organ development in plant embryos. 362

cytoplasm Contents of a cell between the nucleus (nucleoid region of bacteria) and the plasma membrane. 52

cytosine (C) One of four nitrogen-containing bases in the nucleotides composing the structure of DNA and RNA; pairs with guanine. 163

cytoskeleton Internal framework of the cell, consisting of microtubules, actin filaments, and intermediate filaments. 57, 64

cytotoxic T cell T cell that attacks and kills antigen-bearing cells. 468

D

data (sing., datum) Facts or information collected through observation and/or experimentation. 11

day-neutral plant Plant whose flowering is not dependent on day length—e.g., tomato and cucumber. 364

deciduous Shedding leaves annually. 342

decomposer Organism, usually a bacterium or fungus, that breaks down organic matter into inorganic nutrients that can be recycled in the environment. 566

deficiency disorder Disorders caused by the lack of certain vitamins or minerals in the diet. Examples are bleeding gums, rickets (vitamin deficiencies), osteoporosis, and anemia (mineral deficiencies). 442

deforestation Removal of trees from a forest in a way that ever reduces the size of the forest. 584

dehydration reaction Chemical reaction resulting in a covalent bond with the accompanying loss of a water molecule. 34

delayed allergic response Allergic response initiated at the site of the allergen by sensitized T cells, involving macrophages and regulated by cytokines. 471

deletion Change in chromosome structure in which the end of a chromosome breaks off or two simultaneous breaks lead to the loss of an internal segment; often causes abnormalities—e.g., cri du chat syndrome. 202

demographics Study of human populations, their characteristics, and their changes. 545

denatured Loss of an enzyme's normal shape so that it no longer functions; caused by a less than optimal pH and temperature. 43

dendrite Part of a neuron that sends signals toward the cell body. 480

dense fibrous connective tissue Type of connective tissue containing many collagen fibers packed together; found in tendons and ligaments, for example. 386

density-dependent factor Biotic factor, such as disease or competition, that affects population size according to the population's density. 548

density-independent factor Abiotic factor, such as fire or flood, that affects population size independent of the population's density. 548

deoxyribose Pentose sugar found in DNA that has one less hydroxyl group than ribose. 35

desertification Denuding and degrading a once-fertile land, initiating a desert-producing cycle that feeds on itself and causes long-term changes in the soil, climate, and biota of an area. 584

detrital food web Complex pattern of interlocking and crisscrossing food chains that begins with a population of detritivores. 568

detritus Organic matter produced by decomposition of substances such as tissues and animal wastes. 567

deuterostome Coelomate animals in which the second embryonic opening is associated with the mouth; the first embryonic opening, the blastopore, is associated with the anus. 313

diabetes mellitus Condition characterized by a high blood glucose level and the appearance of glucose in the urine, due to a deficiency of insulin production and failure of cells to take up glucose. 494

dialysate Material that passes through the membrane in dialysis. 427

diaphragm In mammals, dome-shaped muscularized sheet separating the thoracic cavity from the abdominal cavity important in inhalation. Also, a birth control device consisting of a soft rubber or latex cup that fits over the cervix. 430, 526

diarrhea Excessively frequent and watery bowel movements. 434

diastole Relaxation period of a heart chamber during the cardiac cycle. 404

diet A person's typical food choices; a balanced diet contains all the nutrients in the right proportions to maintain a healthy body. 443

dietary supplement Nutrients and plant products (i.e., herbal teas, protein supplements) that are used to enhance health; do not need to undergo the same safety and effectiveness testing required for prescription drugs. 456

digestive system Organ system that includes the mouth, esophagus, stomach, small intestine, and large intestine (colon) that receives food and digests it into nutrient molecules. Also has associated organs: teeth, tongue, salivary glands, liver, gallbladder, and pancreas. 390

dihybrid Individual that is heterozygous for two traits; shows the phenotype governed by the dominant alleles but carries the recessive alleles. 149

diploid (2n) number Cell condition in which two of each type of chromosome are present. 131

directional selection Outcome of natural selection in which an extreme phenotype at one end of a population distribution is favored over all other phenotypes, leading to one distinct form. 242

disaccharide Sugar that contains two units of a monosaccharide; e.g., maltose. 35

disruptive selection Outcome of natural selection in which both extreme phenotypes at the ends of a population distribution are favored over the average phenotype, leading to more than one distinct form. 242

distal tubule Final portion of a nephron that joins with a collecting duct; associated with tubular secretion. 425

diversity Amount of each specific species in a community. 559

DNA (deoxyribonucleic acid) Nucleic acid polymer produced from covalent bonding of nucleotide monomers that contain the sugar deoxyribose; the genetic material of nearly all organisms. 44

DNA fingerprinting The use of DNA fragment lengths resulting from restriction enzyme cleavage to identify particular individuals. 177

DNA ligase Enzyme that links DNA fragments; used during production of recombinant DNA to join foreign DNA to vector DNA. 175

DNA microarray Thousands of different single-stranded DNA fragments arranged in an array (grid) on a glass slide; used to detect and measure gene expression. 209

DNA polymerase During replication, an enzyme that joins the nucleotides complementary to a DNA template. 174

DNA replication Synthesis of a new DNA double helix prior to mitosis or meiosis in eukaryotic cells and during prokaryotic fission in prokaryotic cells. 115, 166

domain Largest of the categories, or taxa, used by taxonomists to group species; the three domains are Archaea, Bacteria, and Eukarya. 6, 258

domain Archaea One of the three domains of life; contains prokaryotic cells that often live in extreme habitats and have unique genetic, biochemical, and physiological characteristics; its members are sometimes referred to as *archaea*. 6, 263

domain Bacteria One of the three domains of life; contains prokaryotic cells that differ from archaea because they have their own unique genetic, biochemical, and physiological characteristics. 6, 263

domain Eukarya One of the three domains of life, consisting of organisms with eukaryotic cells and further classified into the kingdoms Protista, Fungi, Plantae, and Animalia. 6, 263

dominant allele Allele that exerts its phenotypic effect in the heterozygote; it masks the expression of the recessive allele. 147

dormancy In plants, a cessation of growth under conditions that seem appropriate for growth. 361

dorsal root ganglion Mass of sensory neuron cell bodies located in the dorsal root of a spinal nerve. 487

double fertilization In flowering plants, one sperm nucleus unites with the egg nucleus, and a second sperm nucleus unites with the polar nuclei of an embryo sac. 299, 369

Down syndrome Trisomy 21; the individual has three copies of chromosome 21 and the following characteristics: mental disabilities of varying degree, short stature, eyelid fold, stubby fingers, a palm crease, etc. 138

Duchenne muscular dystrophy Chronic progressive disease affecting the shoulder and hip (pelvic) girdles, commencing in early childhood; characterized by increasing weakness of the muscles followed by atrophy and a peculiar swaying gait with the legs kept wide apart. Transmitted as an X-linked trait; affected individuals, predominantly males, rarely survive to maturity. Death is usually due to respiratory weakness or heart failure. 208

duodenum First part of the small intestine where chyme enters from the stomach. 434

duplication Change in chromosome structure in which a particular segment is present more than once in the same chromosome. 202

dyad Chromosome composed of two sister chromatids. 132

E

echinoderm Marine invertebrate such as a sea star, sea urchin, and sand dollar; characterized by radial symmetry and a water vascular system. 323

ecological niche Role an organism plays in its community, including its habitat and its interactions with other organisms. 562

ecological pyramid Pictorial graph based on the biomass, number of organisms, or energy content of various trophic levels in a food web—from the producer to the final consumer populations. 569

ecological succession Gradual replacement of communities in an area following a disturbance (secondary succession) or the creation of new soil (primary succession). 560

ecology Study of the interactions of organisms with other organisms and with the physical and chemical environment. 552

ecosystem Biological community together with the associated abiotic environment; characterized by a flow of energy and a cycling of inorganic nutrients. 9, 552, 559

ectoderm Outermost primary tissue layer of an animal embryo; gives rise to the nervous system and the outer layer of the integument. 532

ectothermic Having a body temperature that varies according to the environmental temperature. 328

edema Swelling due to tissue fluid accumulation in the intercellular spaces. 427

electrocardiogram (ECG) Recording of the electrical activity associated with the heartbeat. 404

electron Negative subatomic particle, moving about in an energy level around the nucleus of an atom. 18

electronegativity Ability of an atom to attract electrons toward itself in a chemical bond. 24

electron shell Concentric energy levels in which electrons orbit. 20

electron transport chain Passage of electrons along a series of electron carriers from a higher to lower energy level; the energy released is used for the synthesis of ATP. 90, 101

element Substance that cannot be broken down into substances with different properties; composed of only one type atom. 18

embryo sac Female gametophyte of flowering plants. 368

embryonic stem cell Undifferentiated cell obtained from an embryo that can be manipulated to become a specialized cell type such as a muscle, nerve, or red blood cell. 185

emerging virus Causative agent of a disease that is new or is demonstrating increased prevalence, such as the viruses that cause AIDS, SARS, and avian influenza. 271

emulsification Breaking up of fat globules into smaller droplets by the action of bile salts or any other emulsifier. 432

endocrine gland Ductless organ that secretes hormone(s) into the bloodstream. 434, 489

endocrine system Organ system involved in the coordination of body activities; uses hormones as chemical signals secreted into the bloodstream. 391

endocytosis Process by which substances are moved into the cell from the environment by phagocytosis (cellular eating) or pinocytosis (cellular drinking); includes receptor-mediated endocytosis. 81

endoderm Innermost primary tissue layer of an animal embryo that gives rise to the linings of the digestive tract and associated structures. 532

endodermis Internal plant root tissue forming a boundary between the cortex and the vascular cylinder. 351

endomembrane system Collection of membranous structures involved in transport within the cell. 60

endometrium Mucous membrane lining the interior surface of the uterus. 522

endoplasmic reticulum (ER) System of membranous saccules and channels in the cytoplasm, often with attached ribosomes. 59

endoskeleton Protective internal skeleton, as in vertebrates. 508

endosperm In flowering plants, nutritive storage tissue that is derived from the union of a sperm nucleus and polar nuclei in the embryo sac. 297

endospore Spore formed within a cell; certain bacteria form endospores. 275

endosymbiotic theory Explanation of the evolution of eukaryotic organelles by phagocytosis of prokaryotes. 279

endotherm Animal that generates its own internal heat. 329

energy Capacity to do work and bring about change; occurs in a variety of forms. 3, 72

energy laws Two laws explaining energy, and its relationships and exchanges. The first, also called the "law of conservation," says that energy cannot be created or destroyed but can only be changed from one form to another; the second says that energy cannot be changed from one form to another without a loss of usable energy. 72

energy of activation (E$_a$) Energy that must be added in order for molecules to react with one another. 78

enhancer Element that stimulates transcription of nearby genes. Functions by acting as a binding site for transcription factors. 189

entropy Measure of disorder or randomness. 73

enzyme Organic catalyst, usually a protein, that speeds a reaction in cells due to its particular shape. 40, 77

enzyme inhibition Means by which cells regulate enzyme activity; may be competitive or noncompetitive inhibition. 78

epidermal tissue (epidermis) In plants, tissue that covers roots, leaves, and stems of nonwoody organisms. 345

epididymis Coiled tubule next to the testes where sperm mature and may be stored for a short time. 520

epiglottis Structure that covers the glottis and closes off the air tract during the process of swallowing. 430

epinephrine Hormone secreted by the adrenal medulla in times of stress; adrenaline. 493

epithelial tissue Tissue that lines hollow organs and covers surfaces. 384

equilibrium species Species demonstrating a life history pattern in which members exhibit logistic population growth and the population size remains at or near the carrying capacity. Its members are large in size, slow to mature, have a long life span, have few offspring, and provide much care to offspring (e.g., bears, lions). 550

erection Increase in blood flow to the penis during sexual arousal, causing the penis to stiffen and become erect. 521

esophagus Muscular tube for moving swallowed food from the pharynx to the stomach. 430

essential nutrient Substance in the diet that contributes to good health; must be supplied by the diet because the body either cannot synthesize it or makes insufficient amounts to meet the body's needs. 443

estrogen Female sex hormone that helps maintain sexual organs and secondary sex characteristics. 524

ethylene Plant hormone that causes ripening of fruit and is also involved in abscission. 363

euchromatin Chromatin that is extended and accessible for transcription. 188

eudicot Abbreviation of eudicotyledon. Flowering plant group; members have two embryonic leaves (cotyledons), net-veined leaves, vascular bundles in a ring, flower parts in fours or fives and their multiples, and other characteristics. 344

eugenics Gene manipulation to control desired traits in offspring. 211

eukaryotic cell Type of cell that has a membrane-bounded nucleus and membranous organelles; found in organisms within the domain Eukarya. 52

eutrophication Enrichment of water by inorganic nutrients used by phytoplankton. Often, overenrichment caused by human activities leads to excessive bacterial growth and oxygen depletion. 571

evergreen Land plant that sheds leaves over a long period, so some leaves are always present. 342

evolution Descent of organisms from common ancestors with the development of genetic and phenotypic changes over time that make them more suited to the environment. 5, 217

evolutionary tree Diagram that shows the evolutionary history of groups of organisms. 255

exocrine gland Gland that discharges its secretion into ducts; the pancreas is an exocrine gland when it secretes pancreatic juice into the duodenum. 434

exocytosis Process in which an intracellular vesicle fuses with the plasma membrane so that the vesicle's contents are released outside the cell. 81

exophthalmos An abnormal protrusion of the eyes, often a sign of an hyperthyroidism (Graves disease). 492

exoskeleton Protective external skeleton, as in arthropods. 508

exotic species Species that is new to a community, nonnative. 565

experiment A series of actions undertaken to collect data with which to test a hypothesis. 11

experimental design Artificial situation devised to test a hypothesis. 11

experimental variable In a scientific experiment, a condition of the experiment that is deliberately changed. 11

exponential growth Growth, particularly of a population, in which the increase occurs in the same manner as compound interest. 546

external fertilization Fertilization of an egg by sperm that occurs outside of the body, as in many aquatic animals. 518

extinction Total disappearance of a species or higher group. 14, 551

extracellular matrix (ECM) Meshwork of polysaccharides and proteins that provides support for an animal cell and affects its behavior. 66

extraembryonic membranes Membranes that are not a part of the embryo but are necessary to the continued existence and health of the embryo. 519

ex vivo gene therapy Gene therapy in which infected tissue is removed from an organism, injected with normal genes, and then returned to the organism where the normal genes divide producing normal genes instead of infected ones. 212

F

facial bone Axial skeleton bones including the lower and upper jawbones, cheekbones, and nasal bones. 509

facilitated diffusion Passive transfer of a substance into or out of a cell along a concentration gradient by a process that requires a carrier. 79

family One of the categories, or taxa, used by taxonomists to group species; the taxon above the genus level. 6, 258

fat Organic molecule that contains glycerol and fatty acids and is found in adipose tissue of vertebrates. 37, 445

fatty acid Molecule that contains a hydrocarbon chain and ends with an acid group. 37

feedback inhibition Mechanism for regulating metabolic pathways in which the concentration of the product is kept within a certain range until binding shuts down the pathway, and no more product is produced. 78

female gametophyte In seed plants, the gametophyte that produces an egg; in flowering plants, an embryo sac. 368

femur The thighbone, which is the longest, strongest bone of the body. 509

fermentation Anaerobic breakdown of glucose that results in a gain of 2 ATP and end products such as alcohol and lactate. 109

fern Member of a group of plants that have large fronds; in the sexual life cycle, the independent gametophyte produces flagellated sperm, and the vascular sporophyte produces windblown spores. 292

fertilization Fusion of sperm and egg nuclei, producing a zygote that develops into a new individual. 130, 530

fiber Structure resembling a thread; also plant material that is nondigestible. 444

fibroblast Connective tissue cell that synthesizes fibers and ground substance. 386

fibrous protein Structural protein with only a secondary structure (i.e., keratin, silk, collagen). 43

fibula The outer, more slender bone in a human's lower leg. 509

filament End-to-end chains of cells that form as cell division occurs in only one plane; in plants, the elongated stalk of a stamen. 296, 367

filter feeder Method of obtaining nourishment by certain animals, which strain minute organic particles from the water in a way that deposits them in the digestive tract. 314

filtration Movement of small molecules from a blood capillary into the nephron capsule due to the action of blood pressure. 425

fimbriae (sing., fimbria) In bacteria, small, bristlelike fibers on bacterial cell surface that enable bacteria to adhere to surfaces. 53, 522

fitness The reproductive success of an individual relative to other members of a population. 224

five-kingdom system System of classification that contains the kingdoms Monera, Protista, Plantae, Animalia, and Fungi. 263

flagellum (pl., flagella) Long, slender extension used for locomotion by some bacteria, protozoans, and sperm. 53, 65, 274

flatworm Invertebrate such as a planarian and tapeworm that has a thin body, three branched gastrovascular cavity, and ladder-type nervous system. 315

flower Reproductive organ of a flowering plant, consisting of several kinds of modified leaves arranged in concentric rings and attached to a modified stem called the receptacle. 289, 366

follicle Structure in the ovary of animals that contains an oocyte; site of oocyte production. 523

follicular phase First half of the ovarian cycle, during which the follicle matures and much estrogen (and some progesterone) is produced. 523

food chain Order in which one population feeds on another in an ecosystem, thereby showing the flow of energy from a detritivore (detrital food chain) or a producer (grazing food chain) to the final consumer. 569

food web In ecosystems, a complex pattern of linked and crisscrossing food chains. 568

foraminiferan Protozoan; marine amoeba having a calcium carbonate skeleton, many of which make limestone formations such as the White Cliffs of Dover. 282

formed elements Constituents of blood that are either cellular (red blood cells and white blood cells) or at least cellular in origin (platelets). 409

fossil Any past evidence of an organism that has been preserved in the Earth's crust. 219, 255

fossil fuel Fuel such as oil, coal, and natural gas that is the result of partial decomposition of plants and animals coupled with exposure to heat and pressure for millions of years. 573, 589

fossil record History of life recorded from remains from the past. 226

founder effect The effect of establishing a new population by a small number of individuals, carrying only a small fraction of the original population's genetic variation. The alleles carried by these individuals often, by chance, do not occur in the same frequency as the parent population; an example of genetic drift. 240

fovea Region of the retina consisting of densely packed cones; responsible for the greatest visual acuity. 506

frameshift mutation Alteration in a gene due to deletion of a base, so that the reading "frame" is shifted; can result in a nonfunctional protein. 174

frond Fern leaf. 292

fruit In flowering plants, the structure that forms from an ovary and associated tissues and encloses seeds. 297, 370

functional group Specific cluster of atoms attached to the carbon skeleton of organic molecules that enters into reactions and behaves in a predictable way. 33

fungus (pl., fungi) Saprotrophic decomposer; the body is made up of filaments called hyphae that form a mass called a mycelium. 6, 300

G

gallbladder Organ attached to the liver that serves to store and concentrate bile. 432

gamete Haploid sex cell; e.g., egg and sperm. 528

gametophyte Haploid generation of the alternation of generations life cycle of a plant; produces gametes that unite to form a diploid zygote. 290, 366

ganglion Collection or bundle of neuron cell bodies usually outside the central nervous system. 486

gap junction Junction between cells formed by the joining of two adjacent plasma membranes; it lends strength and allows ions, sugars, and small molecules to pass between cells. 67

gastropod Mollusc with a broad flat foot for crawling (e.g., snails and slugs). 318

gastrula Stage of animal development during which the germ layers form, at least in part, by invagination. 532

gastrulation Formation of a gastrula from a blastula; characterized by an invagination of the cell layers to form a caplike structure. 532

gene Unit of heredity existing as alleles on the chromosomes; in diploid organisms, typically two alleles are inherited—one from each parent. 4

gene flow Sharing of alleles among multiple populations through interbreeding. 238

gene linkage Existence of several alleles on the same chromosome. 156

gene locus Specific location of a particular gene on homologous chromosomes. 147

gene mutation Change in the sequence of bases in a gene. 174

gene pool Total of all the genes of all the individuals in a population. 235

gene therapy Correction of a detrimental mutation by the addition of new DNA and its insertion in a genome. 212

genetically modified plant (GMP) Plant that carries the genes of another organism as a result of DNA technology; also transgenic plant. 375

genetic counseling Prospective parents consult a counselor who determines the genotype of each and whether an unborn child will have a genetic disorder. 200

genetic drift Mechanism of evolution due to random changes in the allelic frequencies of a population;

more likely to occur in small populations or when only a few individuals of a large population reproduce. 238

genetic engineering Alteration of genomes for medical or industrial purposes. 175

genetic marker Abnormality in the sequence of a base at a particular location on a chromosome signifying a disorder. 209

genetic profile An individual's complete genotype, including any possible mutations. 209

genital herpes Viral STD characterized by painful blisters on the genitals. 529

genital warts Viral STD caused by the human papillomaviruses; associated with cervical cancer and penile tumors; most common STD in the U.S. 528

genomics The study of genomes. 178

genotype Alleles of an organism for a particular trait or traits; often designated by letters—for example, *BB* or *Aa*. 147

genus One of the categories, or taxa, used by taxonomists to group species; contains those species that are most closely related through evolution. 6, 258

germinate Beginning of growth of a seed, spore, or zygote, especially after a period of dormancy. 372

germ layer Primary tissue layer of a vertebrate embryo—namely, ectoderm, mesoderm, or endoderm. 312, 532

gibberellin Plant hormone promoting increased stem growth; also involved in flowering and seed germination. 361

global warming Predicted increase in the Earth's temperature due to human activities that promote the greenhouse effect. 573

globular protein A protein (i.e., enzymes) whose polypeptides give it a tertiary structure and a globular shape. 43

glottis Opening for airflow in the larynx. 419

glucagon Hormone secreted by the pancreas, which causes the liver to break down glycogen and raises the blood glucose level. 434, 494

glucocorticoid Type of hormone secreted by the adrenal cortex that influences carbohydrate, fat, and protein metabolism. 493

glucose Six-carbon sugar that organisms degrade as a source of energy during cellular respiration. 35

glycerol Three-carbon carbohydrate with three hydroxyl groups attached; a component of fats and oils. 37

glycogen Storage polysaccharide found in animals; composed of glucose molecules joined in a linear fashion but having numerous branches. 36

glycolysis Anaerobic breakdown of glucose that results in a gain of 2 ATP and the end product pyruvate. 101

golden-brown algae Marine photosynthetic protist with elaborate shells of silica; represented by diatoms. 280

Golgi apparatus Organelle consisting of saccules and vesicles that processes, packages, and distributes molecules about or from the cell. 60

Golgi tendon Proprioceptive sensory receptor located where skeletal muscle inserts into tendons; sensitive to changes in tendon tension. 507

gonadotropic hormones (FSH and LH) Substances secreted by the anterior pituitary that regulate the activity of the ovaries and testes; principally, follicle-

stimulating hormone (FSH) and luteinizing hormone (LH). 490

gonads Organs that produce gametes; the ovary produces eggs, and the testis produces sperm. 518

gonorrhea Bacterial STD that can lead to sterility or infertility. 529

granum (pl., grana) Stack of chlorophyll-containing thylakoids in a chloroplast. 62, 87

gravitational equilibrium Maintenance of balance when the head and body are motionless. 504

gravitropism Growth response of roots and stems of plants to the Earth's gravity; roots demonstrate positive gravitropism, and stems demonstrate negative gravitropism. 364

grazing food web Complex pattern of linked and crisscrossing food chains that begins with a population of photosynthesizers serving as producers. 568

green algae Member of a diverse group of photosynthetic protists that contains chlorophylls *a* and *b* and has other biochemical characteristics like those of plants. 280

greenhouse effect Reradiation of solar heat toward the Earth, caused by gases such as carbon dioxide, methane, nitrous oxide, water vapor, ozone, and nitrous oxide in the atmosphere. 573

greenhouse gases Gases in the atmosphere such as carbon dioxide, methane, nitrous oxide, water vapor, ozone, and nitrous oxide that like the panes of a greenhouse allow the sun's rays to pass through but trap the heat. 589

ground tissue Tissue that constitutes most of the body of a plant; consists of parenchyma, collenchyma, and sclerenchyma cells that function in storage, basic metabolism, and support. 345

growth factor Chemical signal that regulates mitosis and differentiation of cells that have receptors for it; important in such processes as fetal development, tissue maintenance and repair, and hematopoiesis; sometimes a contributing factor in cancer. 193

growth hormone (GH) Substance secreted by the anterior pituitary; controls size of an individual by promoting cell division, protein synthesis, and bone growth. 490

guanine (G) One of four nitrogen-containing bases in nucleotides composing the structure of DNA and RNA; pairs with cytosine. 163

gymnosperm Type of woody seed plant in which the seeds are not enclosed by fruit and are usually borne in cones, such as those of the conifers. 290, 295

H

habitat Place where an organism lives and is able to survive and reproduce. 562

halophile Type of archaea that lives in extremely salty habitats. 278

haploid (n) number Cell condition in which only one of each type of chromosome is present. 131

heart Muscular organ whose contraction causes blood to circulate in the body of an animal. 400

heart attack Damage to the myocardium due to blocked circulation in the coronary arteries; myocardial infarction. 408

heart murmur Clicking or swishy sounds, often due to leaky valves. 403

heat Type of kinetic energy; captured solar energy eventually dissipates as heat in the environment. 73

helper T cell Secretes cytokines, which stimulate all kinds of immune cells. 469

heme Iron-containing group found in hemoglobin. 423

hemocoel Body cavity in arthropods where exchange between hemolymph and tissues occurs. 400

hemodialysis Cleansing of blood by using an artificial membrane that causes substances to diffuse from blood into a dialysis fluid. 427

hemoglobin Iron-containing respiratory pigment occurring in vertebrate red blood cells and in the blood plasma of some invertebrates. 40, 409

hepatitis Viral infection of the liver that can be transmitted sexually and can lead to liver failure and liver cancer. 529

herbaceous Plants that have nonwoody stems. 348

herbivore Primary consumer in a grazing food chain; a plant eater. 429, 566

hermaphrodite Type of animal that has both male and female sex organs. 316, 518

heterochromatin Highly compacted chromatin that is not accessible for transcription. 188

heterotroph Organism that cannot synthesize organic compounds from inorganic substances, and therefore must take in organic food. 566

heterozygous Possessing unlike alleles for a particular trait. 147

hinge joint A synovial joint (elbow and knee joints) that permits movement in one direction only. 512

histamine Substance, produced by basophils in blood and mast cells in connective tissue, that causes capillaries to dilate. 464

histone Protein molecule responsible for packing chromatin. 115

HIV (human immunodeficiency virus) Virus responsible for AIDS. 472, 518

homeostasis Maintenance of normal internal conditions in a cell or an organism by means of self-regulating mechanisms. 3, 392

hominid A family designation for both humans and African apes. 333

hominin A designation that includes humans and extinct species very closely related to humans. 333

Homo erectus An early *Homo* who used fire and migrated out of Africa to Europe and Asia. 335

Homo habilis An early *Homo* dated between 2.0 and 1.9 million years ago who is believed to have been the first tool user. 334

homologous chromosome Member of a pair of chromosomes that are alike and come together in synapsis during prophase of the first meiotic division; a *homologue*. 131

homologous structure In evolution, a structure that is similar in different types of organisms because these organisms are derived from a common ancestor. 228, 259

homologue Member of a pair of chromosomes that are alike and come together in synapsis during prophase of the first meiotic division; a *homologous chromosome*. 131

Homo sapiens Modern humans. 335

homozygous Possessing two identical alleles for a particular trait. 147

humerus Long bone of the arm that articulates with the scapula. 509

Huntington disease Genetic disease marked by progressive deterioration of the nervous system and resulting in neuromuscular disorders. 207

hyaline cartilage Cartilage whose cells lie in lacunae separated by a white translucent matrix containing very fine collagen fibers. 387

hybridization Crossing of different species of organisms, or different varieties of plants. 374

hydrogen (H) bond Weak bond that arises between a slightly positive hydrogen atom of one molecule and a slightly negative atom of another molecule or between parts of the same molecule. 24

hydrogen ion (H⁺) Hydrogen atom that has lost its electron and therefore bears a positive charge (H⁺). 27

hydrolysis reaction Splitting of a compound by the addition of water, with the H⁺ being incorporated in one molecule and the OH⁻ in the other. 34

hydrophilic Type of molecule that interacts with water by dissolving in water and/or by forming hydrogen bonds with water molecules. 25, 33

hydrophobic Type of molecule that does not interact with water because it is nonpolar. 25, 33

hydrostatic skeleton Fluid-filled body compartment that provides support for muscle contraction resulting in movement; seen in cnidarians, flatworms, roundworms, and segmented worms. 509

hydroxide ion (OH⁻) One of two ions that results when a water molecule dissociates; it has gained an electron, and therefore bears a negative charge (OH⁻). 27

hypertension Elevated blood pressure, particularly the diastolic pressure. 407

hypertonic solution Higher solute concentration (less water) than the cytoplasm of a cell; causes cell to lose water by osmosis. 80

hypha Filament of the vegetative body of a fungus. 300

hypothalamic-releasing hormone One of many hormones produced by the hypothalamus that stimulates the secretion of an anterior pituitary hormone. 490

hypothalamus In vertebrates, part of the brain that helps regulate the internal environment of the body—for example, heart rate, body temperature, and water balance. 485, 490

hypothesis Supposition established by reasoning after consideration of available evidence; it can be tested by obtaining more data, often by experimentation. 10

hypotonic solution Lower solute (more water) concentration than the cytoplasm of a cell; causes cell to gain water by osmosis. 80

I

immediate allergic response Allergic response that occurs within seconds of contact with an allergen; caused by the attachment of the allergen to receptors on the surface of mast cells. 471

immune system All the cells in the body that protect the body against foreign organisms and substances and also against cancerous cells. 390, 462

immunity Ability of the body to protect itself from foreign substances and cells, including disease-causing agents. 462

incomplete dominance Inheritance pattern in which the offspring has an intermediate phenotype, as when a red-flowered plant and a white-flowered plant produce pink-flowered offspring. 151

induced fit model Change in the shape of an enzyme's active site that enhances the fit between the active site and its substrate(s). 77

induction Ability of a chemical or a tissue to influence the development of another tissue. 532

inductive reasoning Using specific observations and the process of logic and reasoning to arrive at a hypothesis. 10

industrial melanism Increased frequency of darkly pigmented (melanic) forms in a population when soot and pollution make lightly pigmented forms easier for predators to see against a pigmented background. 236

infertility Inability to have as many children as desired. 526

inflammatory response Tissue response to injury that is characterized by redness, swelling, pain, and heat. 464

inorganic chemistry The study of compounds not having a carbon basis; chemistry of the nonliving world. 32

insect Type of arthropod. The head has antennae, compound eyes, and simple eyes; the thorax has three pairs of legs and often wings; and the abdomen has internal organs. 322

insulin Hormone secreted by the pancreas that lowers the blood glucose level by promoting the uptake of glucose by cells and the conversion of glucose to glycogen by the liver and skeletal muscles. 434, 494

integration Summing up of excitatory and inhibitory signals by a neuron or by some part of the brain. 482

integumentary system Organ system consisting of skin and various organs, such as hair, which are found in skin. 391

intercalated disk Region that holds adjacent cardiac muscle cells together; disks appear as dense bands at right angles to the muscle striations. 388

interkinesis Period of time between meiosis I and meiosis II during which no DNA replication takes place. 135

intermembrane space Space that occurs between the outer and inner membrane of a mitochondrion. 107

interneuron Neuron located within the central nervous system that conveys messages between parts of the central nervous system. 480

internode In vascular plants, the region of a stem between two successive nodes. 343

interphase Stages of the cell cycle (G₁, S, G₂) during which growth and DNA synthesis occur when the nucleus is not actively dividing. 116

intervertebral disks Pads of cartilage that separate vertebrae. 509

intestinal enzyme Enzyme, produced by the epithelial cells on the surface of villi, which functions in the digestion of small organic molecules. 432

intrauterine device (IUD) Birth-control device consisting of a small piece of molded plastic inserted into the uterus, and believed to alter the uterine environment so that fertilization does not occur. 526

inversion Change in chromosome structure in which a segment of a chromosome is turned around 180°; this reversed sequence of genes can lead to altered gene activity and abnormalities. 202

in vivo gene therapy Gene therapy in which normal genes are injected directly into an organism for growth. 212

ion Charged particle that carries a negative or positive charge. 22

ionic bond Chemical bond in which ions are attracted to one another by opposite charges. 22

iris Muscular ring that surrounds the pupil and regulates the passage of light through this opening. 505

isomer Molecules with the same molecular formula but a different structure, and therefore a different shape. 33

isotonic solution Solution that is equal in solute concentration to that of the cytoplasm of a cell; causes cell to neither lose nor gain water by osmosis. 80

isotope Atom of the same element having the same atomic number but a different mass number due to the number of neutrons. 19

J

jawless fishes Type of fish(es) that have no jaws, includes today's hagfishes and lampreys. 326

K

karyotype Chromosomes arranged by pairs according to their size, shape, and general appearance in mitotic metaphase. 200

keystone species Species whose activities significantly affect community structure. 564

kidneys Paired organs of the vertebrate urinary system that regulate the chemical composition of the blood and produce a waste product called urine. 424

kilocalorie Caloric value of food; 1,000 calories. 72

kinase Enzyme that activates another enzyme by adding a phosphate group. 121

kinetic energy Energy associated with motion. 72

kingdom One of the categories, or taxa, used by taxonomists to group species; the taxon above phylum. 6, 258

Klinefelter syndrome Condition caused by the inheritance of XXY chromosomes. 139

L

lacteal Lymphatic vessel in an intestinal villus; aids in the absorption of fats. 432

lacuna Small pit or hollow cavity, as in bone or cartilage, where a cell or cells are located. 387

lancelet Invertebrate chordate with a body that resembles a lancet and has the four chordate characteristics as an adult. 324

large intestine In vertebrates, portion of the digestive tract that follows the small intestine; in humans, consists of the cecum, colon, rectum, and anal canal. 433

larynx Cartilaginous organ located between the pharynx and the trachea; in humans, contains the vocal cords; sometimes called the voice box. 419

lateral bud Site on a stem where lateral branches grow. 343

lateral line Canal system containing sensory receptors that allow fishes and amphibians to detect water currents and pressure waves from nearby objects. 504

law Synonym for principle. 12

law of independent assortment Alleles of unlinked genes assort independently of each other during meiosis so that the gametes contain all possible combinations of alleles. 148

law of segregation Separation of alleles from each other during meiosis so that the gametes contain one from each pair. Each resulting gamete has an equal chance of receiving either allele. 147

leaf Lateral appendage of a stem, highly variable in structure, often containing cells that carry out photosynthesis. 342

leaf vein Vascular tissue within a leaf. 344

leech Blood-sucking annelid, usually found in fresh water, with a sucker at each end of a segmented body. 319

lens Clear, membranelike structure found in the vertebrate eye behind the iris; brings objects into focus. 505

less-developed country (LDC) Country that is becoming industrialized; typically, population growth is expanding rapidly, and the majority of people live in poverty. 541

lichen Symbiotic relationship between certain fungi and algae, in which the fungi possibly provide inorganic food or water and the algae provide organic food. 303

life cycle Recurring pattern of genetically programmed events by which individuals grow, develop, maintain themselves, and reproduce. 131

life history Adaptations in characteristics that influence an organism's biology, such as how many offspring it produces, its survival, and factors such as age and size that determine its reproductive maturity. 550

ligament Tough cord or band of dense fibrous tissue that binds bone to bone at a joint. 387, 512

light reaction Portion of photosynthesis that captures solar energy and takes place in thylakoid membranes of chloroplasts; it produces ATP and NADPH. 88

lignin Chemical that hardens the cell walls of plants. 292

limbic system In humans, functional association of various brain centers, including the amygdala and hippocampus; governs learning and memory and various emotions such as pleasure, fear, and happiness. 486

lineage Evolutionary line of descent. 332

lipase Fat-digesting enzyme secreted by the pancreas. 432

lipid Class of organic compounds that tends to be soluble in nonpolar solvents; includes fats and oils. 37

liver Large, dark red internal organ that produces urea and bile, detoxifies the blood, stores glycogen, and produces the plasma proteins, among other functions. 432

lobe-finned fishes Type of fish(es) with limblike fins. 327

logistic growth Population increase that results in an S-shaped curve; growth is slow at first, steepens, and then levels off due to environmental resistance. 547

long-day plant/short-night plant Plant that flowers when day length is longer and the night is shorter than a critical length; e.g., wheat, barley, clover, and spinach. 364

loose fibrous connective tissue Tissue composed mainly of fibroblasts widely separated by a matrix containing collagen and elastic fibers. 386

lumen Cavity of a tubular organ. 384

lungs Internal respiratory organs containing moist surfaces for gas exchange. 327, 419

luteal phase Second half of the ovarian cycle, during which the corpus luteum develops and much progesterone (and some estrogen) is produced. 523

lycophyte Seedless plant with microphylls and well-developed vascular tissue in roots, stems, and leaves; also called club mosses. 292

lymph Fluid, derived from tissue fluid, that is carried in lymphatic vessels. 407

lymphatic organs Organs other than a lymphatic vessel that are part of the lymphatic system; the lymphatic organs are the lymph nodes, tonsils, spleen, thymus gland, and bone marrow. 462

lymphatic system Organ system consisting of lymphatic vessels and lymphatic organs; transports lymph and lipids, and aids the immune system. 390, 407

lymphatic vessel Vessel that carries lymph. 407

lymph node Mass of lymphatic tissue located along the course of a lymphatic vessel. 463

lymphocyte Specialized white blood cell that functions in specific defense; occurs in two forms—T lymphocytes and B lymphocytes. 410

lysogenic cycle Bacteriophage life cycle in which the virus incorporates its DNA into that of a bacterium; occurs preliminary to the lytic cycle. 269

lysosome Membrane-bounded vesicle that contains hydrolytic enzymes for digesting macromolecules. 60

lytic cycle Bacteriophage life cycle in which the virus takes over the operation of the bacterium immediately upon entering it and subsequently destroys the bacterium. 269

M

macroevolution Large-scale evolutionary change, such as the formation of new species. 248

macronutrient Essential element needed in large amounts by plants and humans. In plants, nitrogen, calcium, and sulfur are needed for plant growth; in humans, carbohydrates, lipids, and proteins supply the body's energy needs. 443

macrophage In vertebrates, large phagocytic cell derived from a monocyte that ingests microbes and debris. 410, 463

major mineral Essential inorganic nutrient (such as calcium, potassium, phosphorus, sodium, chloride, magnesium, or sulfur) required daily by humans to regulate metabolic activities and maintain health. 447

male gametophyte In seed plants, the gametophyte that produces sperm; a pollen grain. 368

malignant Invasive tumor that may spread. 123

Malpighian tubule Blind, threadlike excretory tubule near the anterior end of an insect's hindgut. 426

maltase Enzyme produced in small intestine that breaks down maltose to two glucose molecules. 436

mammal Endothermic vertebrate characterized especially by the presence of hair and mammary glands. 330

Marfan syndrome Congenital disorder of connective tissue characterized by abnormal length of the extremities. 207

marsupial Member of a group of mammals bearing immature young nursed in a marsupium, or pouch—for example, kangaroo and opossum. 330

mass extinction Episode of large-scale extinction in which large numbers of species disappear in a few million years or less. 256

mast cell Connective tissue cell that releases histamine in allergic reactions. 464

matrix Unstructured semifluid substance that fills the space between cells in connective tissues or inside organelles. 63, 386

matter Anything that takes up space and has mass. 18

medulla oblongata In vertebrates, part of the brain stem that is continuous with the spinal cord; controls heartbeat, blood pressure, breathing, and other vital functions. 485

megaphylls Large leaves with complex networks of veins. 288, 292

megaspore One of the two types of spores produced by seed plants; develops into a female gametophyte (embryo sac). 366

meiosis, meiosis I, meiosis II Type of nuclear division that occurs as part of sexual reproduction in which the daughter cells receive the haploid number of chromosomes in varied combinations. 130, 132

memory Capacity of the brain to store and retrieve information about past sensations and perceptions; essential to learning. 486

meniscus Cartilaginous wedges that separate the surfaces of bones in synovial joints. 512

menopause Termination of the ovarian and menstrual cycles in older women. 524

menses Flow of blood during menstruation. 524

menstrual cycle Cycle that runs concurrently with the ovarian cycle; it prepares the uterus to receive a developing zygote. 524

menstruation Periodic shedding of tissue and blood from the inner lining of the uterus in primates. 524

meristem Undifferentiated embryonic tissue in the active growth regions of plants. 345

mesoderm Middle primary tissue layer of an animal embryo that gives rise to muscle, several internal organs, and connective tissue layers. 532

mesophyll Inner, thickest layer of a leaf consisting of palisade and spongy mesophyll; the site of most of photosynthesis. 346

messenger RNA (mRNA) Type of RNA formed from a DNA template and bearing coded information for the amino acid sequence of a polypeptide. 167

metabolic pathway Series of linked reactions, beginning with a particular reactant and terminating with an end product. 77

metabolism All of the chemical reactions that occur in a cell during growth and repair. 3

metaphase Mitotic phase during which chromosomes are aligned at the spindle equator. 117

metastasis Spread of cancer from the place of origin throughout the body; caused by the ability of cancer cells to migrate and invade tissues. 123

methanogen Type of archaea that lives in oxygen-free habitats, such as swamps, and releases methane gas. 278

methemoglobinemia Disorder resulting from the accumulation of methemoglobin in the blood, causing the skin of those affected to appear bluish-purple. 206

microevolution Small measurable evolutionary changes within a population from generation to generation; change in allele frequencies within a population over time. 234

micronutrient Essential element needed in small amounts by plants and humans. In plants, boron, copper, and zinc are needed for plant growth; in humans, vitamins and minerals help regulate metabolism and physiological development. 443

microphylls Small, narrow leaves with single, unbranched veins. 292

microspore One of the two types of spores produced by seed plants; develops into a male gametophyte (pollen grain). 366

microtubule Small, cylindrical organelle composed of tubulin protein around an empty central core; present in the cytoplasm, centrioles, cilia, and flagella. 64

microvillus (pl., microvilli) Cylindrical process that extends from an epithelial cell of a villus and serves to increase the surface area of the cell. 432

midbrain Part of the brain stem located between the diencephalon and the pons. 485

mineral Naturally occurring inorganic substance containing one or more elements; certain minerals are needed in the diet. As raw materials in the Earth, they are nonrenewable resources. 447, 592

mineralocorticoid Hormones secreted by the adrenal cortex that regulate salt and water balance, leading to increases in blood volume and blood pressure. 493

mitochondrion Membrane-bounded organelle in which ATP molecules are produced during the process of cellular respiration. 62

mitosis Process in which a parent nucleus produces two daughter nuclei, each having the same number and kinds of chromosomes as the parent nucleus. 116

model Simulation of a process that aids conceptual understanding until the process can be studied firsthand; a hypothesis that describes how a particular process could possibly be carried out. 11

molecule Union of two or more atoms of the same element; also, the smallest part of a compound that retains the properties of the compound. 21

molting Periodic shedding of the exoskeleton in arthropods or the cuticle in roundworms. 320

monocot Abbreviation of monocotyledon. Flowering plant group; members have one embryonic leaf (cotyledon), parallel-veined leaves, scattered vascular bundles, flower parts in threes or multiples of three, and other characteristics. 344

monocyte Type of a granular leukocyte that functions as a phagocyte, particularly after it becomes a macrophage. 410

monomer Small molecule that is a subunit of a polymer—e.g., glucose is a monomer of starch. 34

monosaccharide Simple sugar; a carbohydrate that cannot be decomposed by hydrolysis—e.g., glucose. 35

monotreme Egg-laying mammal—e.g., duckbill platypus and spiny anteater. 330

more-developed country (MDC) Country that is industrialized; typically, population growth is low, and the people enjoy a good standard of living. 541

morula Spherical mass of cells resulting from cleavage during animal development prior to the blastula stage. 531

mosaic evolution Concept that human characteristics did not evolve at the same rate; for example, some body parts are more humanlike than others in early hominins. 334

motor neuron Nerve cell that conducts nerve impulses away from the central nervous system and innervates effectors (muscle and glands). 480

motor protein Proteins associated with the cytoskeleton that allow movement of transport vesicles and organelles. 64

mouth In humans, organ of the digestive tract where food is chewed and mixed with saliva. 429

mRNA transcript Complementary copy of the sequence of bases in template. 169

multicellular Organism composed of many cells; usually has organized tissues, organs, and organ systems. 2

multifactorial trait Trait controlled by polygenes subject to environmental influences; each dominant allele contributes to the phenotype in an additive and like manner. 152

multiregional continuity hypothesis Proposal that modern humans evolved separately in at least three different places: Asia, Africa, and Europe. 335

muscle dysmorphia Mental state where a person thinks his or her body is underdeveloped, and becomes preoccupied with body-building and diet; affects more men than women. 454

muscle spindle Proprioceptor wrapped around a few muscle cells within a connective tissue sheath; can respond to changes in muscle length. 507

muscle tissue Type of animal tissue composed of fibers that shorten and lengthen to produce movements. 388

muscular system System of muscles that produces movement, both within the body and of its limbs; principal components are skeletal muscle, smooth muscle, and cardiac muscle. 391

mutagen Agent, such as radiation or a chemical, that brings about a mutation in DNA. 174

mutation A change made in the DNA base-pair sequences within a gene due to a copying error or external souces called mutagens. Such changes generate variation within a gene pool. 237

mutualism Symbiotic relationship in which both species benefit in terms of growth and reproduction. 562

mycelium Tangled mass of hyphal filaments composing the vegetative body of a fungus. 300

mycorrhizal fungi, mycorrhiza Mutualistic relationship between fungal hyphae and roots of vascular plants; also called mycorrhizal association. 303

myelin sheath White, fatty material—derived from the membranes of tightly spiraled cells—that forms a covering for nerve fibers. 480

myofibril Specific muscle cell organelle containing a linear arrangement of sarcomeres, which shorten to produce muscle contraction. 511

myosin Muscle protein making up the thick filaments in a sarcomere; it pulls actin to shorten the sarcomere, yielding muscle contraction. 511

N

nasal cavity One of two canals in the nose, separated by a septum. 419

native species Indigenous species that colonize an area without human assistance. 565

natural history The study of how organisms are influenced by factors such as climate, predation, competition, and evolution; uses field observations instead of experimentation. 553

natural killer (NK) cell Lymphocyte that causes an infected or cancerous cell to burst. 465

natural selection Mechanism of evolution caused by environmental selection of organisms most fit to reproduce; results in adaptation to the environment. 7, 223, 240

Neandertal Later *Homo* with a sturdy build that lived during the last Ice Age in Europe and the Middle East; hunted large game and left evidence of being culturally advanced. 336

negative feedback Mechanism of homeostatic response by which the output of a system suppresses or inhibits activity of the system. 393, 491

nematocyst In cnidarians, a capsule that contains a threadlike fiber, the release of which aids in the capture of prey. 314

nephridium Segmentally arranged, paired excretory tubules of many invertebrates, as in the earthworm. 319

nephron Microscopic kidney unit that regulates blood composition by filtration, reabsorption, and secretion. 424

nephron capsule Cuplike structure that is the initial portion of a nephron. 425

nephron loop Portion of a nephron between the proximal and distal tubules; functions in water reabsorption. 425

nerve Bundle of long axons outside the central nervous system. 389, 486

nerve deafness Deafness due to spiral organ damage. 503

nervous system Organ system consisting of the brain, spinal cord, and associated nerves that coordinates the other organ systems of the body. 391

nervous tissue Tissue that contains nerve cells (neurons), which conduct impulses, and neuroglia, which support, protect, and provide nutrients to neurons. 389

neural plate Region of the dorsal surface of the chordate embryo that marks the future location of the neural tube. 532

neural tube Tube formed by closure of the neural groove during development. In vertebrates, the neural tube develops into the spinal cord and brain. 532

neuroglia Nonconducting nerve cells that are intimately associated with neurons and function in a supportive capacity. 389

neuron Nerve cell that characteristically has three parts: dendrites, cell body, and an axon. 389, 480

neurotransmitter Chemical stored at the ends of axons that is responsible for transmission across a synapse. 482

neurulation Development of the central nervous system organs in an embryo. 532

neutron Neutral subatomic particle, located in the nucleus and assigned one atomic mass unit. 18

neutrophil Granular leukocyte that is the most abundant of the white blood cells; first to respond to infection. 410, 464

node In plants, the place where one or more leaves attach to a stem. 343

nodes of Ranvier In the peripheral nervous system, gaps in the myelin sheath encasing long axons. 480

nondisjunction Failure of homologous chromosomes or daughter chromosomes to separate during meiosis I and meiosis II, respectively. 138

nonrandom mating Mating among individuals on the basis of their phenotypic similarities or differences, rather than mating on a random basis. 238

nonrenewable resources Minerals, fossil fuels, and other materials present in essentially fixed amounts (within human time scale) in our environment. 582

nonsister chromatid A tetrad consists of four chromatids. Only the chromatids belonging to a homologue are sister chromatids and the others, belonging to the other homologue, are nonsister chromatids. 133

nonvascular plant Bryophytes such as mosses that have no vascular tissue and either occur in moist locations or have special adaptations for living in dry locations. 291

norepinephrine (NE) Neurotransmitter of the postganglionic fibers in the sympathetic division of the autonomic system; also, a hormone produced by the adrenal medulla. 482, 493

notochord Cartilaginous-like supportive dorsal rod in all chordates sometime in their life cycle; replaced by vertebrae in vertebrates. 324, 532

nuclear envelope Double membrane that surrounds the nucleus in eukaryotic cells and is connected to the endoplasmic reticulum; has pores that allow substances to pass between the nucleus and the cytoplasm. 59

nuclear pore Opening in the nuclear envelope that permits the passage of proteins into the nucleus and ribosomal subunits out of the nucleus. 59

nuclease Enzyme that catalyzes decomposition of nucleic acids. 432

nucleic acid Polymer of nucleotides; both DNA and RNA are nucleic acids. 44

nucleoid Region of prokaryotic cells where DNA is located; it is not bounded by a nuclear envelope. 52, 274

nucleolus Dark-staining, spherical body in the nucleus that produces ribosomal subunits. 58

nucleosome In the nucleus of a eukaryotic cell, a unit composed of DNA wound around a core of eight histone proteins, giving the appearance of a string of beads. 115

nucleotide Monomer of DNA and RNA consisting of a 5-carbon sugar bonded to a nitrogenous base and a phosphate group. 44

nucleus Center of an atom, in which portons and neutrons are found; membrane-bounded organelle within a eukaryotic cell that contains chromosomes and controls the structure and function of the cell. 18, 56

nutrient Chemical substances in foods that are essential to the diet and contribute to good health. 442

nutrition The Council on Food and Nutrition of the American Medical Association defines nutrition as "the science of food; the nutrients and the substances therein; their action, interaction, and balance in relation to health and disease; and the process by which the organism (i.e., body) ingests, digests, absorbs, transports, utilizes, and excretes food substances." 442

O

obesity Characterized by excess adipose tissue; exceeding desirable weight by more than 20%. 445

observation Step in the scientific method by which data are collected before a conclusion is drawn. 10

octet rule States that an atom other than hydrogen tends to form bonds until it has eight electrons in its outer shell; an atom that already has eight electrons in its outer shell does not react and is inert. 21

oil Triglyceride, usually of plant origin, that is composed of glycerol and three fatty acids and is liquid in consistency due to many unsaturated bonds in the hydrocarbon chains of the fatty acids. 37

omnivore Organism in a food chain that feeds on both plants and animals. 429, 566

oncogene Cancer-causing gene. 193

oogenesis Production of eggs in females by the process of meiosis and maturation. 131

open circulatory system Circulatory system, such as that found in a grasshopper, where a tubular heart pumps hemolymph through channels and body cavities. 400

operon Group of structural and regulating genes that function as a single unit. 186

opportunistic population Population demonstrating a life history pattern in which members exhibit exponential population growth. Its members are small in size, mature early, have a short life span, produce many offspring, and provide little or no care to offspring (e.g., dandelions). 550

order One of the categories, or taxa, used by taxonomists to group species; the taxon above the family level. 6, 258

organ Combination of two or more different tissues performing a common function. 2, 382

organelle Small, often membranous structure in the cytoplasm having a specific structure and function. 56

organic Molecule that always contains carbon and hydrogen, and often contains oxygen as well; organic molecules are associated with living things. 32

organic chemistry The study of carbon compounds; chemistry of the living world. 32

organism Individual living thing. 2, 552

organ system Group of related organs working together. 2

orgasm Physiological and psychological sensations that occur at the climax of sexual stimulation. 521

osmosis Diffusion of water through a differentially permeable membrane. 79

osteoblast Bone-forming cell. 510

osteoclast Cell that causes erosion of bone. 510

osteocyte Branched bone cell embedded in a calcium-containing extracellular matrix. 510

osteon Cylindrical-shaped unit containing bone cells that surround an osteonic canal; also called Haversian system. 510

osteoporosis Condition in which bones break easily because calcium is removed from them faster than it is replaced. 510

ostium (pl., ostia) Opening in the heart of an animal with an open circulatory system, which allows blood to fill the relaxed heart. 401

out-of-Africa hypothesis Proposal that modern humans originated only in Africa; then they migrated and supplanted populations of *Homo* in Asia and Europe about 100,000 years ago (see replacement model). 335

ovary Female gonad in animals that produces an egg and female sex hormones; in flowering plants, the enlarged, ovule-bearing portion of the carpel that develops into a fruit. 296, 367, 518

ovulation Bursting of a follicle when a secondary oocyte is released from the ovary; if fertilization occurs, the secondary oocyte becomes an egg. 523

ovule In seed plants, a structure that contains the female gametophyte and has the potential to develop into a seed. 294, 367

oxidation Loss of one or more electrons from an atom or molecule; in biological systems, generally the loss of hydrogen atoms. 97

oxygen deficit Amount of oxygen needed to metabolize lactate, a compound that accumulates during vigorous exercise. 109

oxytocin Hormone released by the posterior pituitary that causes contraction of uterus and milk letdown. 491

P

paleontology Study of fossils that results in knowledge about the history of life. 255

palisade mesophyll Layer of tissue in a plant leaf containing elongated cells with many chloroplasts. 346

pancreas Internal organ that produces digestive enzymes and the hormones insulin and glucagon. 432, 494

pancreatic amylase Enzyme that digests starch to maltose. 432

pancreatic islets Masses of cells that constitute the endocrine portion of the pancreas. 494

parasitism Symbiotic relationship in which one species (the *parasite*) obtains nutrients from another species (the *host*) but does not usually kill the host. 562

parasympathetic division Division of the autonomic system that is active under normal conditions; uses acetylcholine as a neurotransmitter. 488

parathyroid gland Gland embedded in the posterior surface of the thyroid gland; it produces parathyroid hormone. 492

parathyroid hormone (PTH) Hormone secreted by the four parathyroid glands that increases the blood calcium level and decreases the phosphate level. 492

parenchyma Plant tissue composed of the least-specialized of all plant cells; found in all organs of a plant. 345

parthenogenesis Development of an egg cell into a whole organism without fertilization. 518

passive immunity Protection against infection acquired by transfer of antibodies to a susceptible individual. 470

pathogen Disease-causing agent such as viruses, parasitic bacteria, fungi, and animals. 277

pectoral (shoulder) girdle Portion of the vertebrate skeleton that provides support and attachment for the upper (fore) limbs; consists of the scapula and clavicle on each side of the body. 509

pedigree Chart showing the relationships of relatives and which ones have a particular trait. 204

pelvic (hip) girdle Portion of the vertebrate skeleton to which the lower (hind) limbs are attached; consists of the coxal bones. 509

pelvic inflammatory disease (PID) Latent infection of gonorrhea or chlamydia in the vasa deferentia or uterine tubes. 529

penis Male copulatory organ; in humans, the male organ of sexual intercourse. 520

pepsin Enzyme secreted by gastric glands that digests proteins to peptides. 430

peptidase Intestinal enzyme that breaks down short chains of amino acids to individual amino acids that are absorbed across the intestinal wall. 436

peptide Two or more amino acids joined together by covalent bonding. 42

peptide bond Type of covalent bond that joins two amino acids. 41

peptide hormone Type of hormone that is a protein, a peptide, or derived from an amino acid. 490

peptidoglycan Unique molecule found in bacterial cell walls. 274

perennial plant Flowering plant that lives more than one growing season because the underground parts regrow each season. 343

pericycle Layer of cells surrounding the vascular tissue of roots; produces branch roots. 351

peripheral nervous system (PNS) Nerves and ganglia that lie outside the central nervous system. 479

peristalsis Wavelike contractions that propel substances along a tubular structure such as the esophagus. 430

petal Flower part that occurs just inside the sepals; often conspicuously colored to attract pollinators. 296, 367

petiole Part of a plant leaf that connects the blade to the stem. 346

pH Hydrogen ion concentration. 28

phagocytosis Process by which amoeboid-type cells engulf large substances, forming an intracellular vacuole. 81

pharynx In vertebrates, common passageway for both food intake and air movement; located between the mouth and the esophagus. 419, 430

phenotype Visible expression of a genotype—e.g., brown eyes or attached earlobes. 147

phloem Vascular tissue that conducts organic solutes in plants; contains sieve-tube members and companion cells. 292, 344

phospholipid Molecule that forms the *phospholipid bilayer* of plasma membranes; has a polar, hydrophilic head bonded to two nonpolar, hydrophobic tails. 38

photoautotroph Organism able to synthesize organic molecules by using carbon dioxide as the carbon source and sunlight as the energy source. 275

photoperiod Relative lengths of daylight and darkness that affect the physiology and behavior of an organism. 364

photoreceptor Sensory receptor that responds to light stimuli. 505

photosynthesis Process occurring usually within chloroplasts whereby chlorophyll-containing organelles trap solar energy to reduce carbon dioxide to carbohydrate. 3, 86

photosystem Photosynthetic unit where solar energy is absorbed and high-energy electrons are generated; contains a pigment complex and an electron acceptor; occurs as PS (photosystem) I and PS II. 90

phototropism Growth response of plant stems to light; stems demonstrate positive phototropism. 364

photovoltaic (solar) cell Energy-conversion device that captures solar energy and directly converts it to electrical current. 591

pH scale Measurement scale for hydrogen ion concentration. 28

phylogenetic tree Diagram that indicates common ancestors and lines of descent among groups of organisms. 259

phylogeny Evolutionary history of a group of organisms. 258

phylum One of the categories, or taxa, used by taxonomists to group species; the taxon above the class level. 6, 258

phytochrome Photoreversible plant pigment that is involved in photoperiodism and other responses of plants, such as etiolation. 365

pinocytosis Process by which vesicle formation brings macromolecules into the cell. 81

pith Parenchyma tissue in the center of some stems and roots. 348

pituitary gland Small gland that lies just inferior to the hypothalamus; consists of the anterior and posterior pituitary, both of which produce hormones. 490

placebo Treatment that contains no medication but appears to be the same medication as that administered to other test groups in a controlled study. 13

placenta Organ formed during the development of placental mammals from the chorion and the uterine wall; allows the embryo, and then the fetus, to acquire nutrients and rid itself of wastes; produces hormones that regulate pregnancy. 519

placental mammal Mammal characterized by the presence of a placenta during the development of an offspring. 331

planarian Free-living flatworm with a ladderlike nervous system. 315

plant Multicellular, usually photosynthetic, organism belonging to the plant kingdom. 7

plaque Accumulation of soft masses of fatty material, particularly cholesterol, beneath the inner linings of the arteries. 408

plasma In vertebrates, the liquid portion of blood; contains nutrients, wastes, salts, and proteins. 409

plasma membrane Membrane surrounding the cytoplasm that consists of a phospholipid bilayer with embedded proteins; functions to regulate the entrance and exit of molecules from a cell. 54

plasmid Self-duplicating ring of accessory DNA in the cytoplasm of bacteria. 274

plasmodesmata In plants, cytoplasmic strands that extend through pores in the cell wall and connect the cytoplasm of two adjacent cells. 66

plasmolysis Contraction of the cell contents due to the loss of water. 80

platelet Formed element in blood that is necessary to blood clotting. 387, 410

pleiotropy Inheritance pattern in which one gene affects many phenotypic characteristics of the individual. 153

point mutation Alteration in a gene due to a change in a single nucleotide; results of this mutation vary. 174

polar In chemistry, bond in which the sharing of electrons between atoms is unequal. 24

pollen grain In seed plants, structure that is derived from a microspore and develops into a male gametophyte. 294, 368

pollen tube In seed plants, a tube that forms when a pollen grain lands on the stigma and germinates. The tube grows, passing between the cells of the stigma and the style to reach the egg inside an ovule, where fertilization occurs. 297

pollination In gymnosperms, the transfer of pollen from pollen cone to seed cone; in angiosperms, the transfer of pollen from anther to stigma. 294, 369

pollution Any environmental change that adversely affects the lives and health of living things. 582

polygenic inheritance Pattern of inheritance in which a trait is controlled by several allelic pairs; each dominant allele contributes to the phenotype in an additive and like manner. 152

polymer Macromolecule consisting of covalently bonded monomers; for example, a polypeptide is a polymer of monomers called amino acids. 34

polymerase chain reaction (PCR) Technique that uses the enzyme DNA polymerase to produce millions of copies of a particular piece of DNA. 177

polyp Small, abnormal growth that arises from the epithelial lining. 434

polypeptide Polymer of many amino acids linked by peptide bonds. 42

polyribosome String of ribosomes simultaneously translating regions of the same mRNA strand during protein synthesis. 59, 171

polysaccharide Polymer made from sugar monomers; the polysaccharides starch and glycogen are polymers of glucose monomers. 36

pons Part of the brain stem located between the midbrain and the medulla oblongata. 485

population Group of interbreeding individuals of the same species occupying the same area at the same time. 2, 9, 234, 552

population density The number of individuals per unit area or volume living in a particular habitat. 544

population genetics The study of gene frequencies and their changes within a population. 235

portal system Pathway of blood flow that begins and ends in capillaries, such as the portal system located between the small intestine and liver. 407

posterior pituitary Portion of the pituitary gland that stores and secretes oxytocin and antidiuretic hormone produced by the hypothalamus. 491

postzygotic isolating mechanism Anatomical or physiological difference between two species that prevents successful reproduction after mating has taken place. 251

potential energy Stored energy as a result of location or spatial arrangement. 72

predation Interaction in which one organism (the *predator*) uses another (the *prey*) as a food source. 549, 562

prefrontal area Association area in the frontal lobe that receives information from other association areas and uses it to reason and plan actions. 485

preparatory (prep) reaction Reaction that oxidizes pyruvate with the release of carbon dioxide; results in acetyl-CoA and connects glycolysis to the citric acid cycle. 101

pressure-flow model Explanation for phloem transport; osmotic pressure following active transport of sugar into phloem brings a flow of sap from a source to a sink. 354

prezygotic isolating mechanism Anatomical or behavioral difference between two species that prevents the possibility of mating. 250

primary motor area Area in the frontal lobe where voluntary commands begin; each section controls a part of the body. 484

primary productivity Amount of biomass produced primarily by photosynthesizers. 576

primary sensory area Area located in the parietal lobe where sensory information arrives from the skin and skeletal muscles. 484

primary succession Stage in ecological succession, which involves the creation of new soil. 560

principle Theory that is generally accepted by an overwhelming number of scientists; also called a law. 12

prion Infectious particle consisting of protein only and no nucleic acid. 272

producer Photosynthetic organism at the start of a grazing food chain that makes its own food—e.g., green plants on land and algae in water. 566

product Substance that forms as a result of a reaction. 23

progesterone Female sex hormone that helps maintain sexual organs and secondary sex characteristics. 524

prokaryote Organism that lacks the membrane-bounded nucleus and membranous organelles typical of eukaryotes. 272

prokaryotic cell Lacking a membrane-bounded nucleus and organelles; the cell type within the domains Bacteria and Archaea. 52

prolactin (PRL) Hormone secreted by the anterior pituitary that stimulates the production of milk from the mammary glands. 490

promoter In an operon, a sequence of DNA where RNA polymerase binds prior to transcription. 169, 186

prophase Mitotic phase during which chromatin condenses so that chromosomes appear; chromosomes are scattered. 117

proprioceptor Sensory receptor that assists the brain in knowing the position of the limbs. 507

prosimian Group of primates that includes lemurs and tarsiers, and may resemble the first primates to have evolved. 332

prostate gland Gland located around the male urethra below the urinary bladder; adds secretions to semen. 521

protein Molecule consisting of one or more polypeptides; a macronutrient in the diet that is digested to amino acids used by cells to synthesize cellular proteins. 40, 446

proteome Collection of proteins resulting from the translation of genes into proteins. 179

proteomics Study of the structure, function, and interaction of proteins. 179

protist Member of the kingdom Protista. 6, 279

protocell A cell-like structure with an outer membrane, thought to have preceded true cells. 273

proton Positive subatomic particle located in the nucleus and assigned one atomic mass unit. 18

proto-oncogene Normal gene that can become an oncogene through mutation. 193

protostome Group of coelomate animals in which the first embryonic opening (the blastopore) is associated with the mouth. 313

protozoan Chemoheterotrophic, unicellular protist that moves by flagella, cilia, or pseudopodia, or is immobile. 279

proximal tubule Portion of a nephron following the nephron capsule where reabsorption of filtrate occurs. 425

pseudocoelomate Animal, such as a roundworm, with a body cavity incompletely lined by mesoderm. 313

pseudopod Cytoplasmic extension of amoeboid protists; used for locomotion and engulfing food. 282

puberty Period of life when secondary sex changes occur in humans; marked by the onset of menses in females and sperm production in males. 522

pulmonary artery Blood vessel that takes blood away from the heart to the lungs. 403

pulmonary circuit Circulatory pathway between the lungs and the heart. 402, 406

pulmonary trunk Large blood vessel that divides into the pulmonary arteries; takes blood away from the heart to the lungs. 403

pulmonary vein Blood vessel that takes blood to the heart from the lungs. 403

pulse Vibration felt in arterial walls due to expansion of the aorta following ventricle contraction. 404

Punnett square Grid used to calculate the expected results of simple genetic crosses. 146

pupil Opening in the center of the iris of the vertebrate eye. 505

R

radial symmetry Body plan in which similar parts are arranged around a central axis, like spokes of a wheel. 311

radiolarian Protozoan; marine amoeba having a calcium carbonate skeleton, many of which make limestone formations such as the White Cliffs of Dover. 282

radius Thicker and shorter forearm bone that contributes to the easy twisting motion of the forearm. 509

random Spatial distribution of individuals in a population in which individuals have an equal chance of living anywhere within an area. 544

range That portion of the globe where a certain species can be found. 544

ray-finned fishes Group of bony fishes with fins supported by parallel bony rays connected by webs of thin tissue. 326

reabsorption Movement of primarily nutrient molecules and water from the contents of the nephron into blood at the proximal tubule. 425

reactant Substance that participates in a reaction. 23

receptacle Area where a flower attaches to a floral stalk. 366

receptor-mediated endocytosis Selective uptake of molecules into a cell by vacuole formation after they bind to specific receptor proteins in the plasma membrane. 81

recessive allele Allele that exerts its phenotypic effect only in the homozygote; its expression is masked by a dominant allele. 147

recombinant DNA (rDNA) DNA that contains genes from more than one source. 175

recombinant gamete New combination of alleles incorporated into a gamete during crossing-over. 156

red algae Marine photosynthetic protist with a notable abundance of phycobilin pigments; includes coralline algae of coral reefs. 280

red blood cell Erythrocyte; contains hemoglobin and carries oxygen from the lungs or gills to the tissues in vertebrates. 387, 409

red bone marrow Vascularized, modified connective tissue that is sometimes found in the cavities of spongy bone within certain bones; site of blood cell formation. 463, 510

redox reaction Oxidation-reduction reaction; one molecule loses electrons (oxidation) while another molecule simultaneously gains electrons (reduction). 88

reduction Gain of electrons by an atom or molecule with a concurrent storage of energy; in biological systems, the electrons are accompanied by hydrogen ions. 88

reflex Automatic, involuntary response of an organism to a stimulus. 487

regeneration Ability of sponges and echinoderms to "generate" into another organism if separated into pieces. 518

regulatory gene In an operon, a gene that codes for a protein that regulates the expression of other genes. 186

release factor A protein complex that binds to an mRNA codon at the A site of the ribosome, causing termination of the transcription. 172

renal cortex Outer portion of the kidney that appears granular. 424

renal medulla Inner portion of the kidney that consists of renal pyramids. 424

renal pelvis Hollow chamber in the kidney that lies inside the renal medulla and receives freshly prepared urine from the collecting ducts. 424

renewable resources Resources normally replaced or replenished by natural processes; resources not depleted by moderate use. Examples include solar energy, biological resources, such as forests and fisheries, biological organisms, and some biogeochemical cycles. 582

replacement model Proposal that modern humans originated only in Africa; then they migrated and supplanted populations of *Homo* in Asia and Europe about 100,000 years ago (see out-of-Africa hypothesis). 335

repressor In an operon, protein molecule that binds to an operator, preventing transcription of structural genes. 186

reproduce To produce a new individual of the same kind. 3

reproductive cloning Genetically identical to the original individual. 184

reproductive system Organ system that contains male or female organs and specializes in the production of offspring. 391

reptile Terrestrial vertebrate with internal fertilization, scaly skin, and a shelled egg; includes snakes, lizards, turtles, crocodiles, and birds. 328

reservoir In chemical cycling, a source (e.g., fossil fuels, minerals in rocks, and ocean sediments), that is normally unavailable to organisms. 570

resource A necessity for an organism to live and reproduce such as food, water, shelter, nesting sites 544

resource partitioning Mechanism that increases the number of niches by apportioning the supply of a resource such as food or living space between species. 563

respiration Sequence of events that results in gas exchange between the cells of the body and the environment. 418

respiratory system Organ system consisting of the lungs and tubes that bring oxygen into the lungs and take carbon dioxide out. 390

restriction enzyme Bacterial enzyme that stops viral reproduction by cleaving viral DNA; used to cut DNA at specific points during production of recombinant DNA. 175

retina Innermost layer of the vertebrate eyeball containing the photoreceptors—rods and cones. 505

retrovirus RNA virus containing the enzyme reverse transcriptase that carries out RNA/DNA transcription. 270

rhodopsin Light-absorbing molecule in rods and cones that contains a pigment and the protein opsin. 506

rib cage Flexible skeletal structure that protects the heart and lungs, moves when we breathe, and is composed of the ribs and sternum. 509

ribose Pentose sugar found in RNA. 35

ribosomal RNA (rRNA) Type of RNA found in ribosomes that translate messenger RNAs to produce proteins. 167

ribosome RNA and protein in two subunits; site of protein synthesis in the cytoplasm. 53, 167

RNA (ribonucleic acid) Nucleic acid produced from covalent bonding of nucleotide monomers that contain the sugar ribose; occurs in three major forms: messenger RNA, ribosomal RNA, and transfer RNA. 44, 166

RNA polymerase During transcription, an enzyme that joins nucleotides complementary to a DNA template. 169

rod Photoreceptor in vertebrate eyes that responds to dim light. 506

root cap Protective cover of the root tip, whose cells are constantly replaced as they are ground off when the root pushes through rough soil particles. 351

root hair Extension of a root epidermal cell that collectively increases the surface area for the absorption of water and minerals. 343

root nodule Structure on plant root that contains nitrogen-fixing bacteria. 352

root system Includes the main root and any and all of its lateral (side) branches. 342

rotational equilibrium Maintenance of balance when the head and body are suddenly moved or rotated. 504

rotator cuff Tendons and ligaments that stabilize the shoulder joint. 509

rough ER Membranous system of tubules, vesicles, and sacs in cells; has attached ribosomes. 60

roundworm An invertebrate with a cylindrical body covered by a cuticle that molts; also has a complete digestive tract and a pseudocoelom;

some forms are free-living in water and soil, and many are parasitic. 316.

RuBP carboxylase (rubisco) Enzyme that is required for carbon dioxide fixation (atmospheric CO_2 attaches to RuBP) in the Calvin cycle. 92

rule of multiplication The chance of two (or more) independent events occurring together is the product of their chance occurring separately. 149

S

saccule Saclike cavity in the vestibule of the vertebrate inner ear; contains sensory receptors for gravitational equilibrium. 504

sacrum The large, triangular bone at the base of the vertebral column; consists of five vertebrae that are fused together. 509

salinization Process in which mineral salts accumulate in the soil, killing plants; occurs when soils in dry climates are irrigated profusely. 587

salivary amylase In humans, enzyme in saliva that digests starch to maltose. 429

salivary gland In humans, gland associated with the mouth that secretes saliva. 429

salt Compound produced by a reaction between an acid and a base. 22

saltatory conduction Movement of nerve impulses from one node of Ranvier to another along a myelinated axon. 481

saltwater intrusion Movement of saltwater into freshwater aquifers in coastal areas where groundwater is withdrawn faster than it is replenished. 585

SA (sinoatrial) node Small region of neuromuscular tissue that initiates the heartbeat; also called the pacemaker. 404

saprotroph Organism that secretes digestive enzymes and absorbs the resulting nutrients back across the plasma membrane; fungus or bacterium that decomposes the remains of plants, animals, and microbes in the soil. 276, 302

sarcomere One of many identical units, arranged linearly within a myofibril, whose contraction produces muscle contraction. 511

saturated fatty acid Fatty acid molecule that lacks double bonds between the carbons of its hydrocarbon chain. The chain bears the maximum number of hydrogens possible. 38

scapula Shoulder blade. 509

scientific theory Concept supported by a broad range of observations, experiments, and data. 12

sclerenchyma Plant tissue composed of cells with heavily lignified cell walls; functions in support. 345

secondary sex characteristic Trait that is sometimes helpful but not absolutely necessary for reproduction and is maintained by the sex hormones in males and females. 522

secondary succession In ecological succession, the stage which is the gradual replacement of communities in an area following a disturbance. 560

secretion In the cell, release of a substance by exocytosis from a cell that may be a gland or part of a gland; in the urinary system, movement of certain molecules from blood into the distal tubule of a nephron so that they are added to urine. 60, 425

seed Mature ovule that contains an embryo, with stored food enclosed in a protective coat. 288, 370

segmentation Repetition of body units as seen in the earthworm. 313

semen (seminal fluid) Thick, whitish fluid consisting of sperm and secretions from several glands of the male reproductive tract. 521

semiconservative Duplication of DNA resulting in two double helix molecules, each having one parental and one new strand. 166

semilunar valve Valve resembling a half-moon located between the ventricles and their attached vessels. 403

seminal vesicle Convoluted, saclike structure attached to the vas deferens near the base of the urinary bladder in males; adds secretions to semen. 521

seminiferous tubule Long, coiled structure contained within chambers of the testis where sperm are produced. 521

senescence Sum of the processes involving aging, decline, and eventual death of a plant or plant part. 362

sensory neuron Nerve cell that transmits nerve impulses to the central nervous system after a sensory receptor has been stimulated. 480

sepal Outermost leaflike covering of the flower; usually green in color. 296, 367

septum Partition or wall that divides two areas; the septum in the heart separates the right half from the left half. 403

sex chromosome Chromosome that determines the sex of an individual; in humans, females have two X chromosomes, and males have an X and a Y chromosome. 130

sexual selection Changes in males and females, often due to male competition and female selectivity, leading to increased fitness. 238

shoot system Aboveground portion of a plant consisting of the stem, leaves, and flowers. 342

short-day plant/long-night plant Plant that flowers when day length is shorter and the night is longer than a critical length—e.g., cocklebur, poinsettia, and chrysanthemum. 364

sickle cell disease Hereditary disease in which red blood cells are narrow and curved so that they are unable to pass through capillaries and are destroyed; causes chronic anemia. 207

sieve-tube member Member that joins with others in the phloem tissue of plants as a means of transport for nutrient sap. 346

signal Molecule that stimulates or inhibits an event in the cell cycle. 121

signal transduction pathway Activation and inhibition of intracellular targets after binding of growth factors. 191

simple diffusion Movement of molecules or ions from a region of higher to lower concentration; it requires no energy and tends to lead to an equal distribution. 79

simple goiter Condition in which an enlarged thyroid produces low levels of thyroxine. 492

sinus Cavity into which hemolymph flows and baths the organs in an open circulatory system. Also, air-filled spaces in nasal cavities. 400, 419

sister chromatid One of two genetically identical chromosomal units that are the result of DNA replication and are attached to each other at the centromere. 115

skeletal muscle Striated, voluntary muscle tissue that comprises skeletal muscles; also called striated muscle. 388

skeletal system System of bones, cartilage, and ligaments that works with the muscular system to protect the body and provide support for locomotion and movement. 391

skull Protective skeletal structure consisting of the cranium and the facial bones. 509

sliding filament model An explanation for muscle contraction based on the movement of actin filaments in relation to myosin filaments. 511

slime mold Protists that decompose dead material and feed on bacteria by phagocytosis; vegetative state is mobile and amoeboid. 282

small intestine In vertebrates, the portion of the digestive tract that precedes the large intestine; in humans, consists of the duodenum, jejunum, and ileum; responsible for most digestion and absorption. 432

smooth ER Membranous system of tubules, vesicles, and sacs in eukaryotic cells; lacks attached ribosomes. 60

smooth (visceral) muscle Nonstriated, involuntary muscles found in the walls of internal organs. 388

sodium-potassium pump Carrier protein in the plasma membrane that moves sodium ions out of and potassium ions into animal cells; important in nerve and muscle cells. 81

solute Substance that is dissolved in a solvent, forming a solution. 79

solution Fluid (the solvent) that contains a dissolved solid (the solute). 79

solvent Liquid portion of a solution that serves to dissolve a solute. 79

somatic cell Body cell; excludes cells that undergo meiosis and become sperm or egg. 122

somatic system Portion of the peripheral nervous system containing motor neurons that control skeletal muscles. 487

sori (sing., sorus) Clusters of sporangia on the underside of a fern's frond. 292

speciation Origin of new species due to the evolutionary process of descent with modification. 222, 248

species Group of similarly constructed organisms capable of interbreeding and producing fertile offspring; organisms that share a common gene pool; the taxon at the lowest level of classification. 6, 258

species richness List of different species in a community. 559

spermatogenesis Production of sperm in males by the process of meiosis and maturation. 131

spinal cord In vertebrates, the nerve cord that is continuous with the base of the brain and housed within the vertebral column. 484

spinal nerve Nerve that arises from the spinal cord. 487

spindle Microtubule structure that brings about chromosomal movement during nuclear division. 117

spindle equator Disk formed during metaphase in which all of a cell's chromosomes lie in a single plane at right angles to the spindle fibers. 117

spiral organ Structure in the vertebrate inner ear that contains auditory receptors (also called organ of Corti). 502

spleen Large, glandular, lymphatic organ located in the upper left region of the abdomen; stores and purifies blood. 463

sponge Invertebrate animal of the phylum Porifera; pore-bearing filter feeder whose inner body wall is lined by collar cells. 314

spongy bone Type of bone that has an irregular, meshlike arrangement of thin plates of bone. 510

spongy mesophyll Layer of tissue in a plant leaf containing loosely packed cells, increasing the amount of surface area for gas exchange. 346

sporangium Spore-producing plant or fungal structure. 291

spore Asexual reproductive or resting cell capable of developing into a new organism without fusion with another cell, in contrast to a gamete. 290

sporophyte Diploid generation of the alternation of generations life cycle of a plant; produces haploid spores that develop into the haploid generation. 290, 366

sporozoan Spore-forming protist that has no means of locomotion and is typically a parasite with a complex life cycle having both sexual and asexual phases. 282

squamous epithelium Type of epithelial tissue in which the cells are flattened. 384

stabilizing selection Outcome of natural selection in which extreme phenotypes are eliminated and the average phenotype is conserved. 242

stamen In flowering plants, the portion of the flower that consists of a filament and an anther containing pollen sacs where pollen is produced. 296, 367

starch Storage polysaccharide found in plants that is composed of glucose molecules joined in a linear fashion with few side chains. 36

statocyst Gravitational equilibrium organ found in some invertebrates; give information about head position. 504

stem Usually the upright, vertical portion of a plant that transports substances to and from the leaves. 343

sternum Breastbone. 509

steroid Type of lipid molecule having a complex of four carbon rings—e.g., cholesterol, estrogen, progesterone, and testosterone. 38

steroid hormone Type of hormone that has the same complex of four carbon rings, but each one has different side chains. 489

stigma In flowering plants, portion of the carpel where pollen grains adhere and germinate before fertilization can occur. 296, 367

stoma (pl., stomata) Small opening between two guard cells on the underside of leaf epidermis through which gases pass. 87, 345

stomach In vertebrates, muscular sac that mixes food with gastric juices to form chyme, which enters the small intestine. 430

strata Layers of rock or sedimentary material, often containing fossils. 219

striated Having bands; in cardiac and skeletal muscle, alternating light and dark bands produced by the distribution of contractile proteins. 388

stroke Condition resulting when an arteriole in the brain bursts or becomes blocked by an embolism; cerebrovascular accident. 408

stroma Fluid within a chloroplast that contains enzymes involved in the synthesis of carbohydrates during photosynthesis. 62, 87

style Elongated, central portion of the carpel between the ovary and stigma. 296, 367

subsidence Occurs when a portion of the Earth's surface gradually settles downward. 585

subspecies Taxonomic designation of populations within a species, each with somewhat distinctive genotypic and phenotypic characteristics; readily interbreed with other such subspecies. 238

substrate Reactant in a reaction controlled by an enzyme. 77

surface-area-to-volume ratio Ratio of a cell's outside area to its internal volume. 51

survivorship Probability of newborn individuals of a cohort surviving to particular ages. 545

sustainable society Ability of a society or ecosystem to maintain itself while also providing services to human beings. 598

symbiotic Relationship that occurs when two different species live together in a unique way; it may be beneficial, neutral, or detrimental to one and/or the other species. 276

sympathetic division Division of the autonomic system that is active when an organism is under stress; uses norepinephrine as a neurotransmitter. 488

sympatric speciation Origin of new species in populations that overlap geographically. 252

synapse Junction between neurons consisting of the axon membrane, the synaptic cleft, and a dendrite. 482

synapsis Pairing of homologous chromosomes during meiosis I. 132

synaptic cleft Small gap between membranes of a synapse. 482

synovial joint Freely moving joint in which two bones are separated by a cavity. 512

syphilis Bacterial STD with three stages separated by latent periods; may result in blindness, cause birth defects or stillbirth, and affect the cardiovascular and/or nervous system. 529

systematics Study of the diversity of organisms to classify them and determine their evolutionary relationships. 258

systemic circuit Circulatory pathway of blood flow between the tissues and the heart. 402, 406

systole Contraction period of the heart during the cardiac cycle. 404

T

taste buds Structures in the vertebrate mouth containing sensory receptors for taste; in humans, most taste buds are on the tongue. 501

taxon (pl., taxa) Group of organisms that fills a particular classification category. 258

taxonomy Branch of biology concerned with identifying, describing, and naming organisms. 5, 258

T-cell receptor (TCR) Receptor on the T cell surface consisting of two antigen-binding peptide chains; it is associated with a large number of other glycoproteins. Binding of antigen to the TCR, usually in association with MHC, activates the T cell. 468

technology The application of scientific knowledge for a practical purpose. 14

telomere Long, repeating DNA sequence at the ends of chromosomes; functions like a cap and keeps chromosomes from fusing with each other. 122, 194

telophase Mitotic phase during which daughter cells are located at each pole. 117

template Parental strand of DNA that serves as a guide for the complementary daughter strand produced during DNA replication. 166

tendon Strap of fibrous connective tissue that connects skeletal muscle to bone. 387

terminal bud Bud that develops at the apex of a shoot. 342

terrestrial ecosystem Also called biomes, these are the tundra, the taiga, temperate forests, tropical grasslands (savanna), temperate grasslands (prairie), deserts, and tropical rain forests. 574

testcross Cross between an individual with the dominant phenotype and an individual with the recessive phenotype. The resulting phenotypic ratio indicates whether the dominant phenotype is homozygous or heterozygous. 146

test group In a controlled experiment, the group that is subjected to the experimental variable. 11

testis (pl., testes) Male gonad that produces sperm and the male sex hormones. 518, 520

testosterone Male sex hormone that helps maintain sexual organs and secondary sex characteristics. 522

tetany Severe twitching caused by involuntary contraction of the skeletal muscles due to a calcium imbalance. 492

tetrad Homologous chromosomes, each having sister chromatids that are joined during meiosis; also called bivalent. 132

thalamus In vertebrates, the portion of the diencephalon that passes on selected sensory information to the cerebrum. 485

therapeutic cloning Used to create mature cells of various cell types. Also, used to learn about specialization of cells and provide cells and tissue to treat human illnesses. 185

thermoacidophile Type of archaea that lives in hot, acidic, aquatic habitats, such as hot springs or near hydrothermal vents. 278

thoracic cavity Located in the ventral cavity above the abdominal cavity and is separated from the abdominal cavity by the diaphragm; contains the heart and the lungs. 430

three-domain system System of classification that recognizes three domains: Bacteria, Archaea, and Eukarya. 263

thrombin Enzyme that converts fibrinogen to fibrin threads during blood clotting. 411

thylakoid Flattened sac within a granum whose membrane contains chlorophyll and where the light reactions of photosynthesis occur. 62, 87

thymine (T) One of four nitrogen-containing bases in nucleotides composing the structure of DNA; pairs with adenine. 163

thymus gland Lymphatic organ involved in the development and functioning of the immune system; T cells mature in the thymus gland. 463

thyroid gland Large gland in the neck that produces several important hormones, including thyroxine, triiodothyronine, and calcitonin. 492

thyroid-stimulating hormone (TSH) Substance produced by the anterior pituitary that causes the thyroid to secrete thyroxine and triiodothyronine. 490

thyroxine (T$_4$) Hormone secreted from the thyroid gland that promotes growth and development; in general, it increases the metabolic rate in cells. 492

tibia Larger lower leg bone, also called the shinbone. 509

tight junction Junction between cells when adjacent plasma membrane proteins join to form an impermeable barrier. 67

tissue Group of similar cells combined to perform a common function. 2, 382

tissue culture Process of growing tissue artificially, usually in a liquid medium in laboratory glassware. 373

tissue fluid Fluid that surrounds the body's cells; consists of dissolved substances that leave the blood capillaries by filtration and diffusion. 390

T lymphocyte (T cell) Lymphocyte that matures in the thymus and exists in four varieties, one of which kills antigen-bearing cells outright. 463

tonsils Partially encapsulated lymph nodules located in the pharynx. 462

totipotent Cell that has the full genetic potential of the organism, including the potential to develop into a complete organism. 373

tracer Substance having an attached radioactive isotope that allows a researcher to track its whereabouts in a biological system. 20

trace mineral Essential inorganic nutrient (such as zinc, iron, copper, iodine, or selenium) needed daily by humans to regulate metabolic activities and maintain good health. 447

trachea (pl., tracheae) In tetrapod vertebrates, air tube (windpipe) that runs between the larynx and the bronchi. 420

tracheid In vascular plants, type of cell in xylem that has tapered ends and pits through which water and minerals flow. 346

tract Bundles of myelinated axons in the central nervous system. 484

transcription Process whereby a DNA strand serves as a template for the formation of mRNA. 168

transcription activator Protein that speeds transcription. 189

transcription factor In eukaryotes, protein required for the initiation of transcription by RNA polymerase. 188

transduction Exchange of DNA between bacteria by means of a bacteriophage. 275

transfer rate Amount of a substance that moves from one component of the environment to another within a specified period of time. 571

transfer RNA (tRNA) Type of RNA that transfers a particular amino acid to a ribosome during protein synthesis; at one end, it binds to the amino acid, and at the other end it has an anticodon that binds to an mRNA codon. 167

transformation Taking up of extraneous genetic material from the environment by bacteria. 275

transgenic organism Free-living organism in the environment that has had a foreign gene inserted into it. 175

transgenic plant Plant that carries the genes of another organism as a result of DNA technology; also genetically modified plant. 375

translation Process whereby ribosomes use the sequence of codons in mRNA to produce a polypeptide with a particular sequence of amino acids. 168

translocation Movement of a chromosomal segment from one chromosome to another nonhomologous chromosome, leading to abnormalities—e.g., Down syndrome. 172, 194, 203

transpiration Plant's loss of water to the atmosphere, mainly through evaporation at leaf stomata. 353

transport vesicles Small membranous sacs that carry products between endomembrane system organelles and prevent the products from entering the cytoplasm. 57

transposon DNA sequence capable of randomly moving from one site to another in the genome. 174

trichomoniasis Sexually transmitted disease caused by the parasitic protozoan *Trichomonas vaginalis*. 529

triglyceride Neutral fat composed of glycerol and three fatty acids. 37, 445

triplet code During gene expression, each sequence of three nucleotide bases stands for a particular amino acid. 168

trophic level Feeding level of one or more populations in a food web. 569

tropism In plants, a growth response toward or away from a directional stimulus. 364

trypanosome Parasitic zooflagellate that causes severe disease in human beings and domestic animals, including a condition called sleeping sickness. 282

trypsin Protein-digesting enzyme secreted by the pancreas. 432

tumor Cells derived from a single mutated cell that has repeatedly undergone cell division; benign tumors remain at the site of origin, while malignant tumors metastasize. 123

tumor suppressor gene Gene that codes for a protein that ordinarily suppresses cell division; inactivity can lead to a tumor. 193

tunicate Type of primitive invertebrate chordate. 324

Turner syndrome Condition caused by the inheritance of a single X chromosome. 139

U

ulna Larger forearm bone that together with the radius contributes to the easy twisting motion of the forearm. 509

umbilical cord Cord connecting the fetus to the placenta through which blood vessels pass. 533

unicellular Made up of but a single cell, as in the bacteria. 2

uniform Spatial distribution of individuals in a population in which individuals are dispersed uniformly through the area. 544

uniformitarianism Belief espoused by Charles Lyell that geological forces act at a continuous, uniform rate. 220

unsaturated fatty acid Fatty acid molecule that has one or more double bonds between the carbons of its hydrocarbon chain. The chain bears fewer hydrogens than the maximum number possible. 38

uracil (U) Pyrimidine base that replaces thymine in RNA; pairs with adenine. 166

ureter Tubular structure conducting urine from the kidney to the urinary bladder. 424

urethra Tubular structure that receives urine from the bladder and carries it to the outside of the body. 424

urinary bladder Organ where urine is stored. 424

urinary system Organ system consisting of the kidneys and urinary bladder; rids the body of nitrogenous wastes and helps regulate the water-salt balance of the blood. 390

urine Liquid waste product made by the nephrons of the vertebrate kidney through the processes of filtration, reabsorption, and secretion. 424

uterus In mammals, expanded portion of the female reproductive tract through which eggs pass to the environment or in which an embryo develops and is nourished before birth. 522

utricle Saclike cavity in the vestibule of the vertebrate inner ear; contains sensory receptors for gravitational equilibrium. 504

V

vaccine Antigens prepared in such a way that they can promote active immunity without causing disease. 470

vacuole Membrane-bounded sac, larger than a vesicle; usually functions in storage and can contain a variety of substances. In plants, the central vacuole fills much of the interior of the cell. 61

valence shell Outer shell of an atom. 21

vascular bundle In plants, primary phloem and primary xylem enclosed by a bundle sheath. 344

vascular cambium In plants, lateral meristem that produces secondary phloem and secondary xylem. 345

vascular cylinder In eudicots, the tissues in the middle of a root, consisting of the pericycle and vascular tissues. 351

vascular plant Plant that has vascular tissue (xylem and phloem); includes seedless vascular plants (e.g., ferns) and seed plants (gymnosperms and angiosperms). 292

vascular tissue Transport tissue in plants, consisting of xylem and phloem. 292, 345

vas deferens Tube that leads from the epididymis to the urethra in males. 520

vector Piece of DNA that can have a foreign DNA attached to it; a common vector is a plasmid. 175

vein Blood vessel that arises from venules and transports blood toward the heart. 403

vena cava (pl., venae cavae) Large systemic vein that returns blood to the right atrium of the heart in tetrapods; either the superior or inferior vena cava. 403, 407

ventricle Cavity in an organ, such as a lower chamber of the heart or the ventricles of the brain. 403

venule Vessel that takes blood from capillaries to a vein. 405

vertebral column Portion of the vertebrate endoskeleton that houses the spinal cord; consists of many vertebrae separated by intervertebral disks. 509

vertebrate Chordate in which the notochord is replaced by a vertebral column. 324

vesicle Small, membrane-bounded sac that stores substances within a cell. 60

vessel element Cell that joins with others to form a major conducting tube found in xylem. 346

vestigial structure Remains of a structure that was functional in some ancestor but is no longer functional in the organism in question. 228

villus (pl., villi) Small, fingerlike projection of the inner small intestinal wall. 432

viroid Infectious strand of RNA devoid of a capsid and much smaller than a virus. 272

virus Noncellular parasitic agent consisting of an outer capsid and an inner core of nucleic acid. 268

vitamin Essential requirement in the diet, needed in small amounts. Vitamins are often part of coenzymes. 448

vocal cord In humans, folds of tissue within the larynx; create vocal sounds when they vibrate. 419

W

water mold Filamentous organisms having cell walls made of cellulose; typically decomposers of dead freshwater organisms, but some are parasites of aquatic or terrestrial organisms. 282

white blood cell Leukocyte, of which there are several types, each having a specific function in protecting the body from invasion by foreign substances and organisms. 387, 410

wild-type Allele or phenotype that is the most common for a certain gene in a population. 148

wood Secondary xylem that builds up year after year in woody plants and becomes the annual rings. 349

X

X-linked Allele that is located on an X chromosome but may control a trait that has nothing to do with the sex characteristics of an animal. 154

xylem Vascular tissue that transports water and mineral solutes upward through the plant body; it contains vessel elements and tracheids. 292, 344

Z

zooflagellate Nonphotosynthetic protist that moves by flagella; typically zooflagellates enter into symbiotic relationships, and some are parasitic. 282

zygote Diploid cell formed by the union of two gametes; the product of fertilization. 131, 518

Credits

Photographs

Chapter 1
Openers(soil organisms): © Runk/Schoenberger/Grant Heilman Photography; (trees): © Steven P. Lynch; (butterfly): © Stockbyte; (toucan): © IT Stock/PunchStock; (frog): © MedioImages/SuperStock; (orchid): © Max and Bea Hunn/Visuals Unlimited; (bacteria): © David Scharf/SPL/Photo Researchers, Inc.; 1.2a: © age fotostock/SuperStock; 1.2b: © Ed Young/Corbis; 1.3(family): © Vol. 113/Getty RF; 1.3(helix): © David Mack/SPL/Photo Researchers, Inc.; 1.5(main): © Ralph Robinson/Visuals Unlimited; 1.5(inset): © M. Rhode/GBF/SPL/Photo Researchers, Inc.; 1.6(main): © A.B. Dowsett/SPL/Photo Researchers, Inc.; 1.6(inset): © Biophoto Associates/Photo Researchers, Inc.; 1.7: © John D. Cunningham/Visuals Unlimited; 1.8: © Corbis RF; 1.9: © Brand X Pictures/PunchStock RF; 1.10: © Photodisc/PunchStock RF; 1.13a: © Kathy Sloane/Photo Researchers, Inc.; 1.13b: © Edgar Bernstein/Peter Arnold; 1.13c: © Luiz C. Marigo/Peter Arnold; 1.15(students): © image100 Ltd; 1.15(scientists): © Phanie/Photo Researchers, Inc.; 1.16: © The McGraw-Hill Companies, Inc./Jill Braaten, photographer.

Chapter 2
Openers(girders): © Corbis RF; (blood): © PhotoLink/Getty RF; (ocean): © Evelyn Jo Johnson; (sunbather): © Adam Smith/Corbis; (melanoma): © James Stevenson/SPL/Photo Researchers, Inc.; (radiographers): © SPL/Photo Researchers, Inc.; 2.1: © Tim Pannell/Corbis RF; 2.4a: © Biomed Commun./Custom Medical Stock Photo; 2.4b(both): Courtesy National Institutes of Health; Courtesy National Institutes of Health; 2.5a: © Natasja Weitsz/Getty Images; 2.5b: © Geoff Tompkinson/SPL/Photo Researchers, Inc.; 2.7b(salt): © Charles M. Falco/Photo Researchers, Inc.; 2.9: © Fritz Polking/Peter Arnold; 2.10: © Corbis RF; 2.11: © Claude Nuridsany & Marie Perennou/Photo Researchers, Inc.; 2.12a: © The McGraw-Hill Companies, Inc./Jill Braaten, photographer; 2.12b: © SuperStock RF; 2.18a: © Frederica Georgia/Photo Researchers, Inc.; 2.18b: © Ray Pfortner/Peter Arnold.

Chapter 3
Openers(omelet): © Dennis Gray/Cole Group/Getty RF; (fiberCon): © The McGraw-Hill Companies, Inc./John Thoeming, photographer; (woman, eggs): © Getty RF; 3.1a: © Lary Lefever/Grant Heilman Photography; 3.1b: © Photodisc/Getty RF; 3.1c: © H. Pol/CNRI SPL/Photo Researchers, Inc.; 3.6: © The McGraw-Hill Companies, Inc./John Thoeming, photographer; 3.8: © Renee Lynn/Photo Researchers, Inc.; 3.10a: © Jeremy Burgess/SPL/Photo Researchers, Inc.; 3.10b: © Don W. Fawcett/Photo Researchers, Inc.; 3.10c: © J.D. Litvay/Visuals Unlimited; 3.11: © Lightwave Photography, Inc./

Animals Animals; 3.15b-c: © Corbis RF; 3.16a(kidman): © Reuters/Corbis; 3.16a(web): © OSF/Animals Animals; 3.16b: © P. Motta & S. Correr/Photo Researchers, Inc.; 3.16c: © Duomo/Corbis; 3.21: © Eye of Science/Photo Researchers, Inc.

Chapter 4
Openers(bone cell): © Biophoto Associates/Photo Researchers, Inc; (apoptosis): © David McCarthy/SPL/Photo Researchers, Inc.; (canoeing): © Warren Morgan/Corbis; (neuron, red blood cells): © Ed Reschke; (cone cell): © Eye of Science/Photo Researchers, Inc.; (egg cell): © Ed Reschke/Peter Arnold; 4.1(epithelial cells): © Ed Reschke; 4.1(light microscope): © Corbis Images/Jupiter Images; 4.1(electron microscope): © Inga Spence/Visuals Unlimited; 4.1(leaf cells): © Ray F. Evert, University of Wisconsin, Madison; 4.1(lymphocyte): Courtesy Dr. Joseph R. Goodman; 4.1(euglena): © T.E. Adams/Visuals Unlimited; 4.4: © Ralph A. Slepecky/Visuals Unlimited; 4.7a: © Dr. Dennis Kunkel/Visuals Unlimited; 4.8a: © Newcomb/Wergin/BPS; 4.9: © Ron Milligan, Scripps Research Institute; 4.11: © R. Bolender and D. Fawcett/Visuals Unlimited; 4.13a: © Roland/Birke/Peter Arnold; 4.13b: © Newcomb/Wergin/BPS; 4.13c: © The McGraw Hill Companies, Inc./Al Telser, photographer; 4.14: Courtesy Herbert W. Israel, Cornell University; 4.15: Courtesy Keith Porter; 4.19: Courtesy Kent McDonald; 4.20a(cilia): © Manfred Kage/Peter Arnold; 4.20a(sperm): © David M. Phillips/Photo Researchers, Inc.; 4.20b(flagellum, basal body): © William L. Dentler/BPS; 4.21: © E.H. Newcomb/BPS; 4.23: From Douglas E. Kelly, *Journal of Cell Biology,* 28 (1966:51). Reproduced by permission of The Rockefeller University Press.

Chapter 5
Openers(antibiotic): © Tom Pantages/Phototake; (bacteria): © Oliver Meckes/SPL/Photo Researchers, Inc.; (athlete): © age fotostock/SuperStock; (nurse, girl): © Vol. 41/Getty RF; (woman): © GoodShoot/SuperStock RF; 5.2a–b: © Keith Eng, 2008; 5.6(plant): © Comstock/PunchStock RF; 5.6(woman): © Getty RF; 5.7(potato, grain): © The McGraw-Hill Companies, Inc./John Thoeming, photographer; 5.7(fruit): © Getty RF; 5.7(meat): © Foodcollection/Getty RF; 5.16(both): Courtesy Mark Bretscher.

Chapter 6
Openers(light): Courtesy Hydrofarm Horticultural Products; (leaf): © Comstock/PunchStock RF; (dinoflagellates): © SPL/Photo Researchers, Inc.; (diatoms): © Carolina Biological Supply/Phototake; 6.1(trees): © Getty RF; 6.1(sunflower): © Grant Heilman Photography; 6.1(euglena): © T.E. Adams/Visuals Unlimited; 6.1(moss): © Bruce Iverson; 6.1(kelp): © Chuck Davis/Getty Images; 6.1(diatoms): © Ed Reschke/Peter Arnold; 6.1(cyanobacteria): © Sherman Thomas/Visuals Unlimited; 6.2(top): © Manfred Kage/

Peter Arnold; 6.2(bottom): Courtesy Herbert W. Israel, Cornell University; 6.5(both): © Digital Vision/Getty RF; 6.11: © Jim Steinberg/Photo Researchers, Inc.; 6.12: © Charlie Waite; 6.13: © Beverly Factor Photography.

Chapter 7
Openers(soup): © PhotoLink/Getty RF; (scale): © Photodisc/Getty RF; (grapefruit): © Corbis RF; (exhaust): © Getty RF; (athlete): © Bill Robbins/Index Stock Imagery; 7.1: © Corbis RF; 7.5: Courtesy Dr. Keith Porter; 7.10: © C Squared Studios/Getty RF; 7.11(athlete): © Rubber Ball Productions/RF Index Stock Imagery, Inc.; 7.11(wine, bread): © Getty RF.

Chapter 8
Openers(patient): © Antonia Reeve/SPL/Photo Researchers, Inc.; (cells): © VVG/SPL/Photo Researchers, Inc.; (brain): © Simon Fraser/Royal Victoria Infirmary, Newcastle Upon Tyne/SPL/Photo Researchers, Inc.; 8.1a: © Corbis RF; 8.1b(tail): © McDonald Wildlife Photography/Animals Animals; 8.1b(lizard): © Kevin Hanley/Animals Animals; 8.1c: © Biophoto Associates/Photo Researchers, Inc.; 8.1d(left): © Anatomical Travelogue/Photo Researchers, Inc.; 8.1d(right): © Brand X Pictures/PunchStock; 8.2(chromatin): Courtesy O.L. Miller, Jr. and Steve L. McKnight; 8.2(chromosome): © Biophoto Associates/Photo Researchers, Inc.; 8.5(all): Courtesy Dr. Andrew S. Bajer, University of Oregon; 8.6(all): © Ed Reschke; 8.7(both): Copyright by R.G. Kessel and C.Y. Shih, *Scanning Electron Microscopy in Biology: A Students' Atlas on Biological Organization,* 1974, Springer-Verlag, New York; 8.8: © B.A. Palevitz and E.H. Newcomb/BPS/Tom Stack & Associates; 8.11(blebs): Courtesy Klaus Hahn, Cecelia Subauste, and Gary Bikoch from *Science* Vol. 280, April 3, 1998 p. 32; 8.11(apoptotic cell): Courtesy Douglas R. Green, LaJolla Institute for Allergy and Immunology, San Diego; 8.12b: © Breast Scanning Unit, Kings College Hospital, London/SPL/Photo Researchers, Inc.; 8.13: © Nancy Kedersha/Immunogen/SPL/Photo Researchers, Inc.; 8.14(chard): © Roy Morsch/Corbis; 8.14(cabbage, berries, broccoli, oranges): © Corbis RF.

Chapter 9
Openers(child): © Gary Parker/SPL/Photo Researchers, Inc.; (sperm, egg): © Pascal Goetgheluck/SPL/Photo Researchers, Inc.; (couple): © Corbis RF; 9.1: © CNRI/SPL/Photo Researchers, Inc.; 9.10(child): © Jill Cannefax Photography; 9.10(karyotype): © CNRI/SPL/Photo Researchers, Inc.

Chapter 10
Openers(hand): © SPL/Photo Researchers, Inc.; (chromosomes): © Photo Researchers, Inc.; (couple): © SuperStock RF; (smoking): © Corbis RF; 10.1: © Bettmann/Corbis; 10.8(earlobes, both): © The

18.20a: © Steven P. Lynch; 18.20b: © Bill Keogh/Visuals Unlimited; 18.20c: © Gary R. Robinson/Visuals Unlimited; 18.20d: © Corbis RF; 18.21a: © Gary T. Cole/BPS; 18.22(bread): © Precision Graphics/Kim Brucker, Photographer; 18.22(inset): © James W. Richardson/Visuals Unlimited; 18.23b: © Glenn Oliver/Visuals Unlimited; 18.24: © John Taylor/Visuals Unlimited; 18.25: © L. West/Photo Researchers, Inc.; 18.26b(crustose): © Digital Vision/Getty RF; 18.26b(foliose, fruticose): © Steven P. Lynch; 18.27: © R. Roncandore/Visuals Unlimited; 18.28: © The McGraw-Hill Companies, Inc./John Thoeming, photographer; 18.29a: © Kingsley Stern; 18.30a: © Everett S. Beneke/Visuals Unlimited; 18.30b: © John Hadfield/SPL/Photo Researchers, Inc.; 18.30c: © P. Marazzi/SPL/Photo Researchers, Inc.

Chapter 19

Openers(kitten): © Getty RF; (platypus): © D. Parer & E. Parer-Cook/Ardea; (penguin): © Johnny Johnson/Animals Animals; (gull): © Tony Camacho/SPL/Photo Researchers, Inc.; 19.1(adult frog): © Dwight Kuhn; 19.1(all, top row): © CABISCO/Phototake; 19.1(all, bottom row): © Dwight Kuhn; 19.7a: © Andrew J. Martinez/Photo Researchers, Inc.; 19.8a: © CABISCO/Visuals Unlimited; 19.8b: © CABISCO/Phototake; 19.9a: © Runk/Schoenberger/Grant Heilman Photography; 19.9b: © Ron Taylor/Bruce Coleman; 19.9c: © Under Watercolours; 19.11a: © James Webb/Phototake; 19.11b: © NIBSC/SPL/Photo Researchers, Inc.; 19.12c: © John Burbidge/SPL/Photo Researchers, Inc.; 19.15(snail): © Bob Coyle; 19.15(nudibranch): © Kenneth W. Fink/Bruce Coleman; 19.15(octopus): © Alex Kerstitch/Bruce Coleman; 19.15(nautilus): © Douglas Faulkner/Photo Researchers, Inc.; 19.15(scallop): Courtesy Larry S. Roberts; 19.15(mussels): © Rick Harbo; 19.17a: © Heather Angel; 19.17b: © James Carmichael/Nature Photographics; 19.17c: © St. Bartholomews Hospital/Photo Researchers, Inc.; 19.19a–e: © John Shaw/Tom Stack & Associates; 19.20a: © Kim Taylor/Bruce Coleman; 19.20b: © James Carmichael/Nature Photographics; 19.20c: © Natural History Photographic Agency; 19.20d: © Kjell Sandved/Butterfly Alphabet; 19.21a: © C. Allan Morgan/Peter Arnold; 19.21b: © John Bell/Bruce Coleman; 19.21c(ventral): © Daniel Lyons/Bruce Coleman; 19.21c(dorsal): © Ken Lucas/Ardea; 19.21d: © David M. Dennis/Animals Animals; 19.21e: © Geof de Feu/Imagestate; 19.22(scale): © Science VU/Visuals Unlimited; 19.22(beetle): © Kjell Sandved/Bruce Coleman; 19.22(lacewing): © Glenn Oliver/Visuals Unlimited; 19.22(housefly): © L. West/Bruce Coleman; 19.22(walking stick): © Art Wolfe; 19.22(honeybee): Courtesy of The National Human Genome Research Institute; 19.22(flea, true bug): © Edward S. Ross; 19.22(dragonfly): © Creatas Images/PictureQuest RF; 19.22(moth): © Clevland P. Hickman; 19.22(grasshopper): © Gary Meszaros/Visuals Unlimited; 19.22(leaf hopper): © Farley Bridges; 19.23b(sea lily): © P. Danna/Peter Arnold; 19.23b(feather Star): © Carl Roessler/Tom Stack & Associates; 19.23c: © Alex Kerstitch/Visuals Unlimited; 19.23d: © Jim Greenfield/Imagequestmarine.com; 19.23e(sea urchins): © Randy Morse/Animals Animals; 19.23e(sand dollar): © Roger Steene/Imagequestmarine.com; 19.25a: © Rick Harbo; 19.25b, 19.28b: © Heather Angel; 19.28b: © Ingram Publishing/Alamy RF; 19.28c: © Ron & Valerie Taylor/Bruce Coleman; 19.29b: © John Dudak/Phototake; 19.30: © Zig Leszczynski/Animals Animals; 19.31a: © Martin

Harvey/Gallo Images/Corbis; 19.32(both): © Daniel J. Cox; 19.33a: © Kirtley Perkins/Visuals Unlimited; 19.33b: © Thomas Kitchin/Tom Stack & Associates; 19.33c: © Brian Parker/Tom Stack & Associates; 19.34a: © B.G. Thompson/Photo Researchers, Inc.; 19.34b: © Leonard Lee Rue/Photo Researchers; 19.34c: © Fritz Prenzel/Animals Animals; 19.35a: © Leonard Rue; 19.35b: © Stephen J. Krasemann/DRK Photo; 19.35c: © Denise Tackett/Tom Stack & Associates; 19.35d: © Mike Bacon; 19.38a: © Dan Dreyfus and Associates; 19.38b: © John Reader/Photo Researchers, Inc.; 19.40: Courtesy American Museum of Natural History, Neg. # 608.

Chapter 20

Openers(fertilizer): © Evelyn Jo Johnson; (redwoods): © Corbis RF; (carnivorous plant): © Vol. 6/Getty RF; 20.2a: © Michael Gadomski/Photo Researchers, Inc.; 20.2b: © Steven P. Lynch; 20.2c: © Nature Picture Library; 20.2d: © The McGraw Hill Companies, Inc./Carlyn Iverson, photographer; 20.3a: © G.R. Roberts/Natural Sciences Image Library; 20.3b: © Ed Degginger/Color Pic; 20.3c: © David Newman/Visuals Unlimited; 20.5a: © Runk/Schoenberger/Grant Heilman Photography; 20.5b: © J.R. Waaland/BPS; 20.5c, 20.6a–c: © Biophoto Associates/Photo Researchers, Inc.; 20.7a: Courtesy Wilfred A. Cote, SUNY College of Environmental Forestry; 20.7b: © Randy Moore/Visuals Unlimited; 20.9: © Jeremy Burgess/SPL/Photo Researchers, Inc.; 20.10: © J. Robert Waaland/BPS; 20.11a: Courtesy Ray F. Evert, University of Wisconsin, Madison; 20.11b: © CABISCO/Phototake; 20.12: © Ed Reschke/Peter Arnold; 20.13(left): © CABISCO/Phototake; 20.13(right): © Terry Ashley/Tom Stack & Associates; 20.14: © John D. Cunningham/Visuals Unlimited; 20.15a–b: © Ken Wagner/Phototake; 20.16: Courtesy Mary E. Doohan; 20.17a: © Dwight Kuhn; 20.17b: © Runk/Schoenberger/Grant Heilman Photography.

Chapter 21

Openers(plants bend): © Runk/Schoenberger/Grant Heilman Photography; (flower): © PhotoDisc/Getty RF; (crops): © Photodisc Green/Getty RF; 21.2a–b: Courtesy Prof. Malcolm B. Wilkins; 21.3: © Robert E. Lyons/Visuals Unlimited; 21.4(twig): Courtesy John Selier Virginia Tech Forestry Department; 21.4(stoma): © Jeremy Burgess/SPL/Photo Researchers, Inc.; 21.5a–b: © Kingsley Stern; 21.5c: © Anton Vengo/SuperStock RF; 21.6: © Runk/Schoenberger/Grant Heilman Photography; 21.7: © Kingsley Stern; 21.9: © David Young-Wolff/PhotoEdit; 21.12a: Courtesy Timothy J. Fehr; 21.12b: © Ed Reschke; 21.13a: © Arthur C. Smith/Grant Heilman Photography; 21.13b: © Grant Heilman/Grant Heilman Photography; 21.14(pollen grain): Courtesy Graham Kent; 21.14(embryo sac): © Ed Reschke; 21.15a: © Vol. 7/PhotoDisc/Getty RF; 21.15b: © Simko/Visuals Unlimited; 21.15c: © Dwight Kuhn; 21.16: © Jack Bostrack/Visuals Unlimited; 21.17(walnut half): © Jack Star/PhotoLink/Getty RF; 21.17(whole walnut): © Siede Preis/Getty RF; 21.17(peas): © Andrew Thomas/Getty Images; 21.17(peach half): © Peter Fakler/Alamy RF; 21.17(whole peach): © Jupiterimages/Image Source RF; 21.17(apples): © Photolink/Getty RF; 21.18a: © Marie Read/Animals Animals; 21.18b: © Scott Camazine/Photo Researchers, Inc.; 21.18c: © Getty RF; 21.18d: © James Mauseth; 21.22(all): Courtesy Prof. Dr. Hans-Ulrich Koop, from Plant Cell Reports, 17:601-604;

21.23: © Kingsley Stern; 21.24a(all): Courtesy Monsanto; 21.25: © William Freese; 21.26b(both): Courtesy Dr. Eduardo Blumwald; 21.26c: © Getty RF.

Chapter 22

Openers(finger): © PhotoDisc/Getty RF; (insulin): © SIU/Visuals Unlimited; 22.3a–b, d: © Ed Reschke; 22.3c: © Ed Reschke/Peter Arnold; 22.4(left): © Jim Merli/Visuals Unlimited; 22.4(right)-22.5, 22.7–22.8 (all): © Ed Reschke.

Chapter 23

Openers(giving blood): © The McGraw Hill Companies, Inc./Jill Braaten, photographer; (treadmill): © Fotopic/Miles Simons/Phototake; (pacemaker): © Vol. 59/PhotoDisc/Getty RF; 23.11: © Biophoto Associates/Photo Researchers, Inc.; 23.12: © Pascal Goethgheluck/SPL/Photo Researchers, Inc.; 23.14: © Prof. P.M. Motta, and S. Correr/SPL/Photo Researchers, Inc.; 23.16d: © Manfred Kage/Peter Arnold.

Chapter 24

Openers(urine): © Vol. 54/PhotoDisc/Getty RF; (infant): © Ansel Horn/Phototake; 24.3: © Photo Insolite Realite/SPL/Photo Researchers, Inc.; 24.4b: © Ed Reschke; 24.22: © Ardea London Ltd; 24.24c: © Fawcett, Hirokawa, Heuser/SPL/Photo Researchers, Inc.

Chapter 25

Openers(beef): © Corbis RF; (vitamins): © Michael Keller/Corbis; (trans fats): © Reuters/Corbis; 25.1: © Getty RF; 25.2a: © Dr. Ken Greer/Visuals Unlimited, Inc.; 25.2b: © ISM/Phototake; 25.3: © Precision Graphics/Kim Brucker, photographer; 25.4: © The McGraw-Hill Companies, Inc., John Thoeming, photographer; 25.5: © Cole Group/Getty RF; 25.6: © Volume 20/Getty RF; 25.7: © Biophoto Associates/Photo Researchers, Inc.; 25.9a: © Corbis RF; 25.9b: © Lester Lefkowitz/The Image Bank/Getty Images; 25.9c: © Sean Bagshaw/Photo Researchers, Inc.; 25.11: Ryan McVay/Getty RF; 25.12: © Corbis RF; 25.13a: © Tony Freeman/PhotoEdit; 25.13b: © PhotoLink/Getty RF; p. 458: © Precision Graphics/Kim Brucker, photographer.

Chapter 26

Openers(HIV): © NIBSC/Photo Researchers, Inc.; (cyclosporine): © Brad Nelson/Phototake; (nurse): © Vol. 41/Getty RF; 26.2: © Comstock/PunchStock RF; 26.4a: © Dennis Kunkel/Phototake; 26.6b: Courtesy Dr. Arthur J. Olson, Scripps Institute; 26.8a: © Steve Gschmeissner/Photo Researchers, Inc.; 26.9a: © Photodisc/Getty RF; 26.10(pollen): © David Scharf/SPL/Photo Researchers, Inc.; 26.10(girl): © Damien Lovegrove/SPL/Photo Researchers, Inc.; 26.11: © Richard Anderson; 26.12a: © A. Ramey/PhotoEdit.

Chapter 27

Openers(woman, couple): © Corbis RF; (brain): © Neil Bordon/Photo Researchers, Inc.; 27.7a(both): © Science VU/Visuals Unlimited; 27.8: © Reuters/Corbis; 27.11: © Dennis Gray/Cole Group/Getty RF; 27.14: © Tom McHugh/Photo Researchers, Inc.; 27.19: © Biophoto Associates/SPL/Photo Researchers, Inc.; 27.20: © Dr. P. Marazzi/SPL/Photo Researchers, Inc.; 27.21a(both): *Atlas of Pediatric Physical Diagnosis,* 2nd edition by Zitelli and Davis, 1992. Mosby-Wolfe Europe LTd., London, U.K.

Chapter 28

Openers(weightlifter): © Corbis RF; (tongue): © William Radcliffe/Science Faction/Getty Images; (food): © Vol. 67/PhotoDisc/Getty RF; 28.4a(both): Courtesy Robert S. Preston and Joseph E. Hawkins, Kresgke Research Institute, University of Michigan; 28.14a: © Michael Fogden/OSF/Animals Animals; 28.16(center): Courtesy Hugh Huxley; 28.18: © Julie Lemberger/Corbis; 28.20a: © Gerard Vandystadt/Photo Researchers, Inc.

Chapter 29

Openers(sperm): © Dennis Kunkel/Phototake; (infant): © Corbis RF; (herpes virus): © NIAD/NIH/Peter Arnold; 29.1: © Dennis Kunkel/Visuals Unlimited; 29.2: © Hans Pfletschinger/Peter Arnold; 29.3: © Leonard Lee Rue, Jr./Photo Researchers, Inc.; 29.9(implant): © Population Council/Karen Tweedy-Holmes; 29.9(IUD): © The McGraw-Hill Companies, Inc./Vincent Ho, photographer; 29.9(birth control pills, jelly, diaphragm, female condom, Depo-provera): © The McGraw-Hill Companies, Inc./Bob Coyle, photographer; 29.10: © Myrleen Ferguson Cate/PhotoEdit; 29.11: © CC Studio/SPL/Photo Researchers, Inc.; 29.12a(all): © Nicholas Nixon; 29.13: © PhotoAlto/PunchStock; 29.19–29.21: © Lennart Nilsson, *A Child is Born,* Dell Publishing Company; 29.22: © John Watney/Photo Researchers, Inc.; 29.23: © James Stevenson/SPL/Photo Researchers, Inc.; 29.24:Lennart Nilsson, *A Child is Born,* 1990 Delacorte Press, Pg. 111(top frame); 29.26: © Dennis MacDonald/PhotoEdit.

Chapter 30

Openers(babies): © ER Productions/Corbis; (Asian street): © D. Normark/Photo Link/Getty RF; (billboard): © Bohemian Nomad Picturemakers/Corbis; 30.3a: © Jeff Greenberg/Peter Arnold; 30.3b: © Ben Osborne/OSB/Animals Animals; 30.4c: © Still Pictures/Peter Arnold; 30.5(MDC): © Comstock Images/Getty RF; 30.5(LDC): © Alasdair Drysdale RF; 30.6a: © The McGraw Hill Companies, Inc./Evelyn Jo Johnson, photographer; 30.9a: © age fotostock/SuperStock; 30.9b: © Corbis RF; 30.16: © Creatas/PunchStock; 30.17a: © Ted Levin/Animals Animals; 30.17b: © Martin Harvey/Gallo Images/Corbis; 30.19: © Frank & Joyce Burek/Getty RF; 30.20a: © Jeff Henry/National Park Service; 30.20b: © William P. Cunningham.

Chapter 31

Openers(Earth): © Stock Trek/Getty RF; (kelp): © Chuck Davis/Getty Images; (lake): © Michael Gadomski/Animals Animals; 31.1a: © Anthony Mercieca/Photo Researchers, Inc.; 31.1b: Courtesy Dr. Rod Peakall, School of Botany and Zoology, Australian National University; 31.1c: © Merlin D. Tuttle/Bat Conservation International; 31.2a: © Charlie Ott/Photo Researchers, Inc.; 31.2b: © Michael Graybill and Jan Hodder/Biological Photo Service; 31.3a–b: © Vol. 63/Corbis RF; 31.5(all): © Breck P. Kent/Animals Animals; 31.6: © Steve Austin/Papilio/Corbis; 31.10a(all): Courtesy Daniel Janzen; 31.11a: © George Armistead/Vireo; 31.11b: © Jack Wilburn/Animals Animals; 31.11c: © Eastcott/Momatiuk/Animals Animals; 31.12a: © Stockbyte RF; 31.12b: © John Mitchell/Photo Researchers, Inc.; 31.13(diatoms): © Ed Reschke/Peter Arnold; 31.13(tree): © PunchStock RF; 31.14a(giraffe): © George W. Cox; 31.14a(caterpillar): © Corbis RF; 31.14b(osprey): © Joe McDonald/Visuals Unlimited; 31.14b(mantis): © Scott Camazine; 31.15(mushrooms): © IT Stock/age fotostock RF; 31.15(bacteria): © David M. Phillips/Visuals Unlimited; 31.17: © Sylvia S. Mader; 31.23: © Tom McHugh/Photo Researchers, Inc.; 31.25: © Biophoto Associates/Photo Researchers, Inc.; 31.27a: © Vol. 121/Corbis RF; 31.27b: © The McGraw-Hill Companies, Inc./Carlyn Iverson, photographer; 31.27c: © David Hall/Photo Researchers, Inc.; 31.27d–31.28 (tundra): © Vol. 121/Corbis RF; 31.28(taiga, temperate, desert): © Vol. 63/Corbis RF; 31.28(prairie): © Vol. 86/Corbis RF; 31.28(savanna): © Gregory Dimijan/Photo Researchers, Inc.; 31.28(tropical rain forest): © Vol. 121/Corbis RF.

Chapter 32

Openers(deforestation, bulldozer): © PhotoLink/Getty RF; (Biosphere 2): © James Marshall/Corbis; 32.1(land): © Vol. 39/Getty RF; 32.1(water): © Evelyn Jo Johnson; 32.1(food): © The McGraw-Hill Companies, Inc., John Thoeming, photographer; 32.1(energy): © Gerald and Buff Corsi/Visuals Unlimited; 32.1(minerals): © James P. Blair/National Geographic Image Collection; 32.1(population): © D. Normark/Photo Link/Getty RF; 32.2b: © Melvin Zucker/Visuals Unlimited; 32.4b: Courtesy of Carla Montgomery; 32.5a: © Corbis RF; 32.5b: © FoodPix/Getty Images; 32.5c: © Stockbyte/PunchStock RF; 32.6a: © Jodi Jacobson/Peter Arnold; 32.6b: © Peter Essick/Aurora; 32.7a: © Laish Bristo/Visuals Unlimited; 32.7b: © Inga Spence/Visuals Unlimited; 32.7c: Courtesy V. Jane Windsor, Division of Plant Industry, Florida Department of Agriculture and Consumer Services; 32.8: © Corbis RF; 32.9a: © Shane Moore/Animals Animals; 32.9b: © Michael Pole/Corbis; 32.10b: Courtesy Walther C. Japp, Florida Fish and Wildlife Conservation Commission; 32.11a: © David L. Pearson/Visuals Unlimited; 32.11b: © S.K. Patrick/Visuals Unlimited; 32.11c: © Argus Foto Archiv/Peter Arnold; 32.11d: © Gerald and Buff Corsi/Visuals Unlimited; 32.12a: © Matt Stiveson/NREL/DOE; 32.12b: Courtesy of DaimlerChrysler; 32.12c: © Warren Gretz/NREL/DOE; 32.13: © James P. Blair/National Geographic Image Collection; 32.15b: © Gunter Ziesler/Peter Arnold; 32.16(periwinkle): © Kevin Schaefer/Peter Arnold; 32.16(armadillo): © John Cancalosi/Peter Arnold; 32.17a: © Anthony Mercieca/Photo Researchers, Inc.; 32.17b: © Merlin D. Tuttle/Bat Conservation International; 32.17c: © Herve Donnezan/Photo Researchers, Inc.; 32.17d: © Bryn Campbell/Stone/Getty; 32.18(forest): © William M. Smithey Jr.; 32.18(seashore): © Vol. 121/Corbis RF; 32.19: © Stuart Westmorland/Corbis; 32.20a: © Kent Knudson/PhotoLink/Getty RF; 32.20b: © Scenics of America/PhotoLink/Getty RF; 32.21: © Dan McCoy/Rainbow.

Index

Note: Page numbers followed by *f* and *t* indicate figures and tables, respectively.

nucleoid, 52, 274
nucleolus, 58
nucleosome, 115
nucleotide, 44
nucleus, 18, 56, 58–59, 59f
   structure of, 58
nutrient(s), 442
   classes of, 444–448, 444f
nutrient transport, in plants, 353–355
nutrition, 441–460
   and health, 450–454
   planning healthy meals, 455–457
Nutrition Facts panel, 456, 456f

O

obesity, 445, 450
   disorders associated with, 452–453
observation, 10
octet rule, 21
oils, 37
omega-3 unsaturated fatty acids, 453
omnivores, 429, 566
   dentition of, 429f
oncogene, 184, 193
one-trait inheritance, 146
one-trait testcross, 146–147, 146f
oocyte, 530
oogenesis, 131, 523f
open circulatory systems, 400–401, 401f
operon, 186–187, 186f
opportunistic populations, 550, 550f
oral contraception, 525, 525f, 525t
order, 6, 258
organ(s), 2, 382, 390
   formation of, 533
organelles, 56
   energy-related, 62–63
organic chemistry, 32
organic molecules, 31–48, 32f
organisms, 2, 552
   sizes of, 51f
organ rejection, 469
organ systems, 2, 390–391
orgasm, 521
osmosis, 79, 79f
   effects on cells, 80, 80f
ossicles, 502
osteoarthritis, 512
osteoblasts, 510
osteoclasts, 510
osteocytes, 510
osteons, 510
osteoporosis, 447–448, 510
ostium, 401
outgroup, 261
out-of-Africa hypothesis, 335, 335f
ovarian cycle, 523f, 524, 524f

ovary(ies), 296, 296f, 367, 518, 522–523
oviducts, 522
ovoviviparous, 519
ovulation, 523–524
ovules, 367
oxygen deficit, 109
oxytocin, 491–492, 535
oyster fungus, 302, 303f

P

paleontology, 219, 255
palisade mesophyll, 346
pancreas, 432, 432f, 434, 494
pancreatic amylase, 432, 436
pancreatic islets, 494
Pangaea, 256, 257f
Pantherophis obsoleta, 238, 238f
pantothenic acid, 449t
papaya ring spot virus (PRSV), 270
paramecium, 281, 281f, 282
   and competitive exclusion
    principle, 563, 563f
parasitic flatworms, 316, 316f
parasitism, 562, 562t
parasympathetic division, 488
parathyroid glands, 492
parathyroid hormone (PTH), 492
parenchyma, 345, 345f
parent cell, 114
parsimony, 261
parthenogenesis, 518
particulate theory of inheritance, 145
parturition, stages of, 535, 535f
passive immunity, 470
passive transport, 79–80
Patagonian hare, 221, 221f
patents, for plants, 359–360
pathogens, 277
Patua syndrome, ultrasound and, 210
PCD. See programmed cell death
PCR. See polymerase chain reaction
peach leaf curl, 304f
pearls, 319
peas, garden, 144f–145f
   development of, 298f
peat moss, 291
pectoral (shoulder) girdle, 509
pedigree, 204, 204f–205f
pelvic (hip) girdle, 509
pelvic inflammatory disease (PID), 527, 529
penetration, 269
penicillin, 10
penis, 520
pepsin, 430, 436
peptidases, 436
peptide hormones, 490
peptides, 41–42, 41f, 436
peptidoglycan, 274

% daily value, 456, 456f
perennial plants, 343
pericycle, 351
periodic table, 19–20, 19f
peripheral nervous system
   (PNS), 479, 486–488
peristalsis, 430
PET. See positron emission tomography
petals, 296, 296f, 367
petiole, 346
PGD. See preimplantation
   genetic diagnosis
pH, 28
   regulation of, 426
phagosytosis, 81
pharyngeal pouches, 324, 324f
pharynx, 419, 430, 430f
phenotype, 147
   environment and, 152, 152f
   versus genotype, 147, 147t
Philadelphia chromosome, 194, 194f
phloem, 292, 344, 346, 346f
   nutrient transport in, 354–355, 354f
phospholipids, 38, 39f
   in plasma membrane, 54, 54f
phosphorus, 447t
phosphorus cycle, 571, 571f
photoautotrophs, 275, 275f
photoperiodism, 364–365, 365f
photoreceptors, 505–506, 506f
photosynthesis, 3, 85–98
   organisms conducting, 86–87, 86f
   other types of, 94–95
   overview of, 86–88
   process of, 88, 88f
photosystems, 90, 90f
phototropism, 364, 364f
photovoltaic (solar) cell, 591, 591f
pH scale, 28, 28f
phylogenetic tree, 259, 259f
phylogeny, 258
   tracing, 259–261
phylum, 6, 258
phytochrome, 365
PID. See pelvic inflammatory disease
pigments, photosynthetic, 89, 89f
pineal gland, 485
pine trees, 295, 295f
pinocytosis, 81
pinworm, 317
pioneer species, 561
Pisum sativum, 144f–145f
pitch, 503
pitcher plant, 341–342
pith, 348
pituitary gland, 490–492, 491f
placebo, 13
placenta, 534, 534f
placental mammals, 331, 331f, 519
planarian, 315–316, 315f

R